The Tube Amplifier Schematic Bible Volume 2

G-Z

Salvatore Gambino

ISBN: 150058245X
ISBN-13: 978-1500582456

INTRODUCTION

This book of amp schematics was assembled with service and repair in mind. I have always had a very deep respect for the design and performance that tube amps produce. Let's face it, guitar tube amps don't always get the respect that they deserve. Tube amplifiers have always worked hard and should be looked at as a major part of your sound as they inspire you to dig deep into your playing. If you feel somewhat the same way I do about tube amps, then you know each amplifier has their own characteristics and tone. I hope you can use this educational information to understand how tube amps are designed and how they work. Look for my other book

Reading Schematics Made Easy.

CONTENTS

CONTENTS

CONTENTS

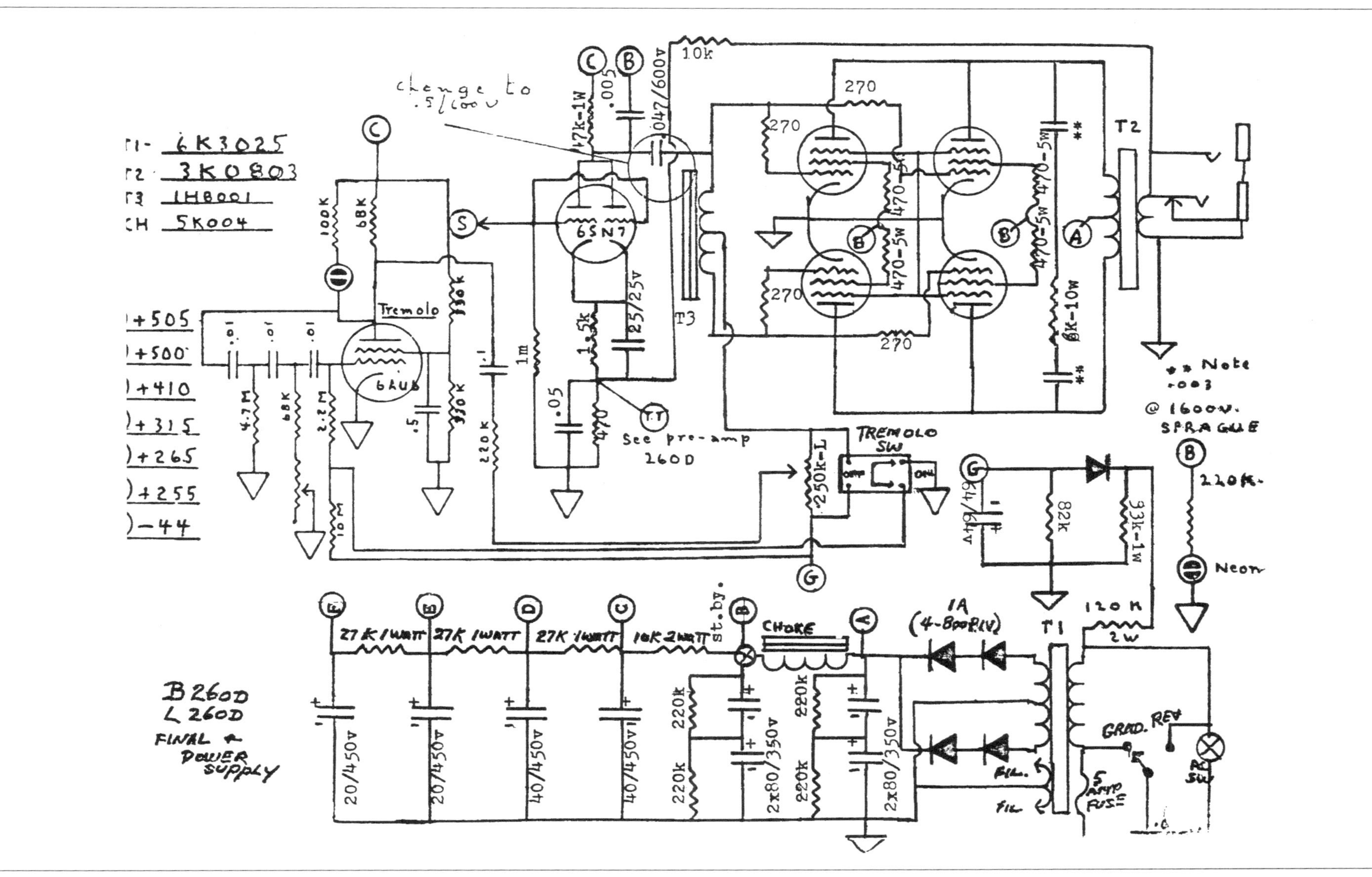

T1- 6K3025
T2 3K0803
T3 1H8001
CH 5K004
+505
+500
+410
+315
+265
+255
-44
Tremolo
6AU6
6SN7
change to .5/600v
47k-1W
.005
.047/600v
10k
270
470-5w
ßK-10w
T2
T3
** Note .003 @ 1600v. SPRAGUE
See pre-amp 260D
250k-L
TREMOLO SW
OFF
ON
64/64v
82k
93k-1w
220K
Neon
120 K 2w
T1
1A (4-800PIV)
CHOKE
st.by.
27K 1WATT
10K 2WATT
20/450v
40/450v
2x80/350v
220k
GRND. REV
AC SW
5 AMP FUSE
FIL.
B260D L260D FINAL & POWER SUPPLY

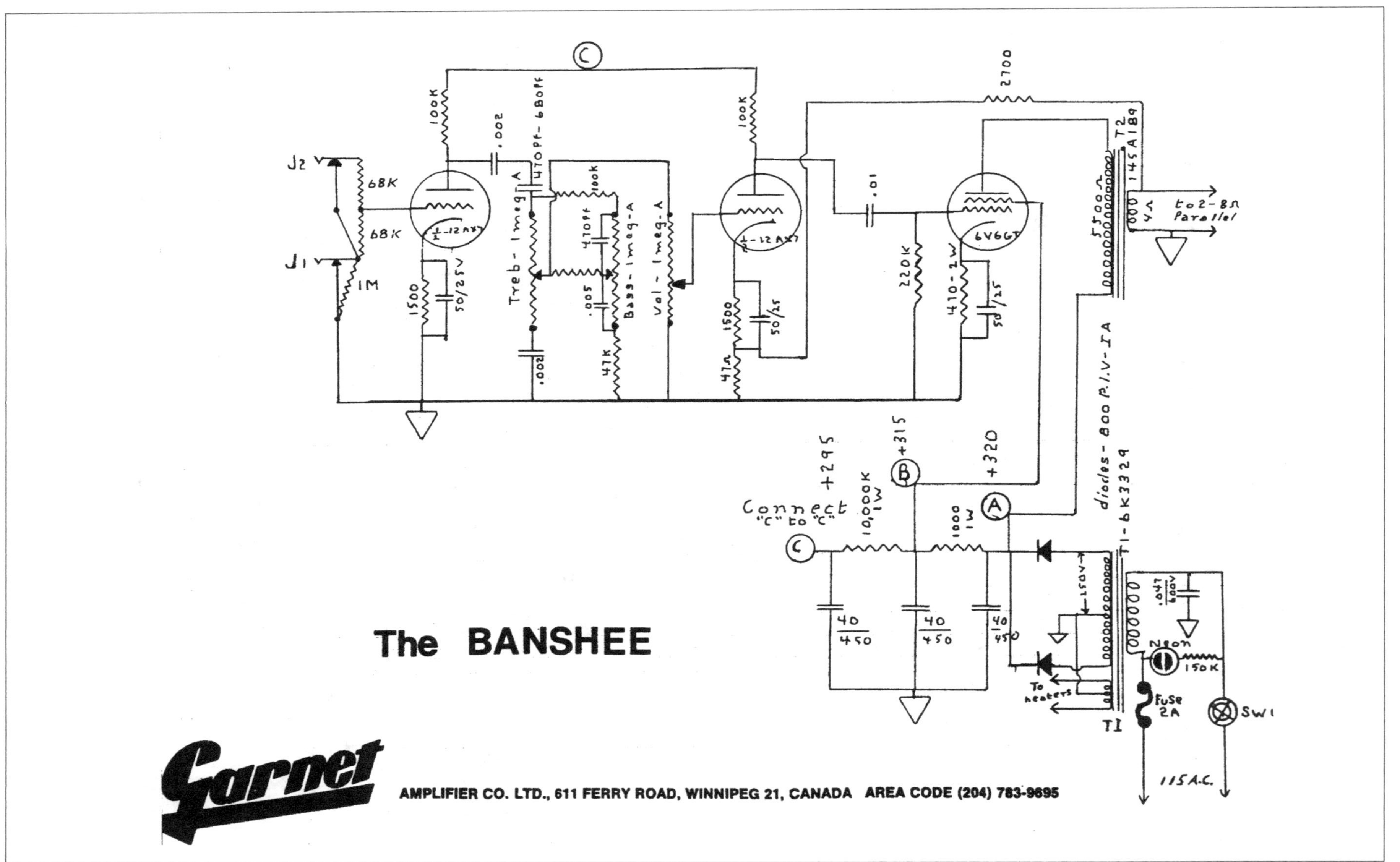
J2
J1
68K
68K
1M
100K
.002
1/2-12AX7
1500
50/25V
Treb - 1meg-A
470PF - 680PF
100K
470PF
.005
Bass - 1meg-A
Vol - 1meg-A
.002
47K
100K
1/2-12AX7
1500
50/25
47Ω
.01
220K
6V6GT
410-2W
50/25
2700
5500Ω
T2
145A189
4Ω
to 2-8Ω
Parallel
diodes - 800 P.I.V - 1A
T1 - 6K3329
+295
+315
+320
Connect
"C" to "C"
10,000K
1W
1000
1W
40
450
40
450
40
450
150V
To heaters
.047
600V
Neon
150K
Fuse
2A
SW1
T1
115A.C.
The BANSHEE
Garnet
AMPLIFIER CO. LTD., 611 FERRY ROAD, WINNIPEG 21, CANADA AREA CODE (204) 783-9695

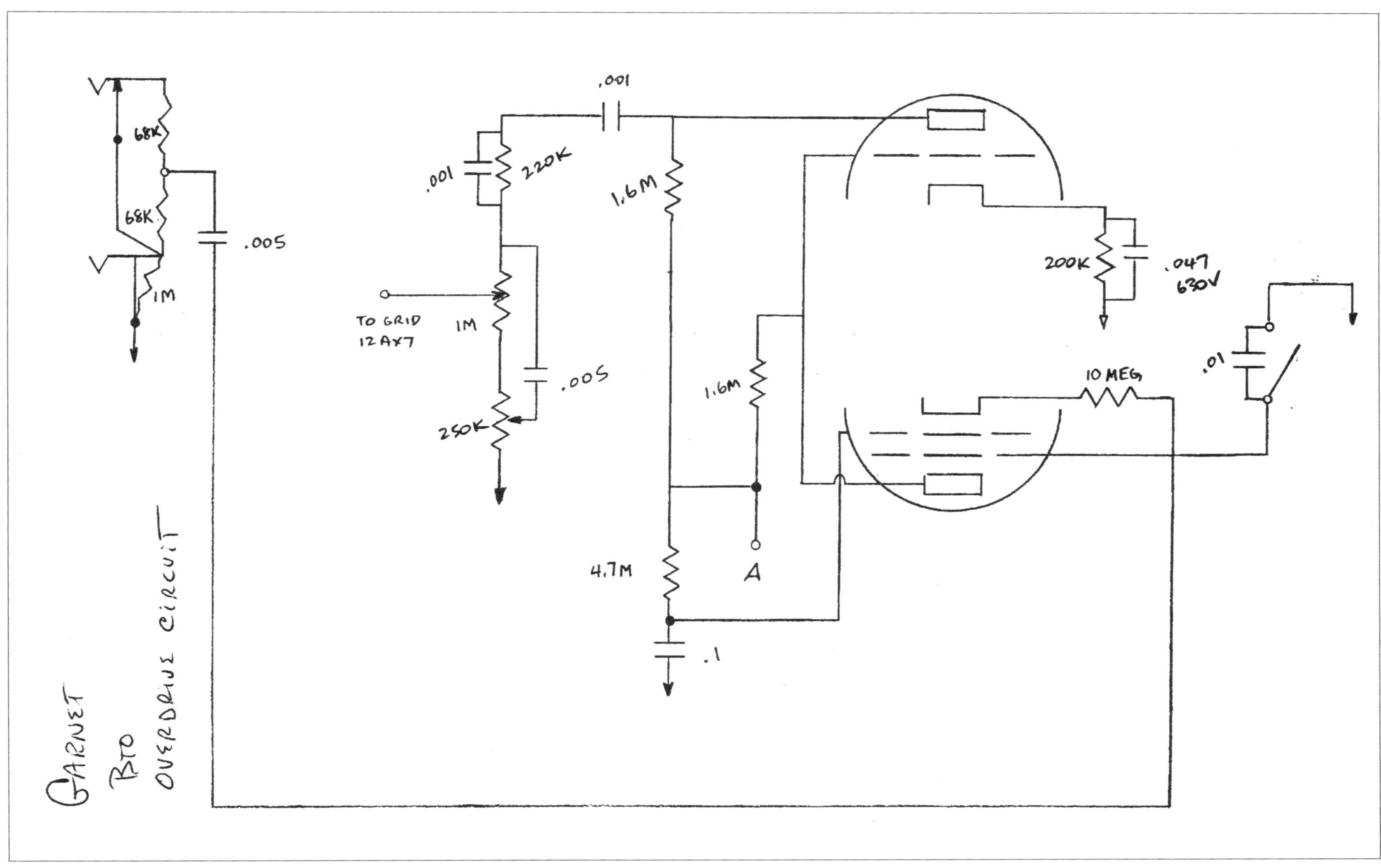
GARNET
BTO
OVERDRIVE CIRCUIT
68K
68K
1M
.005
.001
.001
220K
TO GRID
12AX7
1M
.005
250K
1.6M
1.6M
4.7M
A
.1
200K
.047
630V
10 MEG
.01

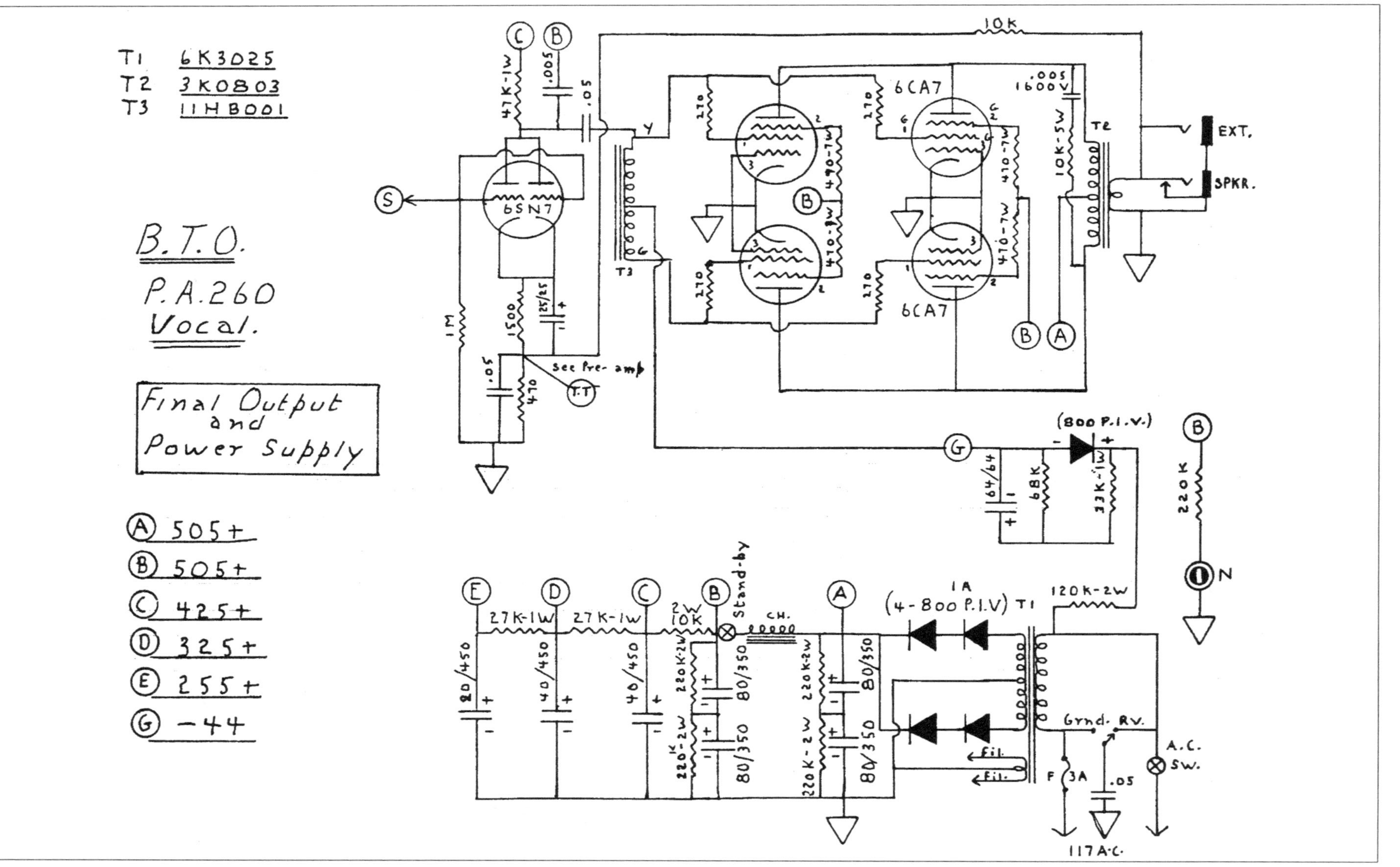
T1 6K3025
T2 3K0803
T3 11HB001
B.T.O.
P.A.260
Vocal.
Final Output and Power Supply
A 505+
B 505+
C 425+
D 325+
E 255+
G −44
6SN7
6CA7
6CA7
47K-1W
.005
.05
1M
1500
25/25
470
See Pre-amp
T.T
10K
270
490-7W
470-7W
470-1W
.005 1600V
10K-5W
T2
T3
EXT.
SPKR.
(800 P.I.V.)
64/64
68K
33K-1W
220K
N
120K-2W
1A
(4-800 P.I.V) T1
Stand-by
CH.
27K-1W
2W 10K
20/450
40/450
80/350
220K-2W
Fil.
Grnd. RV.
F 3A
A.C. SW.
117 A.C.

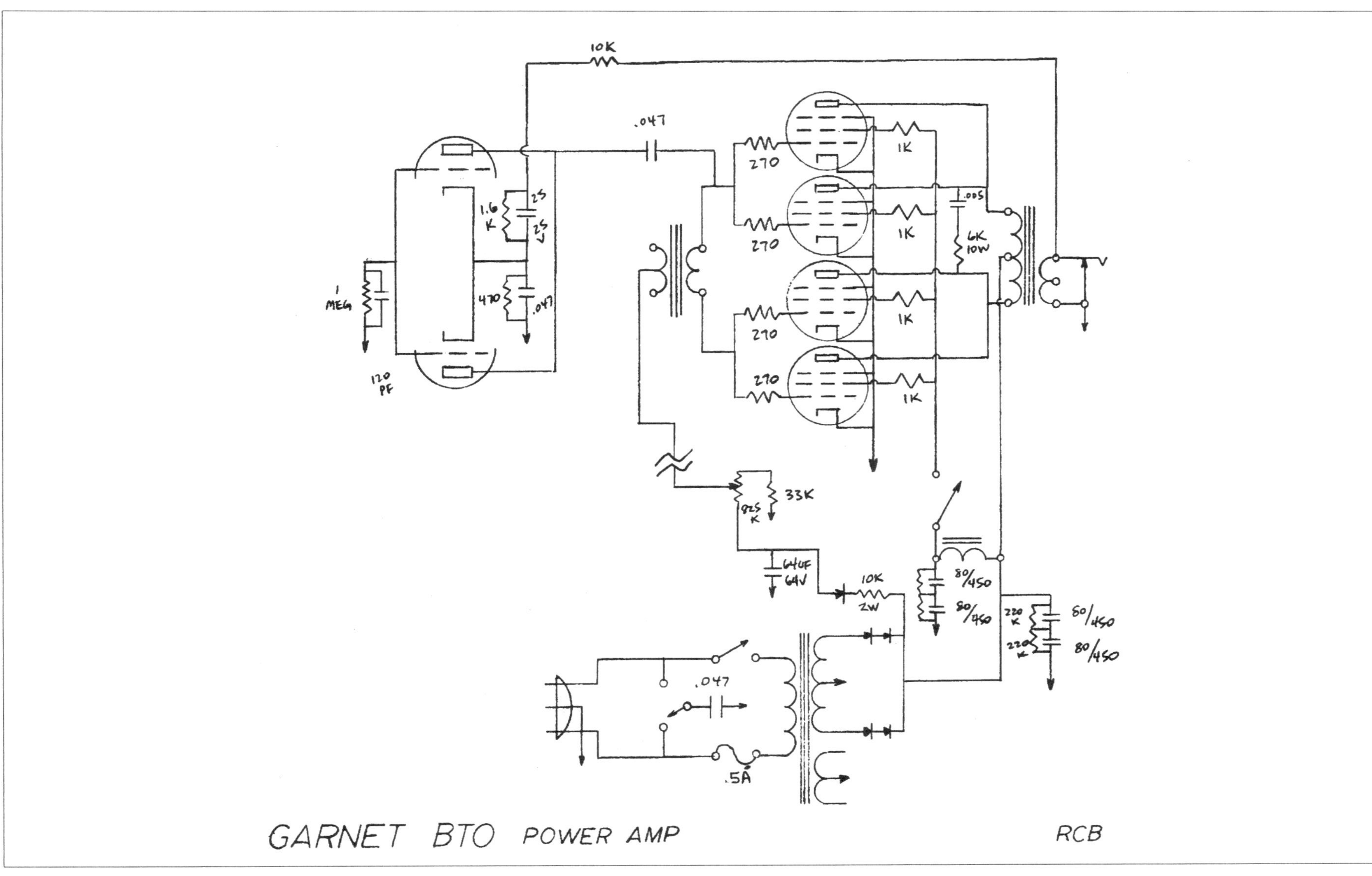
10K
.047
270
270
270
270
1K
1K
1K
1K
.005
6K 10W
1.6 K
25 25 V
1 MEG
470
.047
120 PF
33K
825 K
64UF 64V
10K 2W
80/450
80/450
220 K
220 K
80/450
80/450
.047
.5A
GARNET BTO POWER AMP
RCB

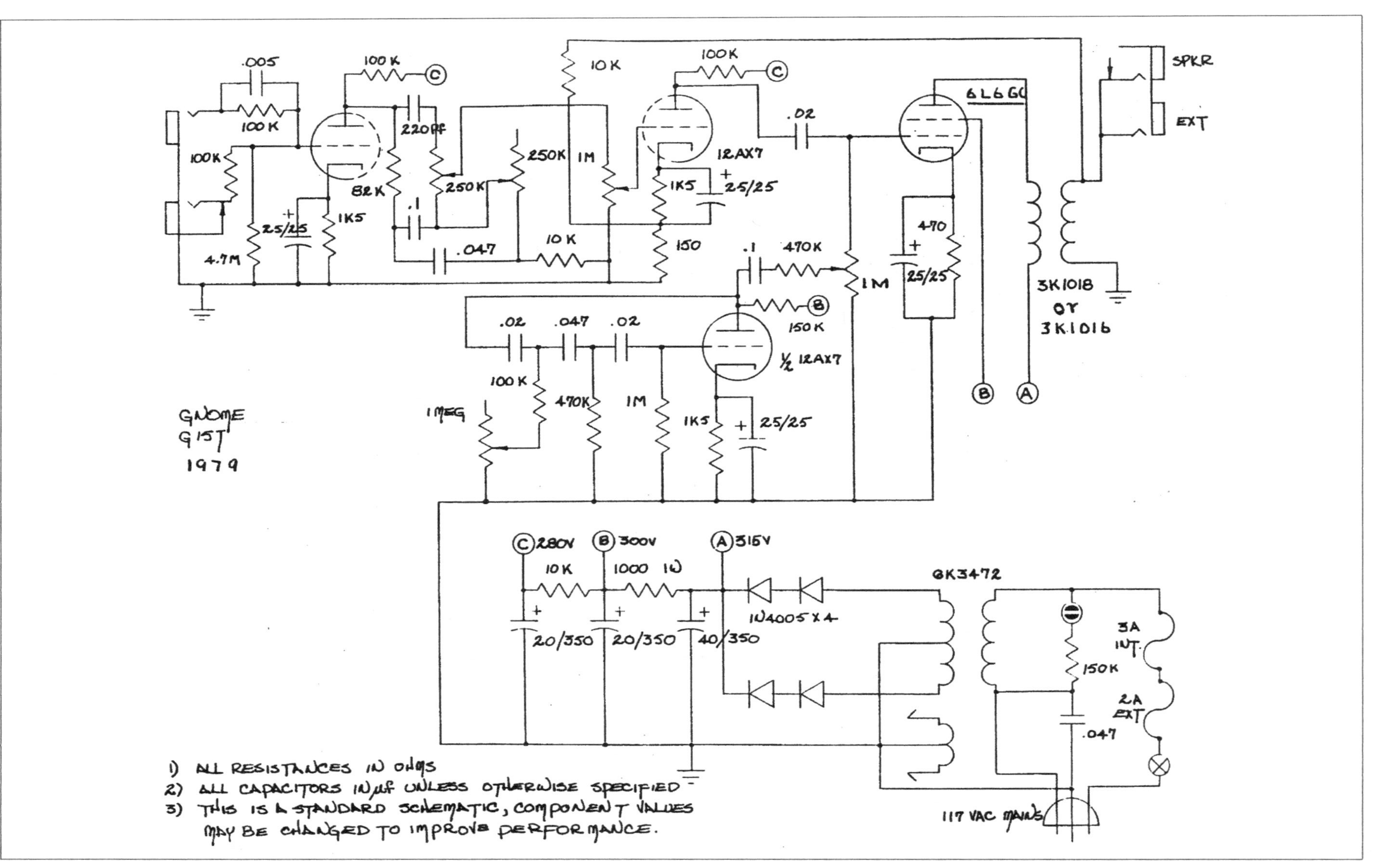
.005
100 K
100 K
100K
4.7M
25/25
1K5
82K
220PF
.1
250K
250K
.047
10 K
10 K
1M
100 K
12AX7
1K5
25/25
150
.02
6L6GC
470
25/25
1M
.1
470K
150K
½ 12AX7
.02
.047
.02
100 K
1MEG
470K
1M
1K5
25/25
SPKR
EXT
3K1018
or
3K1016
GNOME
G15T
1979
C 280V
B 300V
A 315V
10 K
1000 1W
20/350
20/350
40/350
1N4005 X 4
GK3472
150K
3A
INT.
2A
EXT
.047
117 VAC MAINS
1) ALL RESISTANCES IN OHMS
2) ALL CAPACITORS IN µF UNLESS OTHERWISE SPECIFIED
3) THIS IS A STANDARD SCHEMATIC, COMPONENT VALUES
MAY BE CHANGED TO IMPROVE PERFORMANCE.

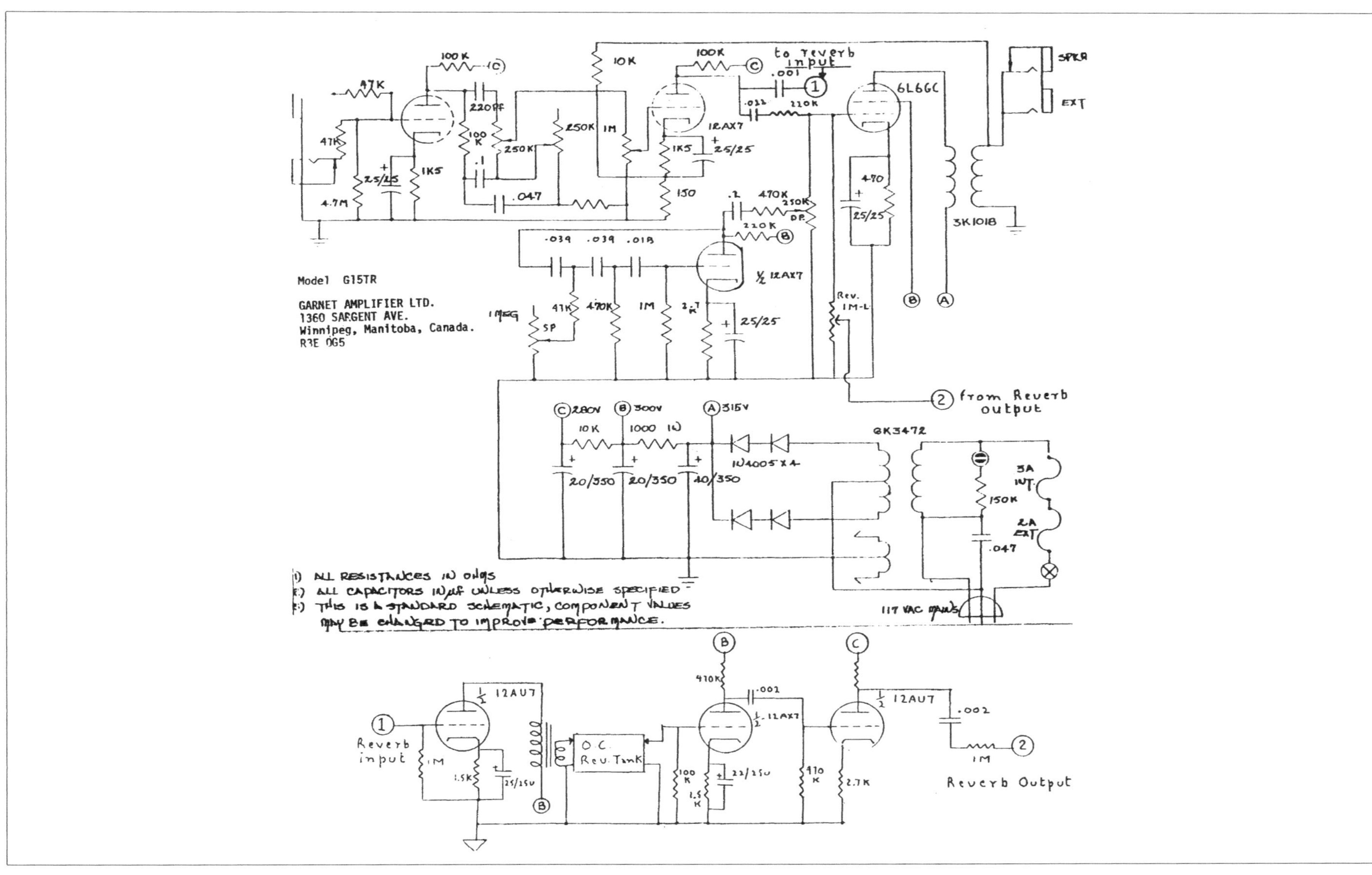

Model G15TR
GARNET AMPLIFIER LTD.
1360 SARGENT AVE.
Winnipeg, Manitoba, Canada.
R3E 0G5
1) ALL RESISTANCES IN OHMS
2) ALL CAPACITORS IN µF UNLESS OTHERWISE SPECIFIED
3) THIS IS A STANDARD SCHEMATIC, COMPONENT VALUES MAY BE CHANGED TO IMPROVE PERFORMANCE.
to reverb input
from Reverb output
Reverb input
Reverb Output
O.C. Rev. Tank
12AX7
½ 12AX7
½ 12AU7
6L6GC
3K101B
GK3472
1N4005 X 4
117 VAC MAINS
SPKR
EXT
280V
300V
315V
5A INT.
2A EXT
Rev. 1M-L
1MEG SP

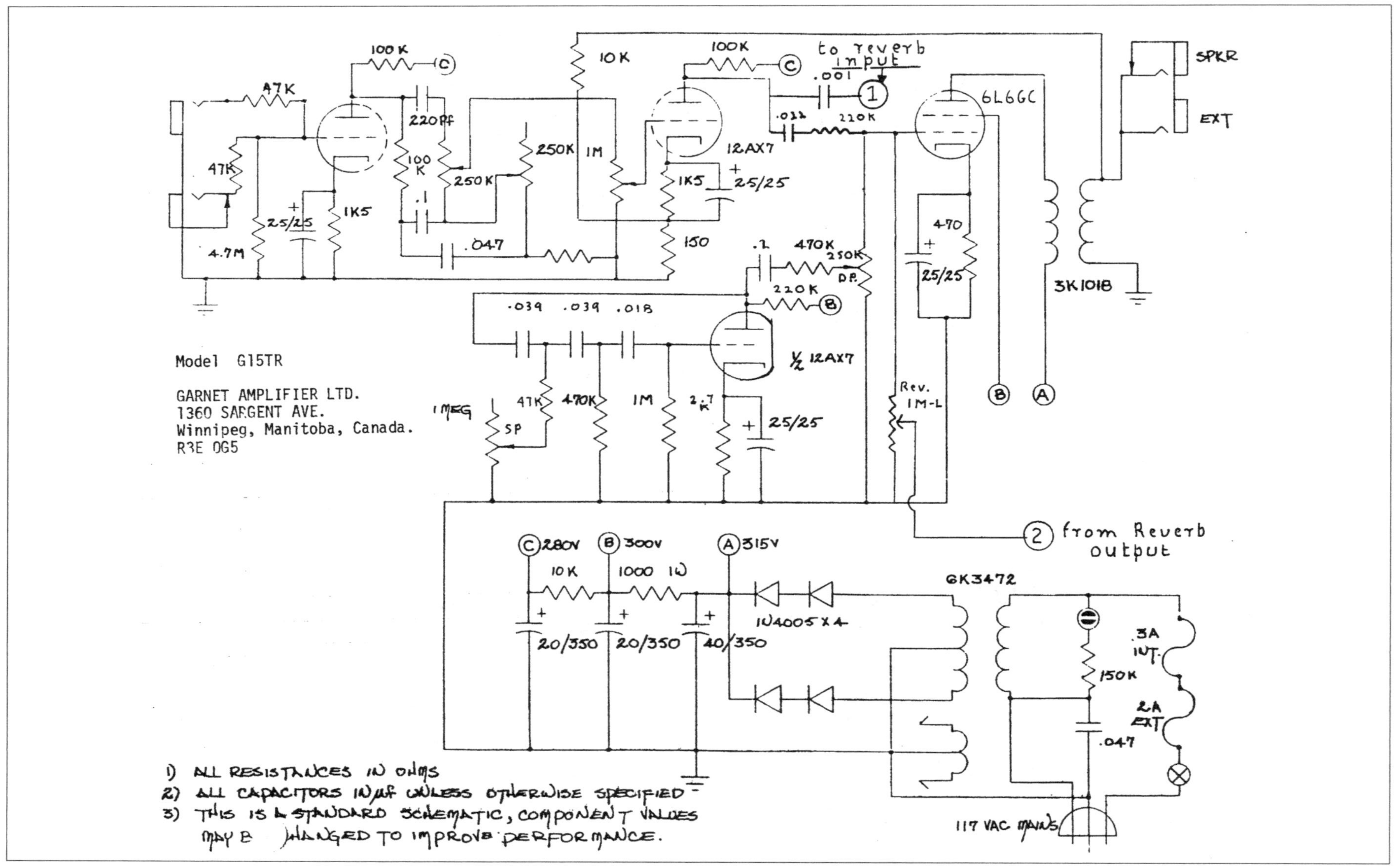

Model G15TR

GARNET AMPLIFIER LTD.
1360 SARGENT AVE.
Winnipeg, Manitoba, Canada.
R3E 0G5

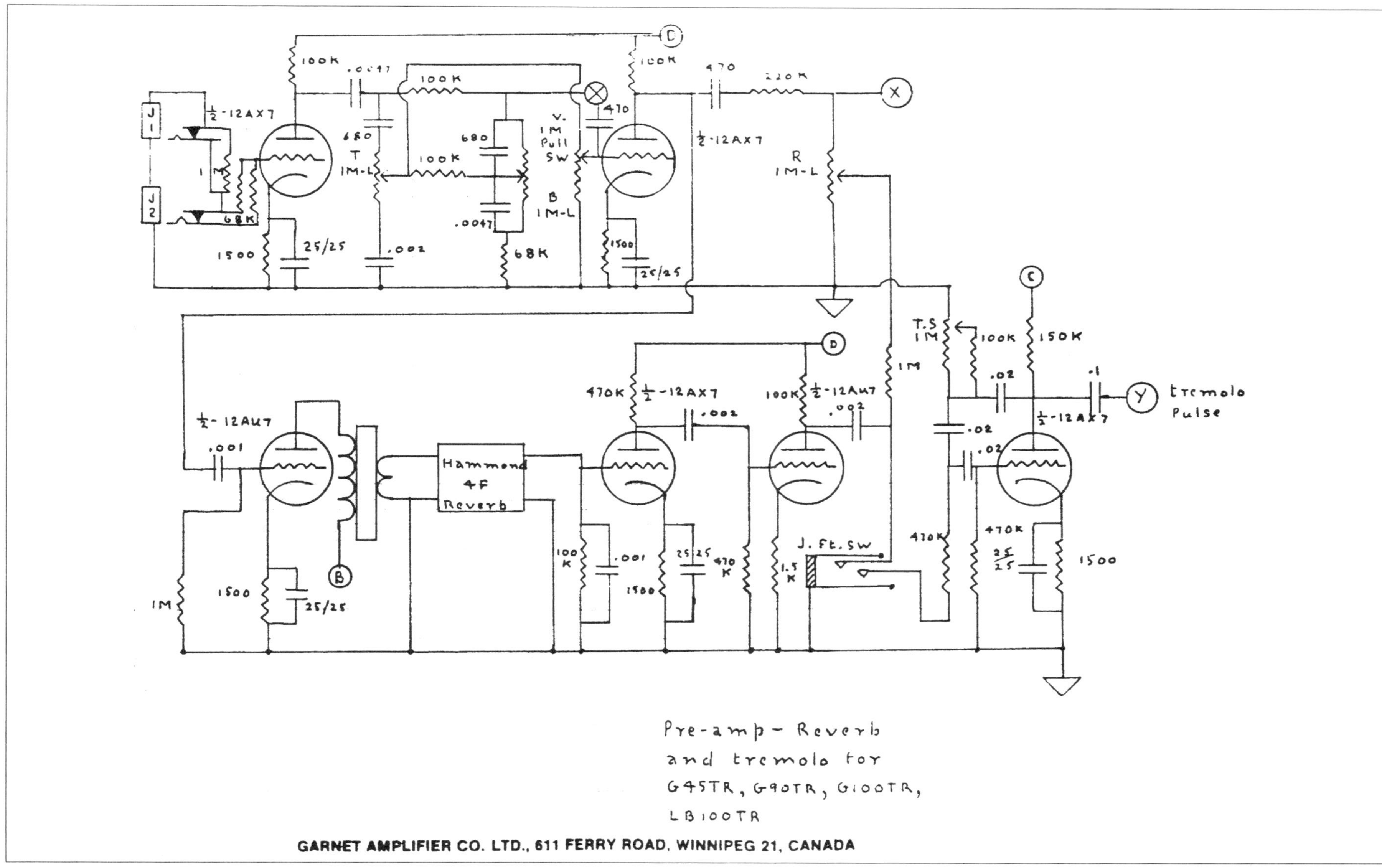

J 1
J 2
½-12AX7
1M
68K
1500
25/25
100K
.0047
680
T
1M-L
.002
100K
680
100K
.0047
68K
V.
1M
Pull
SW
B
1M-L
470
1500
25/25
100K
D
470
220K
X
½-12AX7
R
1M-L
C
T.S
1M
100K
150K
.02
.1
Y
tremolo
Pulse
½-12AX7
.02
.02
470K
470K
25/25
1500
D
470K
½-12AX7
.002
100K
½-12AU7
.002
1M
½-12AU7
.001
Hammond
4F
Reverb
B
1500
25/25
1M
100K
.001
1500
25/25
470K
1.5K
J. Ft. SW
Pre-amp - Reverb
and tremolo for
G45TR, G90TR, G100TR,
LB100TR
GARNET AMPLIFIER CO. LTD., 611 FERRY ROAD, WINNIPEG 21, CANADA

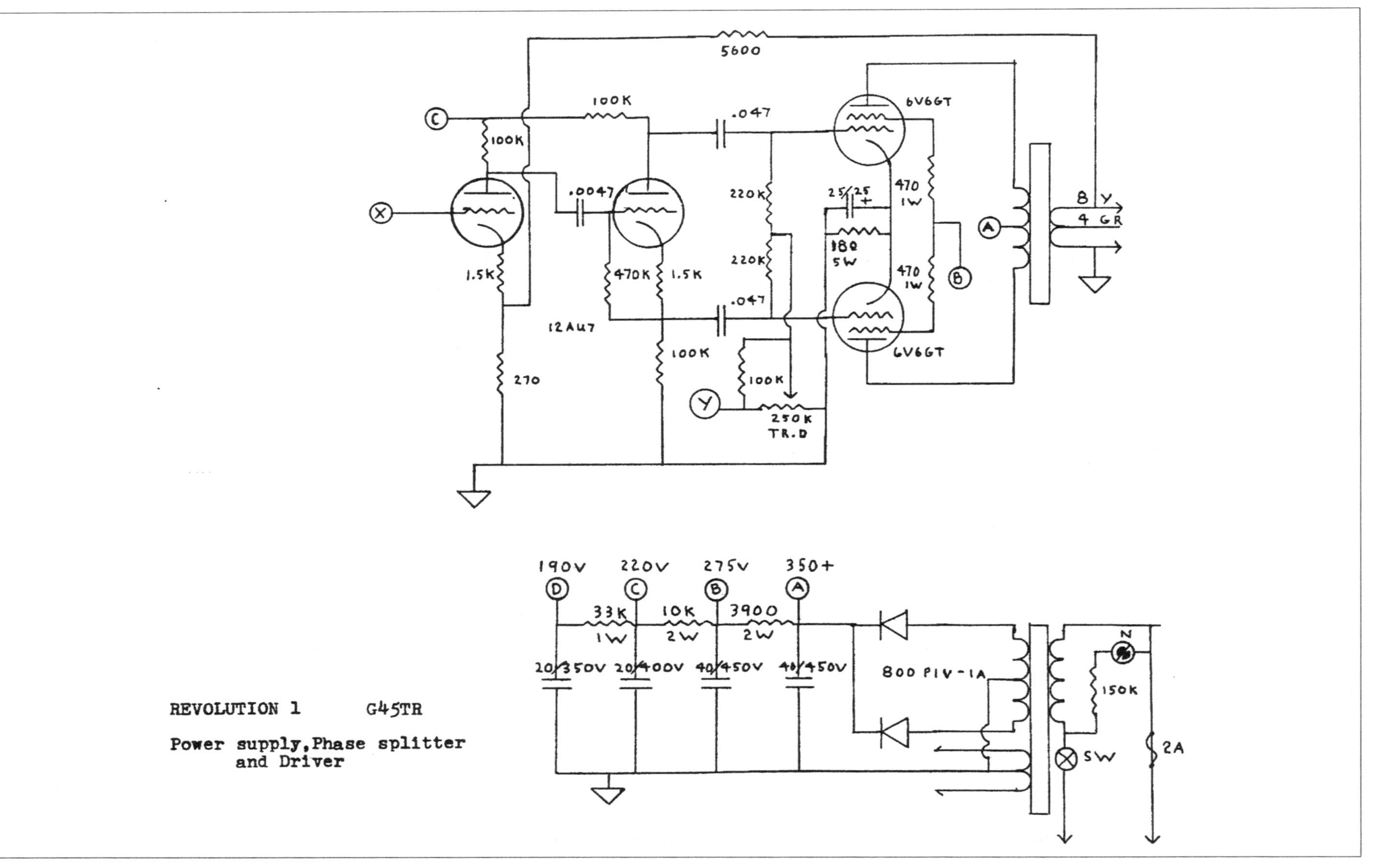
5600
100K
.047
6V6GT
100K
.0047
220K
25/25
470
1W
8
Y
4
GR
180
5W
220K
470
1W
1.5K
470K
1.5K
.047
12AU7
100K
6V6GT
100K
270
250K
TR.D
190V
220V
275V
350+
33K
1W
10K
2W
3900
2W
20/350V
20/400V
40/450V
40/450V
800 PIV-1A
N
150K
SW
2A
REVOLUTION 1
G45TR
Power supply, Phase splitter and Driver

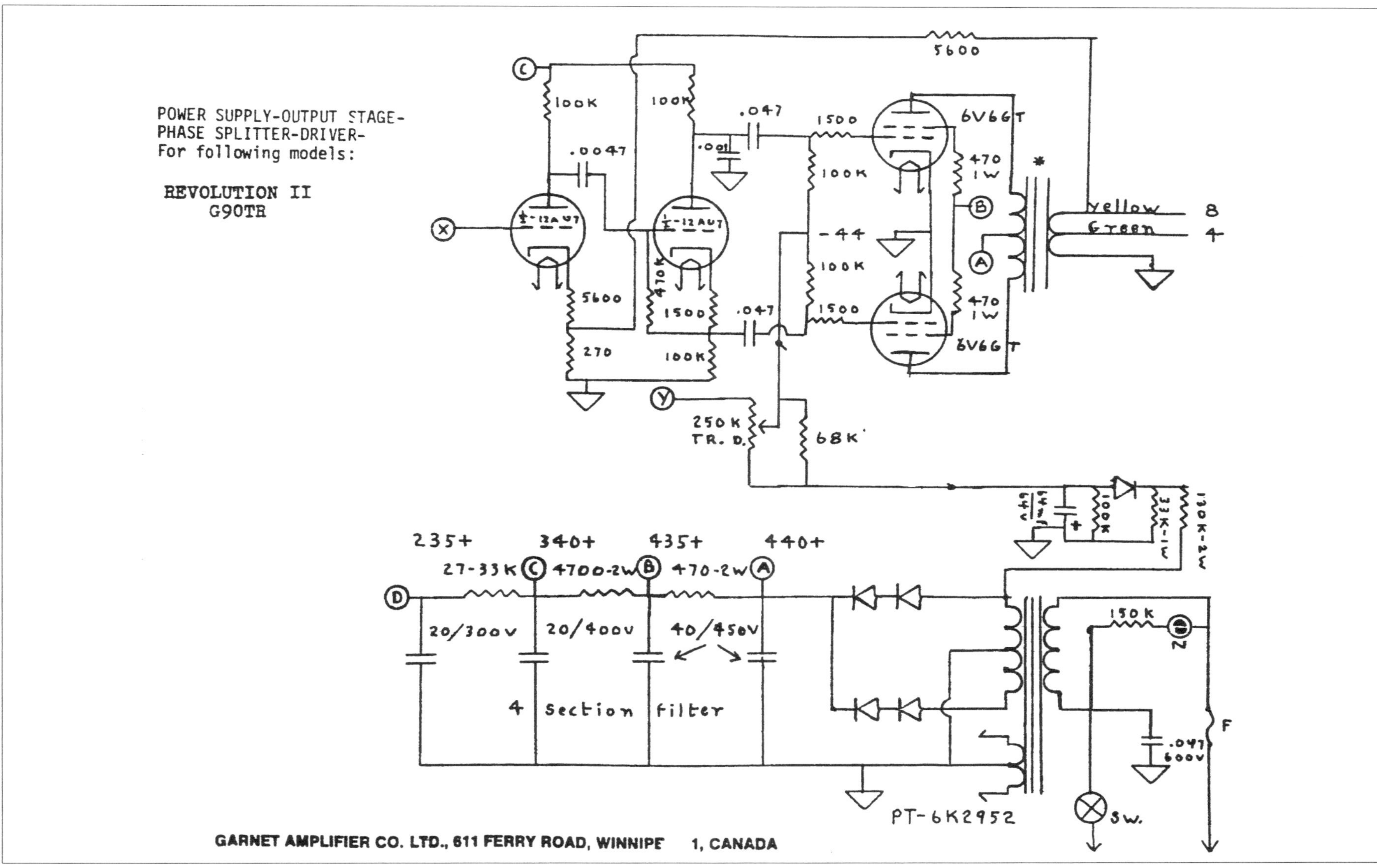
POWER SUPPLY-OUTPUT STAGE-
PHASE SPLITTER-DRIVER-
For following models:
REVOLUTION II
G90TR
100K
100K
.0047
.047
.001
1500
6V6GT
470 1W
½-12AU7
5600
470K
1500
270
100K
.047
-44
100K
1500
5600
yellow
Green
8
4
250K TR. D.
68K
235+
340+
435+
440+
27-33K
4700-2W
470-2W
20/300V
20/400V
40/450V
4 Section filter
33K-1W
130K-2W
150K
.047 600V
F
SW.
PT-6K2952
GARNET AMPLIFIER CO. LTD., 611 FERRY ROAD, WINNIPE 1, CANADA

G100 **PREAMP**

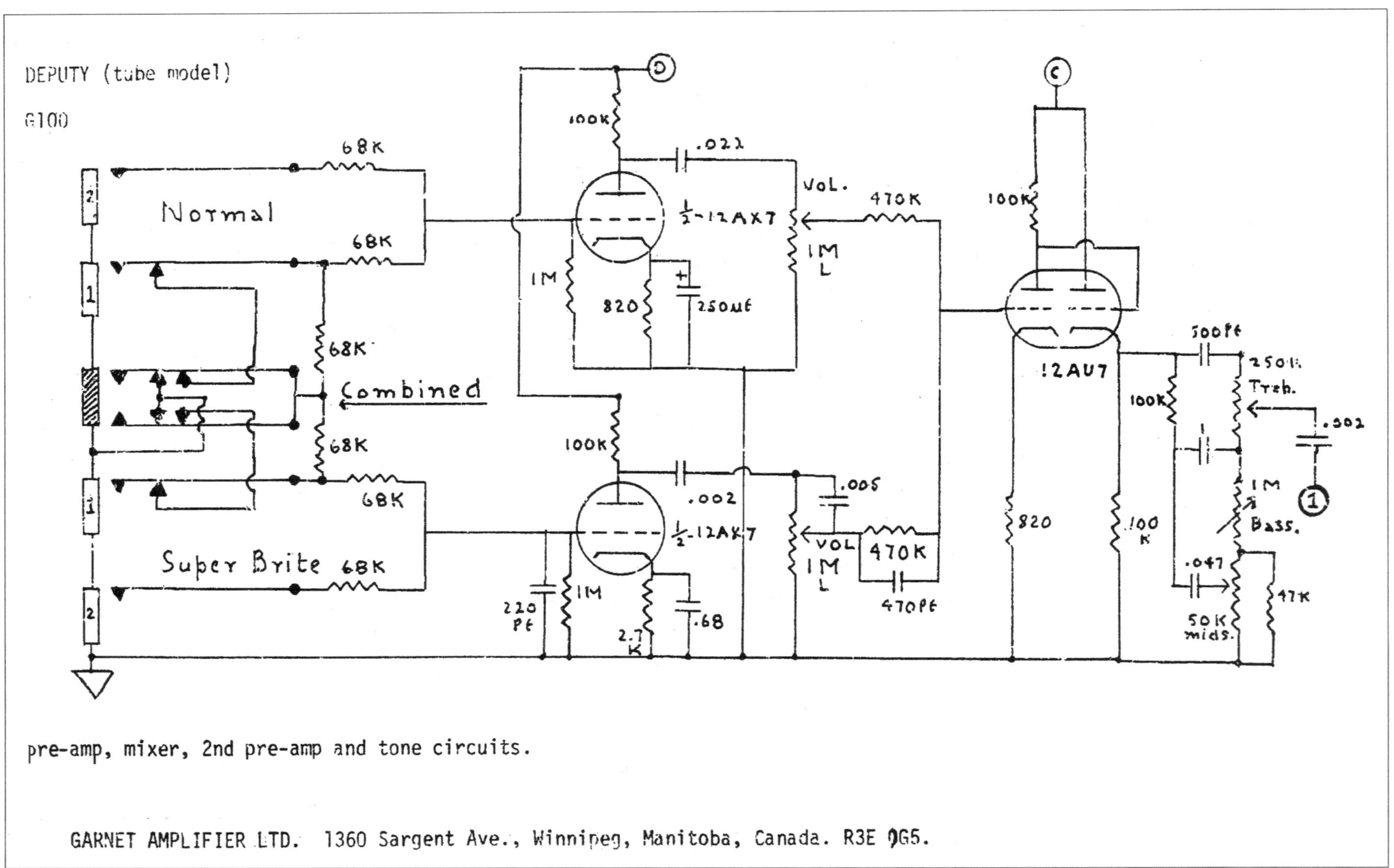

pre-amp, mixer, 2nd pre-amp and tone circuits.

GARNET AMPLIFIER LTD. 1360 Sargent Ave., Winnipeg, Manitoba, Canada. R3E 0G5.

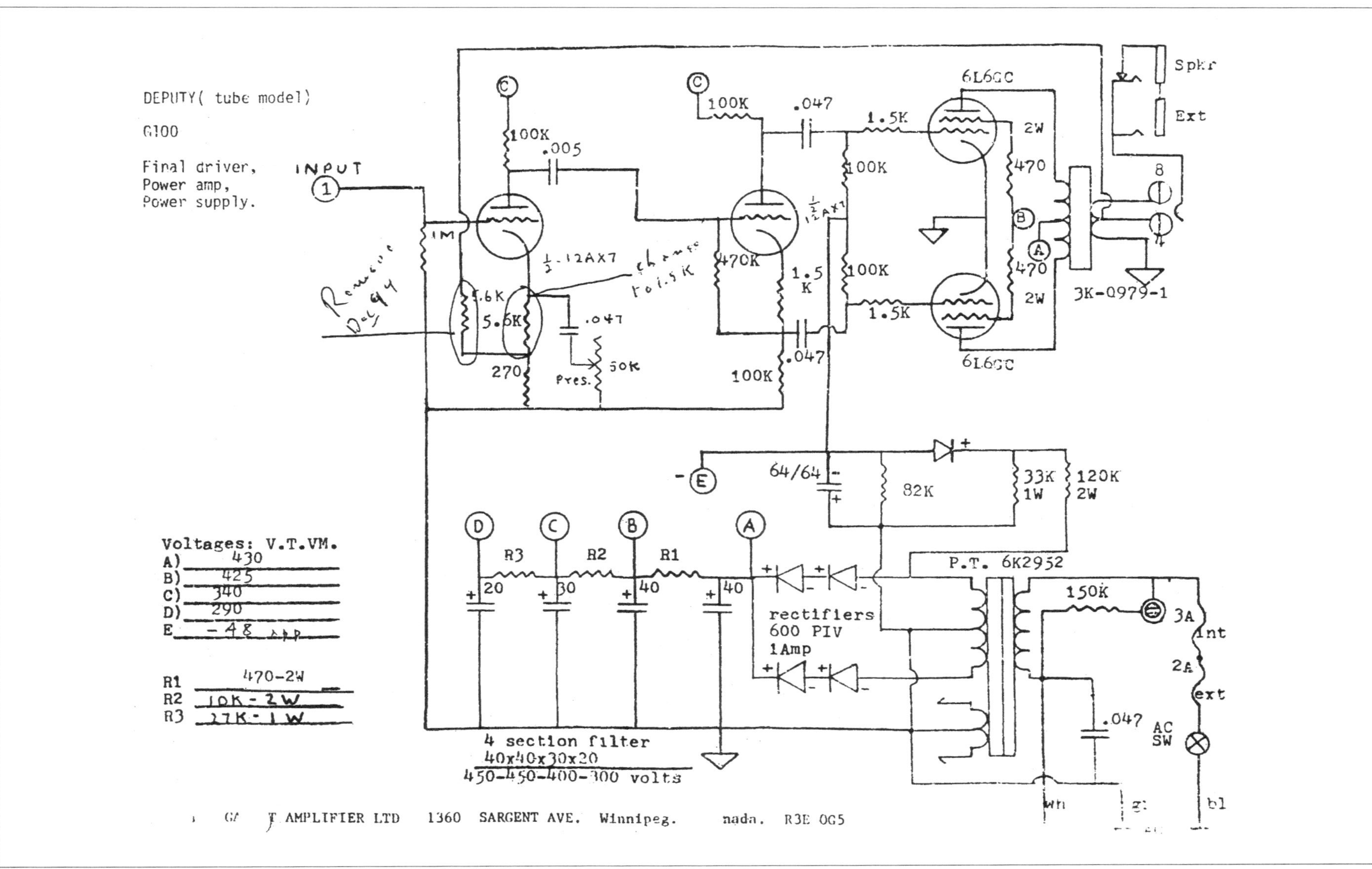

DEPUTY(tube model)
G100
Final driver,
Power amp,
Power supply.
INPUT
1
1M
100K
.005
1/2 12AX7
5.6K
270
.047
50K
Pres.
100K
.047
470K
1.5 K
1/2 12AX7
100K
1.5K
100K
100K
1.5K
6L6GC
6L6GC
470
2W
470
2W
3K-0979-1
Spkr
Ext
8
-E
64/64
82K
33K 1W
120K 2W
Voltages: V.T.VM.
A) 430
B) 425
C) 340
D) 290
E -48
R1 470-2W
R2 10K-2W
R3 27K-1W
R3
R2
R1
20
30
40
40
rectifiers
600 PIV
1Amp
P.T. 6K2952
150K
3A
int
2A
ext
.047
AC SW
wh
bl
4 section filter
40x40x30x20
450-450-400-300 volts
AMPLIFIER LTD 1360 SARGENT AVE. Winnipeg. nada. R3E 0G5

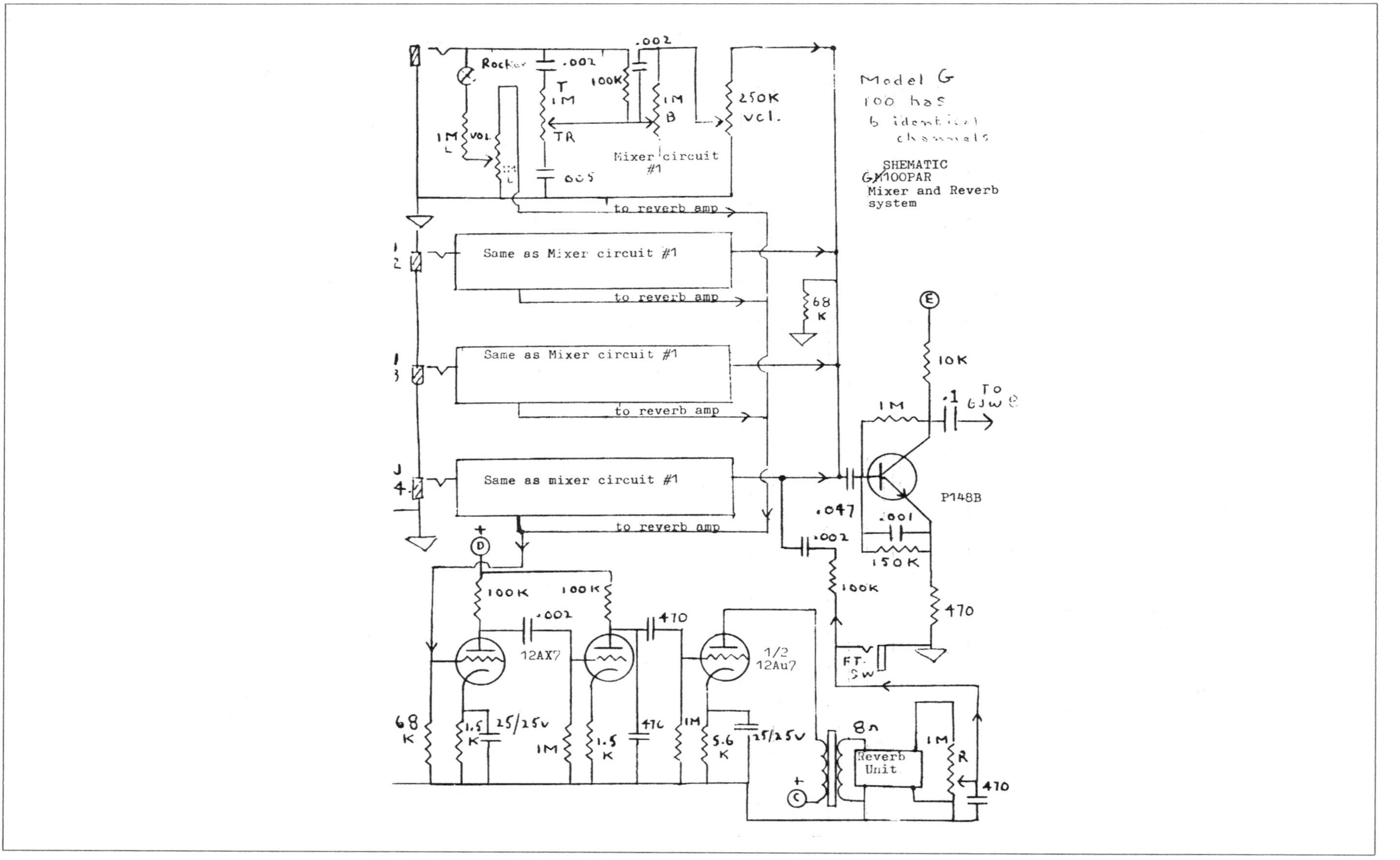
Model G
100 has
6 identical
channels
SHEMATIC
G/M100PAR
Mixer and Reverb
system
Rocker
.002
T
1M
100K
.002
1M
B
250K
vol.
TR
1M
L
VOL
.005
Mixer circuit
#1
to reverb amp
Same as Mixer circuit #1
to reverb amp
Same as Mixer circuit #1
to reverb amp
Same as mixer circuit #1
to reverb amp
68
K
E
10K
.1
To
6JW8
1M
P148B
.047
.001
.002
150K
100K
470
FT.
SW
D
100K
100K
.002
470
12AX7
1/2
12Au7
68
K
1.5
K
25/25v
1M
1.5
K
470
1M
5.6
K
25/25v
8n
Reverb
Unit
1M
R
470
C
J
4

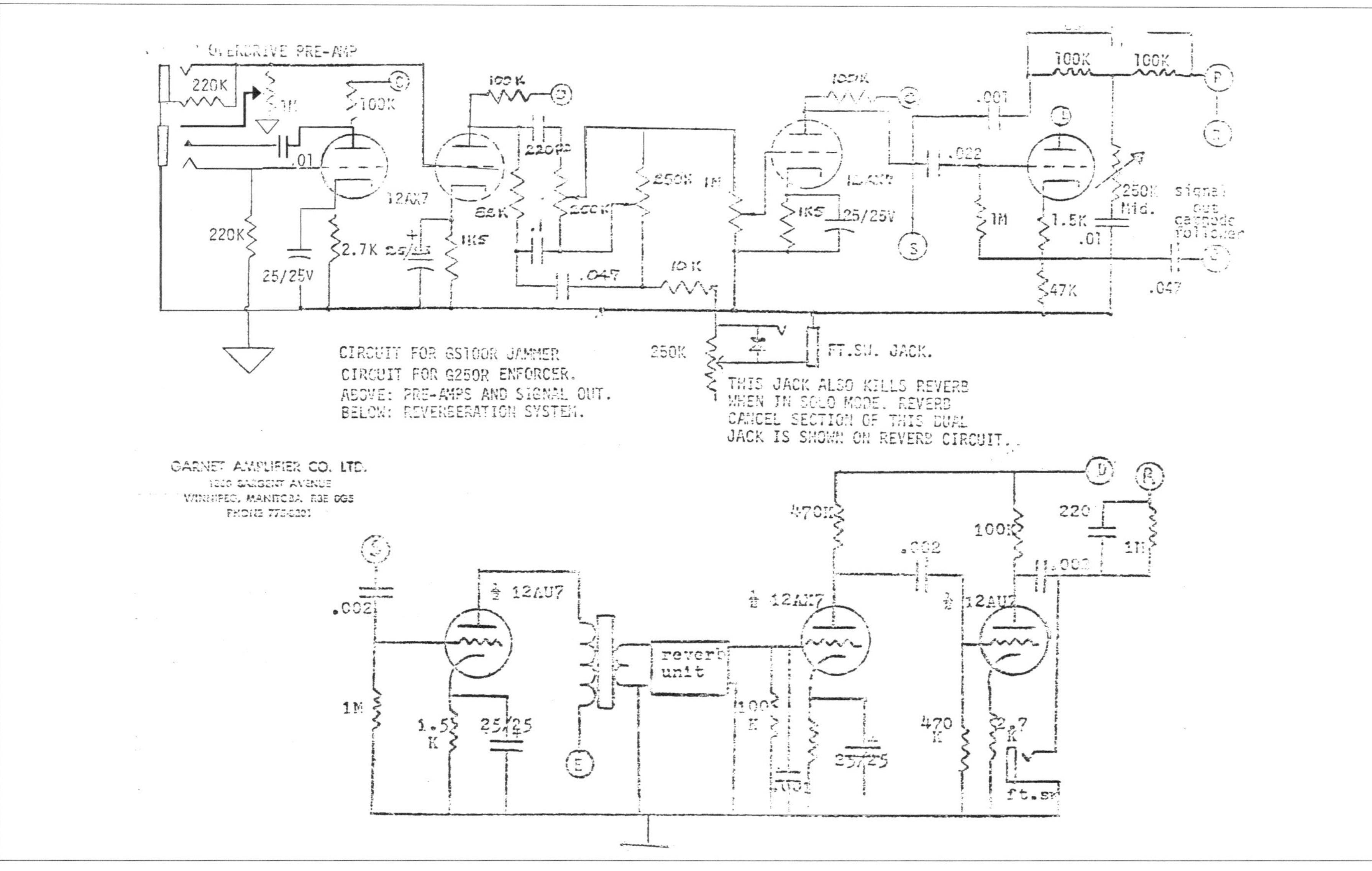
OVERDRIVE PRE-AMP
220K
100K
.01
12AX7
220K
25/25V
2.7K
1K5
100K
220pf
250K
1M
10K
.047
12AX7
1K5
25/25V
.001
.022
100K
100K
1M
1.5K
.01
250K Mid.
signal out cathode follower
47K
.047
CIRCUIT FOR GS100R JAMMER
CIRCUIT FOR G250R ENFORCER.
ABOVE: PRE-AMPS AND SIGNAL OUT.
BELOW: REVERBERATION SYSTEM.
250K
FT.SW. JACK.
THIS JACK ALSO KILLS REVERB
WHEN IN SOLO MODE. REVERB
CANCEL SECTION OF THIS DUAL
JACK IS SHOWN ON REVERB CIRCUIT.
GARNET AMPLIFIER CO. LTD.
.002
½ 12AU7
1M
1.5 K
25/25
reverb unit
470K
½ 12AX7
100 K
25/25
.002
½ 12AU7
100K
220
1M
.002
470 K
2.7 K
ft.sw

GS100R **POWER AMP**

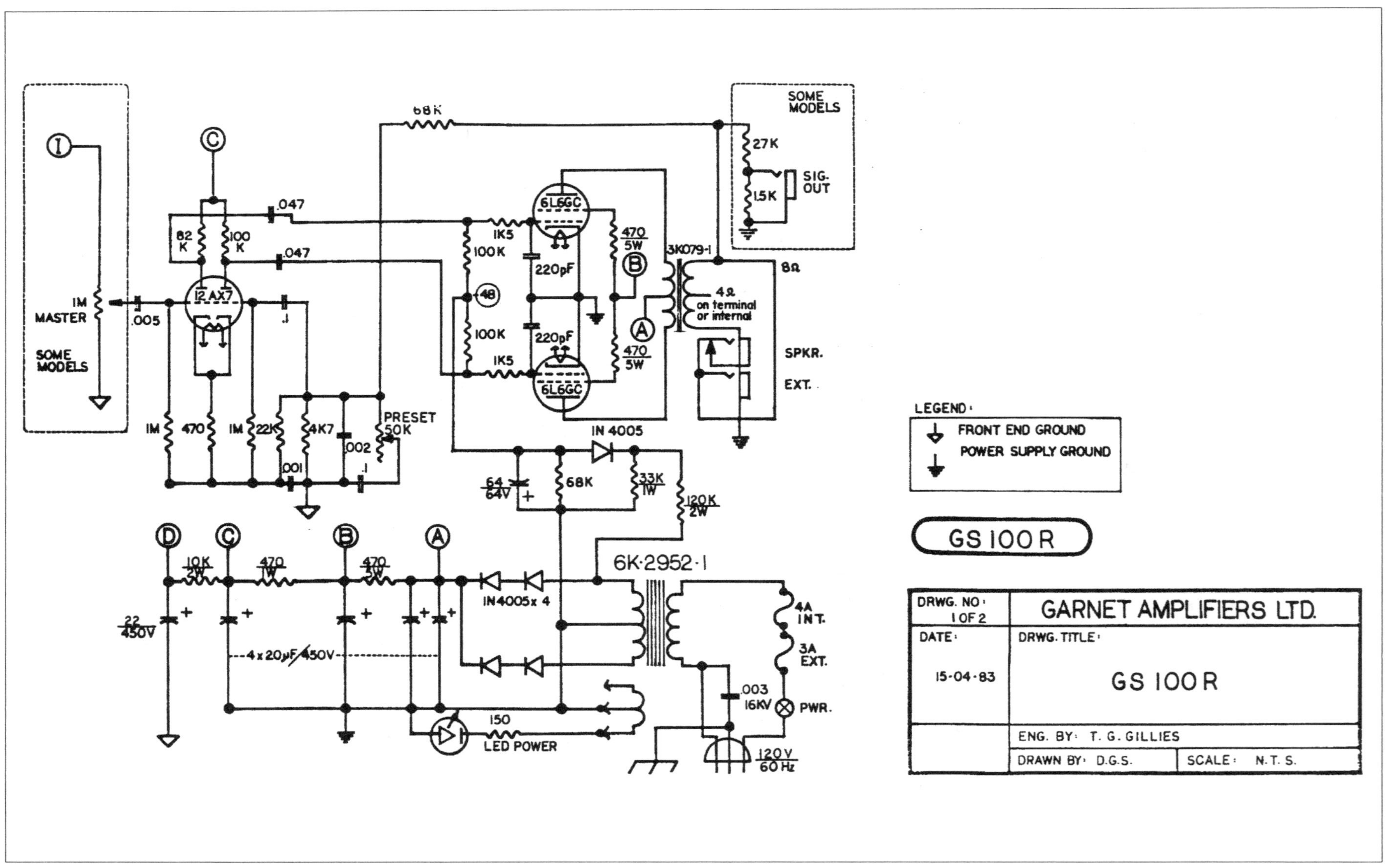

GS100R **PREAMP & REVERB**

Garnet

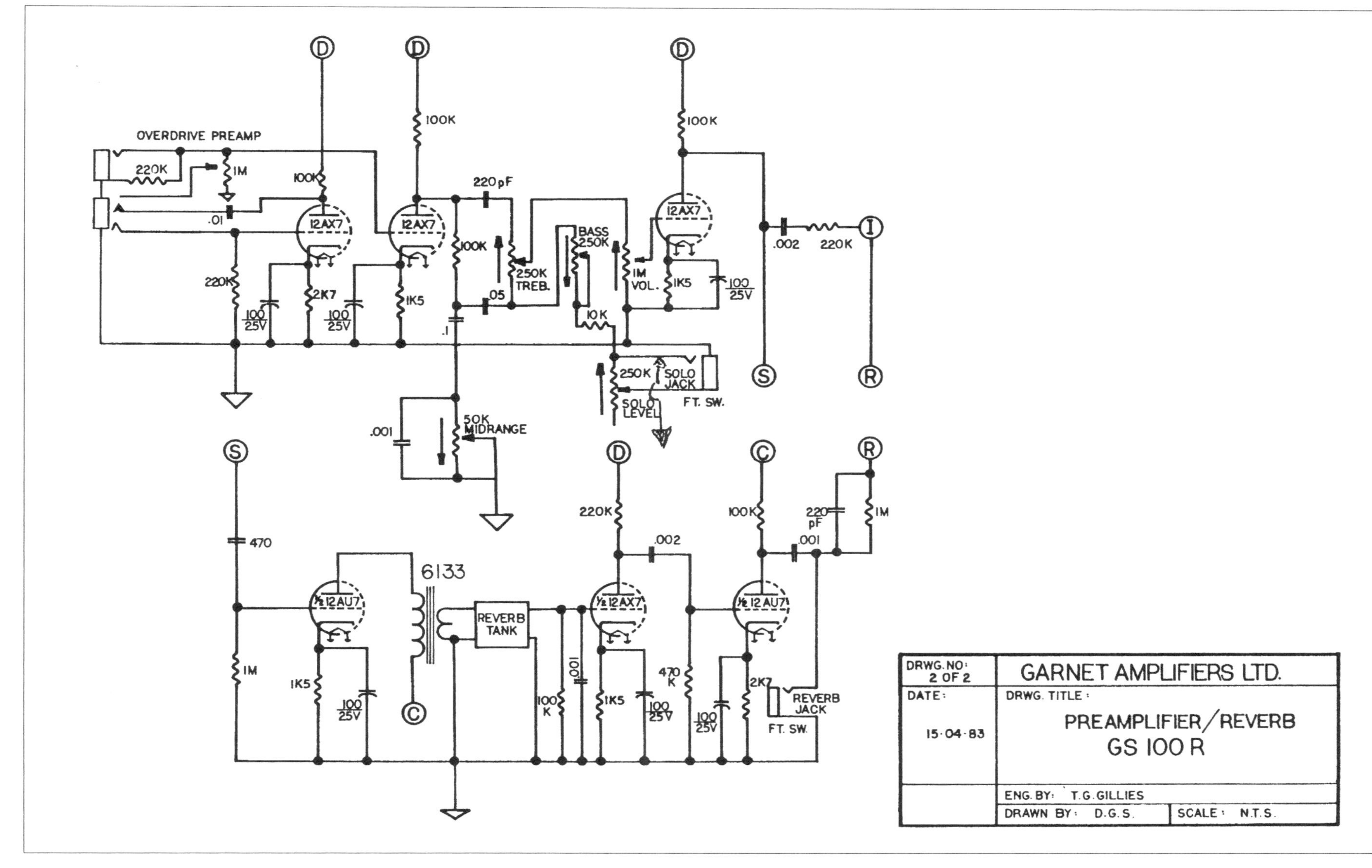

G200R **ENFORCER**

Garnet

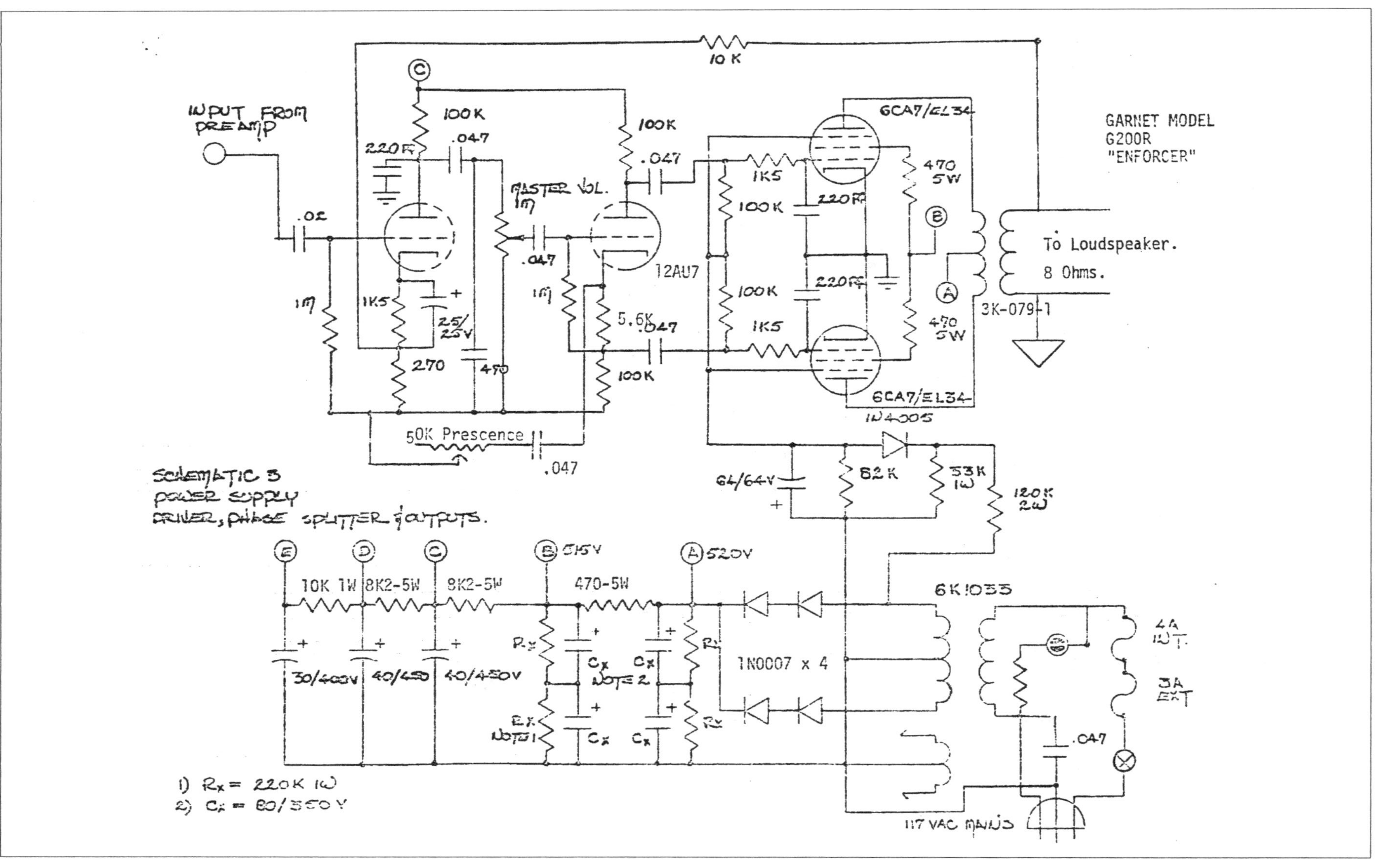

G250FTR & G250D

Garnet

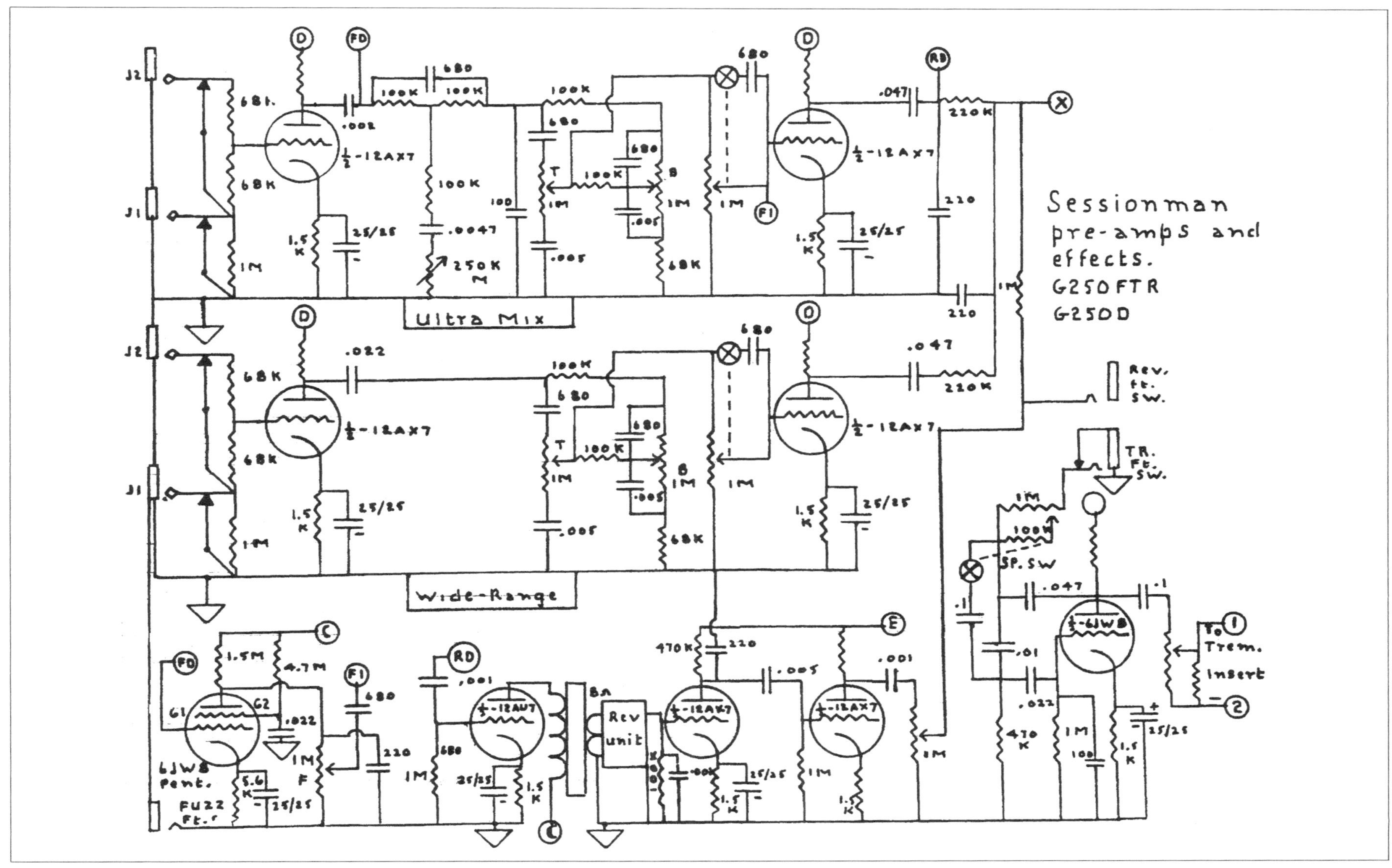

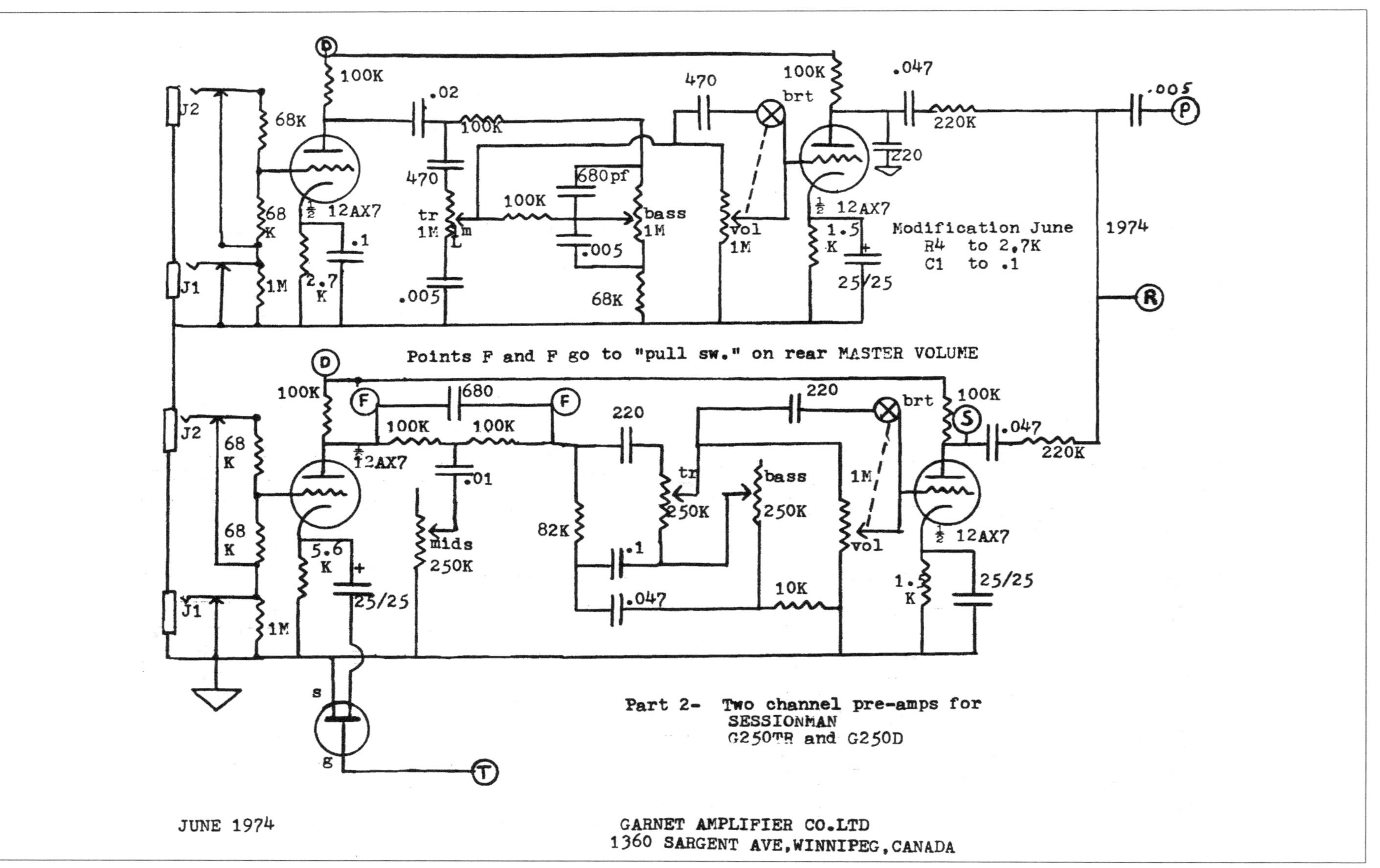
100K
.02
100K
68K
J2
470
tr
1M
100K
680pf
bass
1M
.005
68K
68
K
½ 12AX7
.1
2.7
K
1M
J1
.005
470
brt
vol
1M
100K
.047
220K
.005
P
220
½ 12AX7
1.5
K
25/25
Modification June
R4 to 2,7K
C1 to .1
1974
R
Points F and F go to "pull sw." on rear MASTER VOLUME
D
100K
F
680
F
100K
100K
12AX7
.01
mids
250K
68
K
68
K
J2
J1
1M
5.6
K
25/25
220
82K
tr
250K
.1
.047
220
bass
250K
10K
brt
1M
vol
100K
S
.047
220K
½ 12AX7
1.5
K
25/25
s
g
T
Part 2- Two channel pre-amps for
SESSIONMAN
G250TR and G250D
JUNE 1974
GARNET AMPLIFIER CO.LTD
1360 SARGENT AVE,WINNIPEG,CANADA

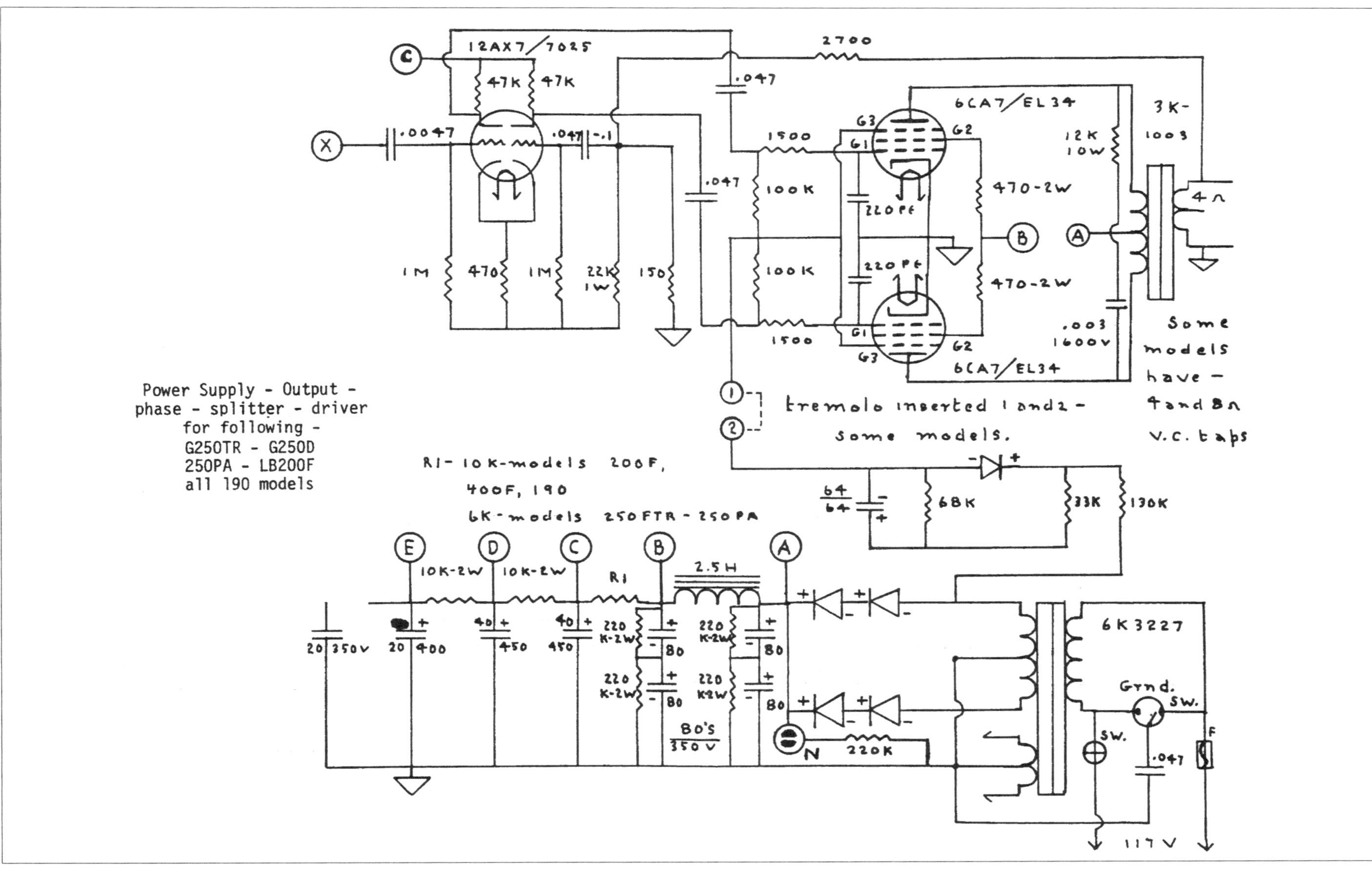
12AX7/7025
47K
47K
.0047
.047
-.1
1M
470
1M
22K 1W
150
2700
.047
.047
1500
100K
100K
1500
220 PF
220 PF
G1
G2
G3
6CA7/EL34
6CA7/EL34
470-2W
470-2W
12K 10W
3K-100Ʃ
4Ω
.003 1600V
Some models have - 4 and 8Ω V.C. taps
tremolo inserted 1 and 2 - some models.
Power Supply - Output - phase - splitter - driver for following - G250TR - G250D 250PA - LB200F all 190 models
R1- 10K-models 200F, 400F, 190
6K-models 250FTR-250PA
10K-2W
10K-2W
R1
2.5H
20 350V
20 400
40 450
40 450
220 K-2W
80
80's 350V
64 64
68K
33K
130K
220K
6K3227
Grnd. SW.
SW.
.047
F
117V

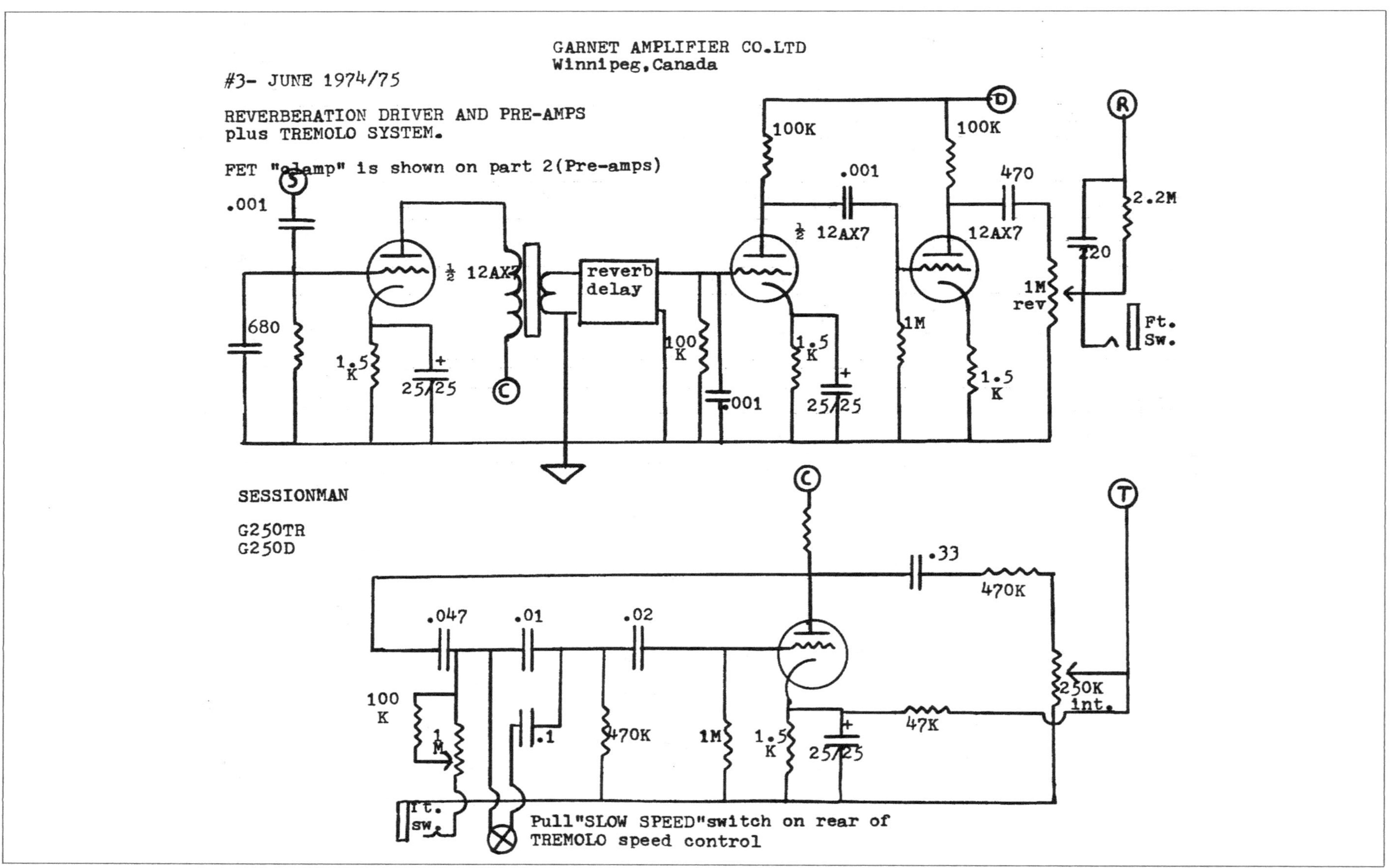
GARNET AMPLIFIER CO.LTD
Winnipeg, Canada
#3- JUNE 1974/75
REVERBERATION DRIVER AND PRE-AMPS
plus TREMOLO SYSTEM.
FET "clamp" is shown on part 2(Pre-amps)
.001
680
1.5 K
25/25
½ 12AX7
reverb
delay
100 K
.001
100K
.001
½ 12AX7
1.5 K
25/25
1M
100K
470
12AX7
1.5 K
1M rev
220
2.2M
Ft. Sw.
SESSIONMAN
G250TR
G250D
.33
470K
.047
.01
.02
100 K
1 M
.1
470K
1M
1.5 K
25/25
47K
250K int.
Ft. sw.
Pull"SLOW SPEED"switch on rear of
TREMOLO speed control

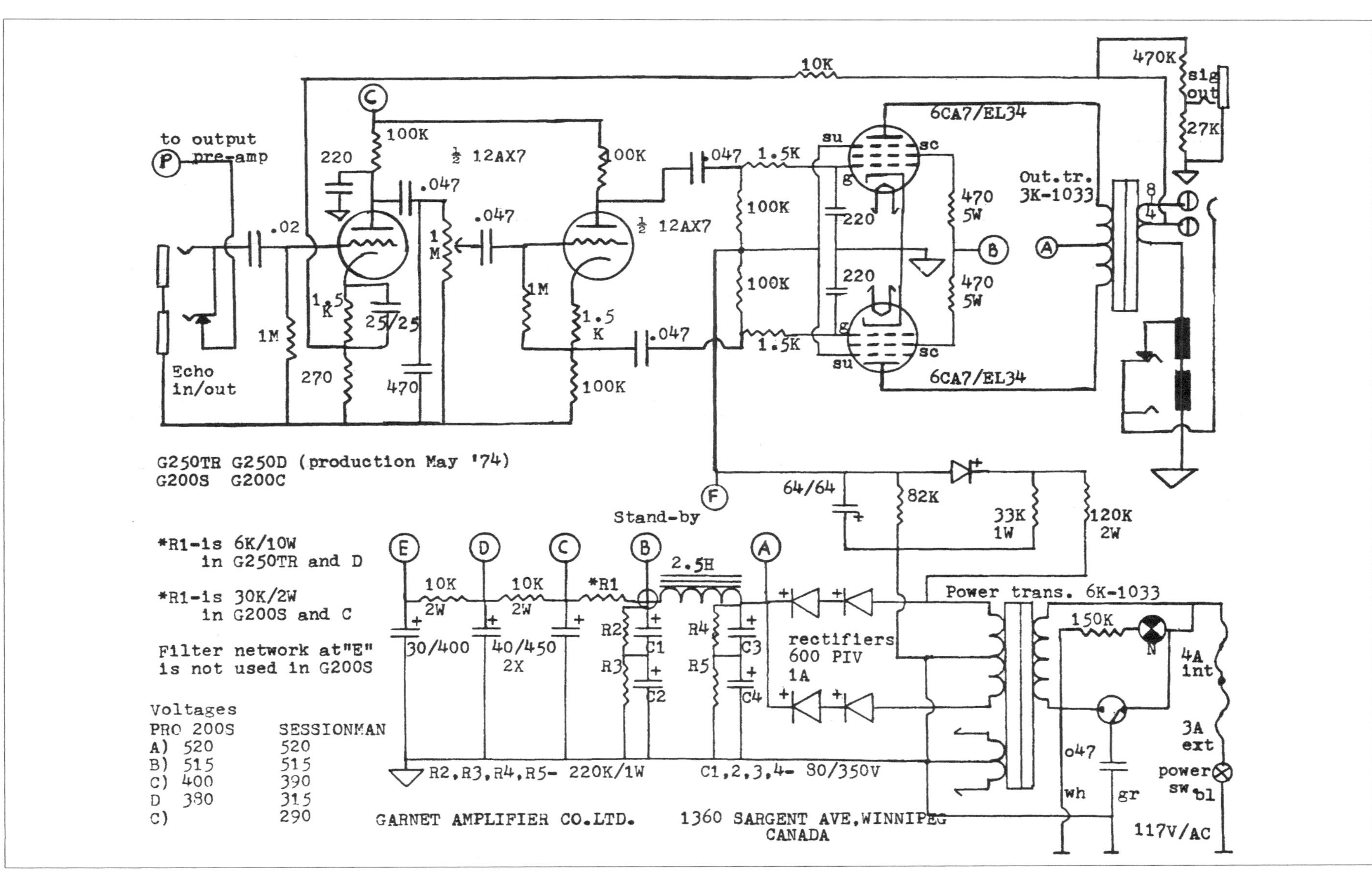

to output pre-amp
Echo in/out
.02
1M
270
220
100K
.047
1.5 K
25/25
470
1 M
½ 12AX7
.047
1M
1.5 K
100K
100K
.047
1.5K
100K
100K
.047
1.5K
220
220
su
sc
g
6CA7/EL34
6CA7/EL34
470 5W
470 5W
Out.tr. 3K-1033
8
4
10K
470K
sig out
27K
G250TR G250D (production May '74)
G200S G200C
Stand-by
64/64
82K
33K 1W
120K 2W
*R1-is 6K/10W
in G250TR and D
*R1-is 30K/2W
in G200S and C
Filter network at"E"
is not used in G200S
10K 2W
10K 2W
*R1
2.5H
30/400
40/450 2X
R2
R3
C1
C2
R4
R5
C3
C4
rectifiers
600 PIV
1A
Power trans. 6K-1033
150K
N
4A int
3A ext
power sw bl
o47
wh
gr
117V/AC
Voltages
PRO 200S SESSIONMAN
A) 520 520
B) 515 515
C) 400 390
D 380 315
C) 290
R2,R3,R4,R5- 220K/1W
C1,2,3,4- 80/350V
GARNET AMPLIFIER CO.LTD. 1360 SARGENT AVE,WINNIPEG
CANADA

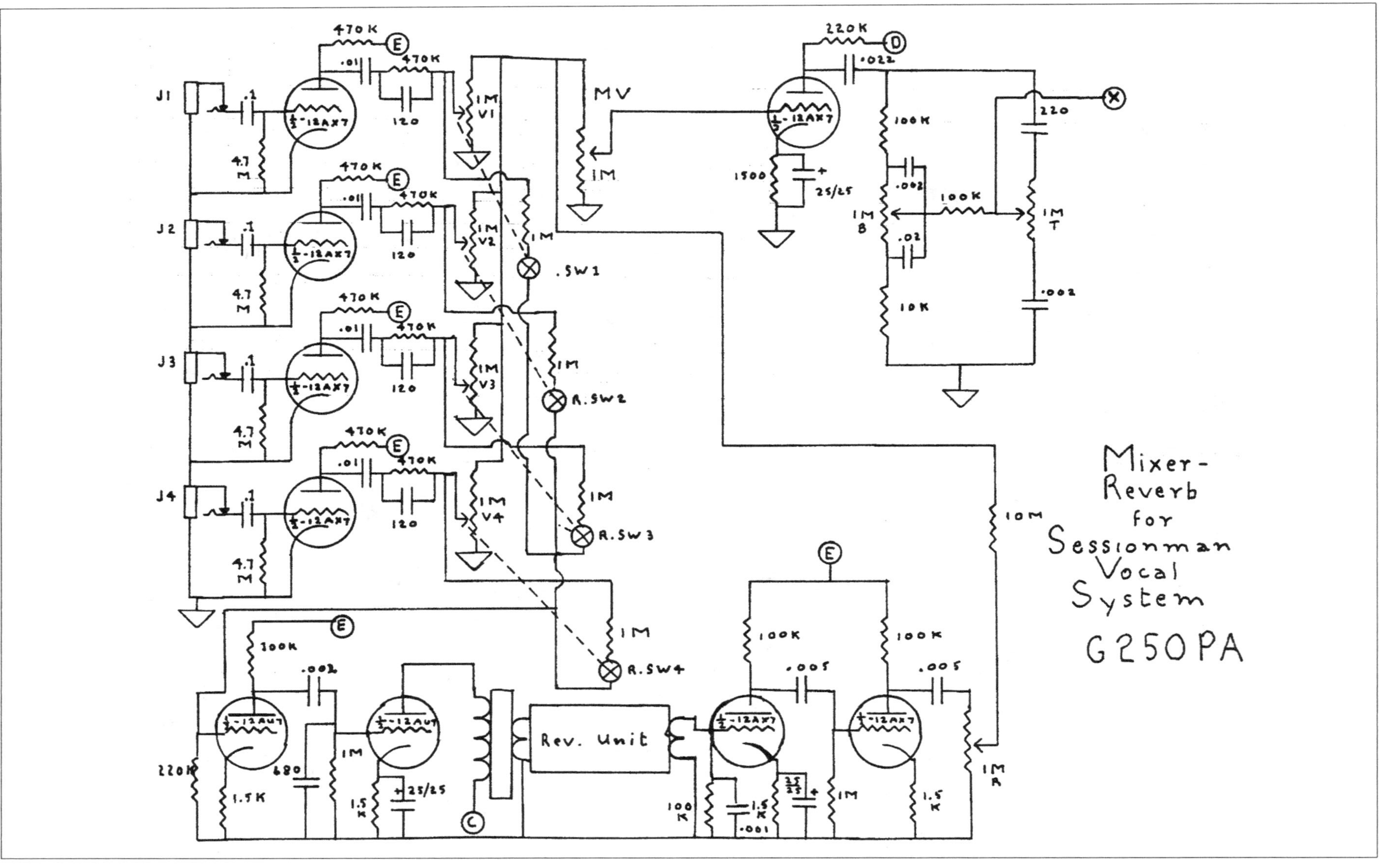
Mixer-
Reverb
for
Sessionman
Vocal
System
G250PA
J1
J2
J3
J4
MV
Rev. Unit
.SW1
R.SW2
R.SW3
R.SW4

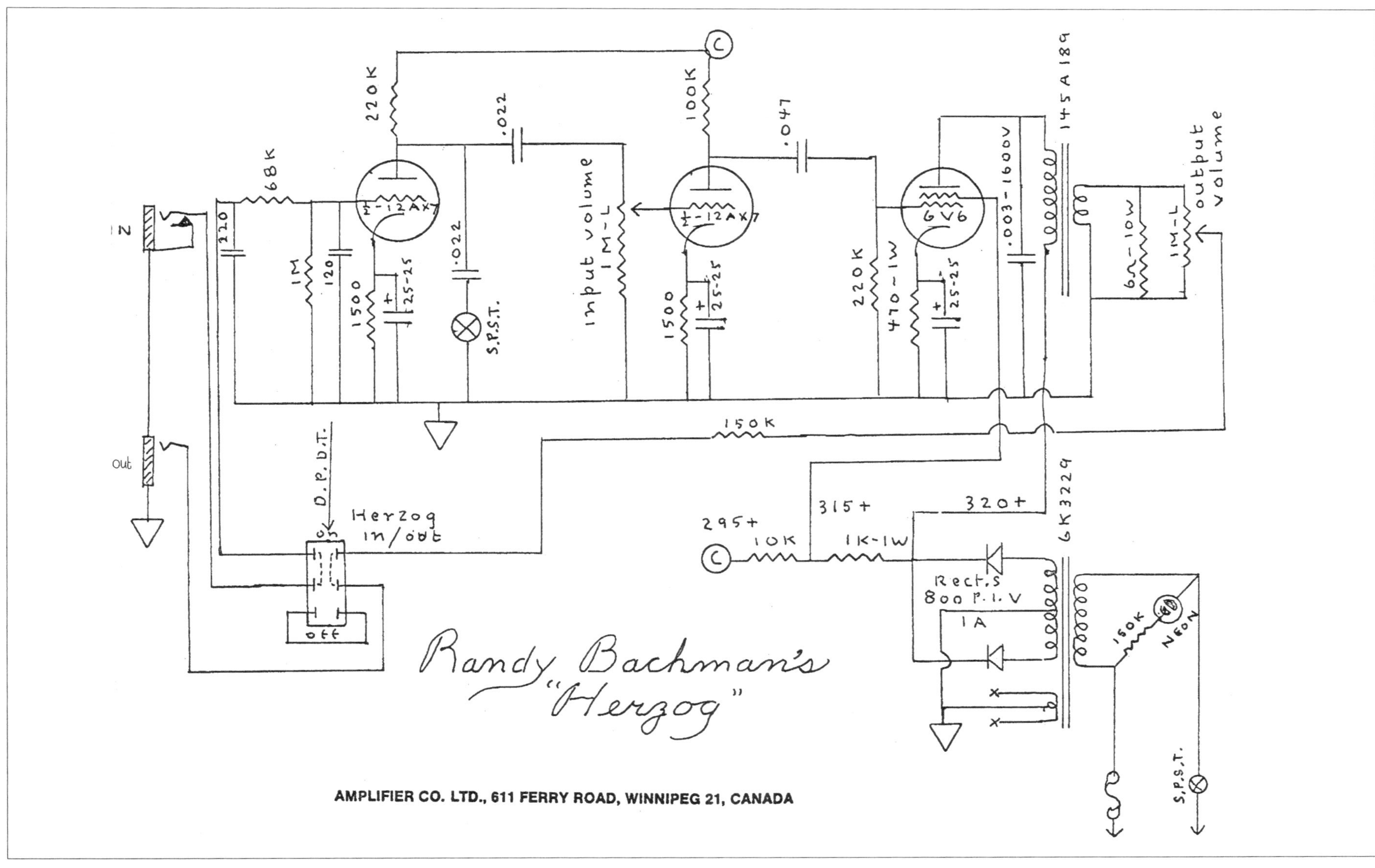
IN
out
220
68K
1M
120
1500
25-25
½-12AX7
220K
.022
S.P.S.T.
.022
input volume
1M-L
100K
1500
25-25
½-12AX7
.047
220K
470-1W
25-25
6V6
.003-1600V
145A189
6Ω-10W
1M-L
output volume
150K
D.P.D.T.
Herzog in/out
on
off
295+
10K
315+
1K-1W
320+
6K3229
Rect.s 800 P.I.V
1A
150K
NEON
S.P.S.T.
Randy Bachman's "Herzog"
AMPLIFIER CO. LTD., 611 FERRY ROAD, WINNIPEG 21, CANADA

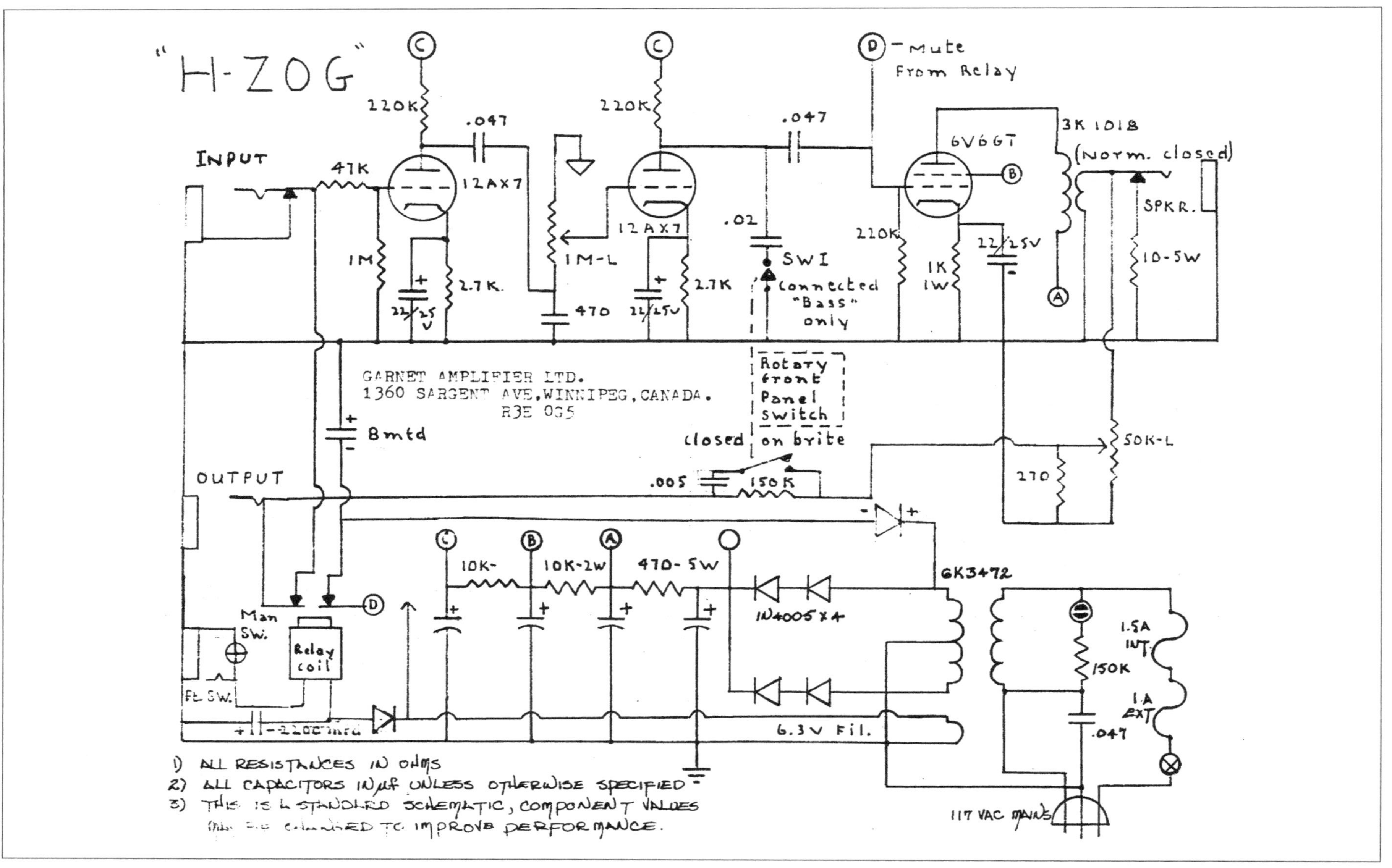
"H-ZOG"
INPUT
OUTPUT
220K
.047
12AX7
47K
1M
2.7K
22/25 V
1M-L
470
22/25V
.02
SW1
Connected "Bass" only
Rotary front Panel switch
closed
on brite
.005
150K
8mfd
6V6GT
3K 1018
(Norm. closed)
SPKR.
10-5W
22/25V
1K 1W
50K-L
270
D - Mute From Relay
GARNET AMPLIFIER LTD.
1360 SARGENT AVE, WINNIPEG, CANADA.
R3E 0G5
10K-
10K-2W
470-5W
1N4005 X 4
GK3472
6.3V Fil.
Man Sw.
Relay coil
Ft. Sw.
1.5A INT.
1A EXT
150K
.047
117 VAC MAINS
1) ALL RESISTANCES IN OHMS
2) ALL CAPACITORS IN µF UNLESS OTHERWISE SPECIFIED
3) THIS IS A STANDARD SCHEMATIC, COMPONENT VALUES
CHANGED TO IMPROVE PERFORMANCE.

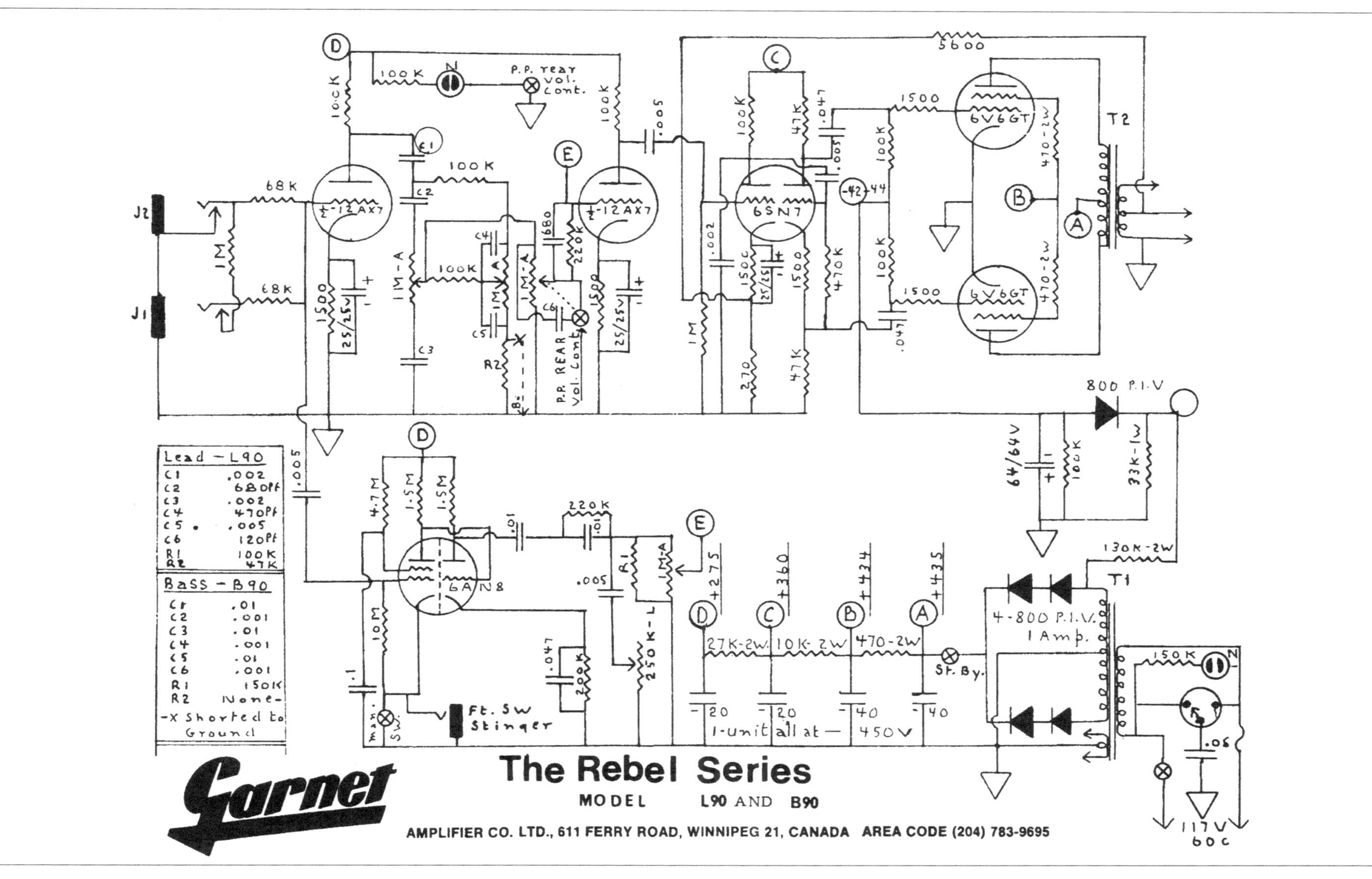
Garnet
The Rebel Series
MODEL L90 AND B90
AMPLIFIER CO. LTD., 611 FERRY ROAD, WINNIPEG 21, CANADA AREA CODE (204) 783-9695
Lead – L90
C1 .002
C2 680Pf
C3 .002
C4 470Pf
C5 .005
C6 120Pf
R1 100K
R2 47K
Bass – B90
C1 .01
C2 .001
C3 .01
C4 .001
C5 .01
C6 .001
R1 150K
R2 None-
-X Shorted to Ground
½-12AX7
6AN8
6SN7
6V6GT
T1
T2
J1
J2
P.P. rear vol. cont.
P.P. REAR Vol. Cont.
Ft. Sw Stinger
+275
+360
+434
+435
27K-2W
10K-2W
470-2W
St. By.
1-unit all at — 450V
4-800 P.I.V. 1 Amp.
800 P.I.V
64/64V
33K-1W
130K-2W
117V 60c

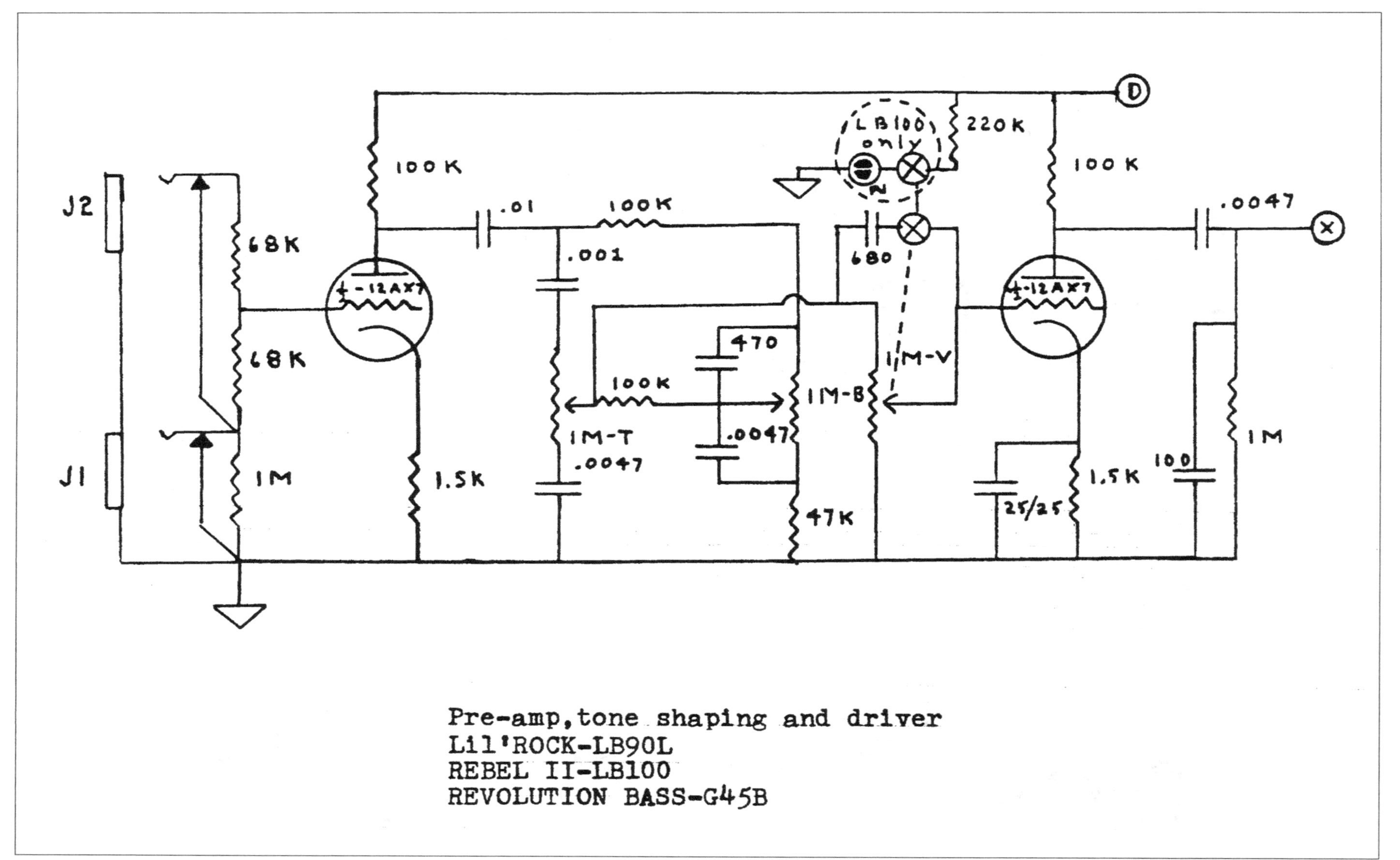

Pre-amp, tone shaping and driver
Lil'ROCK-LB90L
REBEL II-LB100
REVOLUTION BASS-G45B

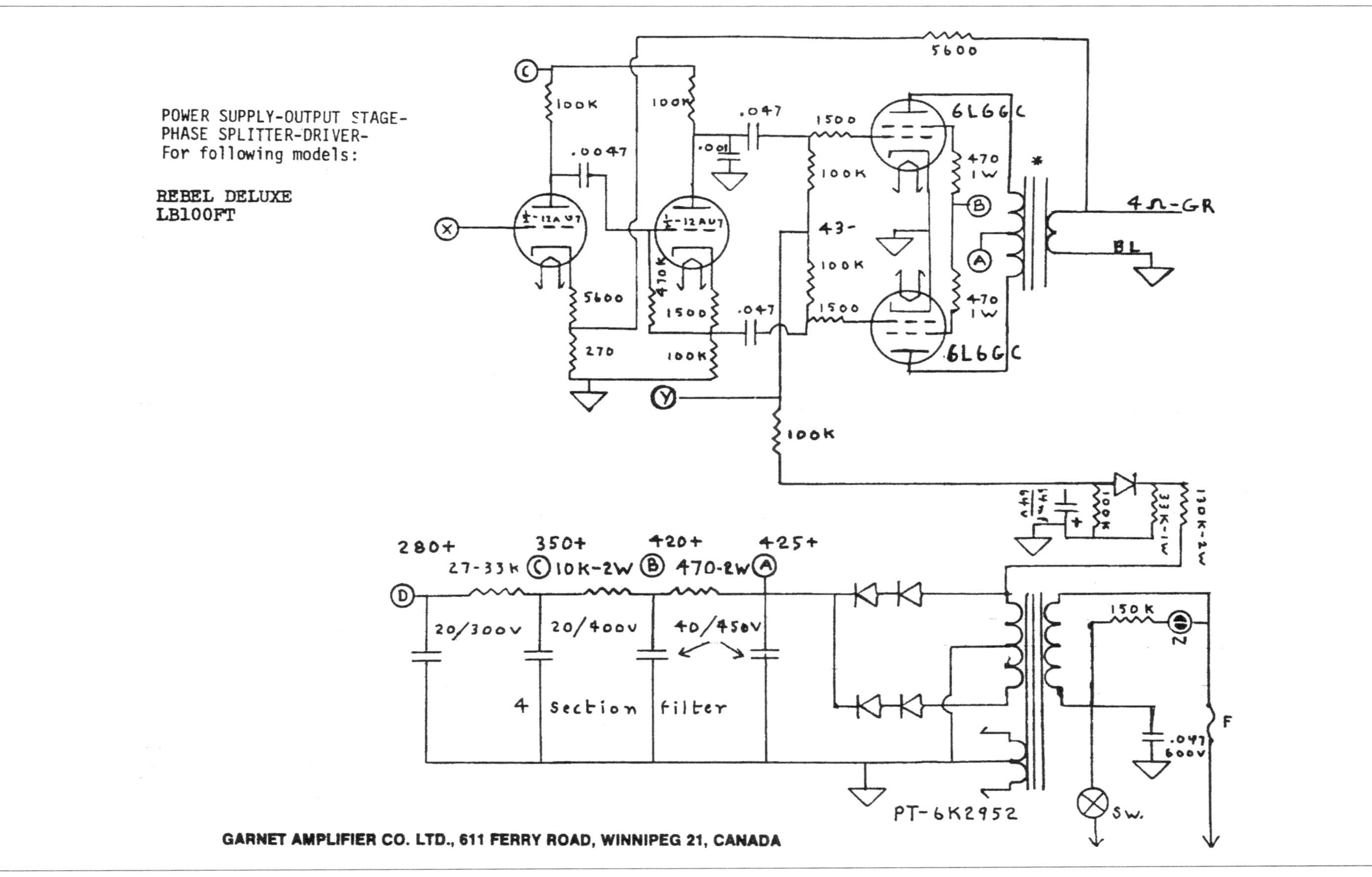
POWER SUPPLY-OUTPUT STAGE-
PHASE SPLITTER-DRIVER-
For following models:
REBEL DELUXE
LB100FT
5600
100K
100K
.047
.0047
.001
1500
6L6GC
470
1W
½-12AU7
½-12AU7
100K
43-
100K
4Ω-GR
BL
5600
470K
1500
.047
1500
270
100K
100K
280+
350+
420+
425+
27-33K
10K-2W
470-2W
20/300V
20/400V
40/450V
4
section
filter
150K
130K-2W
33K-1W
.047
600V
F
PT-6K2952
SW.
GARNET AMPLIFIER CO. LTD., 611 FERRY ROAD, WINNIPEG 21, CANADA

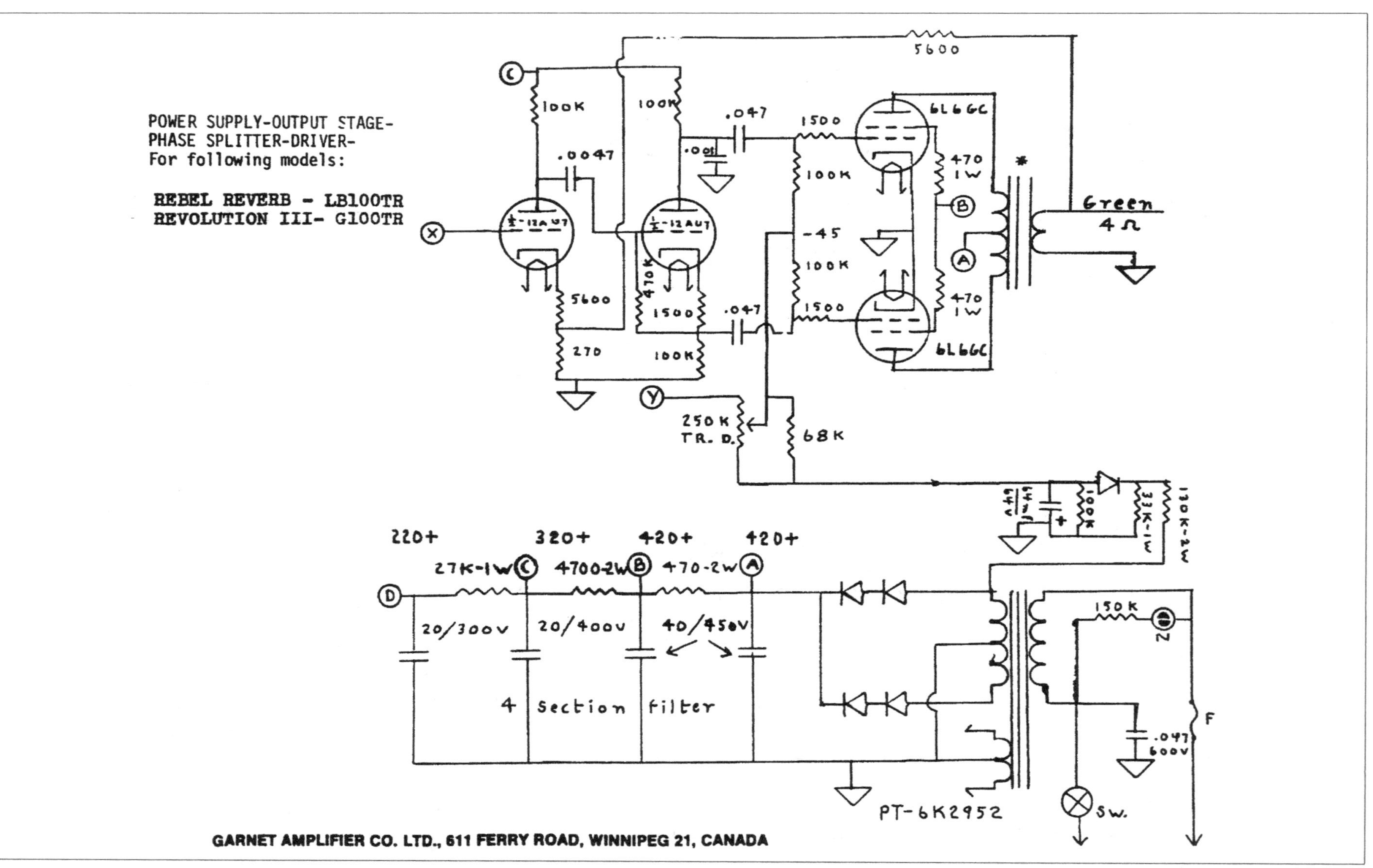
POWER SUPPLY-OUTPUT STAGE-
PHASE SPLITTER-DRIVER-
For following models:
REBEL REVERB - LB100TR
REVOLUTION III- G100TR
100K
100K
.0047
.047
.001
1500
100K
-45
100K
1500
6L6GC
6L6GC
470 1W
470 1W
Green 4Ω
5600
470K
1500
5600
270
100K
.047
250K TR. D.
68K
130K-2W
33K-1W
100K
64/64
220+
320+
420+
420+
27K-1W
4700-2W
470-2W
20/300V
20/400V
40/450V
4 Section filter
150K
.047 600V
F
SW.
PT-6K2952
GARNET AMPLIFIER CO. LTD., 611 FERRY ROAD, WINNIPEG 21, CANADA

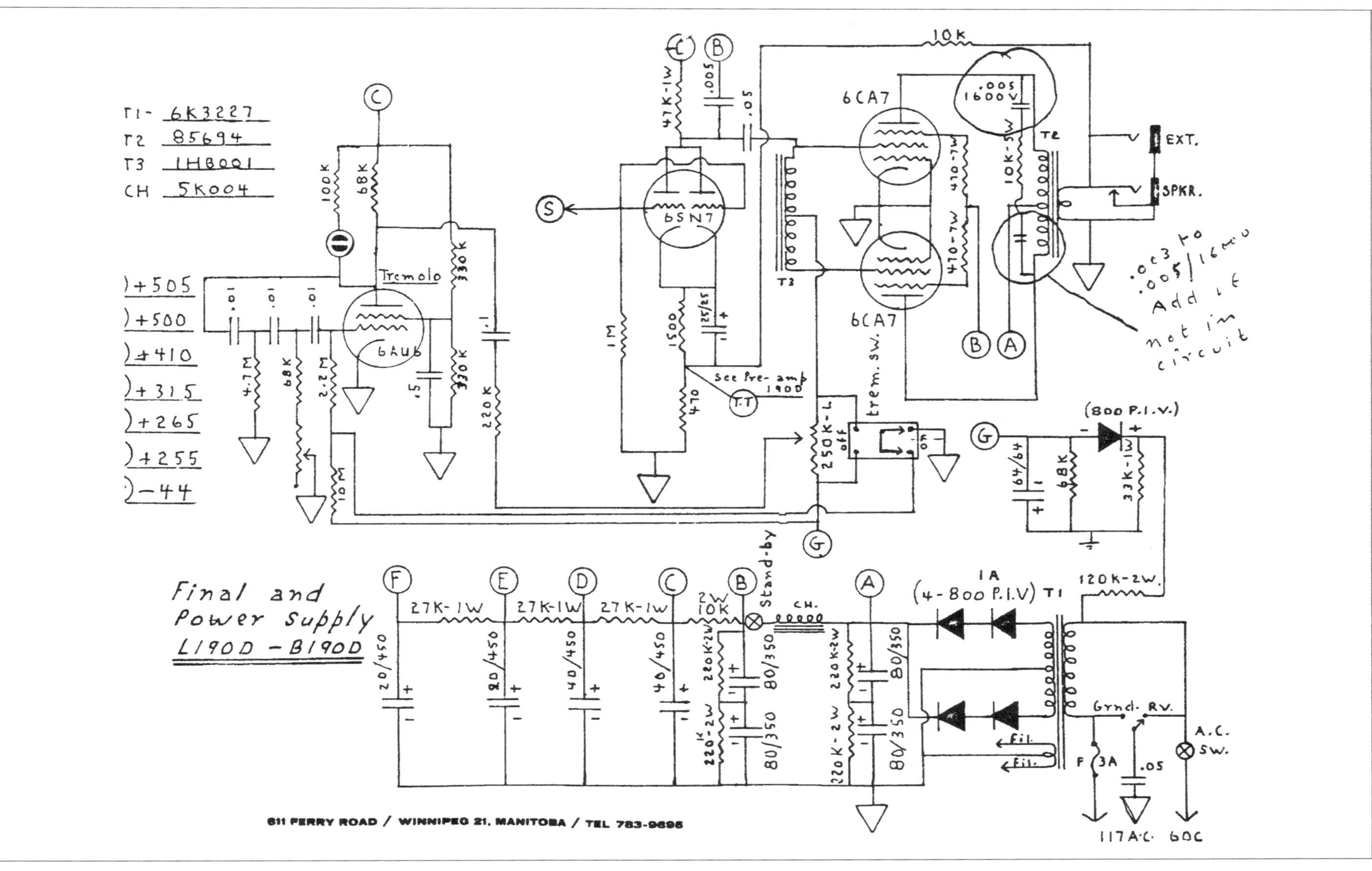
T1- 6K3227
T2 85694
T3 1HB001
CH 5K004
+505
+500
+410
+315
+265
+255
-44
Tremolo
6AU6
6SN7
6CA7
6CA7
.005 1600V
10K-5W
T2
EXT.
SPKR.
.003 to .005/1600 Add if not in circuit
See Pre-amp 190D
trem. sw.
250K-L
(800 P.I.V.)
64/64
68K
33K-1W
Final and Power Supply L190D - B190D
27K-1W
20/450
40/450
2W 10K
Stand-by
80/350
220K-2W
CH.
1A
(4-800 P.I.V.) T1
120K-2W.
Grnd. RV.
A.C. SW.
F 3A
.05
117 A.C. 60C
Fil.
611 FERRY ROAD / WINNIPEG 21, MANITOBA / TEL 783-9695

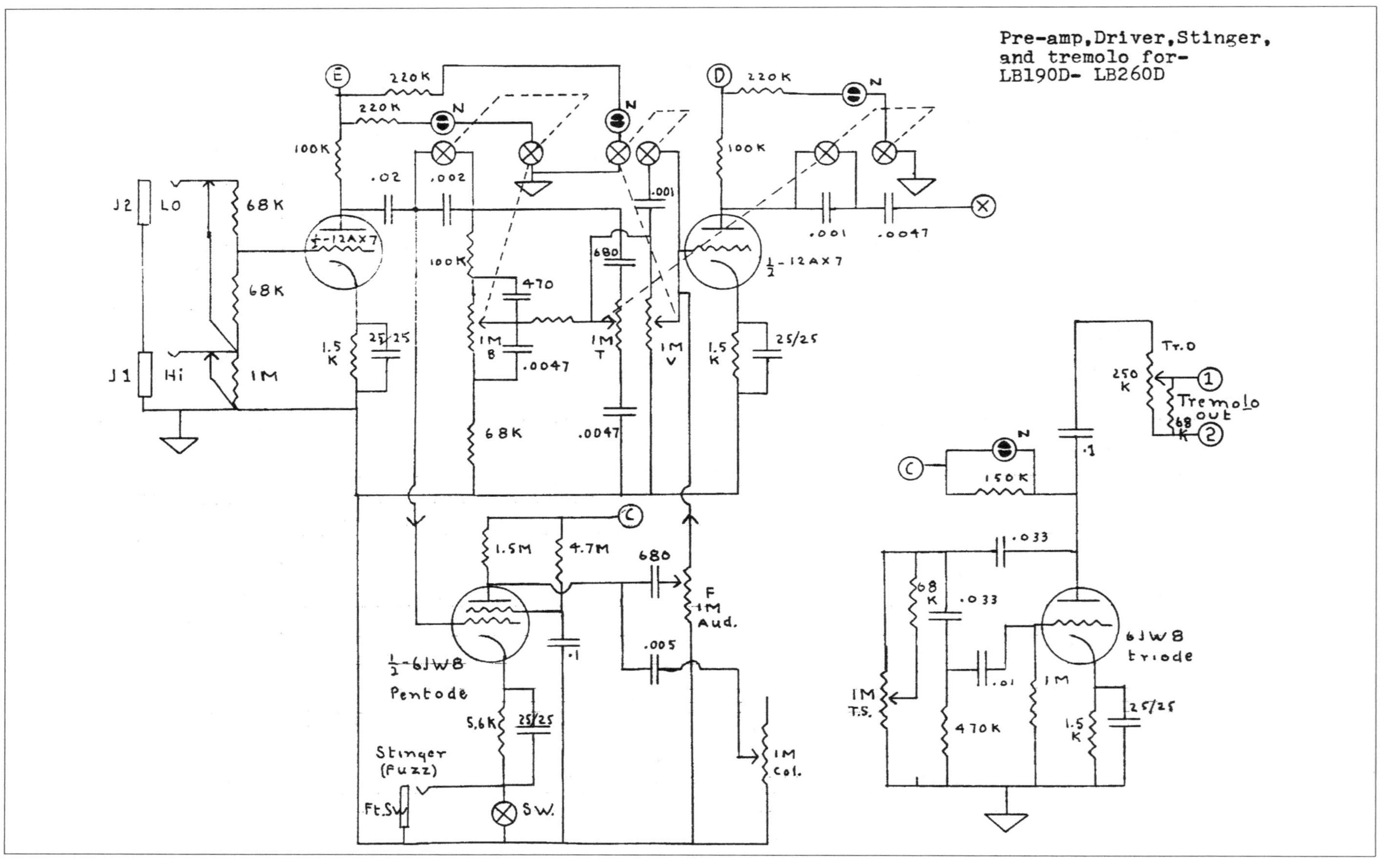
Pre-amp, Driver, Stinger, and tremolo for- LB190D- LB260D
J2
LO
J1
Hi
68K
68K
1M
E
220K
220K
100K
N
1/2-12AX7
.02
.002
1.5 K
25/25
100K
470
1M B
.0047
68K
680
1M T
.0047
1M V
.001
D
220K
100K
1/2-12AX7
1.5 K
25/25
.001
.0047
X
C
1.5M
4.7M
680
F 1M Aud.
.005
1M Col.
.1
1/2-6JW8 Pentode
5.6K
25/25
Stinger (Fuzz)
Ft.Sw
SW.
Tr.O
250 K
1
Tremolo out
68 K
2
150K
.1
.033
68 K
.033
.01
1M
470K
1M T.S.
6JW8 triode
1.5 K
25/25

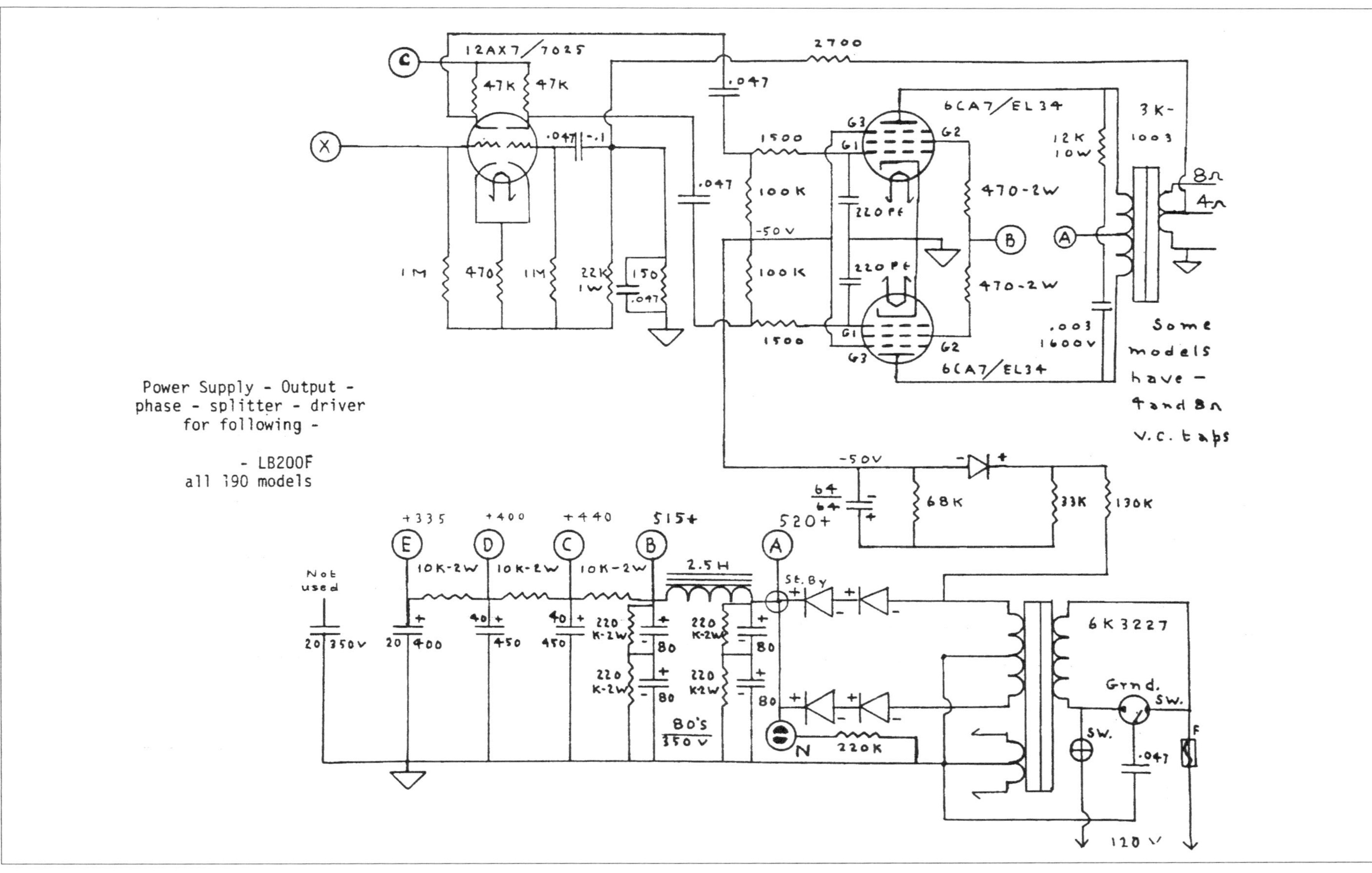
12AX7/7025
47K
47K
.047
.1
1M
470
1M
22K 1W
150
.047
2700
.047
.047
1500
100K
-50V
100K
1500
220pf
220pf
G1
G2
G3
6CA7/EL34
6CA7/EL34
470-2W
470-2W
12K 10W
3K-1003
8Ω
4Ω
.003 1600V
Some models have - 4 and 8Ω V.C. taps
Power Supply - Output - phase - splitter - driver for following -
- LB200F all 190 models
-50V
64 64
68K
33K
130K
+335
+400
+440
515+
520+
10K-2W
10K-2W
10K-2W
2.5H
St.By
Not used
20 350V
20 400
40 450
40 450
220 K-2W
220 K-2W
220 K-2W
220 K2W
80
80
80
80
80's 350V
N
220K
6K3227
Grnd. SW.
SW.
.047
F
120V

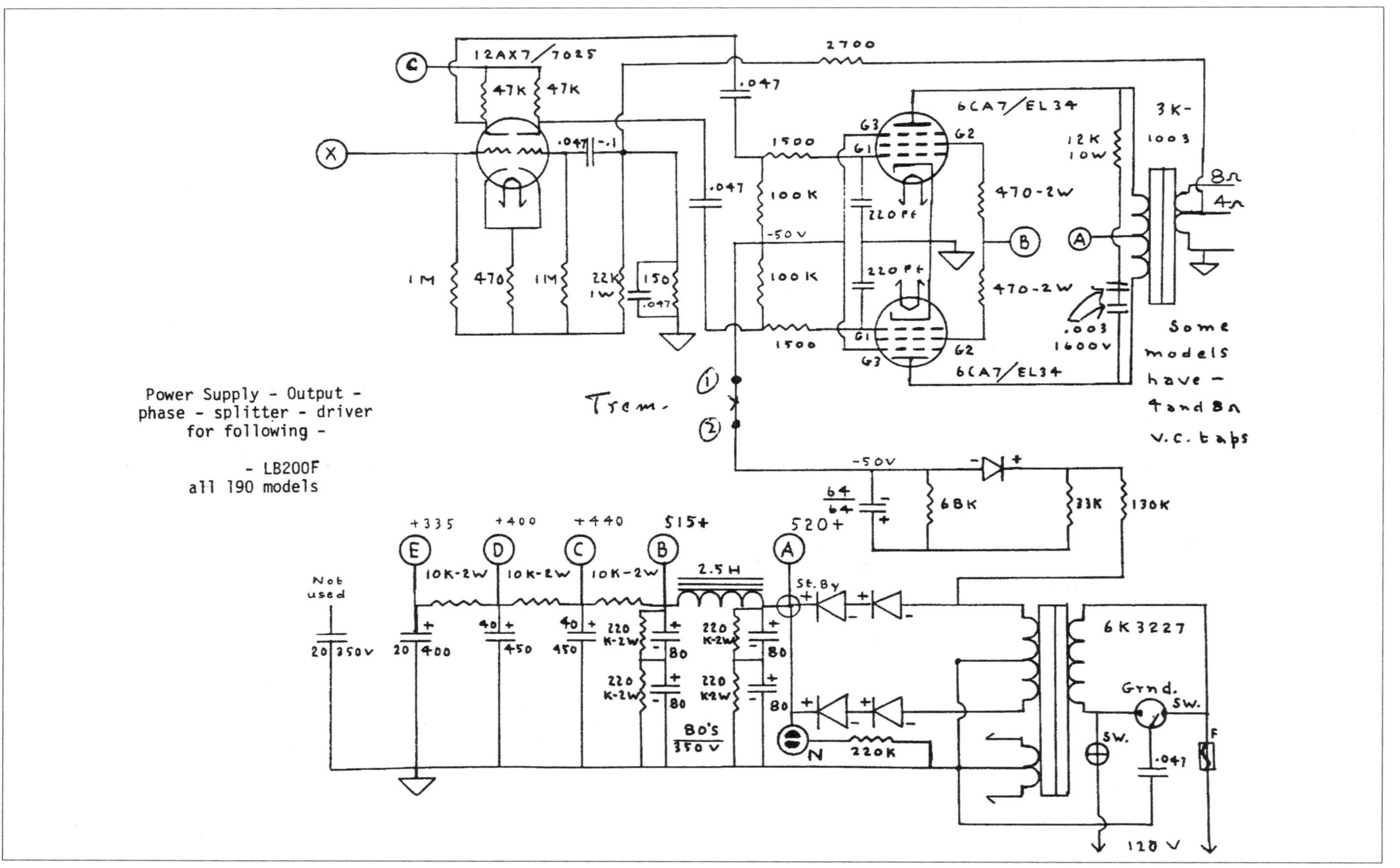

Power Supply - Output - phase - splitter - driver for following -

- LB200F
all 190 models

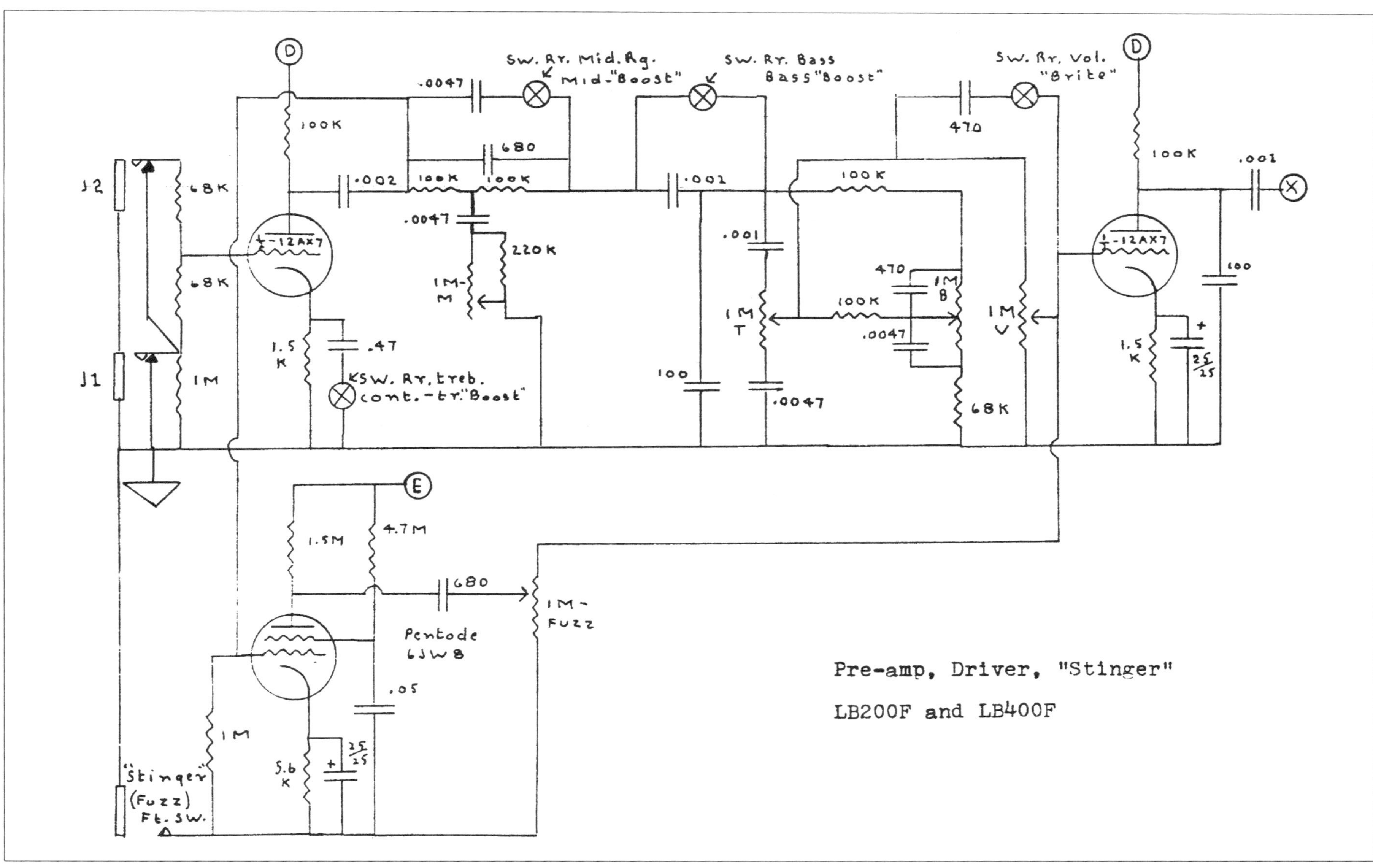

Pre-amp, Driver, "Stinger"

LB200F and LB400F

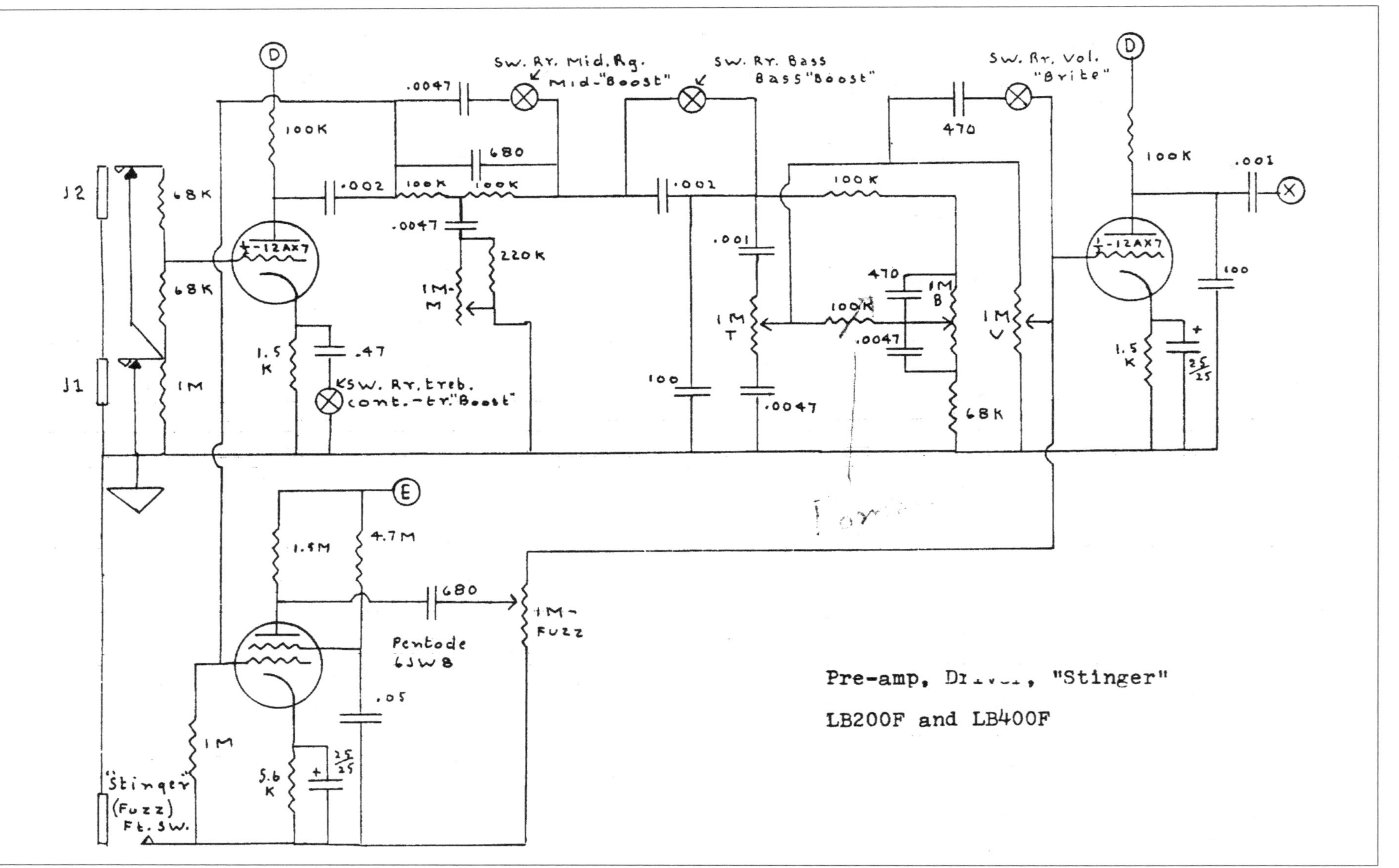

J2
J1
68K
68K
1M
100K
.002
.0047
680
100K
100K
.0047
220K
1M-
M
1.5
K
.47
SW. RY. TREB.
CONT.-TR."BOOST"
SW. RY. MID. RG.
MID-"BOOST"
SW. RY. BASS
BASS "BOOST"
.002
.001
1M
T
100
.0047
100K
470
1M
B
100K
.0047
68K
SW. RY. VOL.
"BRITE"
470
1M
V
100K
.001
1/2-12AX7
100
1.5
K
25/25
E
1.5M
4.7M
680
1M-
FUZZ
Pentode
6JW8
.05
1M
5.6
K
25/25
"Stinger"
(Fuzz)
Ft. Sw.
Pre-amp, Dr..., "Stinger"
LB200F and LB400F

LIL ROCK **B90L**

Garnet

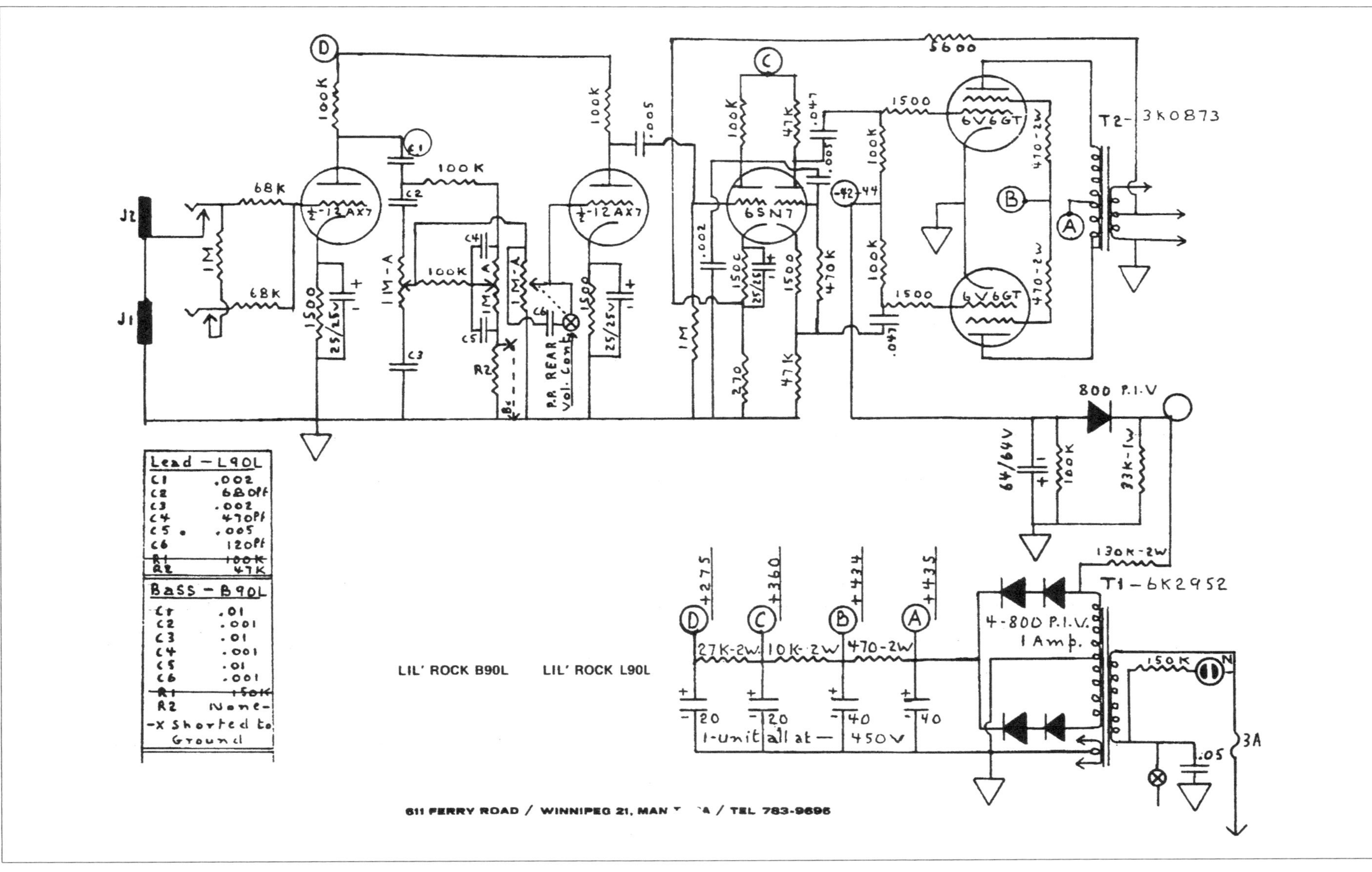

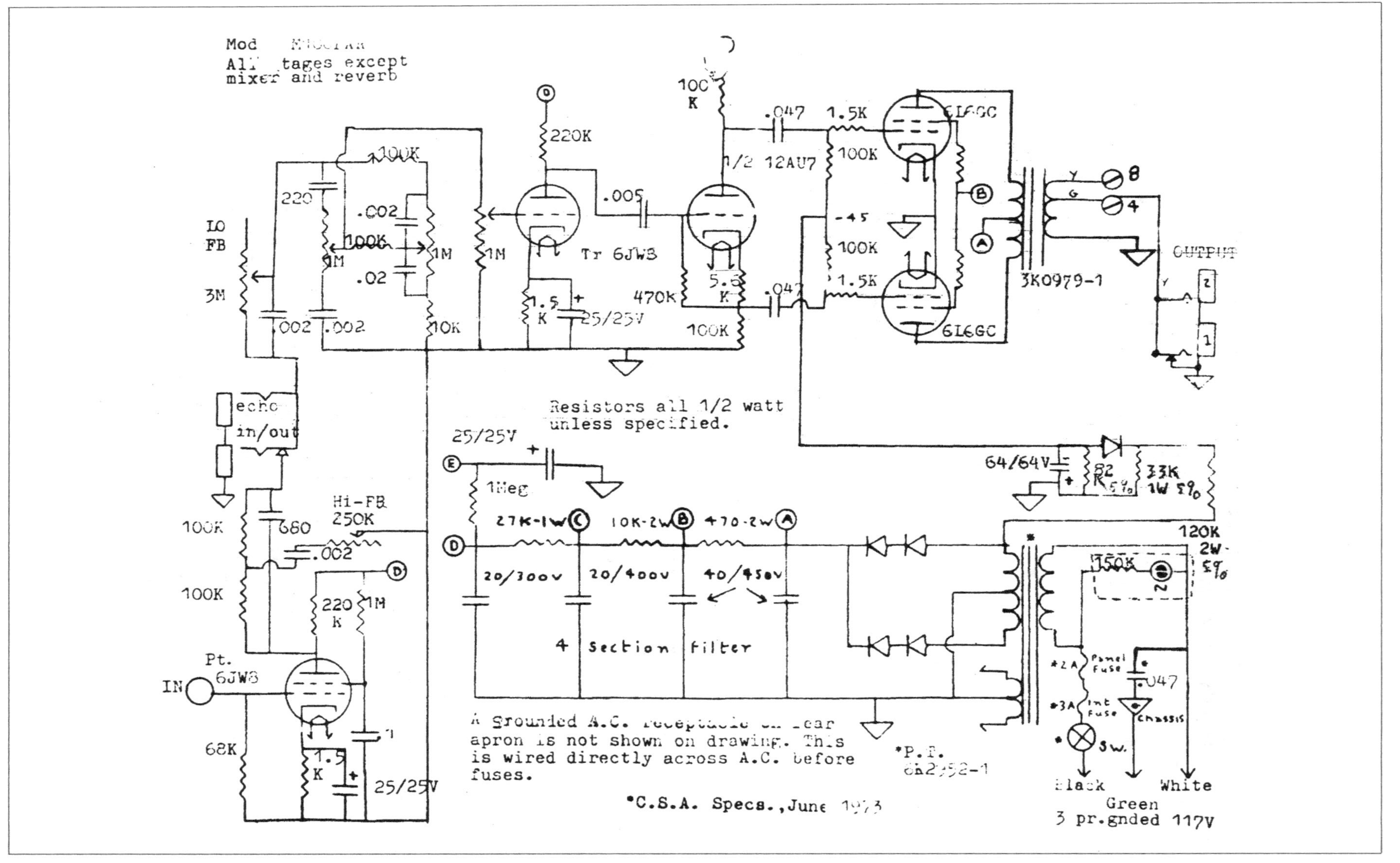
Mod
All tages except mixer and reverb
LO FB
3M
220
.002
.002
100K
.002
100K
1M
.02
1M
10K
220K
1M
Tr 6JW3
1.5 K
25/25V
.005
100 K
1/2 12AU7
470K
5.6 K
100K
.047
1.5K
100K
6L6GC
-45
100K
1.5K
.047
6L6GC
B
A
Y
G
8
4
3K0979-1
OUTPUT
Y
2
1
echo in/out
Hi-FB 250K
680
.002
100K
100K
D
220 K
1M
Pt. 6JW3
IN
68K
1.5 K
.1
25/25V
Resistors all 1/2 watt unless specified.
25/25V
E
1Meg
D
27K-1W
C
10K-2W
B
470-2W
A
20/300v
20/400v
40/450v
4 Section filter
64/64V
82 K 5%
33K 1W 5%
120K 2W 5%
150K
#2A Panel Fuse
#3A Int Fuse
.047
Chassis
Sw.
Black
Green
White
3 pr.gnded 117V
*P.T.
*C.S.A. Specs., June 1973
apron is not shown on drawing. This is wired directly across A.C. before fuses.

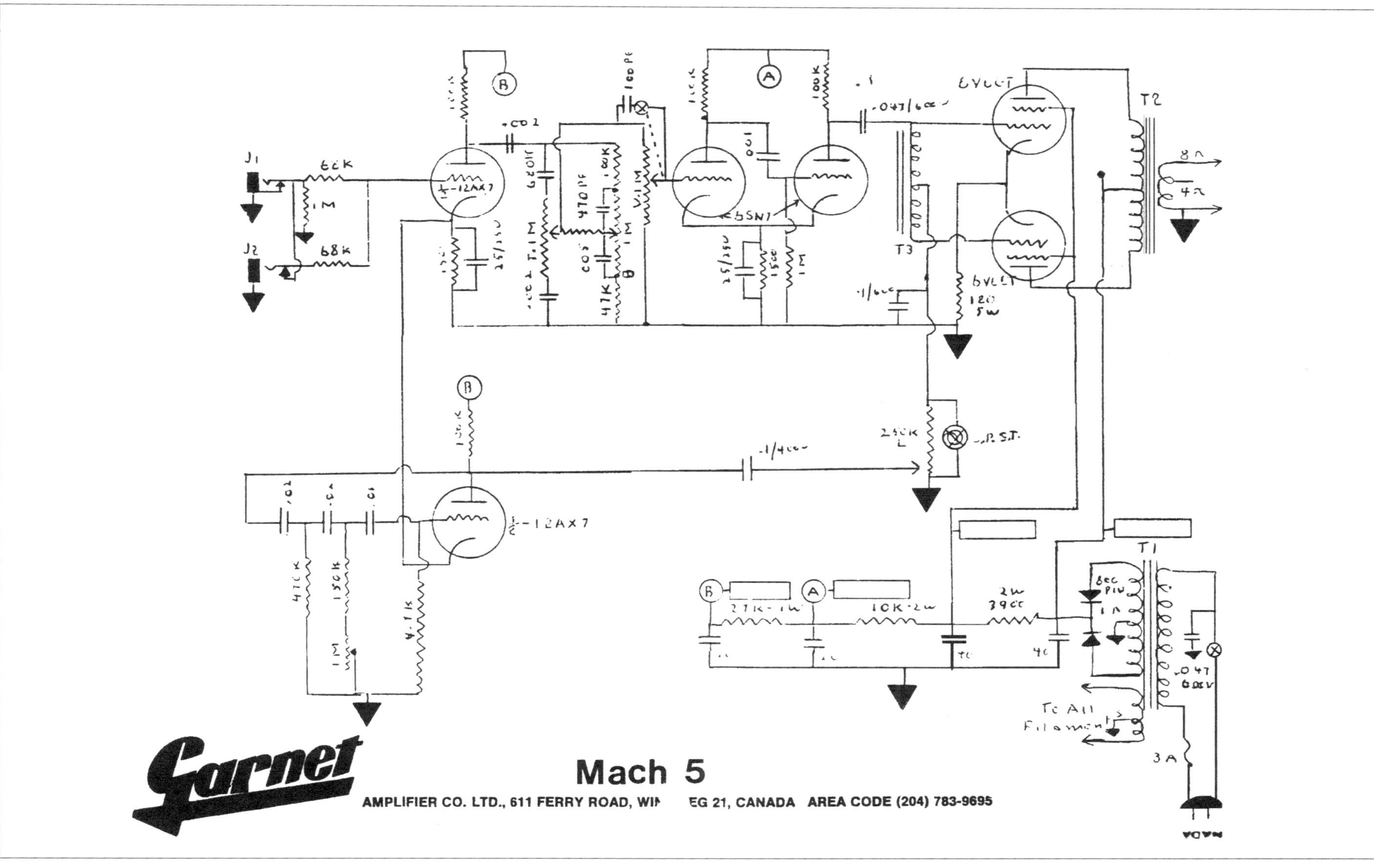
J1
J2
68K
1M
68K
½-12AX7
.002
25/25V
.002
T.1M
.0022
.005
470PF
100K
47K
1M
V.1M
100PF
5SN7
.001
25/25V
1500
1M
100K
.1
.047/600V
6V6GT
T2
8Ω
4Ω
T3
.1/600V
6V6GT
120
5W
250K
L
S.P.S.T.
.1/400V
.02
.02
.01
½-12AX7
470K
150K
1M
4.7K
T1
2W
39CC
8CC
PIV
1A
27K-1W
10K-2W
40
.047
600V
To All
Filaments
3A
Garnet
Mach 5
AMPLIFIER CO. LTD., 611 FERRY ROAD, WI EG 21, CANADA AREA CODE (204) 783-9695

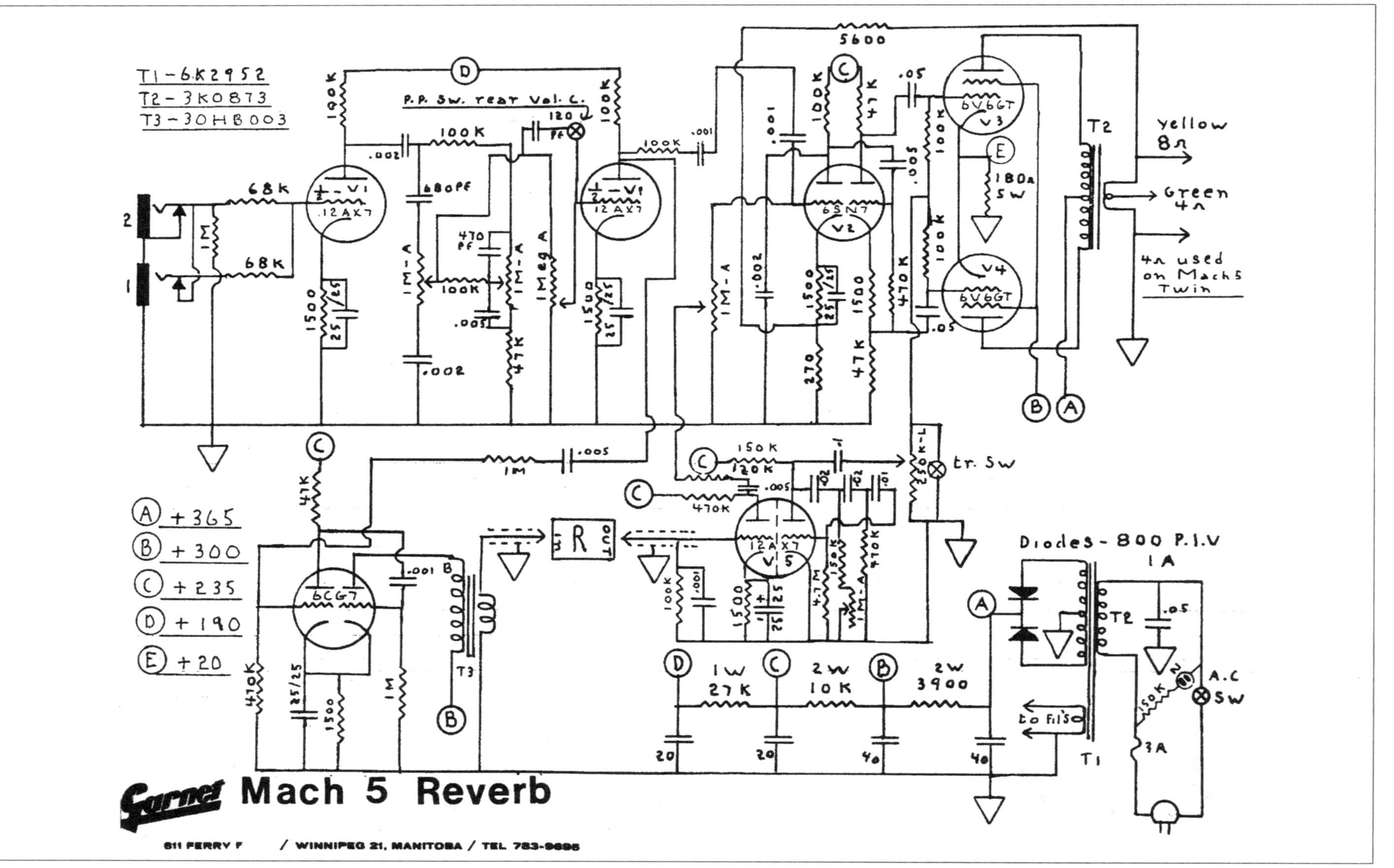
T1 - 6K2952
T2 - 3K0873
T3 - 30HB003
A +365
B +300
C +235
D +190
E +20
12AX7
V1
6SN7
V2
6V6GT
V3
V4
6CG7
V5
T2
T3
T1
yellow 8Ω
Green 4Ω
4Ω used on Mach 5 Twin
P.P. Sw. rear Vol. C.
Diodes - 800 P.I.V 1A
tr. Sw
A.C Sw
to Fil's
R
OUT
IN
Garnet Mach 5 Reverb
611 FERRY F / WINNIPEG 21, MANITOBA / TEL 783-9695

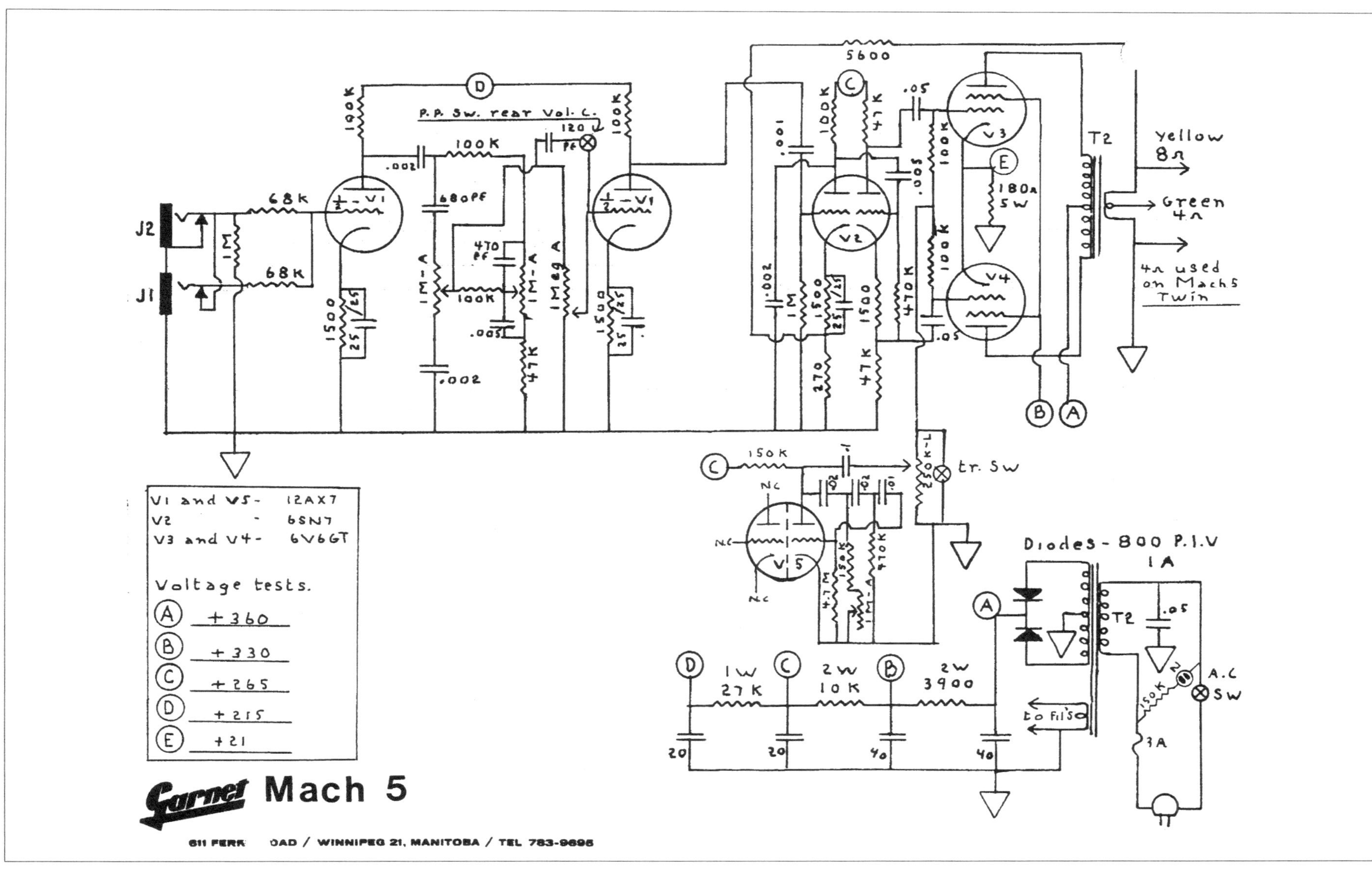
J2
J1
1M
68K
68K
1/2 - V1
1500
25 / 25
100K
.002
1M-A
680PF
100K
470 PF
.005
47K
1M-A
1Meg A
120 PF
P.P. Sw. rear Vol. C.
D
.002
1500
25 / 25
5600
C
100K
47K
.05
.001
.005
100K
V3
V2
V4
E
180Ω
SW
100K
470K
.05
1M
.002
1500
25 / 25
1500
270
47K
T2
Yellow 8Ω
Green 4Ω
4Ω used on Mach5 Twin
B
A
C
150K
.1
.02
.02
.01
250K-L
tr. Sw
NC
V5
4.7M
150K
1M-A
470K
V1 and V5- 12AX7
V2 - 6SN7
V3 and V4- 6V6GT
Voltage tests.
A +360
B +330
C +265
D +215
E +21
Diodes - 800 P.I.V 1A
T2
.05
A.C SW
150K
3A
to Fil's
D
1W 27K
C
2W 10K
B
2W 3900
20
20
40
40
Garnet Mach 5
611 FERR OAD / WINNIPEG 21, MANITOBA / TEL 783-9696

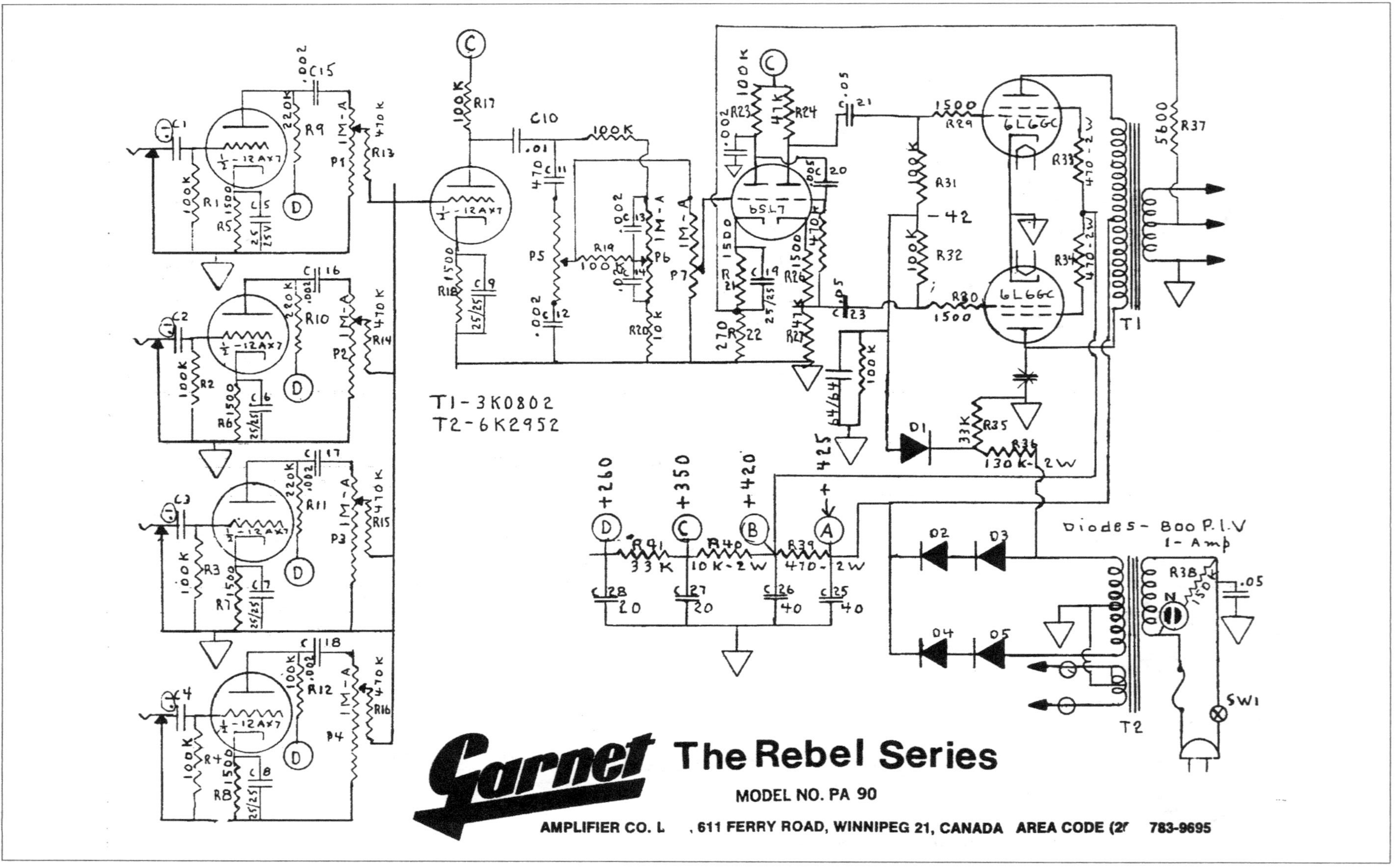
T1-3K0802
T2-6K2952
Diodes- 800 P.I.V
1-Amp
Garnet
The Rebel Series
MODEL NO. PA 90
AMPLIFIER CO. L , 611 FERRY ROAD, WINNIPEG 21, CANADA AREA CODE (2' 783-9695

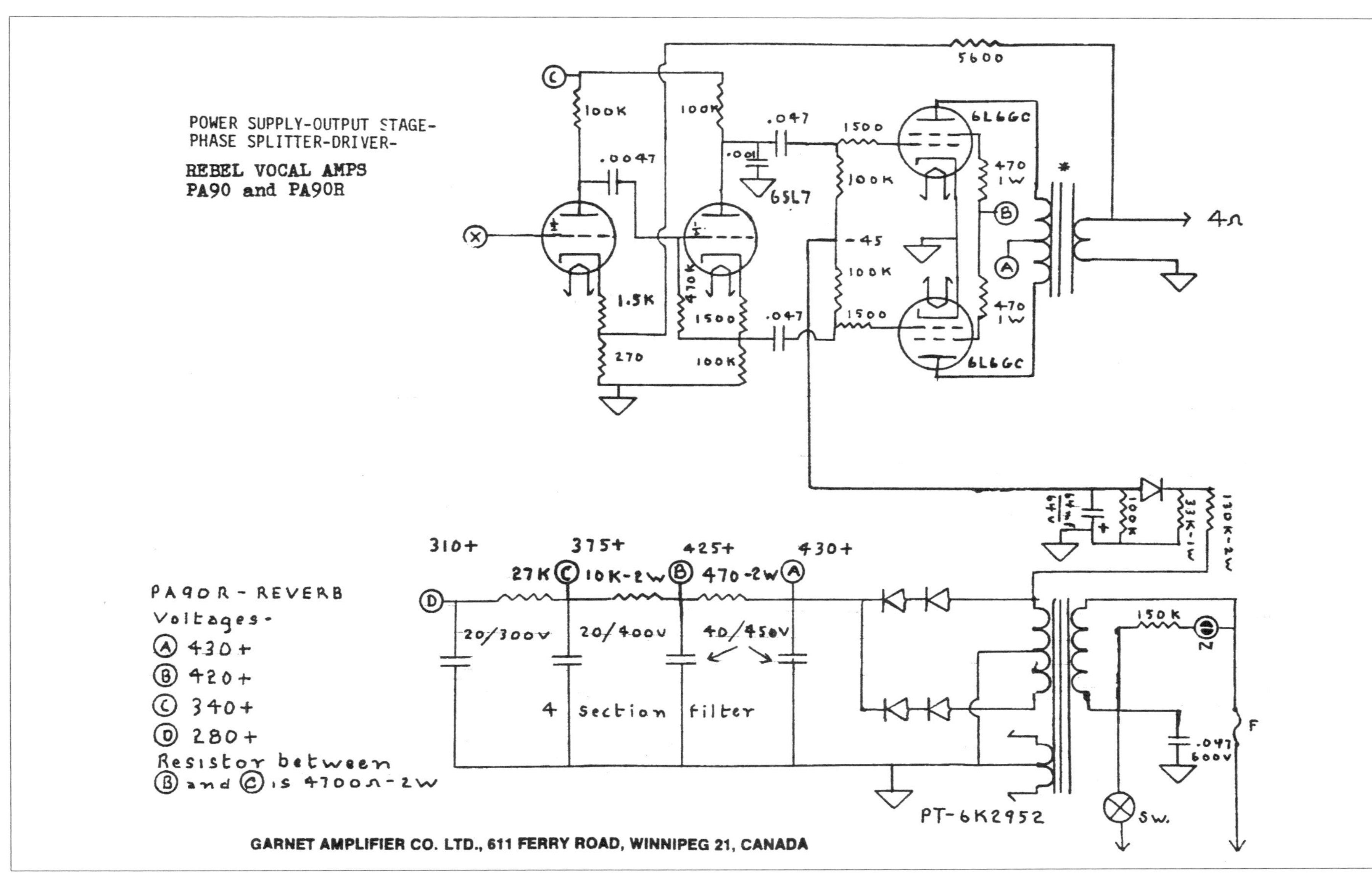
POWER SUPPLY-OUTPUT STAGE-
PHASE SPLITTER-DRIVER-
REBEL VOCAL AMPS
PA90 and PA90R
100K
100K
.0047
.047
.001
6SL7
1500
6L6GC
470
1W
100K
-45
1.5K
470K
1500
270
100K
5600
4Ω
PA90R - REVERB
Voltages -
Ⓐ 430+
Ⓑ 420+
Ⓒ 340+
Ⓓ 280+
Resistor between
Ⓑ and Ⓒ is 4700Ω-2W
310+
375+
425+
430+
27K
10K-2W
470-2W
20/300V
20/400V
40/450V
4 section filter
33K-1W
130K-2W
150K
PT-6K2952
SW.
F
.047
600V
GARNET AMPLIFIER CO. LTD., 611 FERRY ROAD, WINNIPEG 21, CANADA

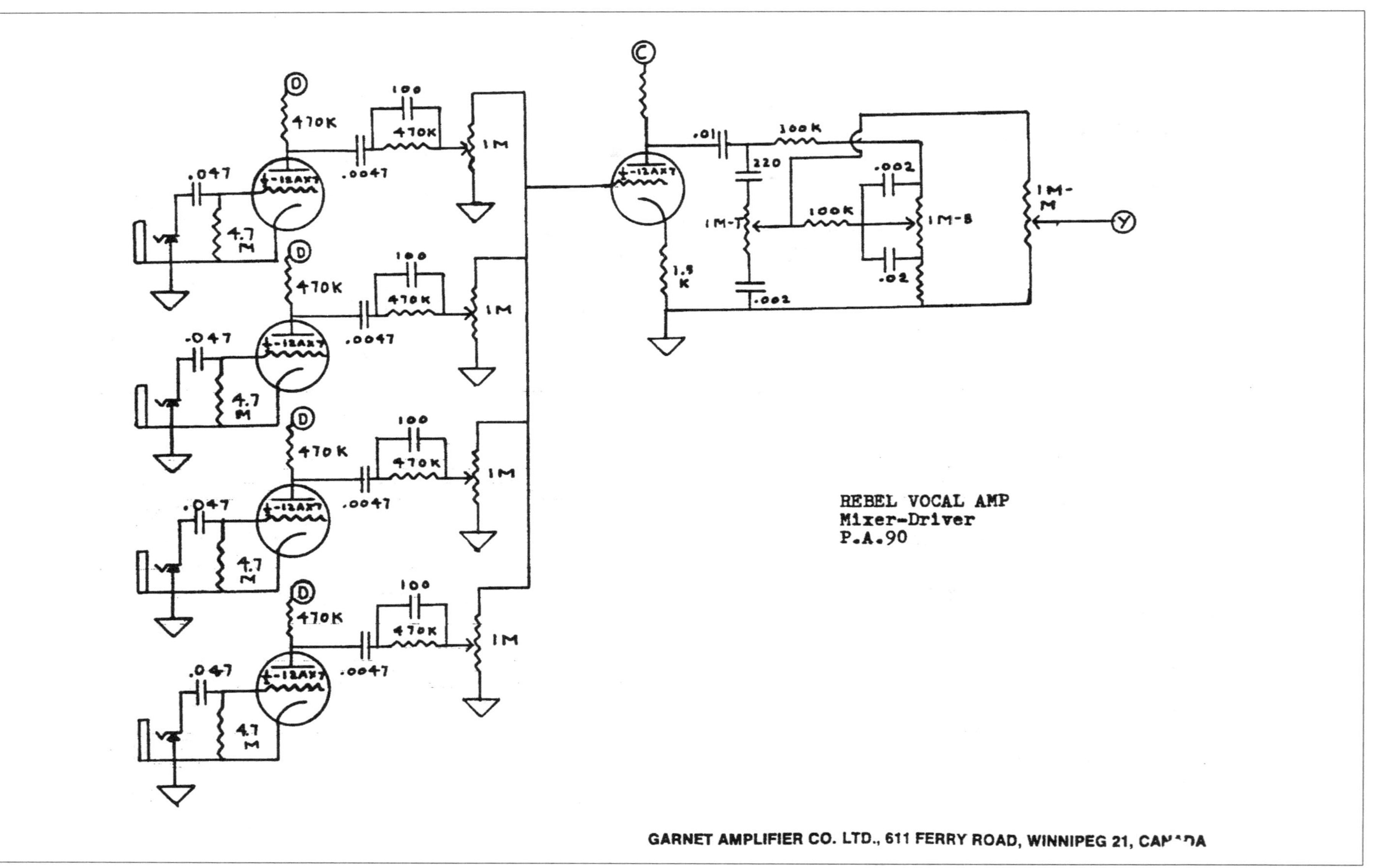
D
470K
100
470K
.0047
.047
4.7 M
1M
C
.01
100K
220
.002
.02
1M-T
100K
1M-B
1M-M
Y
1.5 K
.002
REBEL VOCAL AMP
Mixer-Driver
P.A.90
GARNET AMPLIFIER CO. LTD., 611 FERRY ROAD, WINNIPEG 21, CANADA

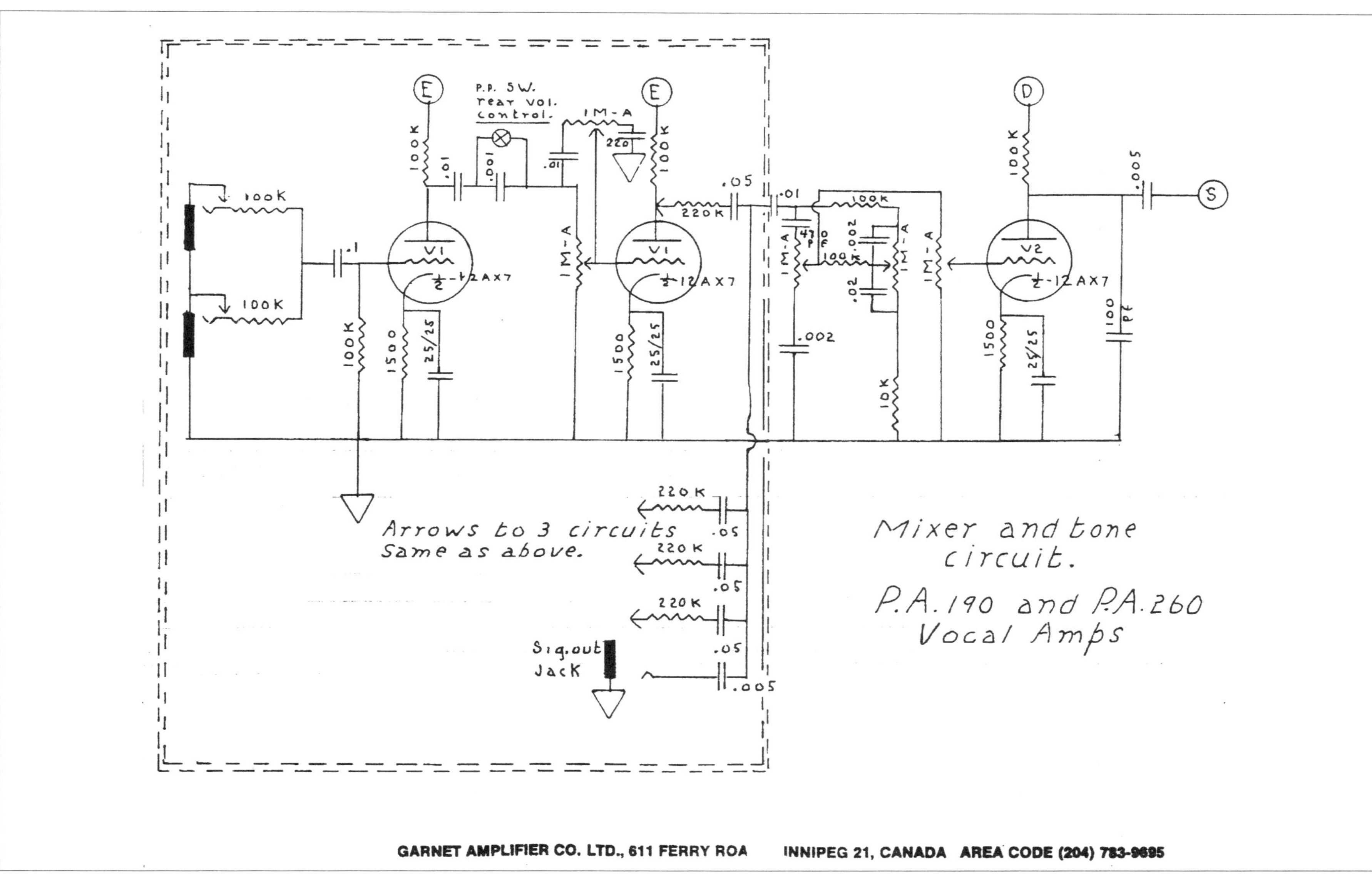

GARNET AMPLIFIER CO. LTD., 611 FERRY ROA INNIPEG 21, CANADA AREA CODE (204) 783-9695

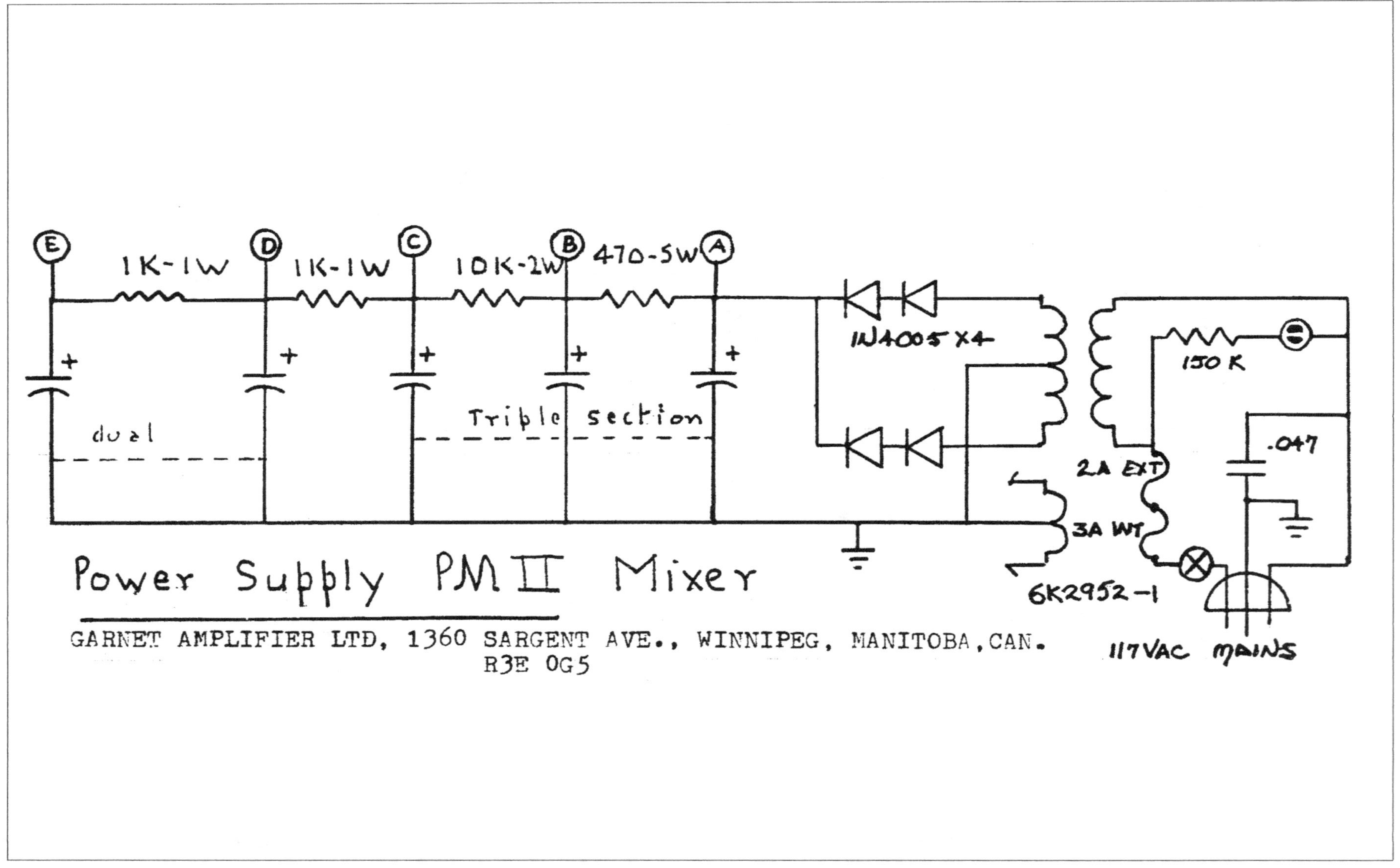
E
D
C
B
A
1K-1W
1K-1W
10K-2W
470-5W
+
+
+
+
+
dual
Triple section
1N4005 X4
150 K
.047
2A EXT
3A WT
6K2952-1
117VAC MAINS
Power Supply PM II Mixer
GARNET AMPLIFIER LTD, 1360 SARGENT AVE., WINNIPEG, MANITOBA, CAN.
R3E 0G5

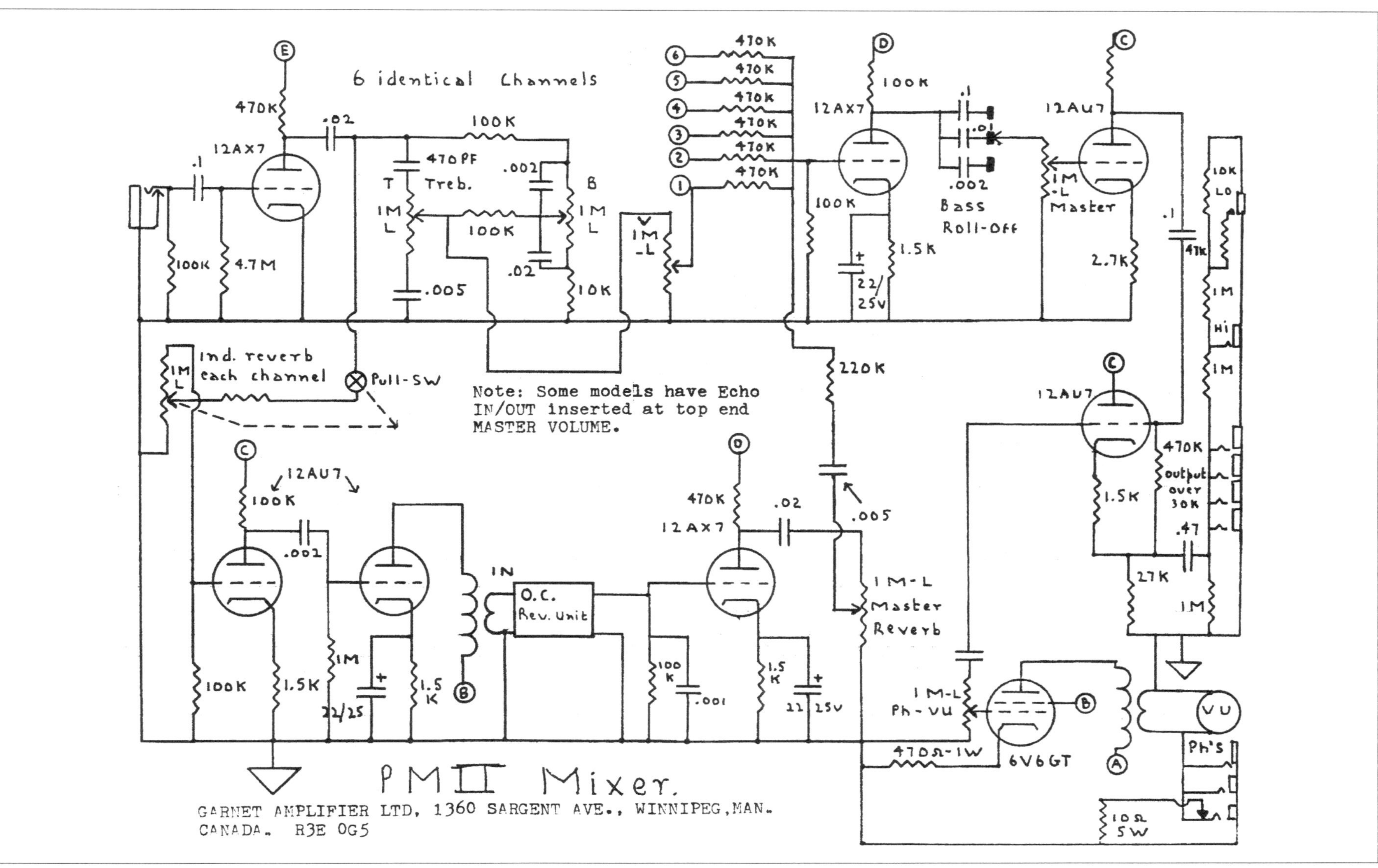
6 identical Channels
12AX7
470K
.02
.1
100K
4.7M
470PF
Treb.
T
1M L
.002
B
100K
.005
.02
10K
1M -L
12AU7
Bass Roll-Off
Master
2.7K
220K
Ind. reverb each channel
Pull-SW
Note: Some models have Echo IN/OUT inserted at top end MASTER VOLUME.
IN
O.C. Rev. Unit
1M-L Master Reverb
1M-L Ph-VU
470Ω-1W
6V6GT
output over 30K
27K
VU
Ph's
10Ω SW
Hi
Lo
PMII Mixer.
GARNET AMPLIFIER LTD, 1360 SARGENT AVE., WINNIPEG, MAN.
CANADA. R3E 0G5

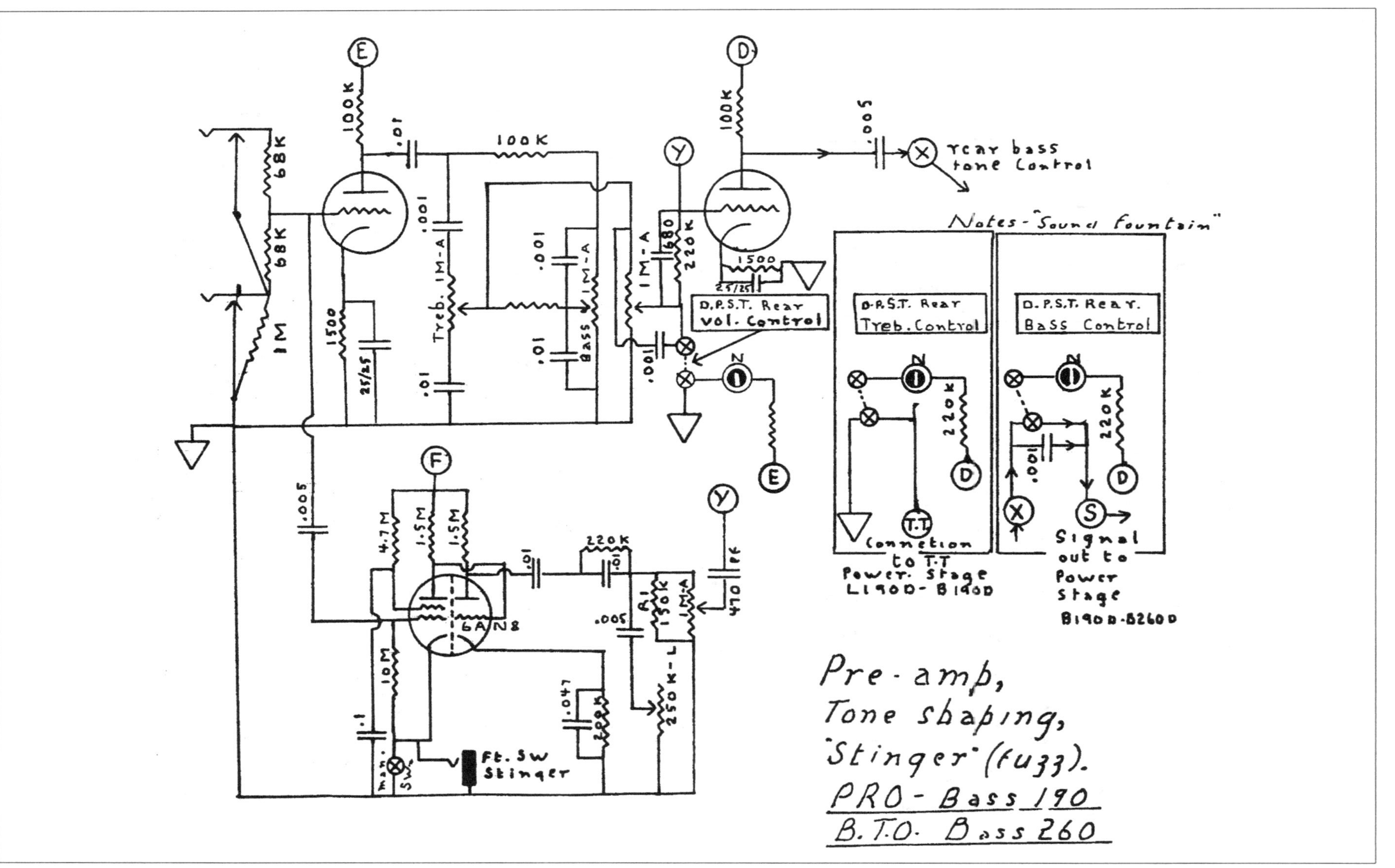
100K
.01
100K
68K
68K
1M
1500
25/25
Treb. 1M-A
.001
.01
Bass 1M-A
1M-A
680
220K
.001
.005
rear bass tone control
1500
25/25
D.P.S.T. Rear Vol. Control
Notes - "Sound Fountain"
D.P.S.T. Rear Treb. Control
220K
Connetion to T.T Power Stage L190D - B190D
D.P.S.T. Rear Bass Control
220K
.001
Signal out to Power Stage B190D-B260D
.005
4.7M
1.5M
1.5M
6AN8
220K
.01
.01
R1 150K
1M-A
470 PF
.005
250K-L
.047
10M
.1
Ft. SW Stinger
Pre-amp,
Tone shaping,
"Stinger" (fuzz).
PRO - Bass 190
B.T.O. Bass 260

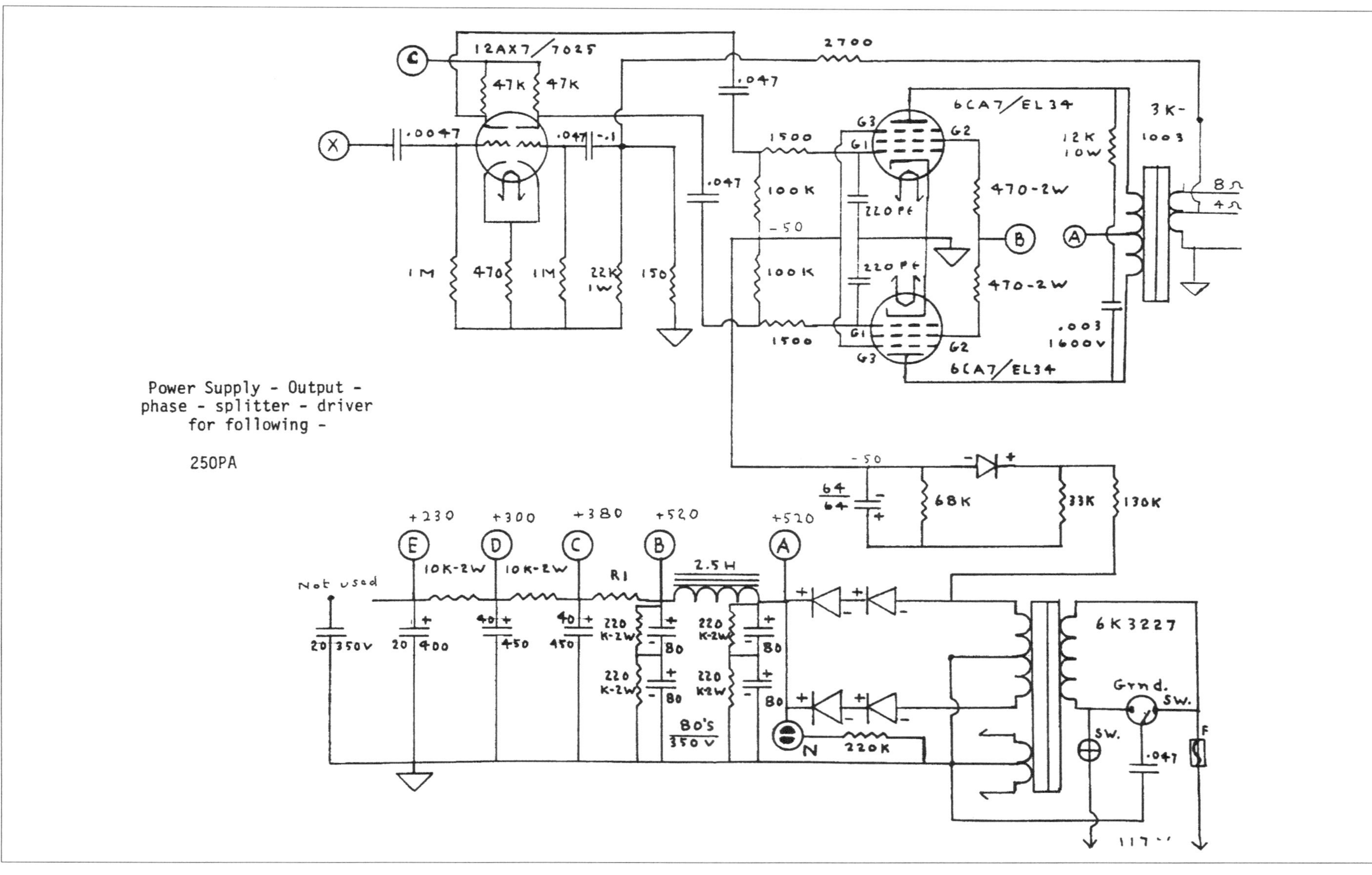

Power Supply - Output - phase - splitter - driver for following -

250PA

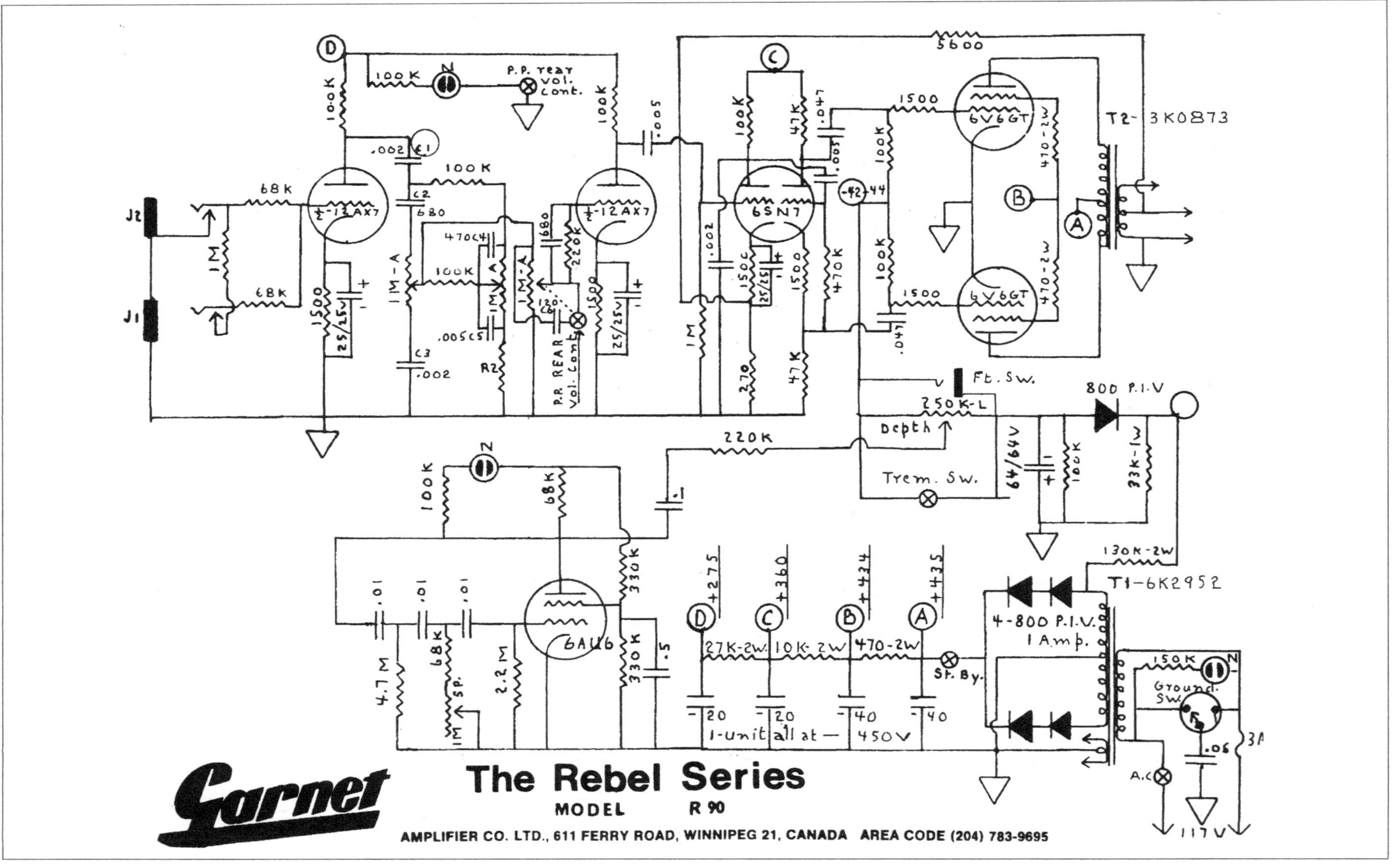
Garnet
The Rebel Series
MODEL R 90
AMPLIFIER CO. LTD., 611 FERRY ROAD, WINNIPEG 21, CANADA AREA CODE (204) 783-9695

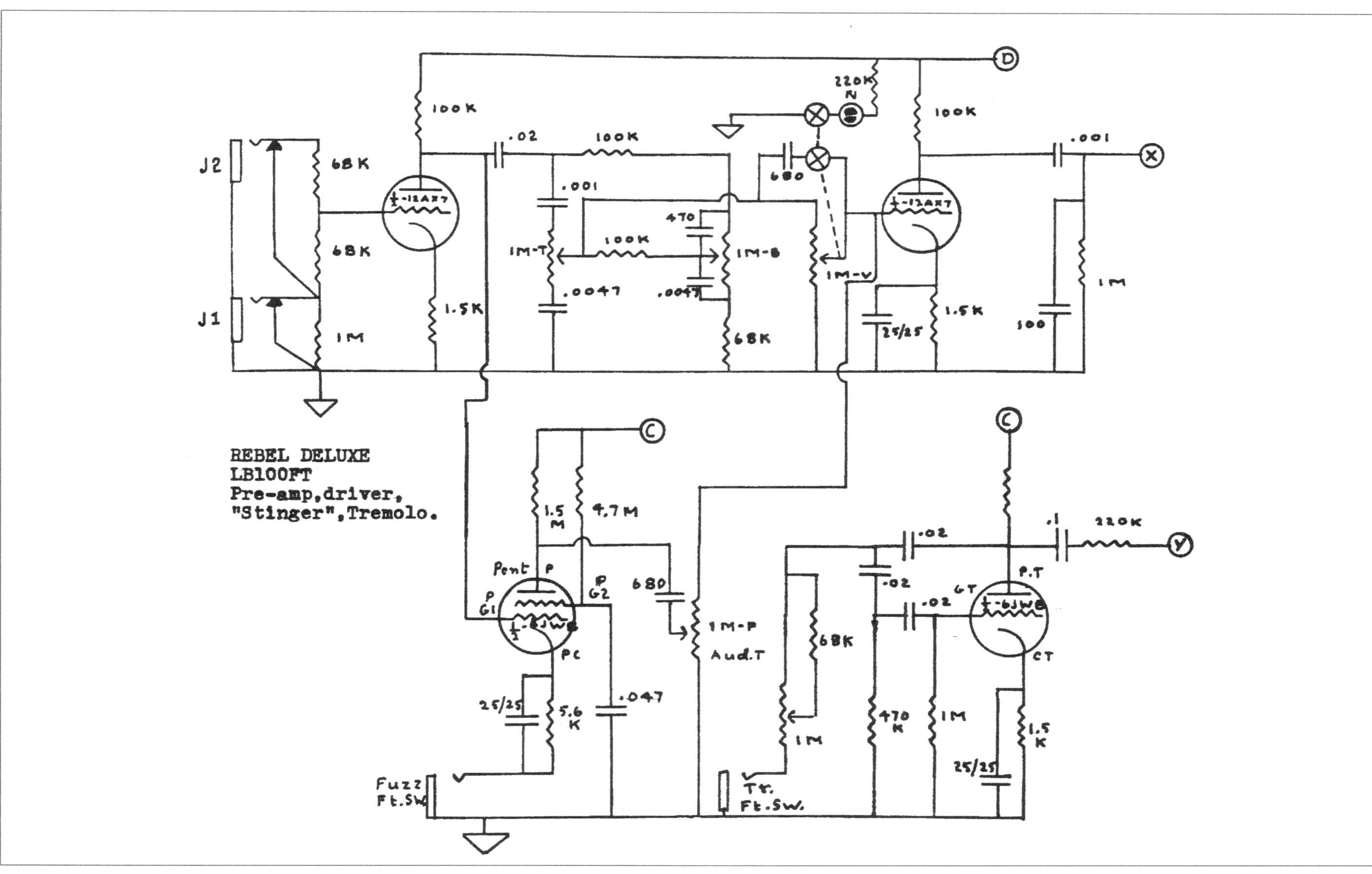
REBEL DELUXE
LB100FT
Pre-amp,driver,
"Stinger",Tremolo.
J2
J1
68K
68K
1M
100K
½-12AX7
1.5K
.02
.001
100K
1M-T
100K
.0047
470
1M-B
.0047
68K
680
220K
N
1M-V
100K
½-12AX7
1.5K
25/25
.001
1M
100
Pent
P
G1
½-6JW8
PC
G2
1.5
M
4.7M
680
1M-P
Aud.T
25/25
5.6
K
.047
Fuzz
Ft.SW
68K
1M
Tr.
Ft.Sw.
.02
.02
.02
470
K
1M
GT
P.T
½-6JW8
CT
1.5
K
25/25
.1
220K

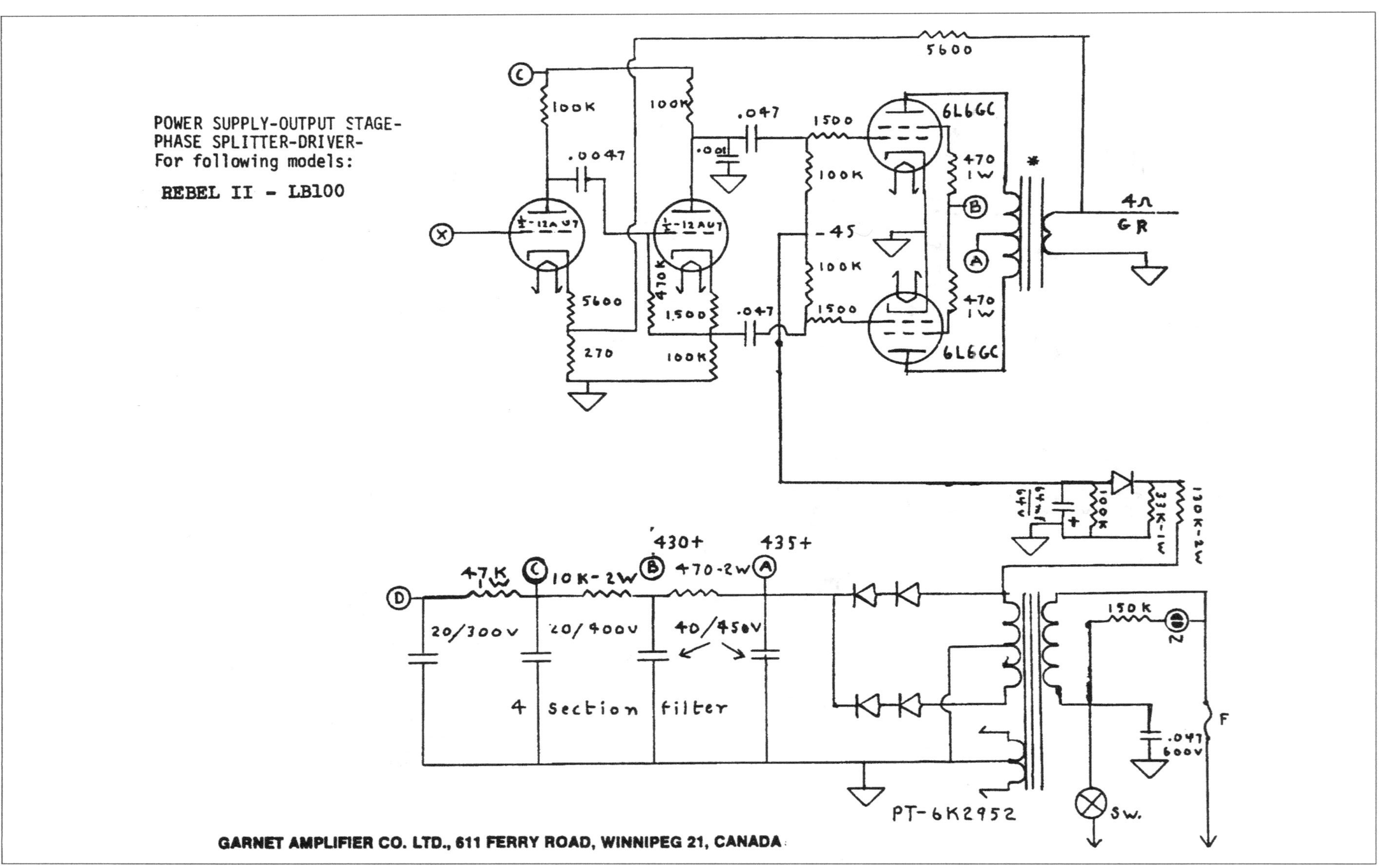
POWER SUPPLY-OUTPUT STAGE-
PHASE SPLITTER-DRIVER-
For following models:
REBEL II - LB100
5600
100K
100K
.047
1500
6L6GC
.0047
.001
470
1W
½-12AU7
½-12AU7
-45
100K
4Ω
GR
5600
470K
1500
.047
1500
6L6GC
270
100K
130K-2W
33K-1W
100K
430+
435+
47K
1W
10K-2W
470-2W
20/300V
20/400V
40/450V
4 Section filter
150K
.047
600V
F
PT-6K2952
SW.
GARNET AMPLIFIER CO. LTD., 611 FERRY ROAD, WINNIPEG 21, CANADA

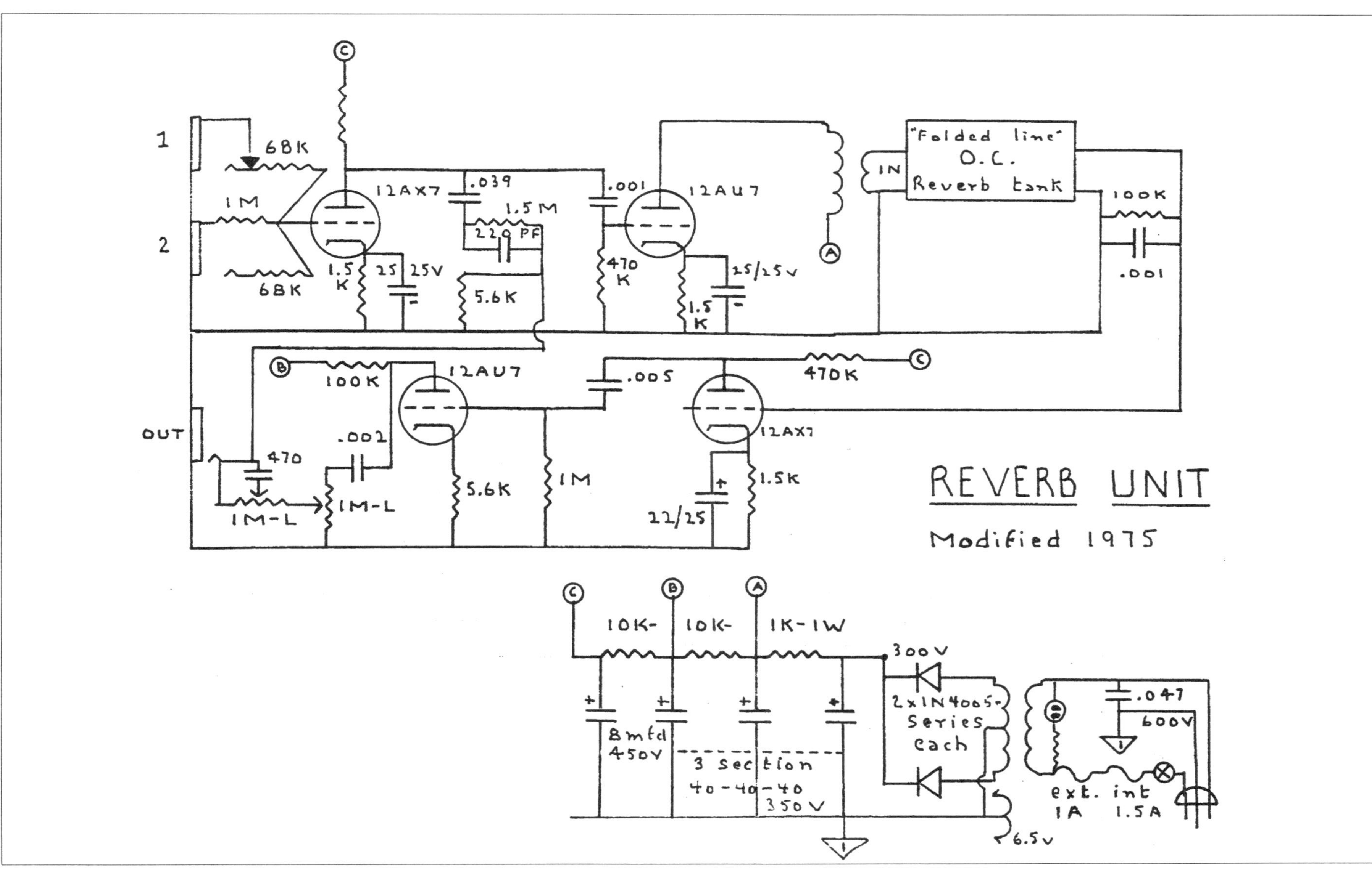
1
2
OUT
68K
1M
68K
1.5 K
25 25V
12AX7
.039
1.5M
220 PF
5.6K
.001
470 K
12AU7
25/25V
1.5 K
"Folded line"
O.C.
Reverb tank
IN
100K
.001
100K
12AU7
.005
470K
470
.002
1M-L
1M-L
5.6K
1M
12AX7
1.5K
22/25
REVERB UNIT
Modified 1975
10K-
10K-
1K-1W
300V
8mfd 450V
3 section
40-40-40
350V
2xIN4005-
Series
each
.047
600V
ext. int
1A 1.5A
6.5V

AMPLIFICADOR MODELO A-102

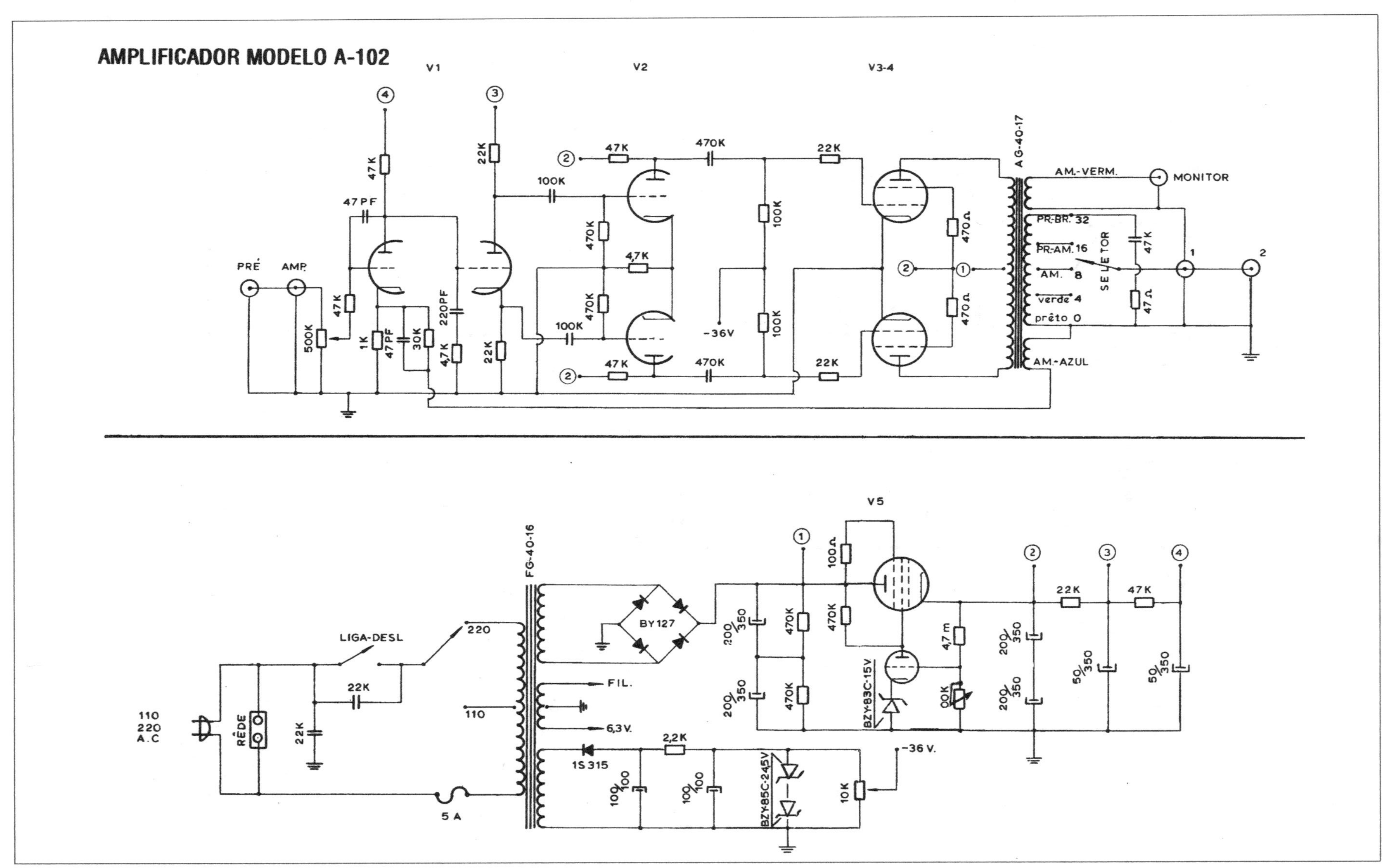

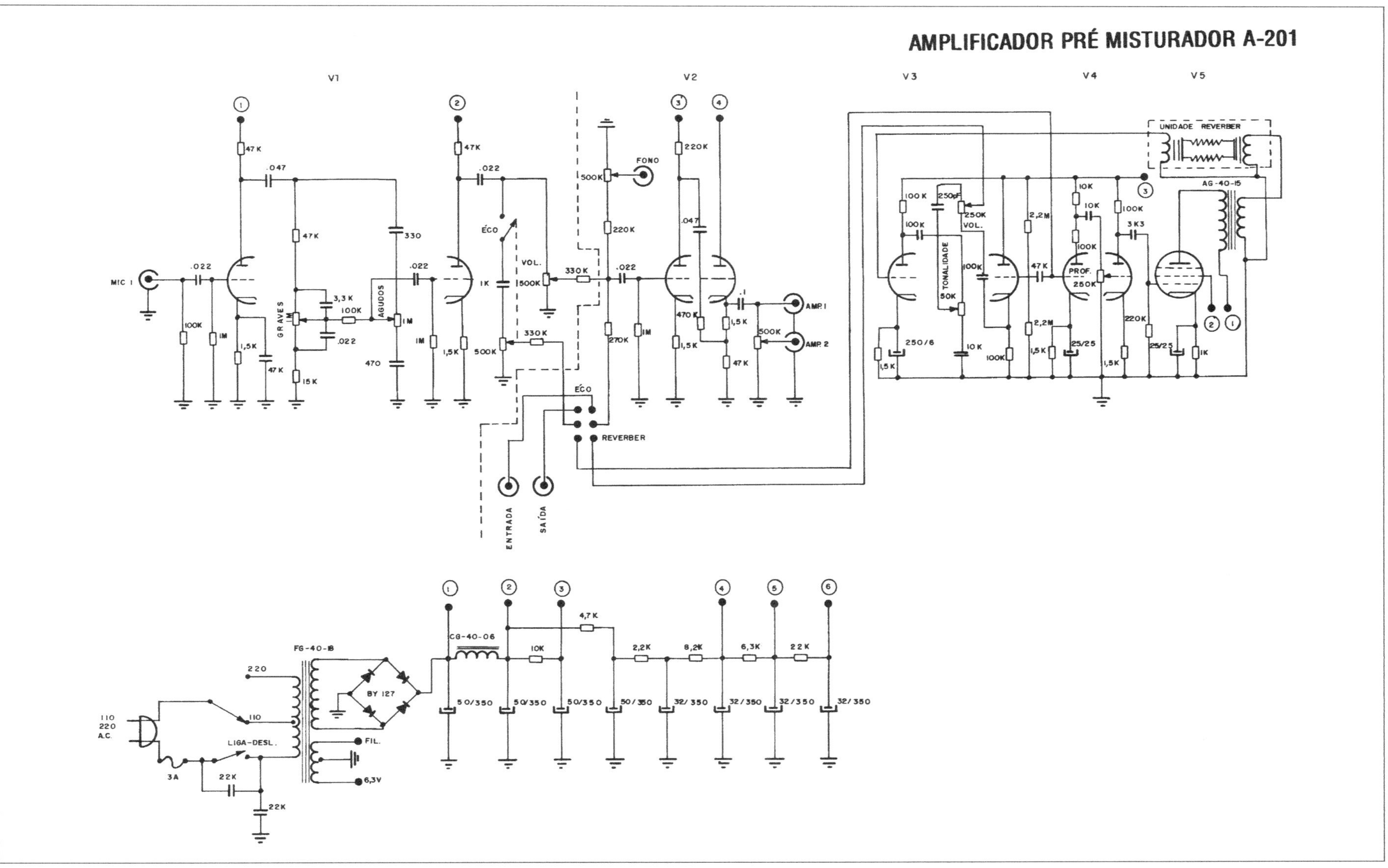
AMPLIFICADOR PRÉ MISTURADOR A-201
V1
V2
V3
V4
V5
MIC 1
GRAVES
AGUDOS
ÉCO
VOL.
FONO
ENTRADA
SAÍDA
REVERBER
AMP 1
AMP 2
TONALIDADE
PROF.
UNIDADE REVERBER
AG-40-15
FG-40-18
CG-40-06
BY 127
LIGA-DESL.
110
220
A.C.
3A
FIL.
6,3V

AMPLIFICADOR A-300

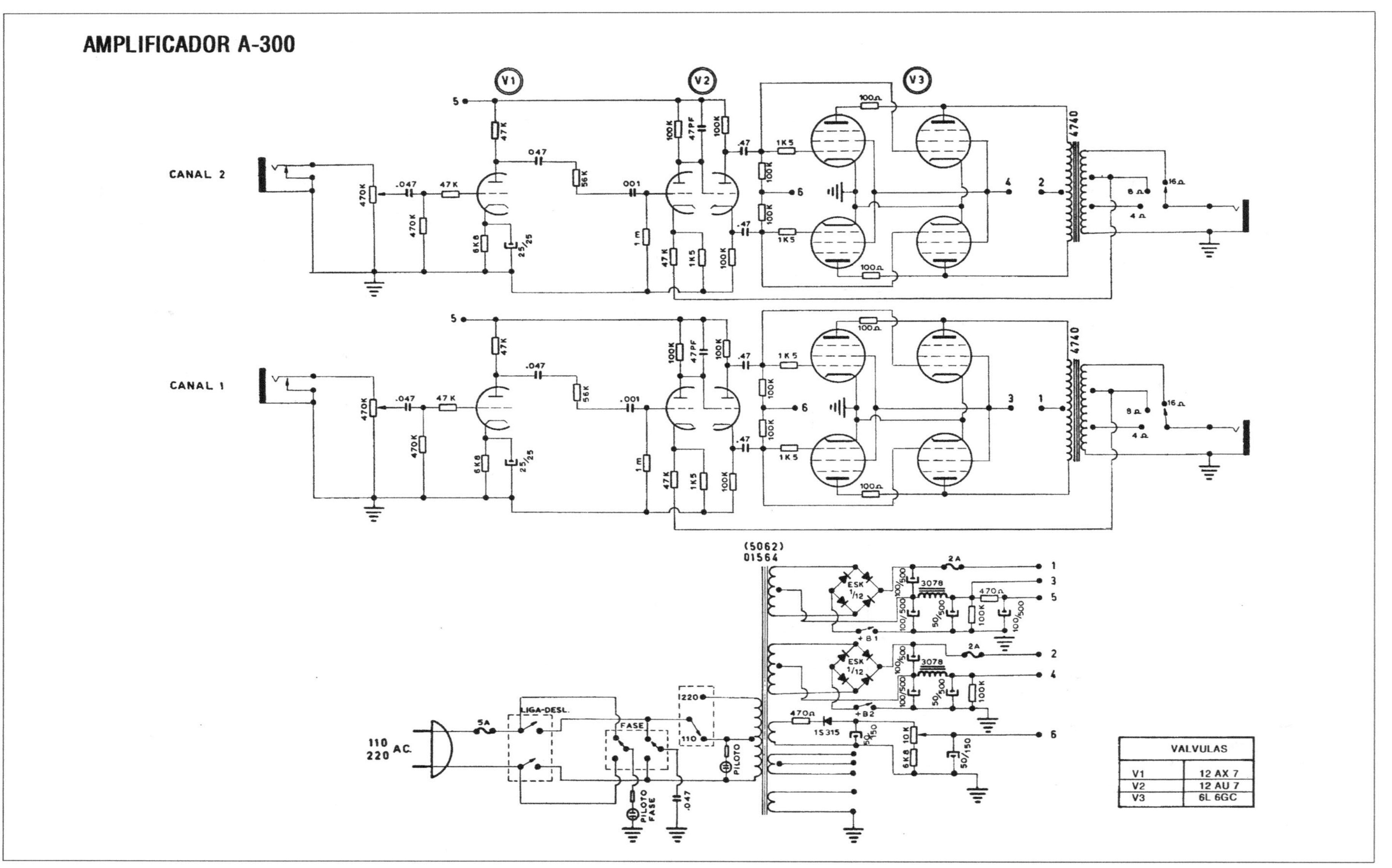

VALVULAS	
V1	12 AX 7
V2	12 AU 7
V3	6L 6GC

AMPLIFICADOR BULLDOG BAIXO

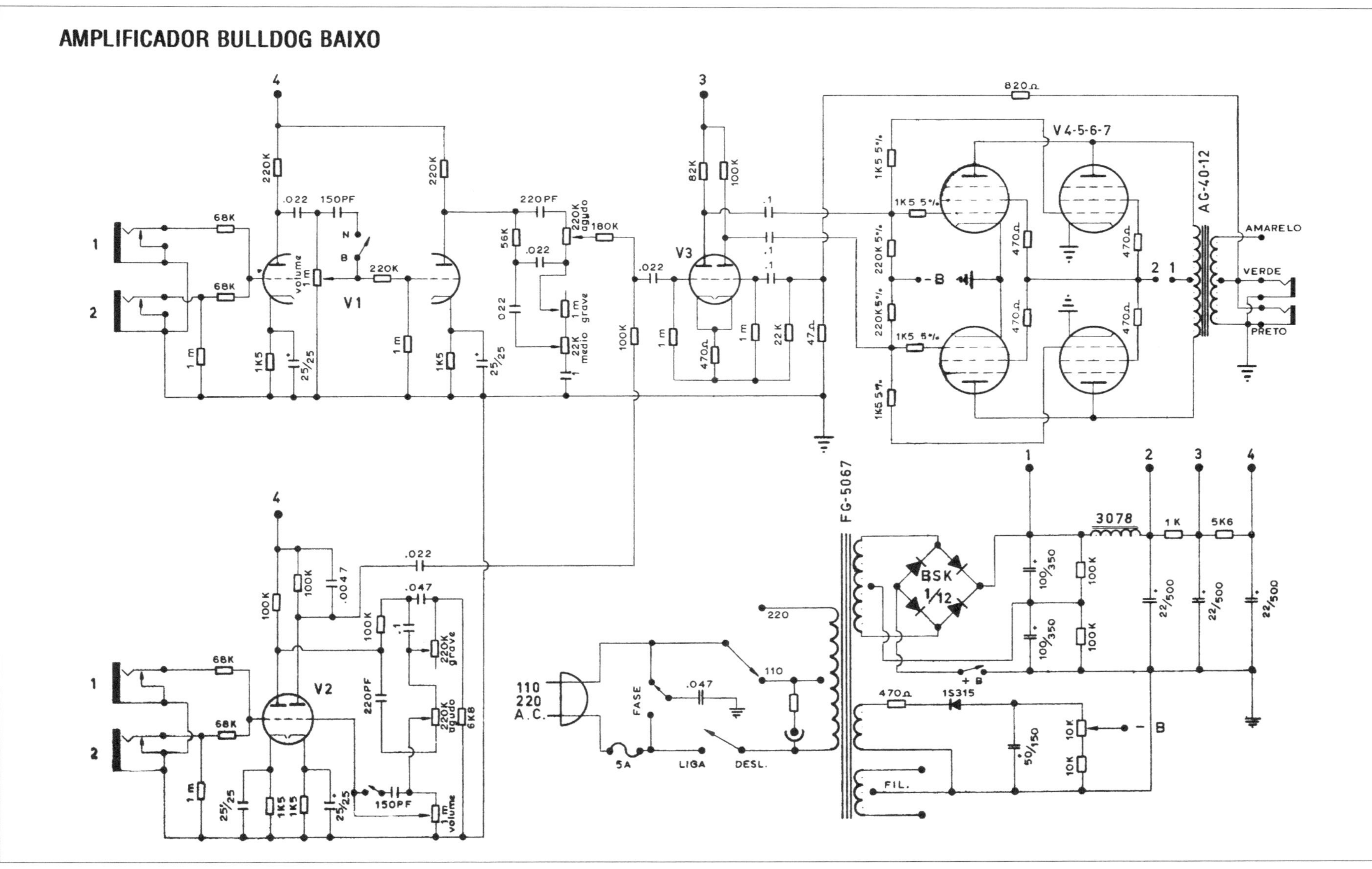

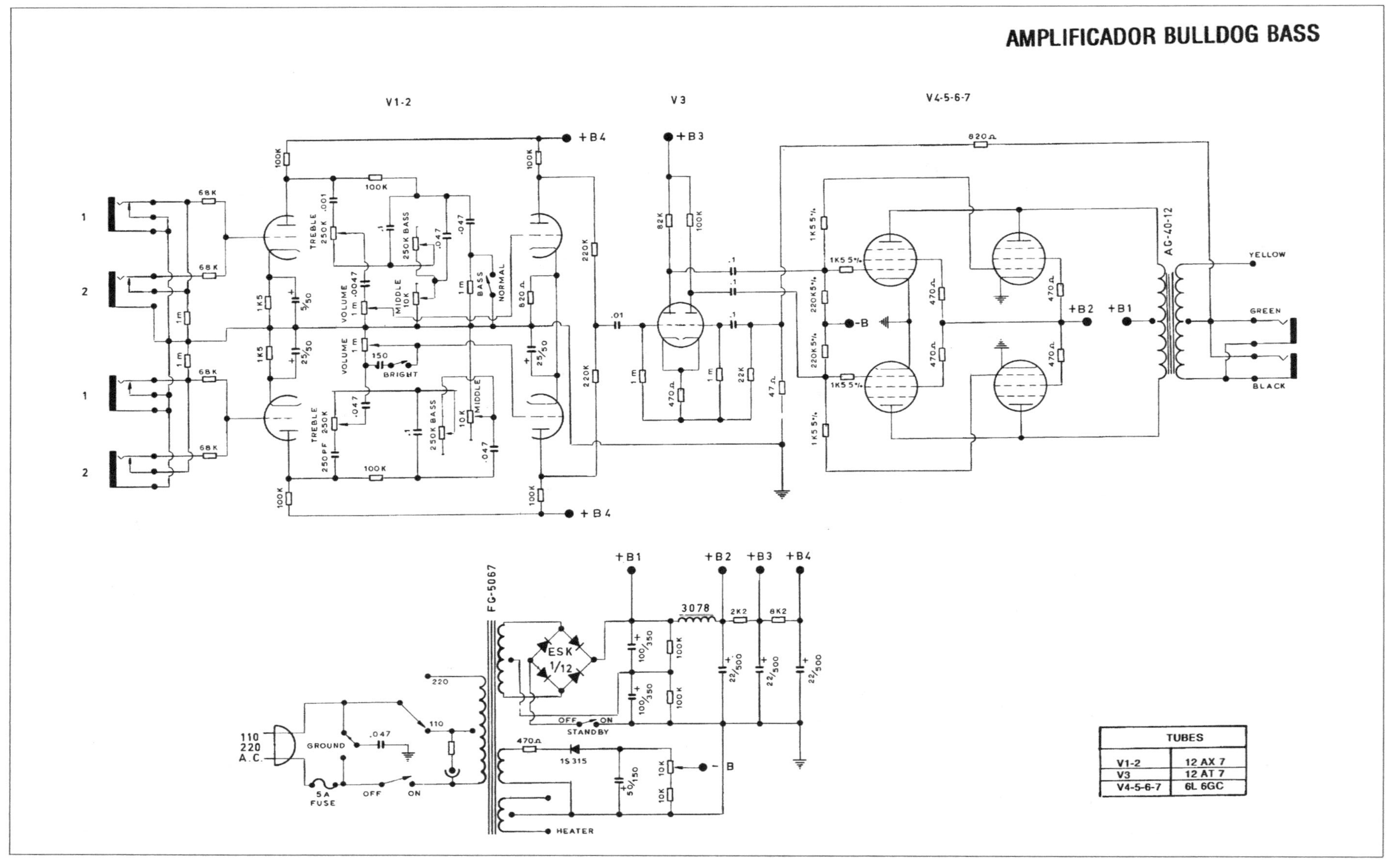

TUBES	
V1-2	12 AX 7
V3	12 AT 7
V4-5-6-7	6L 6GC

AMPLIFICADOR BULLDOG GUITARRA

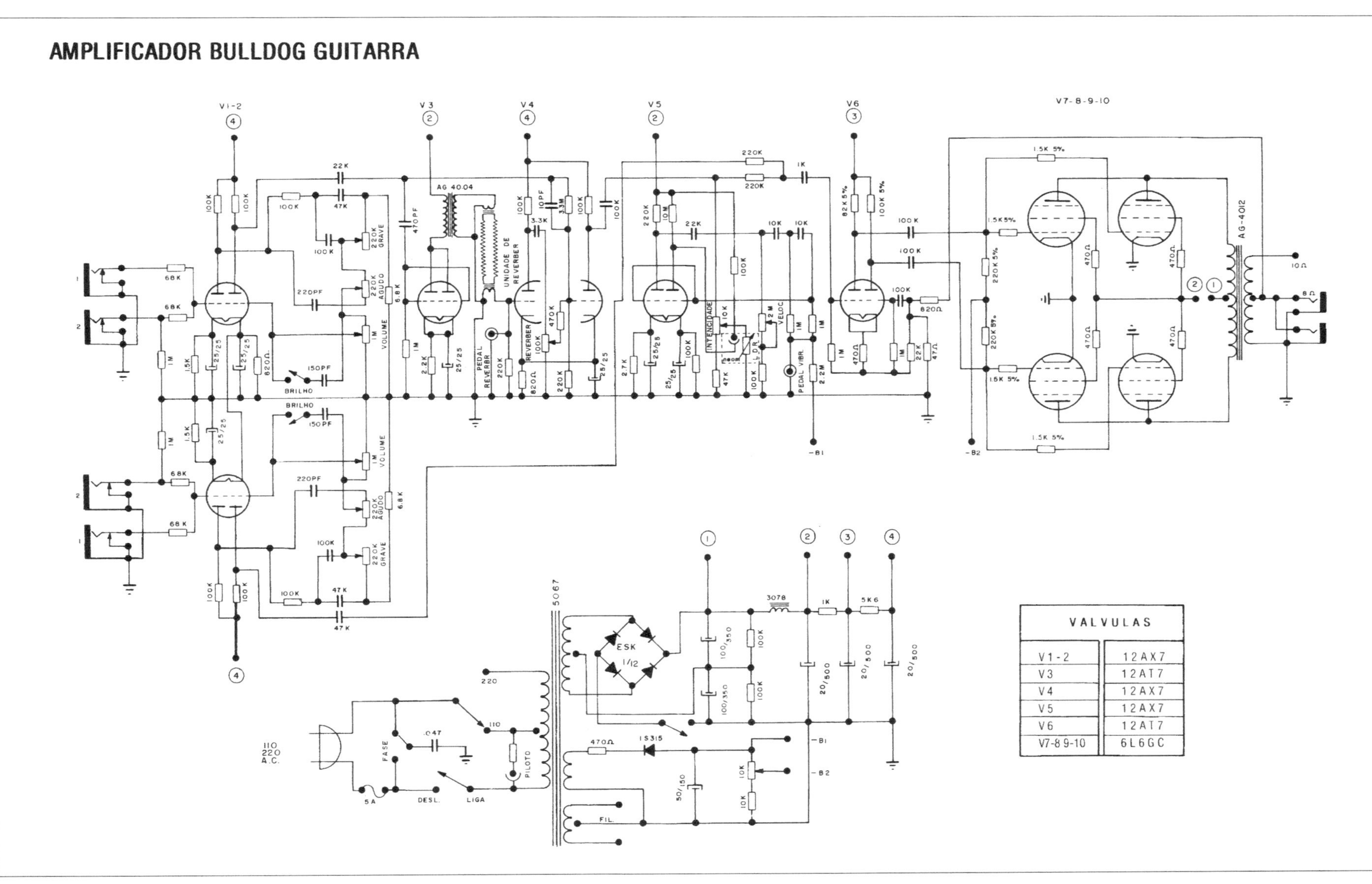

VALVULAS	
V1-2	12AX7
V3	12AT7
V4	12AX7
V5	12AX7
V6	12AT7
V7-8 9-10	6L6GC

AMPLIFICADOR DUOVOX 50B

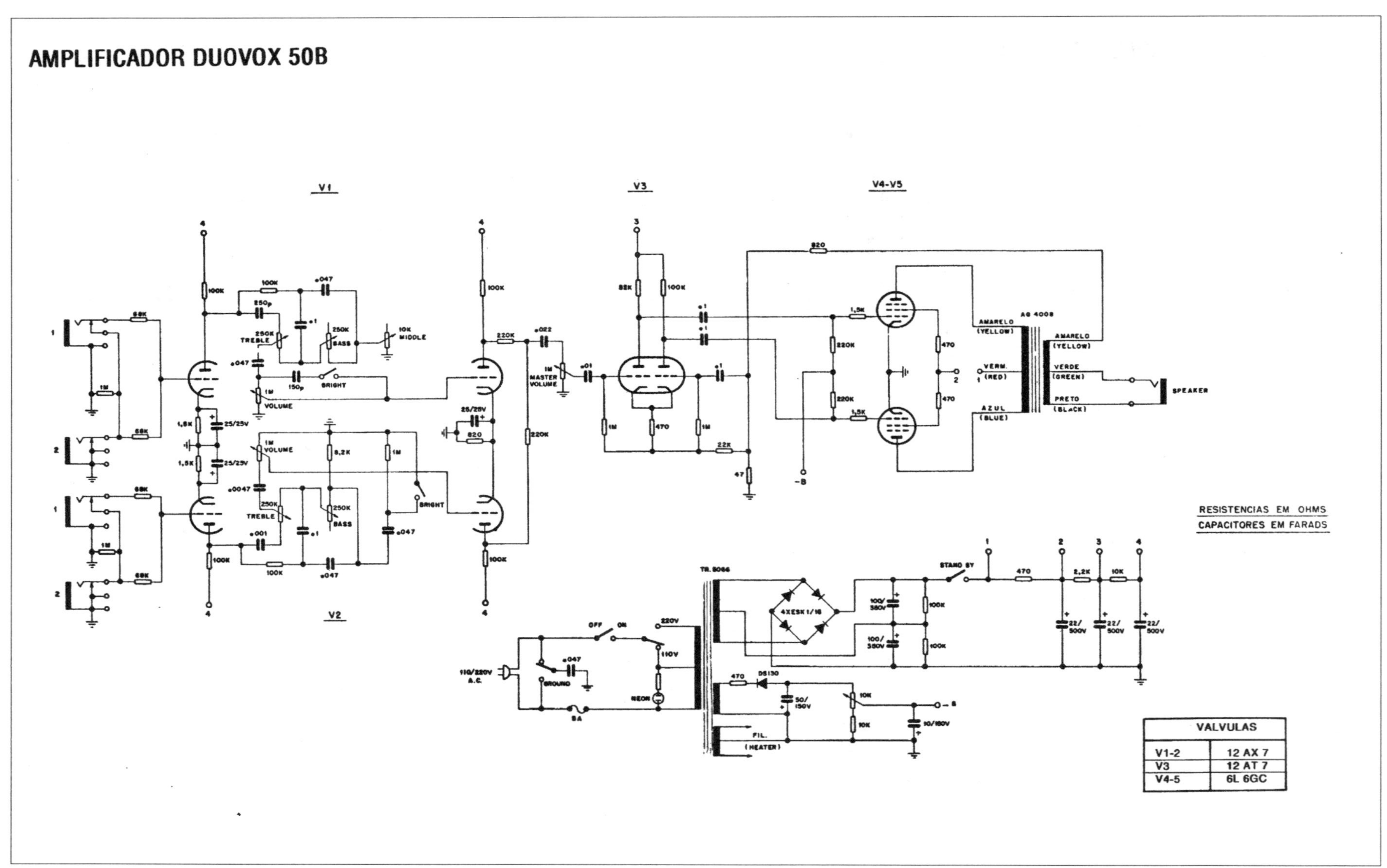

VALVULAS	
V1-2	12 AX 7
V3	12 AT 7
V4-5	6L 6GC

AMPLIFICADOR DUOVOX 50G

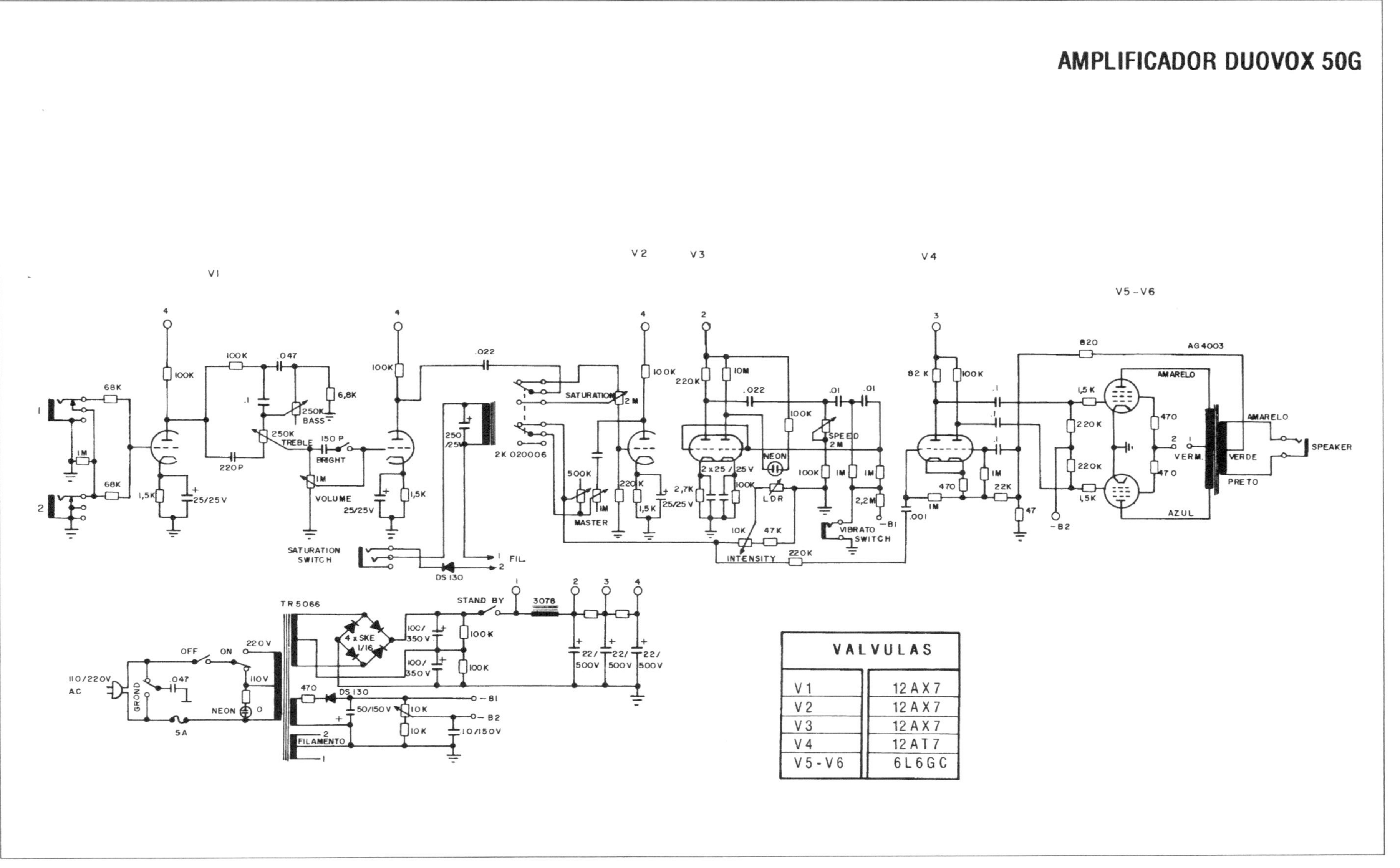

VALVULAS	
V1	12AX7
V2	12AX7
V3	12AX7
V4	12AT7
V5 - V6	6L6GC

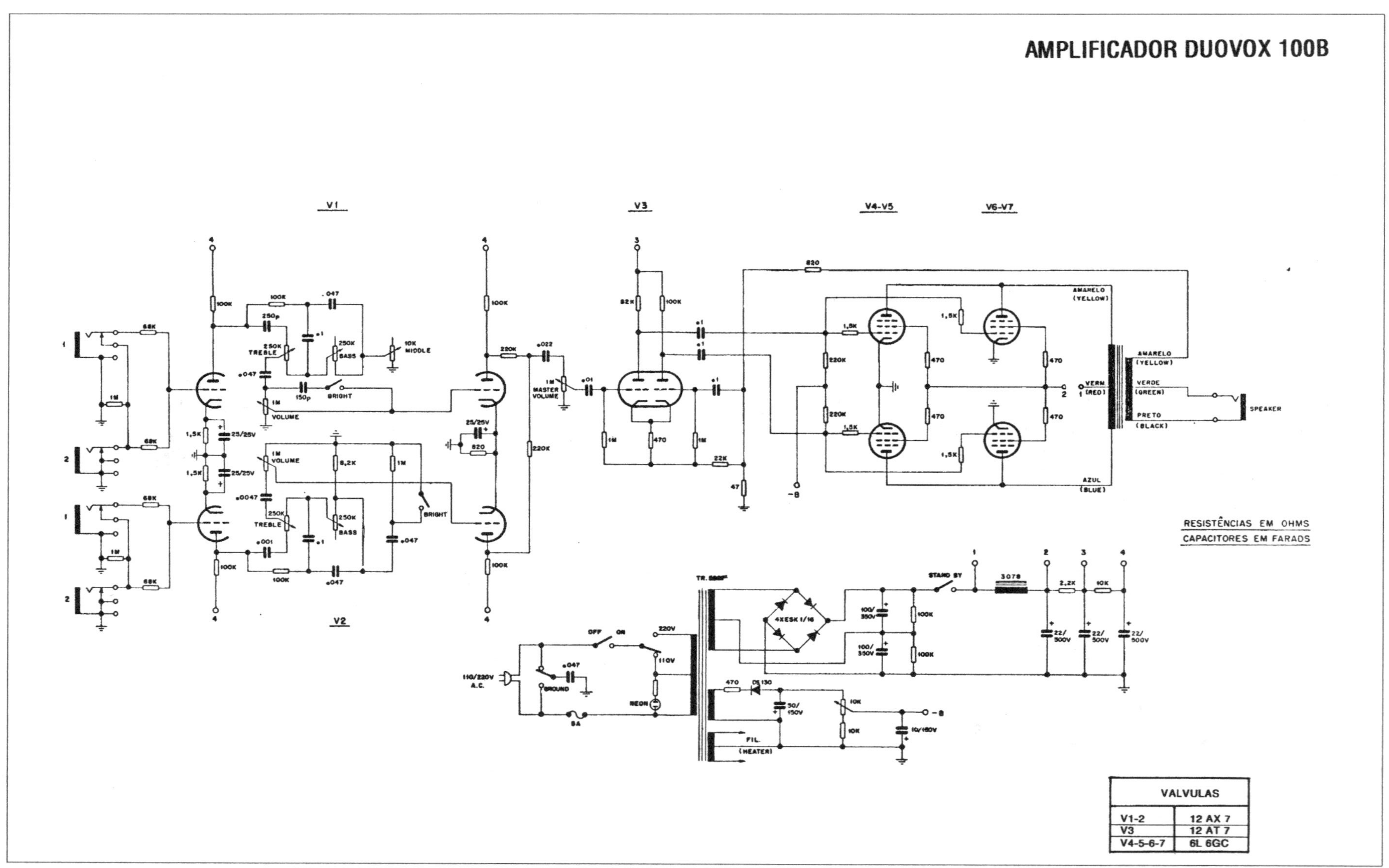
AMPLIFICADOR DUOVOX 100B
V1
V2
V3
V4-V5
V6-V7
AMARELO (YELLOW)
VERDE (GREEN)
PRETO (BLACK)
VERM (RED)
AZUL (BLUE)
SPEAKER
RESISTÊNCIAS EM OHMS
CAPACITORES EM FARADS
VOLUME
TREBLE
BASS
MIDDLE
BRIGHT
MASTER VOLUME
STAND BY
OFF
ON
GROUND
NEON
FIL. (HEATER)
4XESK 1/16
110/220V A.C.
VALVULAS
V1-2 12 AX 7
V3 12 AT 7
V4-5-6-7 6L 6GC

AMPLIFICADOR DUOVOX 100G

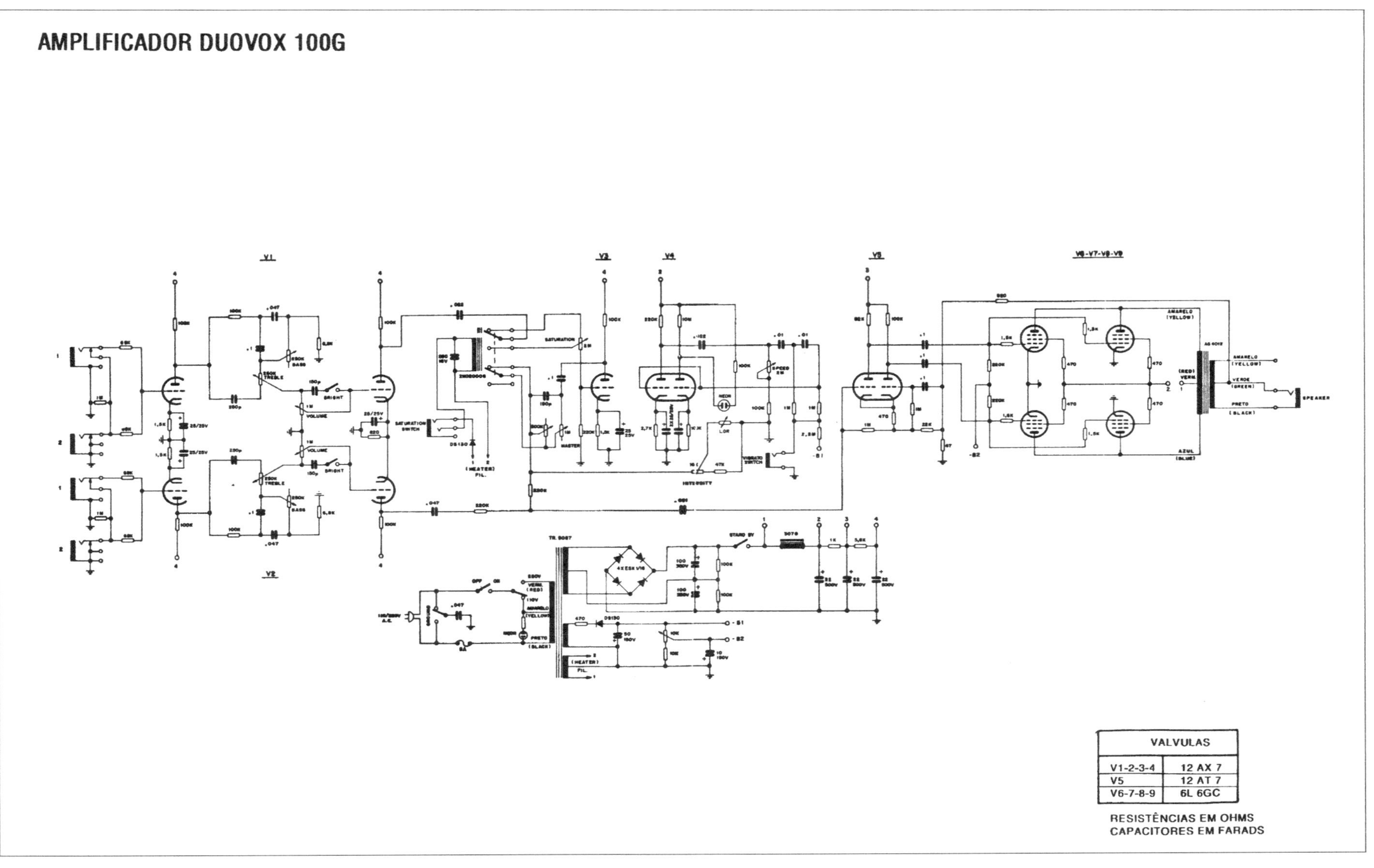

VALVULAS	
V1-2-3-4	12 AX 7
V5	12 AT 7
V6-7-8-9	6L 6GC

RESISTÊNCIAS EM OHMS
CAPACITORES EM FARADS

AMPLIFICADOR DUOVOX 120B

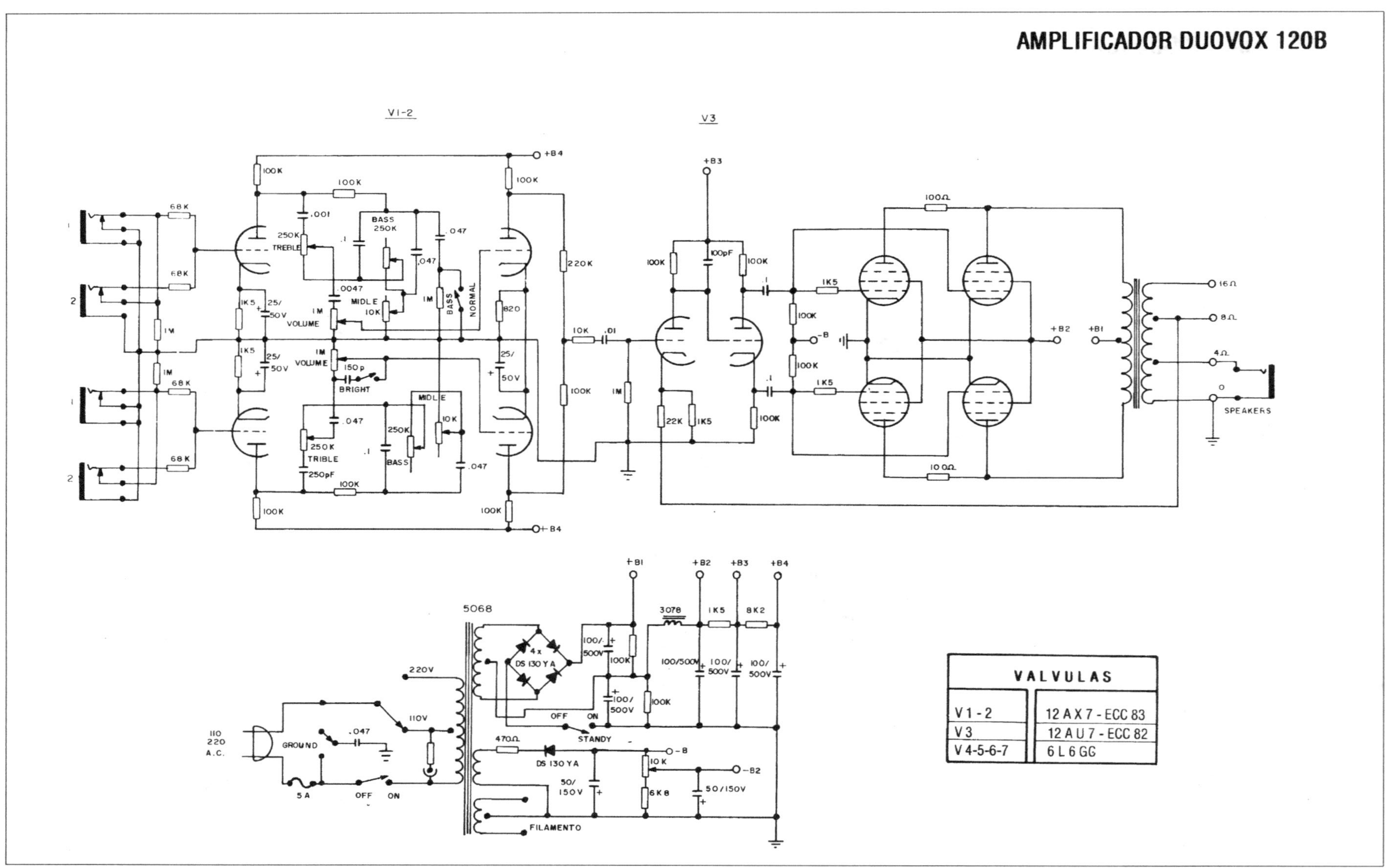

VALVULAS	
V1-2	12AX7 - ECC 83
V3	12AU7 - ECC 82
V 4-5-6-7	6L6GG

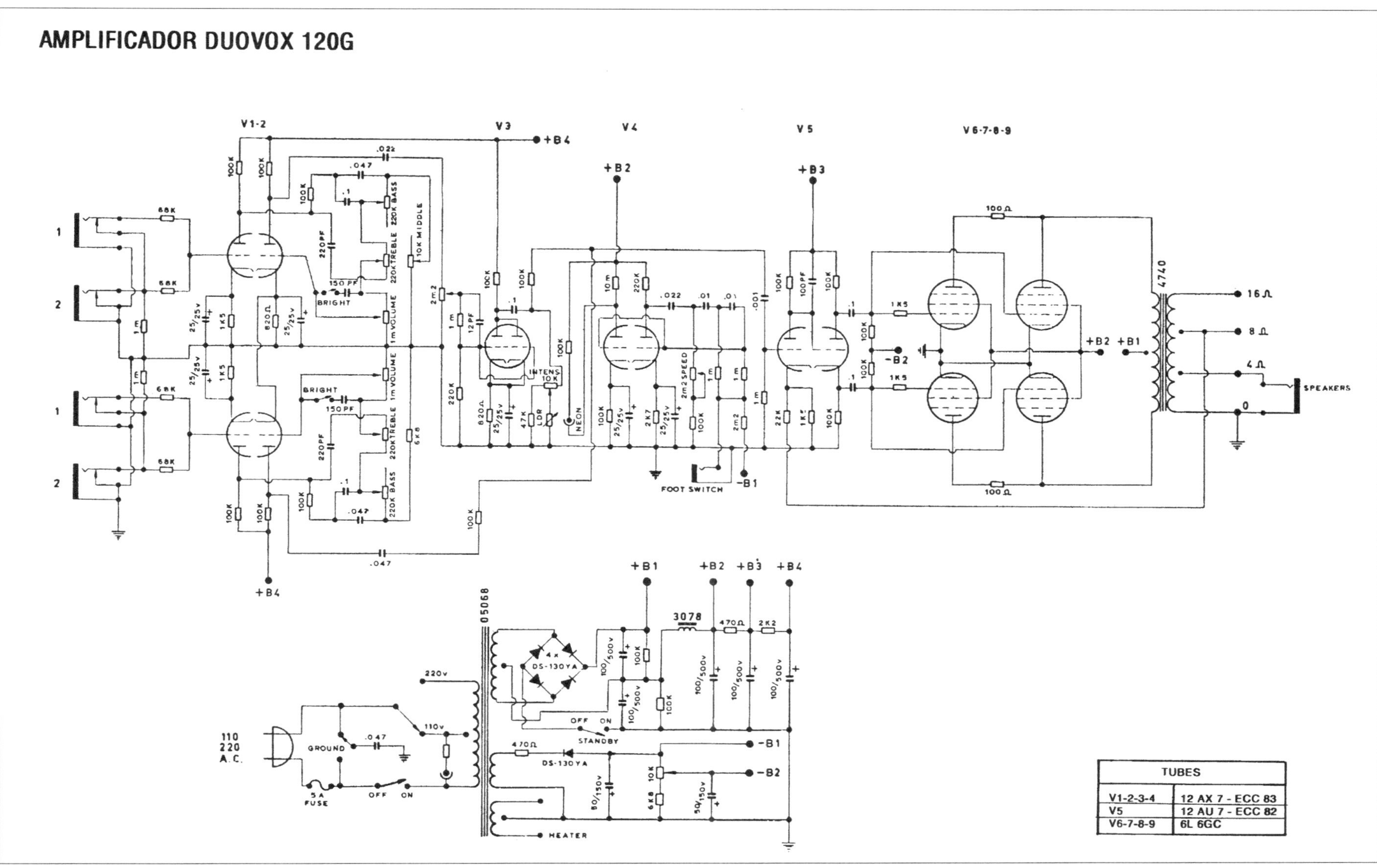

TUBES	
V1-2-3-4	12 AX 7 - ECC 83
V5	12 AU 7 - ECC 82
V6-7-8-9	6L 6GC

AMPLIFICADOR DUOVOX 150B

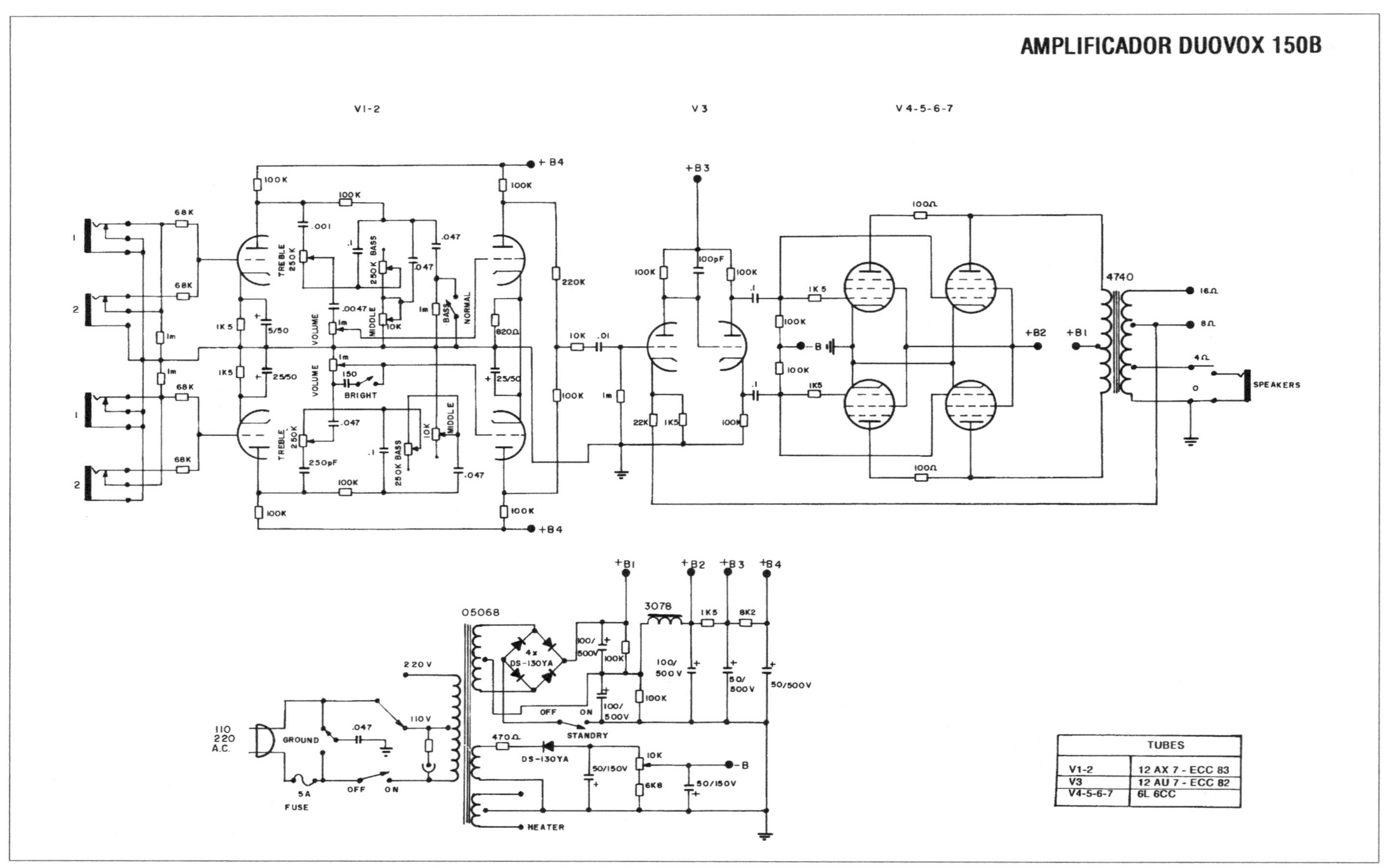

TUBES	
V1-2	12 AX 7 - ECC 83
V3	12 AU 7 - ECC 82
V4-5-6-7	6L 6CC

AMPLIFICADOR DUOVOX 150G

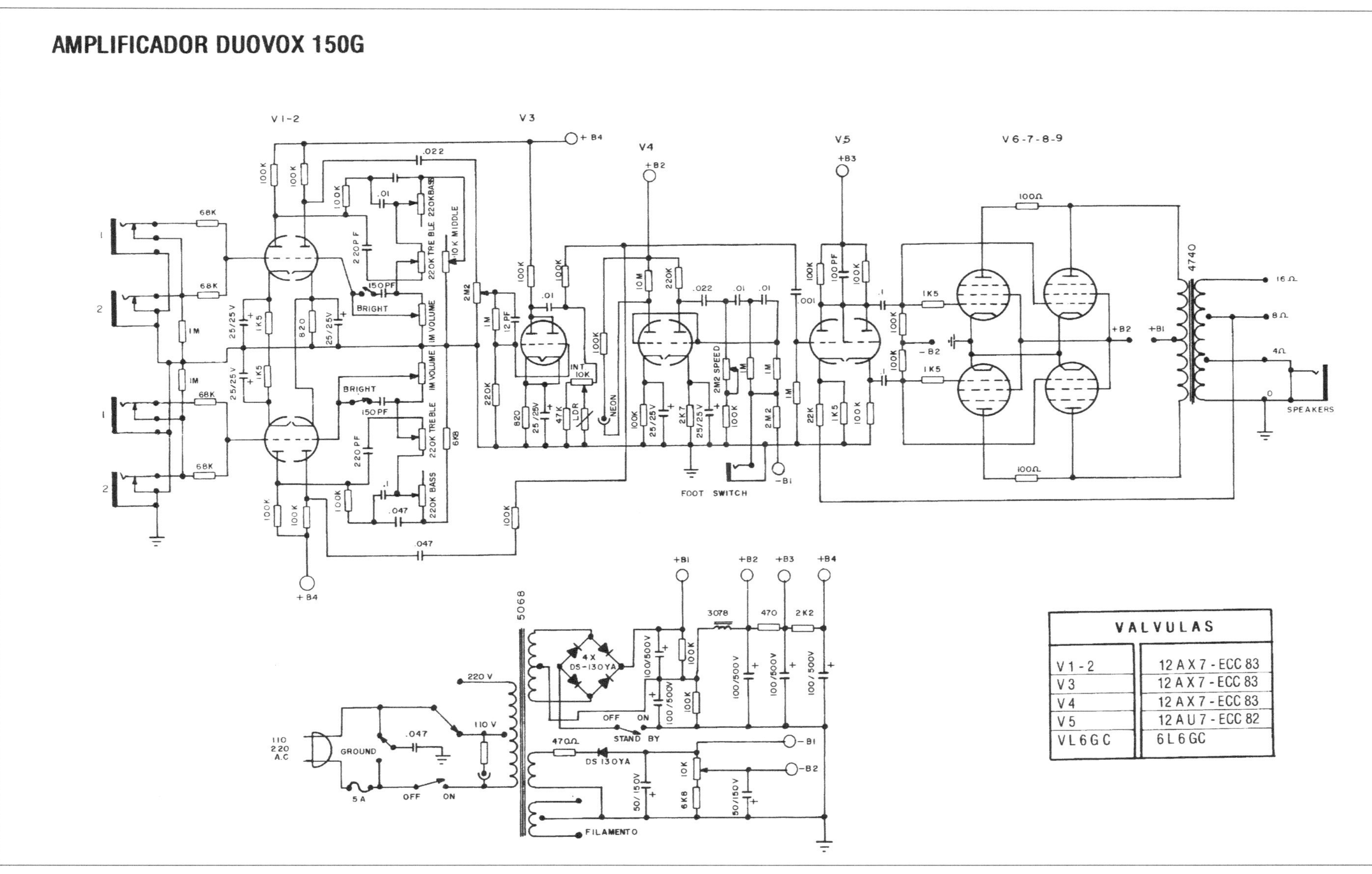

VALVULAS	
V 1 - 2	12 A X 7 - ECC 83
V 3	12 A X 7 - ECC 83
V 4	12 A X 7 - ECC 83
V 5	12 A U 7 - ECC 82
VL 6 G C	6 L 6 GC

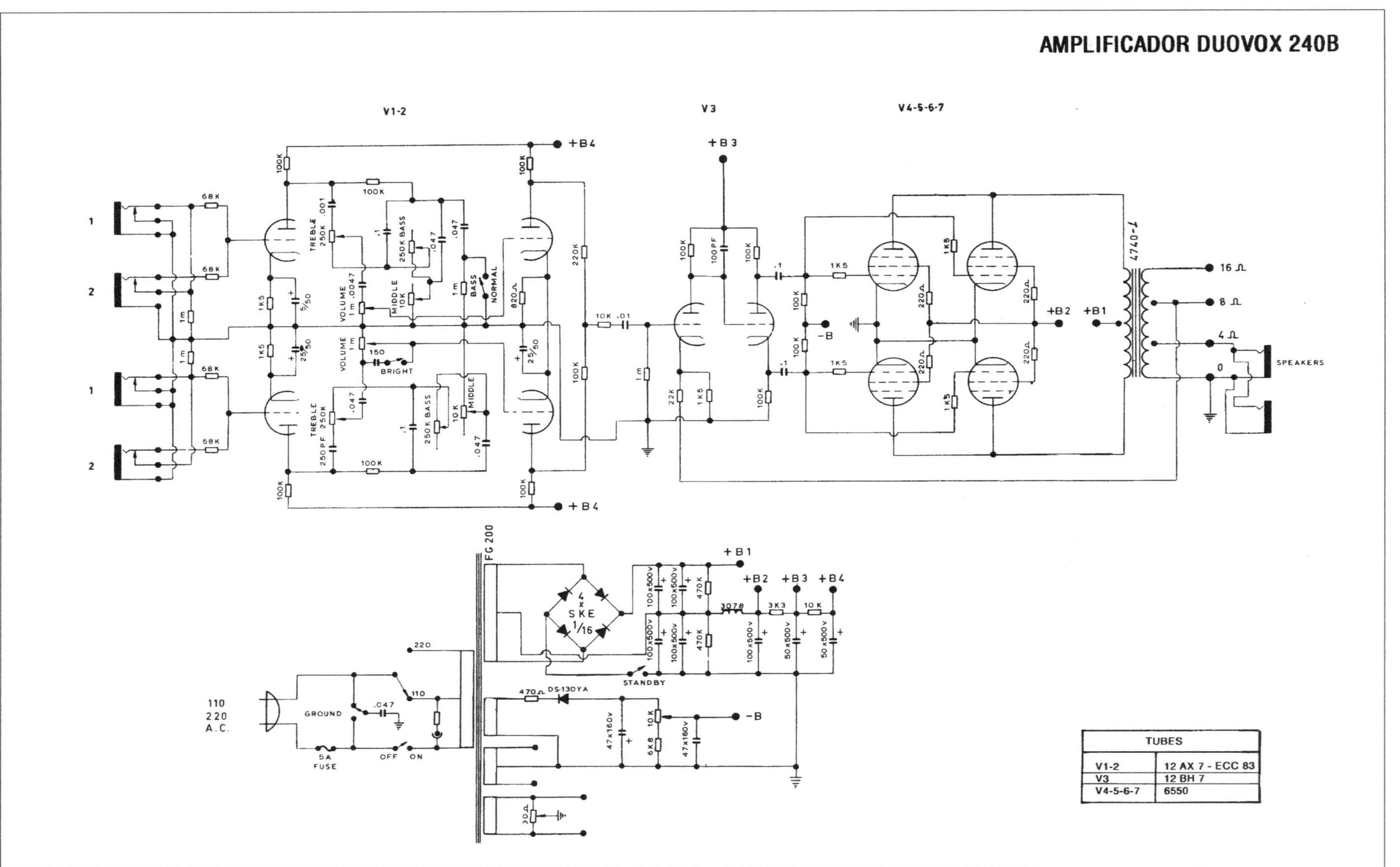

TUBES	
V1-2	12 AX 7 - ECC 83
V3	12 BH 7
V4-5-6-7	6550

AMPLIFICADOR DUOVOX 240G

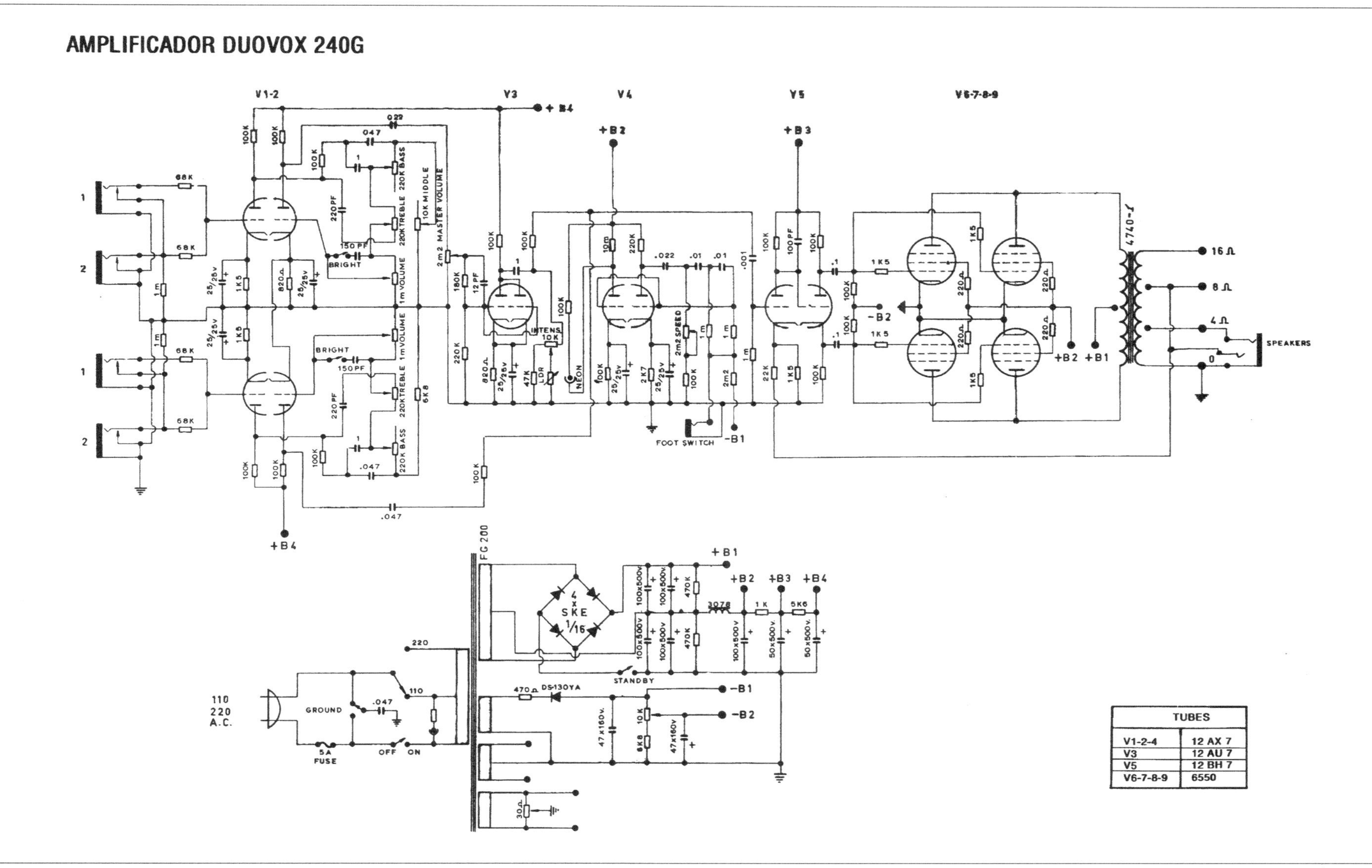

TUBES	
V1-2-4	12 AX 7
V3	12 AU 7
V5	12 BH 7
V6-7-8-9	6550

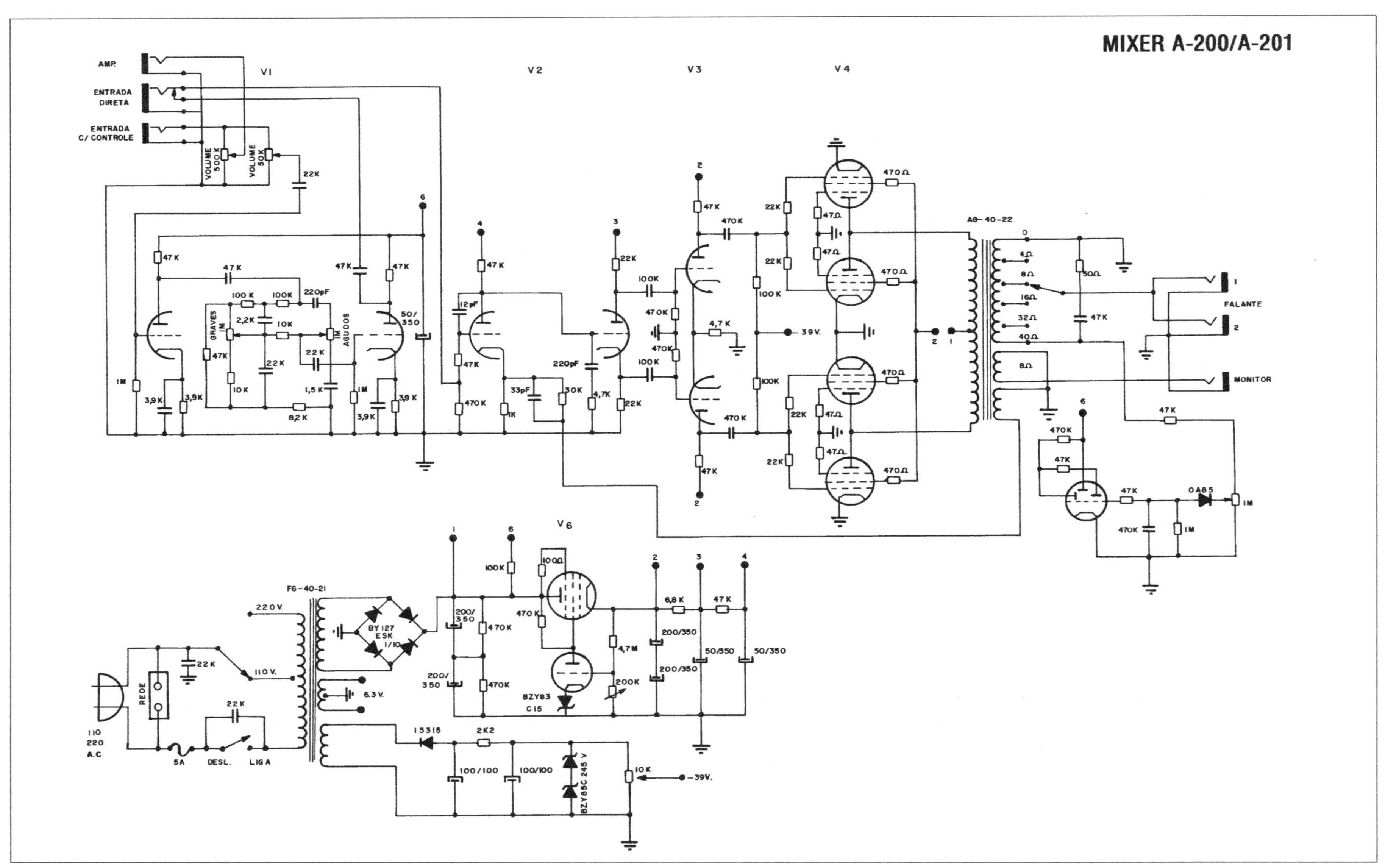
MIXER A-200/A-201
AMP.
ENTRADA DIRETA
ENTRADA C/ CONTROLE
V1
V2
V3
V4
V6
VOLUME 500K
VOLUME 50K
GRAVES 1M
AGUDOS 1M
FALANTE
MONITOR
A8-40-22
F6-40-21
BY 127 ESK 1/10
BZY83 C15
BZY85C 245 V
OA85
REDE
110 220 A.C
5A
DESL.
LIGA
220 V.
110 V.
6.3 V.
-39V.

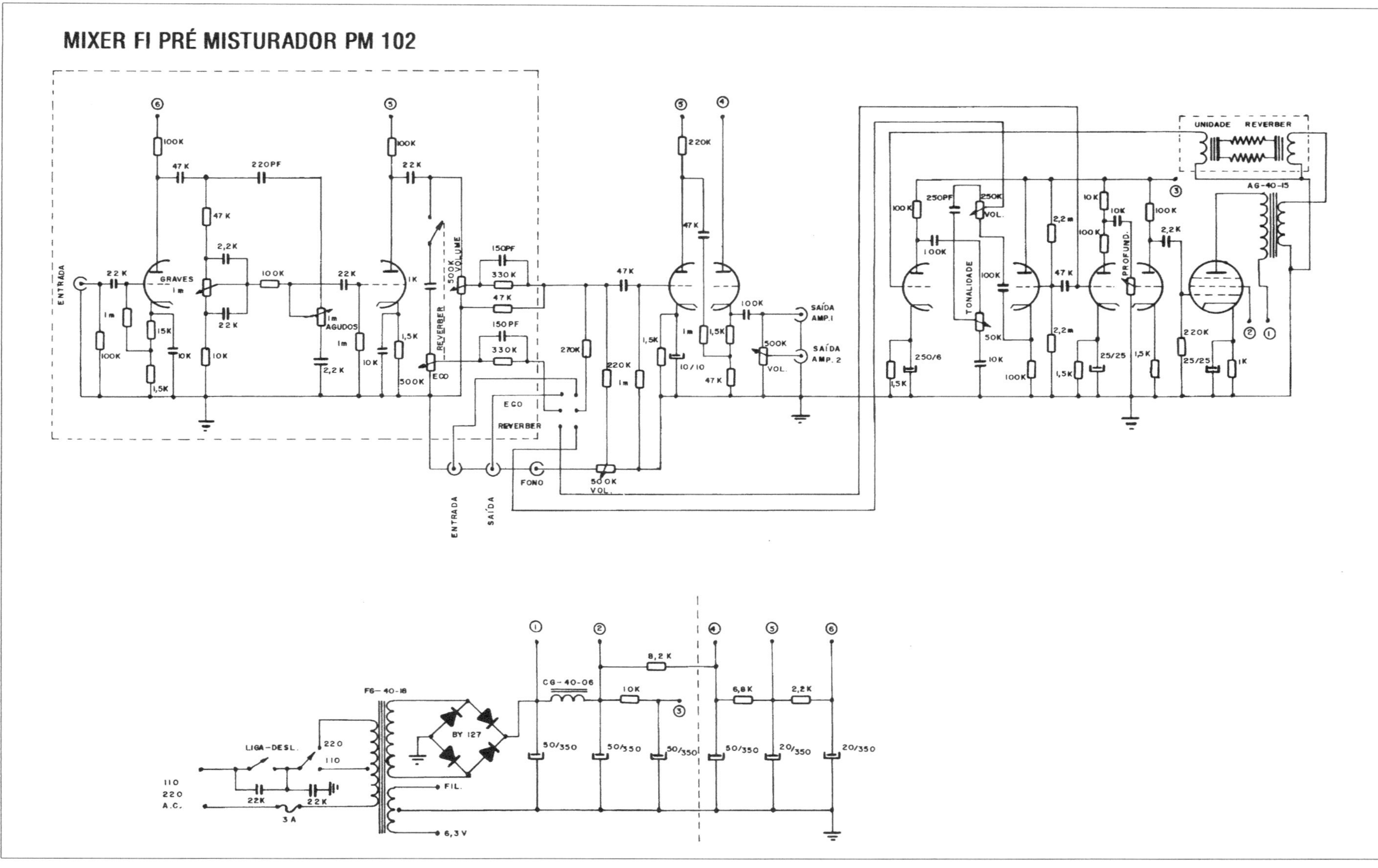
MIXER FI PRÉ MISTURADOR PM 102
ENTRADA
GRAVES
AGUDOS
REVERBER
ECO
VOLUME
FONO
SAÍDA
VOL.
SAÍDA AMP. 1
SAÍDA AMP. 2
TONALIDADE
PROFUND.
UNIDADE REVERBER
AG-40-15
LIGA-DESL.
110
220
A.C.
3 A
F6-40-18
BY 127
CG-40-06
FIL.
6,3 V

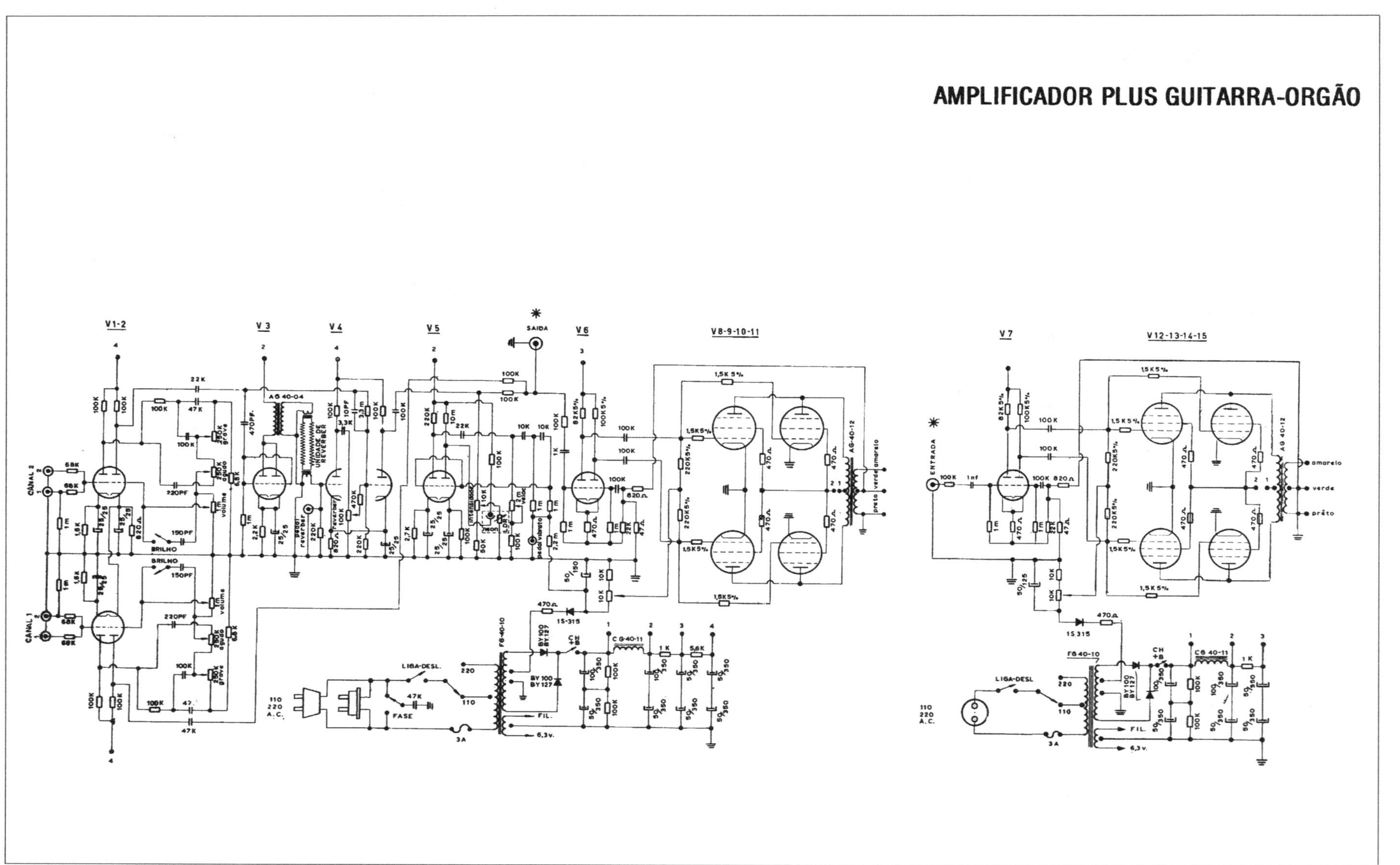
AMPLIFICADOR PLUS GUITARRA-ORGÃO
V1-2
V3
V4
V5
V6
V8-9-10-11
V7
V12-13-14-15
CANAL 2
CANAL 1
SAIDA
ENTRADA
BRILHO
volume
agudo
grave
AG 40-04
UNIDADE DE REVERBER
AG-40-12
AG 40-12
amarelo
verde
prêto
FG-40-10
FG 40-10
CG-40-11
CG 40-11
LIGA-DESL.
FASE
110
220
A.C.
FIL.
6,3v.
3A
1S-315
BY 100
BY 127

AMPLIFICADOR PRÉ SUPER TREMENDÃO

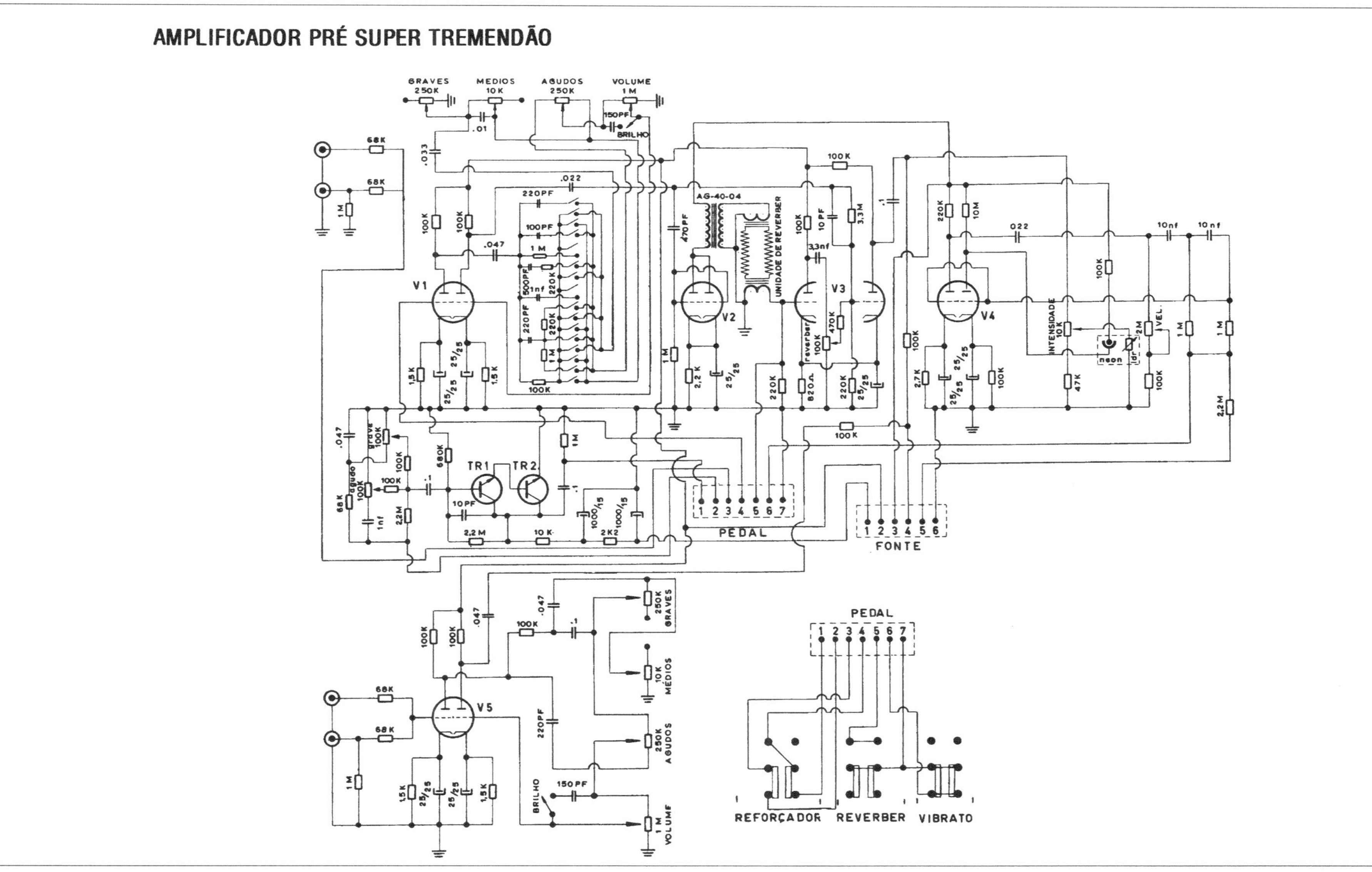

AMPLIFICADOR SUPER TREMENDÃO

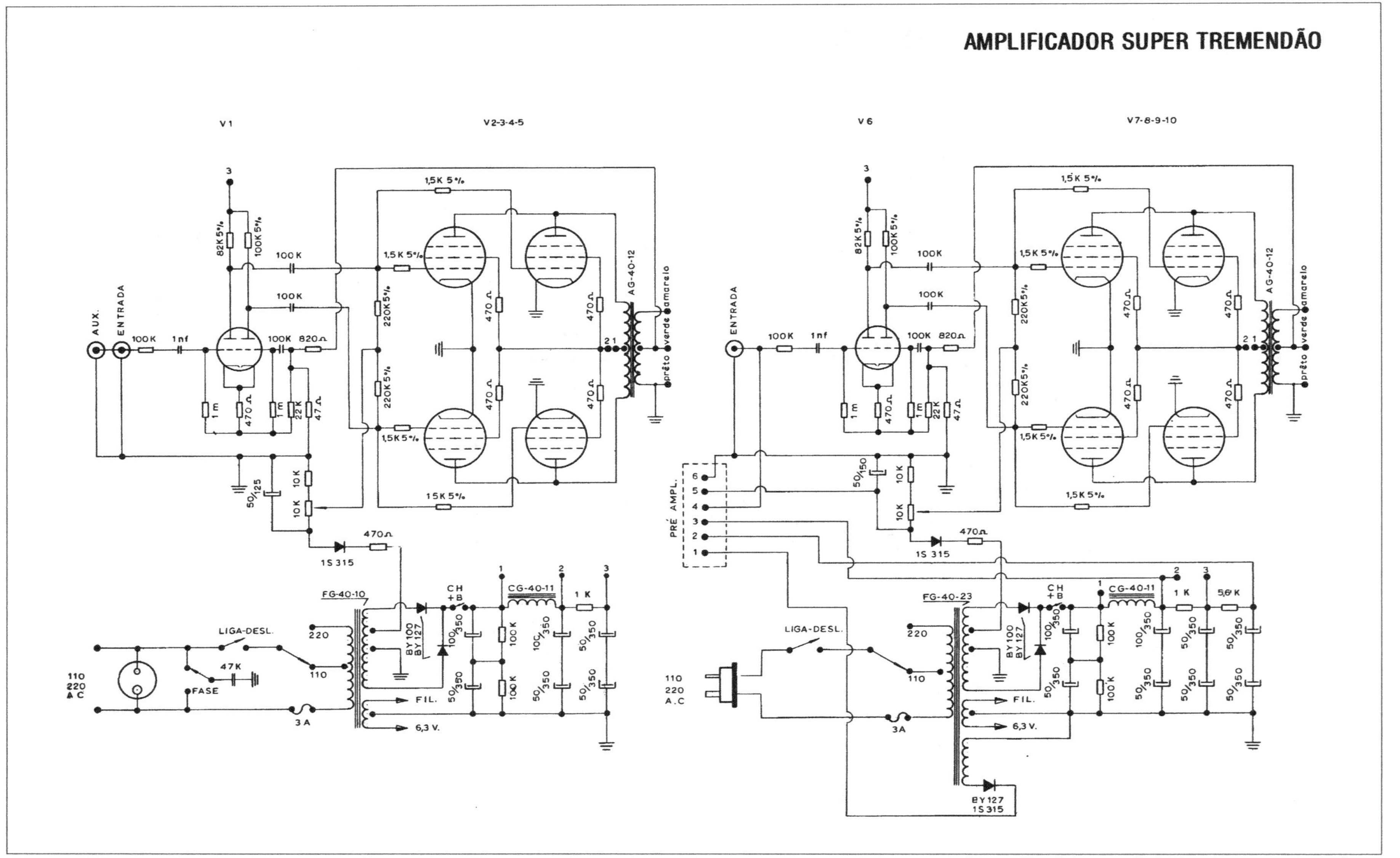

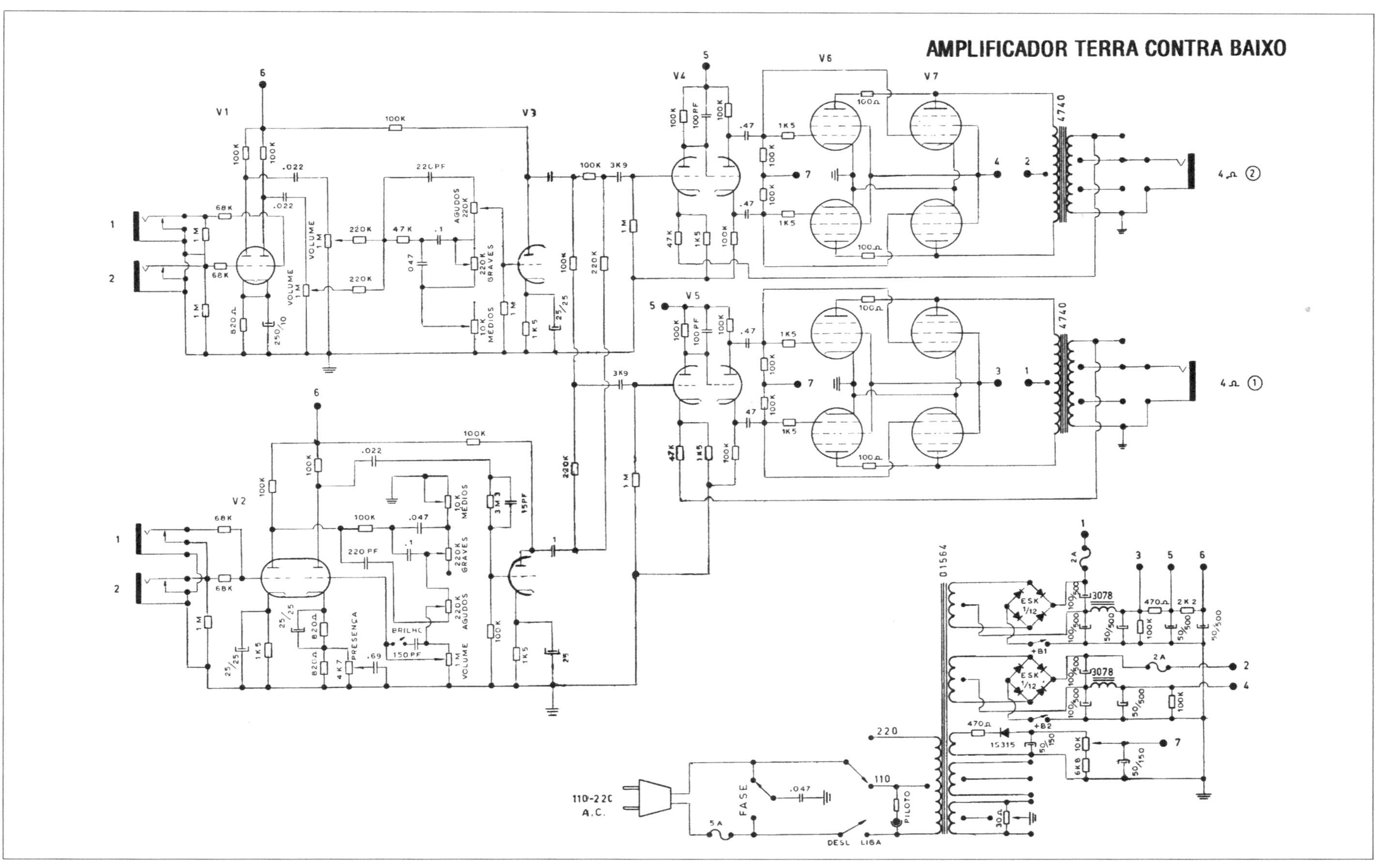
AMPLIFICADOR TERRA CONTRA BAIXO
V1
V2
V3
V4
V5
V6
V7
VOLUME
AGUDOS
GRAVES
MEDIOS
PRESENÇA
BRILHO
FASE
PILOTO
DESL
LIGA
110-220
A.C.
01564
4740
3078
ESK
1S315
+B1
+B2
4 Ω ①
4 Ω ②

AMPLIFICADOR TERRA GUITARRA

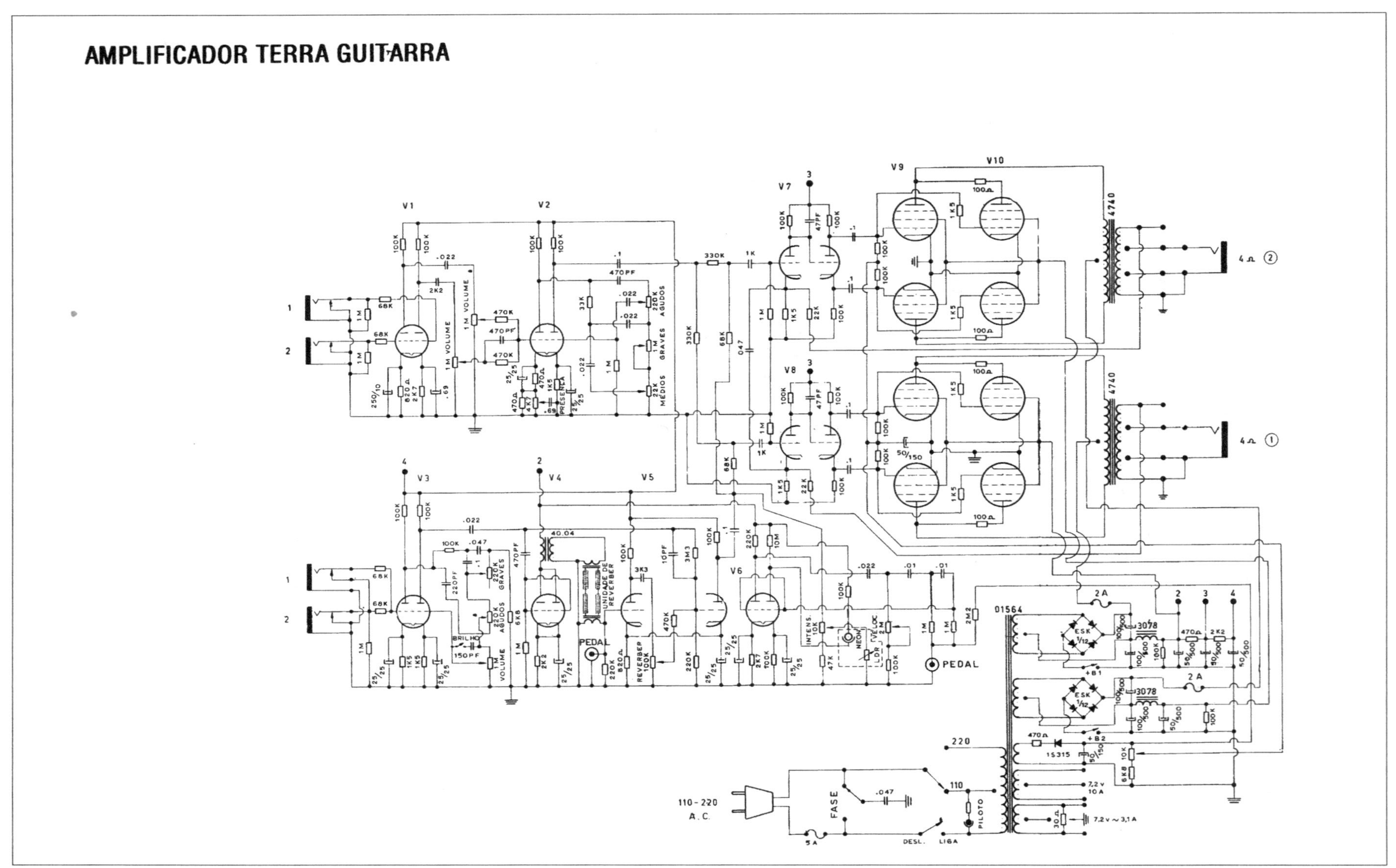

AMPLIFICADOR THOR CONTRA BAIXO

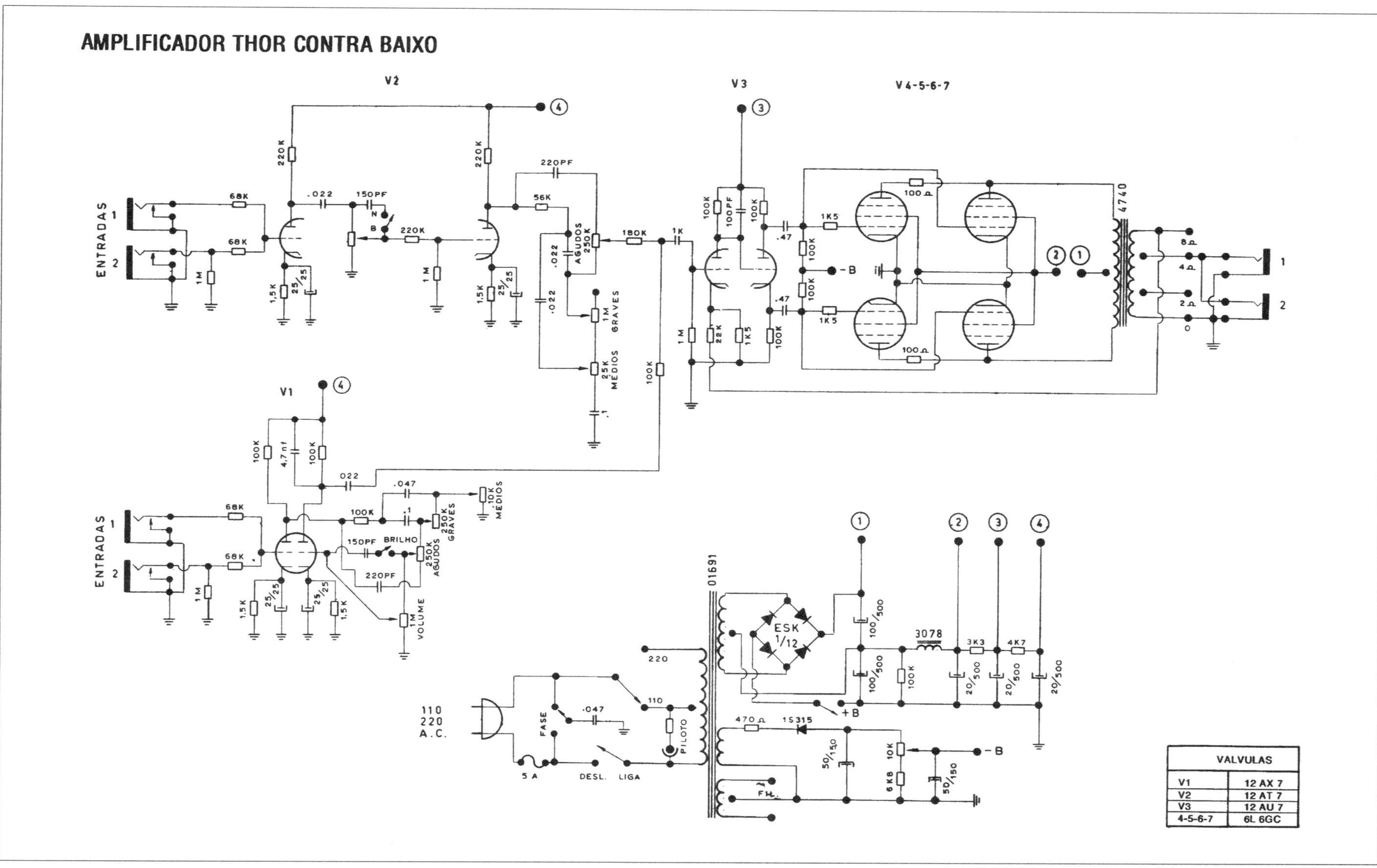

VALVULAS	
V1	12 AX 7
V2	12 AT 7
V3	12 AU 7
4-5-6-7	6L 6GC

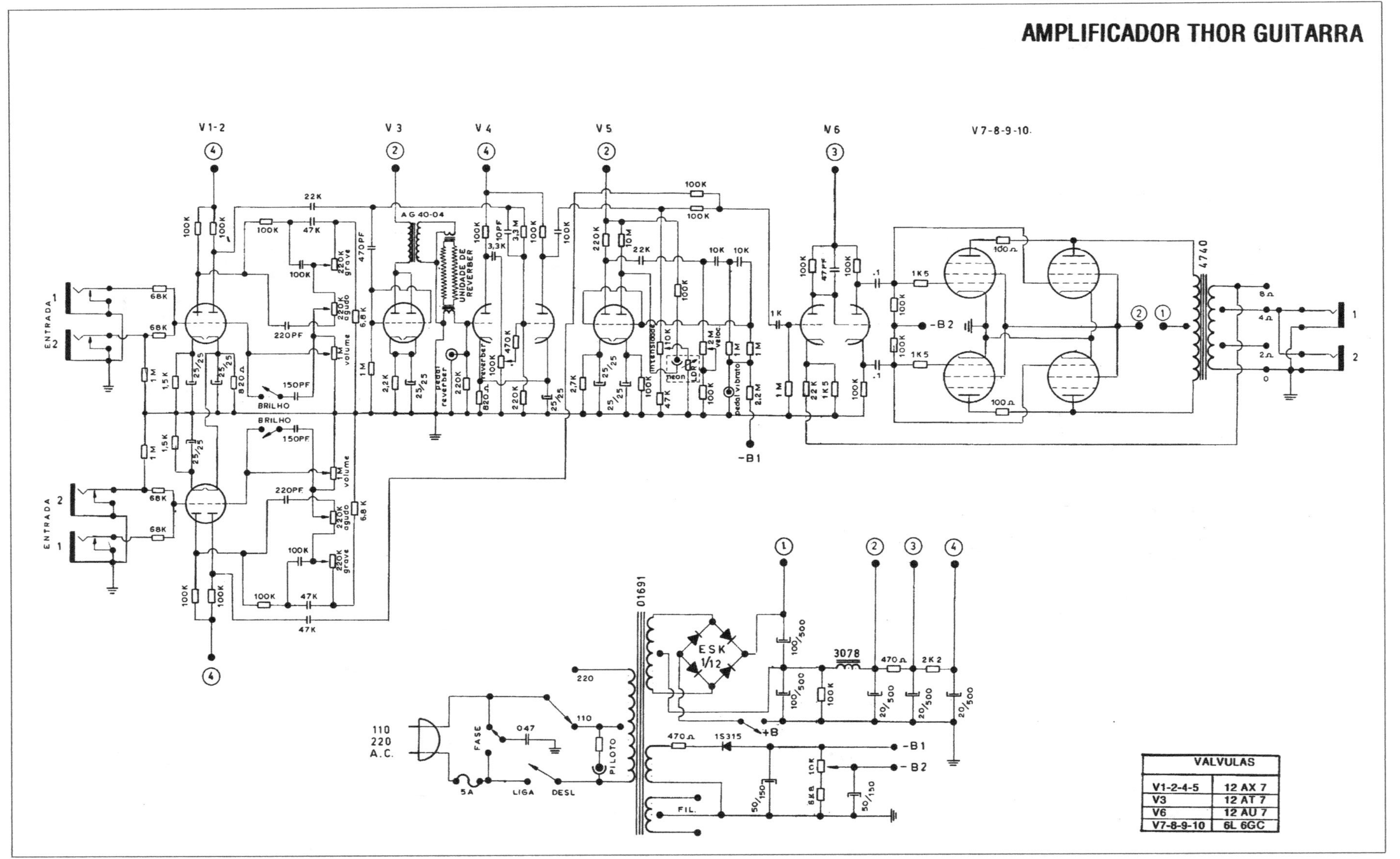
AMPLIFICADOR THOR GUITARRA
V 1-2
V 3
V 4
V 5
V 6
V 7-8-9-10.
ENTRADA
AG 40-04
UNIDADE DE REVERBER
pedal reverber
pedal vibrato
intensidade
veloc
neon
LDR
BRILHO
volume
agudo
grave
4740
01691
ESK 1/12
3078
1S315
FIL.
PILOTO
FASE
LIGA
DESL
5A
110 220 A.C.
-B1
-B2
+B
VALVULAS
V1-2-4-5 12 AX 7
V3 12 AT 7
V6 12 AU 7
V7-8-9-10 6L 6GC

AMPLIFICADOR THUNDER SOUND III (N)

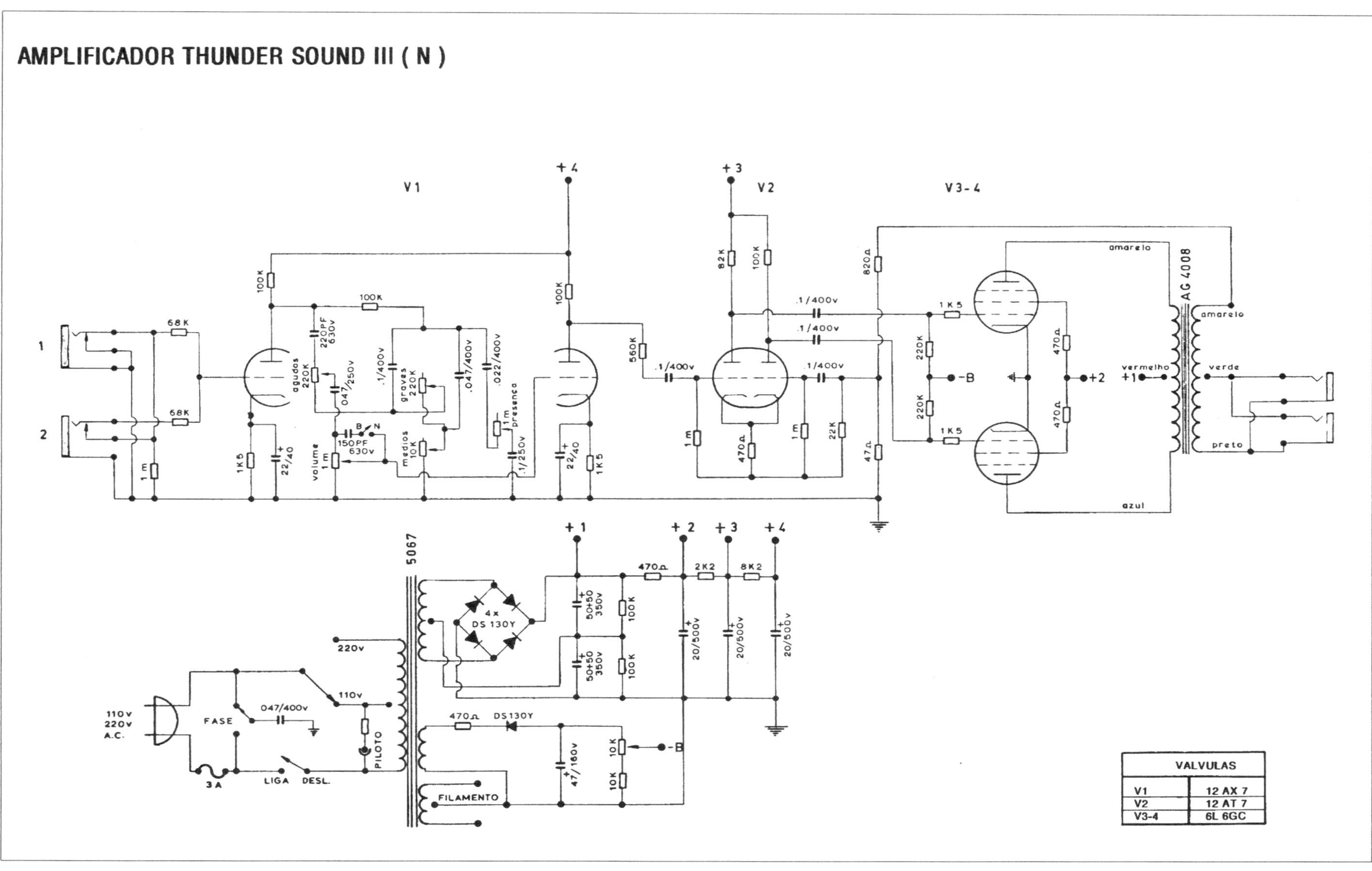

VALVULAS	
V1	12 AX 7
V2	12 AT 7
V3-4	6L 6GC

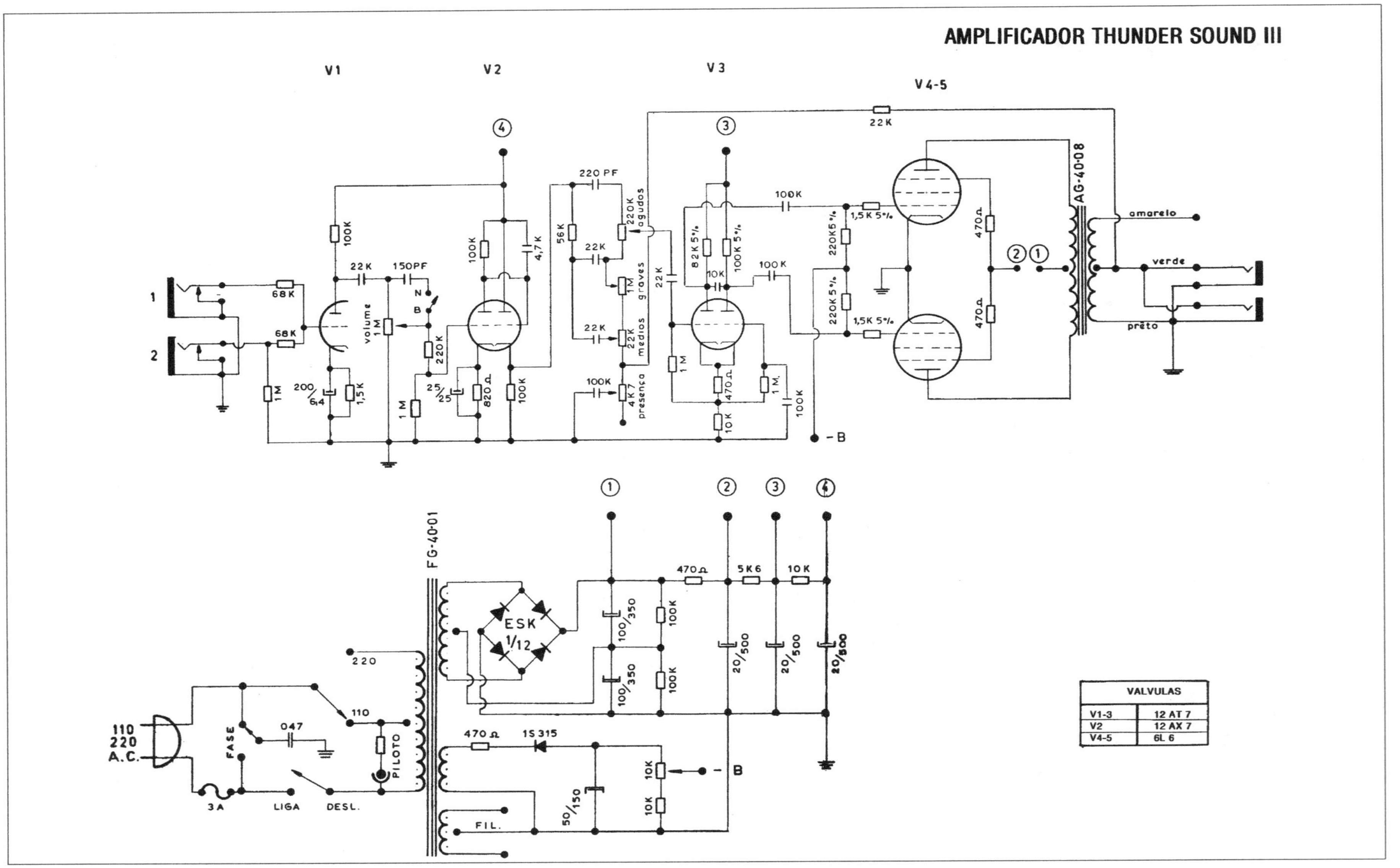
AMPLIFICADOR THUNDER SOUND III
V1
V2
V3
V4-5
1
2
68K
68K
1M
200/6,4
1,5K
100K
22K
150PF
volume 1M
N
B
220K
1M
25/25
820Ω
100K
100K
4,7K
220 PF
56K
22K
22K
220K agudos
1M graves
22K medios
4K7 presença
100K
22K
82K 5%
100K 5%
10K
100K
1M
470Ω
10K
1M
100K
100K
220K 5%
220K 5%
1,5K 5%
1,5K 5%
470Ω
470Ω
22K
AG-40-08
amarelo
verde
prêto
- B
FG-40-01
ESK 1/12
100/350
100/350
100K
100K
470Ω
5K6
10K
20/500
20/500
20/500
470Ω
1S315
50/150
10K
10K
FIL.
220
110
110
220
A.C.
FASE
047
PILOTO
3A
LIGA
DESL.
VALVULAS
V1-3 12 AT 7
V2 12 AX 7
V4-5 6L 6

AMPLIFICADOR TREMENDÃO II

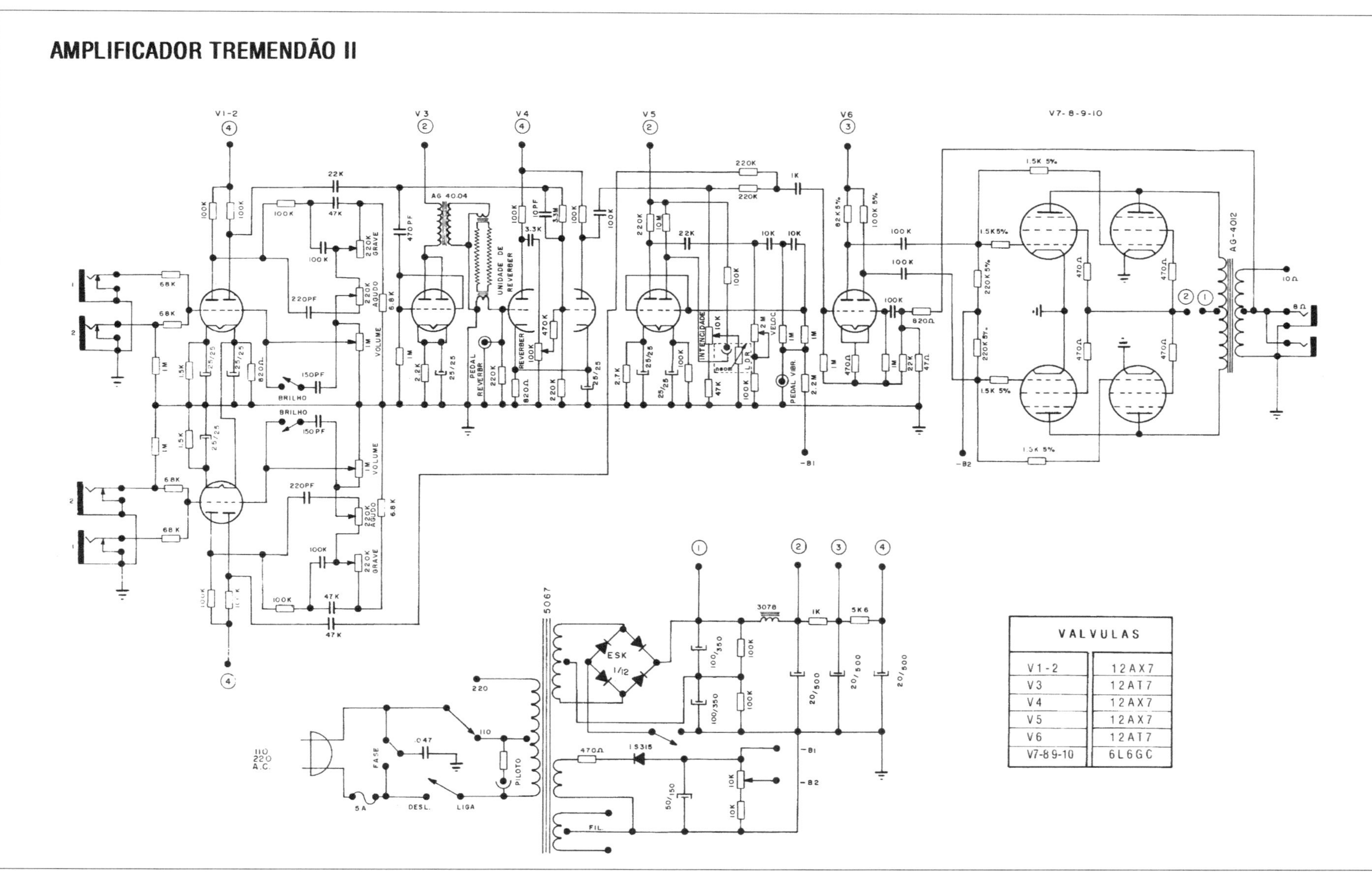

VALVULAS	
V1-2	12AX7
V3	12AT7
V4	12AX7
V5	12AX7
V6	12AT7
V7-8 9-10	6L6GC

AMPLIFICADOR TREMENDÃO

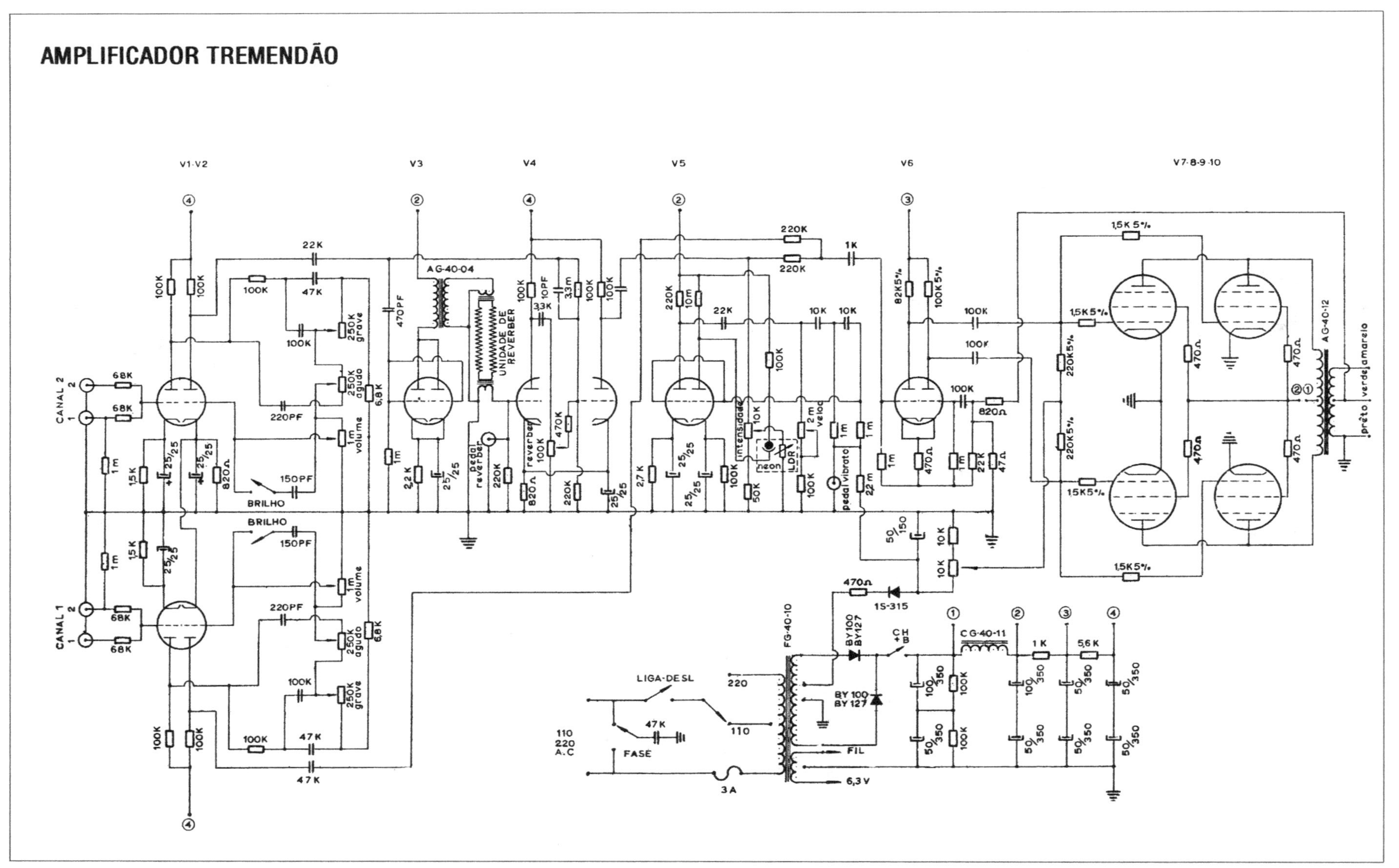

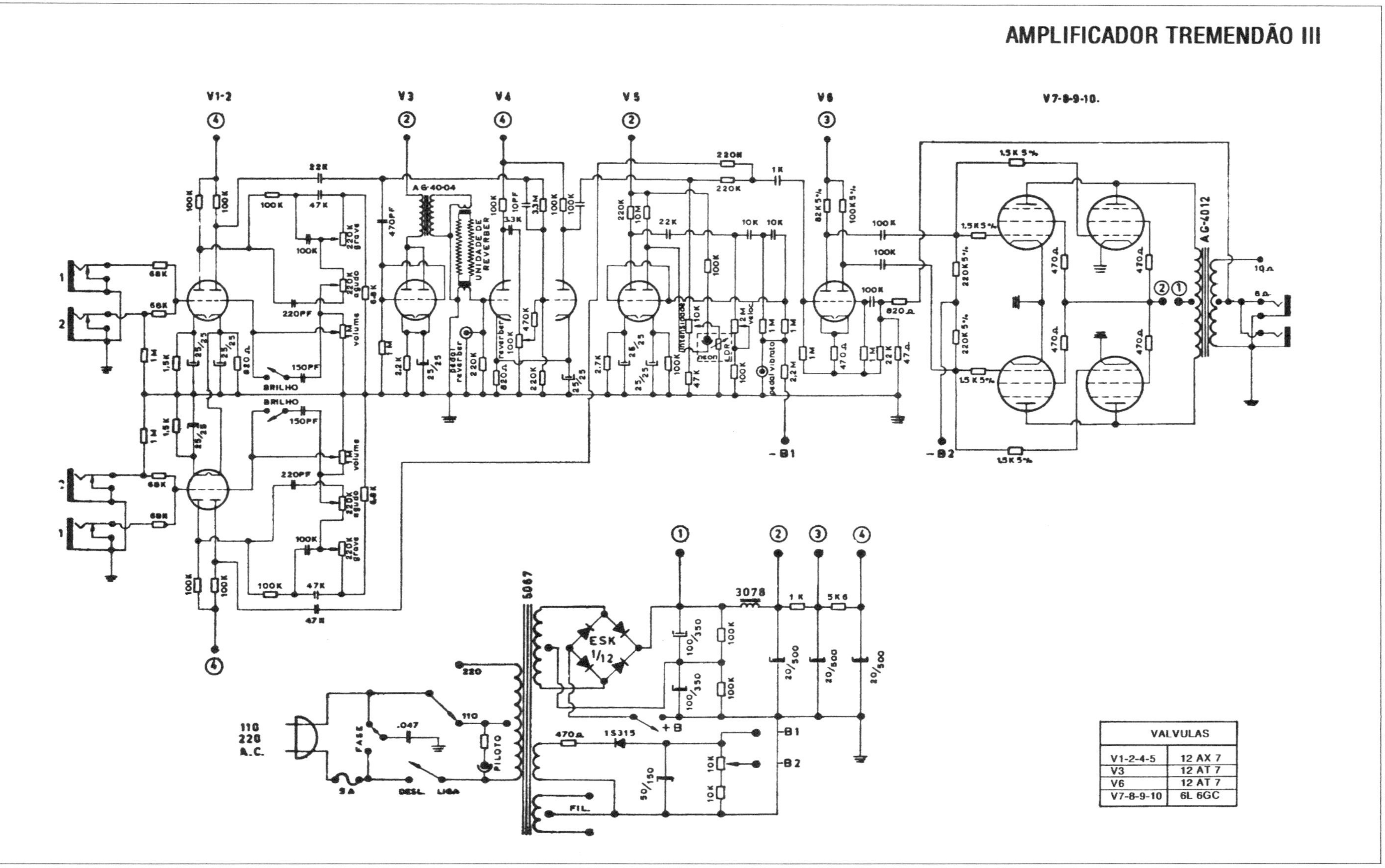

VALVULAS	
V1-2-4-5	12 AX 7
V3	12 AT 7
V6	12 AT 7
V7-8-9-10	6L 6GC

AMPLIFICADOR TREMENDÃO COMPACTO

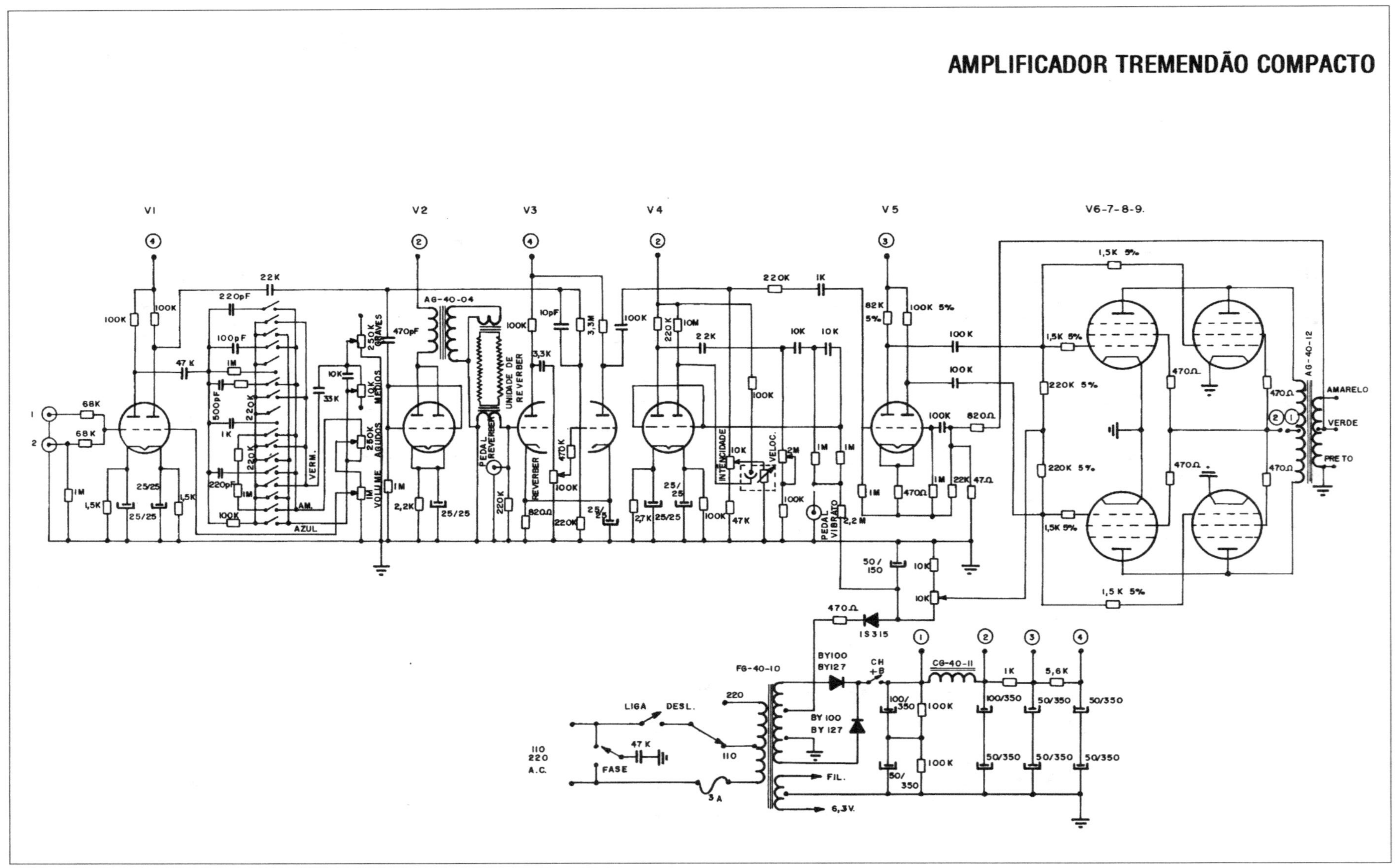

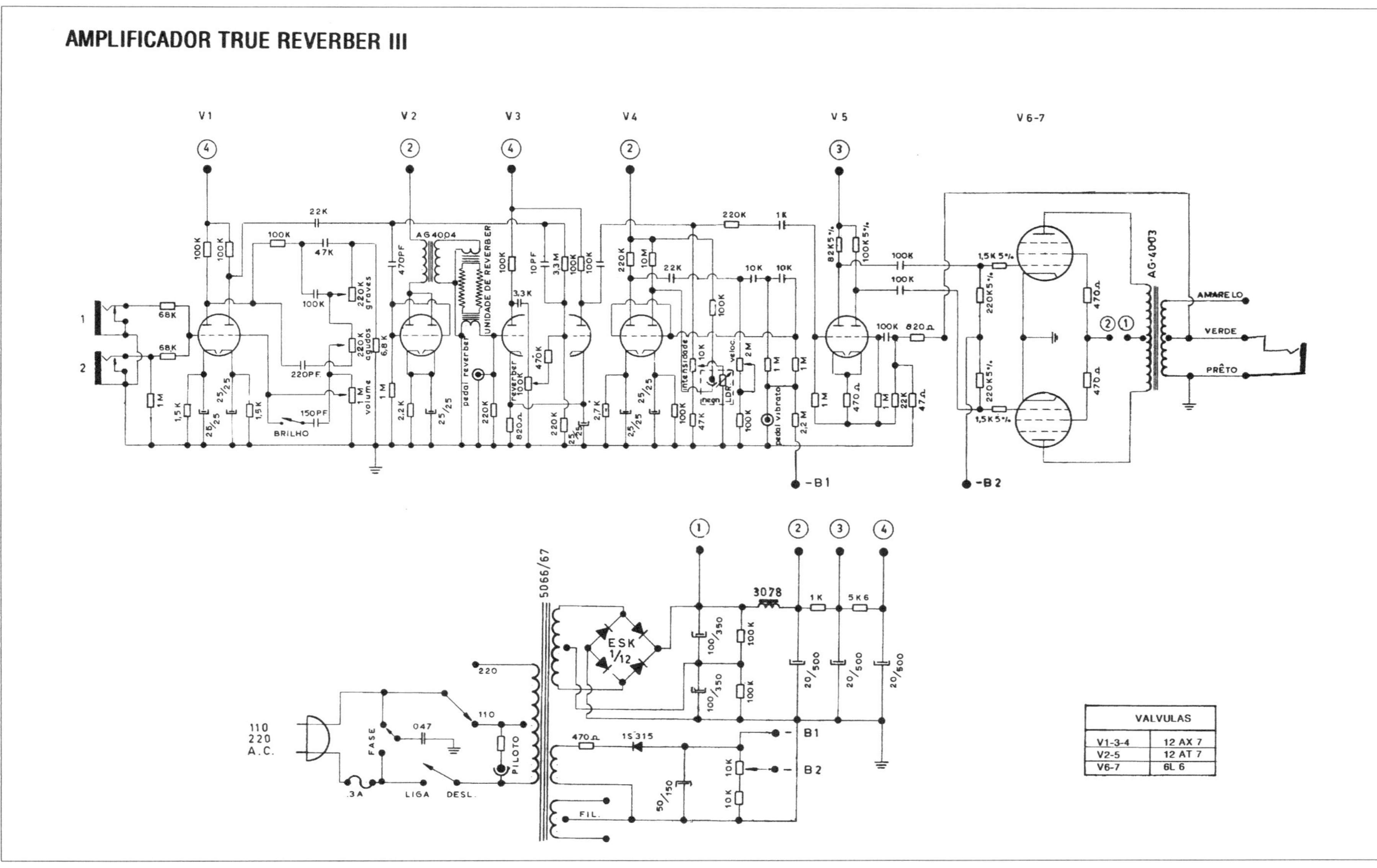

VALVULAS	
V1-3-4	12 AX 7
V2-5	12 AT 7
V6-7	6L 6

AMPLIFICADOR VALIANTE

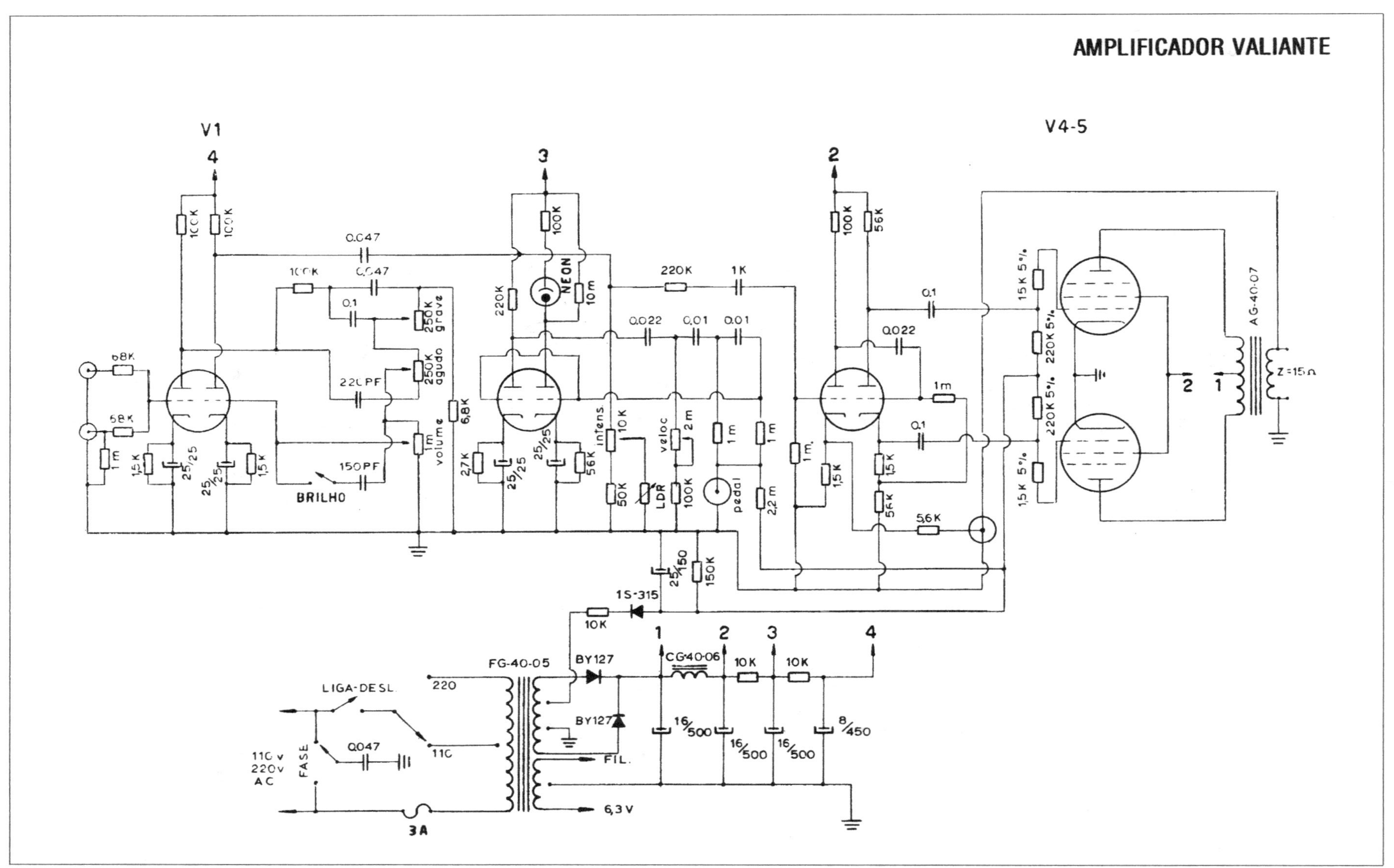

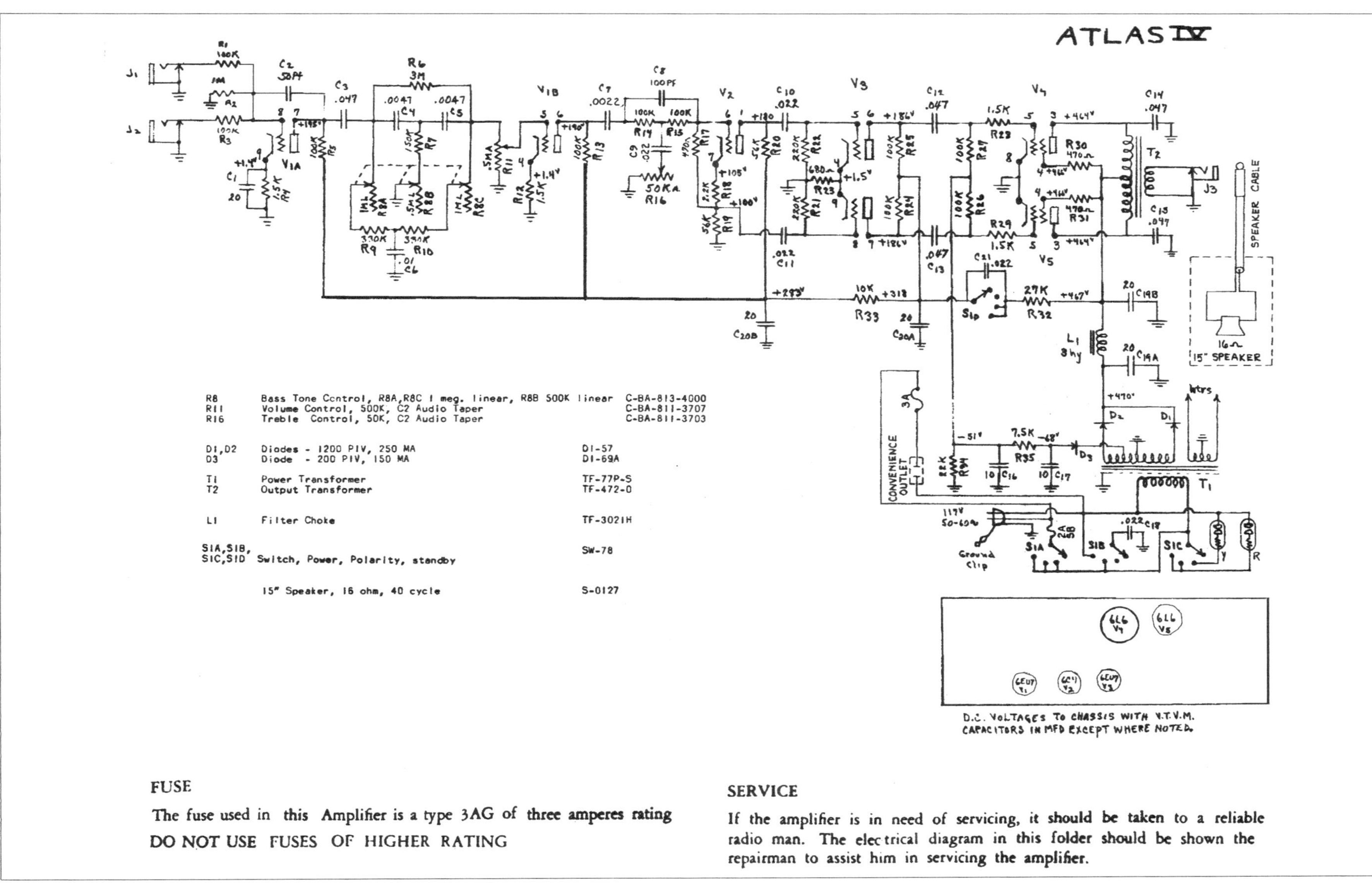

FUSE

The fuse used in this Amplifier is a type 3AG of three amperes rating
DO NOT USE FUSES OF HIGHER RATING

SERVICE

If the amplifier is in need of servicing, it should be taken to a reliable radio man. The electrical diagram in this folder should be shown the repairman to assist him in servicing the amplifier.

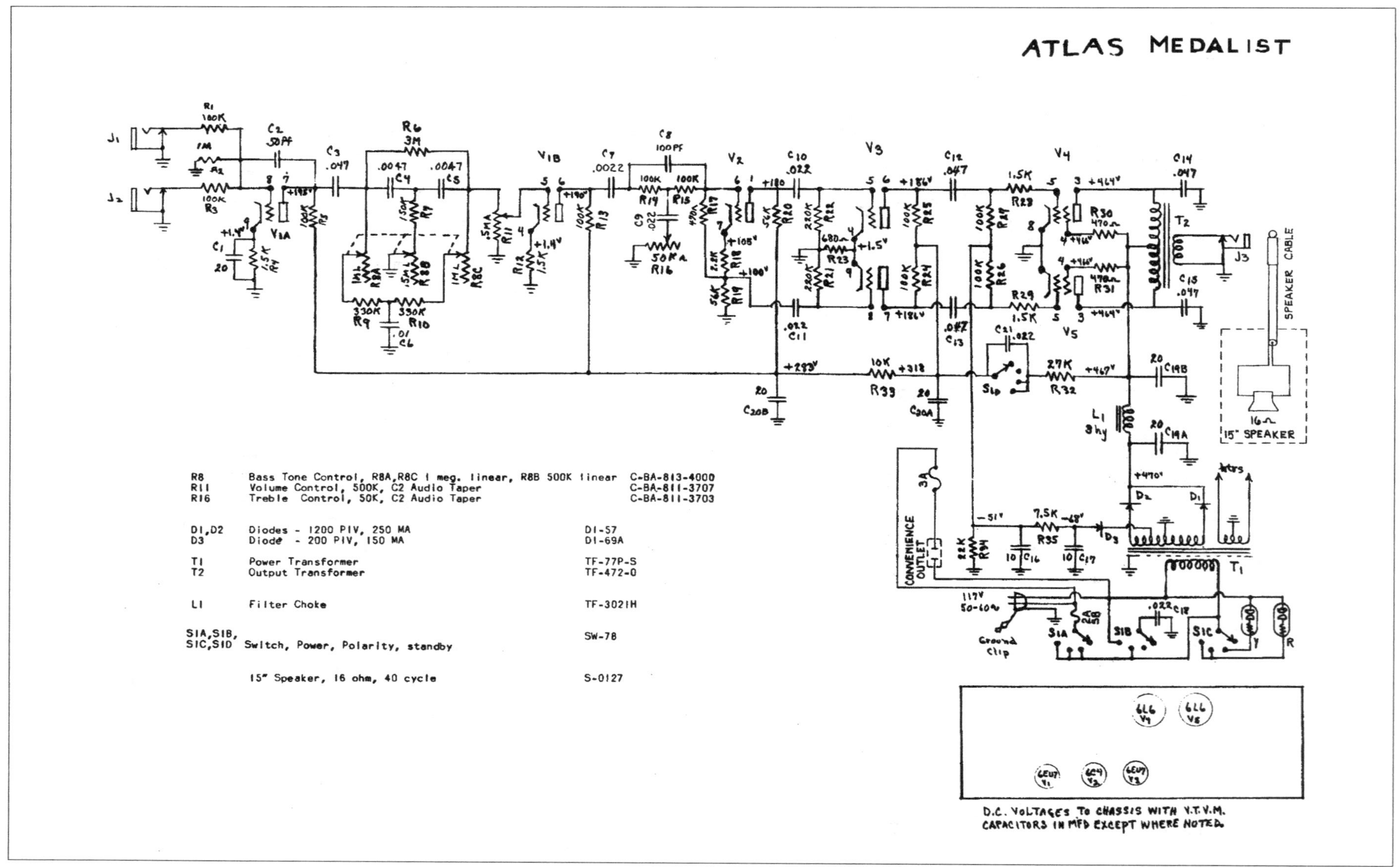
ATLAS MEDALIST
R8 Bass Tone Control, R8A,R8C 1 meg. linear, R8B 500K linear C-BA-813-4000
R11 Volume Control, 500K, C2 Audio Taper C-BA-811-3707
R16 Treble Control, 50K, C2 Audio Taper C-BA-811-3703
D1,D2 Diodes - 1200 PIV, 250 MA D1-57
D3 Diode - 200 PIV, 150 MA D1-69A
T1 Power Transformer TF-77P-S
T2 Output Transformer TF-472-0
L1 Filter Choke TF-3021H
S1A,S1B, S1C,S1D Switch, Power, Polarity, standby SW-78
15" Speaker, 16 ohm, 40 cycle S-0127
SPEAKER CABLE
15" SPEAKER
CONVENIENCE OUTLET
Ground Clip
D.C. VOLTAGES TO CHASSIS WITH V.T.V.M.
CAPACITORS IN MFD EXCEPT WHERE NOTED.

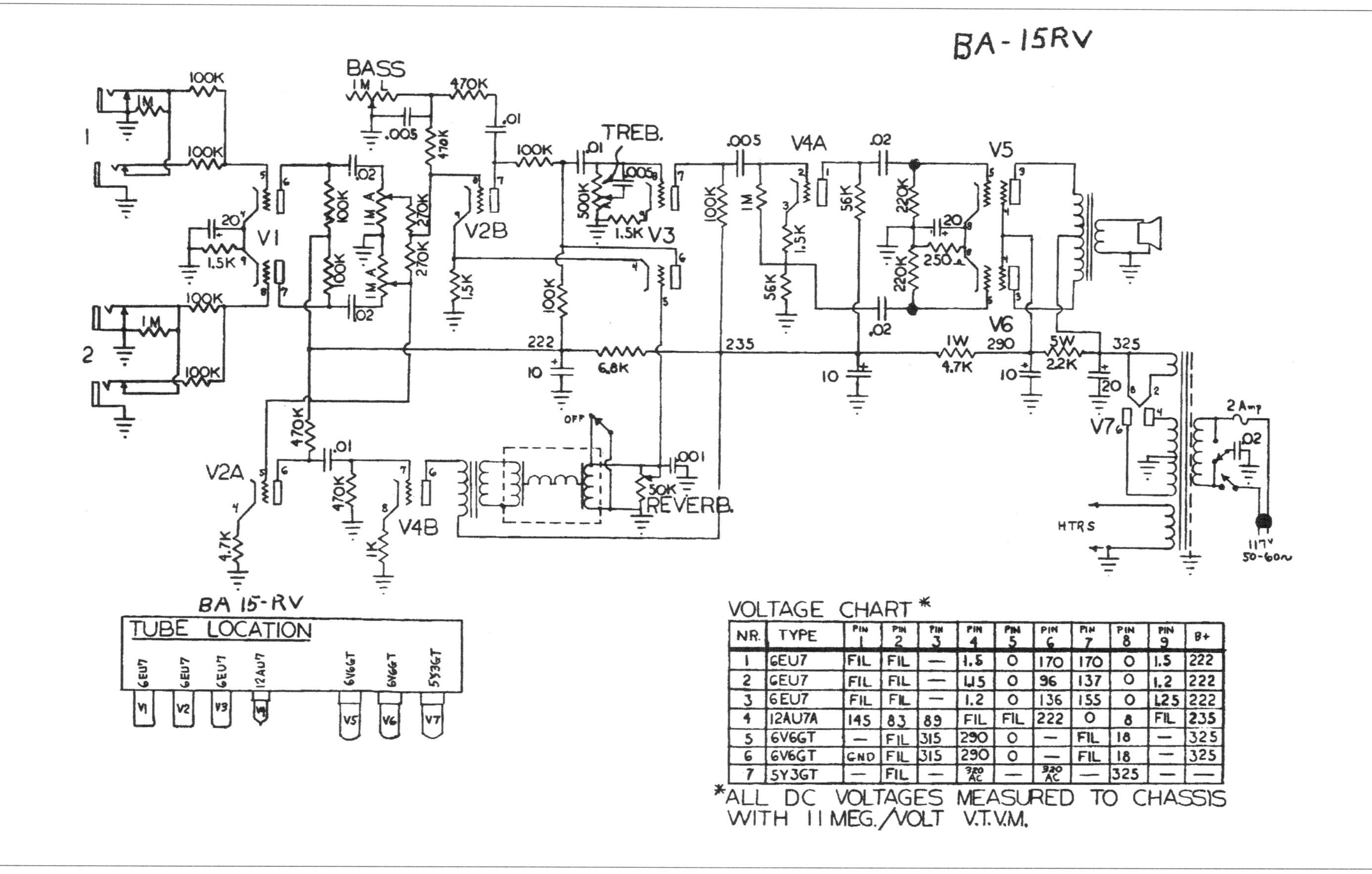

VOLTAGE CHART*

NR.	TYPE	PIN 1	PIN 2	PIN 3	PIN 4	PIN 5	PIN 6	PIN 7	PIN 8	PIN 9	B+
1	6EU7	FIL	FIL	—	1.5	0	170	170	0	1.5	222
2	6EU7	FIL	FIL	—	1.15	0	96	137	0	1.2	222
3	6EU7	FIL	FIL	—	1.2	0	136	155	0	1.25	222
4	12AU7A	145	83	89	FIL	FIL	222	0	8	FIL	235
5	6V6GT	—	FIL	315	290	0	—	FIL	18	—	325
6	6V6GT	GND	FIL	315	290	0	—	FIL	18	—	325
7	5Y3GT	—	FIL	—	320 AC	—	320 AC	—	325	—	—

*ALL DC VOLTAGES MEASURED TO CHASSIS WITH 11 MEG./VOLT V.T.V.M.

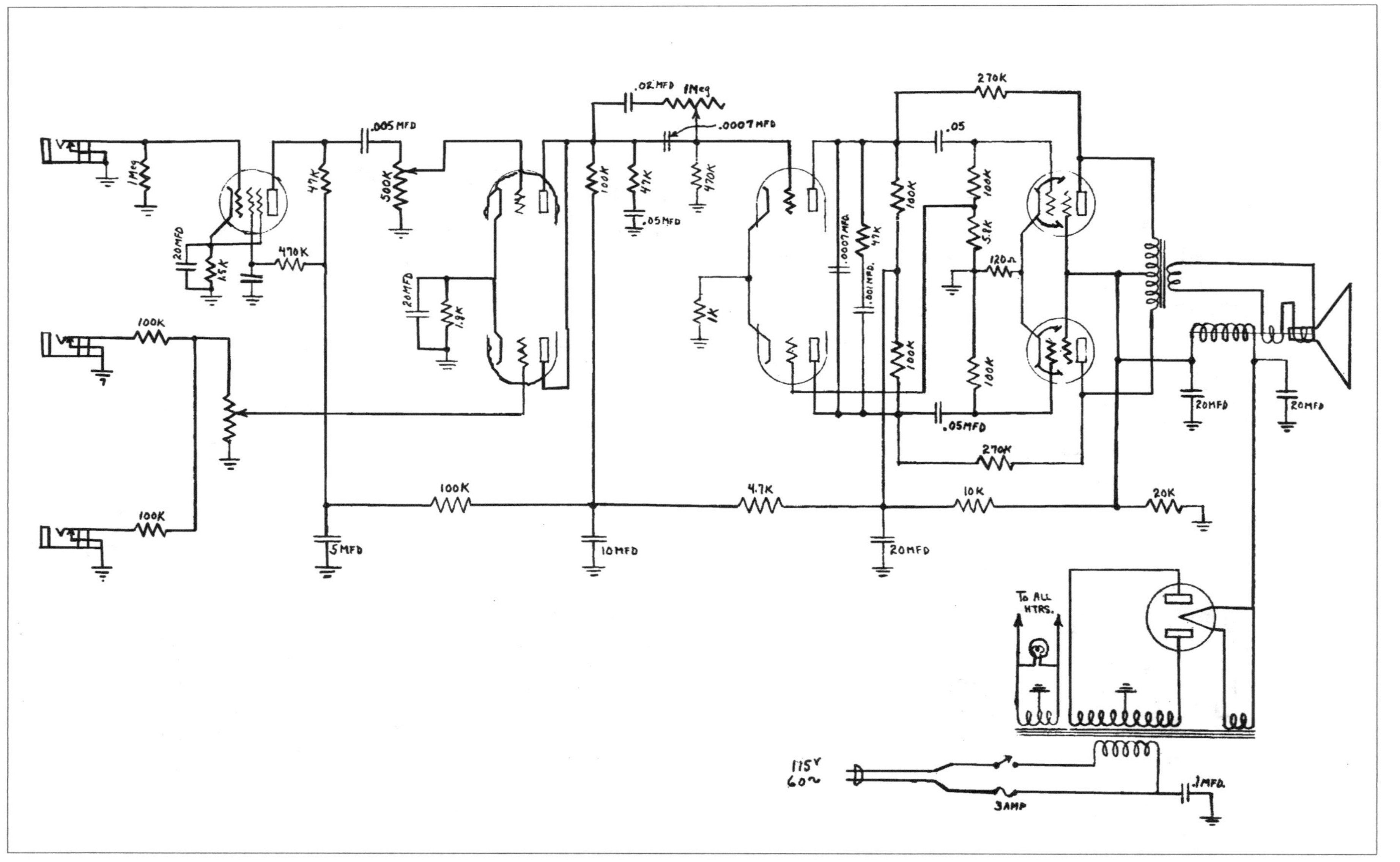
.005 MFD
1Meg
47K
500K
470K
20MFD
1.5K
100K
20MFD
1.9K
5MFD
.02 MFD
.0007 MFD
100K
47K
.05 MFD
470K
1K
10MFD
.0007 MFD
47K
.001 MFD.
270K
.05
100K
5.8K
120Ω
.05MFD
270K
4.7K
10K
20K
20MFD
20MFD
20MFD
To ALL HTRS.
115V 60~
3AMP
.1MFD.

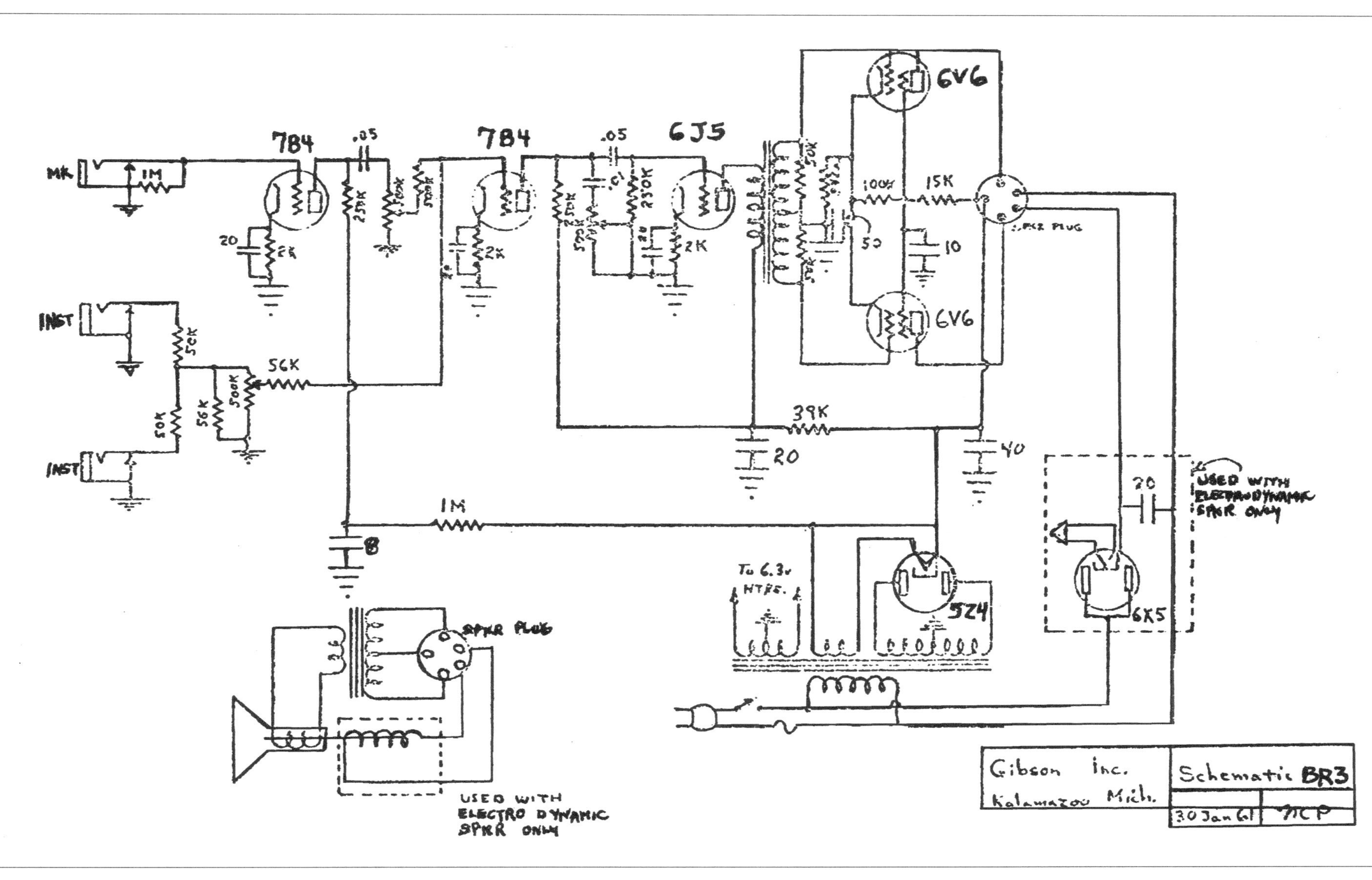
MIK
1M
7B4
.05
250K
500K
20
2K
7B4
.05
6J5
250K
500K
.01
250K
20
2K
INST
50K
56K
56K
500K
50K
INST
6V6
100K
15K
SPKR PLUG
50
10
6V6
39K
20
40
1M
8
To 6.3v HTRS.
5Z4
20
USED WITH ELECTRODYNAMIC SPKR ONLY
6X5
SPKR PLUG
USED WITH ELECTRO DYNAMIC SPKR ONLY
Gibson Inc.
Kalamazoo Mich.
Schematic BR3
30 Jan 61
MCP

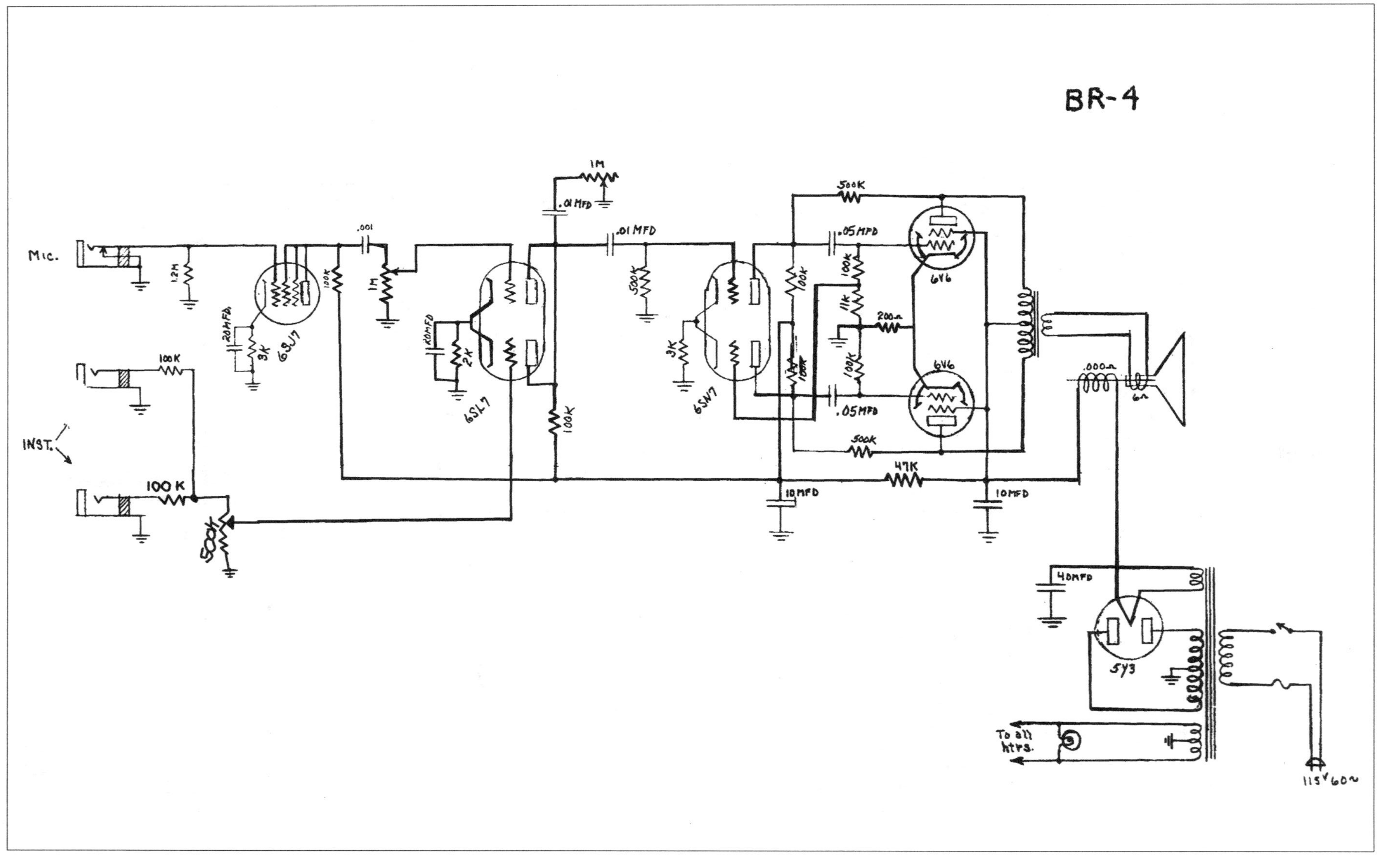
BR-4
MIC.
INST.
1.2M
20MFD
3K
6SJ7
100K
.001
1M
20MFD
2K
6SL7
.01MFD
1M
100K
.01MFD
500K
3K
6SN7
500K
.05MFD
100K
11K
200Ω
.05MFD
6V6
6V6
47K
10MFD
10MFD
.000Ω
6Ω
40MFD
5Y3
To all htrs.
115V 60~
100K
100K
500K

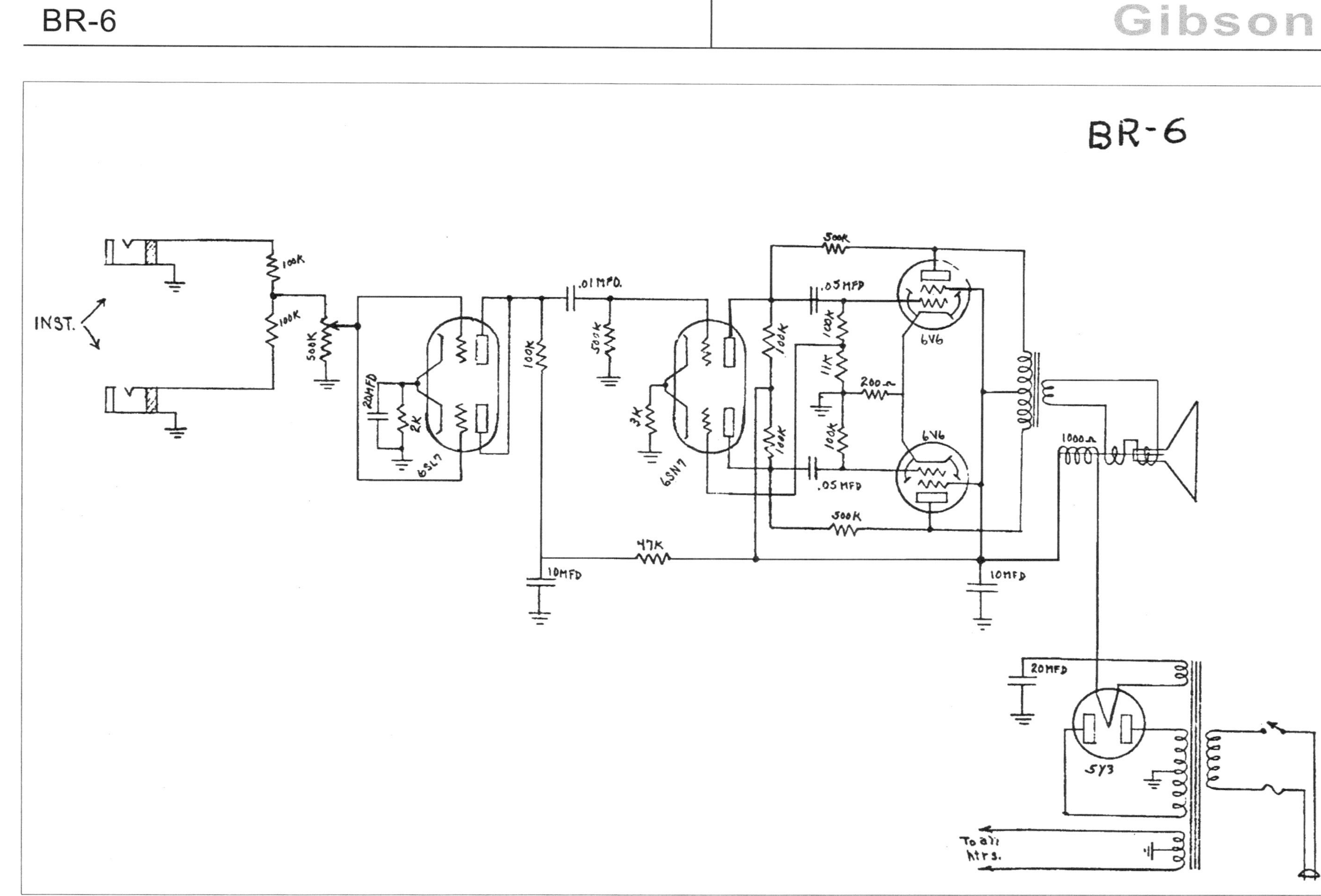

BR-6
INST.
100K
100K
500K
20MFD
2K
6SL7
100K
.01MFD.
500K
3K
6SN7
500K
.05MFD
100K
100K
11K
200Ω
100K
100K
.05MFD
6V6
6V6
500K
47K
10MFD
10MFD
1000Ω
20MFD
5Y3
To all htrs.

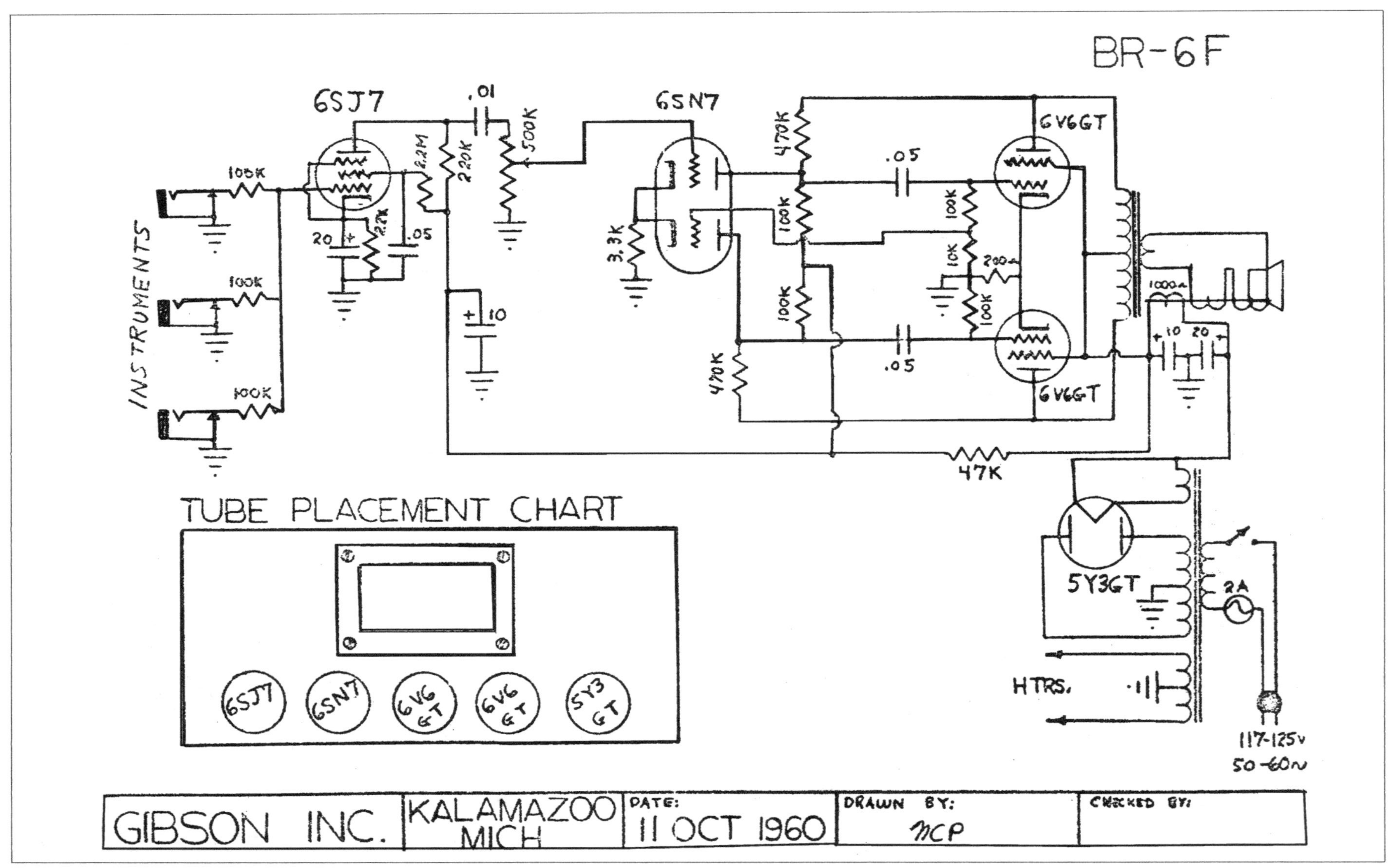
BR-6F
6SJ7
.01
500K
6SN7
470K
6V6GT
.05
100K
2.2M
220K
INSTRUMENTS
20
22K
3.3K
100K
10K
200Ω
1000Ω
10
20
.05
470K
6V6GT
47K
TUBE PLACEMENT CHART
6SJ7
6SN7
6V6 GT
6V6 GT
5Y3 GT
5Y3GT
2A
HTRS.
117-125V
50-60~
GIBSON INC.
KALAMAZOO MICH
DATE: 11 OCT 1960
DRAWN BY: MCP
CHECKED BY:

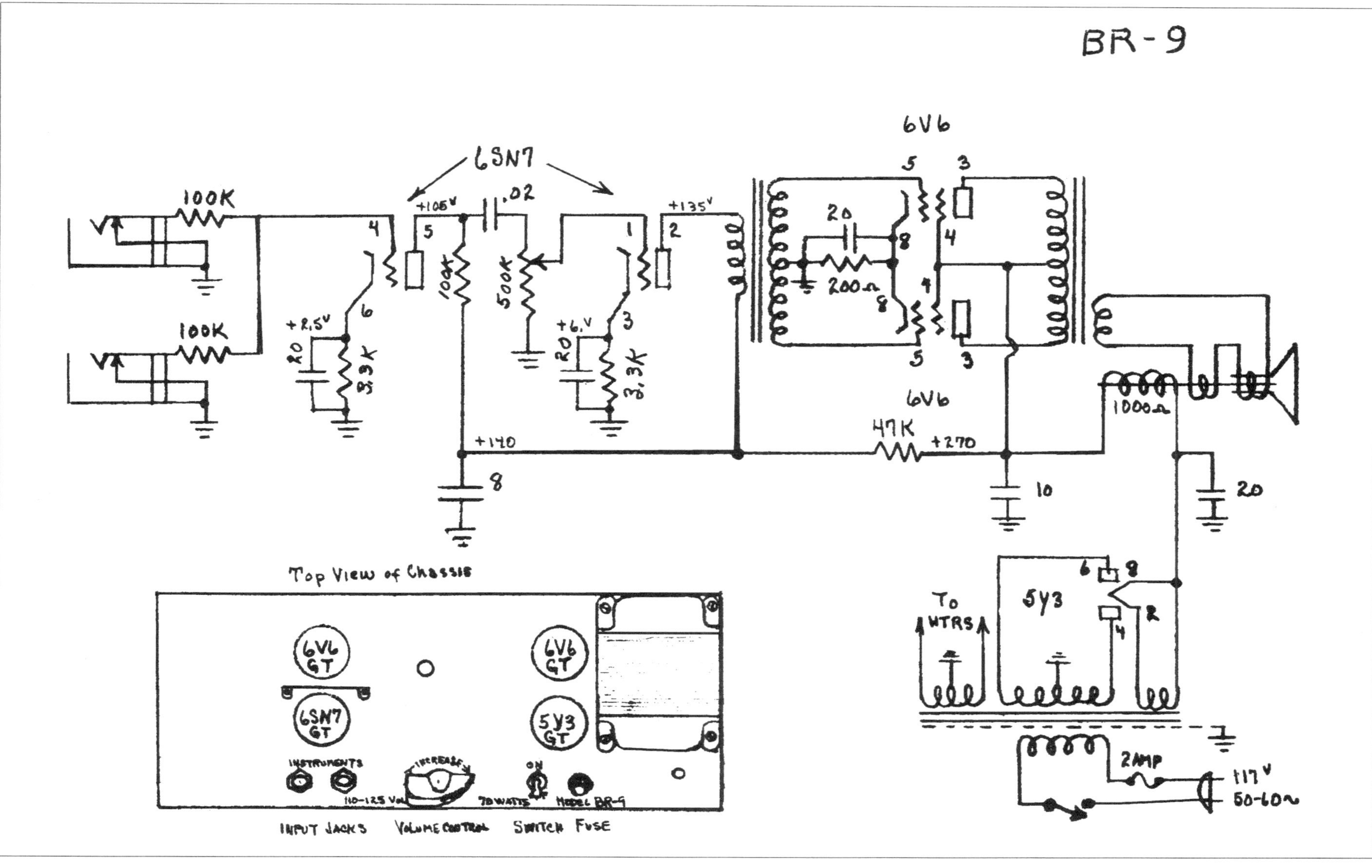
BR-9
6SN7
100K
100K
+105V
.02
+135V
+2.5V
20
3.3K
500K
+6.V
3,3K
+140
8
6V6
20
200Ω
47K
+270
10
1000Ω
20
5Y3
To HTRS
2AMP
117V
50-60~
Top View of Chassis
6V6 GT
6SN7 GT
6V6 GT
5Y3 GT
INSTRUMENTS
INCREASE
110-125 Vol
70 WATTS
ON
MODEL BR-9
INPUT JACKS
VOLUME CONTROL
SWITCH
FUSE

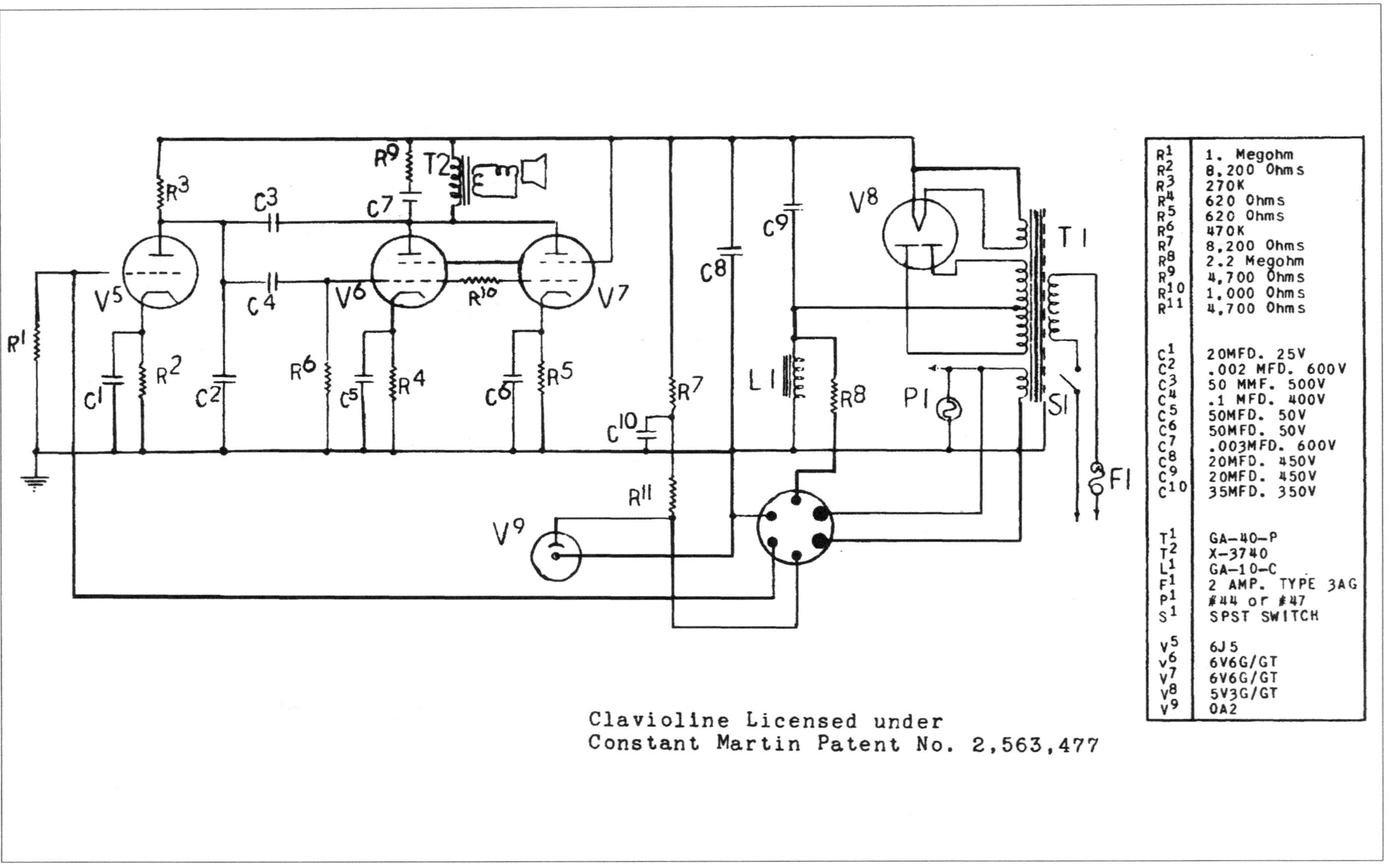

R1	1. Megohm
R2	8,200 Ohms
R3	270K
R4	620 Ohms
R5	620 Ohms
R6	470K
R7	8,200 Ohms
R8	2.2 Megohm
R9	4,700 Ohms
R10	1,000 Ohms
R11	4,700 Ohms
C1	20MFD. 25V
C2	.002 MFD. 600V
C3	50 MMF. 500V
C4	.1 MFD. 400V
C5	50MFD. 50V
C6	50MFD. 50V
C7	.003MFD. 600V
C8	20MFD. 450V
C9	20MFD. 450V
C10	35MFD. 350V
T1	GA-40-P
T2	X-3740
L1	GA-10-C
F1	2 AMP. TYPE 3AG
P1	#44 or #47
S1	SPST SWITCH
V5	6J5
V6	6V6G/GT
V7	6V6G/GT
V8	5V3G/GT
V9	0A2

Clavioline Licensed under
Constant Martin Patent No. 2,563,477

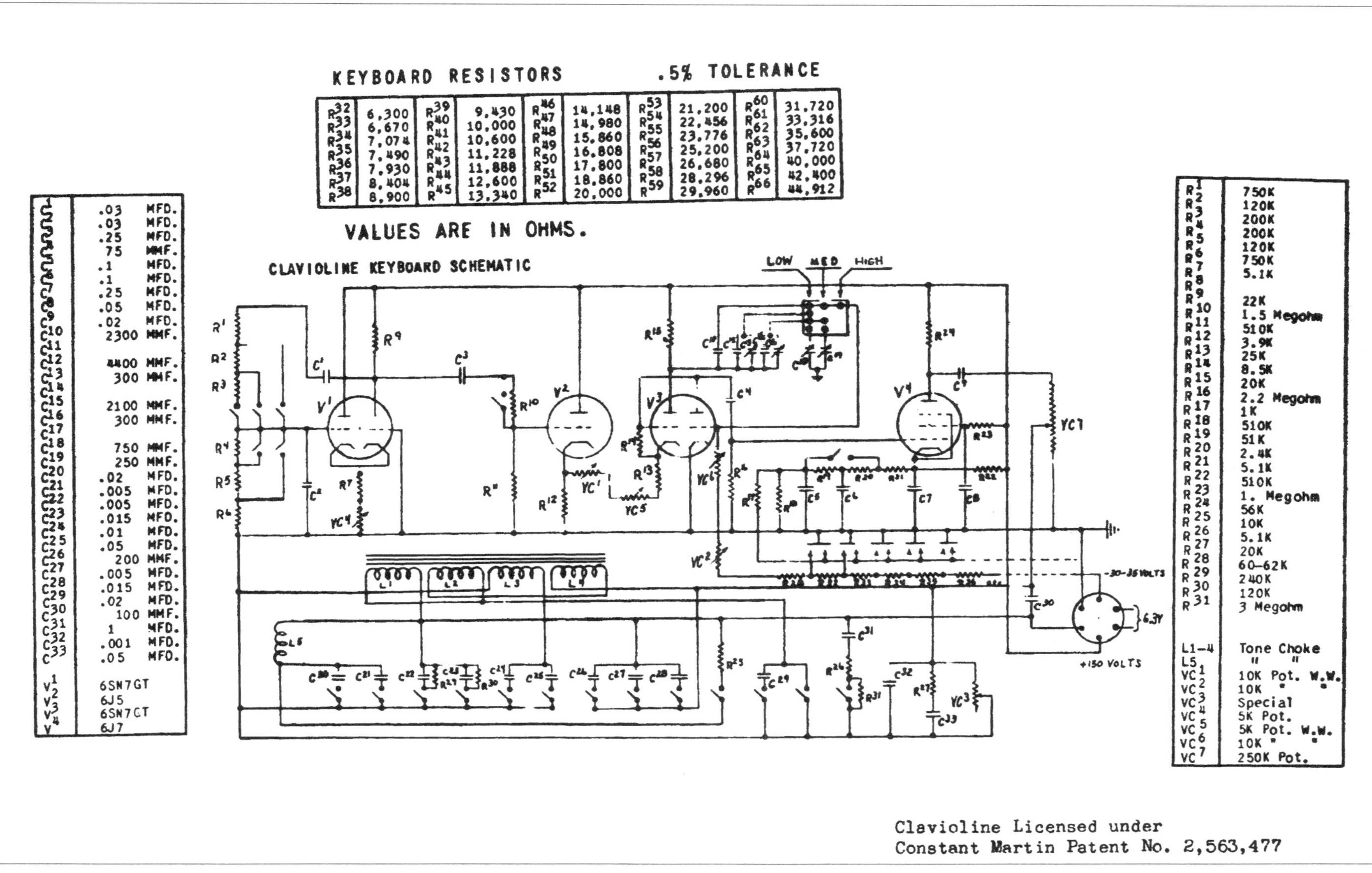

KEYBOARD RESISTORS .5% TOLERANCE

R32	6,300	R39	9,430	R46	14,148	R53	21,200	R60	31,720
R33	6,670	R40	10,000	R47	14,980	R54	22,456	R61	33,316
R34	7,074	R41	10,600	R48	15,860	R55	23,776	R62	35,600
R35	7,490	R42	11,228	R49	16,808	R56	25,200	R63	37,720
R36	7,930	R43	11,888	R50	17,800	R57	26,680	R64	40,000
R37	8,404	R44	12,600	R51	18,860	R58	28,296	R65	42,400
R38	8,900	R45	13,340	R52	20,000	R59	29,960	R66	44,912

VALUES ARE IN OHMS.

CLAVIOLINE KEYBOARD SCHEMATIC

C1	.03 MFD.
C2	.03 MFD.
C3	.25 MFD.
C4	75 MMF.
C5	.1 MFD.
C6	.1 MFD.
C7	.25 MFD.
C8	.05 MFD.
C9	.02 MFD.
C10	2300 MMF.
C11	
C12	4400 MMF.
C13	300 MMF.
C14	
C15	2100 MMF.
C16	300 MMF.
C17	
C18	750 MMF.
C19	250 MMF.
C20	.02 MFD.
C21	.005 MFD.
C22	.005 MFD.
C23	.015 MFD.
C24	.01 MFD.
C25	.05 MFD.
C26	200 MMF.
C27	.005 MFD.
C28	.015 MFD.
C29	.02 MFD.
C30	100 MMF.
C31	1 MFD.
C32	.001 MFD.
C33	.05 MFD.
V1	6SN7GT
V2	6J5
V3	6SN7GT
V4	6J7

R1	750K
R2	120K
R3	200K
R4	200K
R5	120K
R6	750K
R7	5.1K
R8	
R9	22K
R10	1.5 Megohm
R11	510K
R12	3.9K
R13	25K
R14	8.5K
R15	20K
R16	2.2 Megohm
R17	1K
R18	510K
R19	51K
R20	2.4K
R21	5.1K
R22	510K
R23	1. Megohm
R24	56K
R25	10K
R26	5.1K
R27	20K
R28	60–62K
R29	240K
R30	120K
R31	3 Megohm
L1–4	Tone Choke
L5	" "
VC1	10K Pot. W.W.
VC2	10K " "
VC3	Special
VC4	5K Pot.
VC5	5K Pot. W.W.
VC6	10K " "
VC7	250K Pot.

Clavioline Licensed under
Constant Martin Patent No. 2,563,477

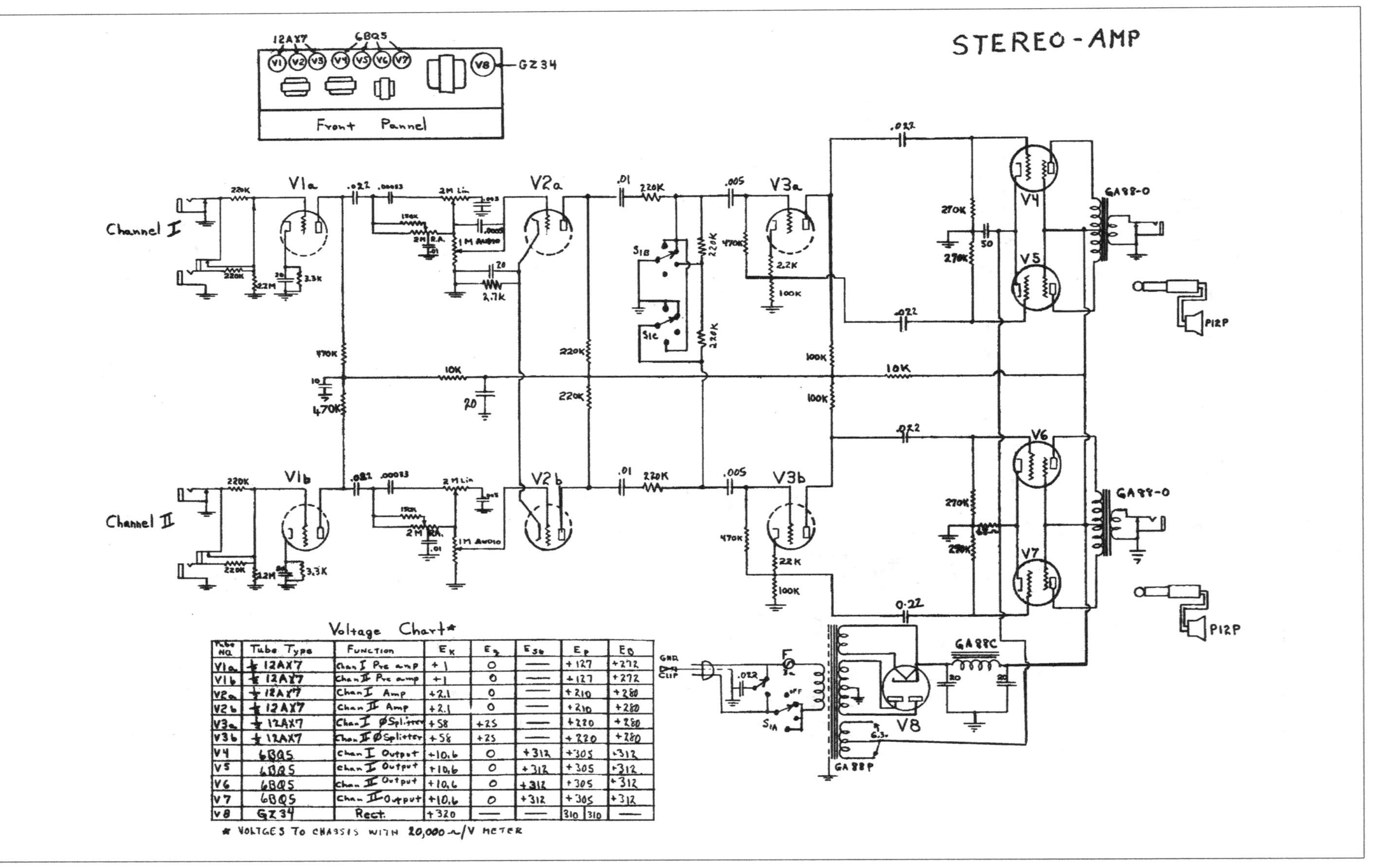

Tube No.	Tube Type	Function	E_K	E_g	E_{sg}	E_p	E_O
V1a	½ 12AX7	Chan I Pre amp	+1	0	—	+127	+272
V1b	½ 12AX7	Chan II Pre amp	+1	0	—	+127	+272
V2a	½ 12AX7	Chan I Amp	+2.1	0	—	+210	+280
V2b	½ 12AX7	Chan II Amp	+2.1	0	—	+210	+280
V3a	½ 12AX7	Chan I ØSplitter	+58	+25	—	+220	+280
V3b	½ 12AX7	Chan II ØSplitter	+58	+25	—	+220	+280
V4	6BQ5	Chan I Output	+10.6	0	+312	+305	+312
V5	6BQ5	Chan I Output	+10.6	0	+312	+305	+312
V6	6BQ5	Chan II Output	+10.6	0	+312	+305	+312
V7	6BQ5	Chan II Output	+10.6	0	+312	+305	+312
V8	GZ34	Rect.	+320	—	—	310 310	—

* VOLTGES TO CHASSIS WITH 20,000 Ω/V METER

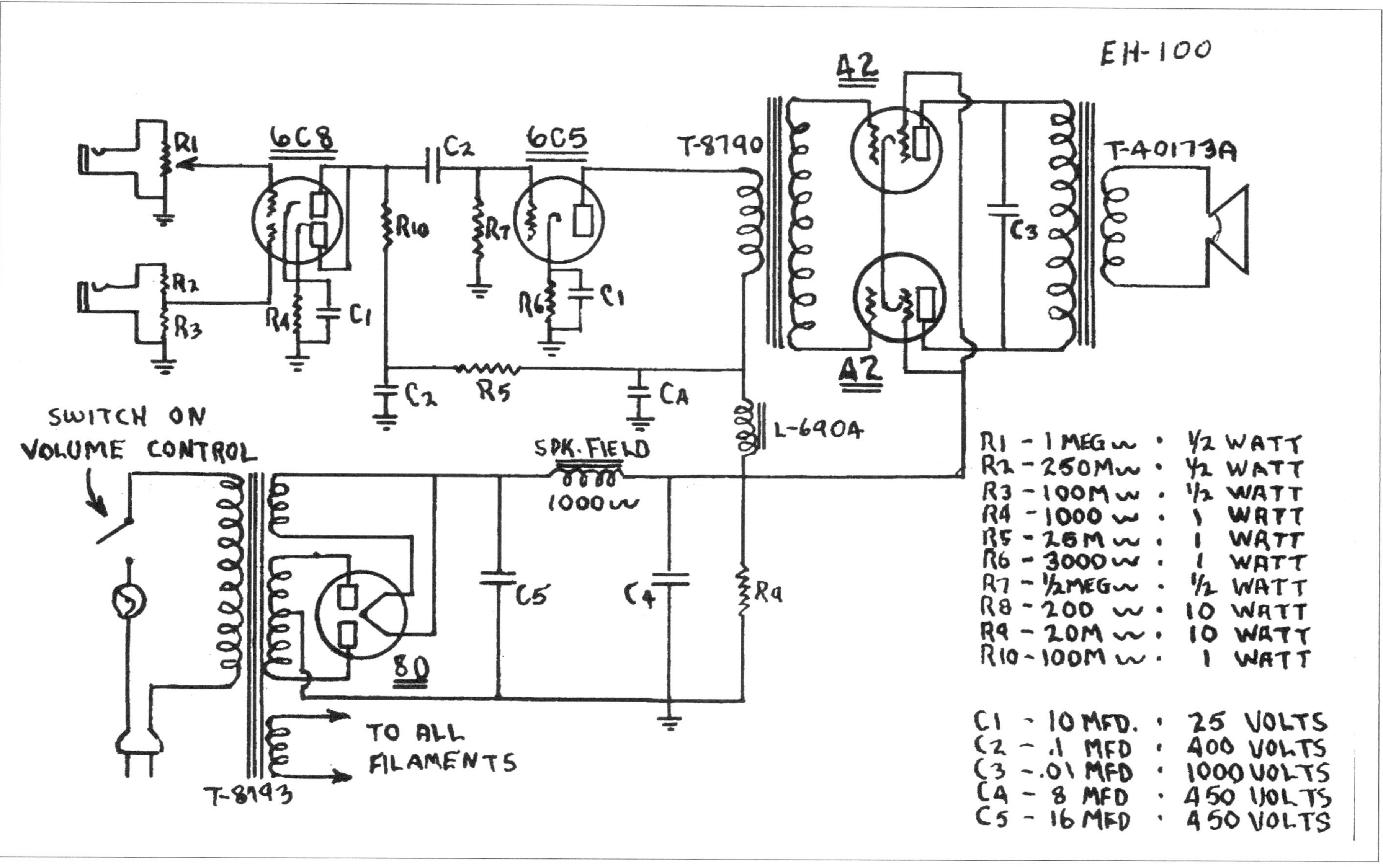

EH-100
6C8
6C5
T-8790
42
42
T-40173A
R1
R2
R3
R4
C1
R10
C2
R7
R6
C1
C3
C2
R5
CA
L-6904
SWITCH ON
VOLUME CONTROL
SPK. FIELD
1000 w
C5
C4
R9
80
T-8193
TO ALL
FILAMENTS
R1 - 1 MEG w · 1/2 WATT
R2 - 250M w · 1/2 WATT
R3 - 100M w · 1/2 WATT
R4 - 1000 w · 1 WATT
R5 - 25M w · 1 WATT
R6 - 3000 w · 1 WATT
R7 - 1/2 MEG w · 1/2 WATT
R8 - 200 w · 10 WATT
R9 - 20M w · 10 WATT
R10 - 100M w · 1 WATT
C1 - 10 MFD. · 25 VOLTS
C2 - .1 MFD · 400 VOLTS
C3 - .01 MFD · 1000 VOLTS
C4 - 8 MFD · 450 VOLTS
C5 - 16 MFD · 450 VOLTS

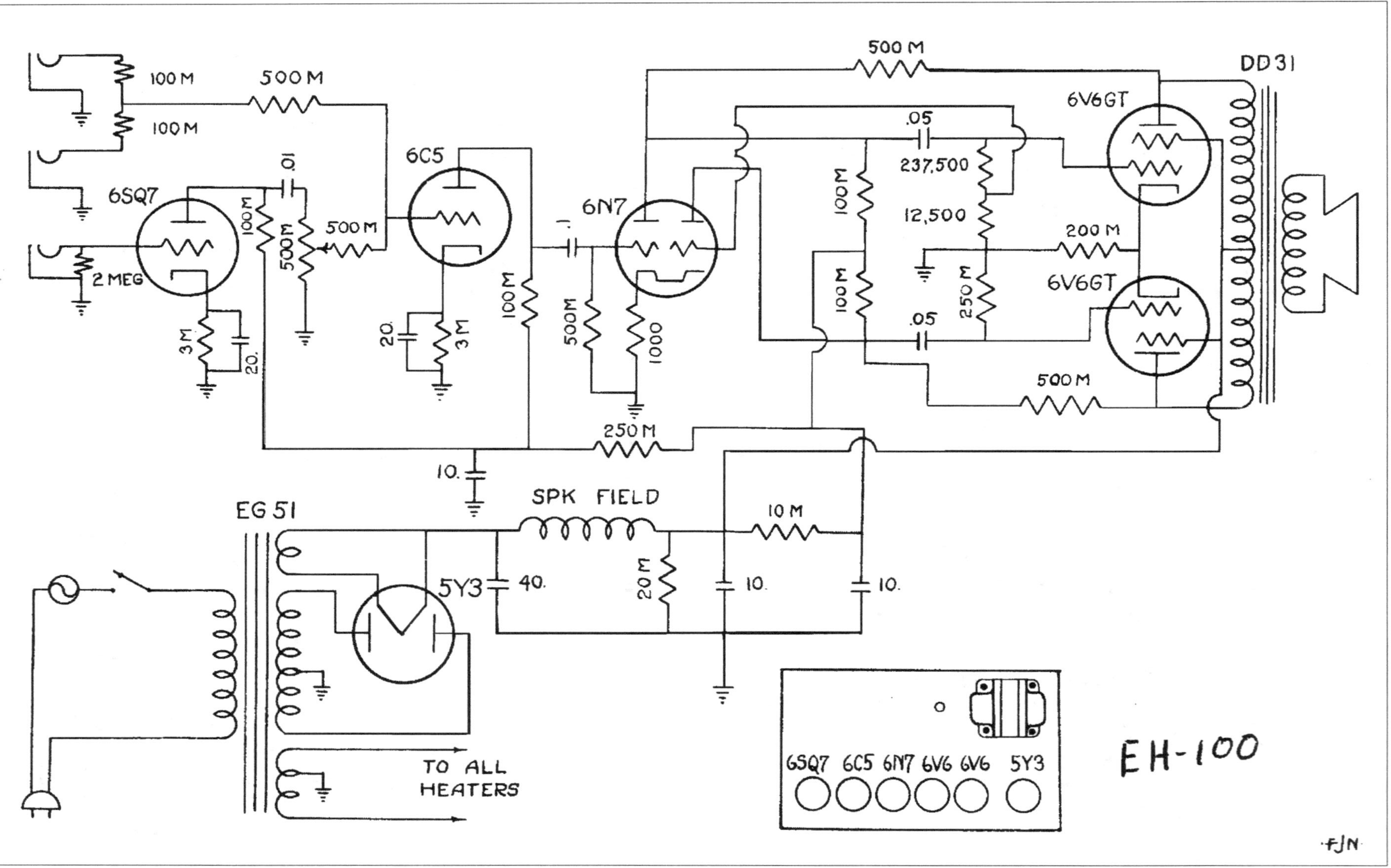
100 M
500 M
100 M
6SQ7
2 MEG
3M
20.
.01
100 M
500 M
500 M
6C5
20.
3 M
100 M
.1
6N7
500M
1000
250 M
10.
500 M
.05
237,500
12,500
100 M
100 M
200 M
250 M
.05
500 M
6V6GT
6V6GT
DD31
SPK FIELD
EG 51
5Y3
40.
20 M
10 M
10.
10.
TO ALL
HEATERS
6SQ7 6C5 6N7 6V6 6V6 5Y3
EH-100
FJN

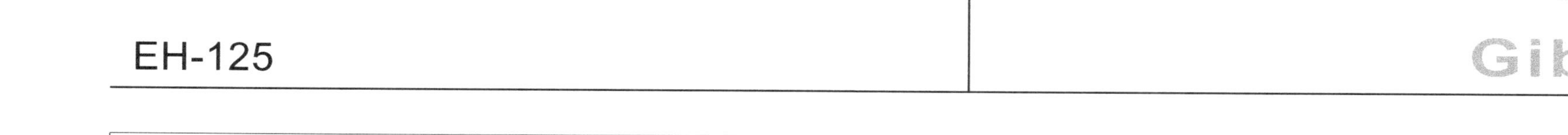

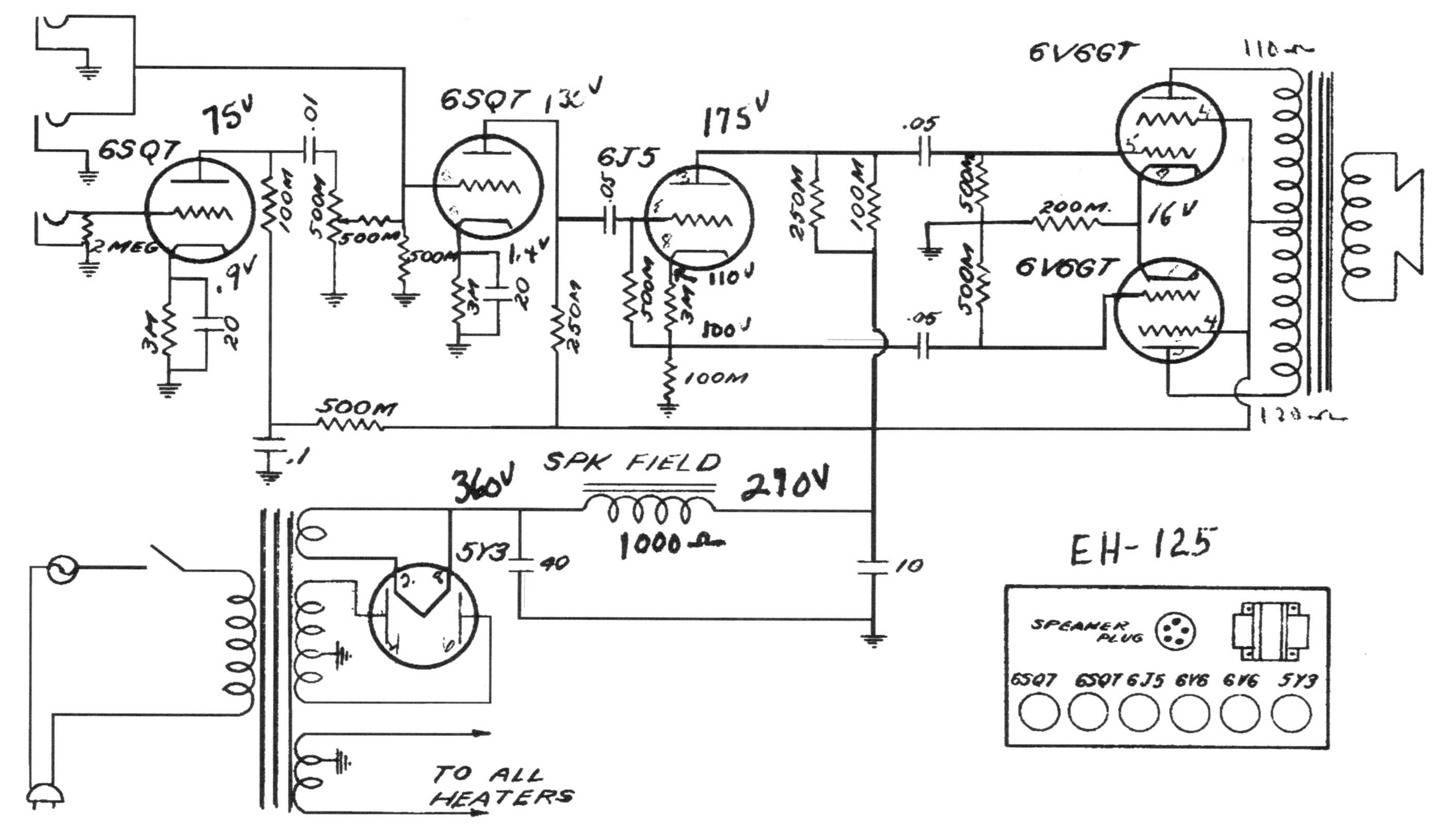

6SQ7
75V
2MEG
.9V
3M
20
100M
.01
500M
500M
500M
6SQ7 130V
1.4V
3M
20
250M
500M
6J5
.05
175V
110V
3M
500M
100V
100M
250M
100M
.05
500M
200M.
16V
500M
.05
6V6GT
6V6GT
110Ω
120Ω
.1
360V
SPK FIELD
1000Ω
270V
5Y3
40
10
TO ALL
HEATERS
EH-125
SPEAKER
PLUG
6SQ7
6SQ7
6J5
6V6
6V6
5Y3

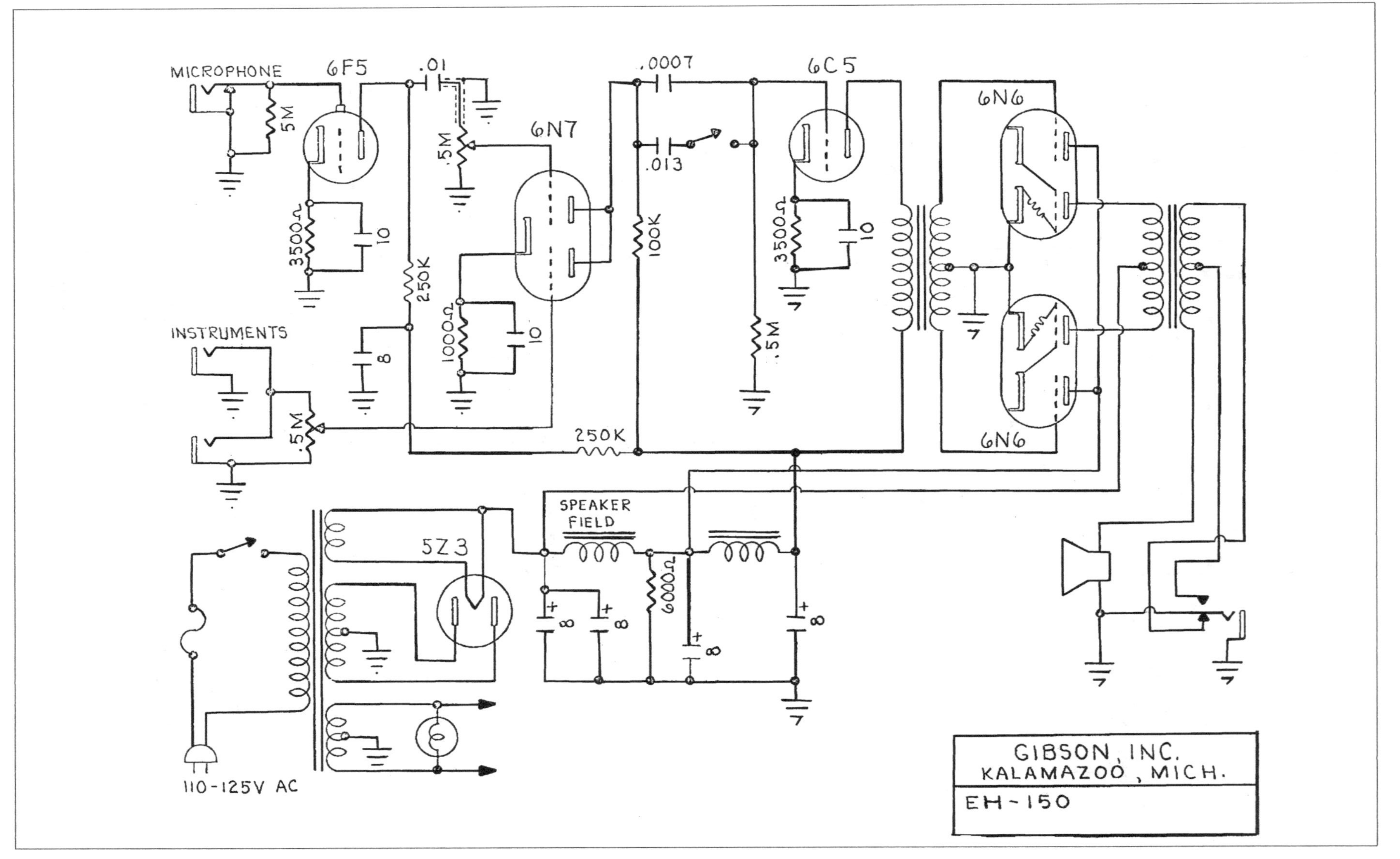
MICROPHONE
6F5
.01
5M
.5M
6N7
3500Ω
10
250K
INSTRUMENTS
8
1000Ω
10
.5M
.0007
.013
100K
250K
6C5
3500Ω
10
.5M
6N6
6N6
SPEAKER
FIELD
5Z3
6000Ω
8
8
8
8
110-125V AC
GIBSON, INC.
KALAMAZOO, MICH.
EH-150

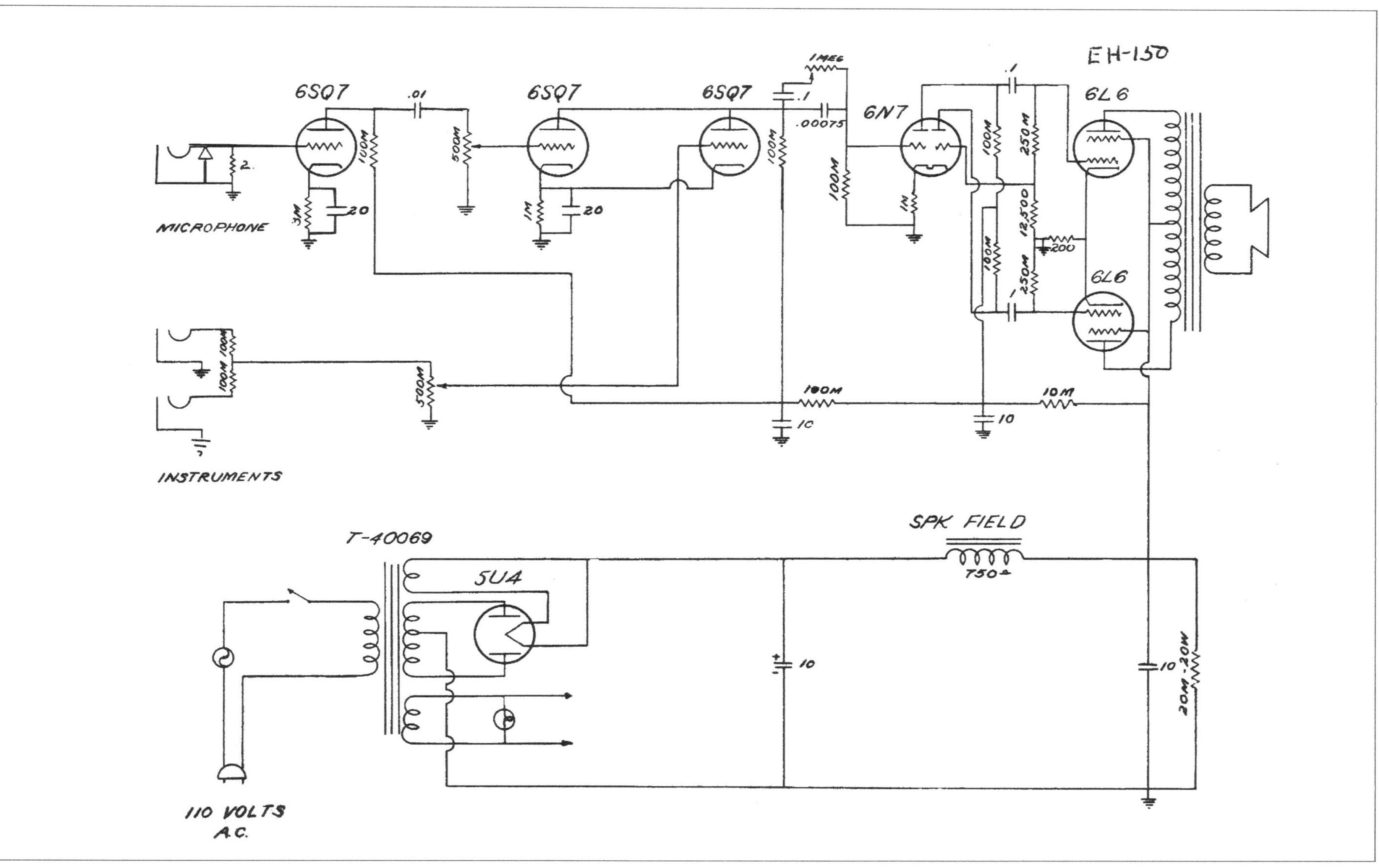
EH-150
6SQ7
6SQ7
6SQ7
6N7
6L6
6L6
.01
.1
.00075
1MEG
100M
500M
2.
3M
20
1M
20
MICROPHONE
INSTRUMENTS
100M
100M
500M
100M
10
10M
10
250M
12,500
250M
200
T-40069
5U4
SPK FIELD
750
10
10
20M-20W
110 VOLTS
A.C.

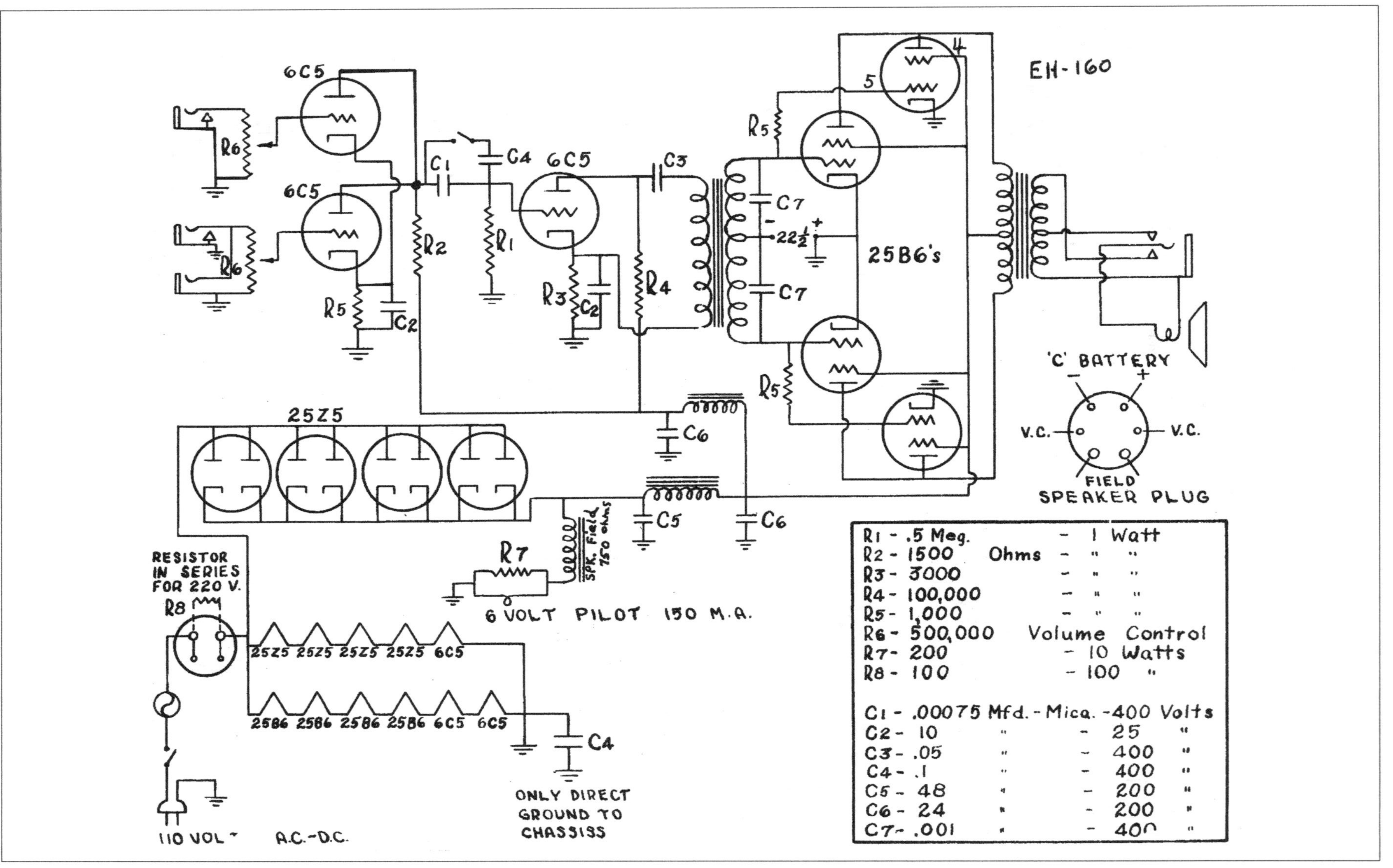

R1 - .5 Meg.		- 1 Watt	
R2 - 1500	Ohms	- " "	
R3 - 3000		- " "	
R4 - 100,000		- " "	
R5 - 1,000		- " "	
R6 - 500,000		Volume Control	
R7 - 200		- 10 Watts	
R8 - 100		- 100 "	
C1 - .00075	Mfd. - Mica.	-400	Volts
C2 - 10	"	- 25	"
C3 - .05	"	- 400	"
C4 - .1	"	- 400	"
C5 - 48	"	- 200	"
C6 - 24	"	- 200	"
C7 - .001	"	- 400	"

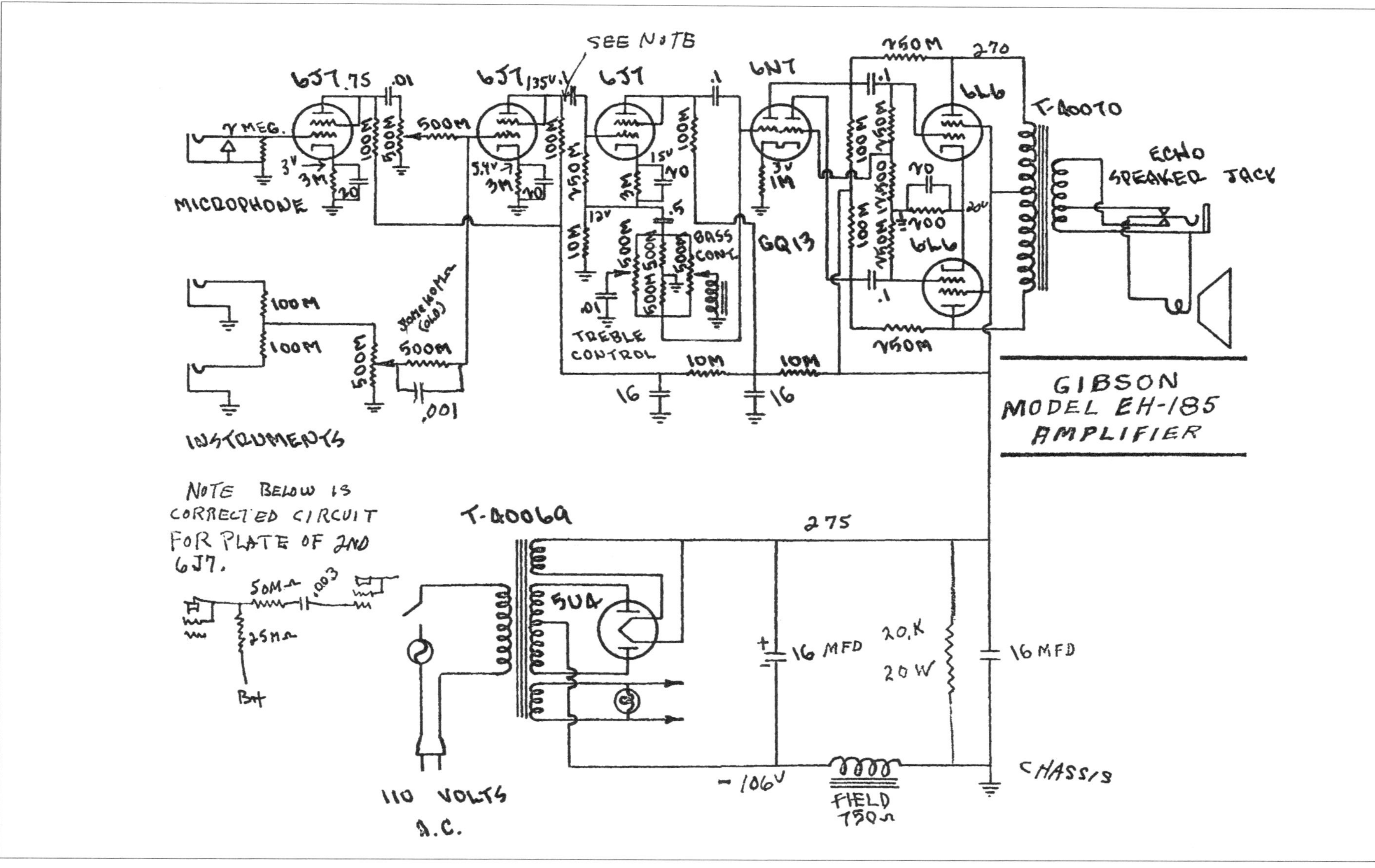

SEE NOTE
6J7
6J7
6J7
6N7
6L6
6L6
T-20070
ECHO
SPEAKER JACK
MICROPHONE
INSTRUMENTS
TREBLE CONTROL
BASS CONT.
GQ13
GIBSON
MODEL EH-185
AMPLIFIER
NOTE BELOW IS CORRECTED CIRCUIT FOR PLATE OF 2ND 6J7.
T-20069
5U4
275
16 MFD
20 K
20 W
16 MFD
CHASSIS
-106V
FIELD 750Ω
110 VOLTS A.C.
B+

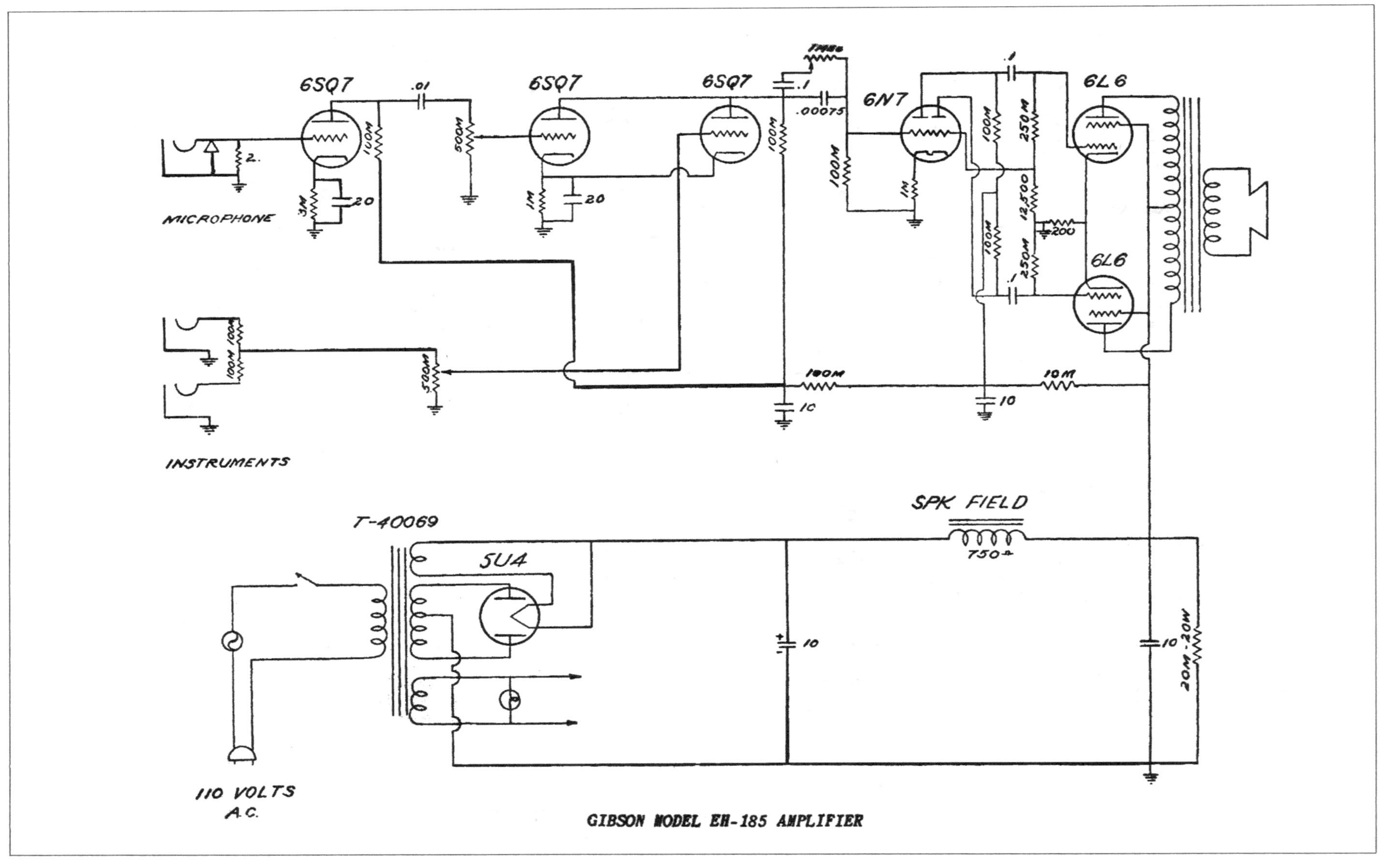

GIBSON MODEL EH-185 AMPLIFIER

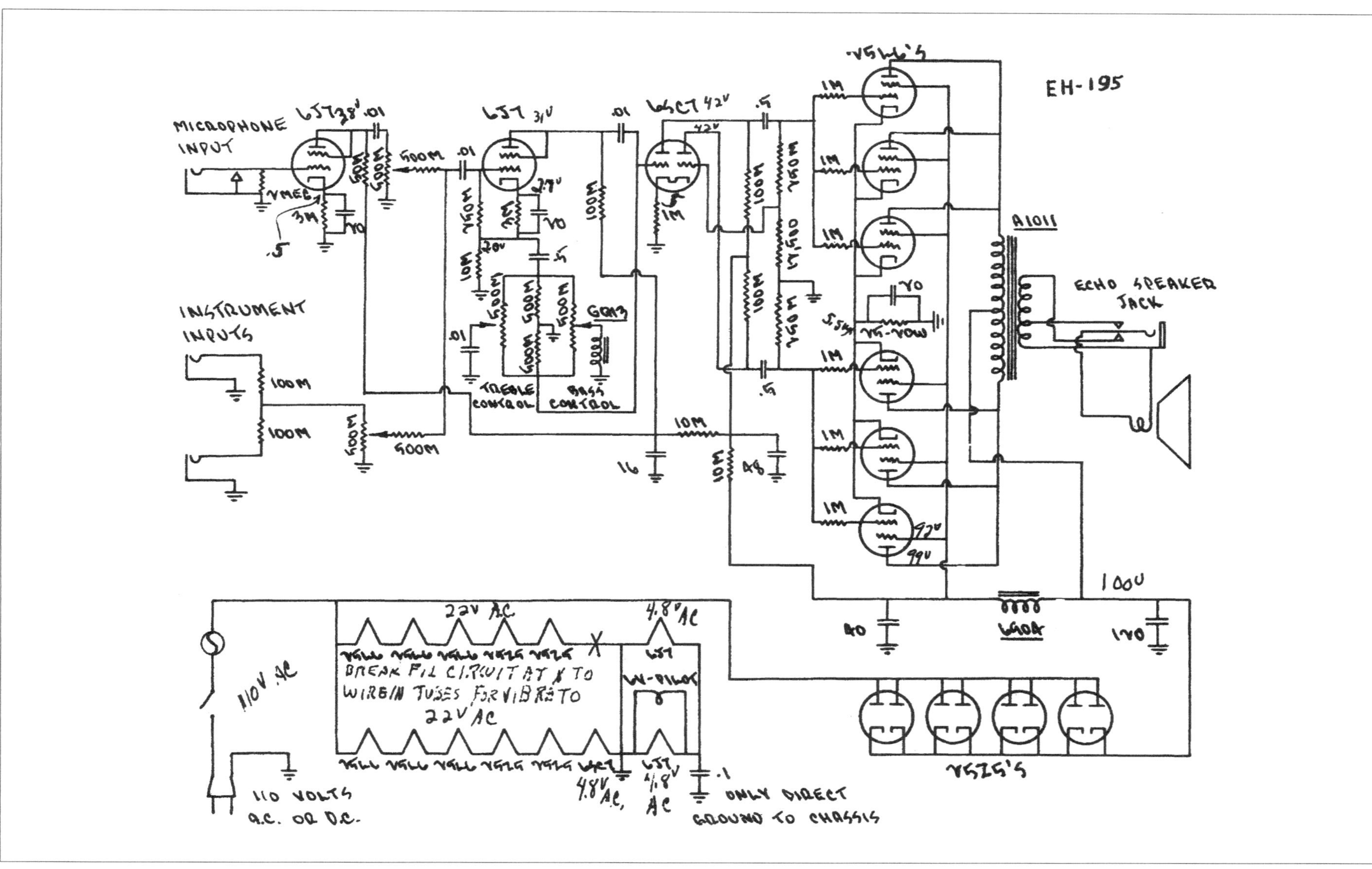
EH-195
MICROPHONE INPUT
INSTRUMENT INPUTS
TREBLE CONTROL
BASS CONTROL
ECHO SPEAKER JACK
110V AC
110 VOLTS A.C. OR D.C.
BREAK FIL CIRCUIT AT X TO WIRE IN TUBES FOR VIBRATO
22V AC
ONLY DIRECT GROUND TO CHASSIS

GIBSON

A.C. - D.C. AMPLIFIER

EH 195

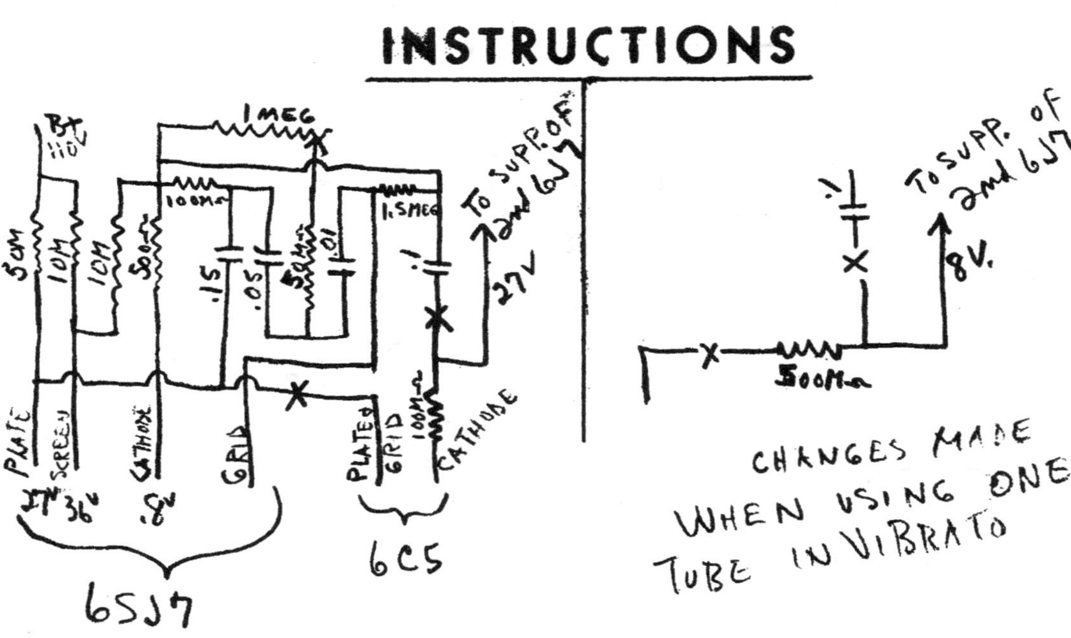

GIBSON INC., KALAMAZOO, MICH.

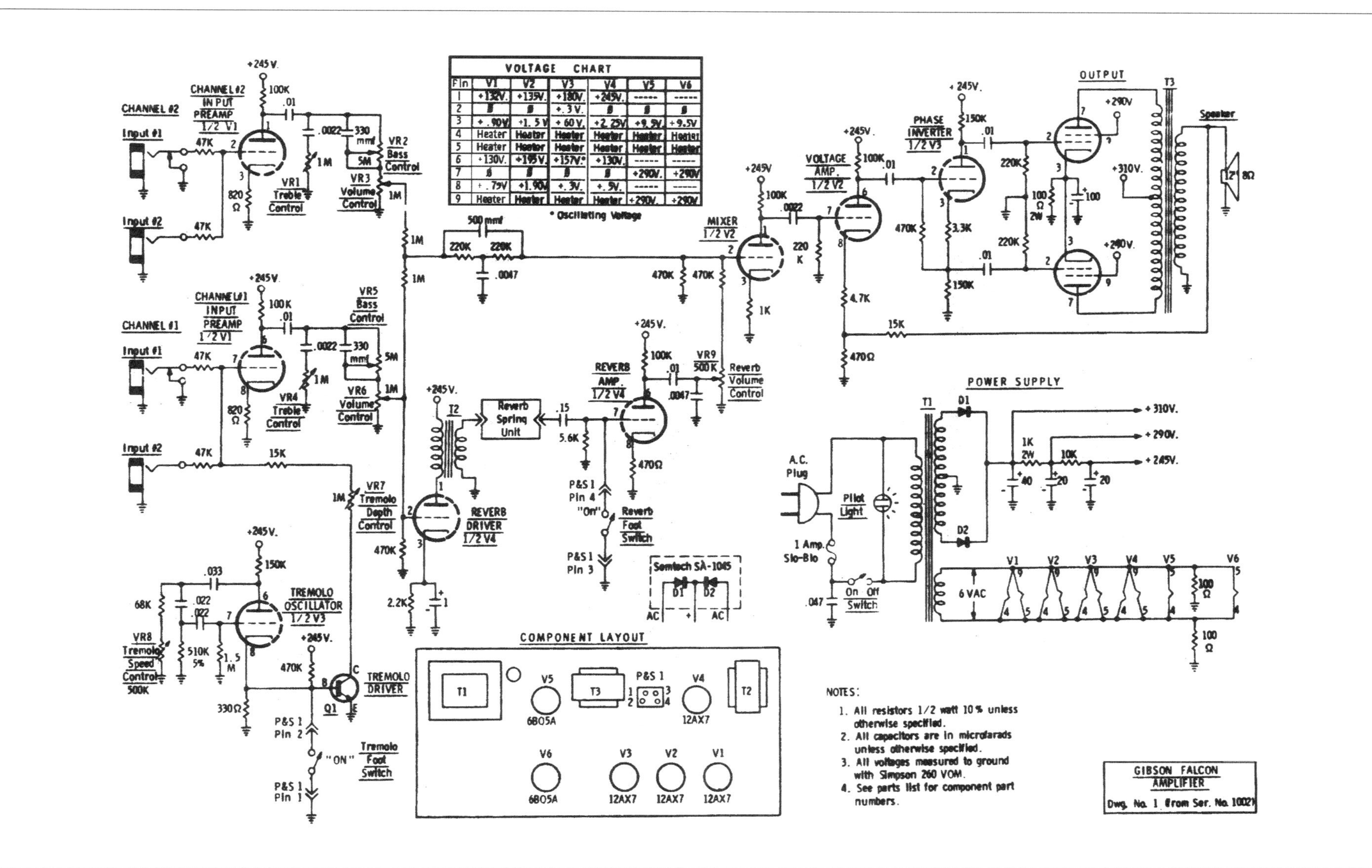

Pin	V1	V2	V3	V4	V5	V6
1	+132V.	+135V.	+180V.	+245V.	-----	-----
2	Ø	Ø	+.3V.	Ø	Ø	Ø
3	+.90V	+1.5V	+60V.	+2.25V	+9.5V.	+9.5V
4	Heater	Heater	Heater	Heater	Heater	Heater
5	Heater	Heater	Heater	Heater	Heater	Heater
6	+130V.	+195V.	+157V.*	+130V.	-----	-----
7	Ø	Ø	Ø	Ø	+290V.	+290V
8	+.79V	+1.90V	+.3V.	+.5V.	-----	-----
9	Heater	Heater	Heater	Heater	+290V.	+290V

* Oscillating Voltage

NOTES:

1. All resistors 1/2 watt 10% unless otherwise specified.
2. All capacitors are in microfarads unless otherwise specified.
3. All voltages measured to ground with Simpson 260 VOM.
4. See parts list for component part numbers.

GIBSON FALCON AMPLIFIER
Dwg. No. 1 (From Ser. No. 1002)

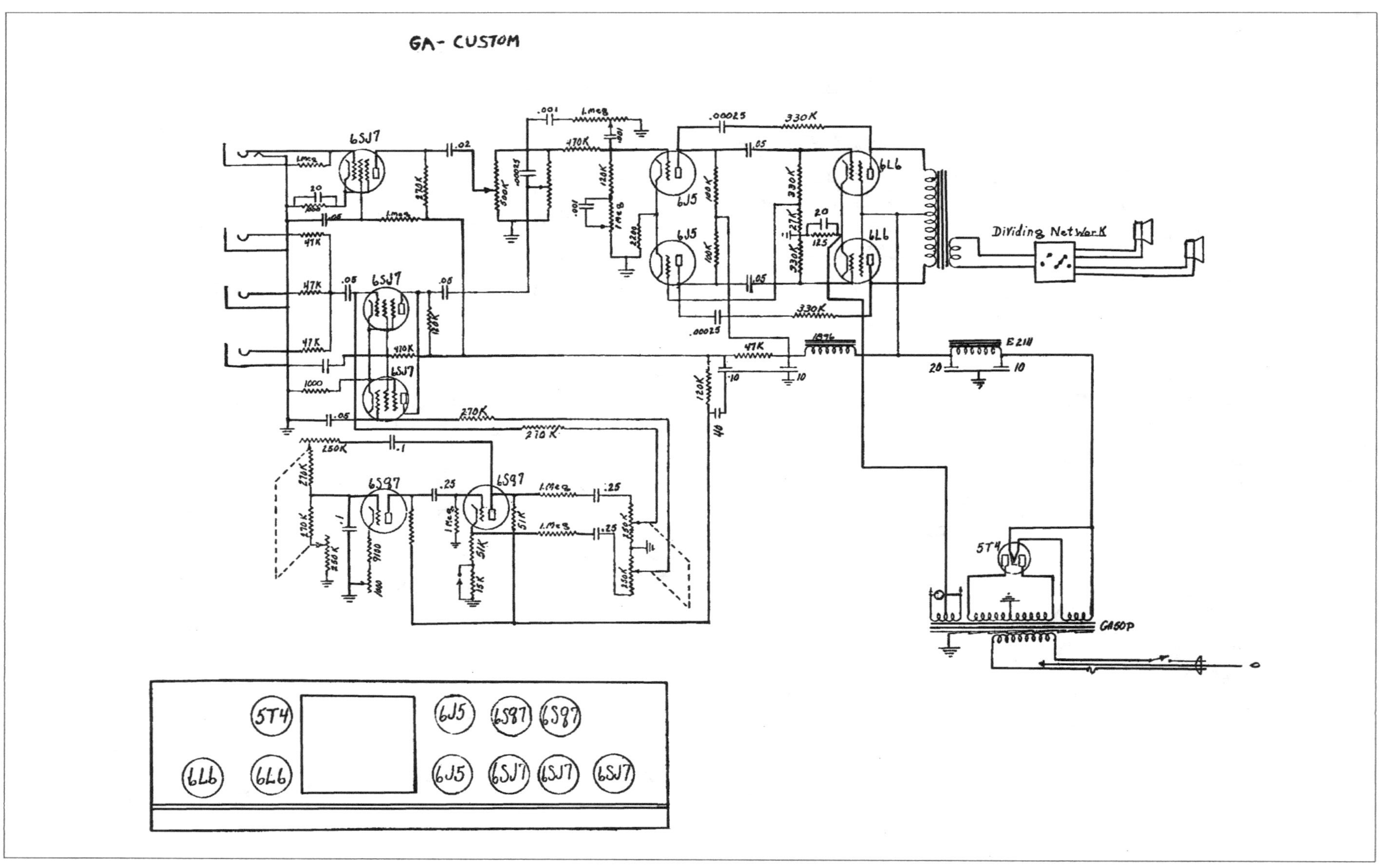
GA- CUSTOM
6SJ7
6SJ7
6SJ7
6J5
6J5
6L6
6L6
6SQ7
6SQ7
5T4
Dividing NetWorK
E211
GA50P
330K
.00025
.05
270K
1.Meg
47K
47K
47K
1000
270K
250K
51K
15K
120K
.25
.1
40
10
20
5T4
6J5
6SQ7
6L6
6SJ7

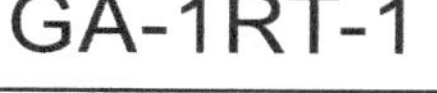

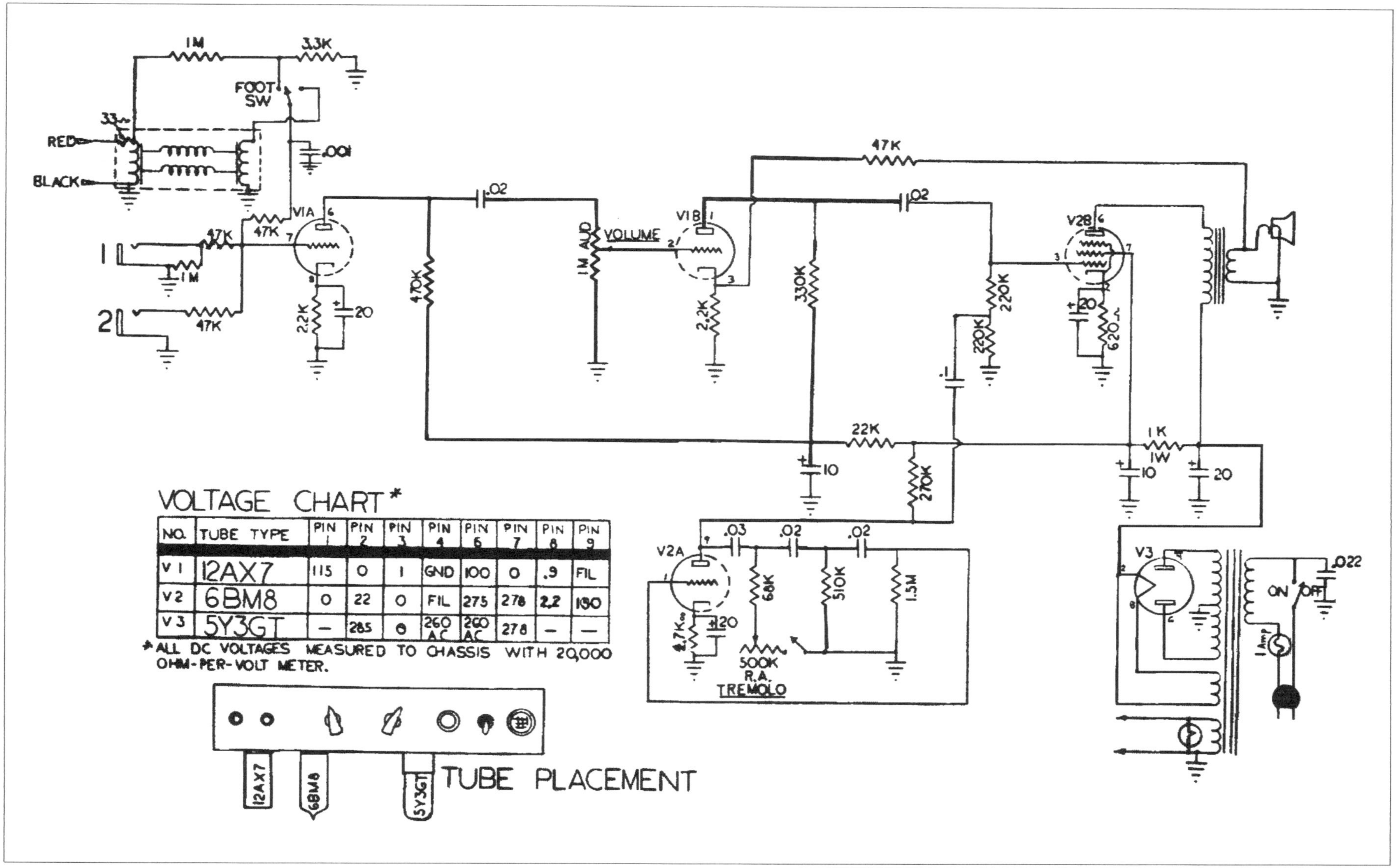

VOLTAGE CHART*

NO.	TUBE TYPE	PIN 1	PIN 2	PIN 3	PIN 4	PIN 6	PIN 7	PIN 8	PIN 9
V1	12AX7	115	0	1	GND	100	0	.9	FIL
V2	6BM8	0	22	0	FIL	275	278	2.2	130
V3	5Y3GT	—	285	0	260 AC	260 AC	278	—	—

*ALL DC VOLTAGES MEASURED TO CHASSIS WITH 20,000 OHM-PER-VOLT METER.

TUBE PLACEMENT

12AX7 6BM8 5Y3GT

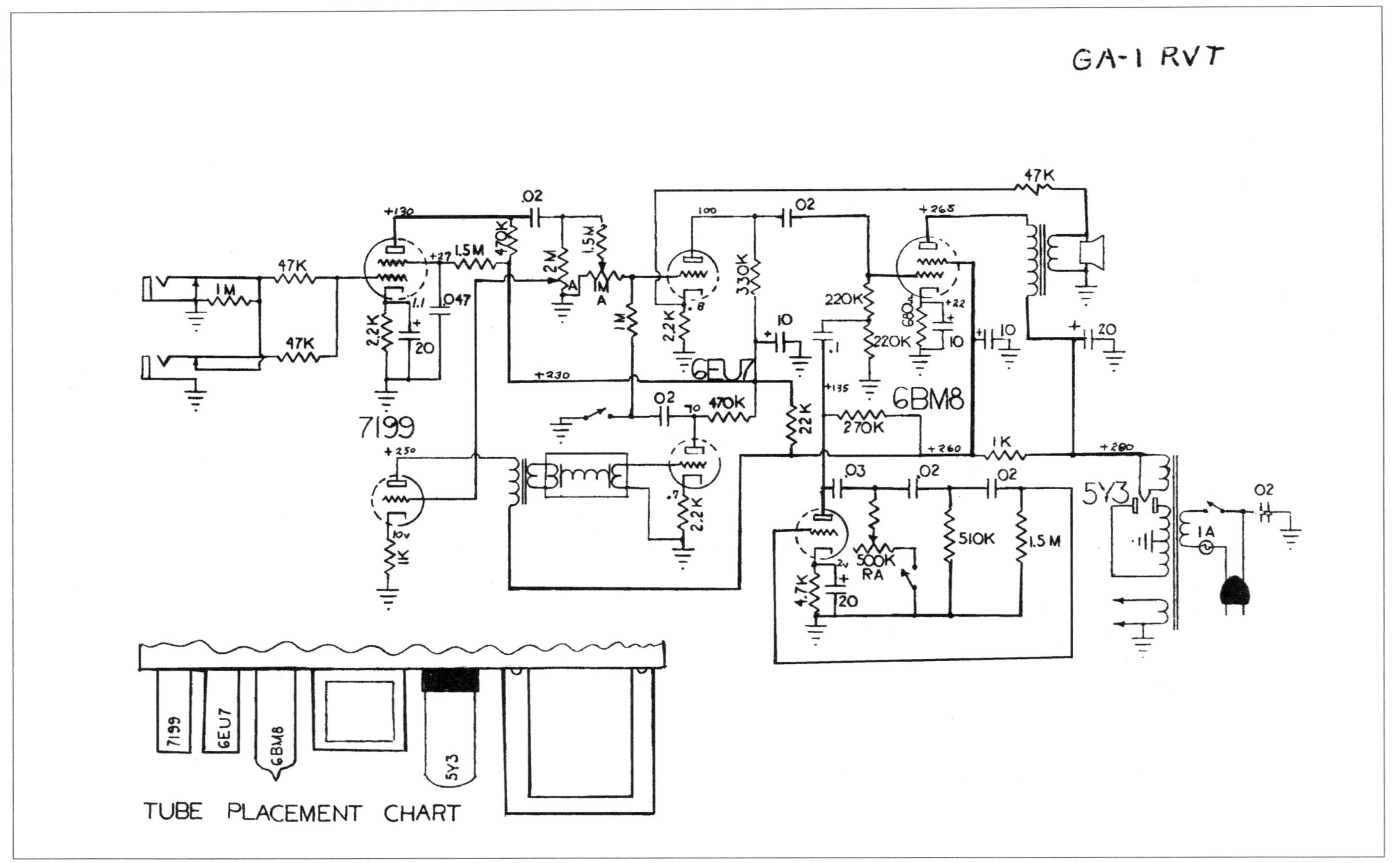
GA-1 RVT
7199
6EU7
6BM8
5Y3
47K
1M
2.2K
20
.047
1.5M
470K
.02
2M
1.5M
1M
+130
+27
+230
+250
1K
10v
2.2K
330K
100
10
.1
+135
220K
270K
22K
470K
+260
+265
680
+22
1K
+280
20
47K
.03
500K
RA
510K
1.5 M
4.7K
2v
1A
TUBE PLACEMENT CHART

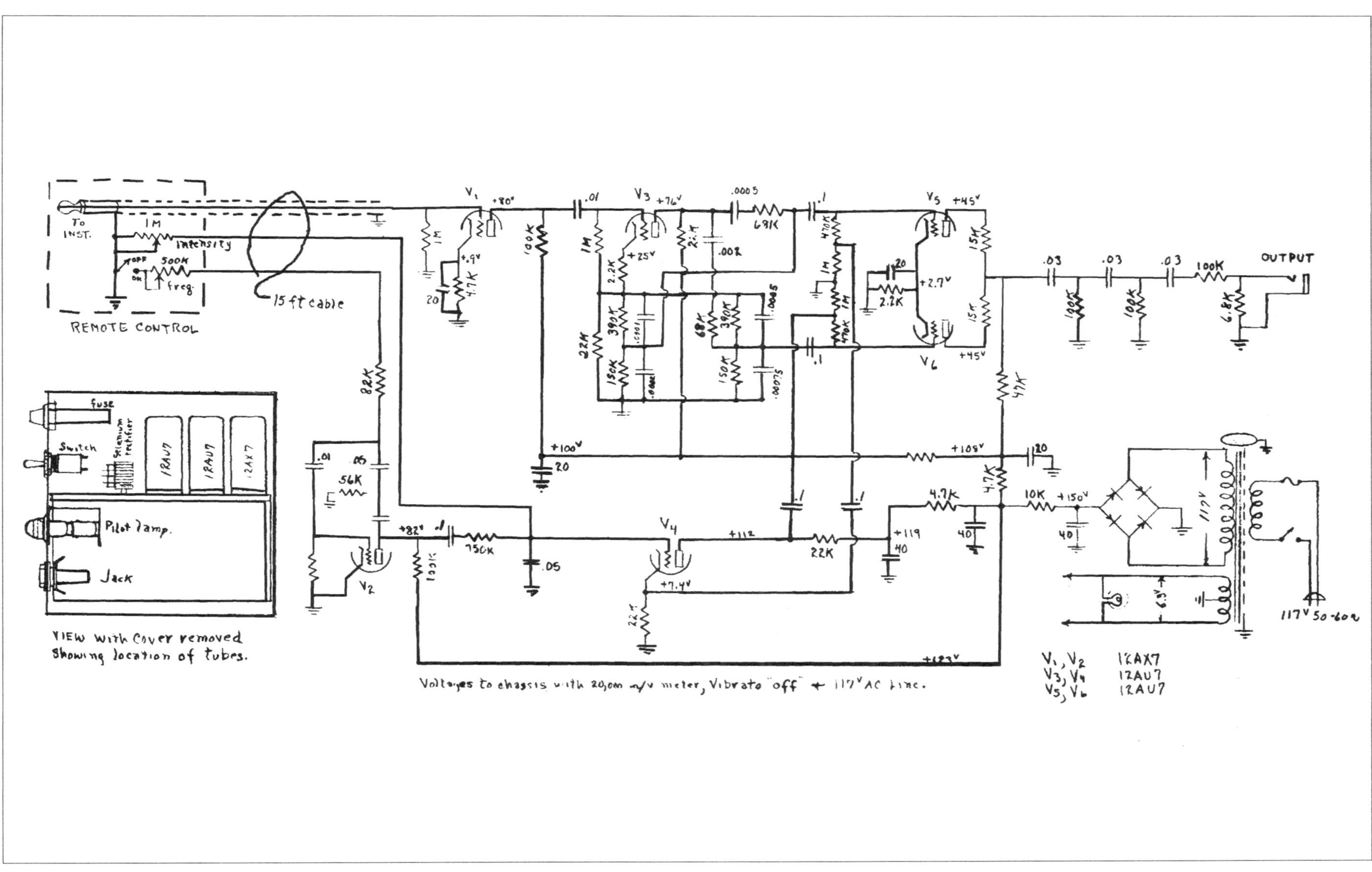
To INST.
1M
intensity
OFF
ON
500K
freq.
REMOTE CONTROL
15 ft cable
fuse
Switch
Selenium rectifier
12AU7
12AU7
12AX7
Pilot lamp.
Jack
VIEW with Cover removed showing location of tubes.
OUTPUT
117V 50-60~
V1, V2 12AX7
V3, V4 12AU7
V5, V6 12AU7
Voltages to chassis with 20,000 Ω/V meter, Vibrato "off" + 117V AC line.

GA-2RVT

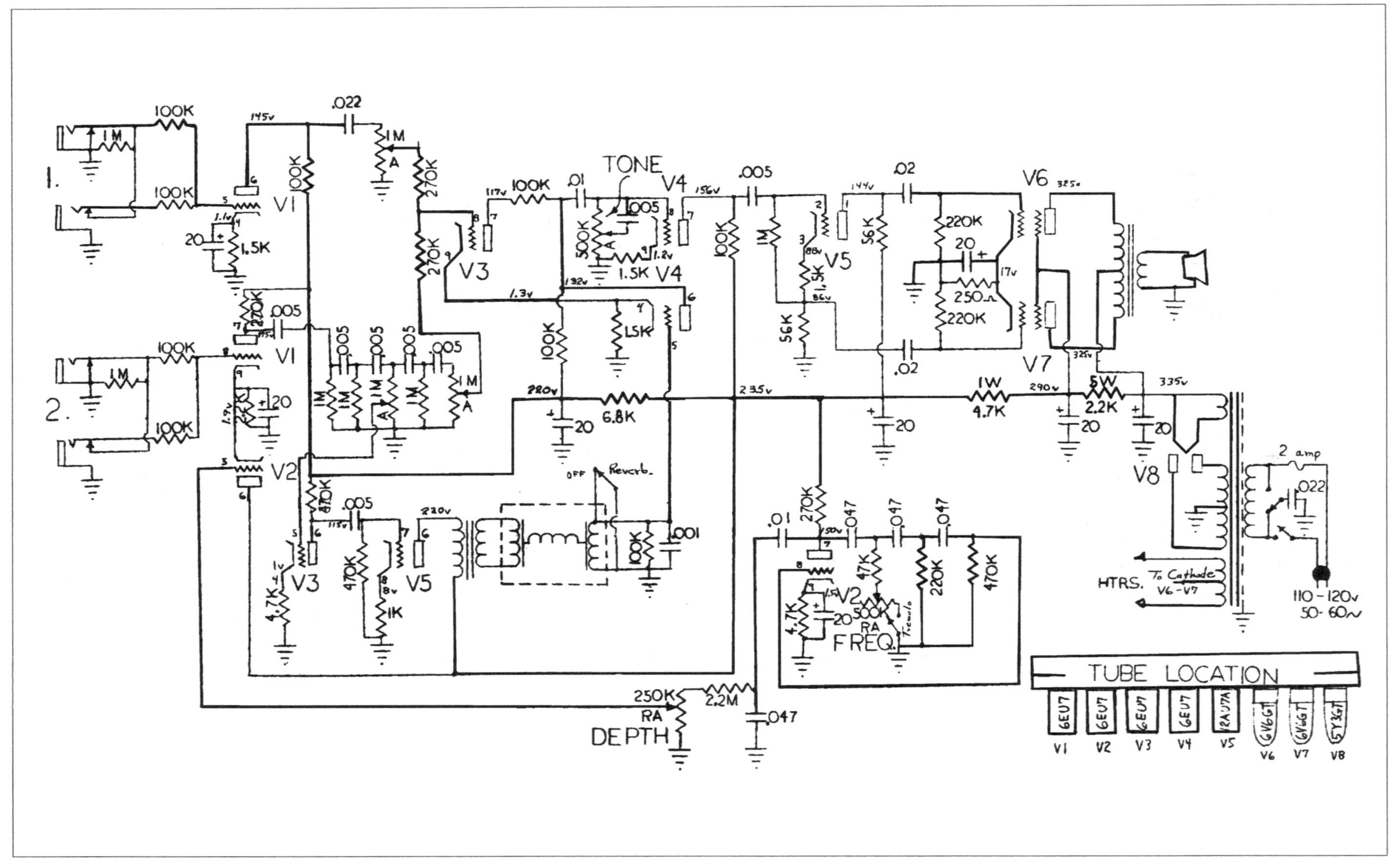

TONE
DEPTH
FREQ.
Reverb.
OFF
Tremolo
2 amp
110 - 120v
50 - 60~
HTRS.
To Cathode V6 - V7
TUBE LOCATION
6EU7
6EU7
6EU7
6EU7
12AU7A
6V6GT
6V6GT
5Y3GT
V1
V2
V3
V4
V5
V6
V7
V8

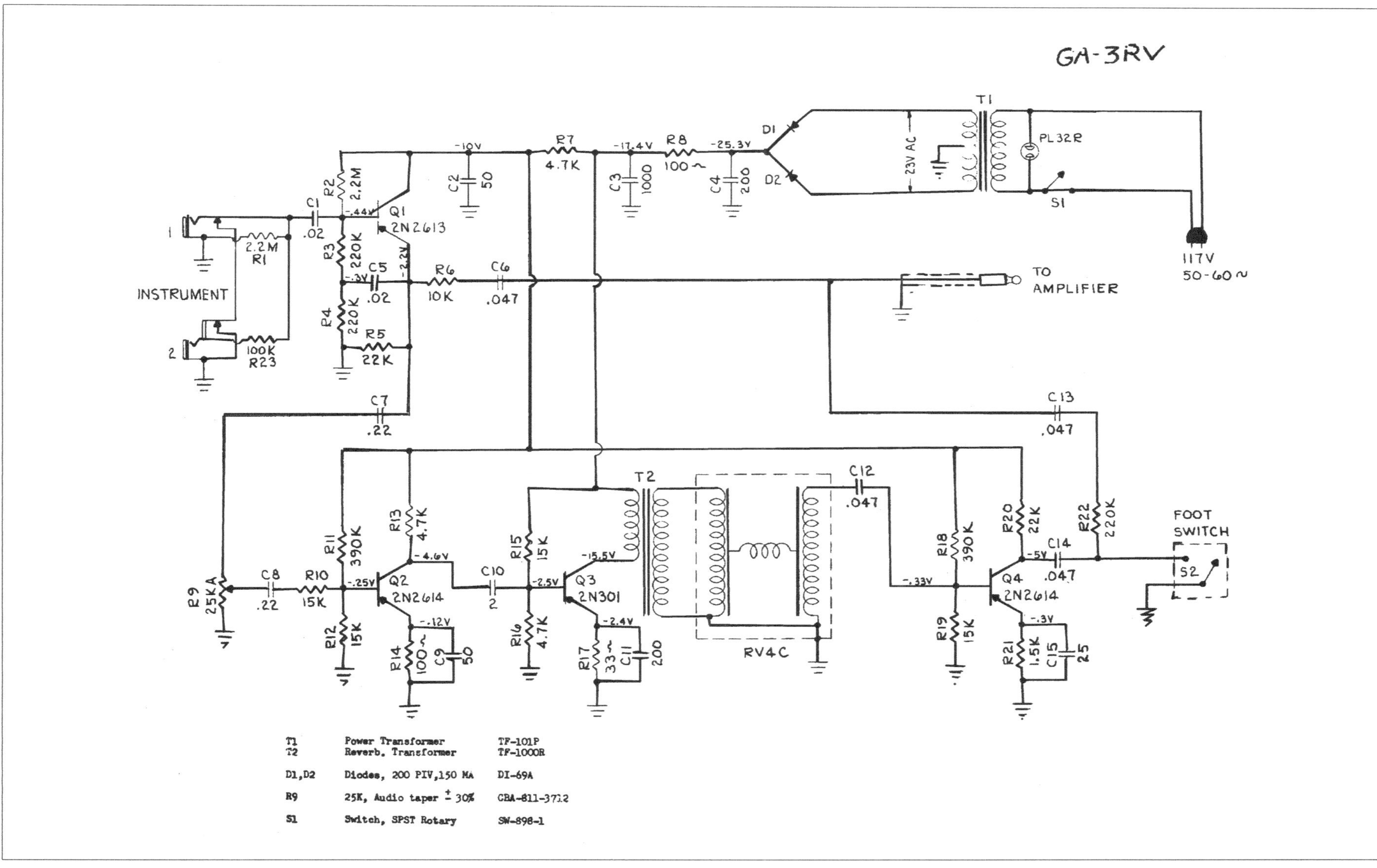
GA-3RV
T1
PL32R
S1
117V
50-60 ~
23V AC
D1
D2
-10V
R7
4.7K
-17.4V
R8
100 ~
-25.3V
C2
50
C3
1000
C4
200
R2
2.2M
C1
.02
-.44V
Q1
2N2613
R3
220K
-2.2V
-.3V
C5
.02
R6
10K
C6
.047
TO
AMPLIFIER
INSTRUMENT
1
2
2.2M
R1
100K
R23
R4
220K
R5
22K
C7
.22
C13
.047
T2
C12
.047
RV4C
R9
25KA
C8
.22
R10
15K
R11
390K
-.25V
Q2
2N2614
R12
15K
R13
4.7K
-4.6V
-.12V
R14
100 ~
C9
50
C10
2
R15
15K
-2.5V
Q3
2N301
-15.5V
R16
4.7K
-2.4V
R17
33 ~
C11
200
R18
390K
-.33V
Q4
2N2614
R19
15K
R20
22K
-5V
C14
.047
-.3V
R21
1.5K
C15
25
R22
220K
FOOT
SWITCH
S2
T1 Power Transformer TF-101P
T2 Reverb. Transformer TF-1000R
D1,D2 Diodes, 200 PIV,150 MA DI-69A
R9 25K, Audio taper ± 30% CBA-811-3712
S1 Switch, SPST Rotary SW-898-1

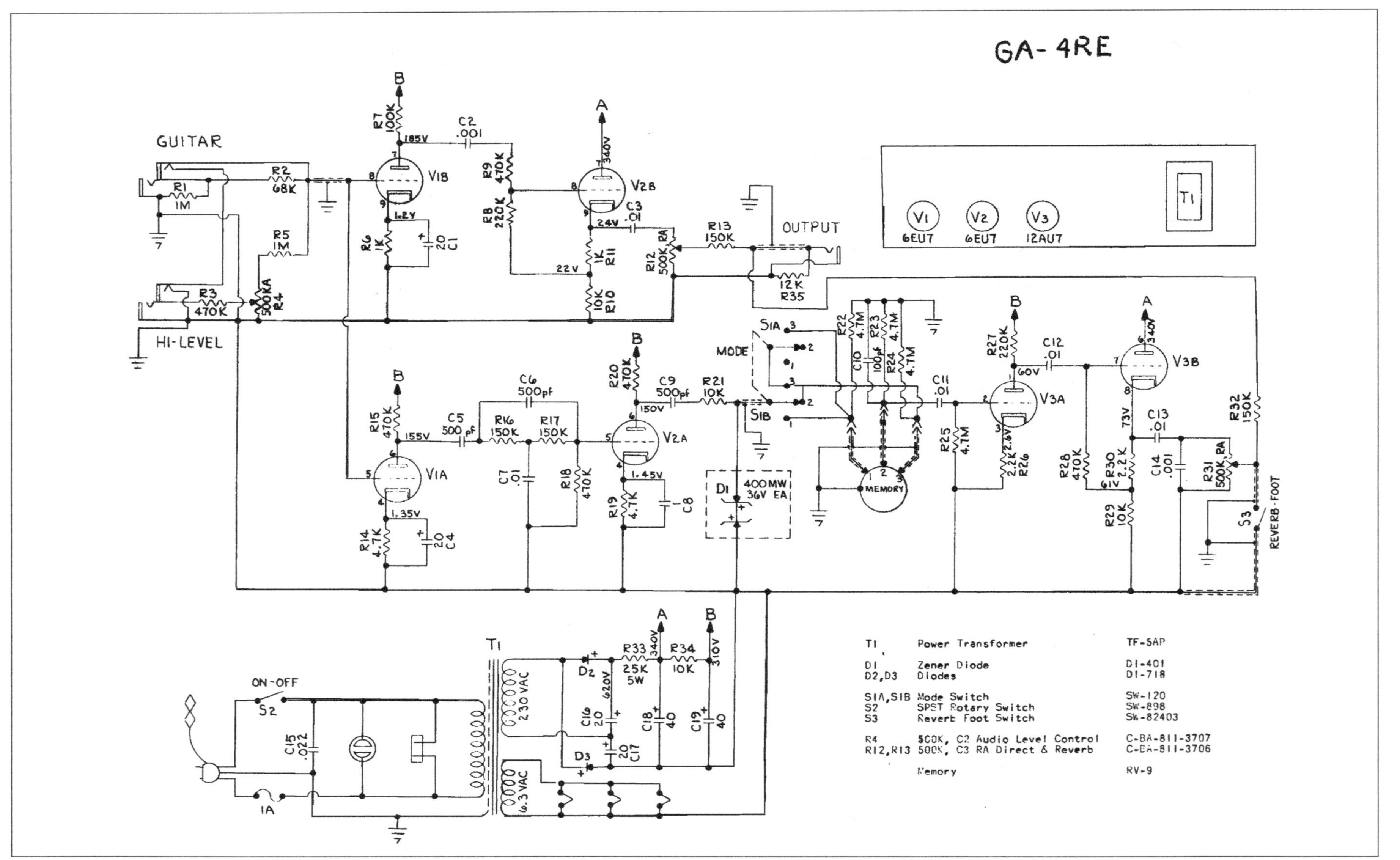
GA-4RE
GUITAR
HI-LEVEL
OUTPUT
MODE
REVERB-FOOT
ON-OFF
MEMORY
V1 6EU7
V2 6EU7
V3 12AU7
T1 Power Transformer TF-5AP
D1 Zener Diode DI-401
D2,D3 Diodes DI-718
S1A,S1B Mode Switch SW-120
S2 SPST Rotary Switch SW-898
S3 Reverb Foot Switch SW-82403
R4 500K, C2 Audio Level Control C-BA-811-3707
R12,R13 500K, C3 RA Direct & Reverb C-BA-811-3706
Memory RV-9

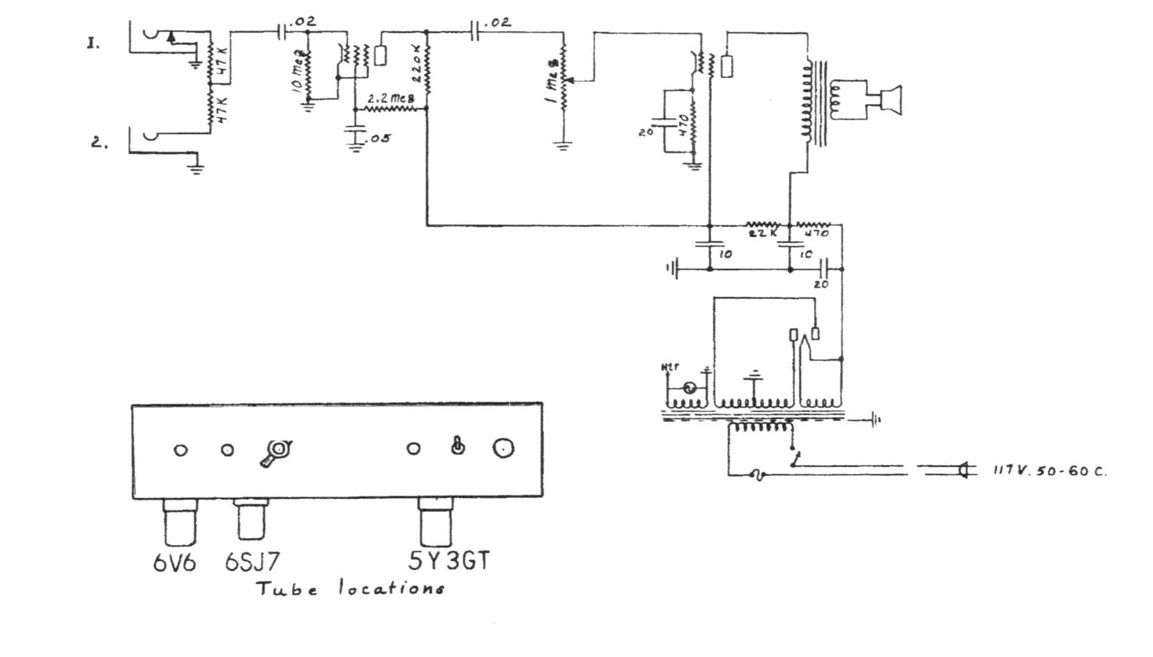

Les Paul

When only one instrument is used plug into #1 input jack.

This amplifier designed for 105-125 volt, 50-60 cycle current. Damage will result if connected to improper power source.

Use the above schematic to facilitate service by a reliable radio man.

Do not use higher rating fuse than one ampere, type 3 A.G.

This amplifier was carefully checked and in good playing condition when shipped. If damaged when received call transportation company immediately and place claim.

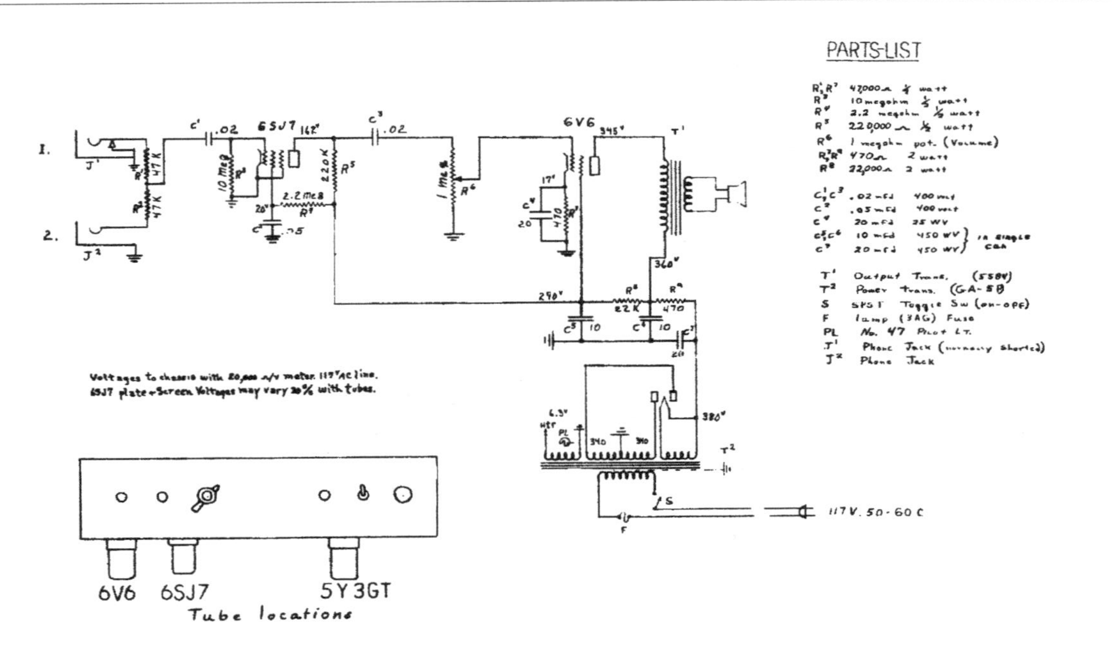

When only one instrument is used plug into #1 input jack.

This amplifier designed for 105-125 volt, 50-60 cycle current. Damage will result if connected to improper power source.

Use the above schematic to facilitate service by a reliable radio man.

Do not use higher rating fuse than one ampere, type 3 A.G.

This amplifier was carefully checked and in good playing condition when shipped. If damaged when received call transportation company immediately and place claim.

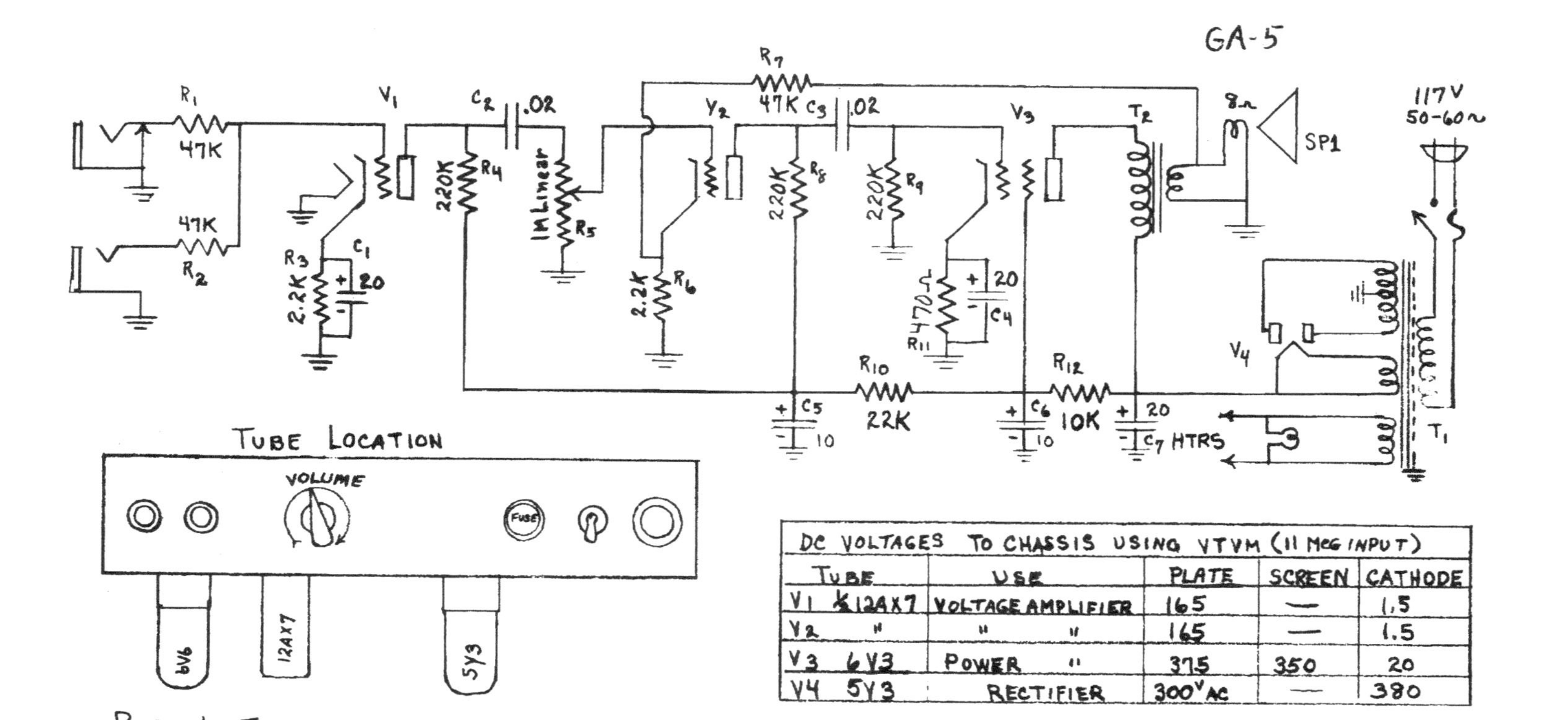

DC VOLTAGES TO CHASSIS USING VTVM (11 MEG INPUT)					
TUBE		USE	PLATE	SCREEN	CATHODE
V1	½12AX7	VOLTAGE AMPLIFIER	165	—	1.5
V2	"	" "	165	—	1.5
V3	6V3	POWER "	375	350	20
V4	5Y3	RECTIFIER	300V AC	—	380

PARTS LIST

$R_1 R_2 R_7$ 47K ½W, $R_3 R_6$ 2.2K ½W, $R_4 R_8 R_9$ 220K ½W, R_{10} 22K 1W, R_{11} 470Ω 1W, R_{12} 10K 1W, R_5 1MEG LINEAR TAPER

$C_1 C_4$ 20MFD 25 VOLT, $C_2 C_3$ .02 MFD 400VOLT, $C_5 C_6$ 10MFD 450VOLT, C_7 20MFD 450VOLT

T_1 POWER TRANSFORMER # GA5P T_2 OUTPUT TRANSFORMER # GA5-O   SP1 SPEAKER, 8" dia, 8Ω VC, Type 8J11.

When only one instrument is used plug into #1 input jack.

This amplifier designed for 105-125 volt, 50-60 cycle current. Damage will result if connected to improper power source.

Use the above schematic to facilitate service by a reliable radio man.

Do not use higher rating fuse than one ampere, type 3 A.G.

This amplifier was carefully checked and in good playing condition when shipped. If damaged when received call transportation company immediately and place claim.

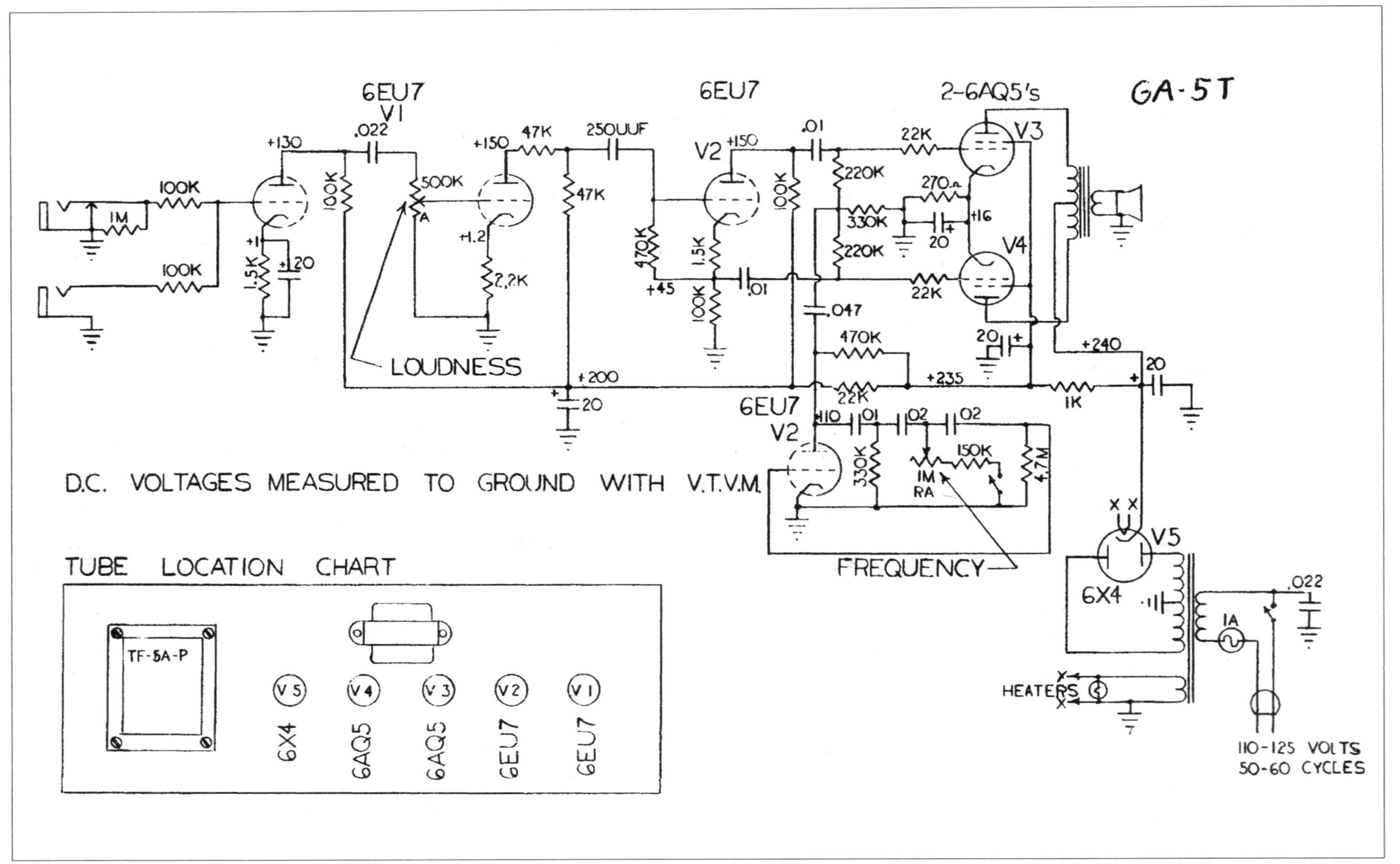
GA-5T
6EU7
V1
6EU7
V2
2-6AQ5's
V3
V4
6EU7
V2
V5
6X4
LOUDNESS
FREQUENCY
HEATERS
110-125 VOLTS
50-60 CYCLES
D.C. VOLTAGES MEASURED TO GROUND WITH V.T.V.M.
TUBE LOCATION CHART
TF-5A-P
V5 6X4
V4 6AQ5
V3 6AQ5
V2 6EU7
V1 6EU7

GA-6

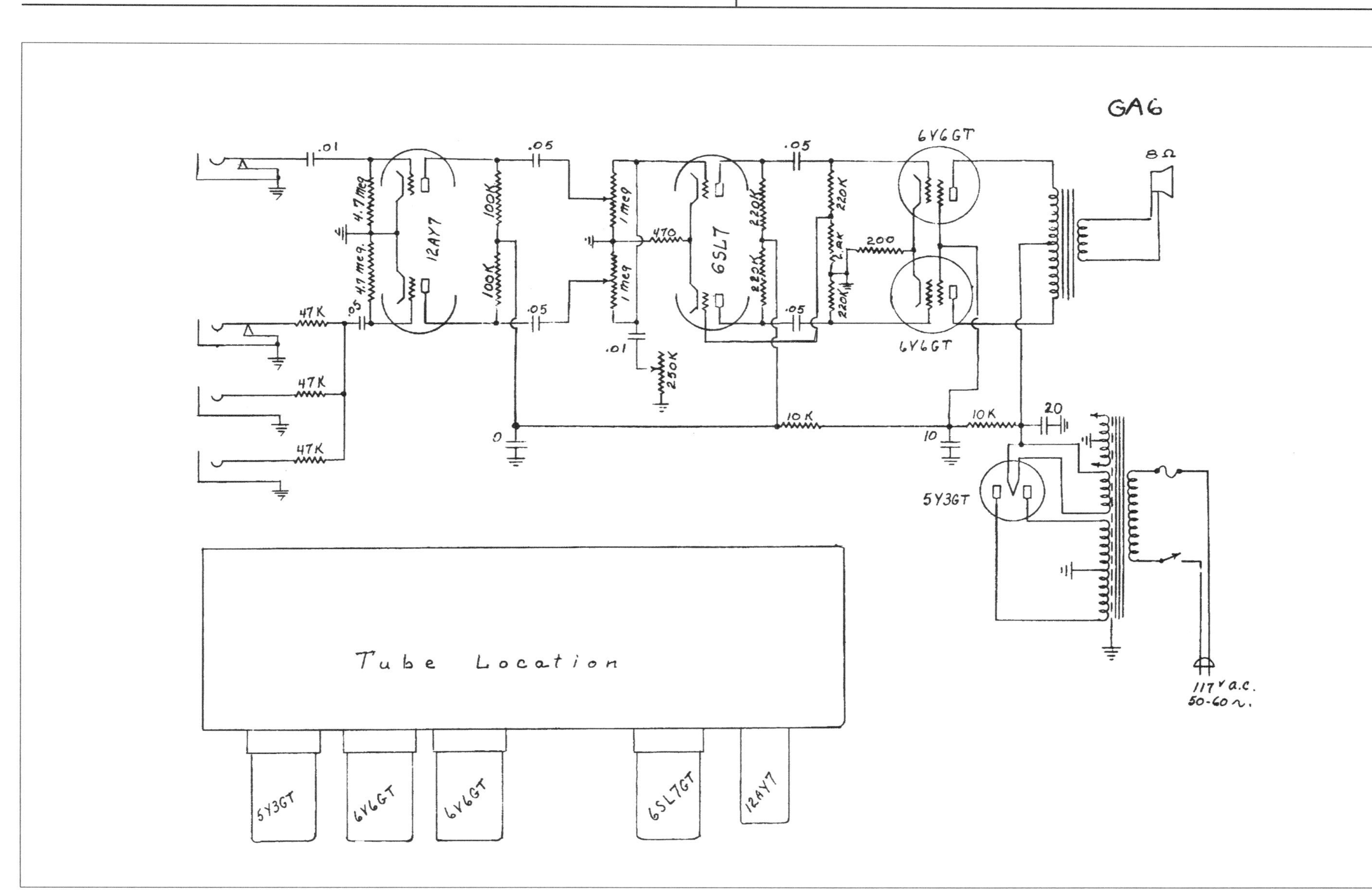

GA6
.01
4.7 meg
4.7 meg.
12AY7
47K
47K
47K
.05
100K
100K
.05
.05
1 meg
1 meg
470
6SL7
.01
250K
220K
220K
.05
.05
220K
2.2K
220K
200
6V6GT
6V6GT
8Ω
0
10K
10K
10
20
5Y3GT
117V A.C.
50-60~.
Tube Location
5Y3GT
6V6GT
6V6GT
6SL7GT
12AY7

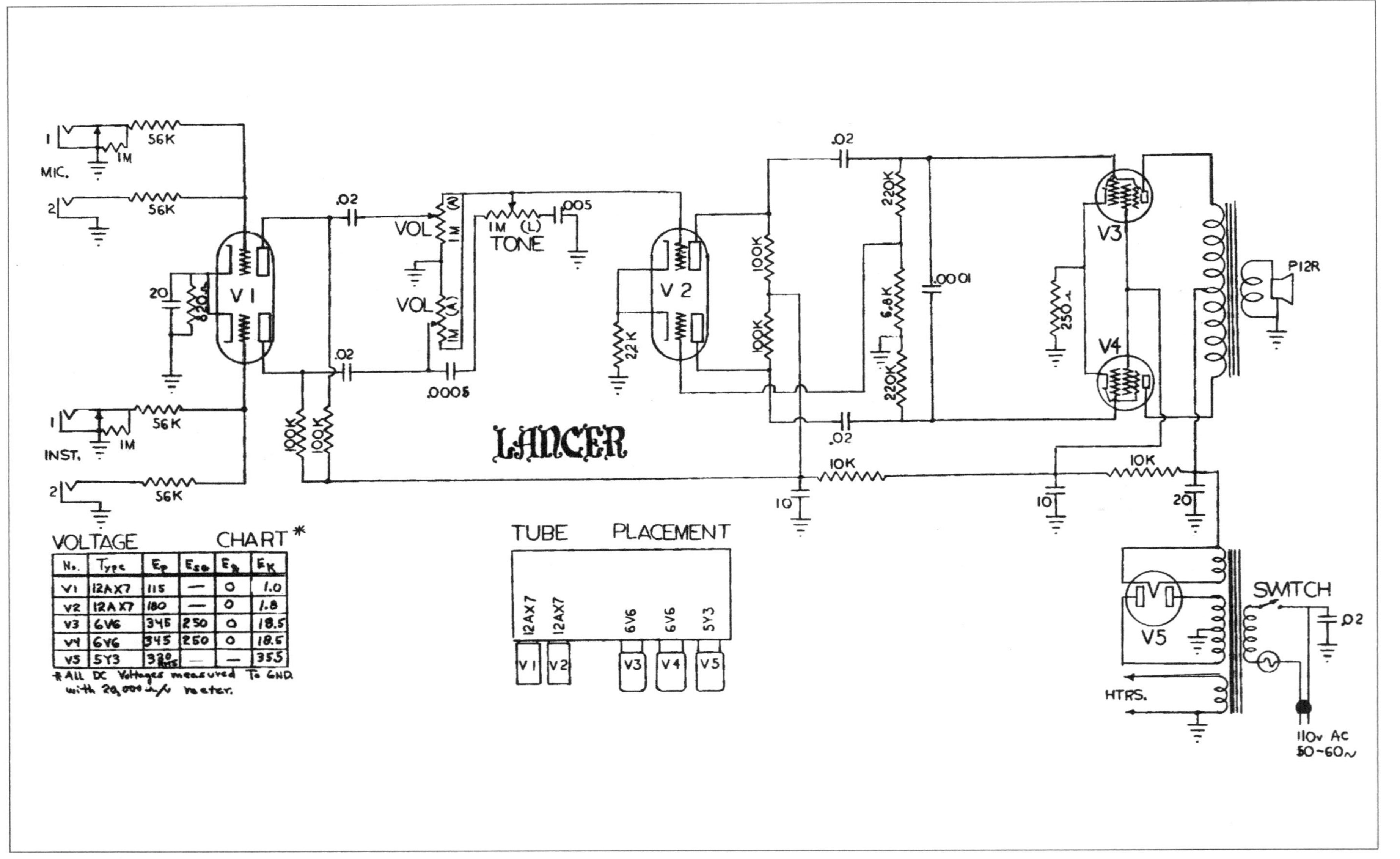

VOLTAGE CHART *

No.	Type	E_p	E_{sg}	E_g	E_K
V1	12AX7	115	—	0	1.0
V2	12AX7	180	—	0	1.8
V3	6V6	345	250	0	18.5
V4	6V6	345	250	0	18.5
V5	5Y3	330 RMS	—	—	355

* All DC Voltages measured To GND with 20,000 Ω/v meter.

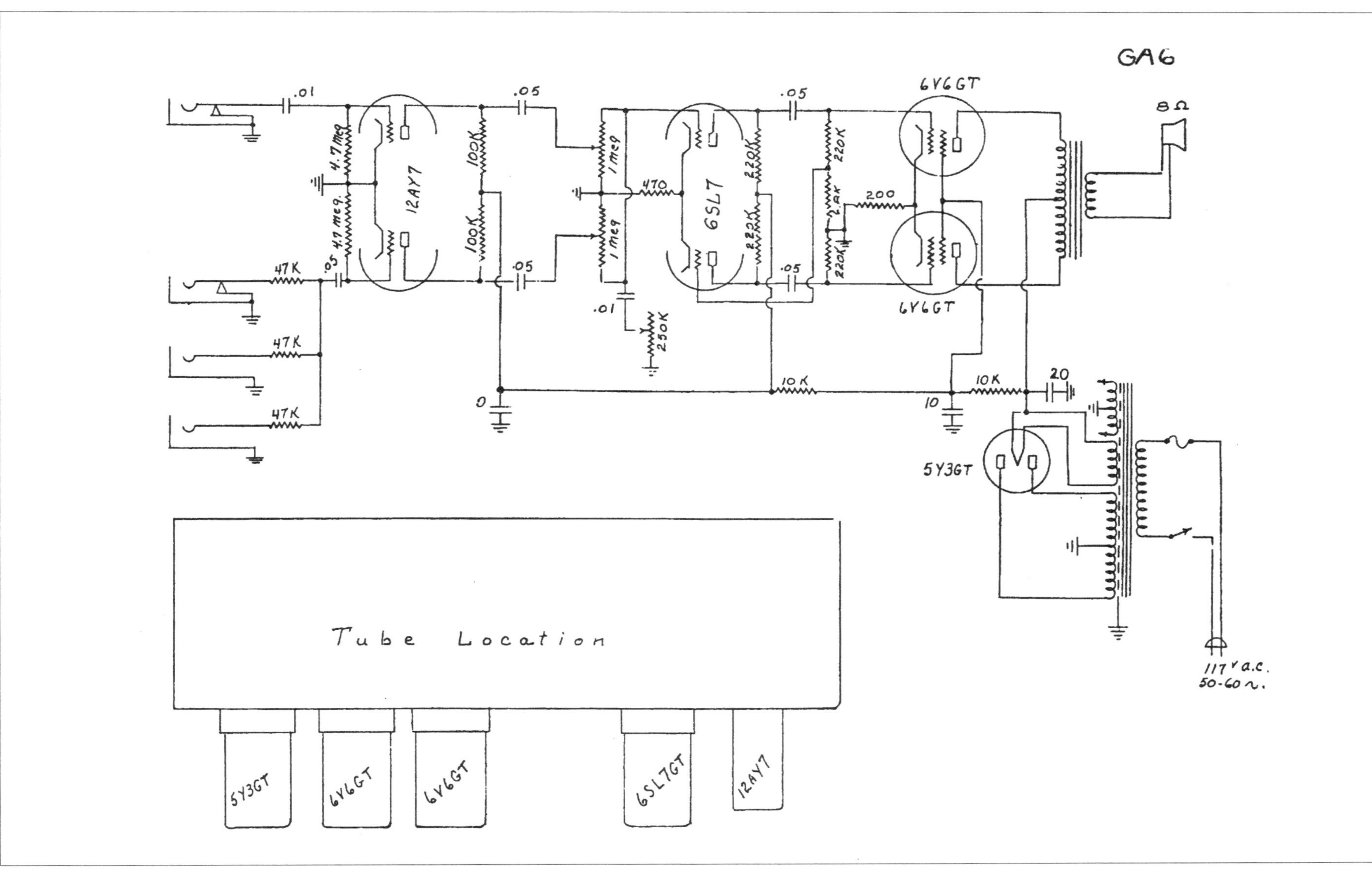
GA6
.01
4.7 meg
4.7 meg.
12AY7
.05
100K
100K
.05
1 meg
1 meg
470
6SL7
.01
250K
.05
220K
220K
.05
220K
2.2K
220K
200
6V6GT
6V6GT
8 Ω
47K
47K
47K
0
10K
10
10K
20
5Y3GT
117 V a.c.
50-60 ~.
Tube Location
5Y3GT
6V6GT
6V6GT
6SL7GT
12AY7

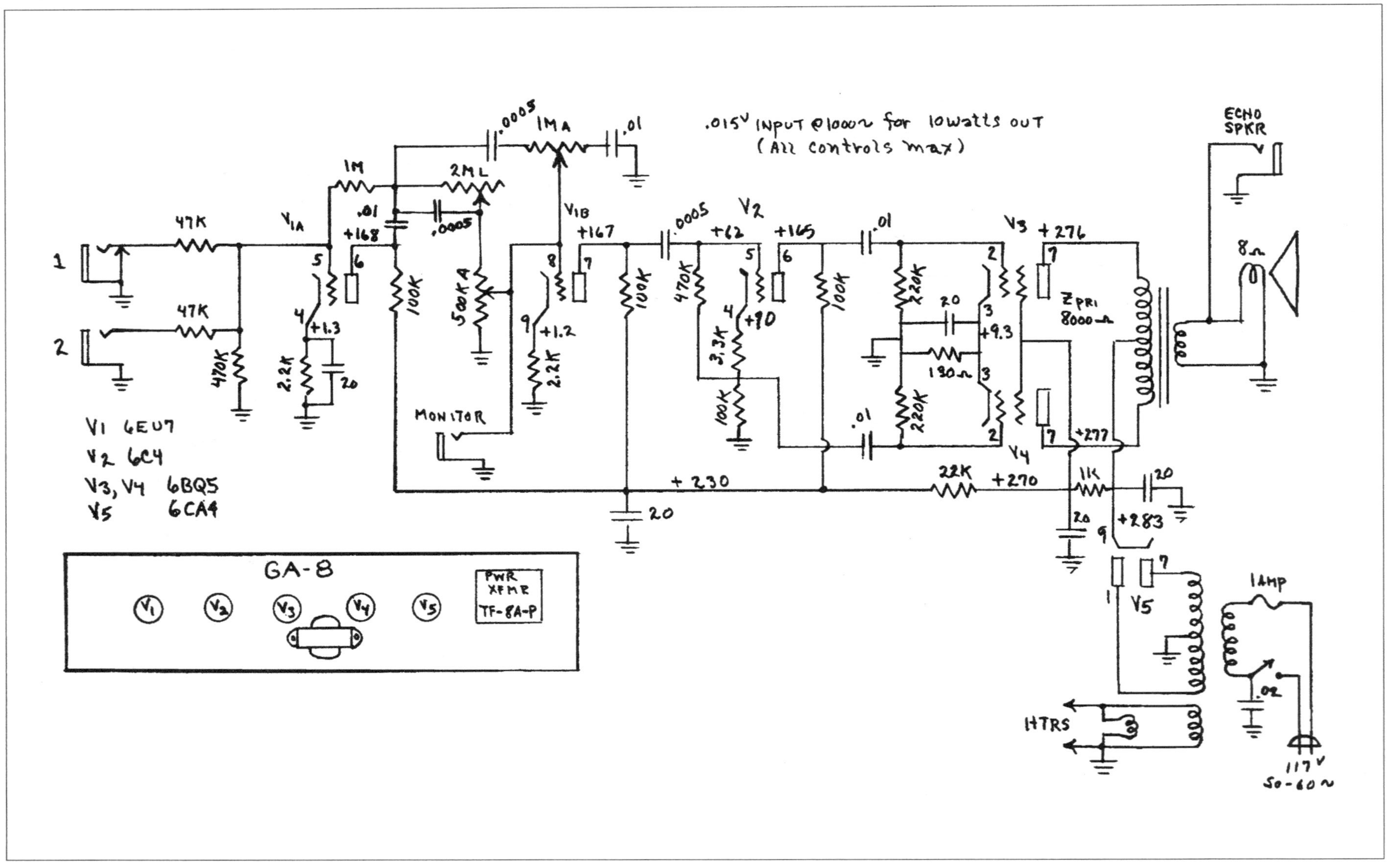
.015V INPUT @1000Ω for 10watts OUT
(All controls max)
ECHO SPKR
8Ω
1
2
47K
47K
470K
V1A
V1B
V2
V3
V4
V5
1M
.01
+168
2ML
.0005
.0005
1MA
.01
500KA
100K
2.2K
+1.3
20
+1.2
+167
+62
+165
+90
3.3K
470K
220K
130Ω
+9.3
+276
+277
ZPRI 8000Ω
22K
+270
1K
+283
+230
MONITOR
HTRS
1AMP
.02
117V
50-60~
V1 6EU7
V2 6C4
V3, V4 6BQ5
V5 6CA4
GA-8
PWR XFMR TF-8A-P

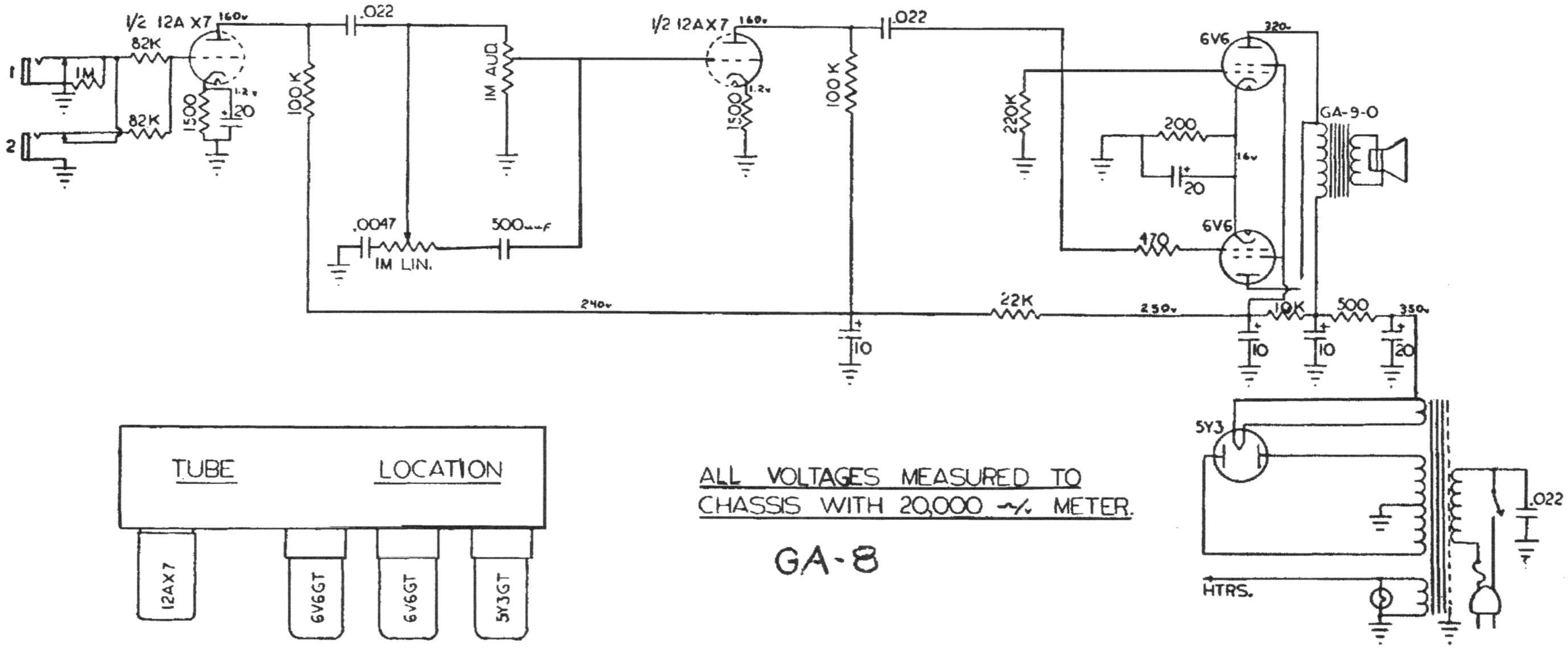

1/2 12AX7
160v
82K
82K
1M
1
2
1500
1.2v
20
.022
100 K
1M AUD.
.0047
1M LIN.
500μμf
1/2 12AX7
160v
1500
1.2v
100 K
.022
220K
6V6
320v
200
16v
20
GA-9-0
6V6
470
240v
22K
250v
10K
500
350v
10
10
10
20
5Y3
.022
HTRS.
TUBE
LOCATION
12AX7
6V6GT
6V6GT
5Y3GT
ALL VOLTAGES MEASURED TO
CHASSIS WITH 20,000 Ω/V METER.
GA-8

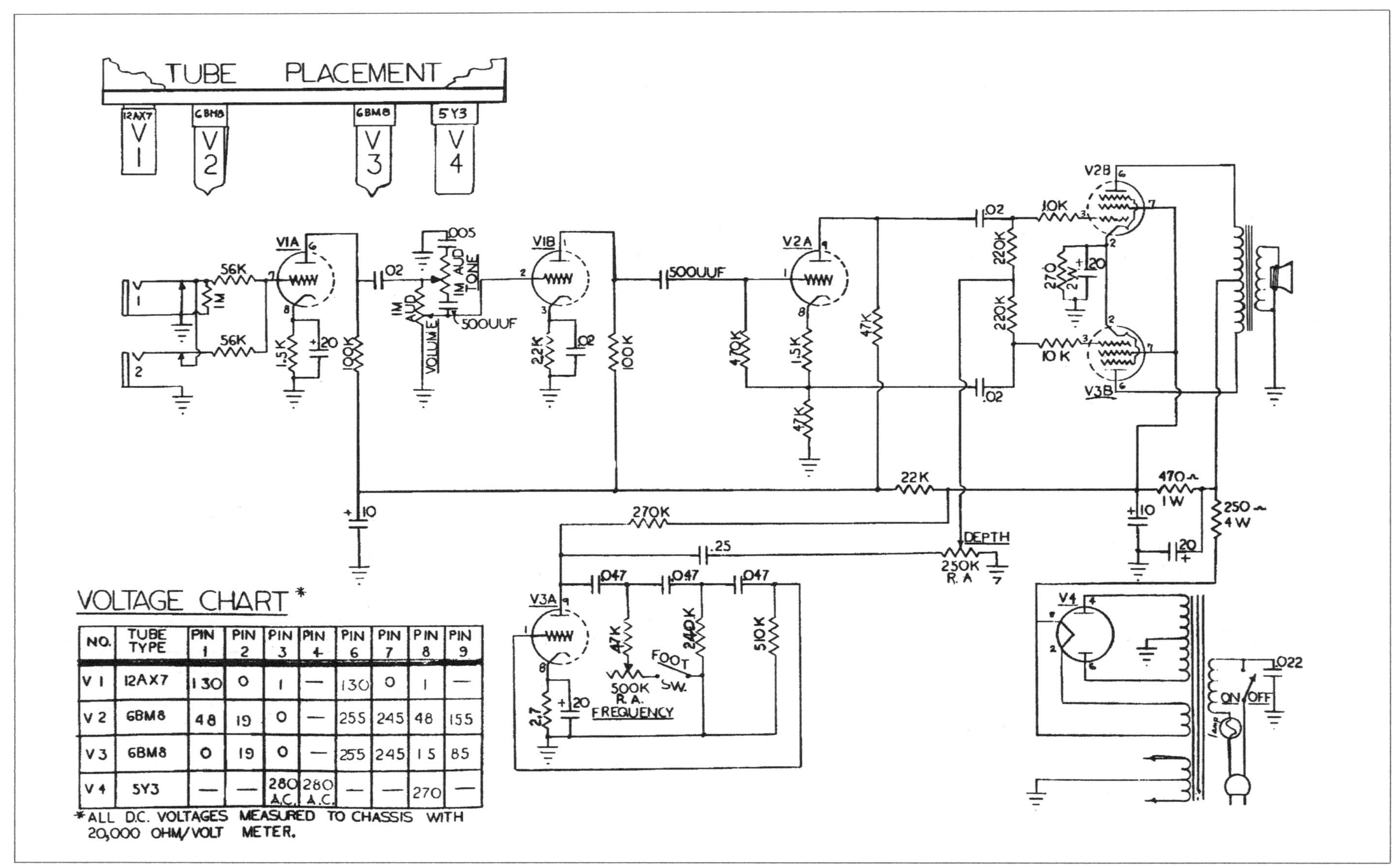

VOLTAGE CHART *

NO.	TUBE TYPE	PIN 1	PIN 2	PIN 3	PIN 4	PIN 6	PIN 7	PIN 8	PIN 9
V 1	12AX7	130	0	1	—	130	0	1	—
V 2	6BM8	48	19	0	—	255	245	48	155
V 3	6BM8	0	19	0	—	255	245	15	85
V 4	5Y3	—	—	280 A.C.	280 A.C.	—	—	270	—

*ALL D.C. VOLTAGES MEASURED TO CHASSIS WITH 20,000 OHM/VOLT METER.

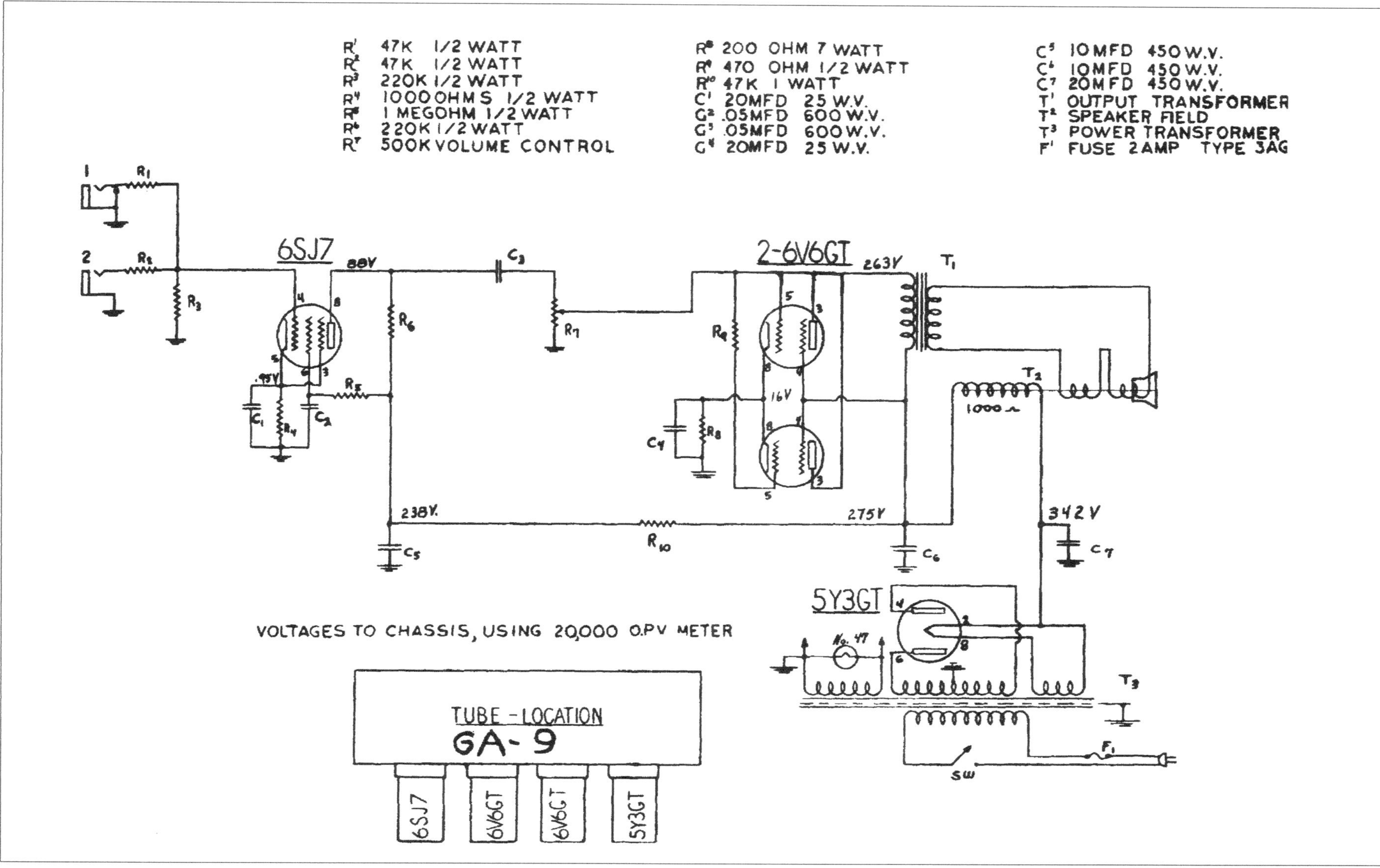
R1 47K 1/2 WATT
R2 47K 1/2 WATT
R3 220K 1/2 WATT
R4 1000 OHMS 1/2 WATT
R5 1 MEGOHM 1/2 WATT
R6 220K 1/2 WATT
R7 500K VOLUME CONTROL
R8 200 OHM 7 WATT
R9 470 OHM 1/2 WATT
R10 47K 1 WATT
C1 20MFD 25 W.V.
C2 .05MFD 600 W.V.
C3 .05MFD 600 W.V.
C4 20MFD 25 W.V.
C5 10MFD 450 W.V.
C6 10MFD 450 W.V.
C7 20MFD 450 W.V.
T1 OUTPUT TRANSFORMER
T2 SPEAKER FIELD
T3 POWER TRANSFORMER
F1 FUSE 2AMP TYPE 3AG
6SJ7
88V
2-6V6GT
263V
16V
238V
275V
342V
1000 Ω
5Y3GT
No. 47
SW
VOLTAGES TO CHASSIS, USING 20,000 O.P.V METER
TUBE - LOCATION
GA-9
6SJ7
6V6GT
6V6GT
5Y3GT

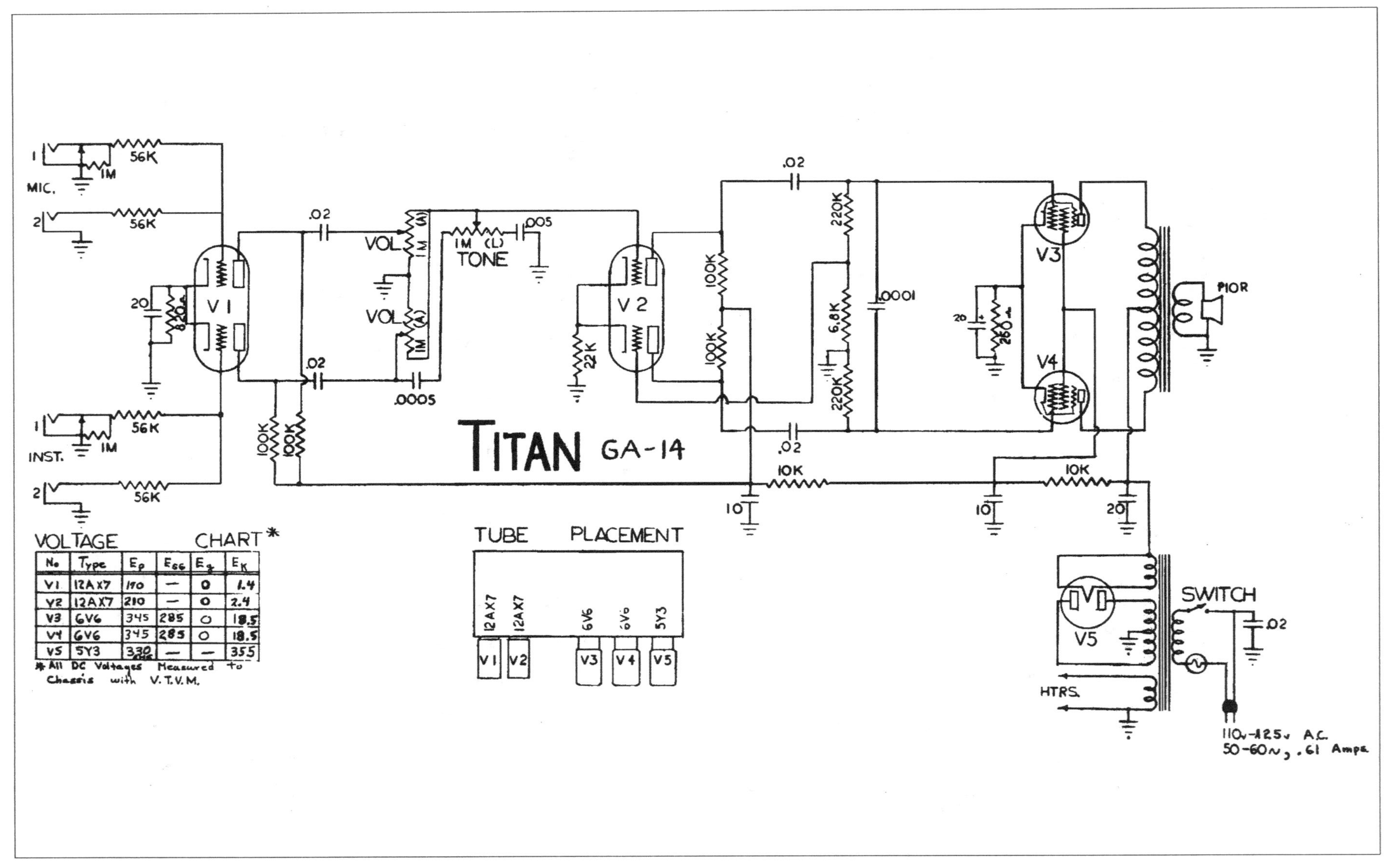

VOLTAGE CHART*

No	Type	E_p	E_{sg}	E_g	E_K
V1	12AX7	190	—	0	1.4
V2	12AX7	210	—	0	2.4
V3	6V6	345	285	0	18.5
V4	6V6	345	285	0	18.5
V5	5Y3	330	—	—	355

*All DC Voltages Measured to Chassis with V.T.V.M.

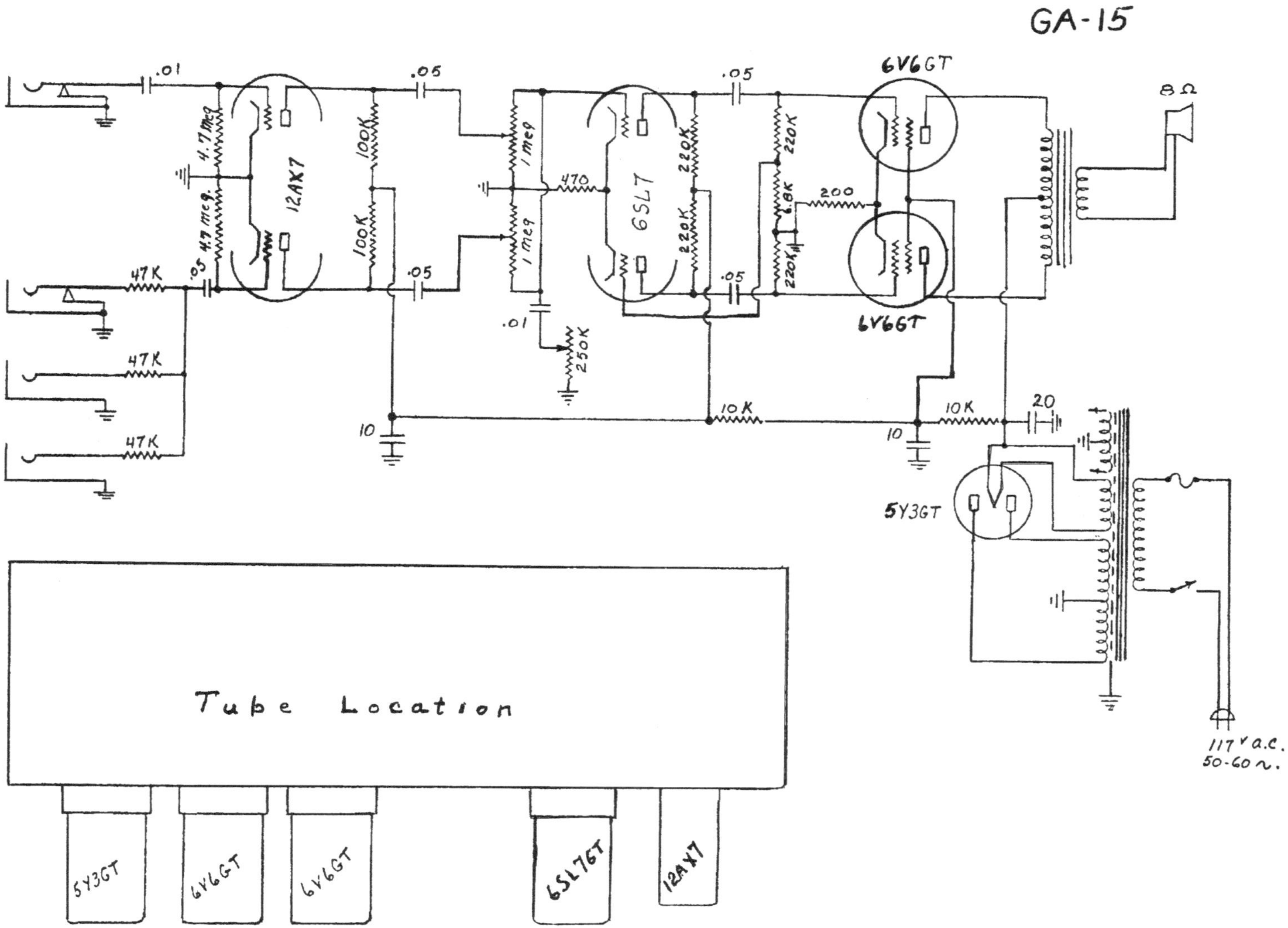

GA-15

.01
4.7 meg.
4.7 meg.
12AX7
.05
47K
47K
47K
100K
100K
.05
.05
1 meg
1 meg
470
.01
250K
6SL7
220K
220K
.05
.05
220K
6.8K
220K
200
6V6GT
6V6GT
8 Ω
10
10K
10K
10
20
5Y3GT
117 v a.c.
50-60 ~.

Tube Location

5Y3GT
6V6GT
6V6GT
6SL7GT
12AX7

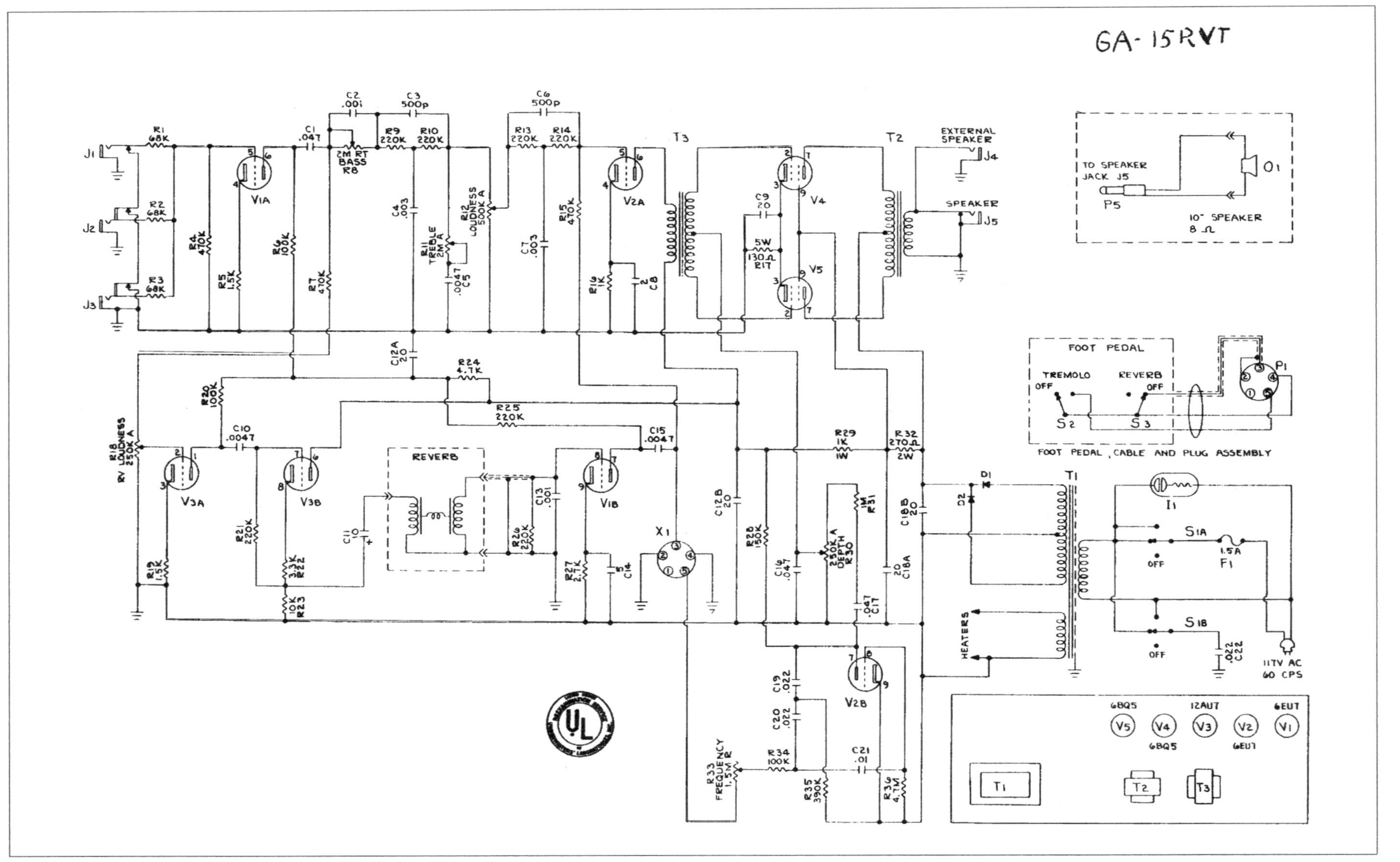
GA-15RVT
J1
J2
J3
R1 68K
R2 68K
R3 68K
R4 470K
R5 1.5K
V1A
C1 .047
R6 100K
R7 470K
C2 .001
2M RT BASS R8
R9 220K
C3 500p
R10 220K
C4 .003
R11 TREBLE 2M A
.0047 C5
R12 LOUDNESS 500K A
C6 500p
R13 220K
R14 220K
C7 .003
R15 470K
V2A
R16 1K
2 C8
T3
C9 20
5W 130 Ω R17
V4
V5
T2
EXTERNAL SPEAKER
J4
SPEAKER
J5
TO SPEAKER JACK J5
P5
O1
10" SPEAKER 8 Ω
C12A 20
R24 4.7K
R25 220K
R20 100K
C10 .0047
R18 RV LOUDNESS 250K A
V3A
V3B
R19 1.5K
R21 220K
3.3K R22
10K R23
C11 10
REVERB
R26 220K
C13 .001
V1B
R27 2.7K
5 C14
C15 .004T
X1
C12B 20
R28 150K
C16 .047
250K A DEPTH R30
1M R31
.047 C17
R29 1K 1W
R32 270 Ω 2W
C18B 20
20 C18A
C19 .022
C20 .022
V2B
R33 FREQUENCY 1.5M R
R34 100K
R35 390K
C21 .01
R36 4.7M
FOOT PEDAL
TREMOLO
REVERB
OFF
OFF
S2
S3
P1
FOOT PEDAL, CABLE AND PLUG ASSEMBLY
D1
D2
T1
I1
S1A
1.5A
F1
OFF
S1B
OFF
.022 C22
117V AC 60 CPS
HEATERS
6BQ5
12AU7
6EU7
V5
V4
V3
V2
V1
6BQ5
6EU7
T1
T2
T3

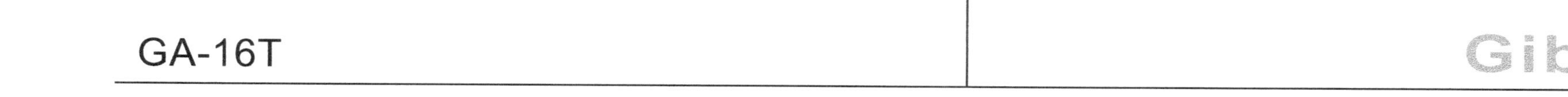

GA-16T

TUBE PLACEMENT

REAR OF CHASSIS

12AX7	12AX7	6V6	6V6	5Y3
V 1	V 2	V 3	V 4	V 5

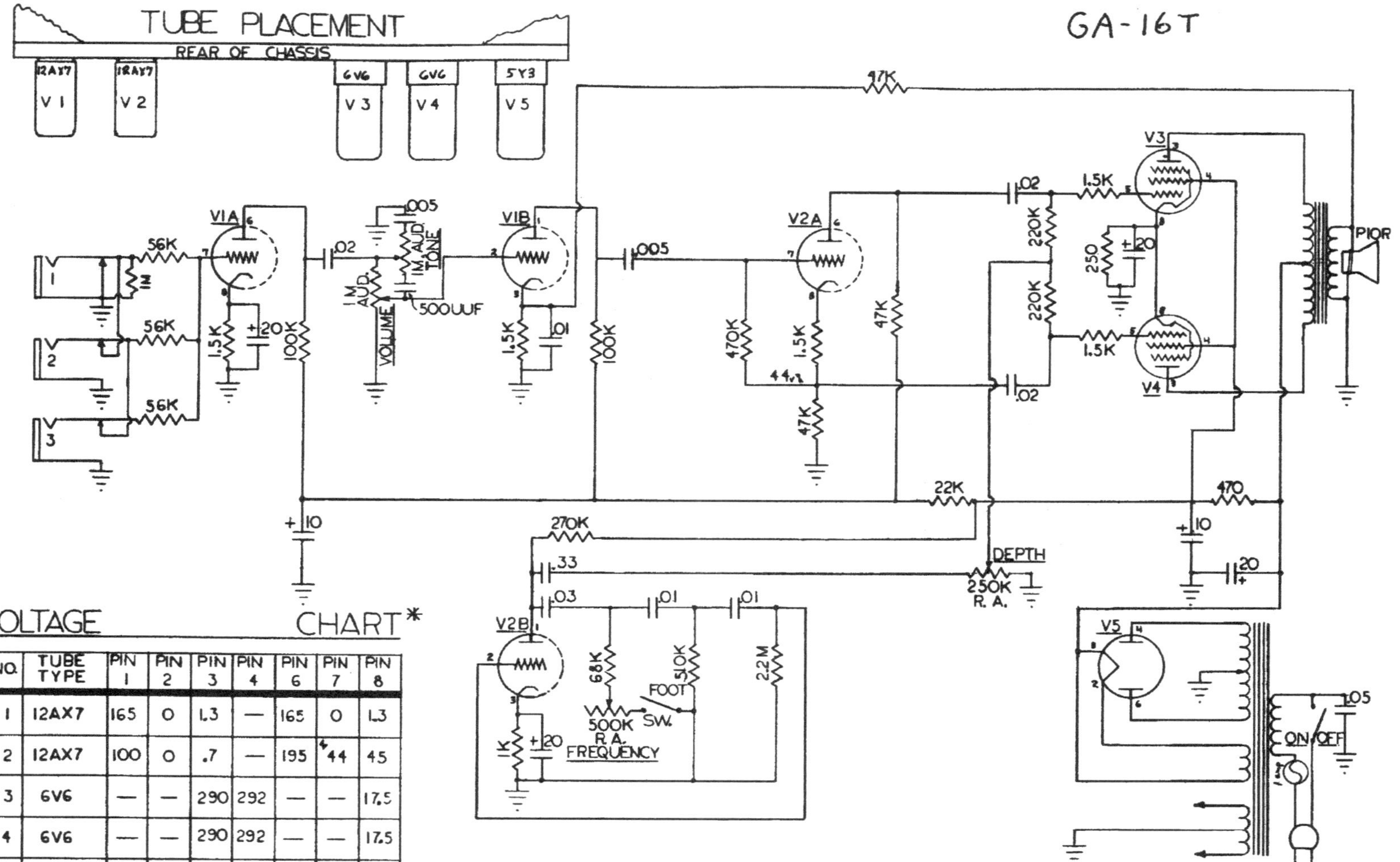

VOLTAGE CHART*

NO.	TUBE TYPE	PIN 1	PIN 2	PIN 3	PIN 4	PIN 6	PIN 7	PIN 8
V 1	12AX7	165	0	1.3	—	165	0	1.3
V 2	12AX7	100	0	.7	—	195	44	45
V 3	6V6	—	—	290	292	—	—	17.5
V 4	6V6	—	—	290	292	—	—	17.5
V 5	5Y3	—	5.2 AC	—	280 AC	280 AC	—	305

*ALL D.C. VOLTAGES MEASURED TO CHASSIS WITH 20,000 OHM/VOLT METER.

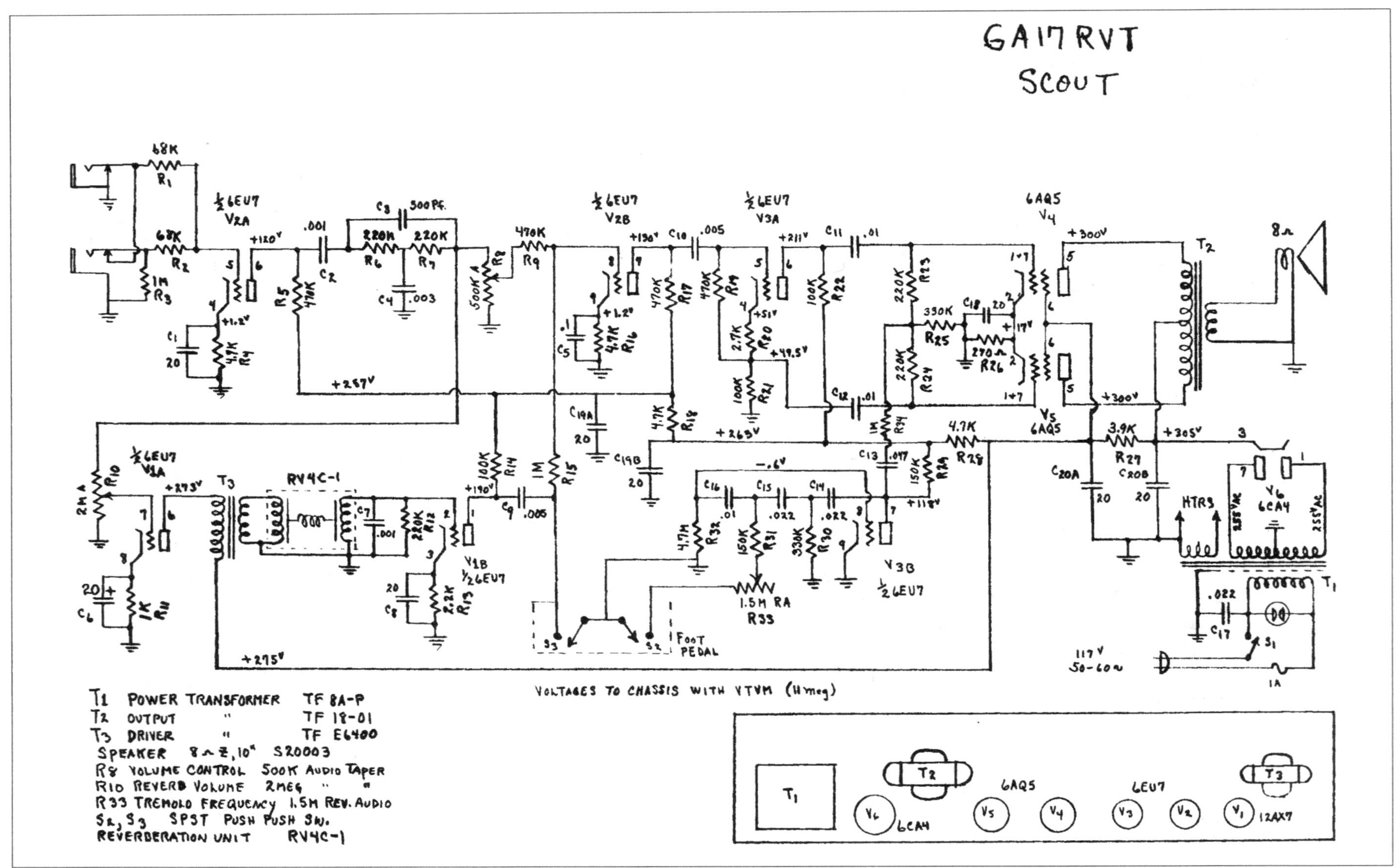
GA17RVT
SCOUT
VOLTAGES TO CHASSIS WITH VTVM (11 meg)
T1 POWER TRANSFORMER TF 8A-P
T2 OUTPUT " TF 18-01
T3 DRIVER " TF E6400
SPEAKER 8Ω Z, 10" S20003
R8 VOLUME CONTROL 500K AUDIO TAPER
R10 REVERB VOLUME 2MEG " "
R33 TREMOLO FREQUENCY 1.5M REV. AUDIO
S2, S3 SPST PUSH PUSH SW.
REVERBERATION UNIT RV4C-1
FOOT PEDAL
117V 50-60~
6AQ5
6EU7
12AX7
6CA4

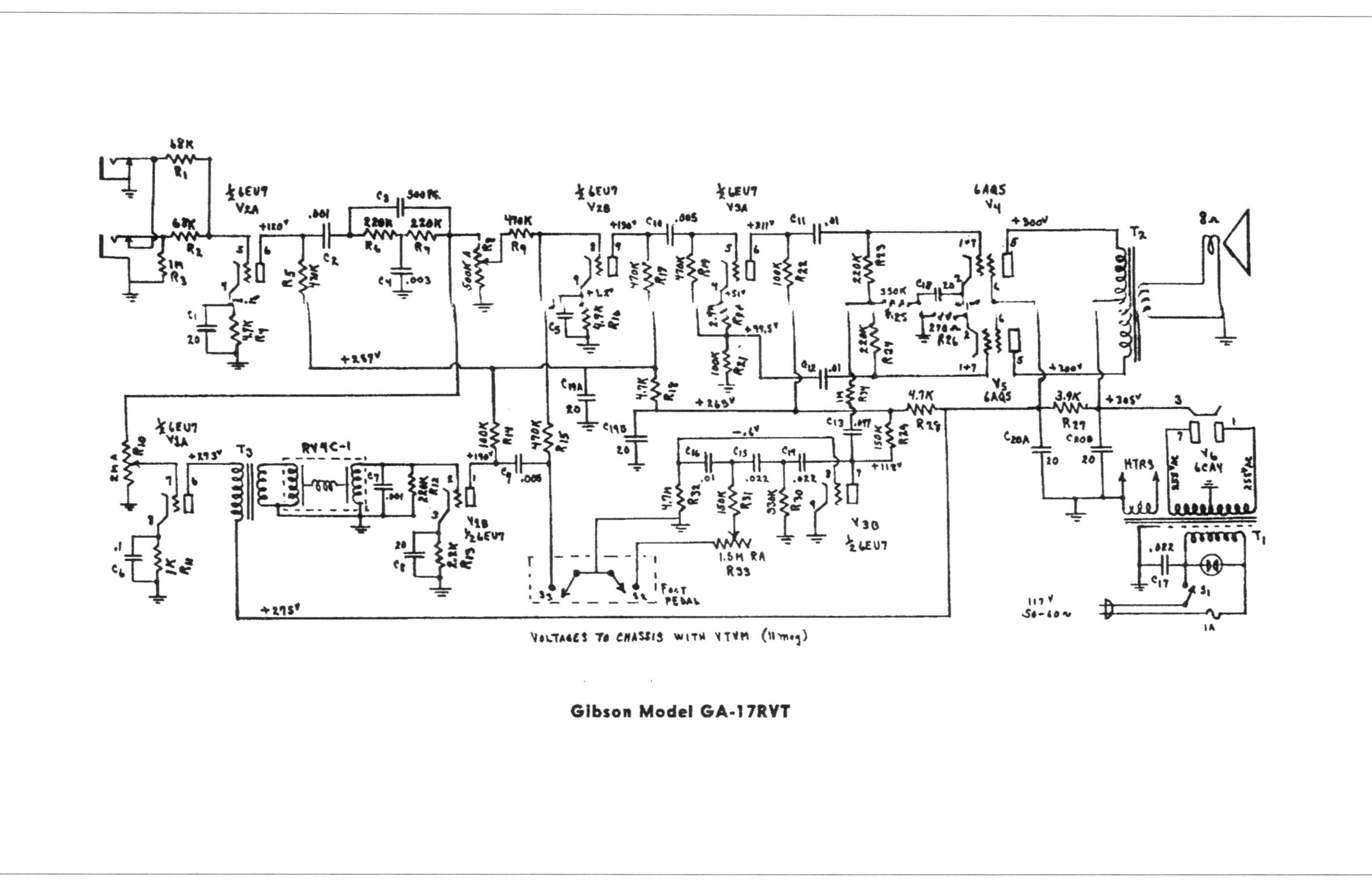

Gibson Model GA-17RVT

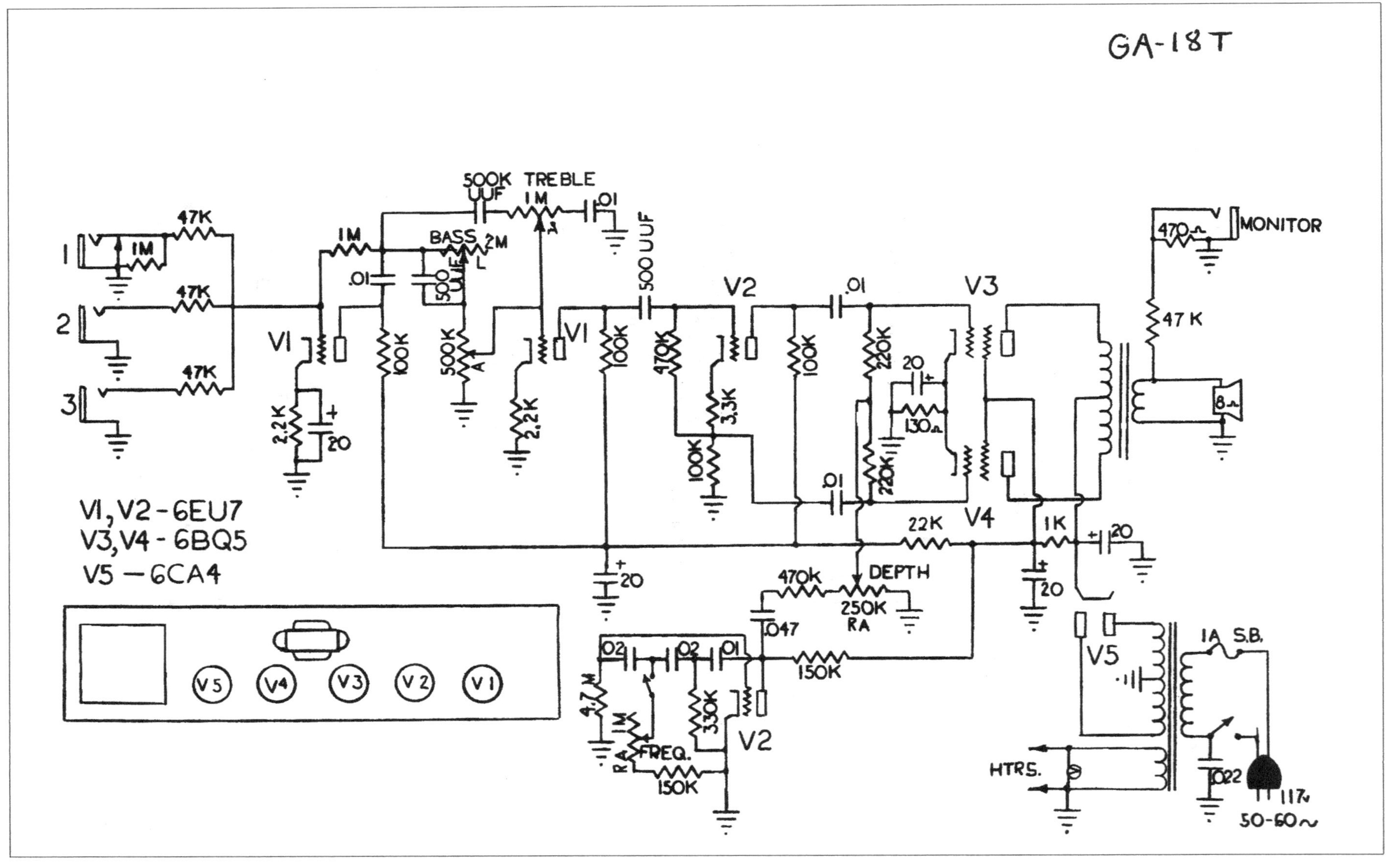
GA-18T
500K UUF
TREBLE 1M
.01
500 UUF
BASS 2M
MONITOR
470Ω
47K
8Ω
1M
.01
500 UUF
V2
V3
V1
500K
100K
470K
3.3K
220K
20
130Ω
2.2K
20
V4
22K
1K
DEPTH
250K RA
.047
150K
.02
.01
4.7M
FREQ.
330K
V5
1A S.B.
HTRS.
.022
117V 50-60~
V1, V2 - 6EU7
V3, V4 - 6BQ5
V5 — 6CA4

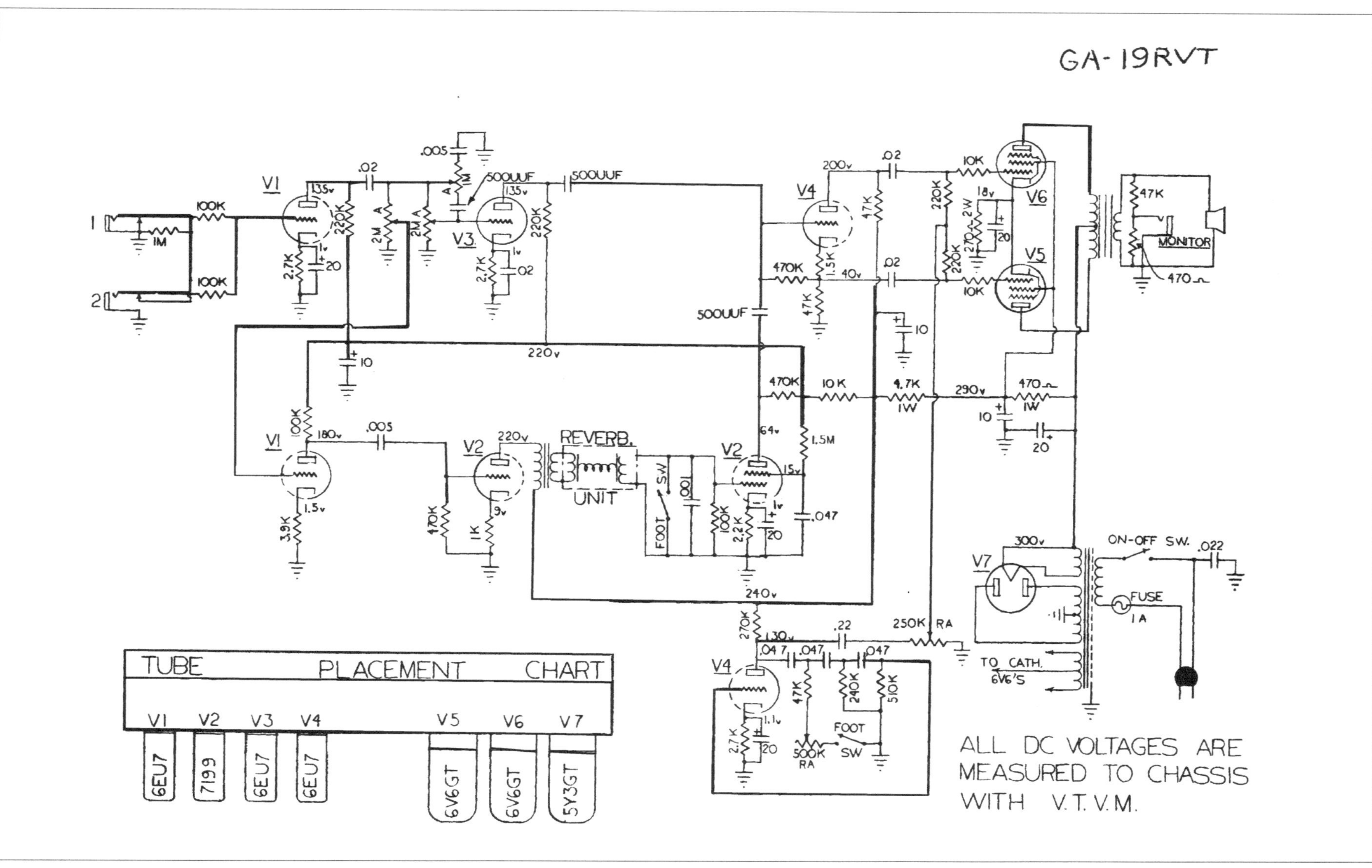
GA-19RVT
REVERB. UNIT
FOOT SW
MONITOR
ON-OFF SW.
FUSE 1A
TO CATH. 6V6'S
TUBE PLACEMENT CHART
V1 6EU7
V2 7199
V3 6EU7
V4 6EU7
V5 6V6GT
V6 6V6GT
V7 5Y3GT
ALL DC VOLTAGES ARE MEASURED TO CHASSIS WITH V.T.V.M.

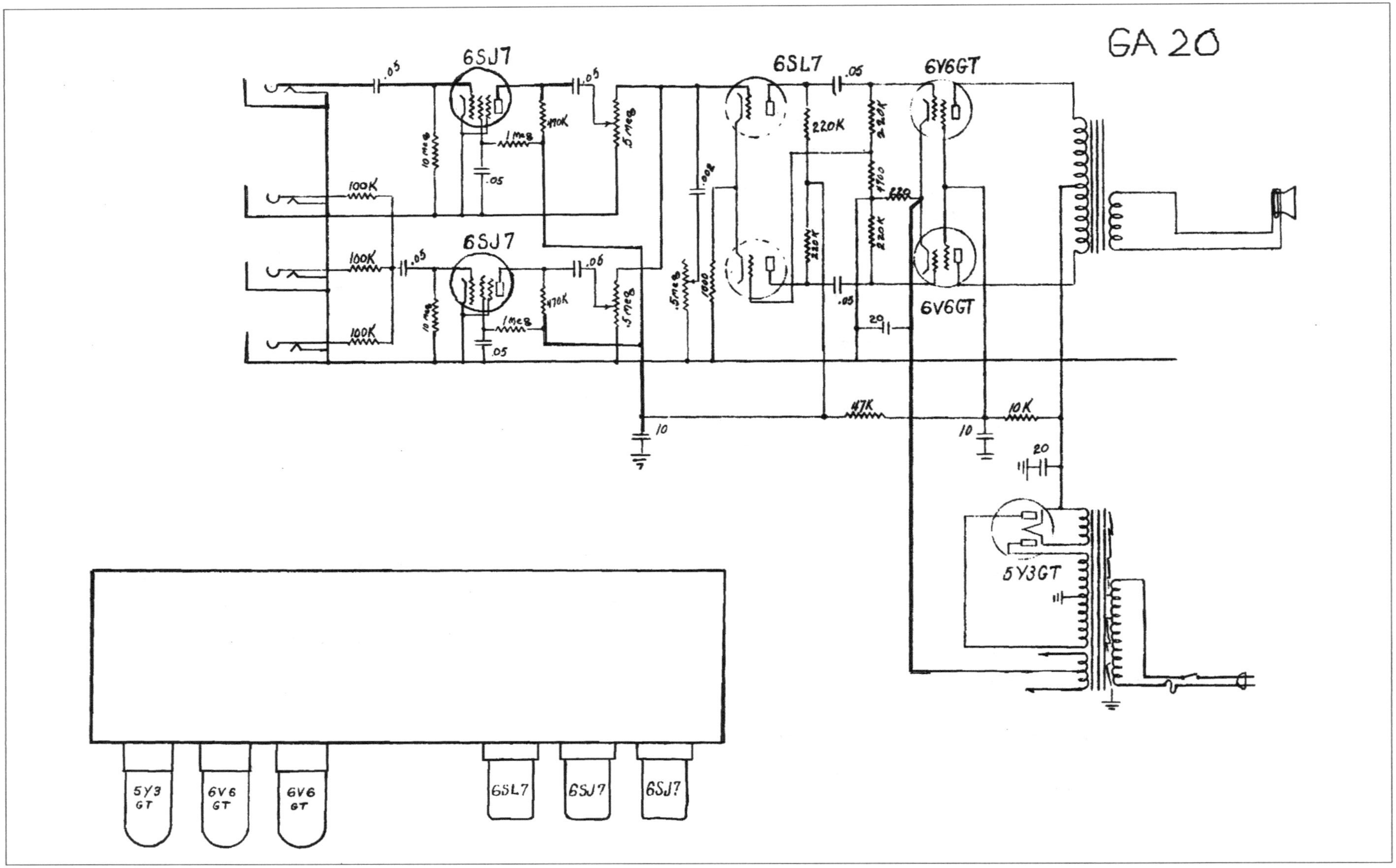

GA 20

6SJ7 6SJ7 6SL7 6V6GT 6V6GT 5Y3GT

.05 .05 .05 .05 .06 .05 .05 .05 .002 10 10 20 20

100K 100K 100K 10 meg 10 meg 1 meg 1 meg 470K 470K .5 meg .5 meg .5 meg 1000 220K 220K 220K 220K 4700 680 47K 10K

5Y3 GT 6V6 GT 6V6 GT 6SL7 6SJ7 6SJ7

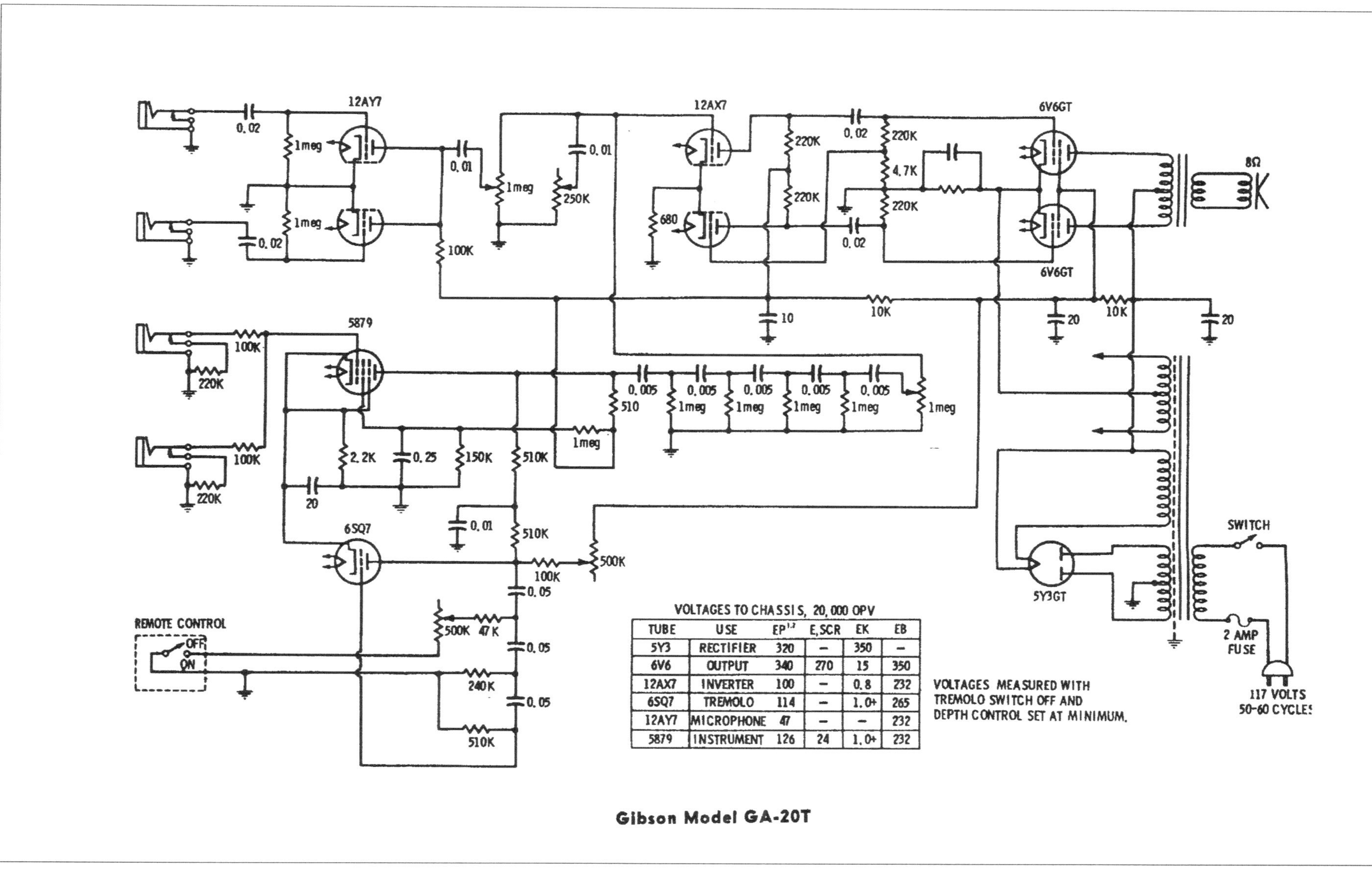

VOLTAGES TO CHASSIS, 20,000 OPV

TUBE	USE	EP[1,2]	E,SCR	EK	EB
5Y3	RECTIFIER	320	–	350	–
6V6	OUTPUT	340	270	15	350
12AX7	INVERTER	100	–	0.8	232
6SQ7	TREMOLO	114	–	1.0+	265
12AY7	MICROPHONE	47	–	–	232
5879	INSTRUMENT	126	24	1.0+	232

VOLTAGES MEASURED WITH TREMOLO SWITCH OFF AND DEPTH CONTROL SET AT MINIMUM.

Gibson Model GA-20T

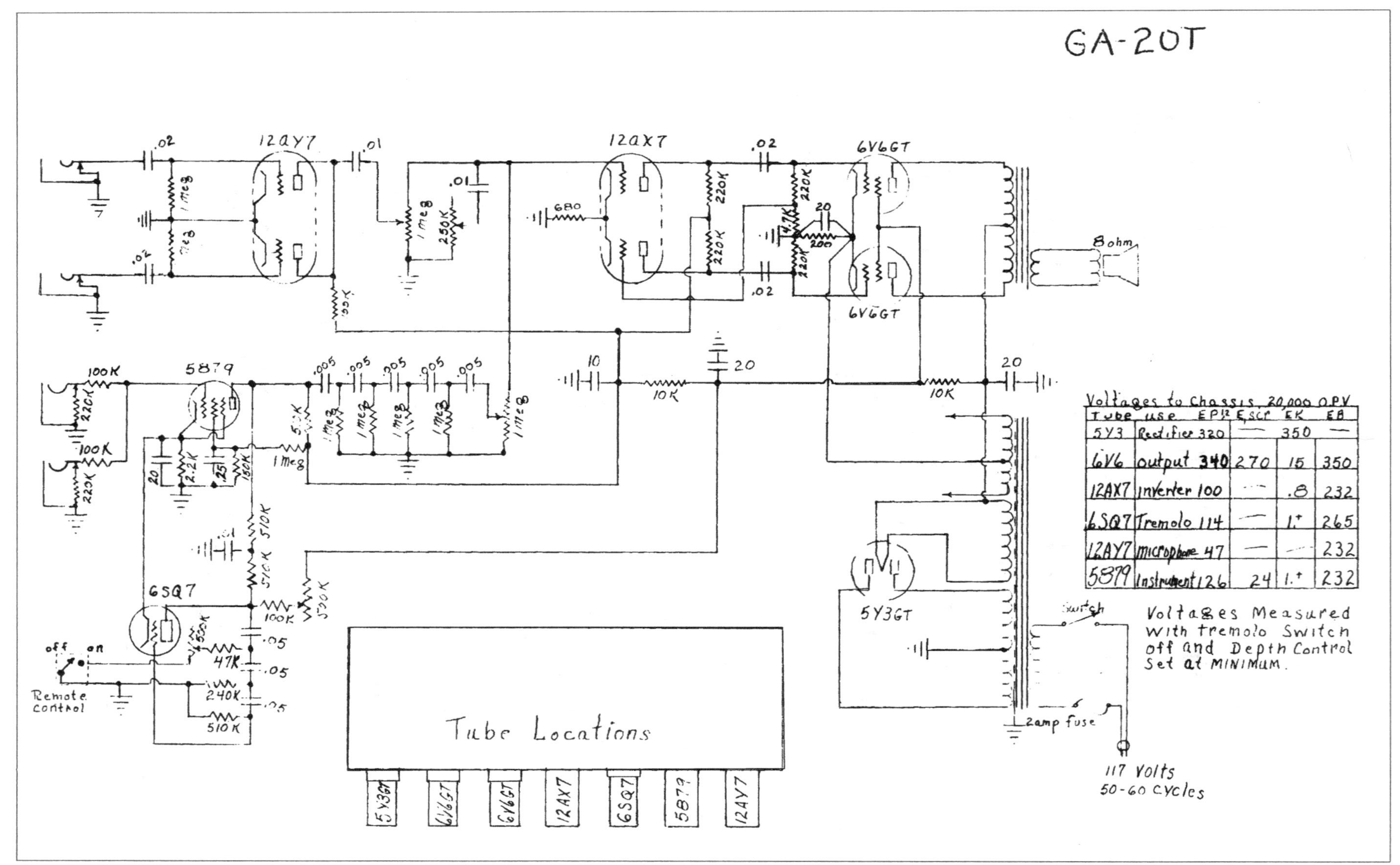

Voltages to Chassis 20,000 OPV

Tube	use	EP½	Escr	EK	EB
5Y3	Rectifier	320	—	350	—
6V6	output	340	270	15	350
12AX7	inverter	100	—	.8	232
6SQ7	Tremolo	114	—	1.+	265
12AY7	microphone	47	—	—	232
5879	Instrument	126	24	1.+	232

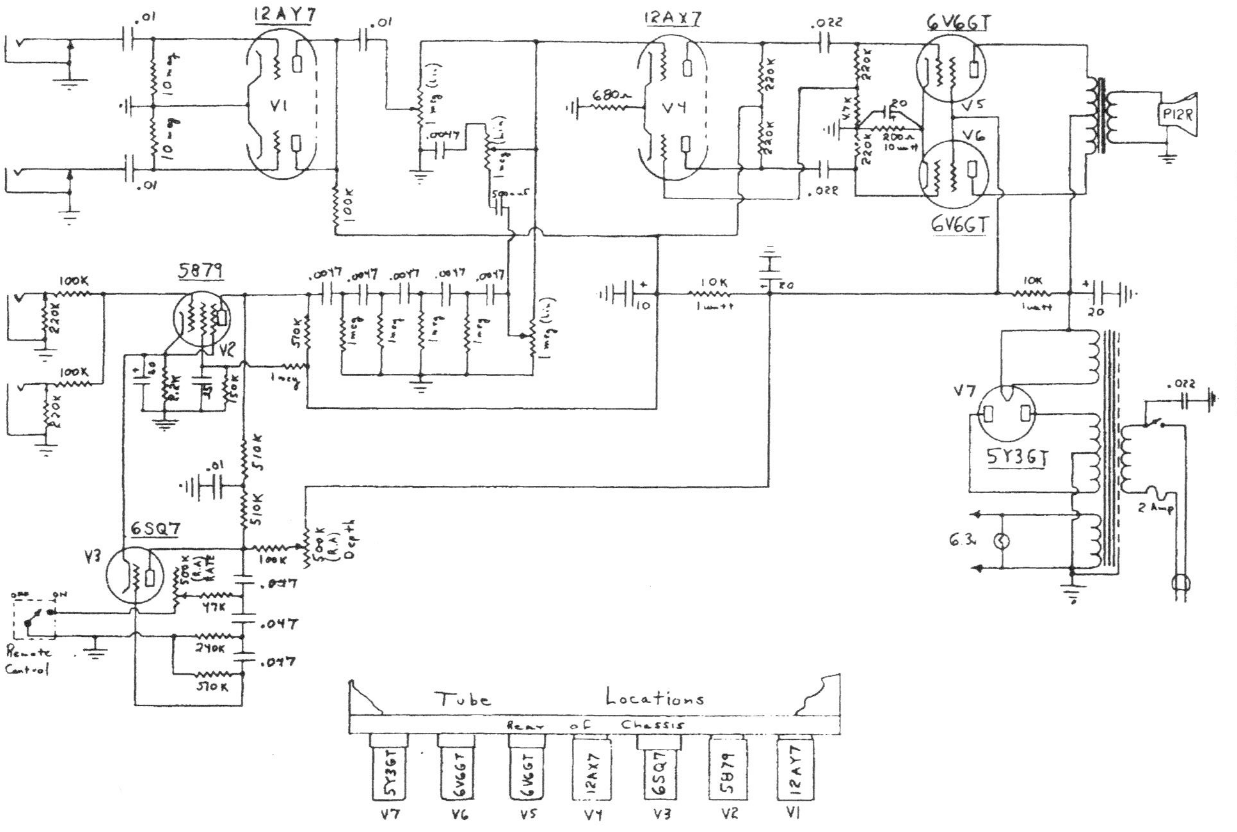

Voltages To Chassis, 20,000 Ω/V

No	Type	E_p	E_{sg}	E_g	E_k
V1	12AY7	55	—	-.8 **	0
V2	5879	115	24	0	1.2
V3	6SQ7	105	—	0	1.2
V4	12AX7	105	—	0	.85
V5-6	6V6GT	350	260	0	15
V7	5Y3GT	320 RMS	—	—	355

Voltages measured with Tremolo Switch off and Depth Control set at minimum.

** measured with VTVM

Tube Locations

Rear of Chassis

5Y3GT	6V6GT	6V6GT	12AX7	6SQ7	5879	12AY7
V7	V6	V5	V4	V3	V2	V1

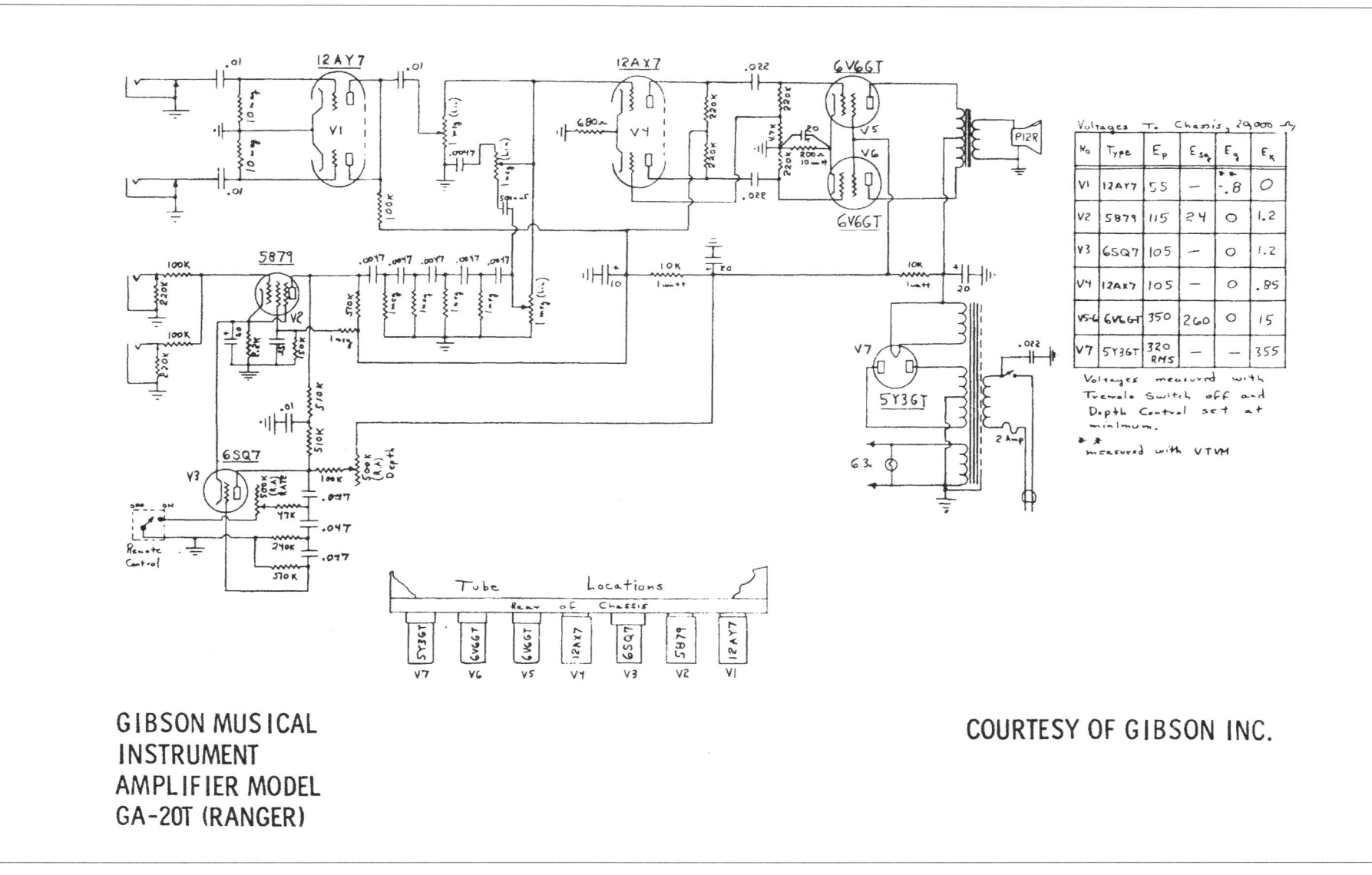

GIBSON MUSICAL
INSTRUMENT
AMPLIFIER MODEL
GA-20T (RANGER)

COURTESY OF GIBSON INC.

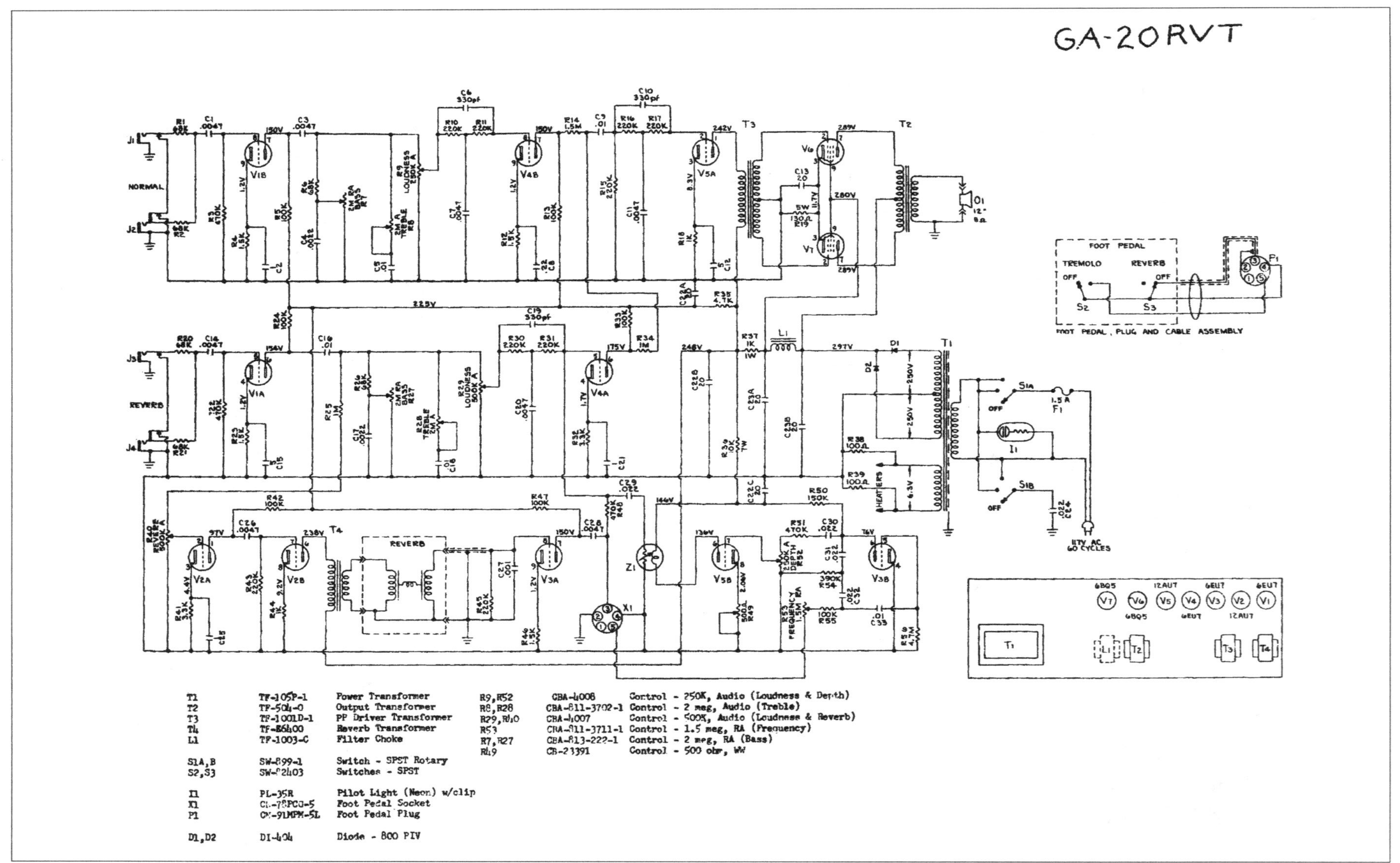
GA-20RVT
NORMAL
REVERB
FOOT PEDAL
TREMOLO
REVERB
OFF
FOOT PEDAL, PLUG AND CABLE ASSEMBLY
117V AC
60 CYCLES
HEATERS
T1 TF-105P-1 Power Transformer
T2 TF-504-0 Output Transformer
T3 TF-1001D-1 PP Driver Transformer
T4 TF-86400 Reverb Transformer
L1 TF-1003-C Filter Choke
S1A,B SW-899-1 Switch - SPST Rotary
S2,S3 SW-82403 Switches - SPST
I1 PL-35R Pilot Light (Neon) w/clip
X1 CL-78PCG-5 Foot Pedal Socket
P1 CM-91MPM-5L Foot Pedal Plug
D1,D2 DI-404 Diode - 800 PIV
R9,R52 CBA-4008 Control - 250K, Audio (Loudness & Depth)
R8,R28 CBA-811-3702-1 Control - 2 meg, Audio (Treble)
R29,R40 CBA-4007 Control - 500K, Audio (Loudness & Reverb)
R53 CBA-811-3711-1 Control - 1.5 meg, RA (Frequency)
R7,R27 CBA-813-222-1 Control - 2 meg, RA (Bass)
R49 CB-23391 Control - 500 ohm, WW

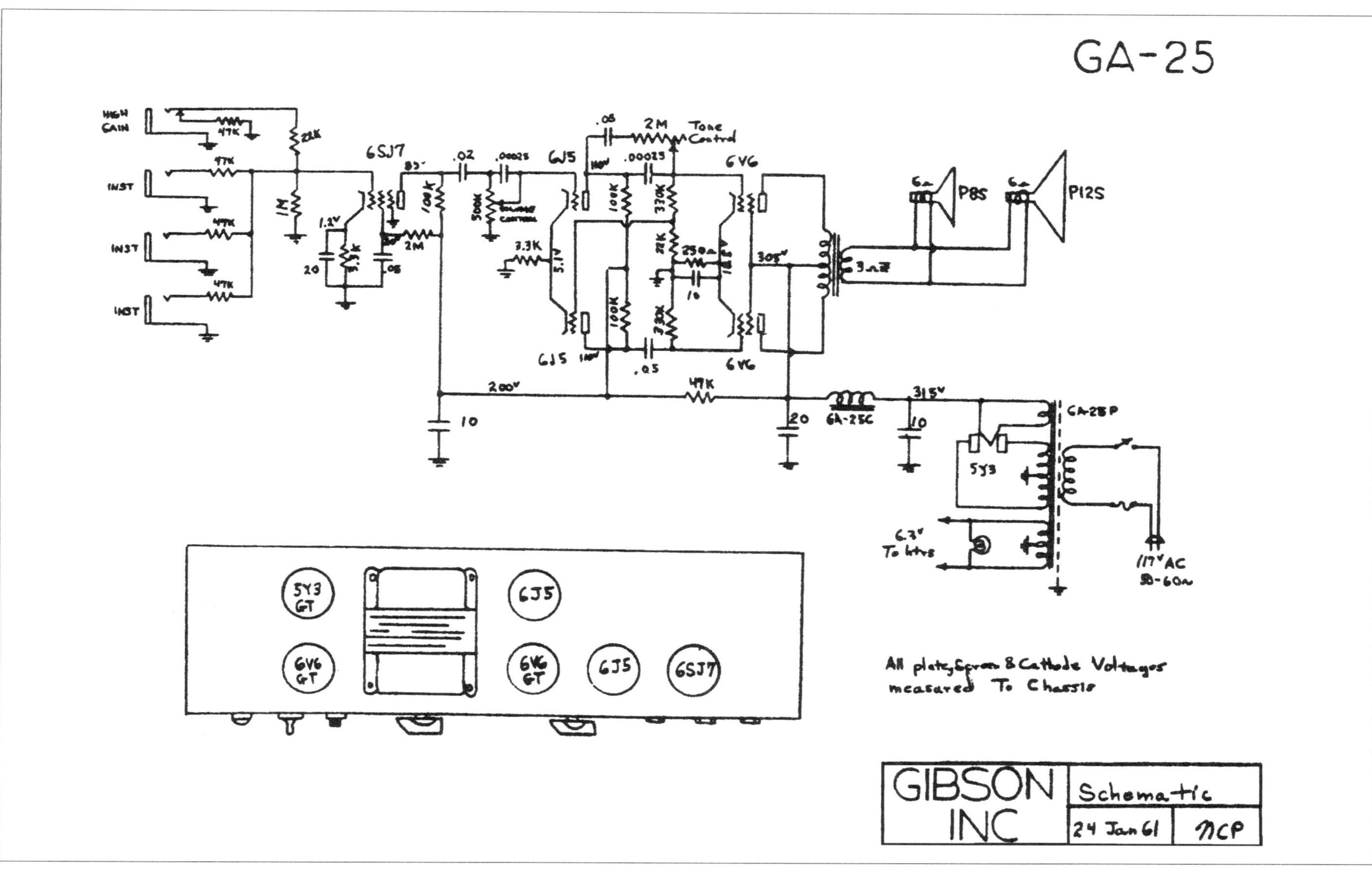
GA-25
HIGH GAIN
INST
INST
INST
47K
22K
1M
6SJ7
1.2V
3.3K
20
.05
2M
100K
85V
.02
.00025
500K
VOLUME CONTROL
3.3K
5.1V
6J5
110V
.05
2M
Tone Control
.00025
100K
330K
22K
250Ω
18.5V
10
6V6
6V6
305V
3Ω
6Ω
P8S
6Ω
P12S
6J5
110V
.05
200V
47K
10
20
GA-25C
315V
10
GA-28P
5Y3
6.3V
To htrs
117V AC
50-60~
5Y3 GT
6J5
6V6 GT
6V6 GT
6J5
6SJ7
All plate, Screen & Cathode Voltages measured To Chassis
GIBSON INC
Schematic
24 Jan 61
MCP

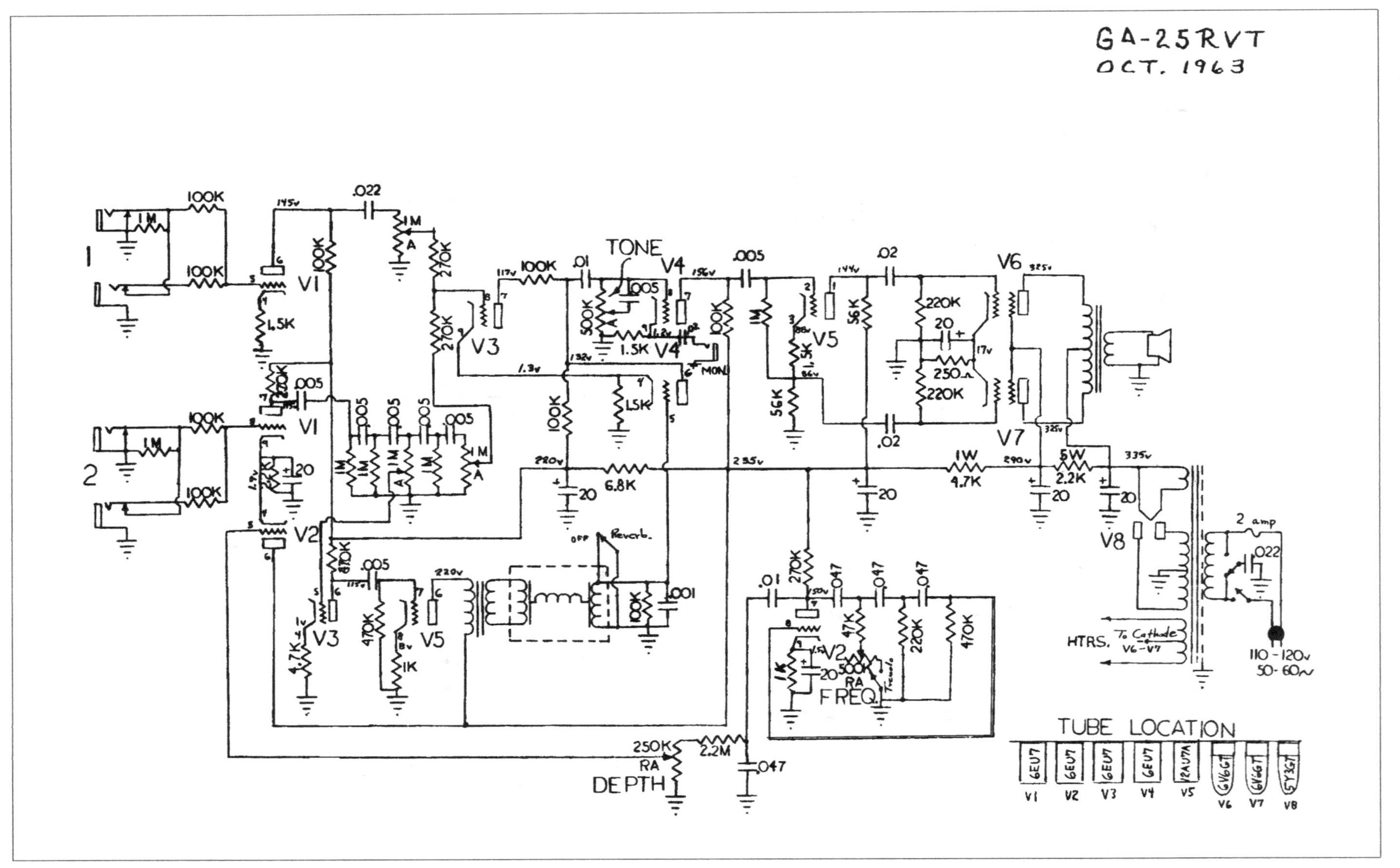
GA-25RVT
OCT. 1963
TONE
DEPTH
FREQ.
Reverb.
MON.
HTRS.
To Cathode V6-V7
110-120v
50-60~
2 amp
TUBE LOCATION
6EU7
6EU7
6EU7
6EU7
12AU7A
6V6GT
6V6GT
5Y3GT
V1
V2
V3
V4
V5
V6
V7
V8

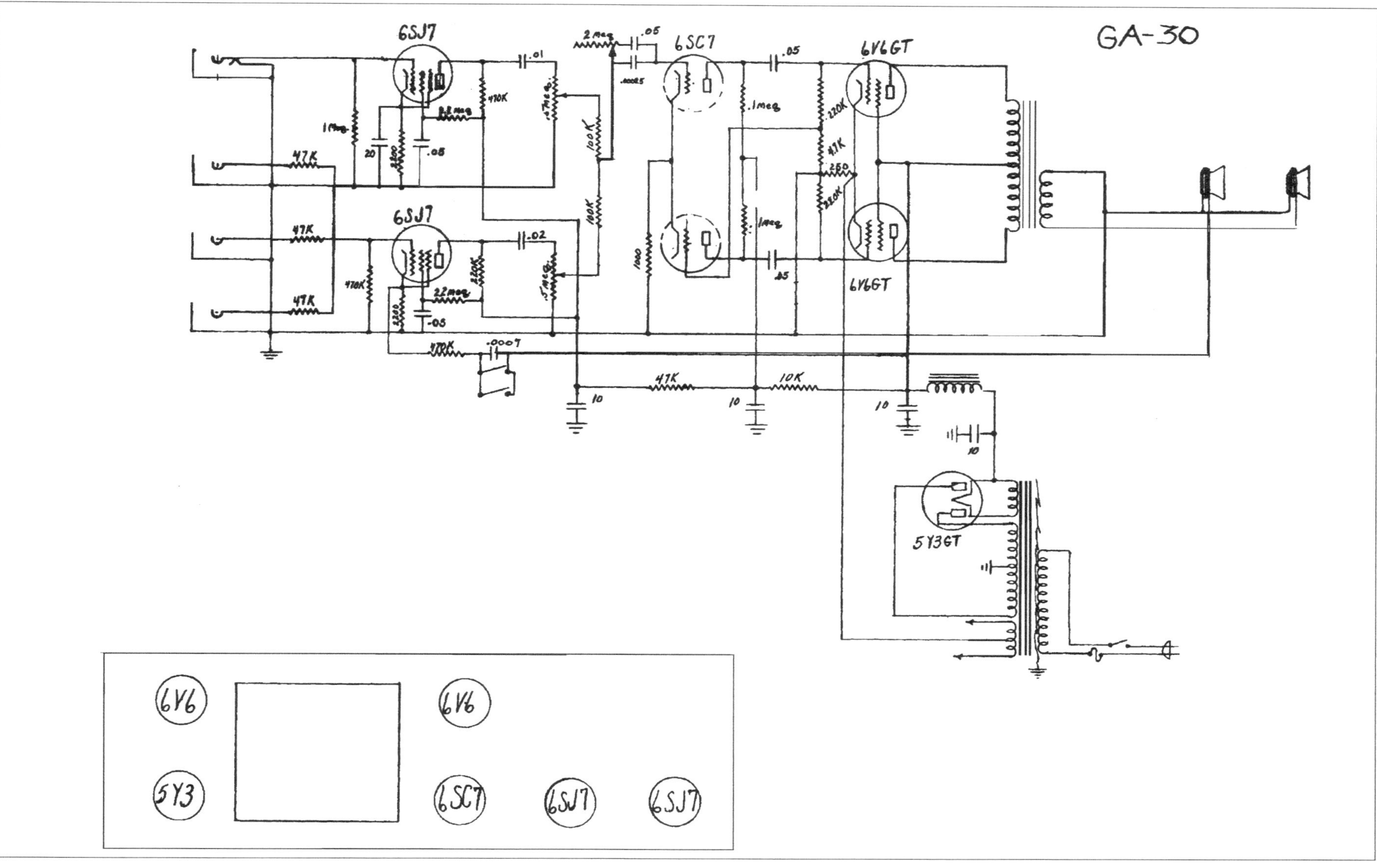
GA-30
6SJ7
6SJ7
6SC7
6V6GT
6V6GT
5Y3GT
47K
47K
47K
1Meg
470K
470K
220K
220K
100K
100K
1000
.1meg
.1meg
220K
47K
250
220K
470K
.0007
47K
10K
10
10
10
10
.01
.02
.05
.05
.05
.05
.05
.0005
2 meg
.5 meg
.5 meg
22 meg
22 meg
2200
2200
20
6V6
6V6
5Y3
6SC7
6SJ7
6SJ7

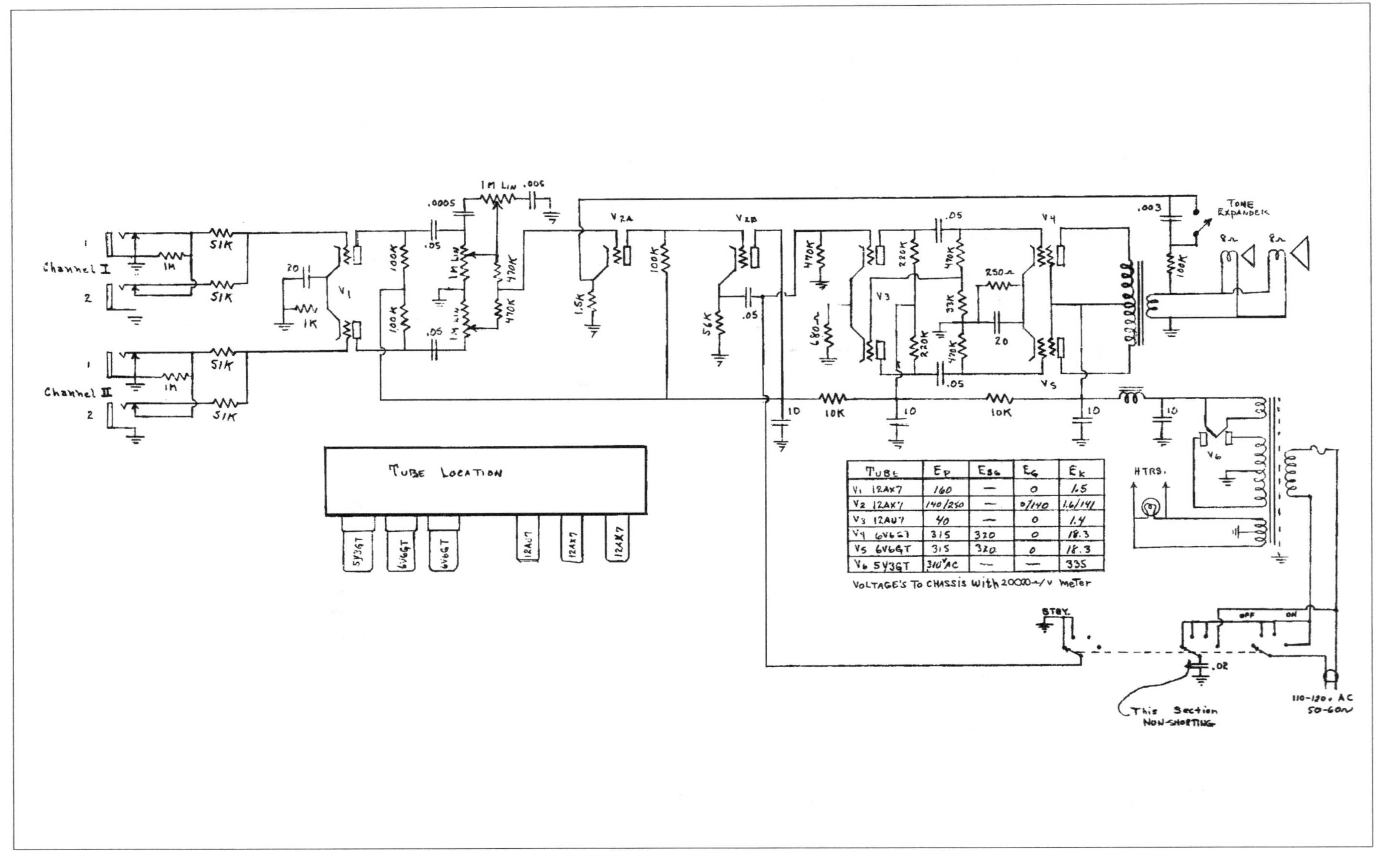

TUBE	Ep	Esg	Eg	Ek
V1 12AX7	160	—	0	1.5
V2 12AX7	140/250	—	0/140	1.6/141
V3 12AU7	40	—	0	1.4
V4 6V6GT	315	320	0	18.3
V5 6V6GT	315	320	0	18.3
V6 5Y3GT	310VAC	—	—	335

VOLTAGE'S TO CHASSIS with 20000Ω/v meter

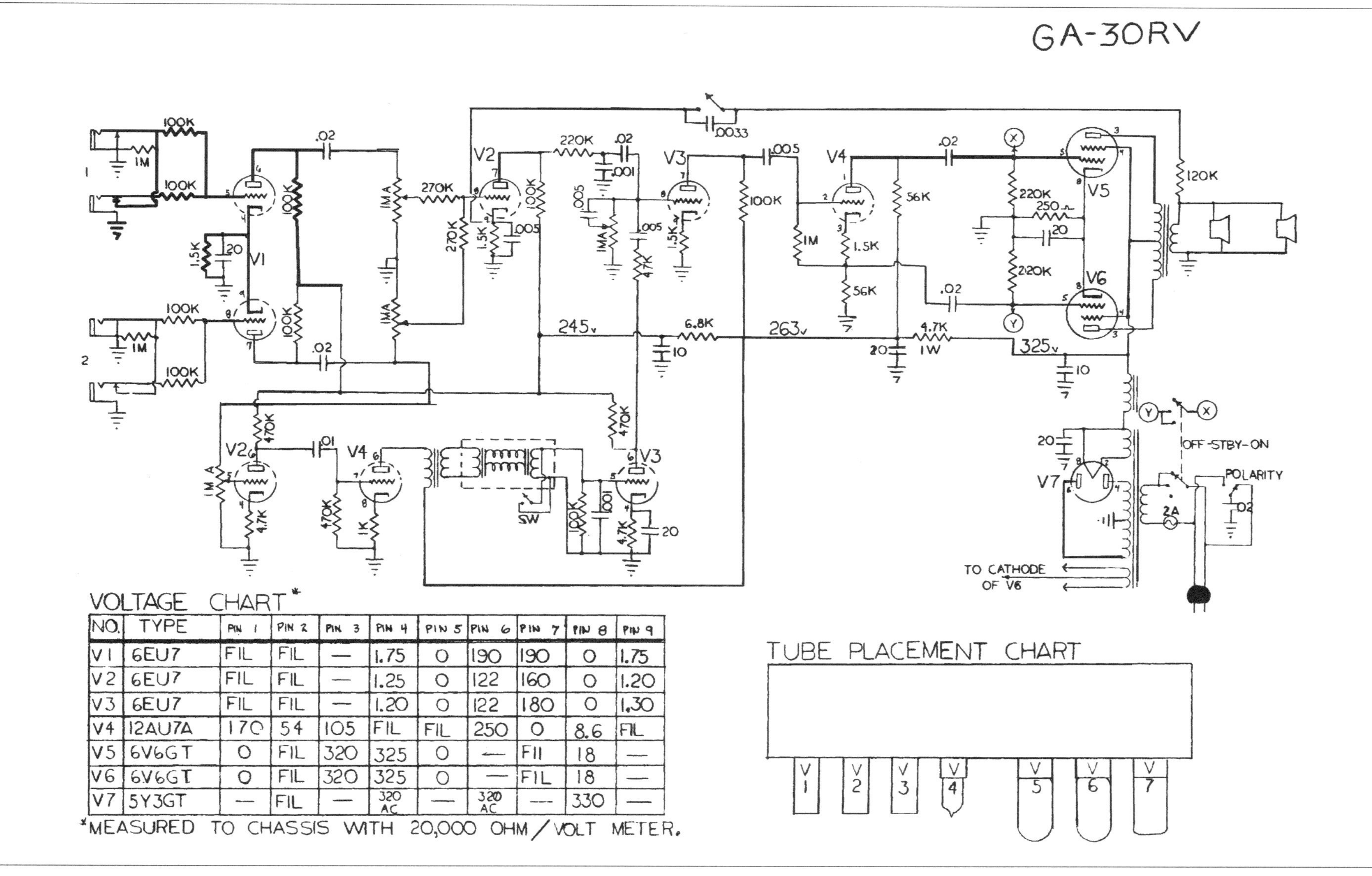

VOLTAGE CHART*

NO.	TYPE	PIN 1	PIN 2	PIN 3	PIN 4	PIN 5	PIN 6	PIN 7	PIN 8	PIN 9
V1	6EU7	FIL	FIL	—	1.75	0	190	190	0	1.75
V2	6EU7	FIL	FIL	—	1.25	0	122	160	0	1.20
V3	6EU7	FIL	FIL	—	1.20	0	122	180	0	1.30
V4	12AU7A	170	54	105	FIL	FIL	250	0	8.6	FIL
V5	6V6GT	0	FIL	320	325	0	—	FIl	18	—
V6	6V6GT	0	FIL	320	325	0	—	FIL	18	—
V7	5Y3GT	—	FIL	—	320 AC	—	320 AC	—	330	—

*MEASURED TO CHASSIS WITH 20,000 OHM/VOLT METER.

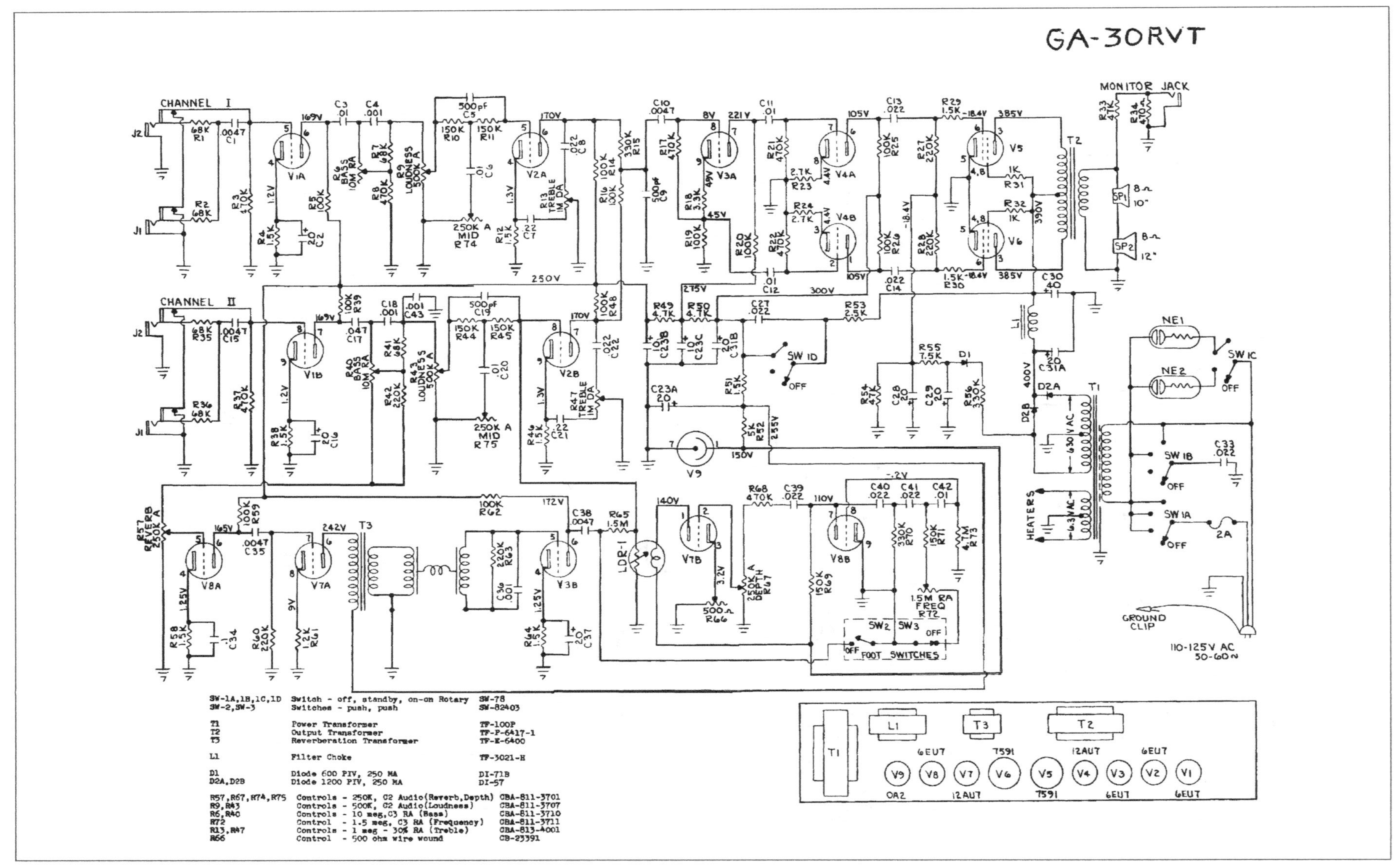
GA-30RVT
CHANNEL I
CHANNEL II
MONITOR JACK
HEATERS
GROUND CLIP
110-125V AC 50-60~
FOOT SWITCHES
SW-1A,1B,1C,1D Switch - off, standby, on-on Rotary SW-78
SW-2,SW-3 Switches - push, push SW-82403
T1 Power Transformer TF-100P
T2 Output Transformer TF-P-6417-1
T3 Reverberation Transformer TF-K-6400
L1 Filter Choke TF-3021-H
D1 Diode 600 PIV, 250 MA DI-71B
D2A,D2B Diode 1200 PIV, 250 MA DI-57
R57,R67,R74,R75 Controls - 250K, C2 Audio(Reverb,Depth) CBA-811-3701
R9,R43 Controls - 500K, C2 Audio(Loudness) CBA-811-3707
R6,R40 Controls - 10 meg,C3 RA (Bass) CBA-811-3710
R72 Control - 1.5 meg, C3 RA (Frequency) CBA-811-3711
R13,R47 Controls - 1 meg - 30% RA (Treble) CBA-813-4001
R66 Control - 500 ohm wire wound CB-23391

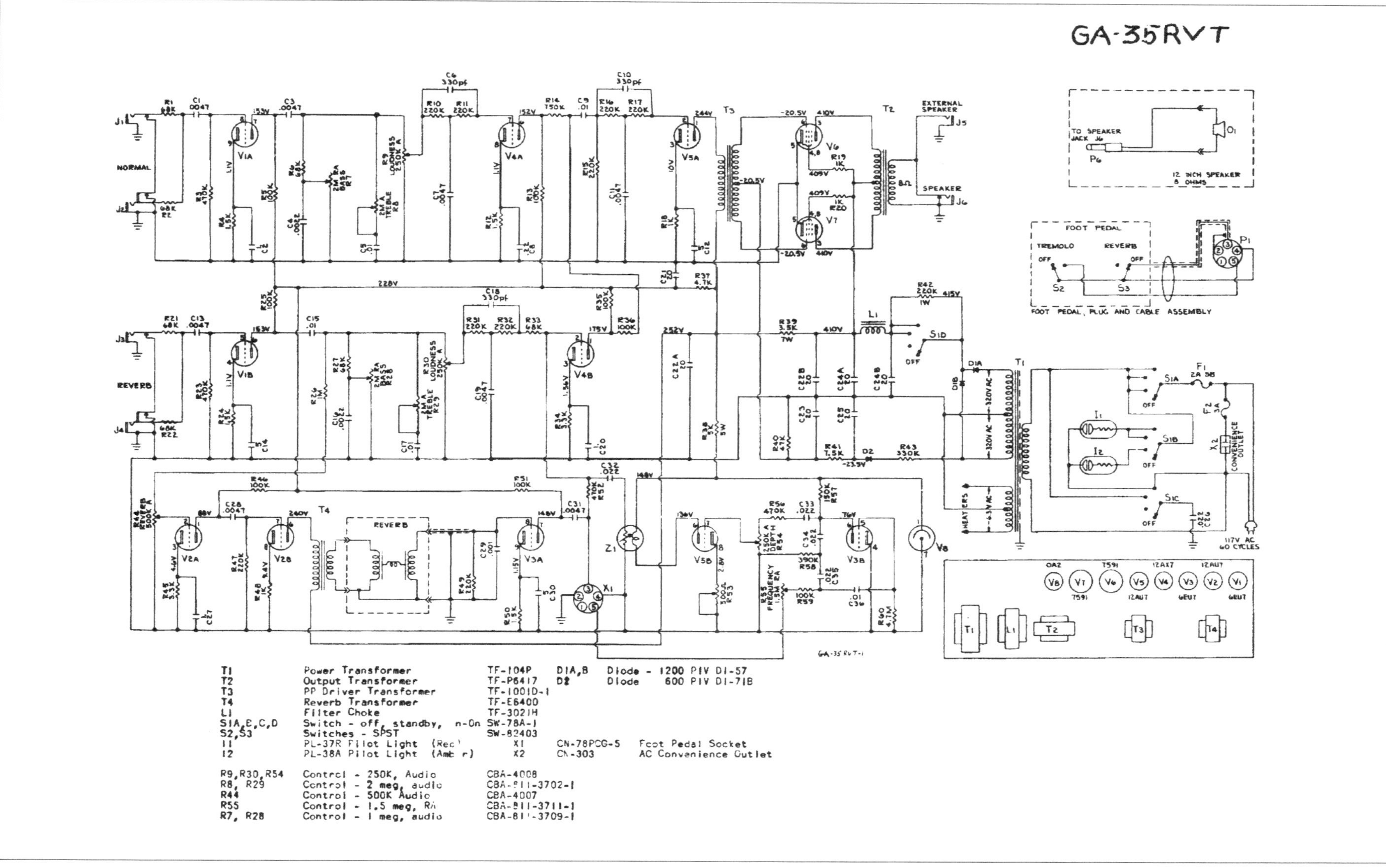
GA-35RVT
NORMAL
REVERB
EXTERNAL SPEAKER
SPEAKER
TO SPEAKER JACK J6
12 INCH SPEAKER 8 OHMS
FOOT PEDAL
TREMOLO
REVERB
FOOT PEDAL, PLUG AND CABLE ASSEMBLY
CONVENIENCE OUTLET
117V AC 60 CYCLES
HEATERS
T1 Power Transformer TF-104P
T2 Output Transformer TF-P6417
T3 PP Driver Transformer TF-1001D-1
T4 Reverb Transformer TF-E6400
L1 Filter Choke TF-3021H
S1A,B,C,D Switch - off, standby, n-On SW-78A-1
S2,S3 Switches - SPST SW-82403
I1 PL-37R Pilot Light (Red) X1 CN-78PCG-5 Foot Pedal Socket
I2 PL-38A Pilot Light (Amber) X2 CN-303 AC Convenience Outlet
D1A,B Diode - 1200 PIV DI-57
D2 Diode 600 PIV DI-71B
R9,R30,R54 Control - 250K, Audio CBA-4008
R8, R29 Control - 2 meg, audio CBA-811-3702-1
R44 Control - 500K Audio CBA-4007
R55 Control - 1.5 meg, RA CBA-811-3711-1
R7, R28 Control - 1 meg, audio CBA-811-3709-1

LES PAUL GA-40

Gibson

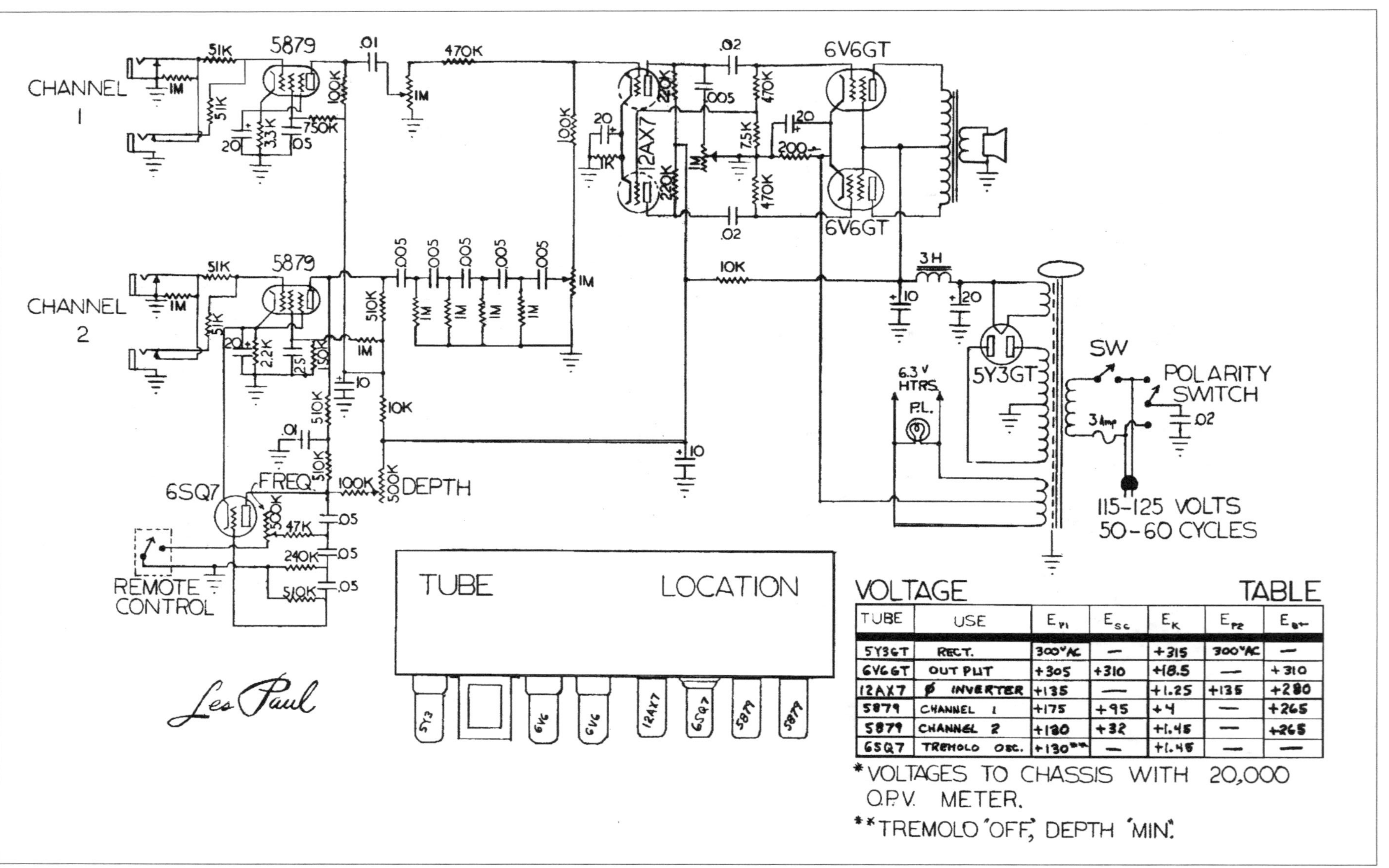

VOLTAGE TABLE

TUBE	USE	E_{P1}	E_{SC}	E_K	E_{P2}	E_{B+}
5Y3GT	RECT.	300V AC	—	+315	300V AC	—
6V6GT	OUT PUT	+305	+310	+18.5	—	+310
12AX7	Ø INVERTER	+135	—	+1.25	+135	+280
5879	CHANNEL 1	+175	+95	+4	—	+265
5879	CHANNEL 2	+180	+32	+1.45	—	+265
6SQ7	TREMOLO OSC.	+130**	—	+1.45	—	—

*VOLTAGES TO CHASSIS WITH 20,000 O.P.V. METER.

**TREMOLO "OFF", DEPTH "MIN."

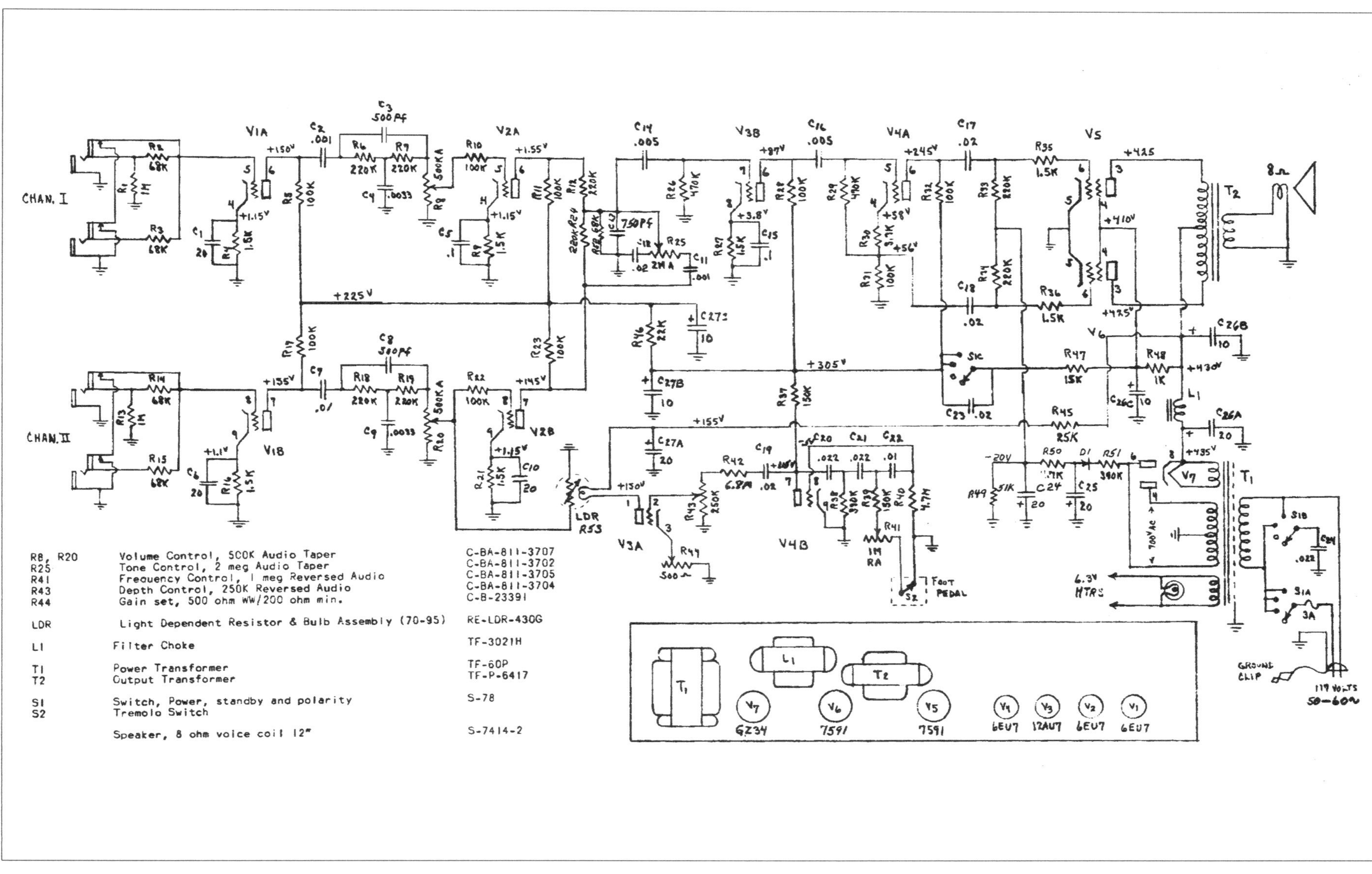
CHAN. I
CHAN. II
R8, R20 Volume Control, 500K Audio Taper C-BA-811-3707
R25 Tone Control, 2 meg Audio Taper C-BA-811-3702
R41 Frequency Control, 1 meg Reversed Audio C-BA-811-3705
R43 Depth Control, 250K Reversed Audio C-BA-811-3704
R44 Gain set, 500 ohm WW/200 ohm min. C-B-23391
LDR Light Dependent Resistor & Bulb Assembly (70-95) RE-LDR-430G
L1 Filter Choke TF-3021H
T1 Power Transformer TF-60P
T2 Output Transformer TF-P-6417
S1 Switch, Power, standby and polarity S-78
S2 Tremolo Switch
Speaker, 8 ohm voice coil 12" S-7414-2
V7 GZ34
V6 7591
V5 7591
V4 6EU7
V3 12AU7
V2 6EU7
V1 6EU7
FOOT PEDAL
6.3V HTRS
GROUND CLIP
117 VOLTS 50-60~

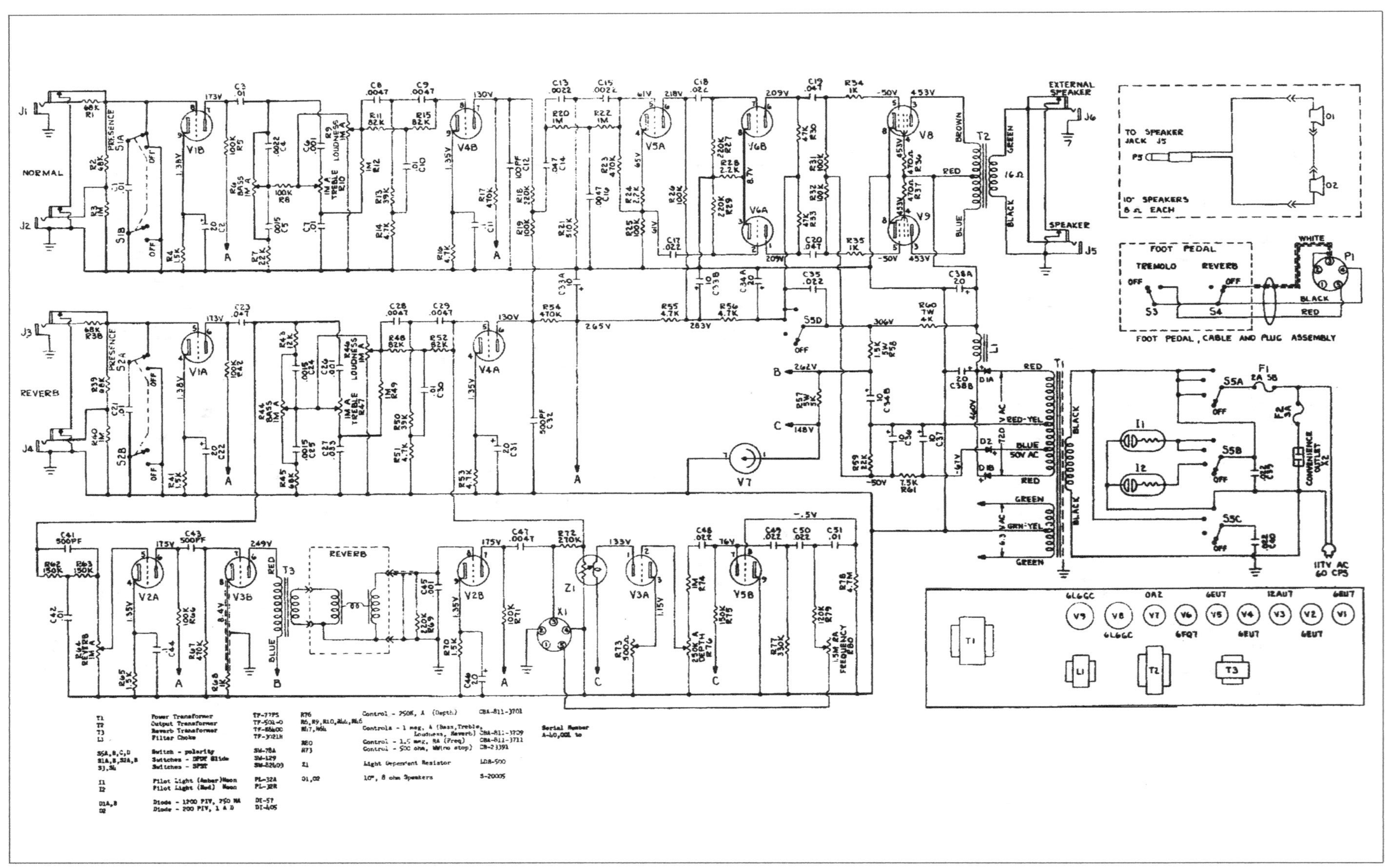
NORMAL
REVERB
PRESENCE
BASS
TREBLE
LOUDNESS
REVERB
DEPTH
FREQUENCY
EXTERNAL SPEAKER
SPEAKER
TO SPEAKER JACK J5
10" SPEAKERS 8 Ω EACH
FOOT PEDAL
TREMOLO
REVERB
FOOT PEDAL, CABLE AND PLUG ASSEMBLY
CONVENIENCE OUTLET X2
117V AC 60 CPS
6L6GC
0A2
6EU7
12AU7
6FQ7
T1 Power Transformer TF-77PS
T2 Output Transformer TF-501-0
T3 Reverb Transformer TF-86400
L1 Filter Choke TF-3021H
S5A,B,C,D Switch - polarity SW-78A
S1A,B,S2A,B Switches - DPDT Slide SW-129
S3,S4 Switches - SPST SW-82409
I1 Pilot Light (Amber) Neon PL-32A
I2 Pilot Light (Red) Neon PL-32R
D1A,B Diode - 1200 PIV, 250 MA DI-57
D2 Diode - 200 PIV, 1 A D DI-405
R76 Control - 250K, A (Depth) CBA-811-3701
R6,R9,R10,R44,R46 R47,R64 Controls - 1 meg, A (Bass,Treble, Loudness, Reverb) CBA-811-3709
R80 Control - 1.5 meg, RA (Freq) CBA-811-3711
R73 Control - 500 ohm, WW(no stop) CB-23391
Z1 Light Dependent Resistor LDR-500
O1,O2 10", 8 ohm Speakers S-20005
Serial Number A-40,001 to

GA-45RVT **SATURN**

Gibson

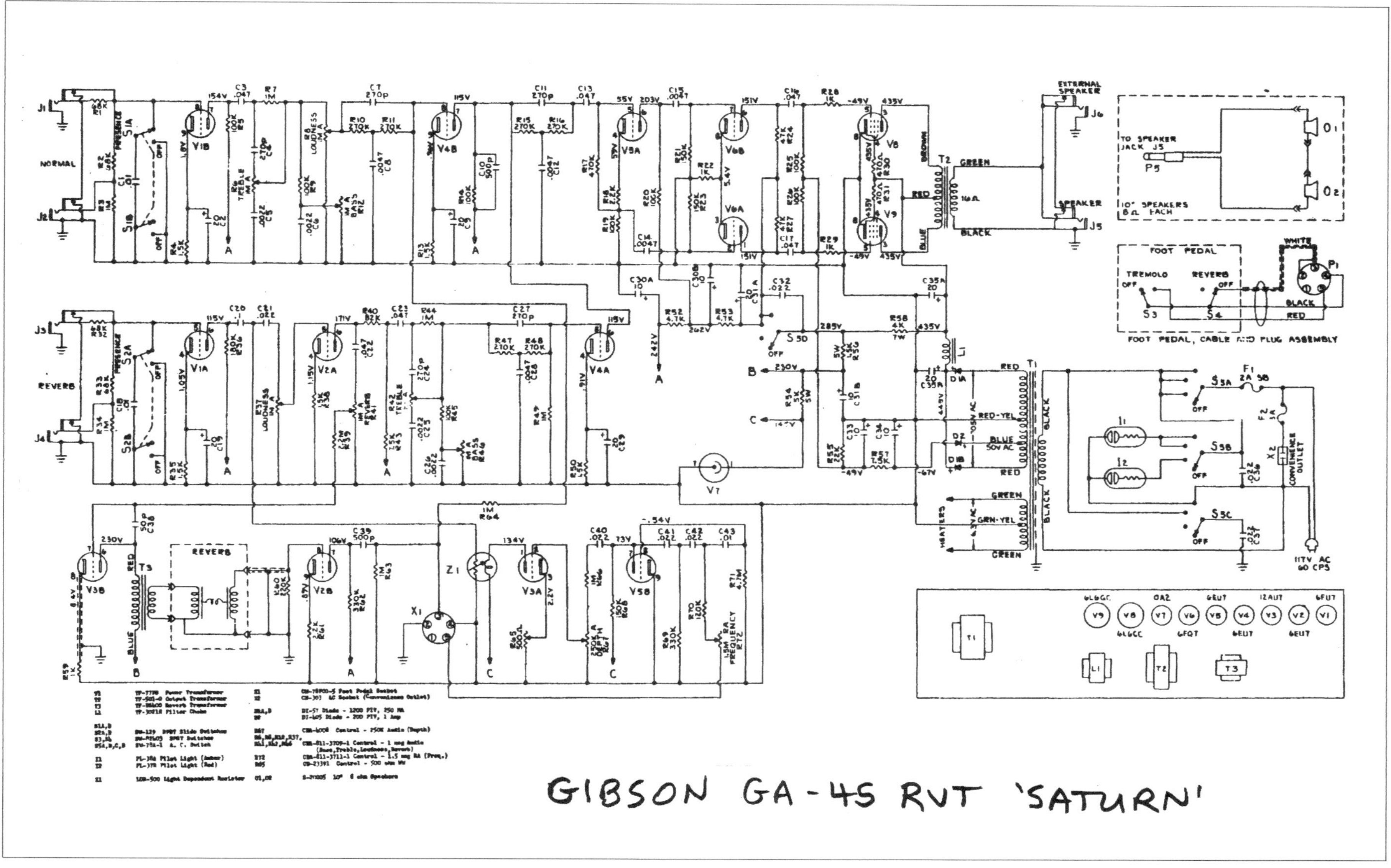

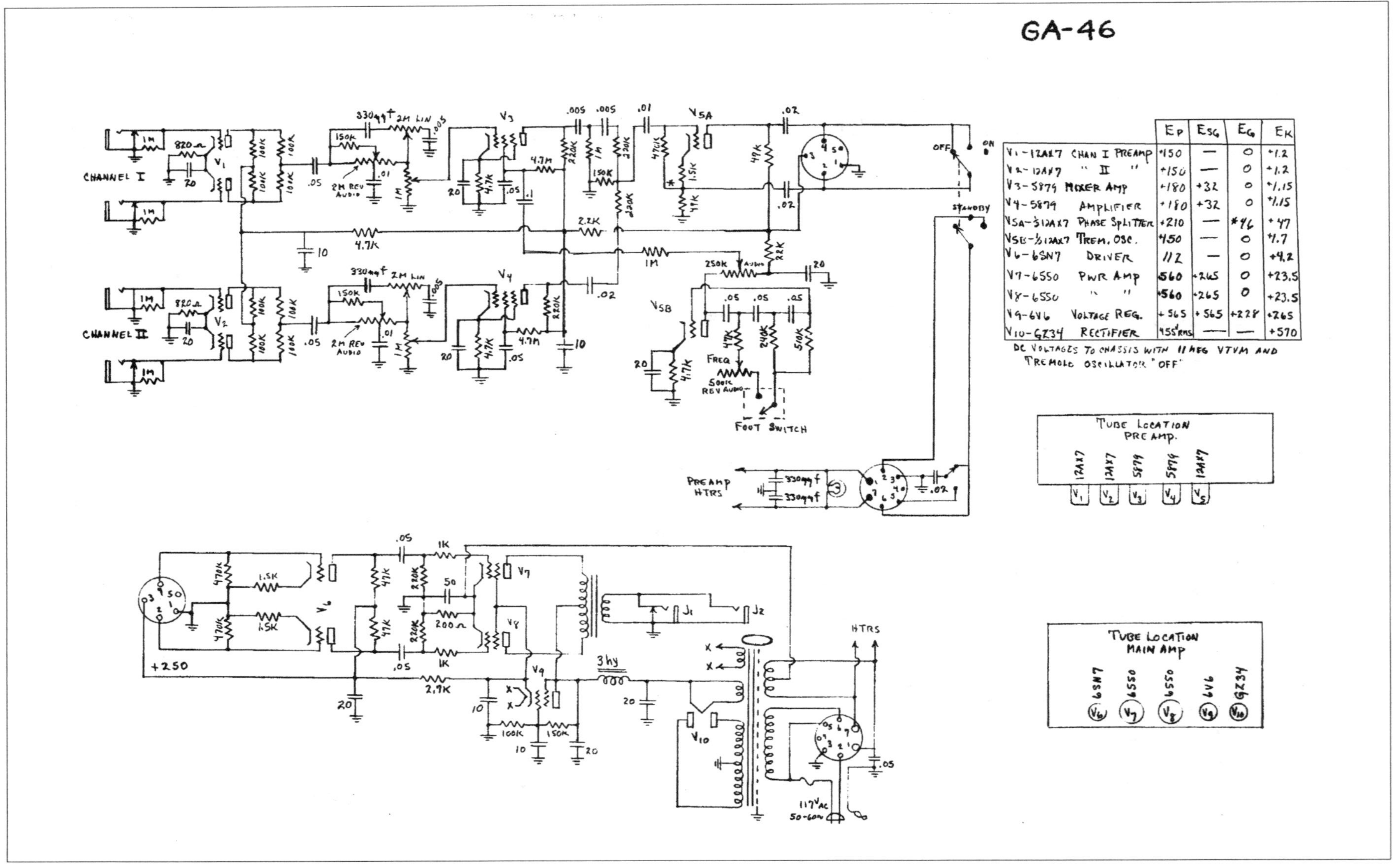

	EP	ESG	EG	EK
V1-12AX7 CHAN I PREAMP	+150	—	0	+1.2
V2-12AX7 " II "	+150	—	0	+1.2
V3-5879 MIXER AMP	+180	+32	0	+1.15
V4-5879 AMPLIFIER	+180	+32	0	+1.15
V5A-½12AX7 PHASE SPLITTER	+210	—	*46	+47
V5B-½12AX7 TREM. OSC.	+150	—	0	+1.7
V6-6SN7 DRIVER	112	—	0	+4.2
V7-6550 PWR AMP	560	+265	0	+23.5
V8-6550 " "	+560	+265	0	+23.5
V9-6V6 VOLTAGE REG.	+565	+565	+228	+265
V10-GZ34 RECTIFIER	455 RMS	—	—	+570

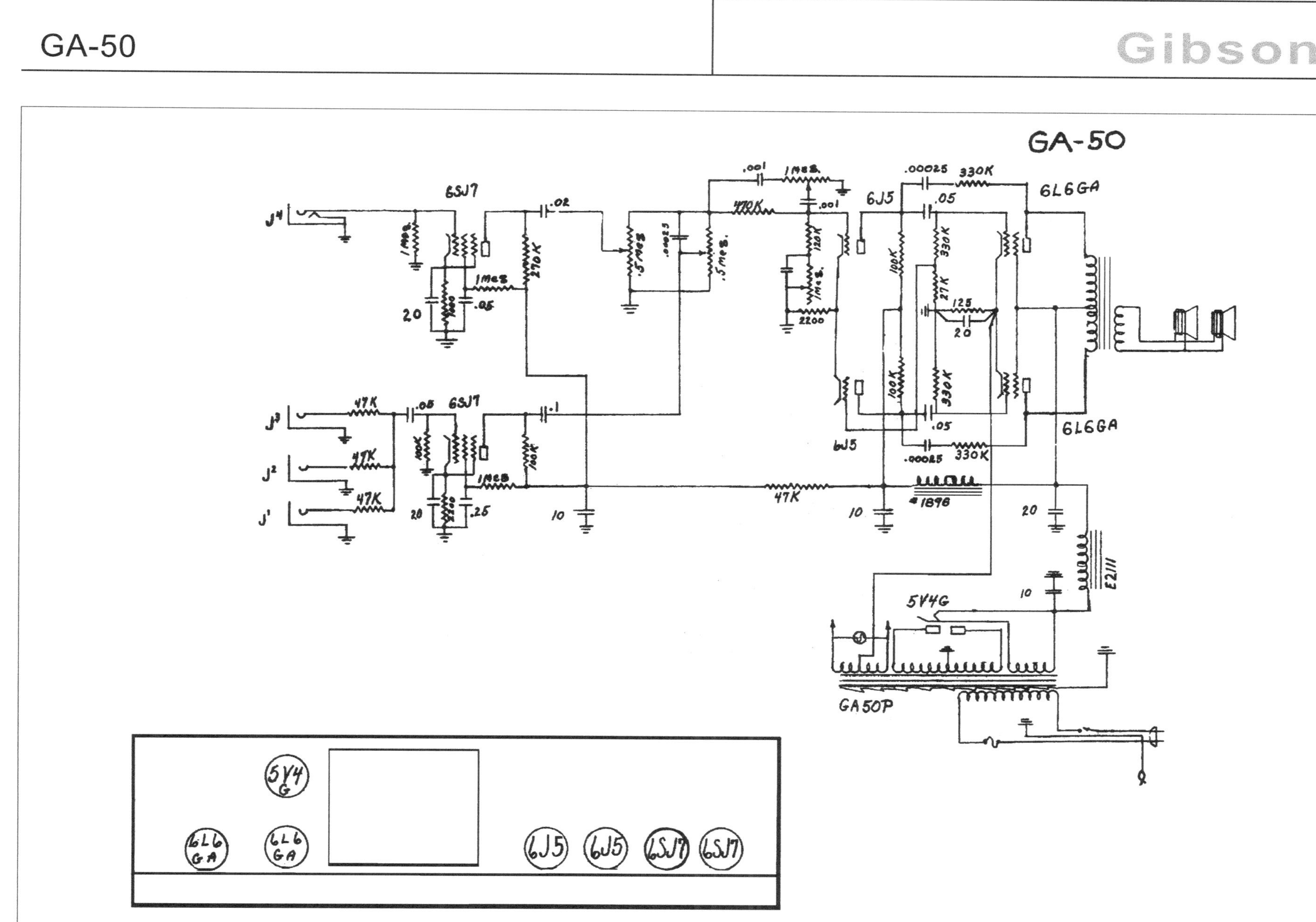
GA-50
6SJ7
6SJ7
6J5
6J5
6L6GA
6L6GA
5Y4G
GA50P
E2111
1898
J1
J2
J3
J4
47K
100K
270K
470K
330K
27K
120K
1 Meg.
.5 meg.
.00025
2200
20
10

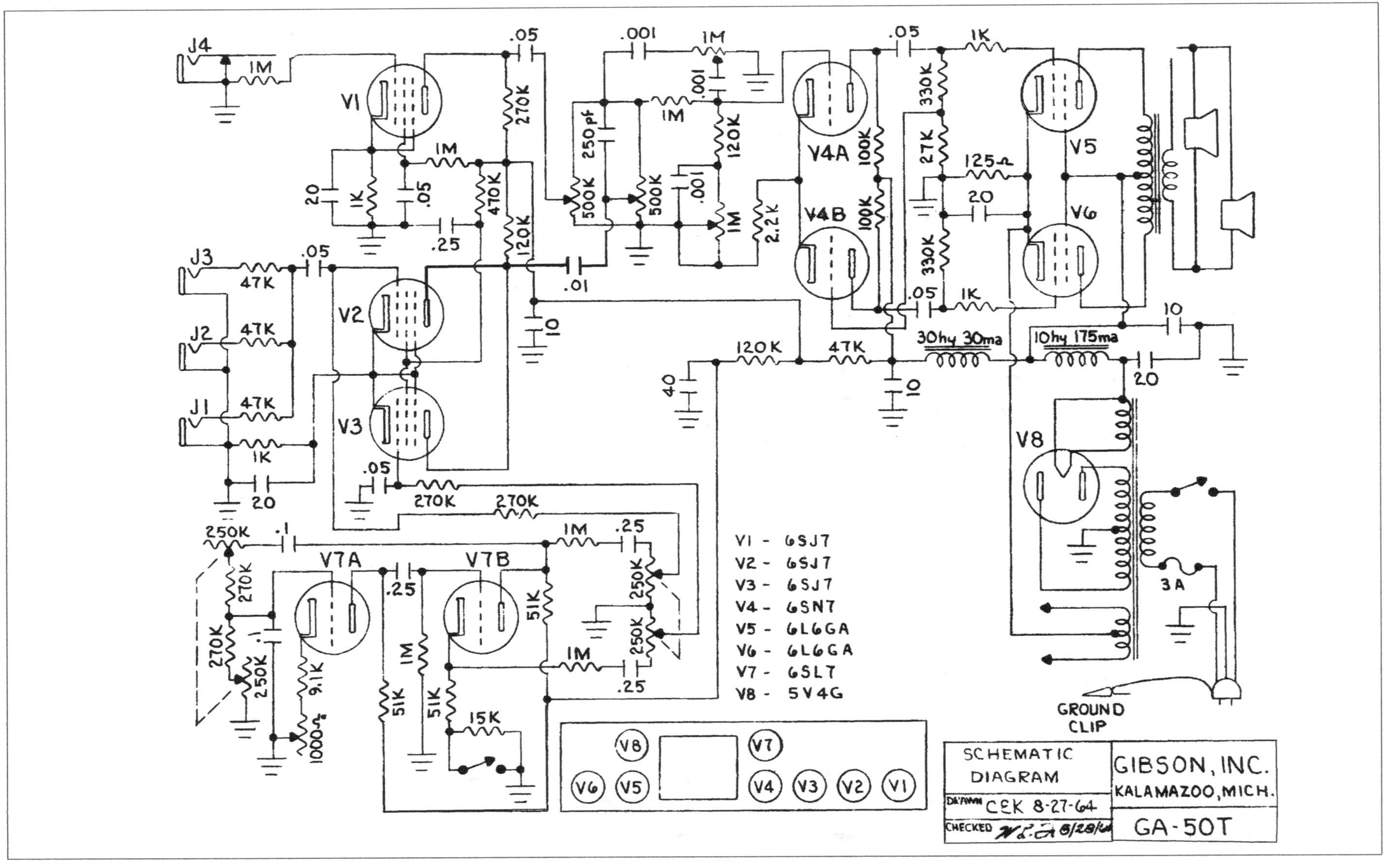
V1 - 6SJ7
V2 - 6SJ7
V3 - 6SJ7
V4 - 6SN7
V5 - 6L6GA
V6 - 6L6GA
V7 - 6SL7
V8 - 5V4G
GROUND CLIP
SCHEMATIC DIAGRAM
DRAWN CEK 8-27-64
CHECKED 8/28/64
GIBSON, INC.
KALAMAZOO, MICH.
GA-50T

GSS-50

Gibson

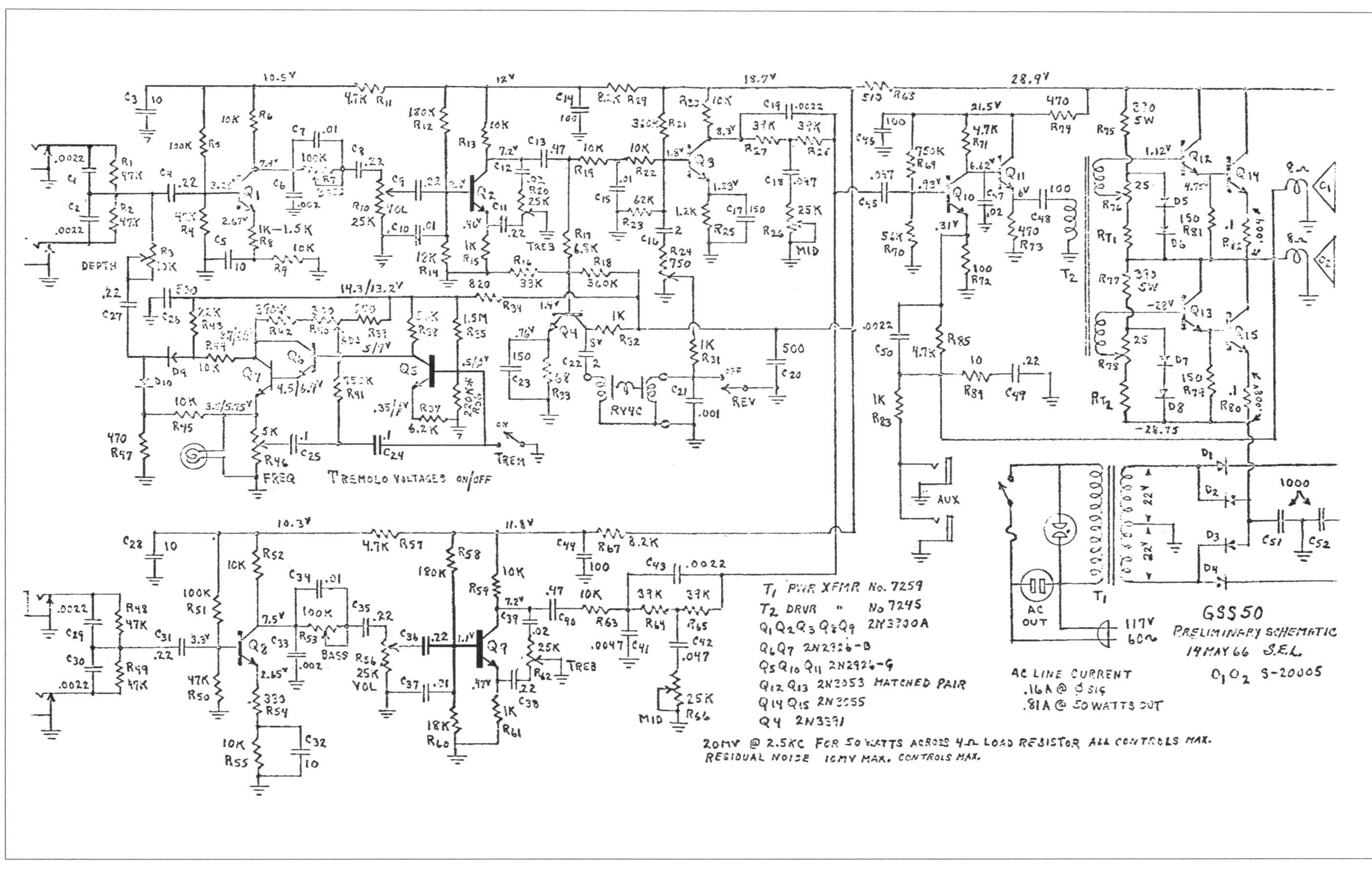

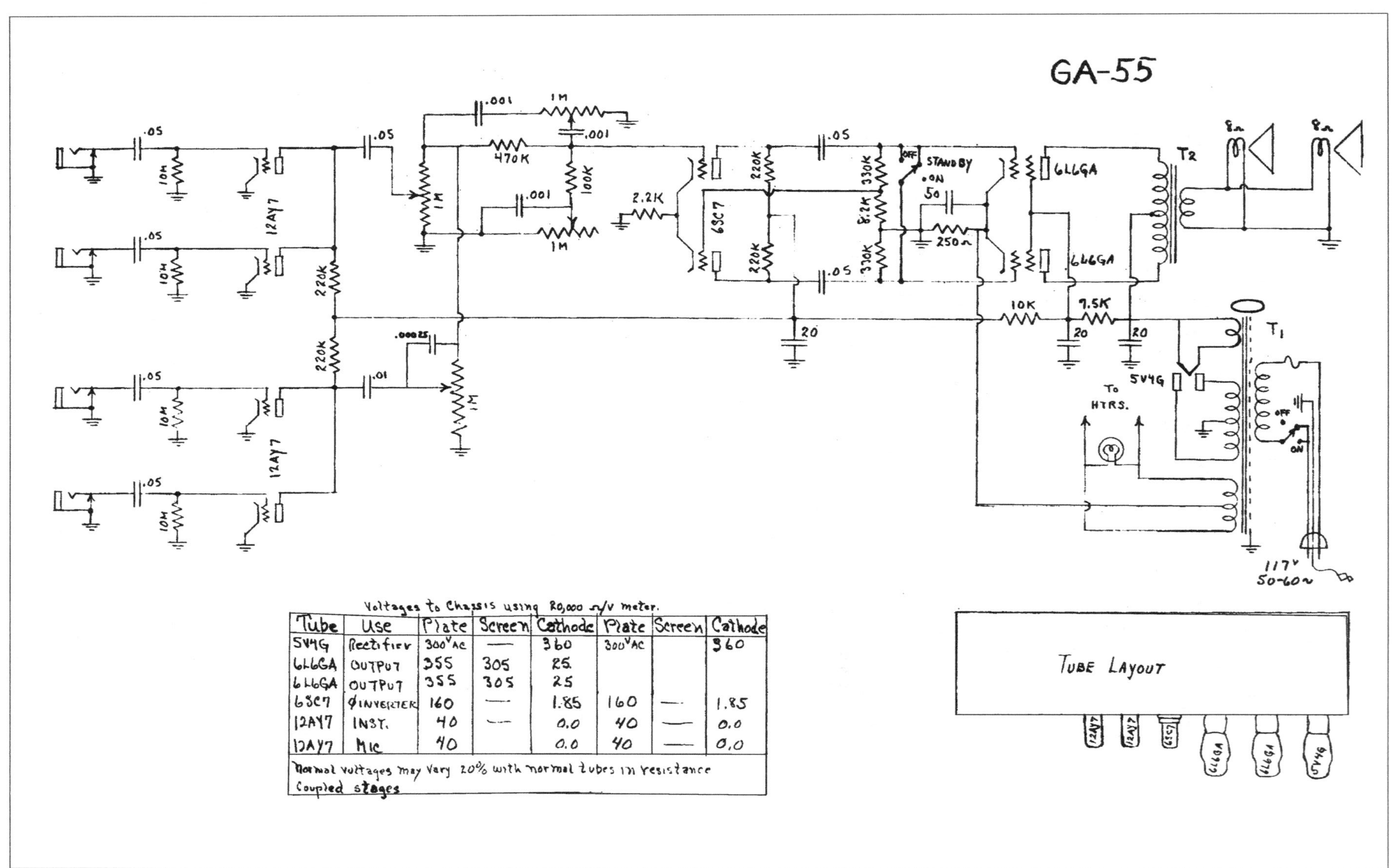

Voltages to Chassis using 20,000 Ω/v meter.

Tube	Use	Plate	Screen	Cathode	Plate	Screen	Cathode
5V4G	Rectifier	300V AC	—	360	300V AC		360
6L6GA	OUTPUT	355	305	25			
6L6GA	OUTPUT	355	305	25			
6SC7	ϕ INVERTER	160	—	1.85	160	—	1.85
12AY7	INST.	40	—	0.0	40	—	0.0
12AY7	Mic	40		0.0	40	—	0.0

Normal voltages may vary 20% with normal tubes in resistance coupled stages

Gibson

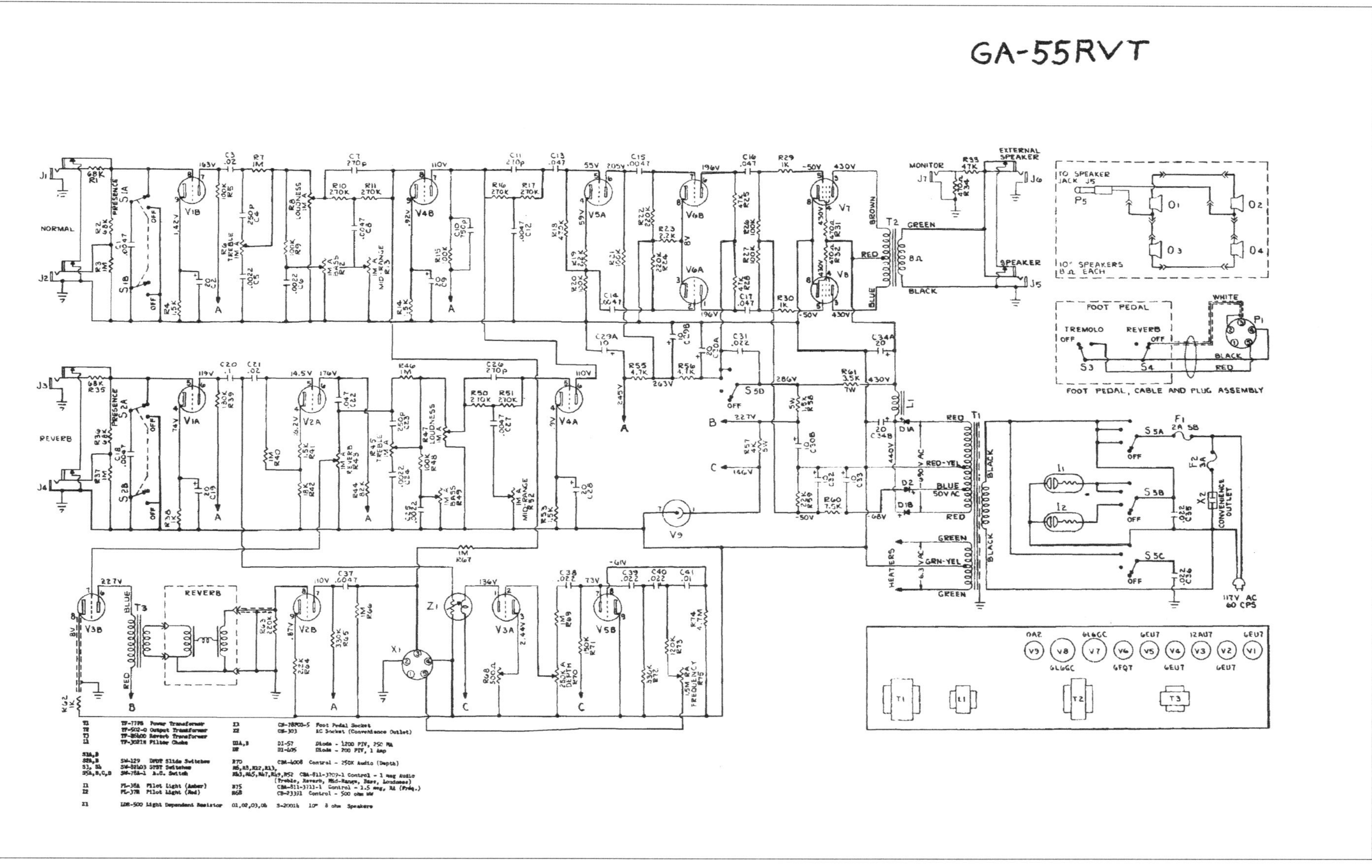

GA-55RVT
NORMAL
REVERB
PRESENCE
TREBLE
LOUDNESS
BASS
MID-RANGE
DEPTH
FREQUENCY
MONITOR
EXTERNAL SPEAKER
SPEAKER
TO SPEAKER JACK J5
10" SPEAKERS 8 Ω EACH
FOOT PEDAL
TREMOLO
FOOT PEDAL, CABLE AND PLUG ASSEMBLY
HEATERS
CONVENIENCE OUTLET
117V AC 60 CPS

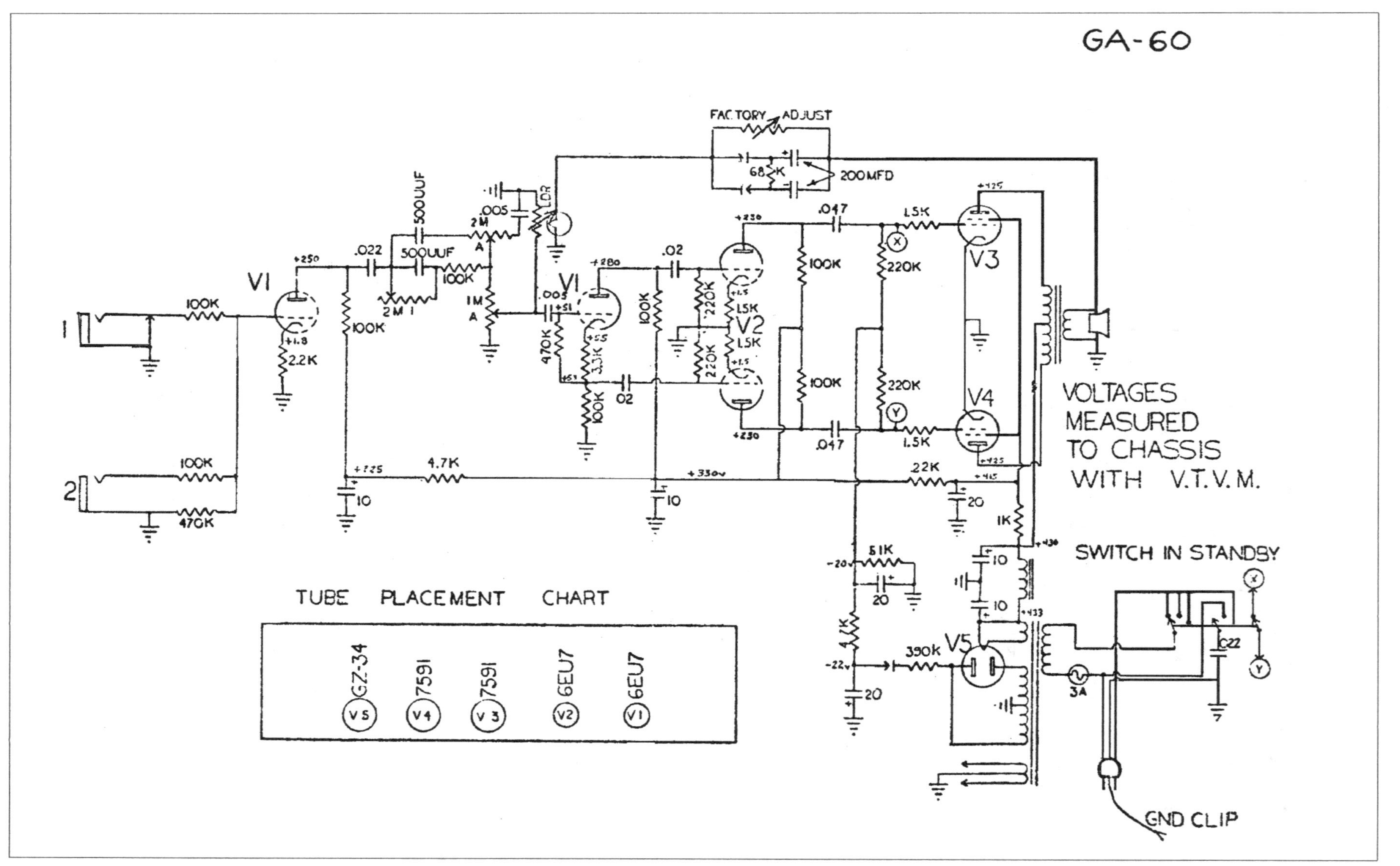
GA-60
VOLTAGES MEASURED TO CHASSIS WITH V.T.V.M.
SWITCH IN STANDBY
FACTORY ADJUST
200 MFD
GND CLIP
TUBE PLACEMENT CHART
V5 GZ-34
V4 7591
V3 7591
V2 6EU7
V1 6EU7

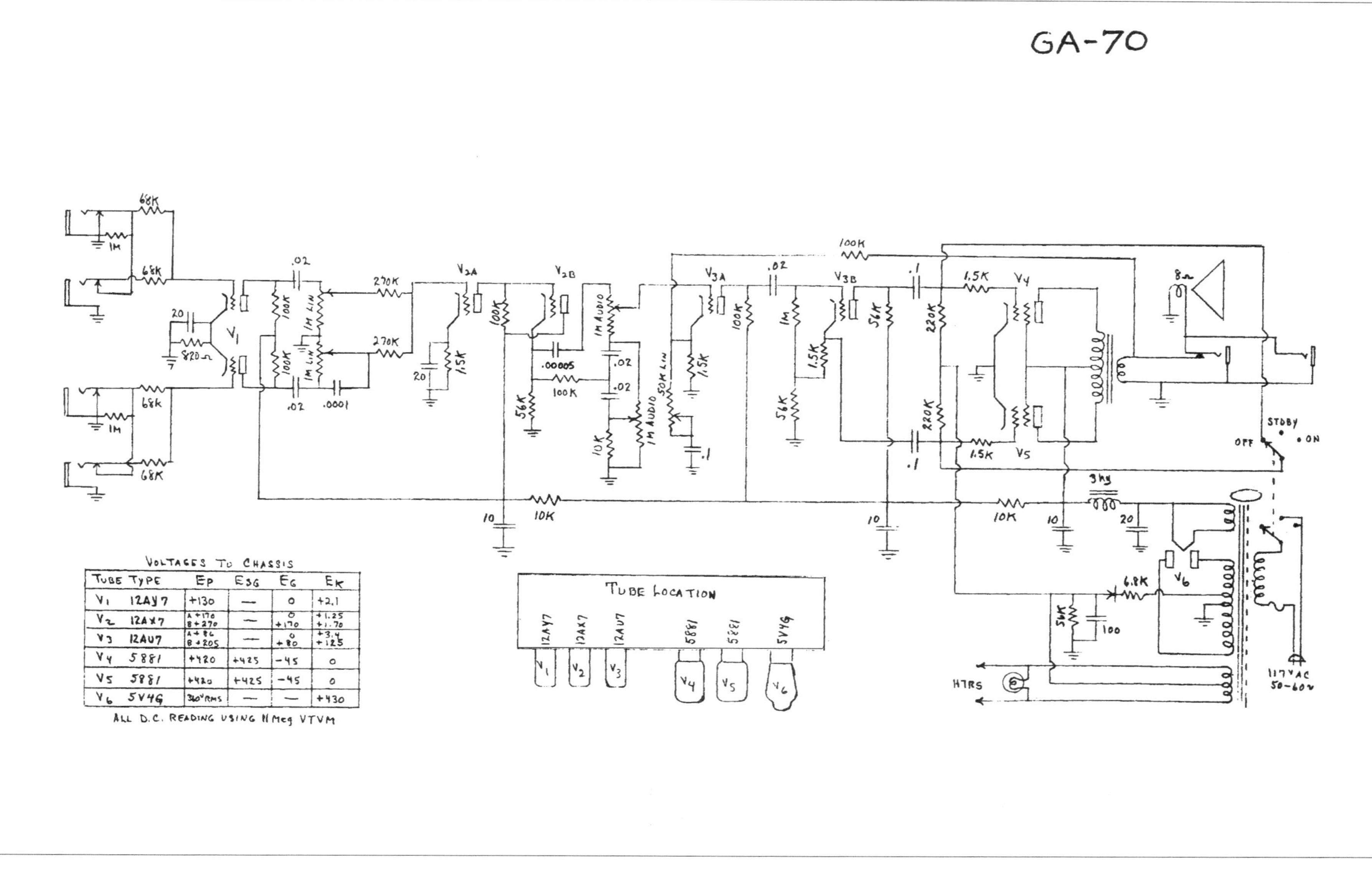

VOLTAGES TO CHASSIS

TUBE	TYPE	EP	ESG	EG	EK
V1	12AY7	+130	—	0	+2.1
V2	12AX7	A +170 B +270	—	0 +170	+1.25 +1.70
V3	12AU7	A +86 B +205	—	0 +80	+3.4 +125
V4	5881	+420	+425	-45	0
V5	5881	+420	+425	-45	0
V6	5V4G	360V RMS	—	—	+430

ALL D.C. READING USING 11 Meg VTVM

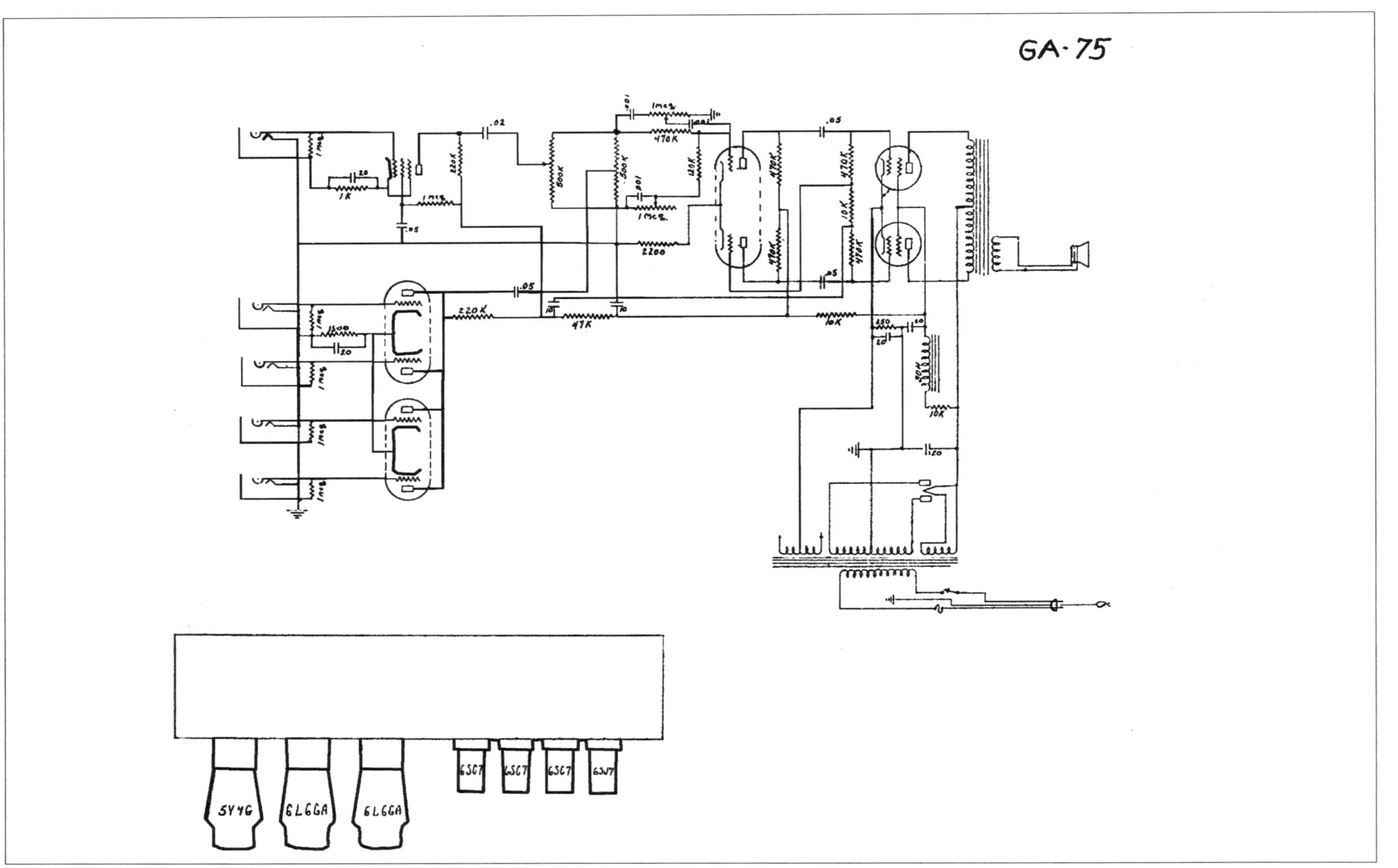
GA-75
5Y4G
6L6GA
6L6GA
6SC7
6SC7
6SC7
6SJ7
470K
10K
220K
47K
500K
2200
120K
1MEG.
.05
.02
.001
1K
1500
20

GA-75 **LATER**

Gibson

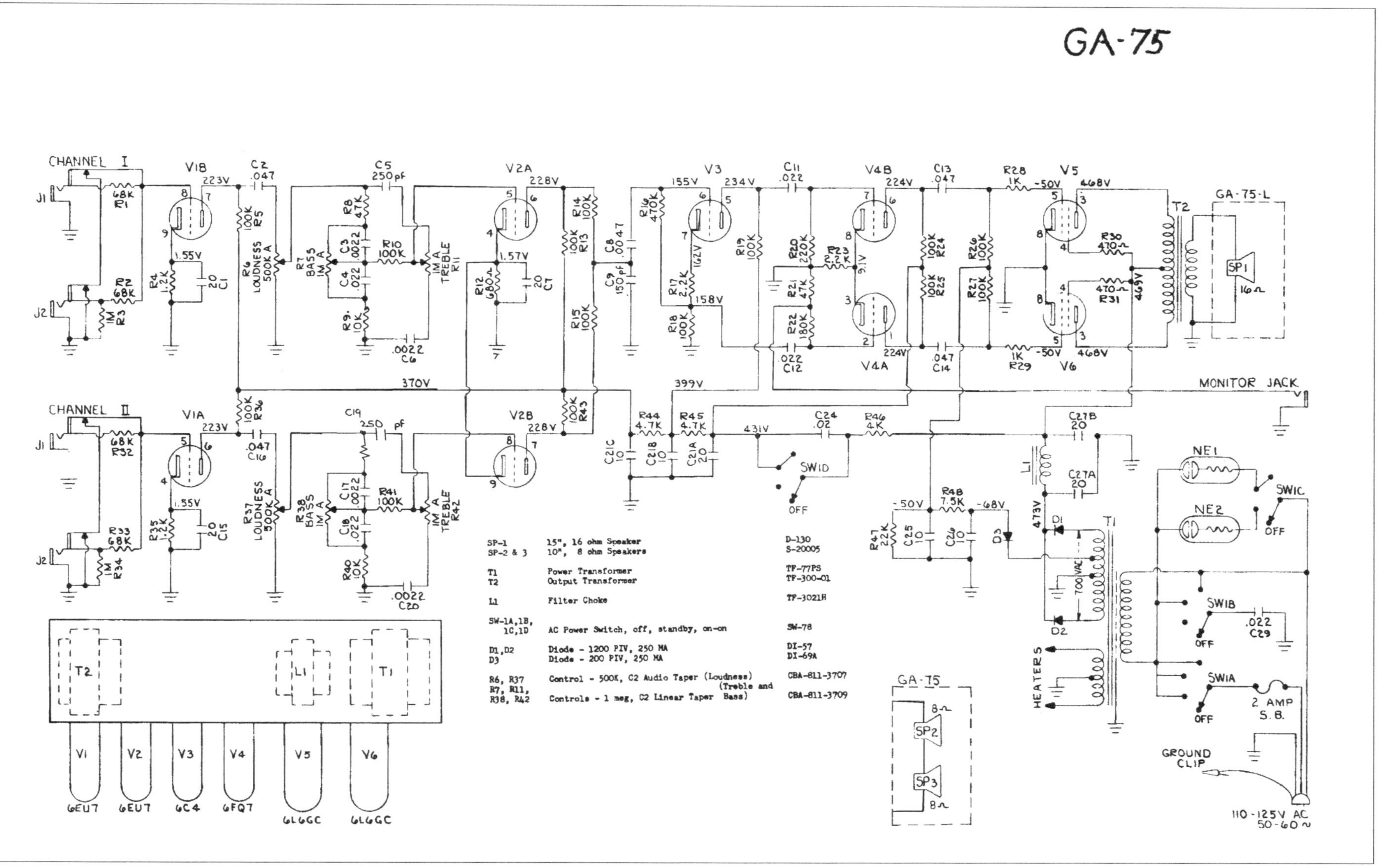

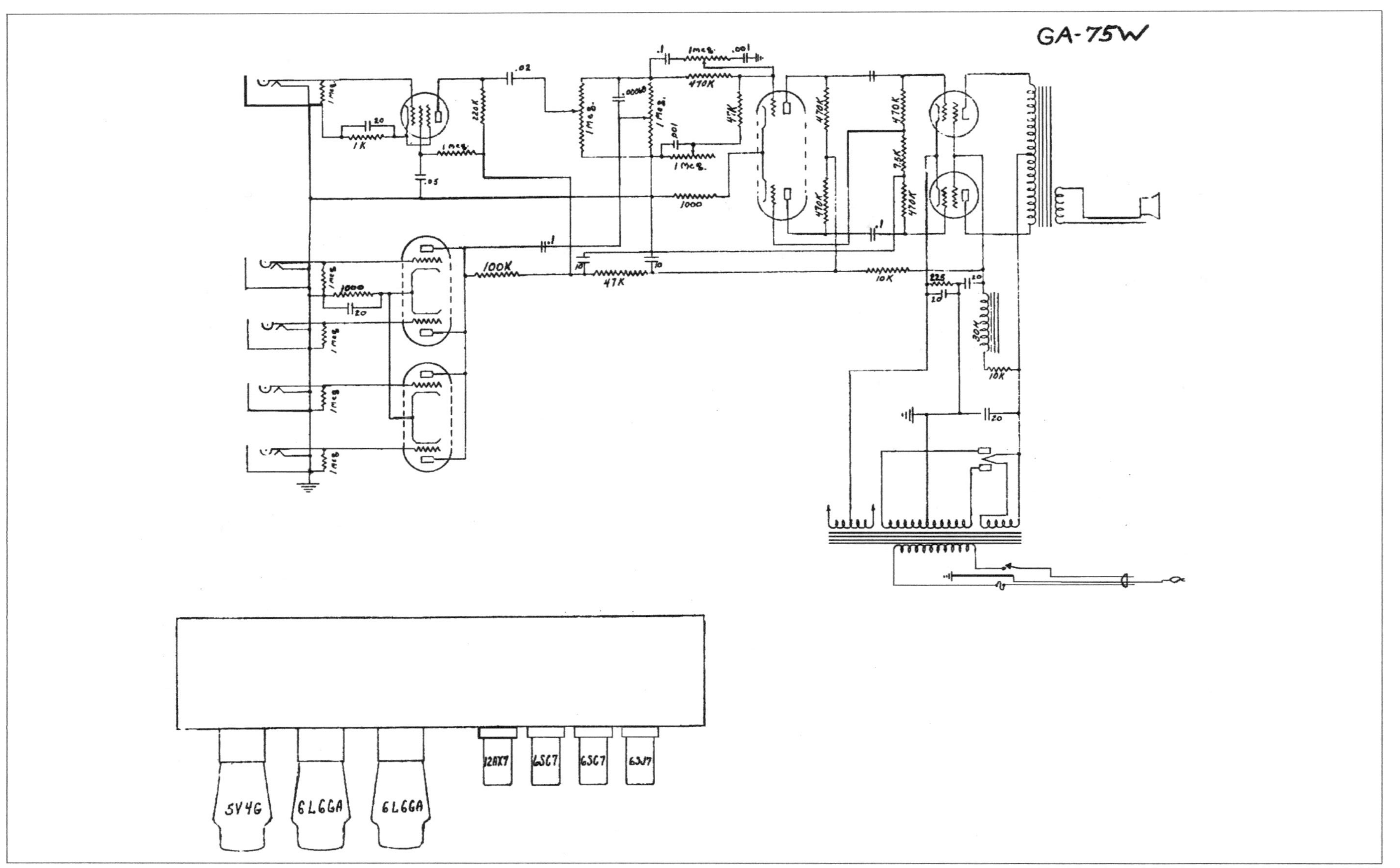
GA-75W
.02
.1
1meg.
.001
470K
47K
220K
1K
20
1meg
.05
1meg.
.00068
.001
1meg.
1000
470K
75K
470K
.1
100K
10
47K
10
10K
225
20
20
30K
10K
20
1000
20
5Y4G
6L6GA
6L6GA
12AX7
6SC7
6SC7
6SJ7

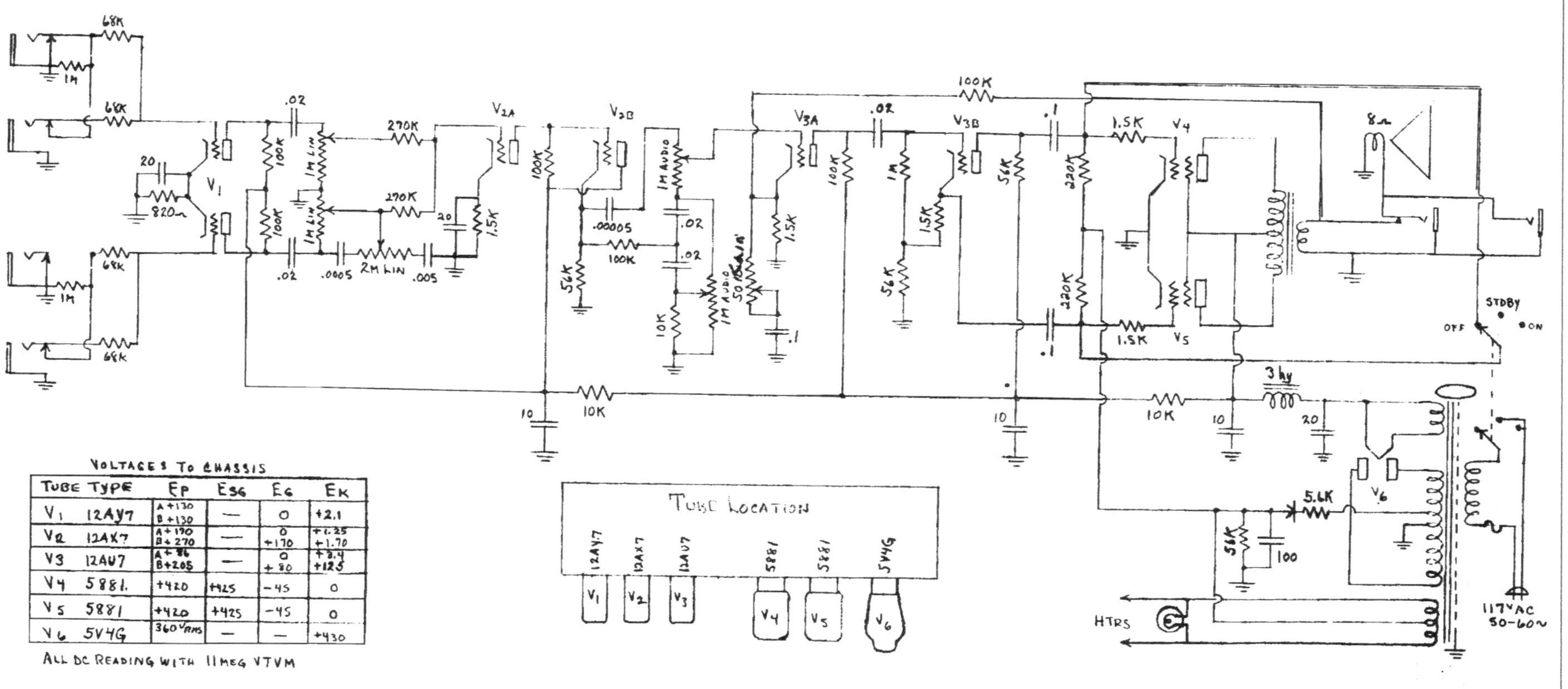

VOLTAGES TO CHASSIS

TUBE	TYPE	EP	ESG	EG	EK
V1	12AY7	A +130 B +130	—	0	+2.1
V2	12AX7	A +170 B +270	—	0 +170	+1.25 +1.70
V3	12AU7	A +86 B +205	—	0 +80	+3.4 +125
V4	5881	+420	+425	−45	0
V5	5881	+420	+425	−45	0
V6	5V4G	360 V RMS	—	—	+430

ALL DC READING WITH 11 MEG VTVM

TUBE LOCATION

V1 12AY7 · V2 12AX7 · V3 12AU7 · V4 5881 · V5 5881 · V6 5V4G

GA-77 Amplifier
1953/54
25W

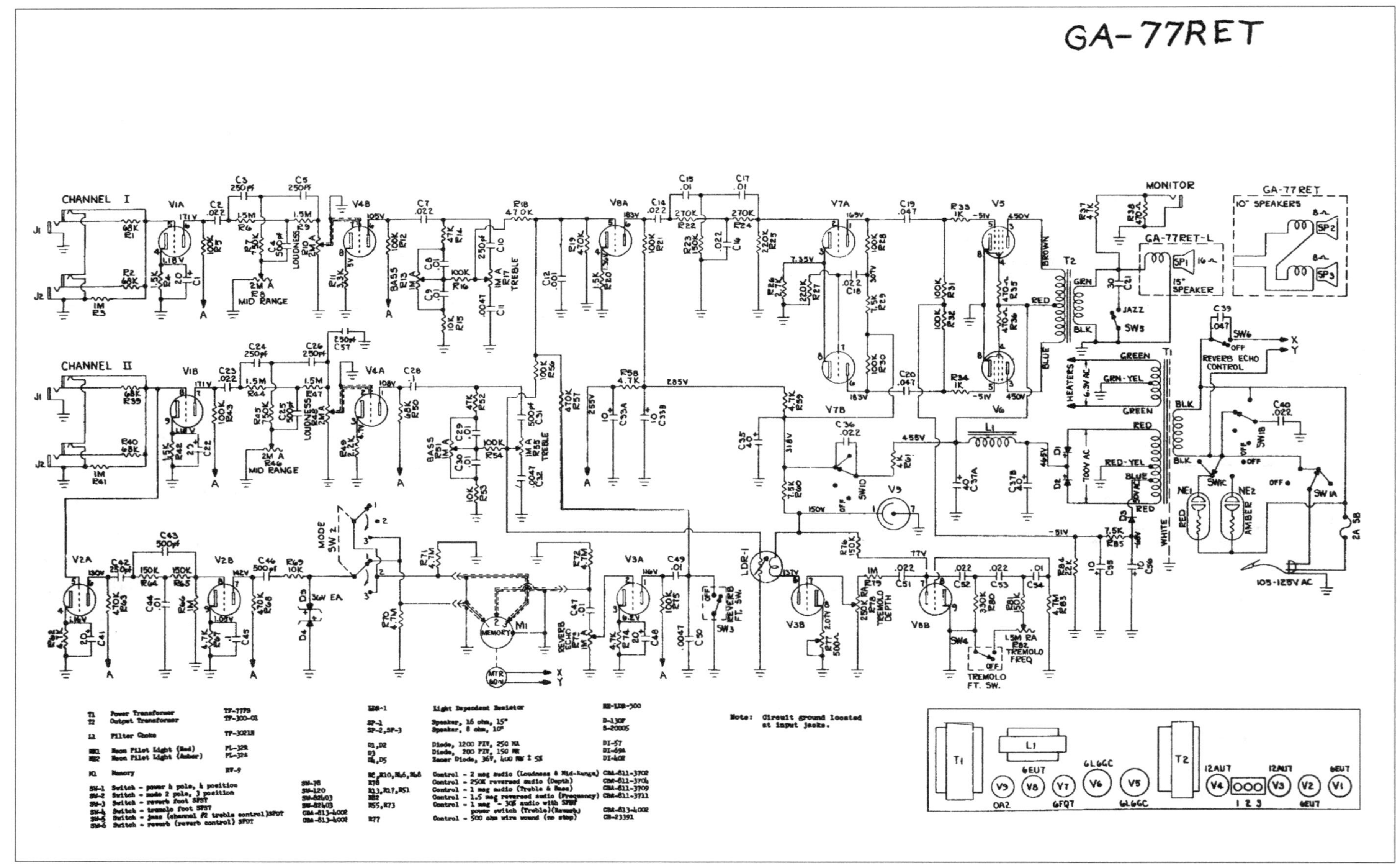
GA-77RET
CHANNEL I
CHANNEL II
MONITOR
GA-77RET-L
15" SPEAKER
GA-77RET
10" SPEAKERS
MID RANGE
LOUDNESS
BASS
TREBLE
MODE
SW 2
REVERB ECHO
CONTROL
TREMOLO
DEPTH
TREMOLO
FREQ
TREMOLO
FT. SW.
REVERB
FT. SW.
MEMORY
HEATERS
105-125V AC

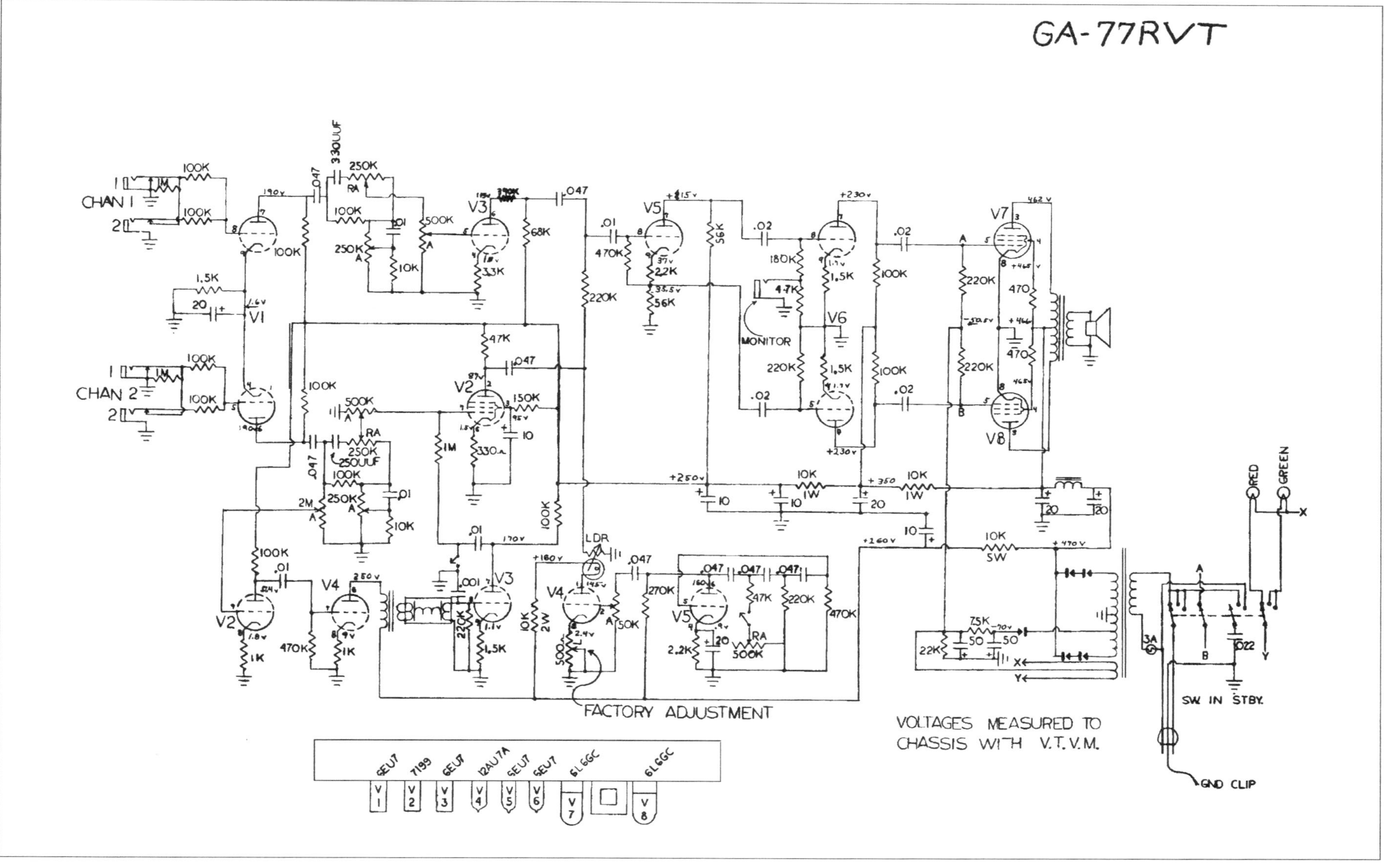
GA-77RVT
CHAN 1
CHAN 2
V1
V2
V3
V4
V5
V6
V7
V8
MONITOR
LDR
FACTORY ADJUSTMENT
VOLTAGES MEASURED TO CHASSIS WITH V.T.V.M.
SW. IN STBY.
RED
GREEN
GND CLIP
6EU7
7199
6EU7
12AU7A
6EU7
6EU7
6L6GC
6L6GC

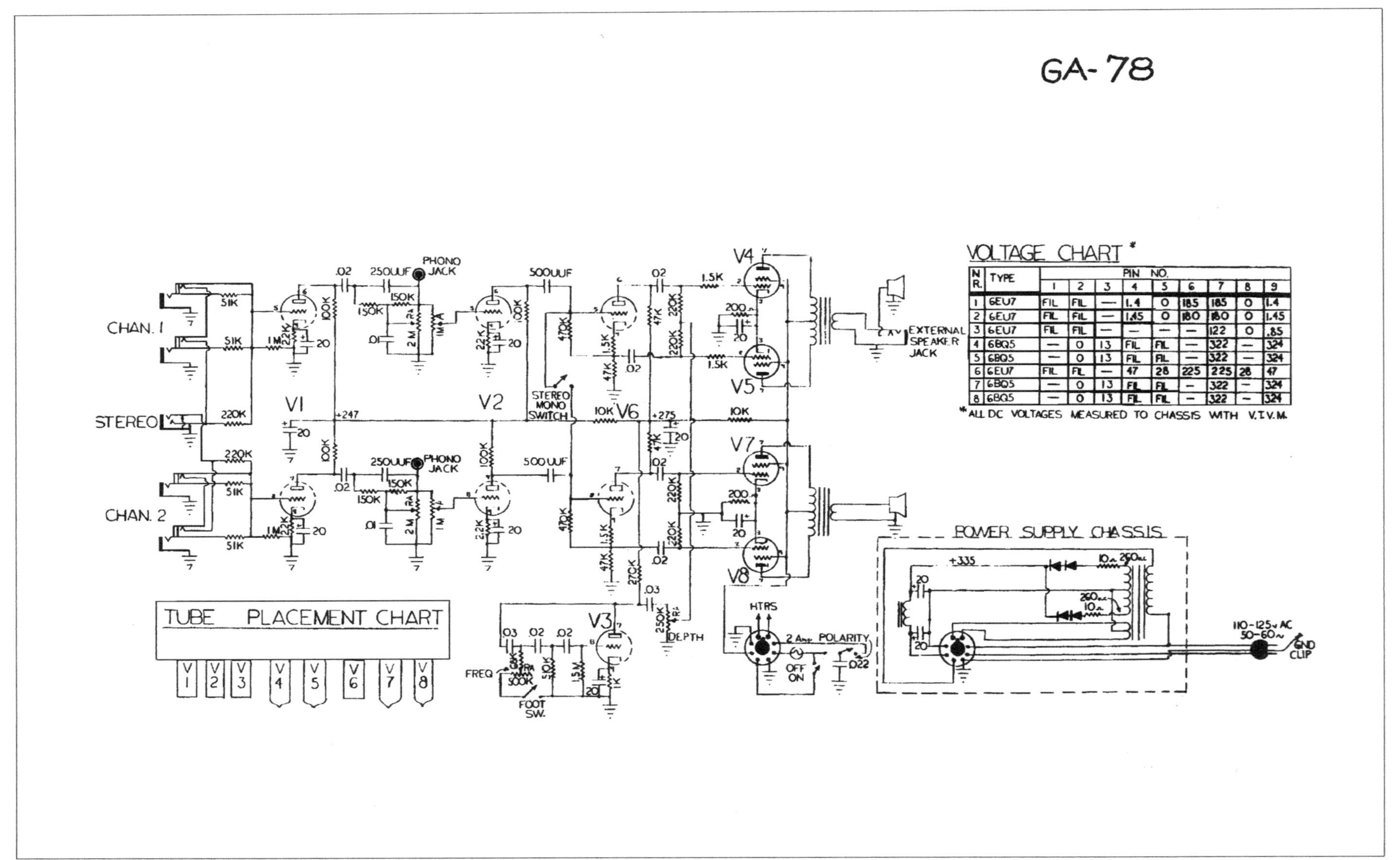

VOLTAGE CHART *

NR.	TYPE	PIN NO. 1	2	3	4	5	6	7	8	9
1	6EU7	FIL	FIL	—	1.4	0	185	185	0	1.4
2	6EU7	FIL	FIL	—	1.45	0	180	180	0	1.45
3	6EU7	FIL	FIL	—	—	—	—	122	0	.85
4	6BQ5	—	0	13	FIL	FIL	—	322	—	324
5	6BQ5	—	0	13	FIL	FIL	—	322	—	324
6	6EU7	FIL	FIL	—	47	28	225	225	28	47
7	6BQ5	—	0	13	FIL	FIL	—	322	—	324
8	6BQ5	—	0	13	FIL	FIL	—	322	—	324

* ALL DC VOLTAGES MEASURED TO CHASSIS WITH V.T.V.M.

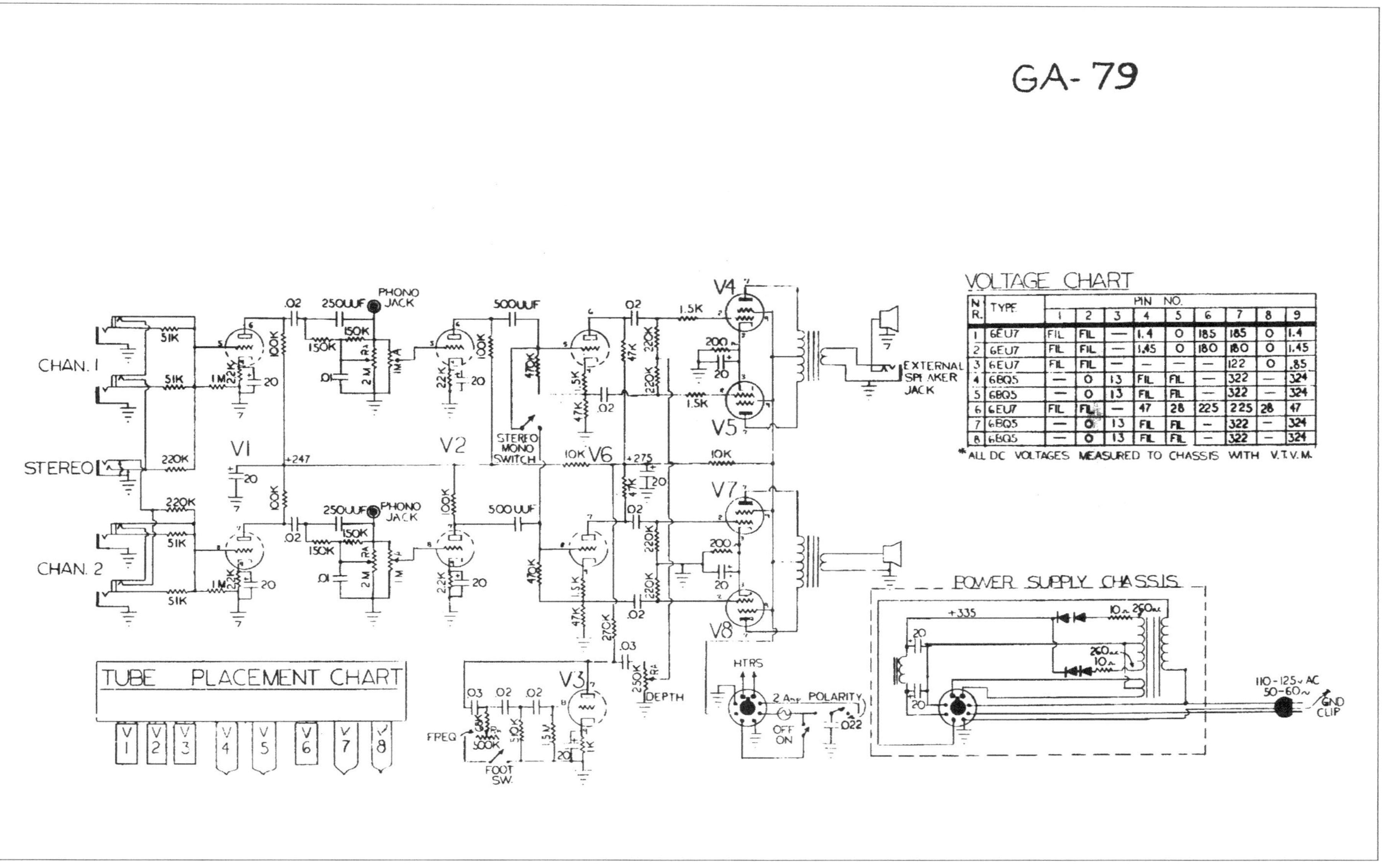

VOLTAGE CHART

N R.	TYPE	PIN NO.								
		1	2	3	4	5	6	7	8	9
1	6EU7	FIL	FIL	—	1.4	0	185	185	0	1.4
2	6EU7	FIL	FIL	—	1.45	0	180	180	0	1.45
3	6EU7	FIL	FIL	—	—	—	—	122	0	.85
4	6BQ5	—	0	13	FIL	FIL	—	322	—	324
5	6BQ5	—	0	13	FIL	FIL	—	322	—	324
6	6EU7	FIL	FIL	—	47	28	225	225	28	47
7	6BQ5	—	0	13	FIL	FIL	—	322	—	324
8	6BQ5	—	0	13	FIL	FIL	—	322	—	324

*ALL DC VOLTAGES MEASURED TO CHASSIS WITH V.T.V.M.

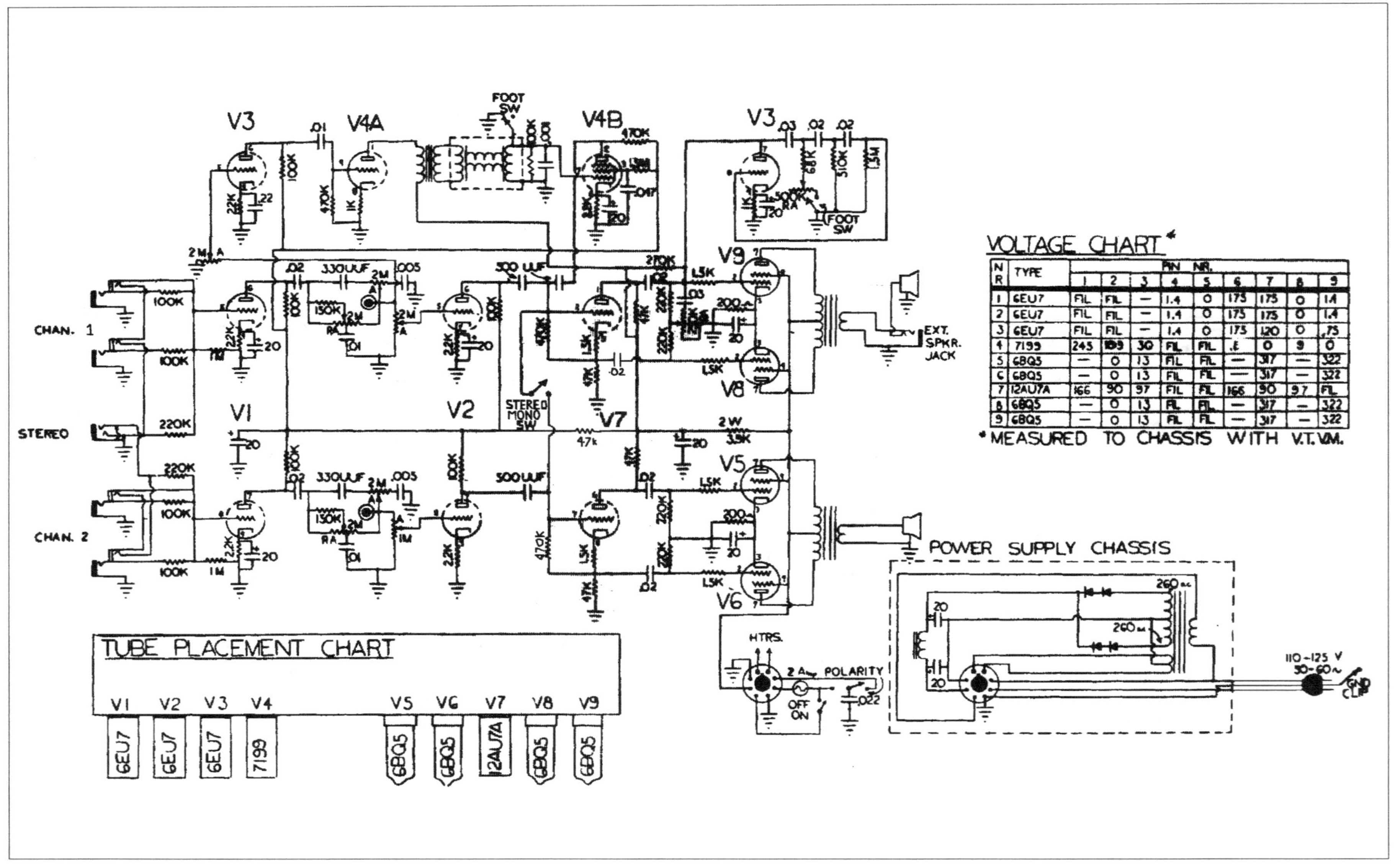

VOLTAGE CHART*

NR	TYPE	PIN NR. 1	2	3	4	5	6	7	8	9
1	6EU7	FIL	FIL	—	1.4	0	175	175	0	1.4
2	6EU7	FIL	FIL	—	1.4	0	175	175	0	1.4
3	6EU7	FIL	FIL	—	1.4	0	175	120	0	.75
4	7199	245	189	30	FIL	FIL	.8	0	9	0
5	6BQ5	—	0	13	FIL	FIL	—	317	—	322
6	6BQ5	—	0	13	FIL	FIL	—	317	—	322
7	12AU7A	166	90	97	FIL	FIL	166	90	97	FIL
8	6BQ5	—	0	13	FIL	FIL	—	317	—	322
9	6BQ5	—	0	13	FIL	FIL	—	317	—	322

*MEASURED TO CHASSIS WITH V.T.V.M.

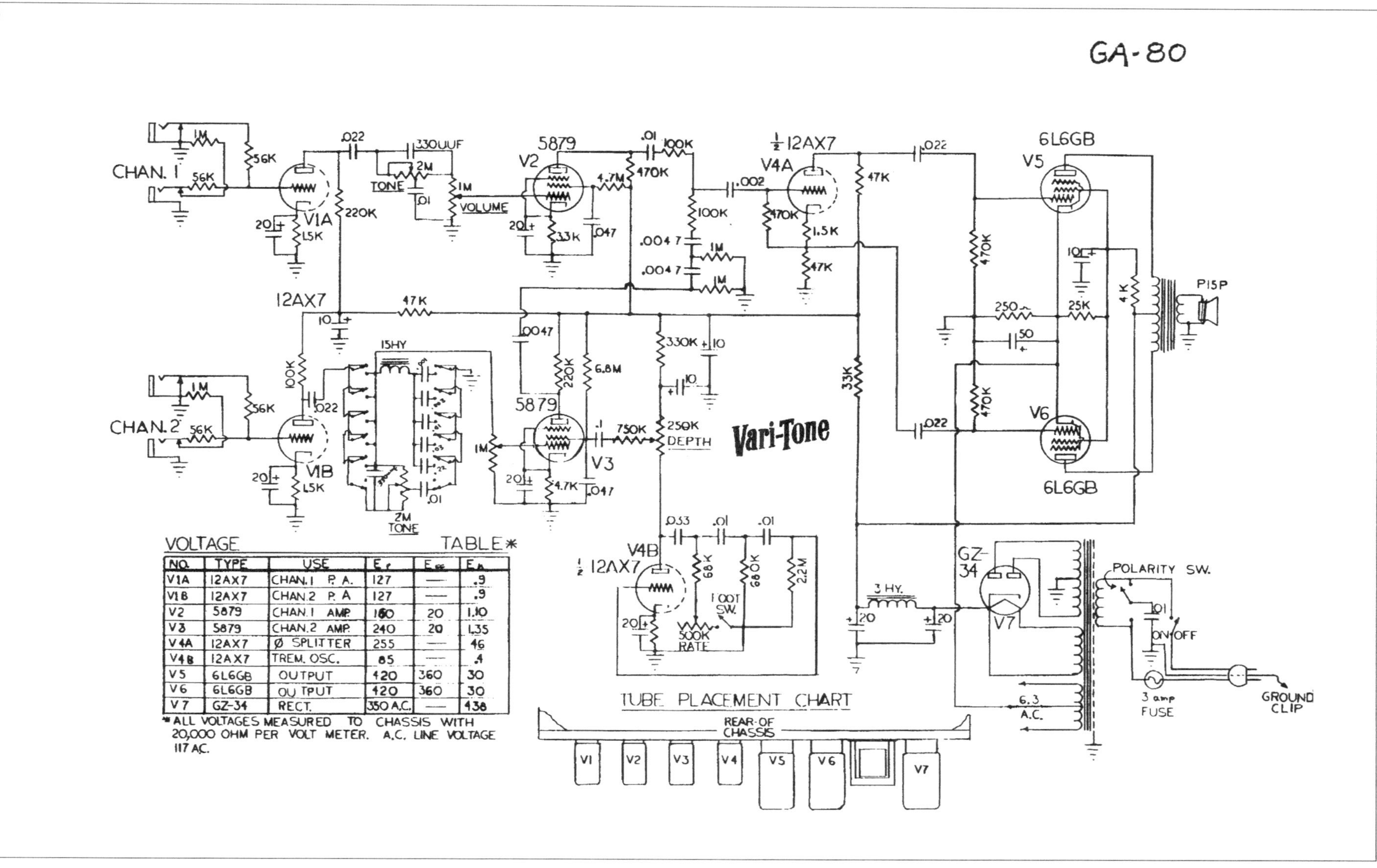

VOLTAGE TABLE*

NO.	TYPE	USE	E_f	E_{sg}	E_A
V1A	12AX7	CHAN.1 P.A.	127	—	.9
V1B	12AX7	CHAN.2 P.A.	127	—	.9
V2	5879	CHAN.1 AMP.	160	20	1.10
V3	5879	CHAN.2 AMP.	240	20	1.35
V4A	12AX7	Ø SPLITTER	255	—	46
V4B	12AX7	TREM. OSC.	85	—	.4
V5	6L6GB	OUTPUT	420	360	30
V6	6L6GB	OUTPUT	420	360	30
V7	GZ-34	RECT.	350 A.C.	—	436

*ALL VOLTAGES MEASURED TO CHASSIS WITH 20,000 OHM PER VOLT METER. A.C. LINE VOLTAGE 117 AC.

Gibson

GA-83S STEREO-VIB (PREAMP)

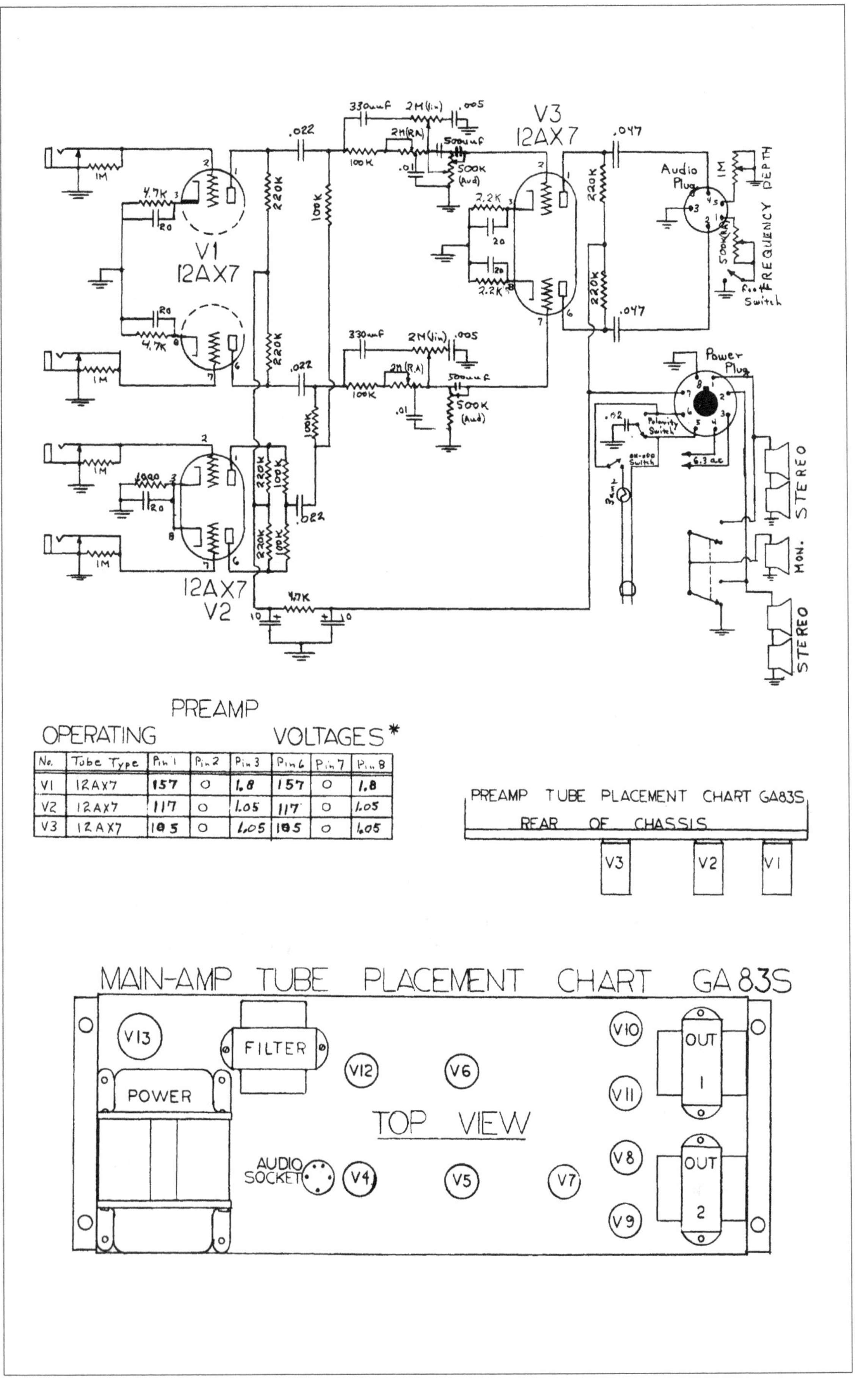

No.	Tube Type	Pin 1	Pin 2	Pin 3	Pin 6	Pin 7	Pin 8
V1	12AX7	157	0	1.8	157	0	1.8
V2	12AX7	117	0	1.05	117	0	1.05
V3	12AX7	105	0	1.05	105	0	1.05

GA-83S **STEREO VIB**

Gibson

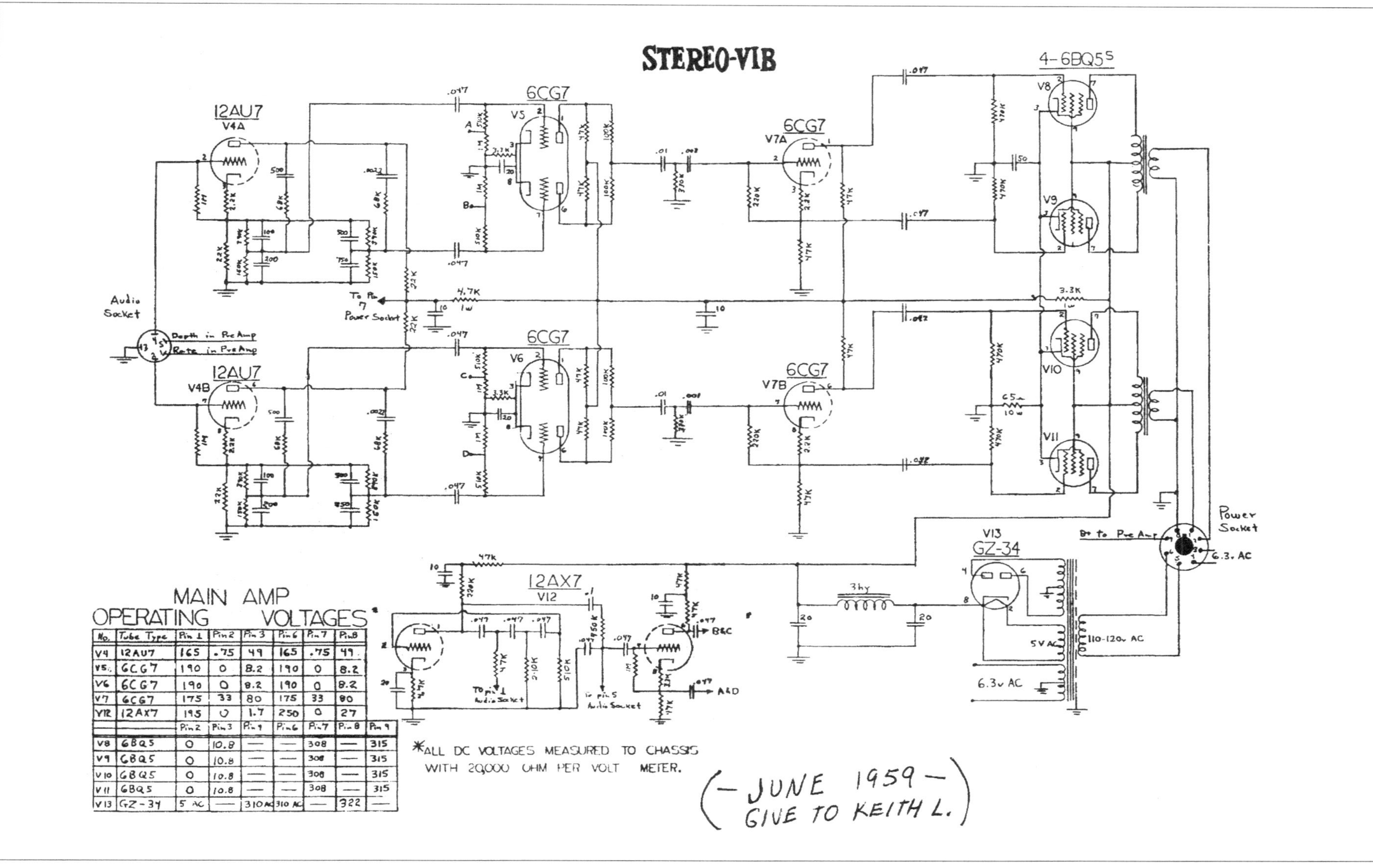

MAIN AMP OPERATING VOLTAGES*

No.	Tube Type	Pin 1	Pin 2	Pin 3	Pin 6	Pin 7	Pin 8	
V4	12AU7	165	.75	49	165	.75	49	
V5	6CG7	190	0	8.2	190	0	8.2	
V6	6CG7	190	0	8.2	190	0	8.2	
V7	6CG7	175	33	80	175	33	80	
V12	12AX7	195	0	1.7	250	0	27	
		Pin 2	Pin 3	Pin 4	Pin 6	Pin 7	Pin 8	Pin 9
V8	6BQ5	0	10.8	—	—	308	—	315
V9	6BQ5	0	10.8	—	—	308	—	315
V10	6BQ5	0	10.8	—	—	308	—	315
V11	6BQ5	0	10.8	—	—	308	—	315
V13	GZ-34	5 AC	—	310 AC	310 AC	—	322	—

*ALL DC VOLTAGES MEASURED TO CHASSIS WITH 20,000 OHM PER VOLT METER.

(—JUNE 1959— GIVE TO KEITH L.)

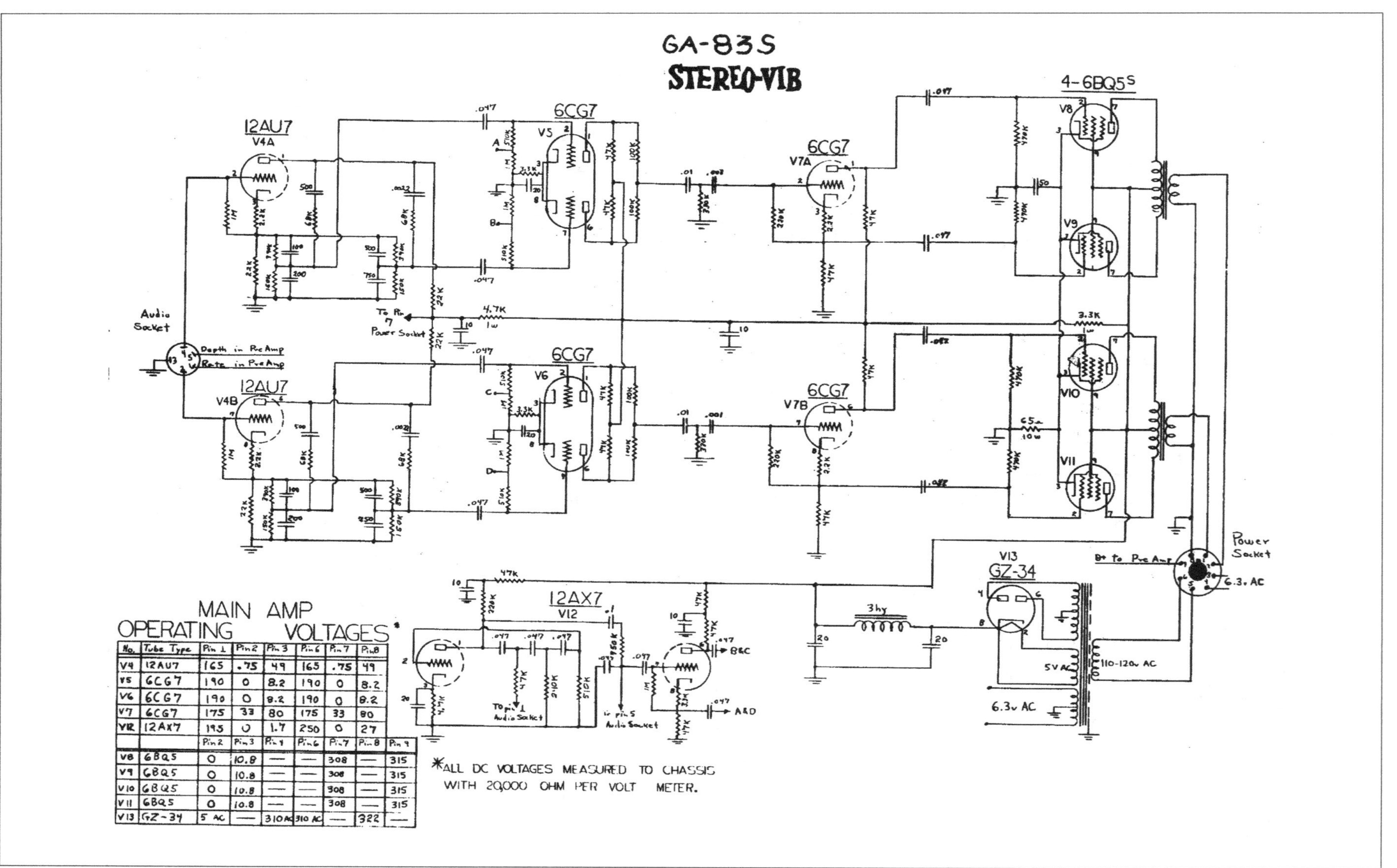

No.	Tube Type	Pin 1	Pin 2	Pin 3	Pin 6	Pin 7	Pin 8	
V4	12AU7	165	.75	49	165	.75	49	
V5	6CG7	190	0	8.2	190	0	8.2	
V6	6CG7	190	0	8.2	190	0	8.2	
V7	6CG7	175	33	80	175	33	80	
V12	12AX7	195	0	1.7	250	0	27	
		Pin 2	Pin 3	Pin 4	Pin 6	Pin 7	Pin 8	Pin 9
V8	6BQ5	0	10.8	—	—	308	—	315
V9	6BQ5	0	10.8	—	—	308	—	315
V10	6BQ5	0	10.8	—	—	308	—	315
V11	6BQ5	0	10.8	—	—	308	—	315
V13	GZ-34	5 AC	—	310 AC	310 AC	—	322	—

*ALL DC VOLTAGES MEASURED TO CHASSIS WITH 20,000 OHM PER VOLT METER.

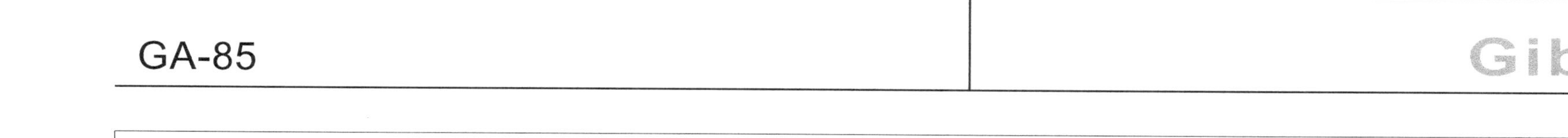

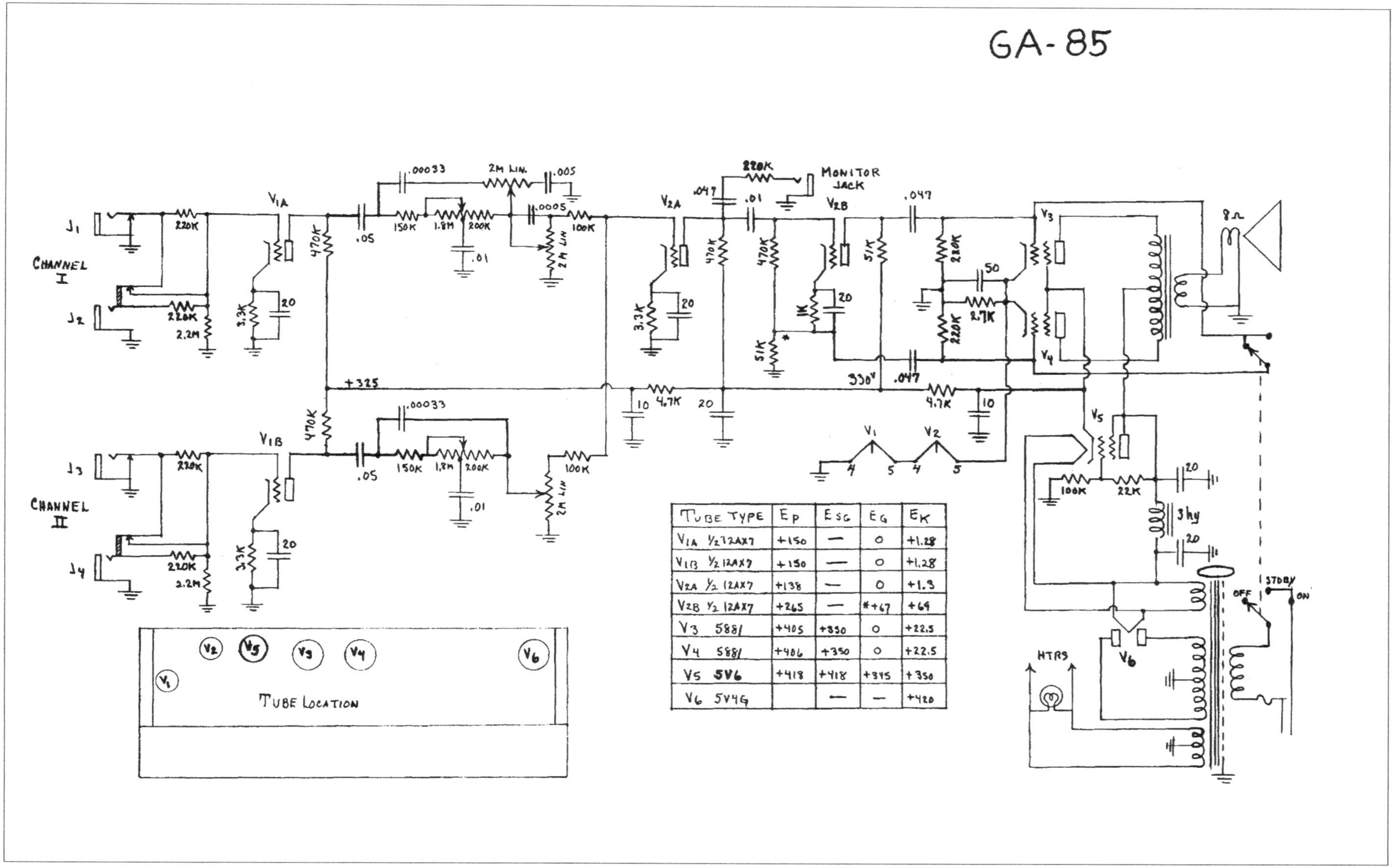

GA-85

CHANNEL I: J1, J2
CHANNEL II: J3, J4
MONITOR JACK

TUBE TYPE	EP	ESG	EG	EK
V1A ½ 12AX7	+150	—	0	+1.28
V1B ½ 12AX7	+150	—	0	+1.28
V2A ½ 12AX7	+138	—	0	+1.3
V2B ½ 12AX7	+265	—	*+67	+69
V3 5881	+405	+330	0	+22.5
V4 5881	+406	+350	0	+22.5
V5 5V6	+418	+418	+345	+350
V6 5V4G		—	—	+420

TUBE LOCATION: V1, V2, V5, V3, V4, V6

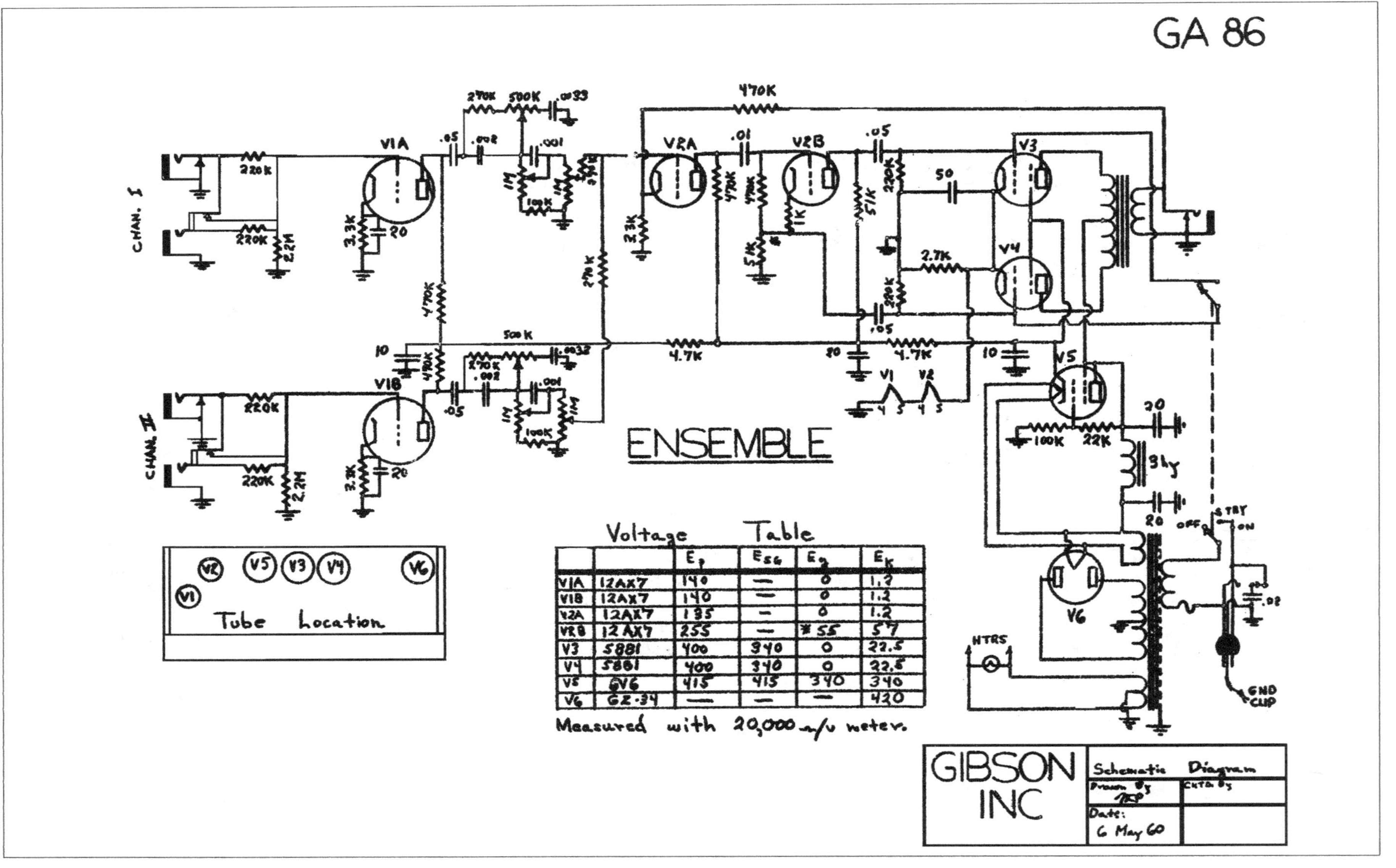

Voltage Table

		E_p	E_{SG}	E_g	E_k
V1A	12AX7	140	—	0	1.2
V1B	12AX7	140	—	0	1.2
V2A	12AX7	135	—	0	1.2
V2B	12AX7	255	—	*55	57
V3	5881	400	340	0	22.5
V4	5881	400	340	0	22.5
V5	6V6	415	415	340	340
V6	GZ-34	—	—	—	420

Measured with 20,000 Ω/v meter.

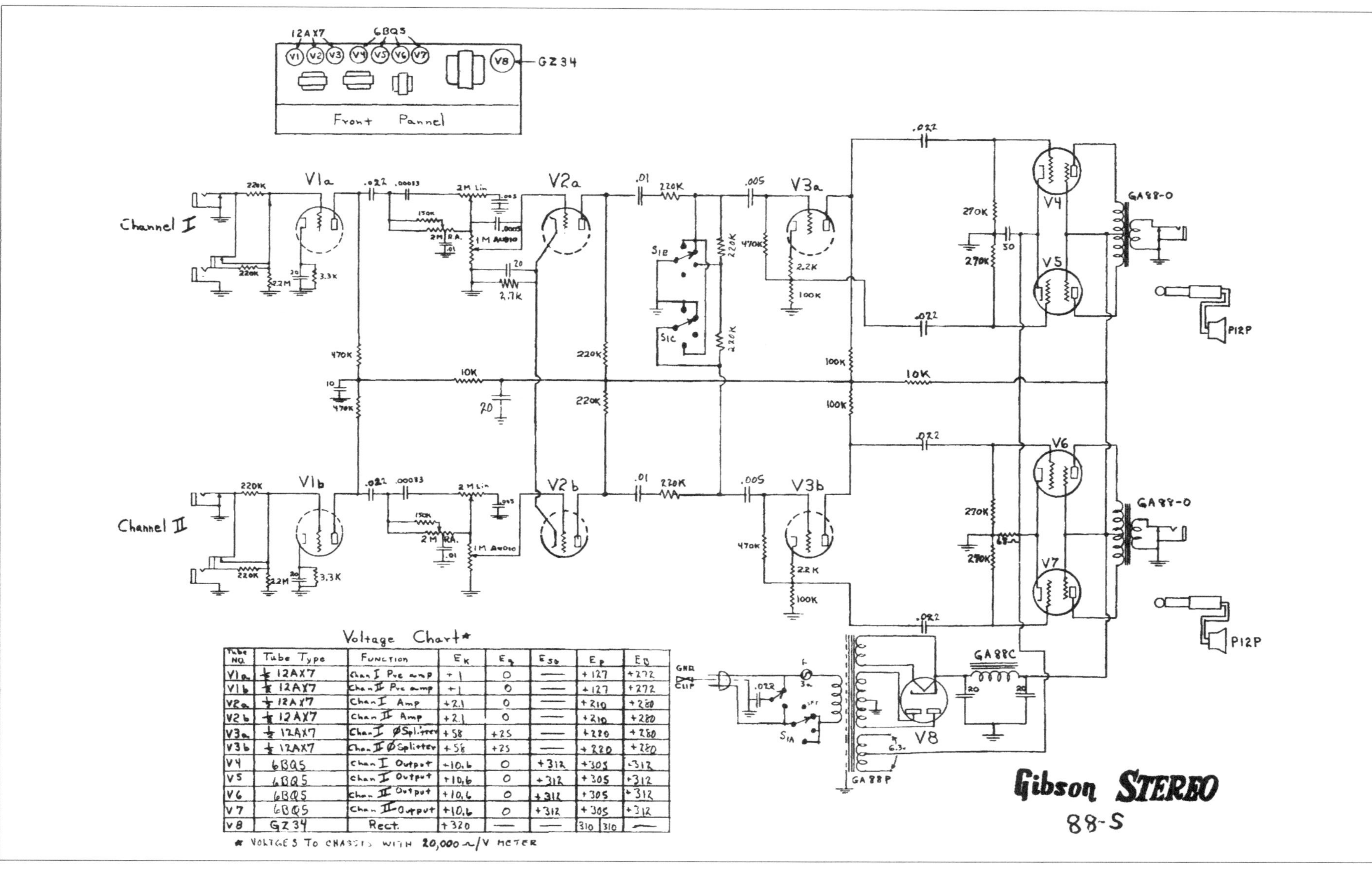

Voltage Chart*

Tube NO.	Tube Type	Function	E_K	E_g	E_{sg}	E_p	E_B
V1a	½ 12AX7	Chan I Pre amp	+1	0	—	+127	+272
V1b	½ 12AX7	Chan II Pre amp	+1	0	—	+127	+272
V2a	½ 12AX7	Chan I Amp	+2.1	0	—	+210	+280
V2b	½ 12AX7	Chan II Amp	+2.1	0	—	+210	+280
V3a	½ 12AX7	Chan I ØSplitter	+58	+25	—	+220	+280
V3b	½ 12AX7	Chan II ØSplitter	+58	+25	—	+220	+280
V4	6BQ5	Chan I Output	+10.6	0	+312	+305	+312
V5	6BQ5	Chan I Output	+10.6	0	+312	+305	+312
V6	6BQ5	Chan II Output	+10.6	0	+312	+305	+312
V7	6BQ5	Chan II Output	+10.6	0	+312	+305	+312
V8	GZ34	Rect.	+320	—	—	310 310	—

* VOLTGES TO CHASSIS WITH 20,000 Ω/V METER

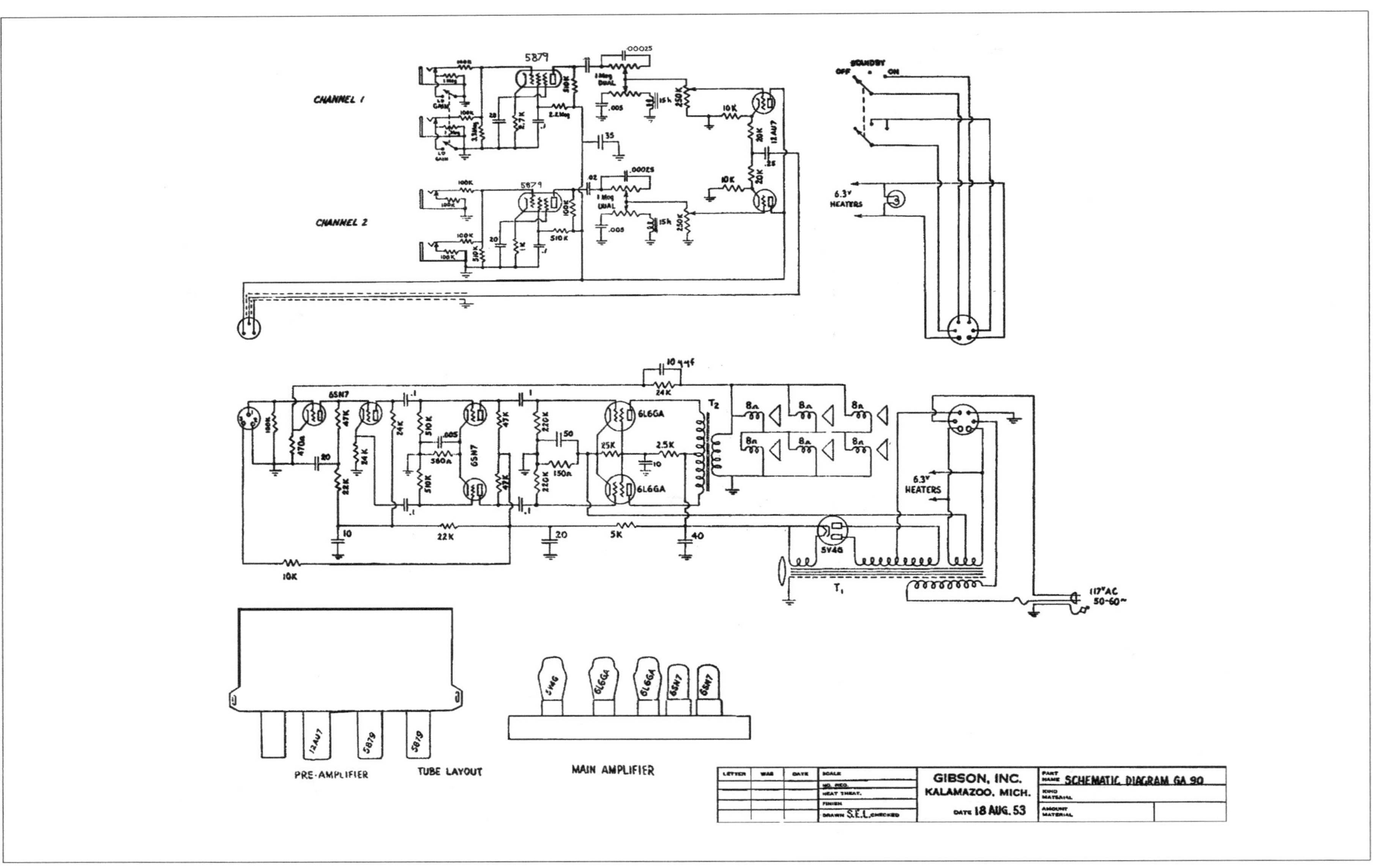

CHANNEL 1
CHANNEL 2
5879
12AU7
STANDBY
OFF
ON
6.3V HEATERS
6SN7
6L6GA
T2
T1
5V4G
117V AC 50-60~
PRE-AMPLIFIER
TUBE LAYOUT
MAIN AMPLIFIER
GIBSON, INC.
KALAMAZOO, MICH.
DATE 18 AUG. 53
SCHEMATIC DIAGRAM GA 90

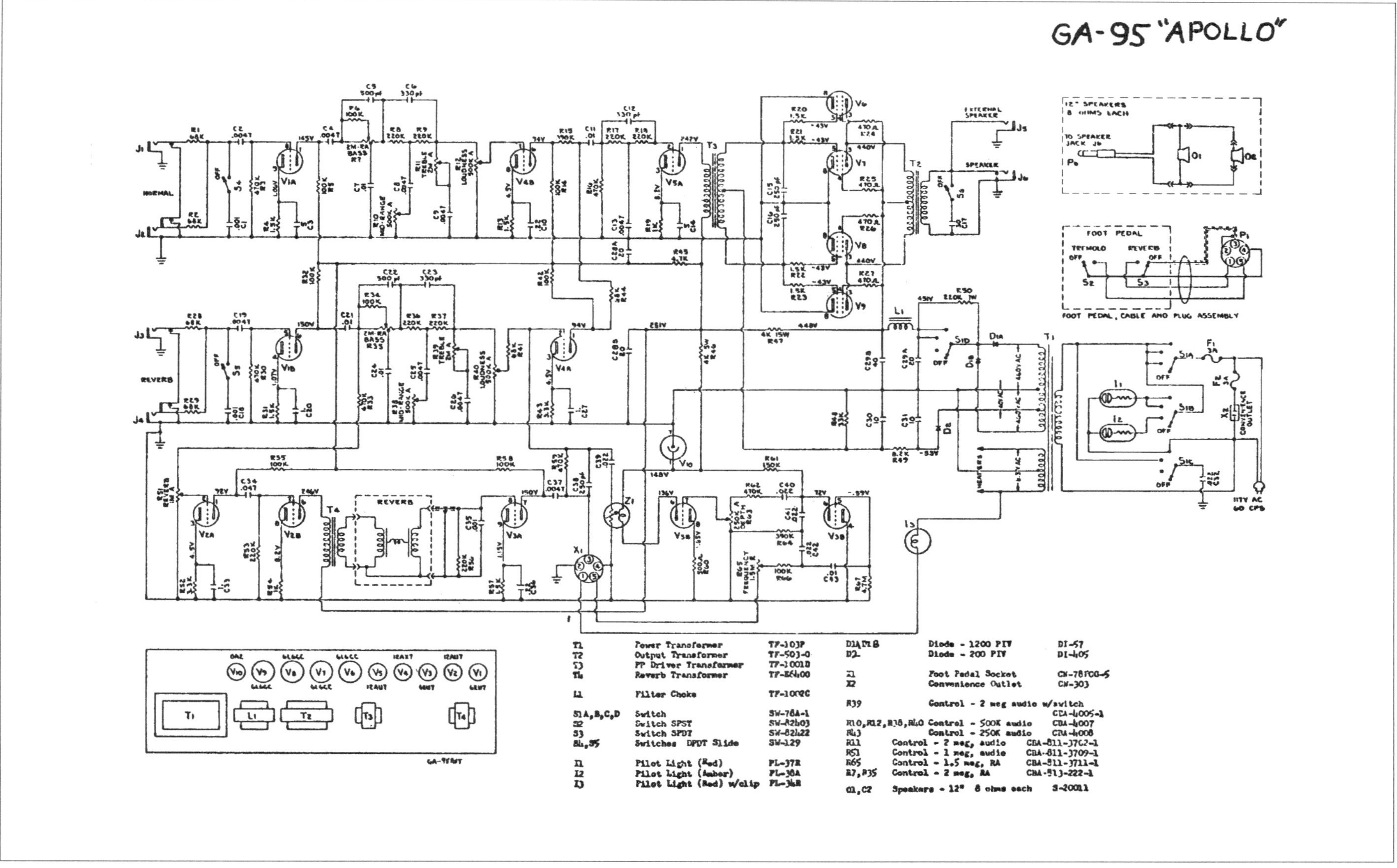
GA-95 "APOLLO"
NORMAL
REVERB
REVERB
EXTERNAL SPEAKER
SPEAKER
12" SPEAKERS 8 OHMS EACH
TO SPEAKER JACK J6
FOOT PEDAL
TREMOLO
REVERB
FOOT PEDAL, CABLE AND PLUG ASSEMBLY
CONVENIENCE OUTLET
117V AC 60 CPS
HEATERS
T1 Power Transformer TF-103P
T2 Output Transformer TF-503-0
T3 PP Driver Transformer TF-1001D
T4 Reverb Transformer TF-E6400
L1 Filter Choke TF-1002C
S1A,B,C,D Switch SW-78A-1
S2 Switch SPST SW-82403
S3 Switch SPDT SW-82422
S4,S5 Switches DPDT Slide SW-129
I1 Pilot Light (Red) PL-37R
I2 Pilot Light (Amber) PL-38A
I3 Pilot Light (Red) w/clip PL-34R
D1A,D1B Diode - 1200 PIV DI-57
D2 Diode - 200 PIV DI-405
X1 Foot Pedal Socket CN-78FC0-5
X2 Convenience Outlet CN-303
R39 Control - 2 meg audio w/switch CDA-4005-1
R10,R12,R38,R40 Control - 500K audio CDA-4007
R43 Control - 250K audio CDA-4008
R11 Control - 2 meg, audio CBA-811-3702-1
R51 Control - 1 meg, audio CBA-811-3709-1
R65 Control - 1.5 meg, RA CBA-811-3711-1
R7,R35 Control - 2 meg, RA CBA-813-222-1
O1,O2 Speakers - 12" 8 ohms each S-20011
GA-95RVT

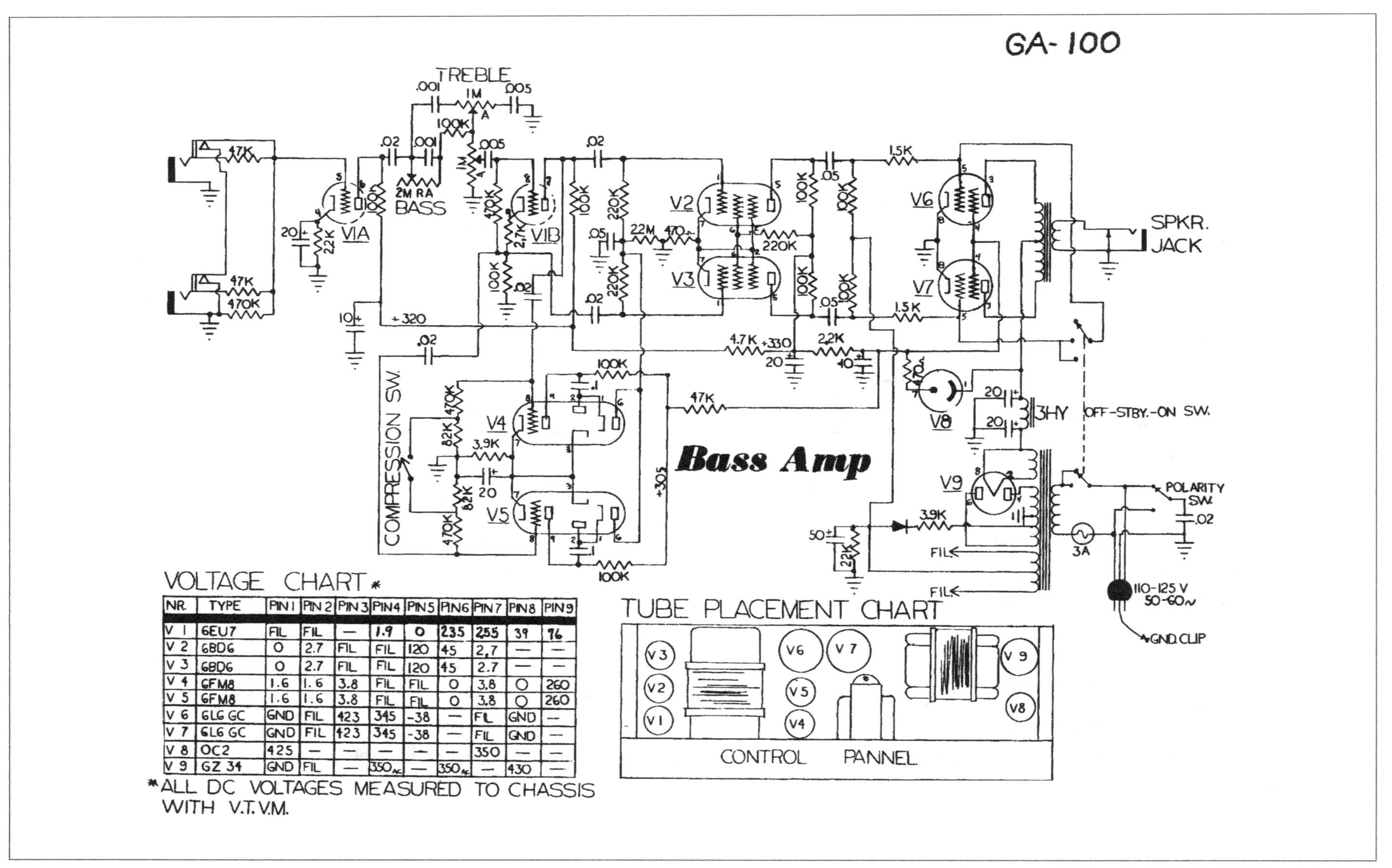

VOLTAGE CHART *

NR.	TYPE	PIN 1	PIN 2	PIN 3	PIN 4	PIN 5	PIN 6	PIN 7	PIN 8	PIN 9
V 1	6EU7	FIL	FIL	—	1.9	0	235	255	39	76
V 2	6BD6	0	2.7	FIL	FIL	120	45	2.7	—	—
V 3	6BD6	0	2.7	FIL	FIL	120	45	2.7	—	—
V 4	6FM8	1.6	1.6	3.8	FIL	FIL	0	3.8	0	260
V 5	6FM8	1.6	1.6	3.8	FIL	FIL	0	3.8	0	260
V 6	6L6 GC	GND	FIL	423	345	-38	—	FIL	GND	—
V 7	6L6 GC	GND	FIL	423	345	-38	—	FIL	GND	—
V 8	OC2	425	—	—	—	—	—	350	—	—
V 9	GZ 34	GND	FIL	—	350 AC	—	350 AC	—	430	—

*ALL DC VOLTAGES MEASURED TO CHASSIS WITH V.T.V.M.

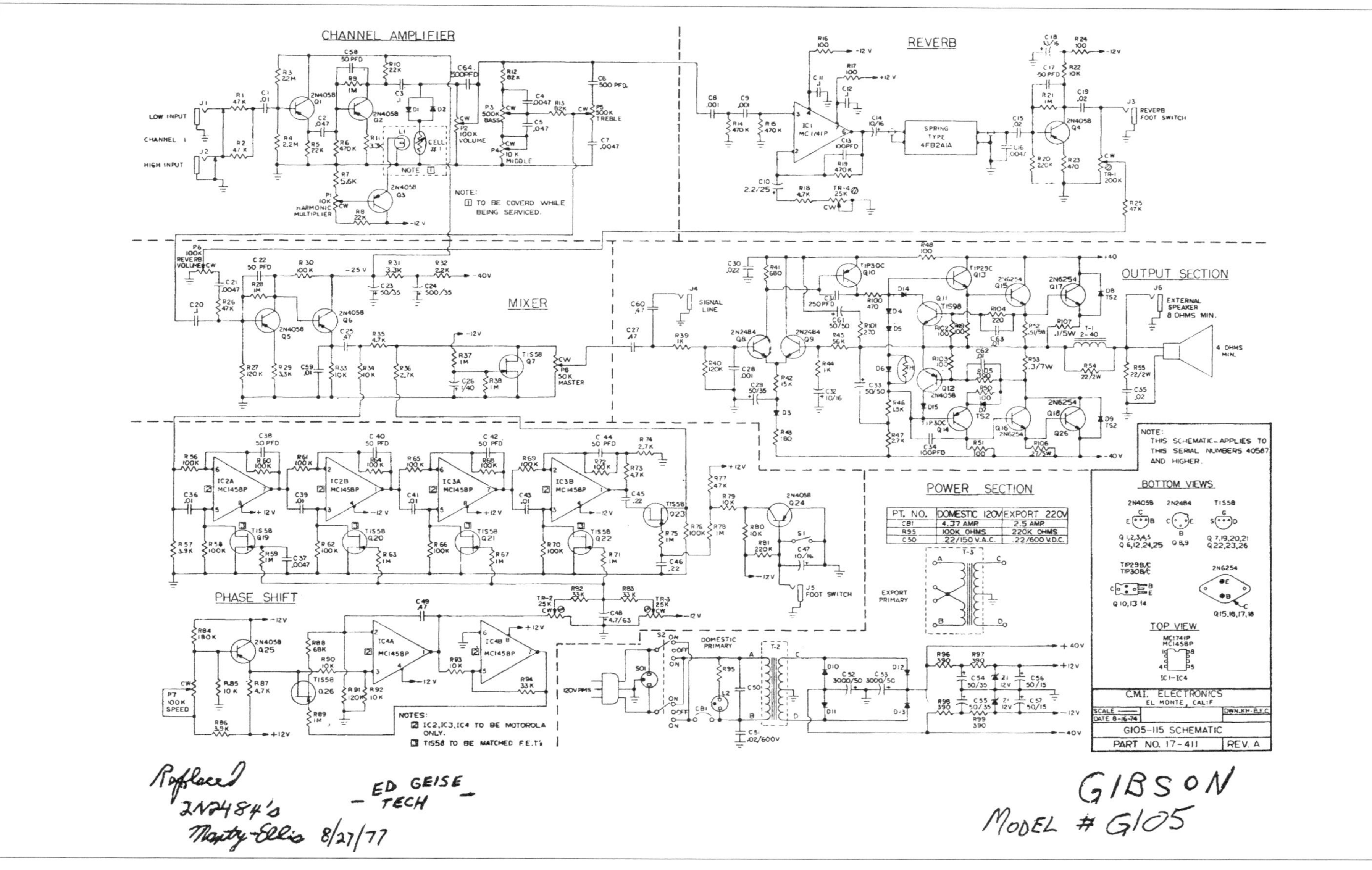

PT. NO.	DOMESTIC 120V	EXPORT 220V
CB1	4.37 AMP	2.5 AMP
R95	100K OHMS	220K OHMS
C50	.22/150 V.A.C.	.22/600 V.D.C.

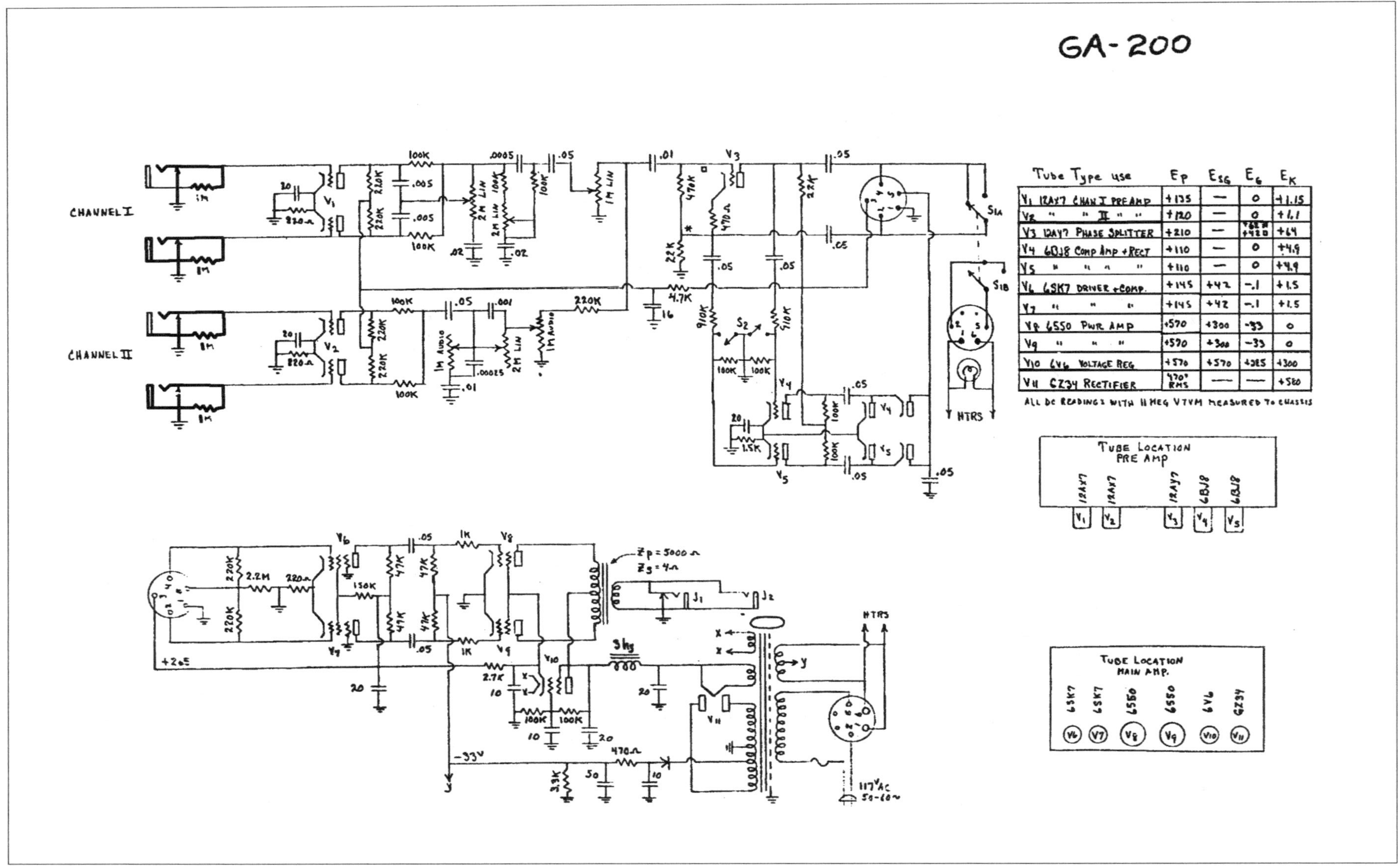

Tube Type use	Ep	Esg	EG	EK
V1 12AY7 CHAN I PREAMP	+135	—	0	+1.15
V2 " " II " "	+120	—	0	+1.1
V3 12AY7 PHASE SPLITTER	+210	—	+62* +42□	+64
V4 6BJ8 Comp Amp + Rect	+110	—	0	+4.9
V5 " " " "	+110	—	0	+4.9
V6 6SK7 DRIVER + COMP.	+145	+42	-.1	+1.5
V7 " " "	+145	+42	-.1	+1.5
V8 6550 PWR AMP	+570	+300	-33	0
V9 " " "	+570	+300	-33	0
V10 6V6 VOLTAGE REG	+570	+570	+285	+300
V11 GZ34 RECTIFIER	470V RMS	—	—	+580

ALL DC READINGS WITH 11 MEG VTVM MEASURED TO CHASSIS

TUBE LOCATION PRE AMP

12AX7	12AX7	12AY7	6BJ8	6BJ8
V1	V2	V3	V4	V5

TUBE LOCATION MAIN AMP.

6SK7	6SK7	6550	6550	6V6	GZ34
V6	V7	V8	V9	V10	V11

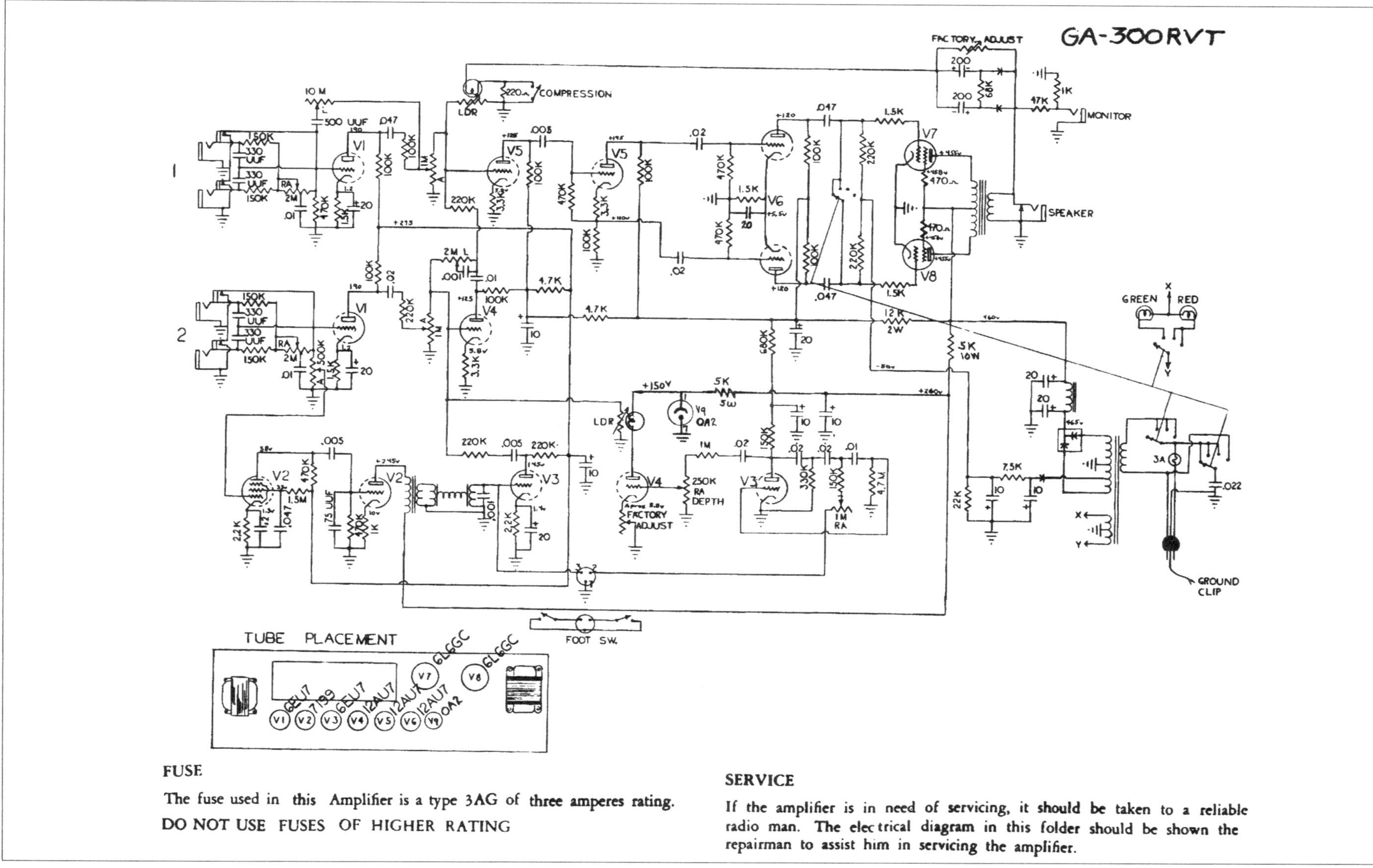

FUSE

The fuse used in this Amplifier is a type 3AG of three amperes rating.
DO NOT USE FUSES OF HIGHER RATING

SERVICE

If the amplifier is in need of servicing, it should be taken to a reliable radio man. The electrical diagram in this folder should be shown the repairman to assist him in servicing the amplifier.

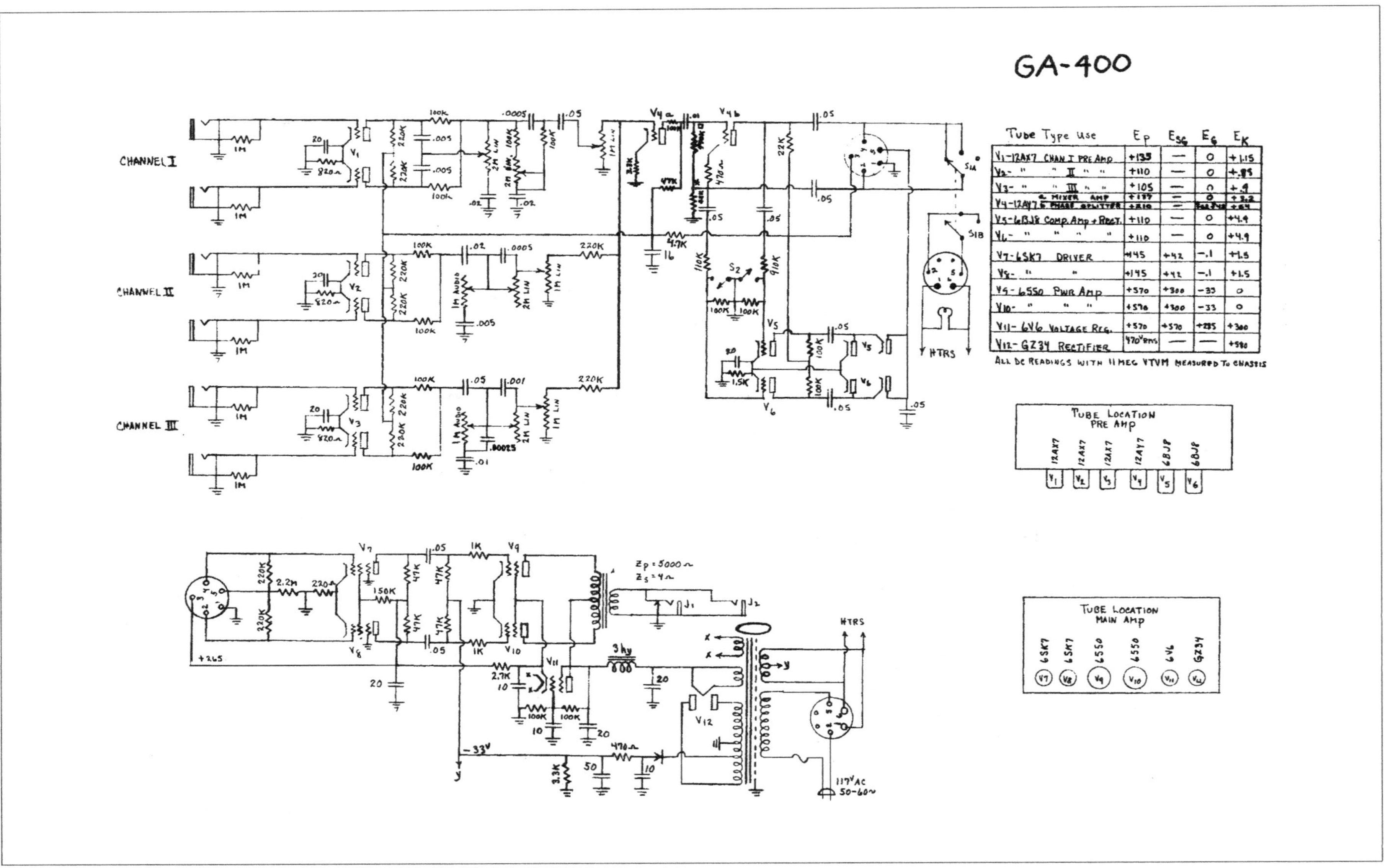
GA-400
CHANNEL I
CHANNEL II
CHANNEL III
Tube Type Use | Ep | Esg | Eg | Ek
V1-12AX7 CHAN I PRE AMP | +135 | — | 0 | +1.15
V2- " " II " " | +110 | — | 0 | +.85
V3- " " III " " | +105 | — | 0 | +.9
V4-12AY7 a MIXER AMP | +137 | — | 0 | +3.2
b PHASE SPLITTER | +210 | — | | +64
V5-6BJ8 COMP. AMP + RECT. | +110 | — | 0 | +4.9
V6- " " " " | +110 | — | 0 | +4.9
V7-6SK7 DRIVER | +145 | +42 | -.1 | +1.5
V8- " " | +145 | +42 | -.1 | +1.5
V9-6550 PWR AMP | +570 | +300 | -33 | 0
V10- " " " | +570 | +300 | -33 | 0
V11-6V6 VOLTAGE REG. | +570 | +570 | +285 | +300
V12-GZ34 RECTIFIER | 470V RMS | — | — | +580
ALL DC READINGS WITH 11 MEG VTVM MEASURED TO CHASSIS
TUBE LOCATION PRE AMP
12AX7 V1
12AX7 V2
12AX7 V3
12AY7 V4
6BJ8 V5
6BJ8 V6
TUBE LOCATION MAIN AMP
6SK7 V7
6SK7 V8
6550 V9
6550 V10
6V6 V11
GZ34 V12
Zp = 5000 Ω
Zs = 4 Ω
J1
J2
HTRS
117V AC 50-60~
S1A
S1B
S2
3 hy
-33V
+265

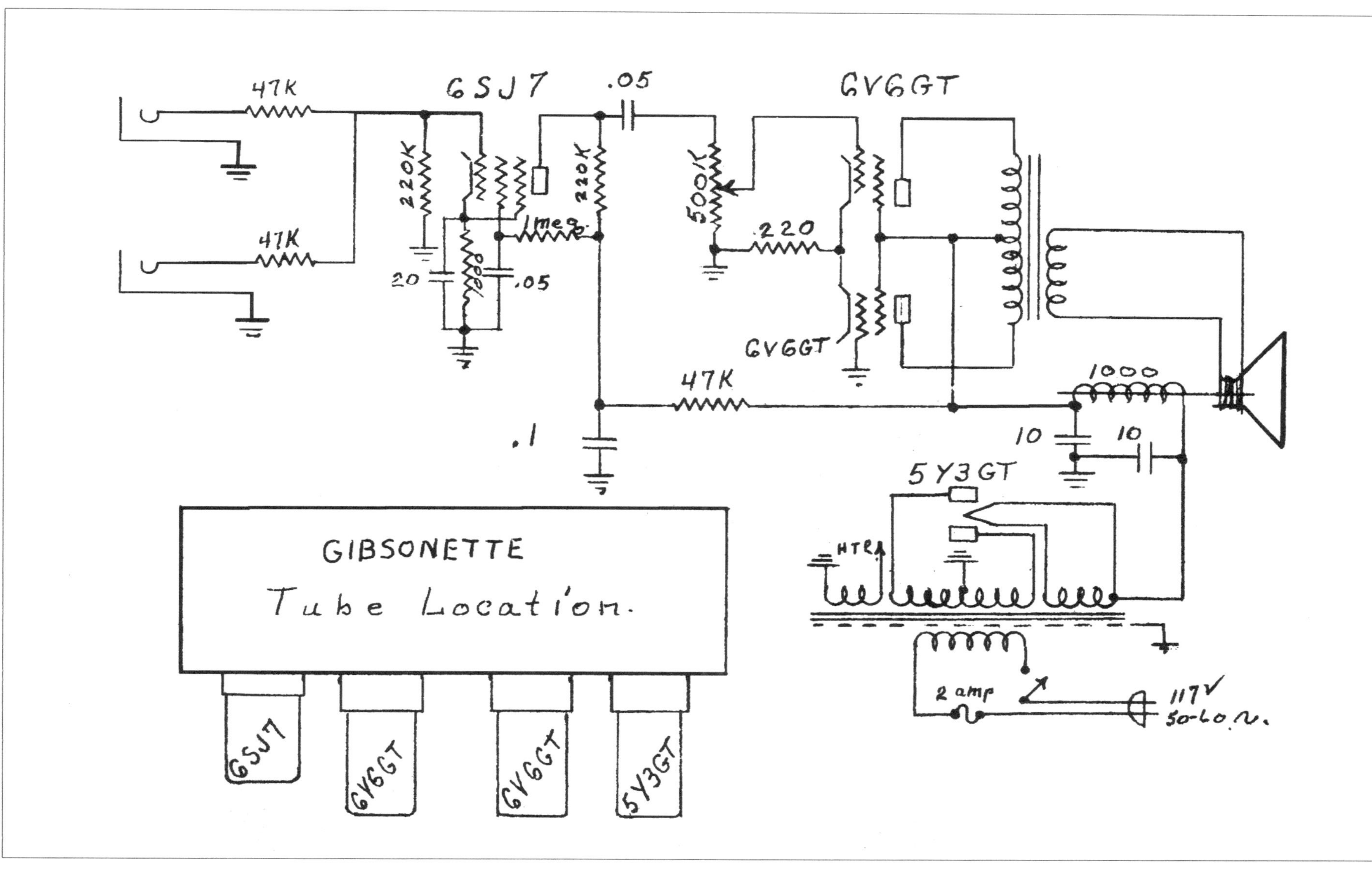
47K
6SJ7
.05
6V6GT
220K
220K
500K
1meg.
47K
20
.05
.220
6V6GT
47K
1000
.1
10
10
5Y3GT
HTR
GIBSONETTE
Tube Location.
2 amp
117V
50-60 ~.
6SJ7
6V6GT
6V6GT
5Y3GT

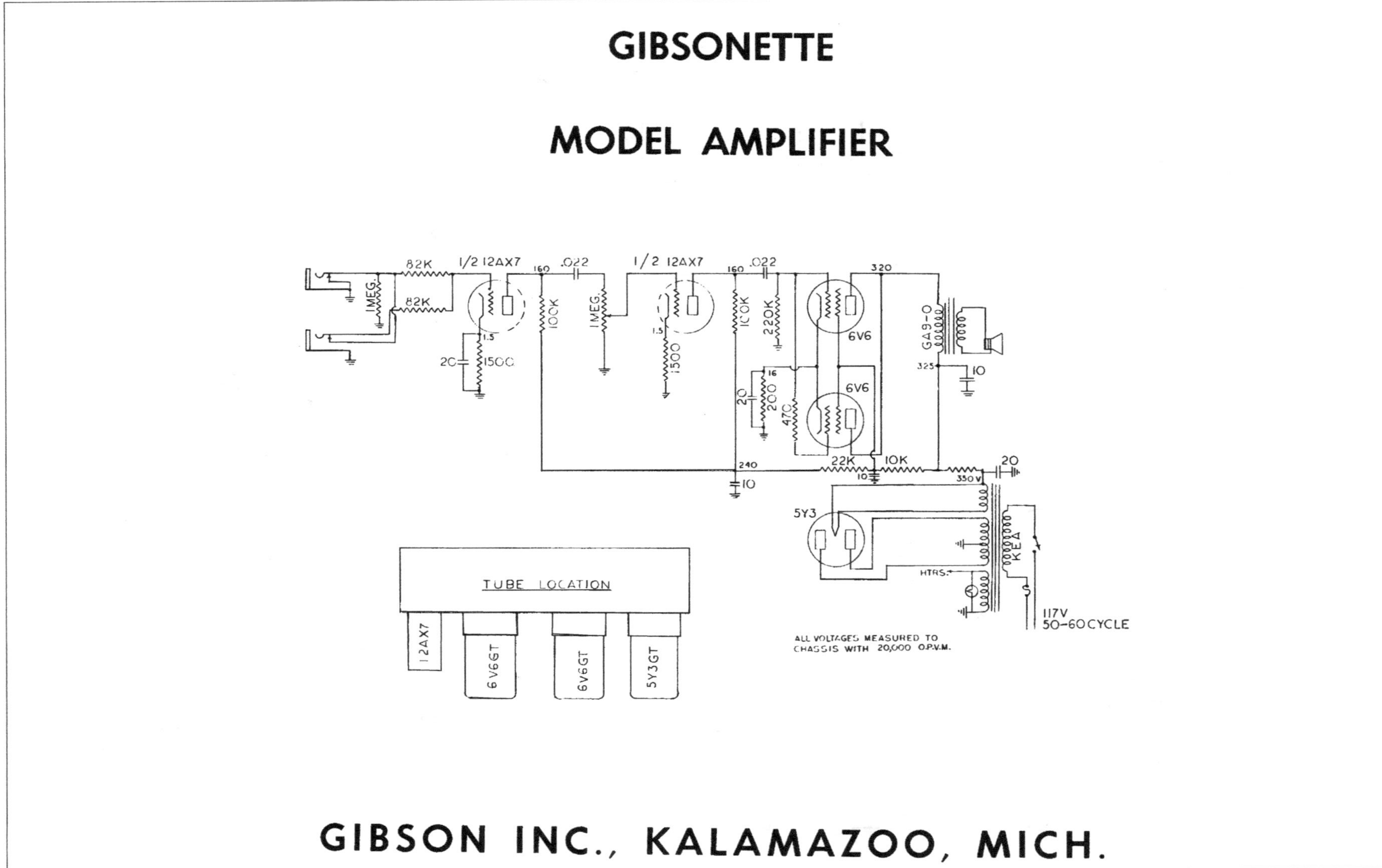
GIBSONETTE
MODEL AMPLIFIER
82K
82K
1MEG.
1/2 12AX7
1.5
20
1500
160
.022
100K
1MEG.
1/2 12AX7
1.5
1500
160
.022
100K
220K
16
20
200
470
6V6
6V6
320
GA9-O
325
10
240
10
22K
10
10K
350V
20
5Y3
HTRS.
KEA
117V
50-60CYCLE
ALL VOLTAGES MEASURED TO
CHASSIS WITH 20,000 O.P.V.M.
TUBE LOCATION
12AX7
6V6GT
6V6GT
5Y3GT
GIBSON INC., KALAMAZOO, MICH.

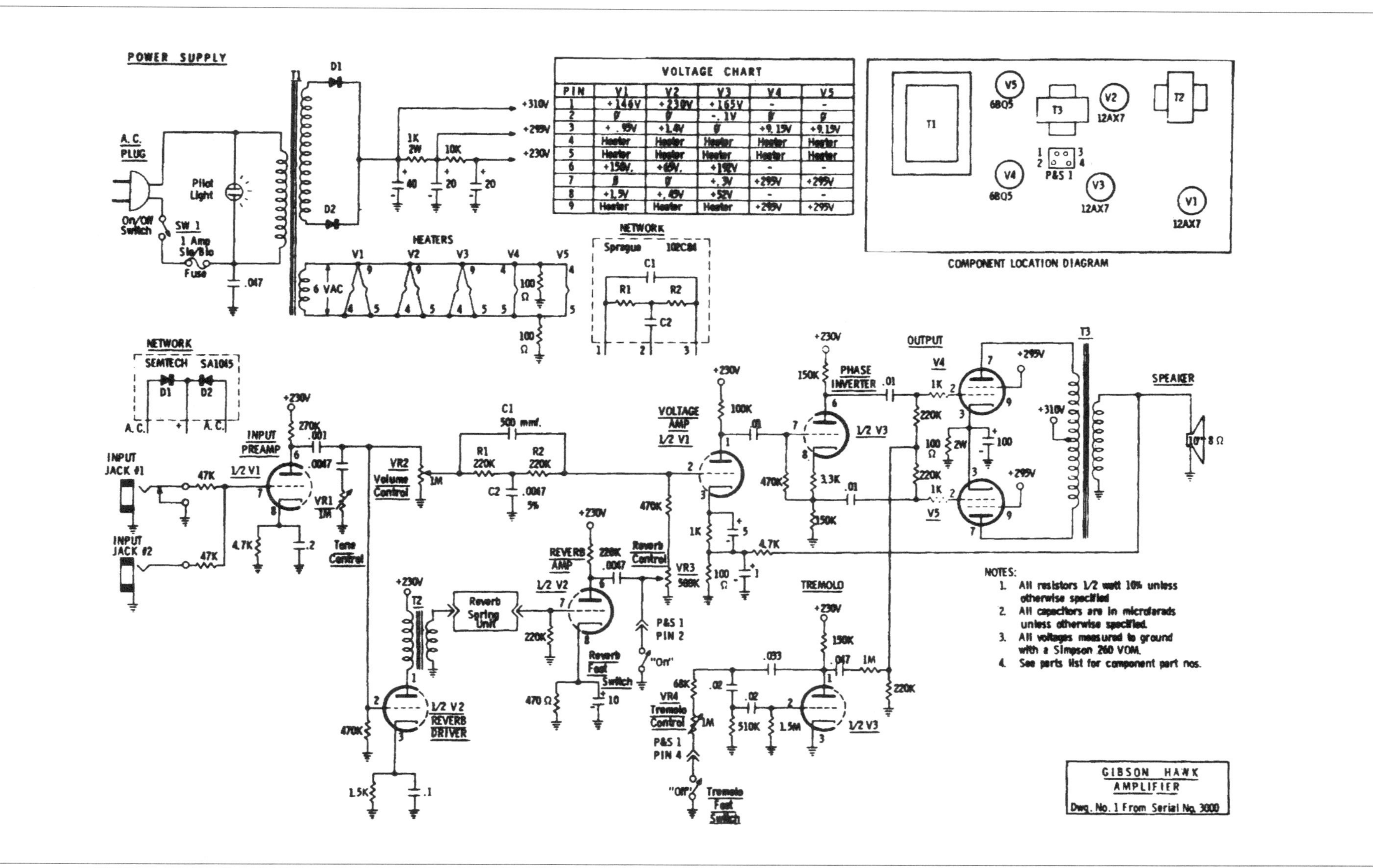

VOLTAGE CHART

PIN	V1	V2	V3	V4	V5
1	+146V	+230V	+165V	-	-
2	Ø	Ø	-.1V	Ø	Ø
3	+.95V	+1.4V	Ø	+9.15V	+9.15V
4	Heater	Heater	Heater	Heater	Heater
5	Heater	Heater	Heater	Heater	Heater
6	+150V.	+65V.	+1[illegible]2V	-	-
7	Ø	Ø	+.3V	+295V	+295V
8	+1.5V	+.45V	+52V	-	-
9	Heater	Heater	Heater	+295V	+295V

NOTES:

1. All resistors 1/2 watt 10% unless otherwise specified
2. All capacitors are in microfarads unless otherwise specified.
3. All voltages measured to ground with a Simpson 260 VOM.
4. See parts list for component part nos.

GIBSON HAWK AMPLIFIER
Dwg. No. 1 From Serial No. 3000

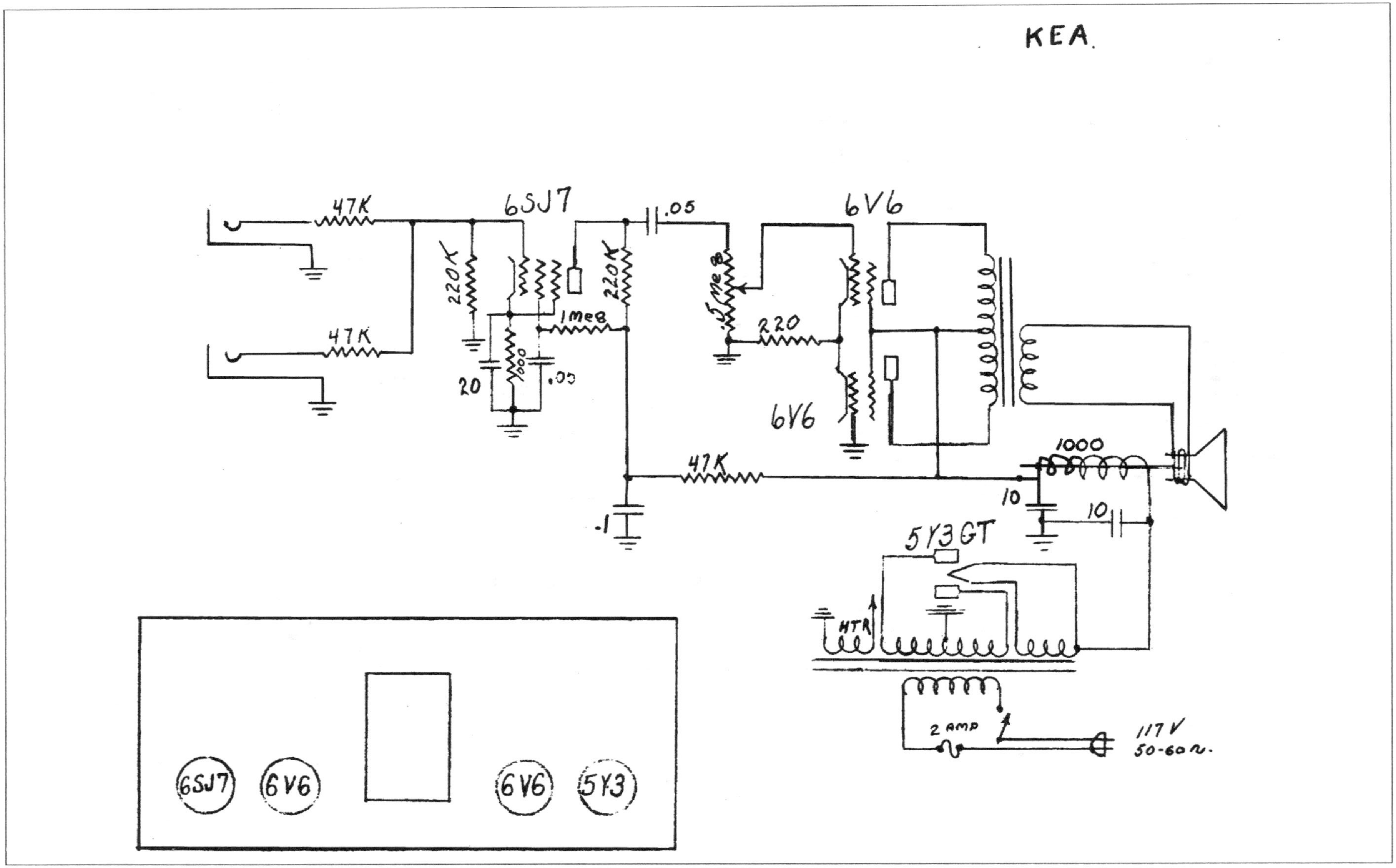

KEA.

47K
47K
220K
6SJ7
20
.00
1meg
220K
.05
.5 meg
220
6V6
6V6
47K
.1
1000
10
10
5Y3GT
HTR
2 AMP
117V
50-60~
6SJ7
6V6
6V6
5Y3

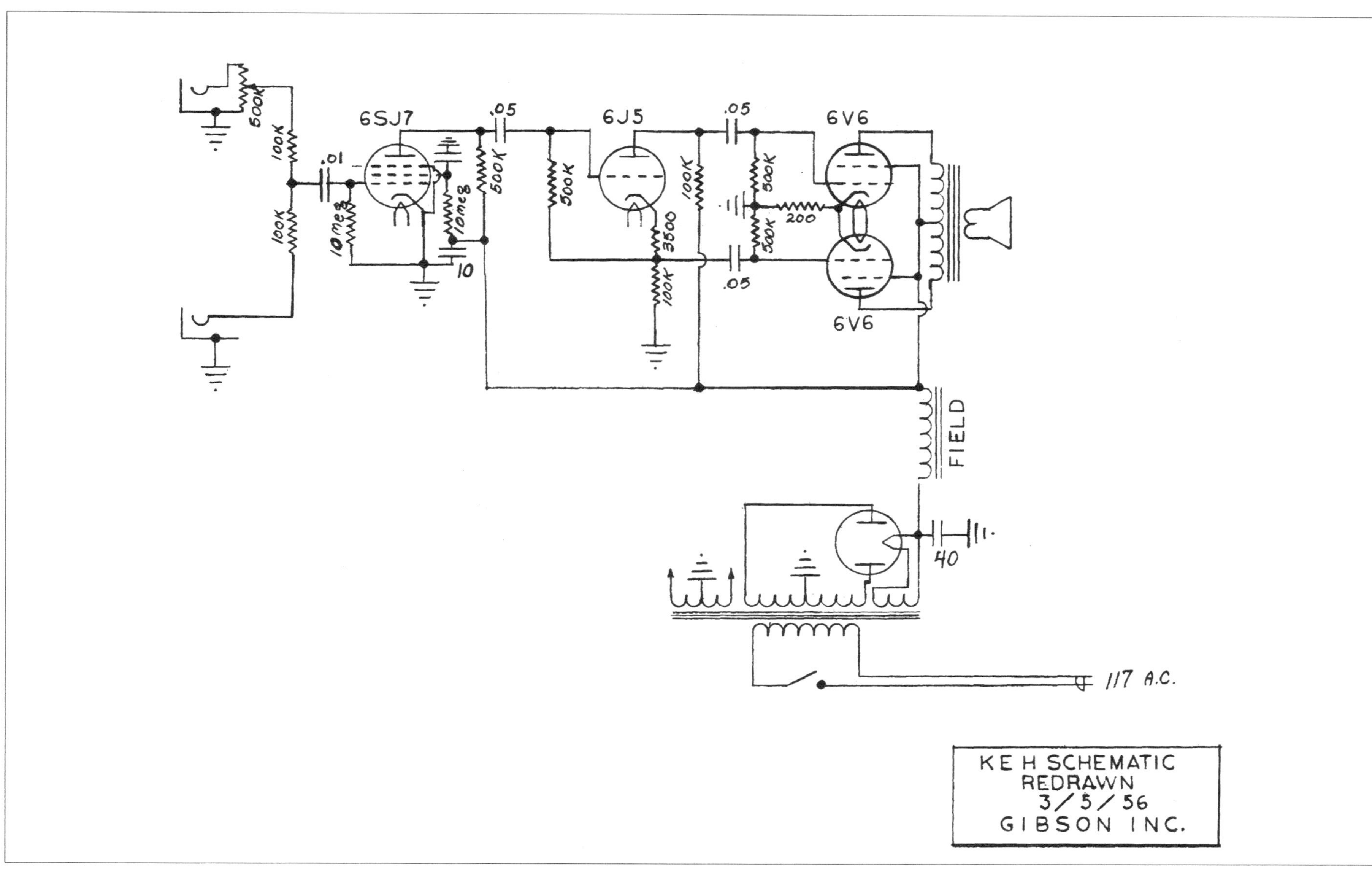
500K
100K
100K
.01
10meg
6SJ7
10meg
10
.05
500K
6J5
500K
3500
100K
100K
.05
.05
500K
500K
200
6V6
6V6
FIELD
40
117 A.C.
KE H SCHEMATIC
REDRAWN
3/5/56
GIBSON INC.

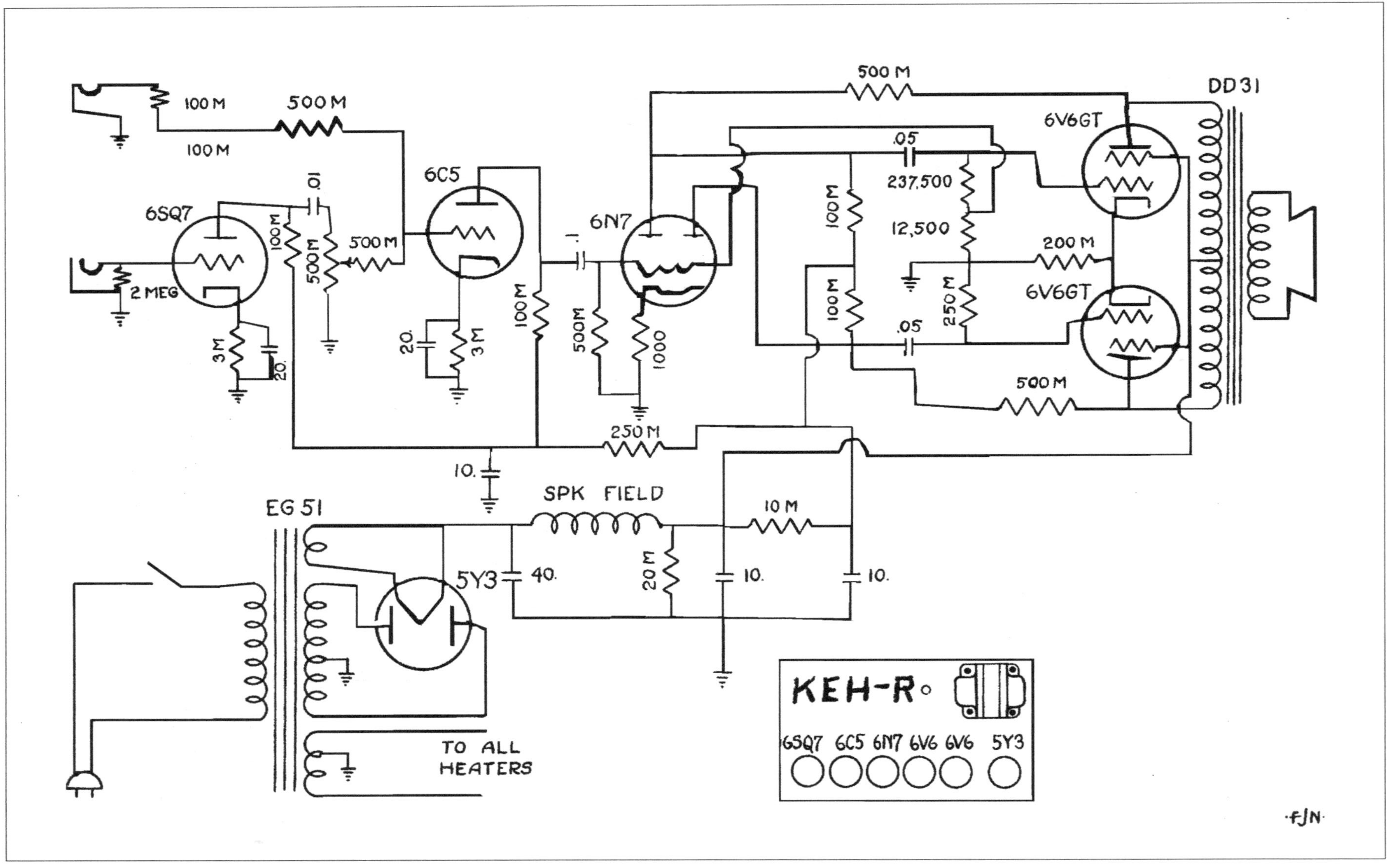
100 M
500 M
100 M
6SQ7
2 MEG
3 M
20.
.01
100 M
500 M
500 M
6C5
20.
3 M
100 M
.1
6N7
500 M
1000
250 M
10.
500 M
.05
237,500
12,500
100 M
100 M
200 M
250 M
.05
500 M
6V6GT
6V6GT
DD 31
EG 51
SPK FIELD
10 M
5Y3
40.
20 M
10.
10.
TO ALL
HEATERS
KEH-R
6SQ7 6C5 6N7 6V6 6V6 5Y3
·FJN·

DUO-MEDALIST "A"

Gibson

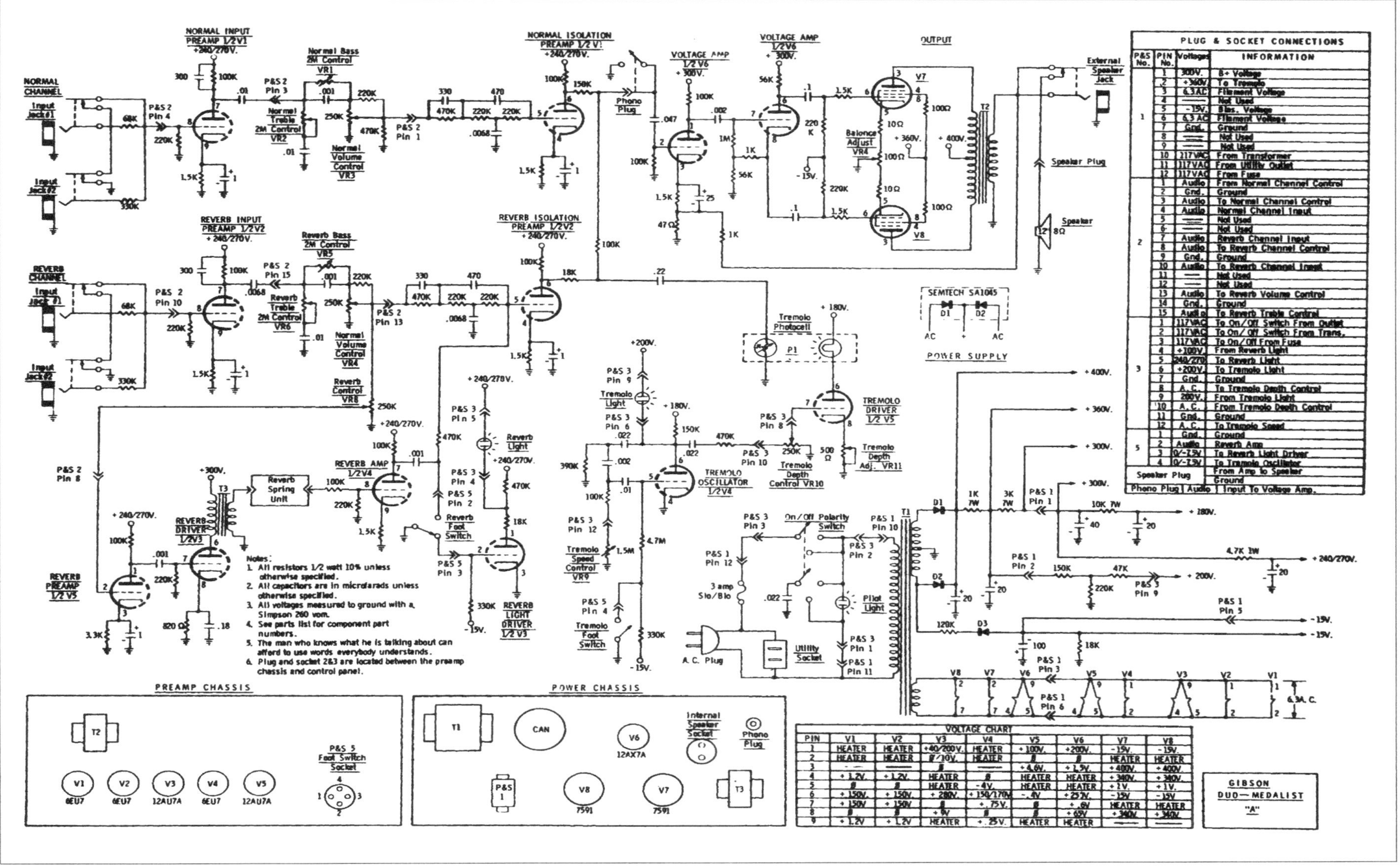

Notes:
1. All resistors 1/2 watt 10% unless otherwise specified.
2. All capacitors are in microfarads unless otherwise specified.
3. All voltages measured to ground with a Simpson 260 vom.
4. See parts list for component part numbers.
5. The man who knows what he is talking about can afford to use words everybody understands.
6. Plug and socket 2&3 are located between the preamp chassis and control panel.

PLUG & SOCKET CONNECTIONS

P&S No.	PIN No.	Voltages	INFORMATION
1	1	300V.	B+ Voltage
	2	+360V.	To Tremolo
	3	6.3AC	Filament Voltage
	4	——	Not Used
	5	-15V.	Bias. Voltage
	6	6.3AC	Filament Voltage
	7	Gnd.	Ground
	8	——	Not Used
	9	——	Not Used
	10	117VAC	From Transformer
	11	117VAC	From Utility Outlet
	12	117VAC	From Fuse
2	1	Audio	From Normal Channel Control
	2	Gnd.	Ground
	3	Audio	To Normal Channel Control
	4	Audio	Normal Channel Input
	5	——	Not Used
	6	——	Not Used
	7	Audio	Reverb Channel Input
	8	Audio	To Reverb Channel Control
	9	Gnd.	Ground
	10	Audio	To Reverb Channel Input
	11	——	Not Used
	12	——	Not Used
	13	Audio	To Reverb Volume Control
	14	Gnd.	Ground
	15	Audio	To Reverb Treble Control
3	1	117VAC	To On/Off Switch From Outlet
	2	117VAC	To On/Off Switch From Trans.
	3	117VAC	To On/Off From Fuse
	4	+100V.	From Reverb Light
	5	240/270	To Reverb Light
	6	+200V.	To Tremolo Light
	7	Gnd.	Ground
	8	A.C.	To Tremolo Depth Control
	9	200V.	From Tremolo Light
	10	A.C.	From Tremolo Depth Control
	11	Gnd.	Ground
	12	A.C.	To Tremolo Speed
5	1	Gnd.	Ground
	2	Audio	Reverb Amp
	3	0/-7.5V	To Reverb Light Driver
	4	0/-7.5V	To Tremolo Oscillator
Speaker Plug			From Amp to Speaker
			Ground
Phono Plug		Audio	Input To Voltage Amp.

VOLTAGE CHART

PIN	V1	V2	V3	V4	V5	V6	V7	V8
1	HEATER	HEATER	+40/200V.	HEATER	+100V.	+200V.	-15V.	-15V.
2	HEATER	HEATER	Ø/10V.	HEATER	Ø	Ø	HEATER	HEATER
3	——	——	Ø	——	+4.6V.	+1.5V.	+400V.	+400V.
4	+1.2V.	+1.2V.	HEATER	Ø	HEATER	HEATER	+340V.	+340V.
5	Ø	Ø	HEATER	-4V.	HEATER	HEATER	+1V.	+1V.
6	+150V.	+150V.	+280V.	+150/170V.	-.4V	+257V.	-15V	-15V
7	+150V	+150V	Ø	+.75V.	Ø	+.6V	HEATER	HEATER
8	Ø	Ø	+9V	Ø	Ø	+65V	+340V	+340V
9	+1.2V	+1.2V	HEATER	+.25V.	HEATER	HEATER	——	——

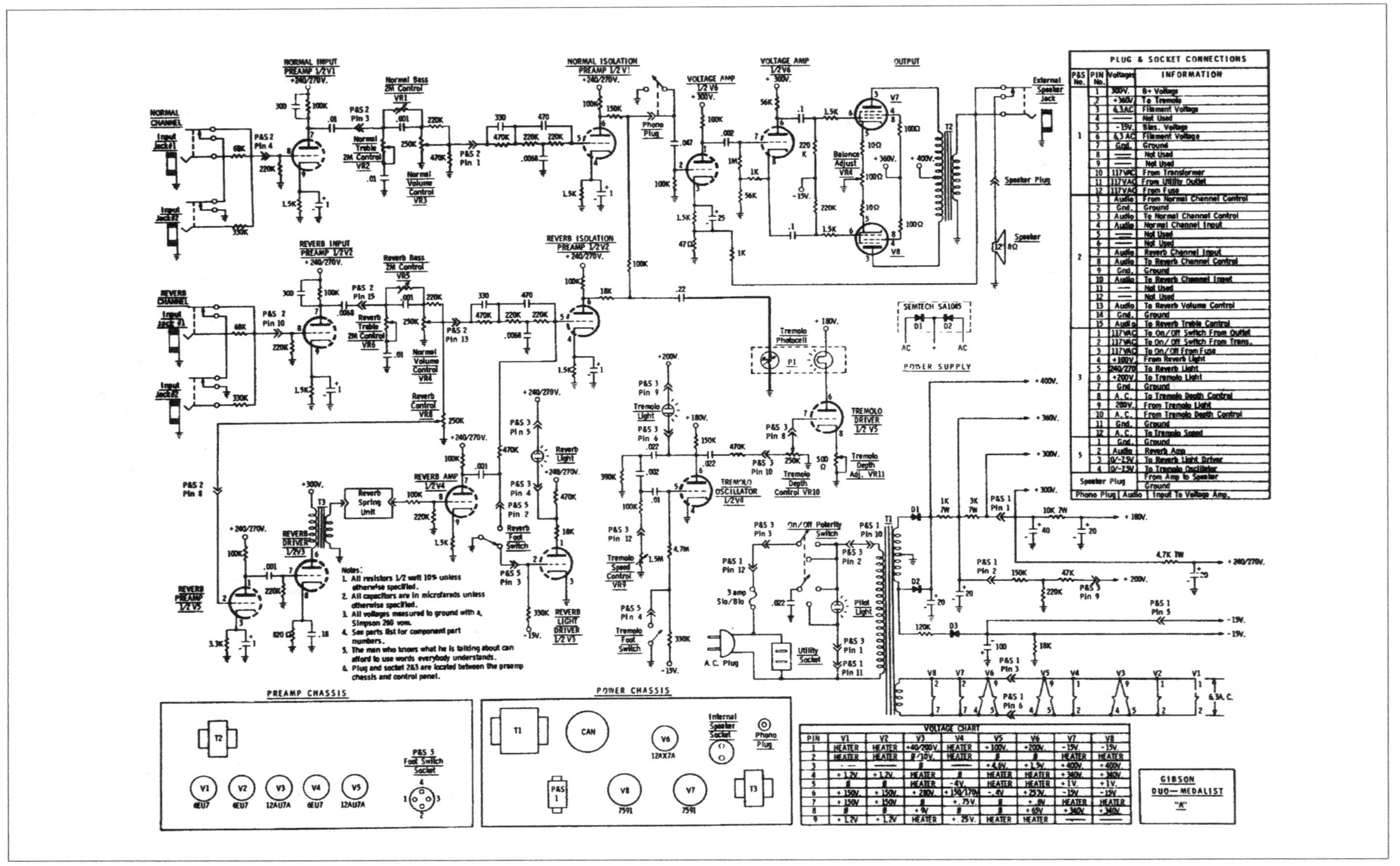
NORMAL INPUT PREAMP 1/2 V1
NORMAL ISOLATION PREAMP 1/2 V1
VOLTAGE AMP 1/2 V6
OUTPUT
REVERB INPUT PREAMP 1/2 V2
REVERB ISOLATION PREAMP 1/2 V2
REVERB AMP 1/2 V4
REVERB DRIVER 1/2 V3
REVERB PREAMP 1/2 V5
REVERB LIGHT DRIVER 1/2 V3
TREMOLO DRIVER 1/2 V5
TREMOLO OSCILLATOR 1/2 V4
PLUG & SOCKET CONNECTIONS
SEMTECH SA1045
POWER SUPPLY
VOLTAGE CHART
PREAMP CHASSIS
POWER CHASSIS
GIBSON
DUO—MEDALIST
"A"
Notes:
1. All resistors 1/2 watt 10% unless otherwise specified.
2. All capacitors are in microfarads unless otherwise specified.
3. All voltages measured to ground with a, Simpson 260 vom.
4. See parts list for component part numbers.
5. The man who knows what he is talking about can afford to use words everybody understands.
6. Plug and socket 2&3 are located between the preamp chassis and control panel.

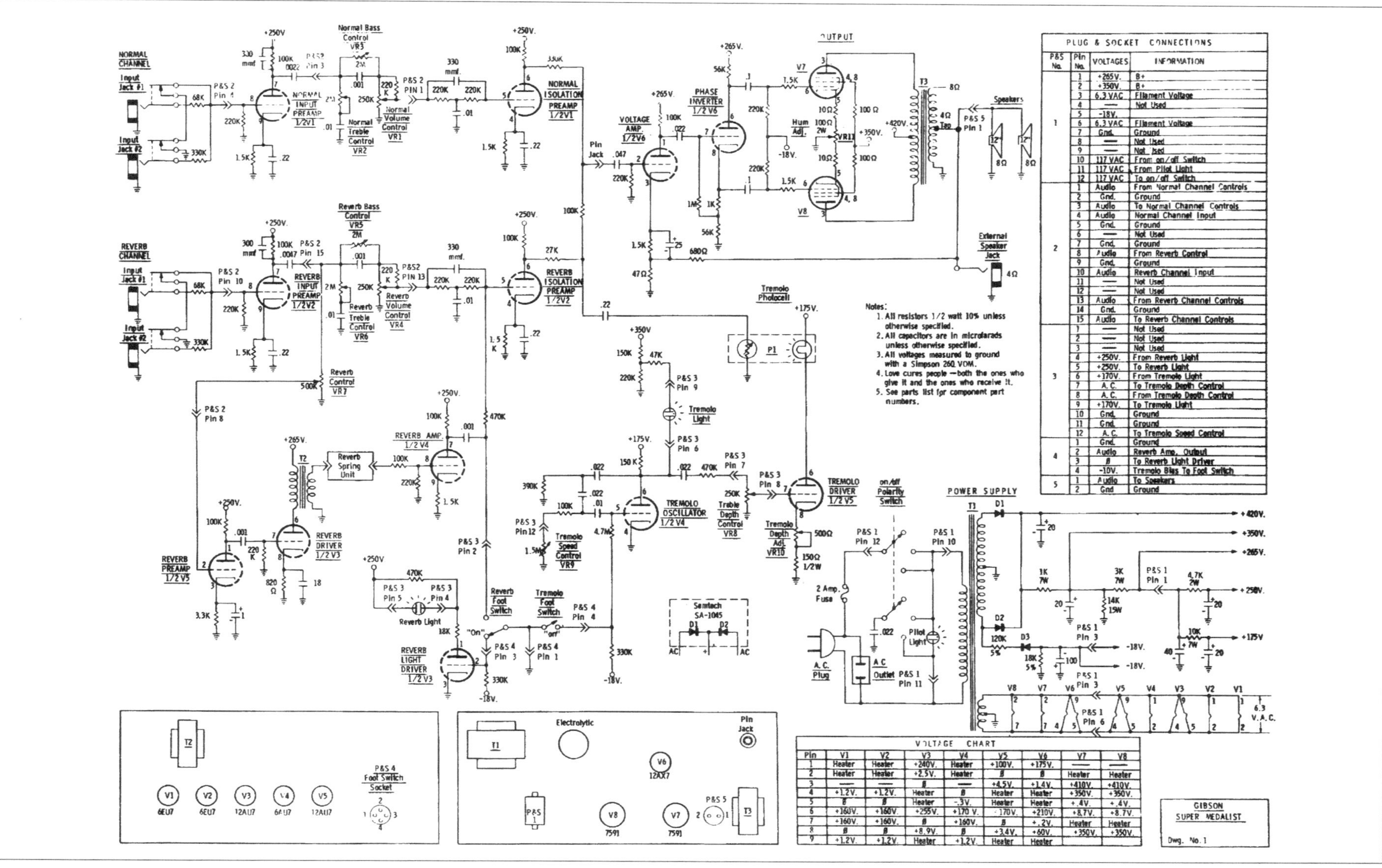

PLUG & SOCKET CONNECTIONS			
P&S No.	Pin No.	VOLTAGES	INFORMATION
1	1	+265V.	B+
	2	+350V.	B+
	3	6.3 VAC	Filament Voltage
	4	—	Not Used
	5	-18V.	
	6	6.3 VAC	Filament Voltage
	7	Gnd.	Ground
	8	—	Not Used
	9	—	Not Used
	10	117 VAC	From on/off Switch
	11	117 VAC	From Pilot Light
	12	117 VAC	To on/off Switch
2	1	Audio	From Normal Channel Controls
	2	Gnd.	Ground
	3	Audio	To Normal Channel Controls
	4	Audio	Normal Channel Input
	5	Gnd.	Ground
	6	—	Not Used
	7	Gnd.	Ground
	8	Audio	From Reverb Control
	9	Gnd.	Ground
	10	Audio	Reverb Channel Input
	11	—	Not Used
	12	—	Not Used
	13	Audio	From Reverb Channel Controls
	14	Gnd.	Ground
	15	Audio	To Reverb Channel Controls
3	1	—	Not Used
	2	—	Not Used
	3	—	Not Used
	4	+250V.	From Reverb Light
	5	+250V.	To Reverb Light
	6	+170V.	From Tremolo Light
	7	A.C.	To Tremolo Depth Control
	8	A.C.	From Tremolo Depth Control
	9	+170V.	To Tremolo Light
	10	Gnd.	Ground
	11	Gnd.	Ground
	12	A.C.	To Tremolo Speed Control
4	1	Gnd.	Ground
	2	Audio	Reverb Amp. Output
	3	Ø	To Reverb Light Driver
	4	-10V.	Tremolo Bias To Foot Switch
5	1	Audio	To Speakers
	2	Gnd	Ground

Notes:
1. All resistors 1/2 watt 10% unless otherwise specified.
2. All capacitors are in microfarads unless otherwise specified.
3. All voltages measured to ground with a Simpson 260 VOM.
4. Love cures people —both the ones who give it and the ones who receive it.
5. See parts list for component part numbers.

VOLTAGE CHART								
Pin	V1	V2	V3	V4	V5	V6	V7	V8
1	Heater	Heater	+240V.	Heater	+100V.	+175V.	—	—
2	Heater	Heater	+2.5V.	Heater	Ø	Ø	Heater	Heater
3	—	—	Ø	—	+4.5V.	+1.4V.	+410V.	+410V.
4	+1.2V.	+1.2V.	Heater	Ø	Heater	Heater	+350V.	+350V.
5	Ø	Ø	Heater	-.3V.	Heater	Heater	+.4V.	+.4V.
6	+160V.	+160V.	+255V.	+170 V.	170V.	+210V.	+8.7V.	+8.7V.
7	+160V.	+160V.	Ø	+160V.	Ø	+.2V.	Heater	Heater
8	Ø	Ø	+8.9V.	Ø	+3.4V.	+60V.	+350V.	+350V.
9	+1.2V.	+1.2V.	Heater	+1.2V.	Heater	Heater		

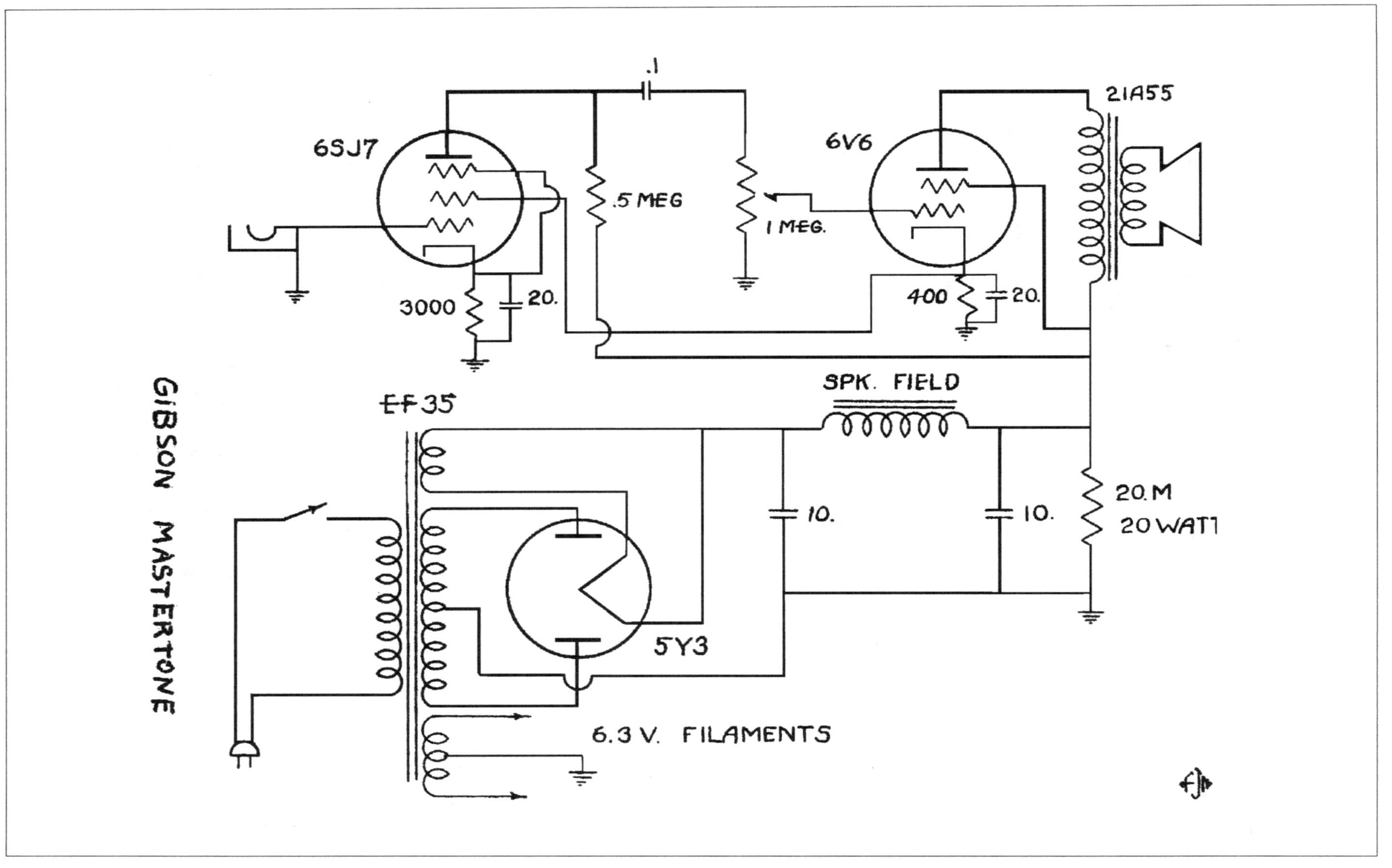
6SJ7
.1
6V6
21A55
.5 MEG
1 MEG.
3000
20.
400
20.
EF35
SPK. FIELD
10.
10.
20.M
20 WATT
5Y3
6.3 V. FILAMENTS
GIBSON MASTERTONE

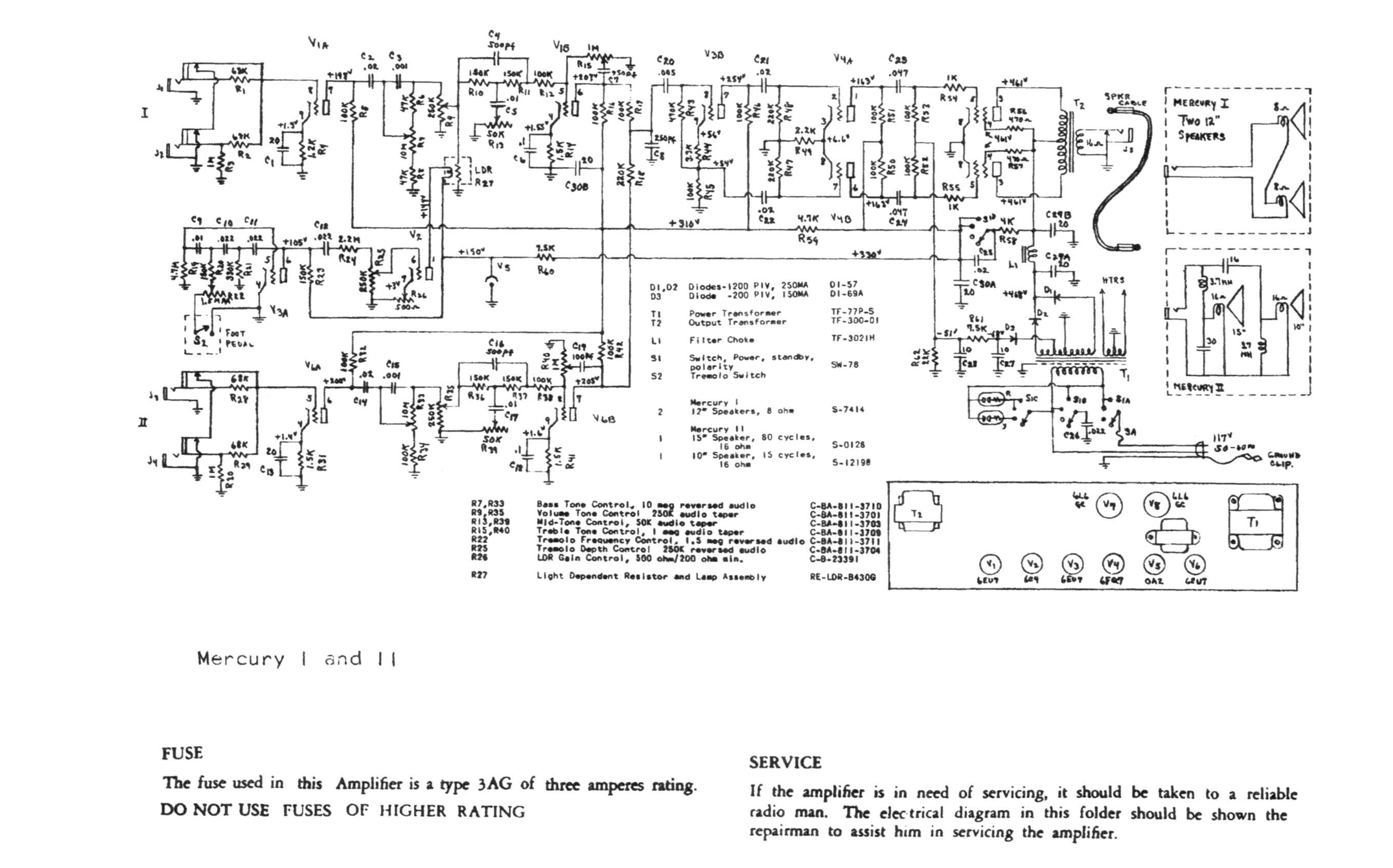

D1, D2	Diodes-1200 PIV, 250MA	D1-57
D3	Diode -200 PIV, 150MA	D1-69A
T1	Power Transformer	TF-77P-S
T2	Output Transformer	TF-300-01
L1	Filter Choke	TF-3021H
S1	Switch, Power, standby, polarity	SW-78
S2	Tremolo Switch	
2	Mercury I 12" Speakers, 8 ohm	S-7414
	Mercury II	
1	15" Speaker, 80 cycles, 16 ohm	S-0128
1	10" Speaker, 15 cycles, 16 ohm	S-12198

R7, R33	Bass Tone Control, 10 meg reversed audio	C-BA-811-3710
R9, R35	Volume Tone Control 250K audio taper	C-BA-811-3701
R13, R39	Mid-Tone Control, 50K audio taper	C-BA-811-3703
R15, R40	Treble Tone Control, 1 meg audio taper	C-BA-811-3709
R22	Tremolo Frequency Control, 1.5 meg reversed audio	C-BA-811-3711
R25	Tremolo Depth Control 250K reversed audio	C-BA-811-3704
R26	LDR Gain Control, 500 ohm/200 ohm min.	C-B-23391
R27	Light Dependent Resistor and Lamp Assembly	RE-LDR-B430G

Mercury I and II

FUSE

The fuse used in this Amplifier is a type 3AG of three amperes rating.
DO NOT USE FUSES OF HIGHER RATING

SERVICE

If the amplifier is in need of servicing, it should be taken to a reliable radio man. The electrical diagram in this folder should be shown the repairman to assist him in servicing the amplifier.

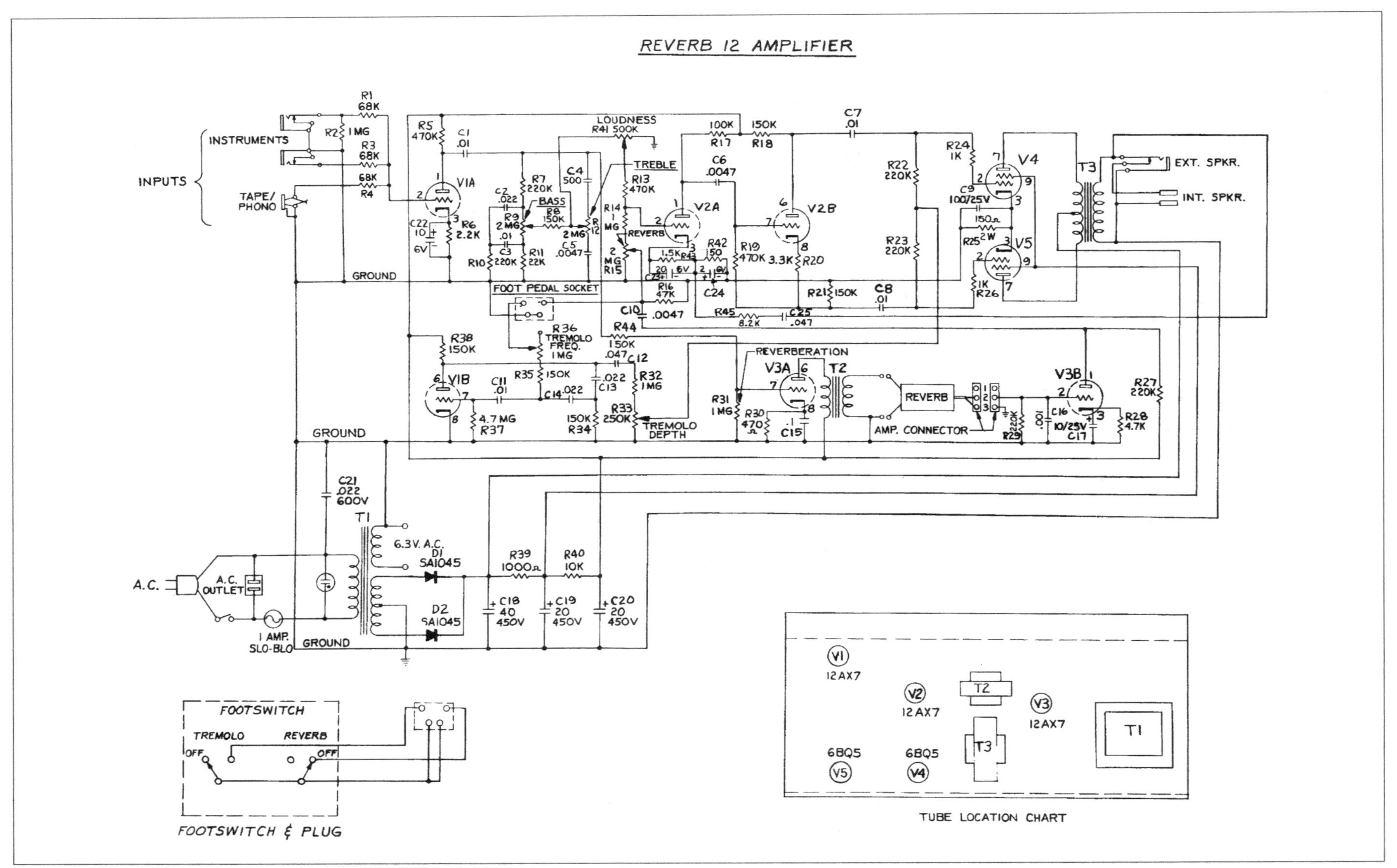
REVERB 12 AMPLIFIER
INPUTS
INSTRUMENTS
TAPE/
PHONO
R1
68K
R2 1MG
R3
68K
68K
R4
GROUND
R5
470K
C1
.01
V1A
C22
10
6V
R6
2.2K
C2
.022
R7
220K
R9
2MG
BASS
R8
150K
C3
.01
R10 220K
R11
22K
C4
500
2MG
R12
C5
.0047
LOUDNESS
R41 500K
TREBLE
R13
470K
R14
1
MG
REVERB
2
MG
R15
FOOT PEDAL SOCKET
100K
R17
150K
R18
C6
.0047
V2A
R42
150
1.5K R43
20
C23
6V
C24
R16
47K
C10
.0047
V2B
R19
470K
3.3K
R20
R21 150K
C8
.01
R45
8.2K
C25
.047
C7
.01
R22
220K
R23
220K
R24
1K
V4
C9
100/25V
150Ω
R25 2W
V5
1K
R26
T3
EXT. SPKR.
INT. SPKR.
R36
TREMOLO
FREQ.
1MG
R38
150K
V1B
C11
.01
R35 150K
C14 .022
4.7MG
R37
150K
R34
R44
150K
.047 C12
.022
C13
R32
1MG
R33
250K
TREMOLO
DEPTH
REVERBERATION
V3A
T2
R31
1MG
R30
470Ω
C15
REVERB
AMP. CONNECTOR
R29
.001
V3B
C16
10/25V
C17
R27
220K
R28
4.7K
GROUND
C21
.022
600V
T1
6.3V. A.C.
D1
SA1045
D2
SA1045
A.C.
A.C.
OUTLET
1 AMP.
SLO-BLO
GROUND
R39
1000Ω
R40
10K
C18
40
450V
C19
20
450V
C20
20
450V
FOOTSWITCH
TREMOLO
REVERB
OFF
OFF
FOOTSWITCH & PLUG
V1
12AX7
V2
12AX7
V3
12AX7
6BQ5
V5
6BQ5
V4
T2
T3
T1
TUBE LOCATION CHART

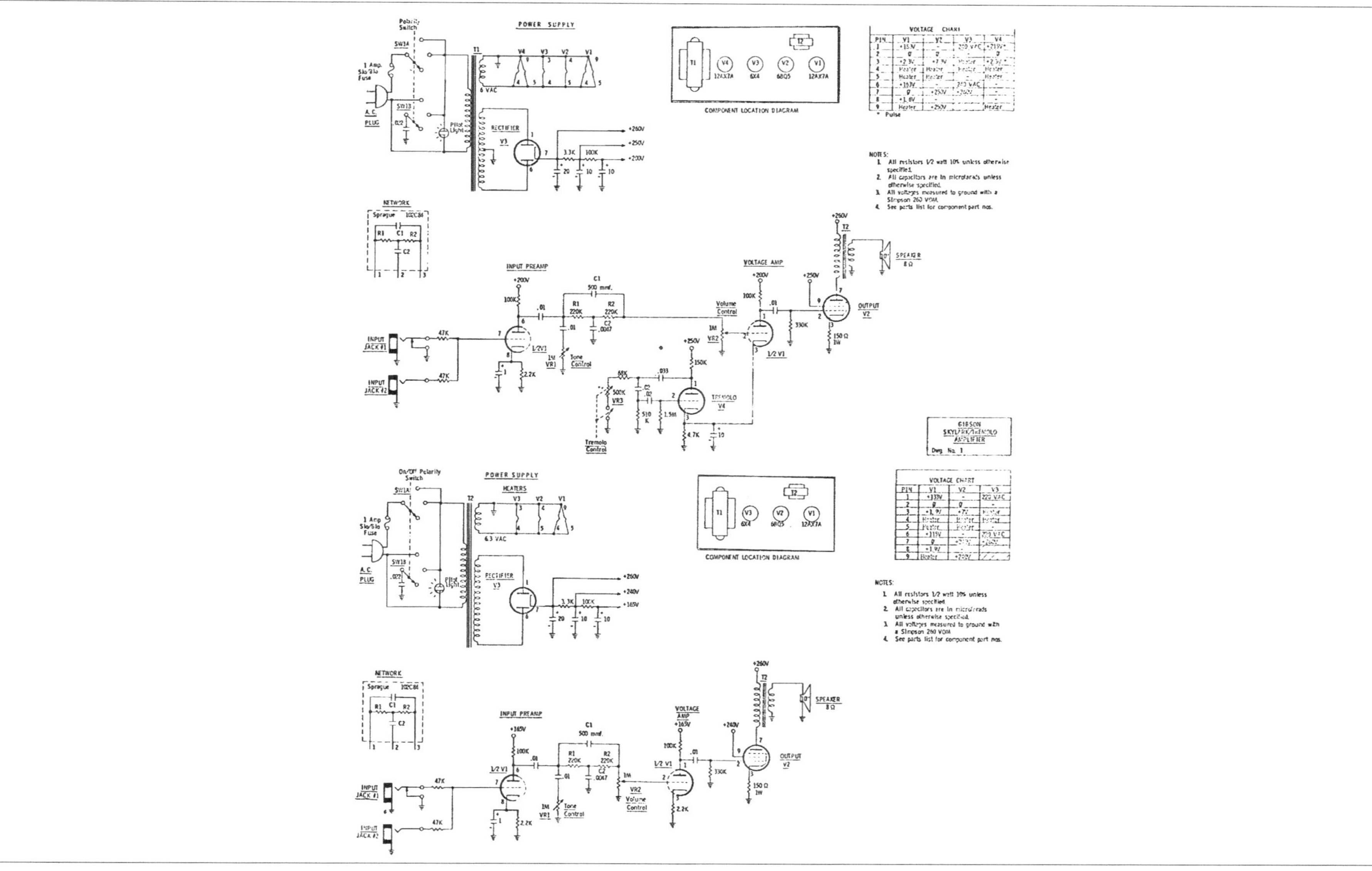
POWER SUPPLY
COMPONENT LOCATION DIAGRAM
NETWORK
INPUT PREAMP
VOLTAGE AMP
OUTPUT
SPEAKER
TREMOLO
GIBSON
SKYLARK/TREMOLO
AMPLIFIER
Dwg. No. 1
NOTES:
1. All resistors 1/2 watt 10% unless otherwise specified.
2. All capacitors are in microfarads unless otherwise specified.
3. All voltages measured to ground with a Simpson 260 VOM.
4. See parts list for component part nos.

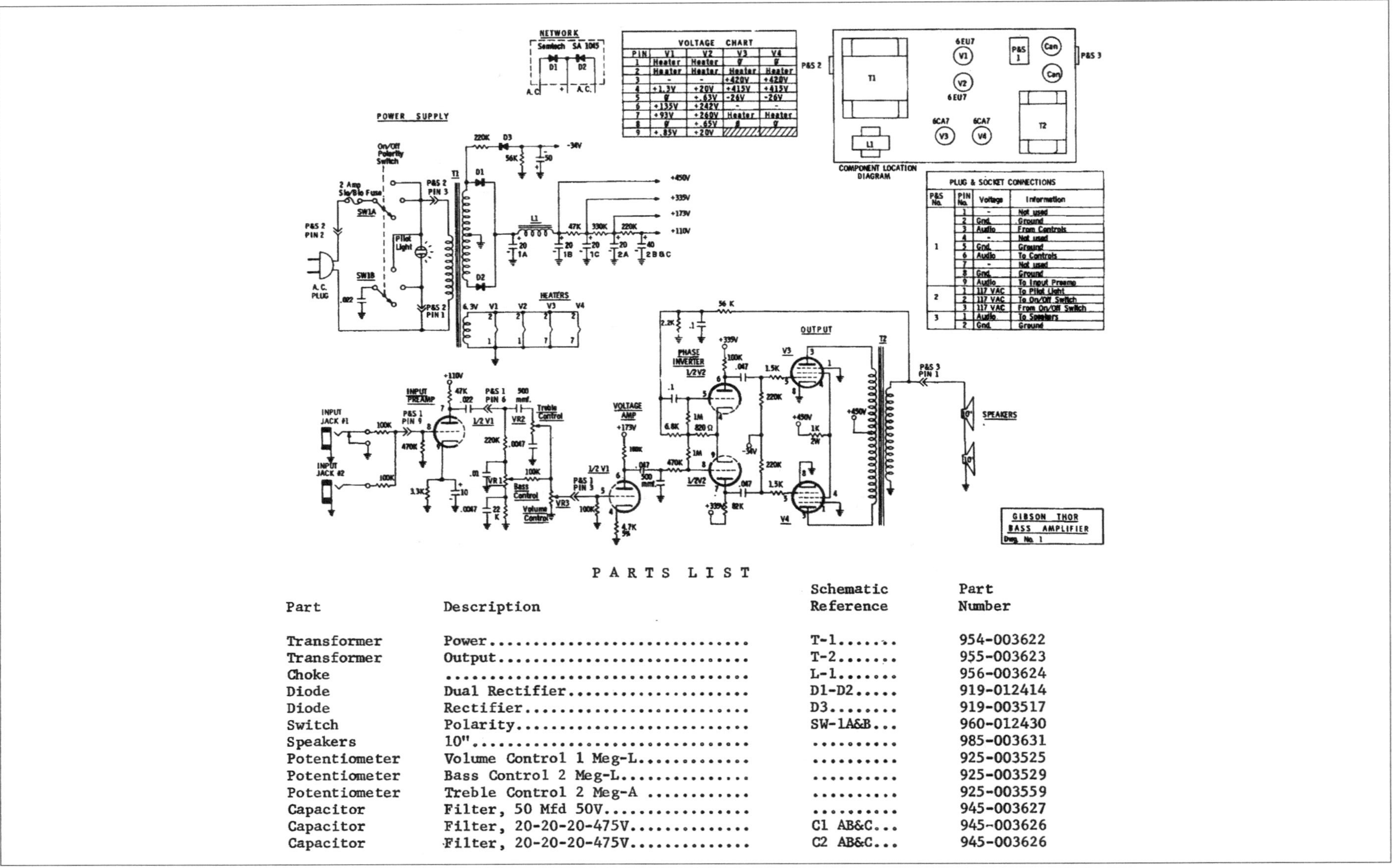

PIN	V1	V2	V3	V4
1	Heater	Heater	∅	∅
2	Heater	Heater	Heater	Heater
3	-	-	+420V	+420V
4	+1.3V	+20V	+415V	+415V
5	∅	+.63V	-26V	-26V
6	+135V	+242V	-	-
7	+93V	+260V	Heater	Heater
8	∅	+.65V	∅	∅
9	+.85V	+20V		

PLUG & SOCKET CONNECTIONS

P&S No.	PIN No.	Voltage	Information
1	1	-	Not used
	2	Gnd.	Ground
	3	Audio	From Controls
	4	-	Not used
	5	Gnd.	Ground
	6	Audio	To Controls
	7	-	Not used
	8	Gnd.	Ground
	9	Audio	To Input Preamp
2	1	117 VAC	To Pilot Light
	2	117 VAC	To On/Off Switch
	3	117 VAC	From On/Off Switch
3	1	Audio	To Speakers
	2	Gnd.	Ground

P A R T S L I S T

Part	Description	Schematic Reference	Part Number
Transformer	Power..............................	T-1.......	954-003622
Transformer	Output.............................	T-2.......	955-003623
Choke	....	L-1.......	956-003624
Diode	Dual Rectifier.....................	D1-D2.....	919-012414
Diode	Rectifier..........................	D3........	919-003517
Switch	Polarity...........................	SW-1A&B...	960-012430
Speakers	10"................................	....	985-003631
Potentiometer	Volume Control 1 Meg-L.............	....	925-003525
Potentiometer	Bass Control 2 Meg-L...............	....	925-003529
Potentiometer	Treble Control 2 Meg-A	....	925-003559
Capacitor	Filter, 50 Mfd 50V.................	....	945-003627
Capacitor	Filter, 20-20-20-475V..............	C1 AB&C...	945-003626
Capacitor	Filter, 20-20-20-475V..............	C2 AB&C...	945-003626

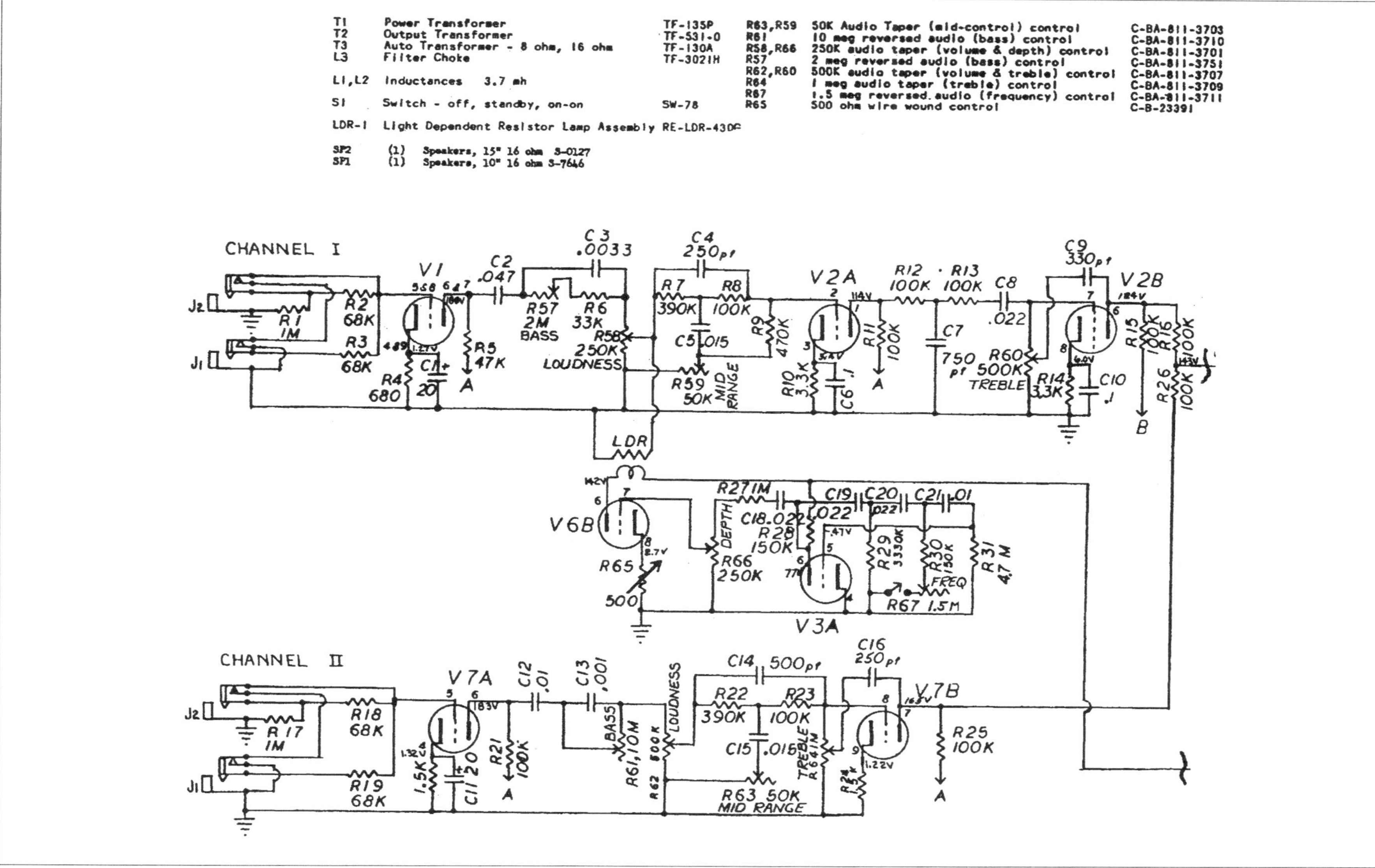
T1 Power Transformer TF-135P
T2 Output Transformer TF-531-0
T3 Auto Transformer - 8 ohm, 16 ohm TF-130A
L3 Filter Choke TF-3021H
L1,L2 Inductances 3.7 mh
S1 Switch - off, standby, on-on SW-78
LDR-1 Light Dependent Resistor Lamp Assembly RE-LDR-430C
SP2 (1) Speakers, 15" 16 ohm S-0127
SP1 (1) Speakers, 10" 16 ohm S-7646
R63,R59 50K Audio Taper (mid-control) control C-BA-811-3703
R61 10 meg reversed audio (bass) control C-BA-811-3710
R58,R66 250K audio taper (volume & depth) control C-BA-811-3701
R57 2 meg reversed audio (bass) control C-BA-811-3751
R62,R60 500K audio taper (volume & treble) control C-BA-811-3707
R64 1 meg audio taper (treble) control C-BA-811-3709
R67 1.5 meg reversed audio (frequency) control C-BA-811-3711
R65 500 ohm wire wound control C-B-23391
CHANNEL I
CHANNEL II
LDR
V1
V2A
V2B
V6B
V3A
V7A
V7B
LOUDNESS
BASS
MID RANGE
TREBLE
DEPTH
FREQ

111

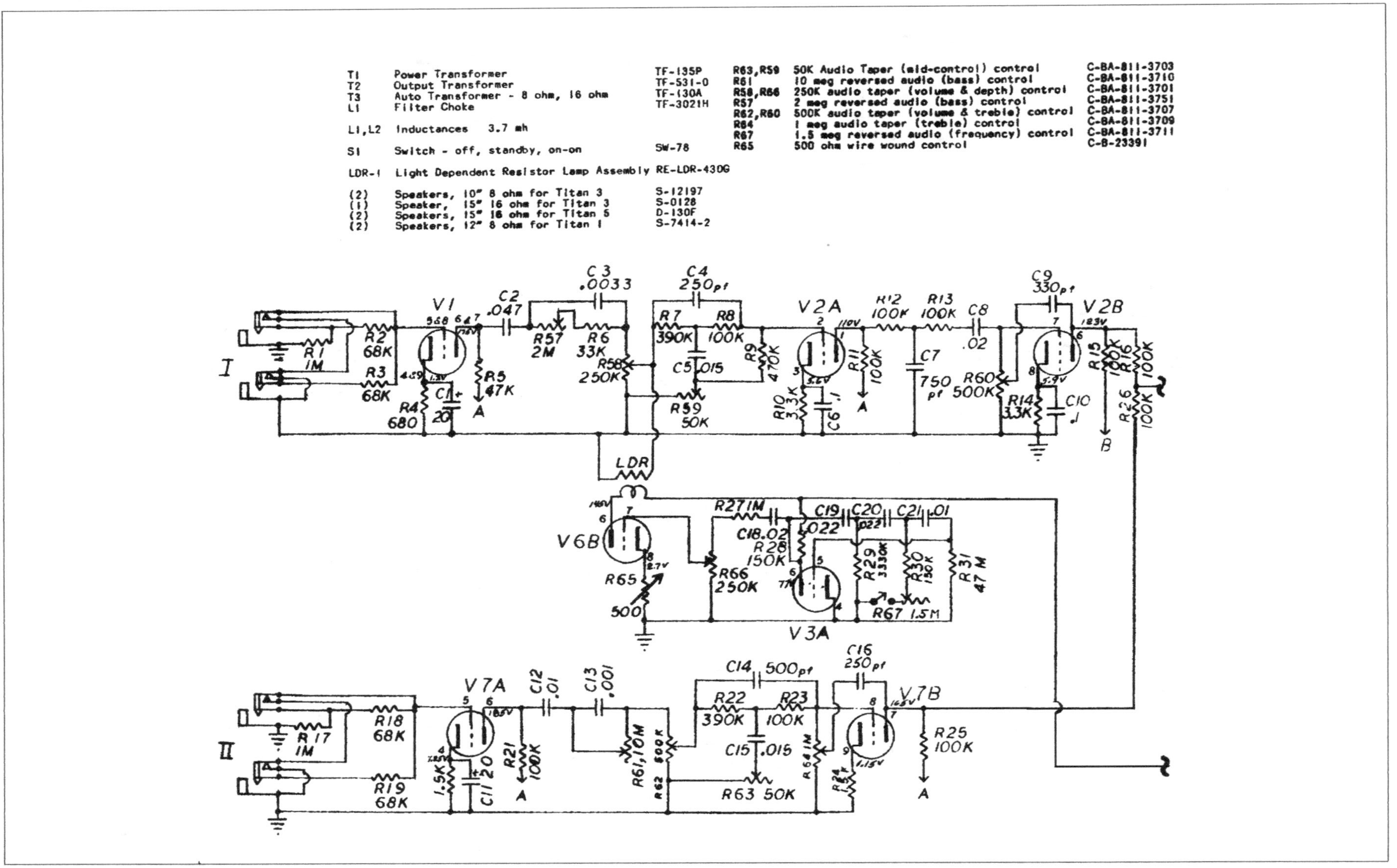
T1 Power Transformer TF-135P
T2 Output Transformer TF-531-0
T3 Auto Transformer - 8 ohm, 16 ohm TF-130A
L1 Filter Choke TF-3021H
L1,L2 Inductances 3.7 mh
S1 Switch - off, standby, on-on SW-78
LDR-1 Light Dependent Resistor Lamp Assembly RE-LDR-430G
(2) Speakers, 10" 8 ohm for Titan 3 S-12197
(1) Speaker, 15" 16 ohm for Titan 3 S-0128
(2) Speakers, 15" 16 ohm for Titan 5 D-130F
(2) Speakers, 12" 8 ohm for Titan 1 S-7414-2
R63,R59 50K Audio Taper (mid-control) control C-BA-811-3703
R61 10 meg reversed audio (bass) control C-BA-811-3710
R58,R66 250K audio taper (volume & depth) control C-BA-811-3701
R57 2 meg reversed audio (bass) control C-BA-811-3751
R62,R60 500K audio taper (volume & treble) control C-BA-811-3707
R64 1 meg audio taper (treble) control C-BA-811-3709
R67 1.5 meg reversed audio (frequency) control C-BA-811-3711
R65 500 ohm wire wound control C-B-23391
I
II
V1
V2A
V2B
V3A
V6B
V7A
V7B
LDR

TITAN I-111 & V **PWR/OUTPUT**

Gibson

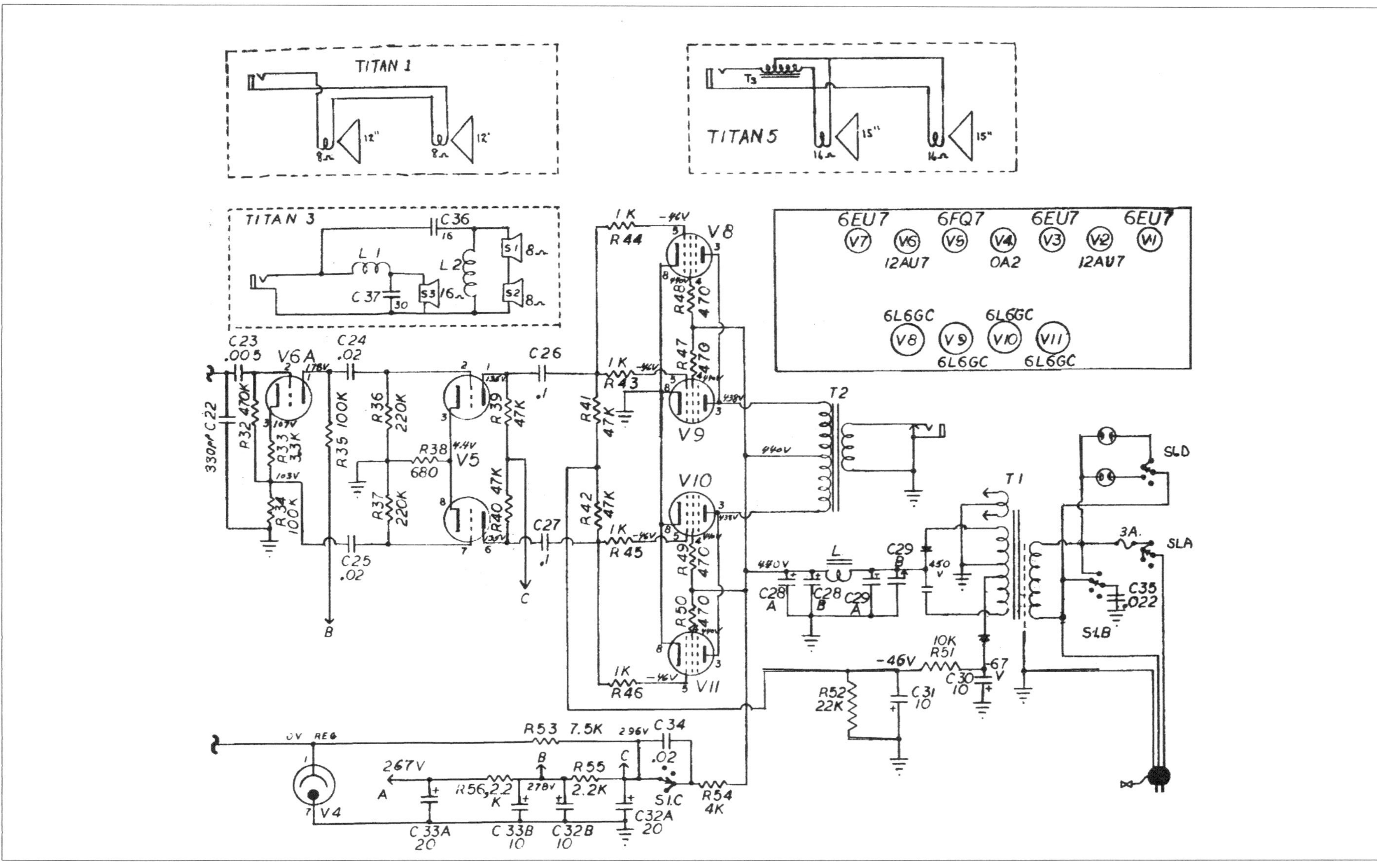

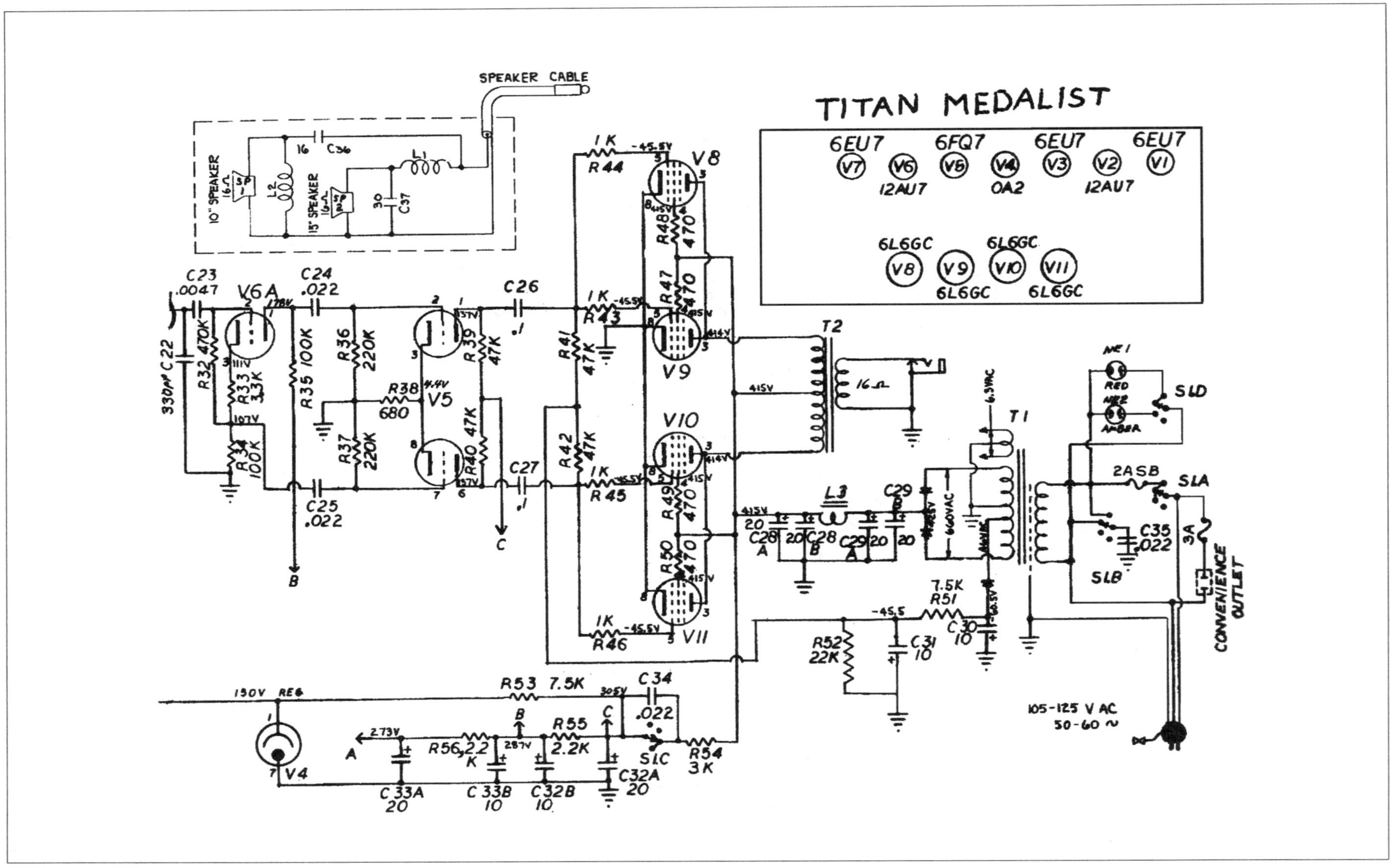
TITAN MEDALIST
6EU7 V7
V6 12AU7
6FQ7 V5
V4 OA2
6EU7 V3
V2 12AU7
6EU7 V1
6L6GC V8
V9 6L6GC
6L6GC V10
V11 6L6GC
SPEAKER CABLE
10" SPEAKER 16Ω
15" SPEAKER 16Ω
L1
L2
16 C36
30 C37
C23 .0047
V6A
330PF C22
R32 470K
R33 3.3K
R34 100K
R35 100K
C24 .022
C25 .022
R36 220K
R37 220K
R38 680
V5
R39 47K
R40 47K
C26 .1
C27 .1
R41 47K
R42 47K
1K R44
1K R43
1K R45
1K R46
R47 470
R48 470
R49 470
R50 470
V8
V9
V10
V11
T2
16Ω
T1
NE1
RED
AMBER
SLD
2ASB
SLA
C35 .022
SLB
3A
CONVENIENCE OUTLET
105-125 V AC 50-60 ~
L3
C28 A 20
C28 B 20
C29 A 20
C29 B 20
7.5K R51
C30 10
C31 10
R52 22K
150V REG
V4
R53 7.5K
C34 .022
R54 3K
S1C
R55 2.2K
R56 2.2 K
C33A 20
C33B 10
C32B 10
C32A 20
A
B
C

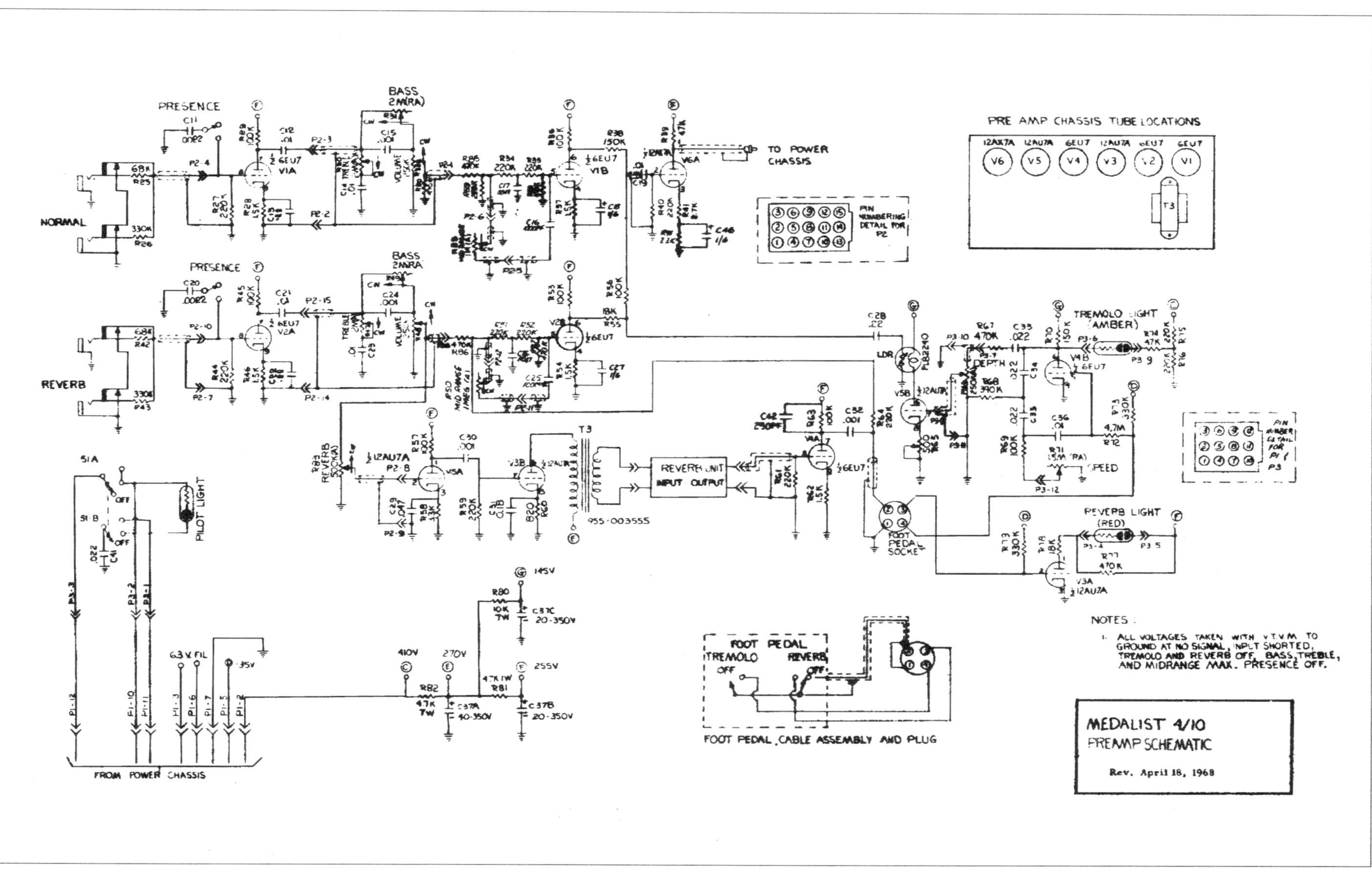
PRESENCE
NORMAL
REVERB
BASS 2M(RA)
TREBLE
VOLUME
V1A
V1B
V2A
V2B
V6A
V5A
V3B
V4A
V5B
V4B
V3A
½ 6EU7
½ 12AU7A
½ 12AX7A
R50 MID RANGE 1MEG (A)
R83 REVERB 500K(A)
TO POWER CHASSIS
PIN NUMBERING DETAIL FOR P2
PRE AMP CHASSIS TUBE LOCATIONS
12AX7A
12AU7A
6EU7
12AU7A
6EU7
6EU7
V6
V5
V4
V3
V2
V1
T3
955-003555
REVERB UNIT
INPUT OUTPUT
LDR
PLB2240
DEPTH
SPEED
TREMOLO LIGHT (AMBER)
REVERB LIGHT (RED)
FOOT PEDAL SOCKET
PIN NUMBER DETAIL FOR P1 / P3
S1A
S1B
PILOT LIGHT
6.3 V FIL
FROM POWER CHASSIS
410V
270V
255V
145V
-35V
FOOT PEDAL
TREMOLO
REVERB
OFF
FOOT PEDAL, CABLE ASSEMBLY AND PLUG
NOTES:
1. ALL VOLTAGES TAKEN WITH V.T.V.M. TO GROUND AT NO SIGNAL, INPUT SHORTED, TREMOLO AND REVERB OFF, BASS, TREBLE, AND MIDRANGE MAX. PRESENCE OFF.
MEDALIST 4/10
PREAMP SCHEMATIC
Rev. April 18, 1968

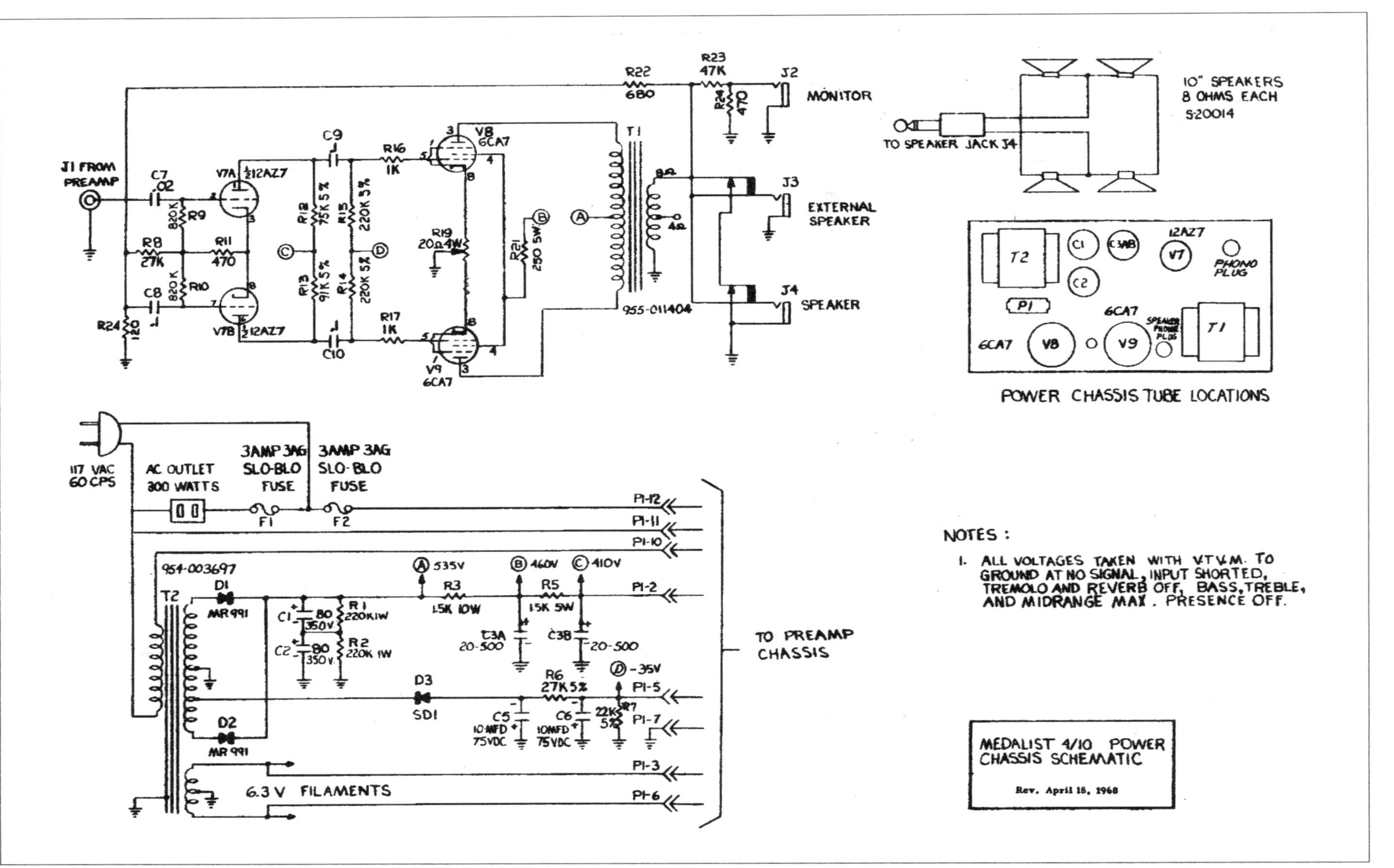
J1 FROM PREAMP
C7 .02
V7A ½12AZ7
V7B ½12AZ7
R9 820K
R10 820K
R8 27K
R11 470
C8 .1
R24 10Ω
C9
C10
R12 75K 5%
R13 91K 5%
R15 220K 5%
R14 220K 5%
R16 1K
R17 1K
V8 6CA7
V9 6CA7
R19 20Ω 4W
R21 250 5W
R22 680
R23 47K
R24 470
T1
955-011404
8Ω
4Ω
J2 MONITOR
J3 EXTERNAL SPEAKER
J4 SPEAKER
10" SPEAKERS
8 OHMS EACH
5-20014
TO SPEAKER JACK J4
T2
C1
C3A/B
12AZ7
V7
PHONO PLUG
C2
P1
6CA7
V8
V9
SPEAKER PHONE PLUG
T1
POWER CHASSIS TUBE LOCATIONS
117 VAC
60 CPS
AC OUTLET
300 WATTS
3AMP 3AG SLO-BLO FUSE
F1
3AMP 3AG SLO-BLO FUSE
F2
954-003697
D1 MR991
D2 MR991
C1 80 350V
C2 80 350V
R1 220K 1W
R2 220K 1W
A 535V
R3 1.5K 10W
B 460V
R5 15K 5W
C 410V
C3A 20-500
C3B 20-500
D3 SD1
C5 10 MFD 75VDC
R6 27K 5%
C6 10MFD 75VDC
D -35V
R7 22K 5%
6.3 V FILAMENTS
P1-12
P1-11
P1-10
P1-2
P1-5
P1-7
P1-3
P1-6
TO PREAMP CHASSIS
NOTES :
1. ALL VOLTAGES TAKEN WITH V.T.V.M. TO GROUND AT NO SIGNAL, INPUT SHORTED, TREMOLO AND REVERB OFF, BASS, TREBLE, AND MIDRANGE MAX. PRESENCE OFF.
MEDALIST 4/10 POWER CHASSIS SCHEMATIC
Rev. April 18, 1968

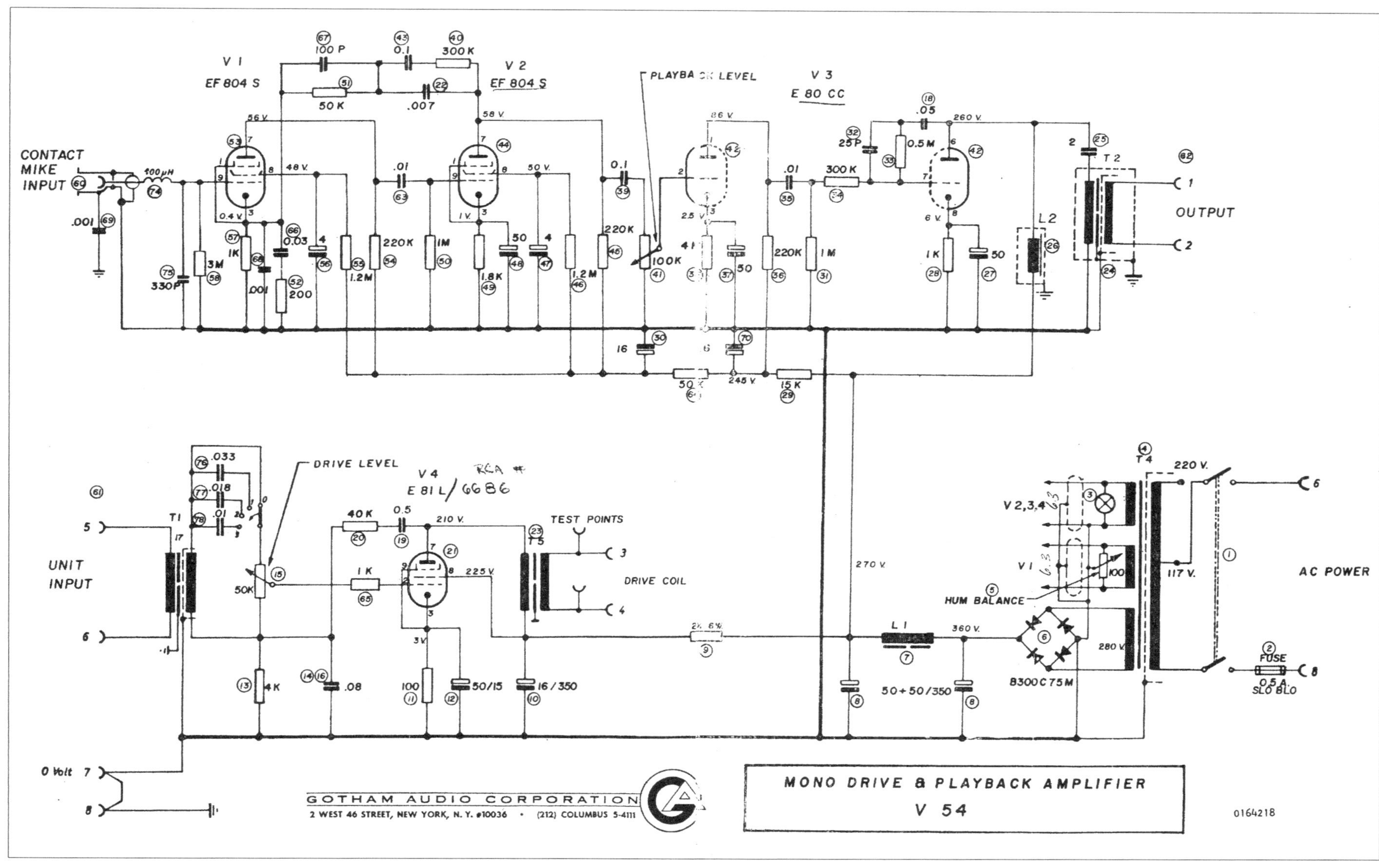
CONTACT MIKE INPUT
V 1
EF 804 S
V 2
EF 804 S
PLAYBACK LEVEL
V 3
E 80 CC
OUTPUT
UNIT INPUT
DRIVE LEVEL
V 4
E 81 L/6686
TEST POINTS
DRIVE COIL
V 2,3,4
V 1
HUM BALANCE
B300C75M
AC POWER
FUSE
0.5 A SLO BLO
0 Volt
GOTHAM AUDIO CORPORATION
2 WEST 46 STREET, NEW YORK, N. Y. #10036 • (212) COLUMBUS 5-4111
MONO DRIVE & PLAYBACK AMPLIFIER
V 54
0164218

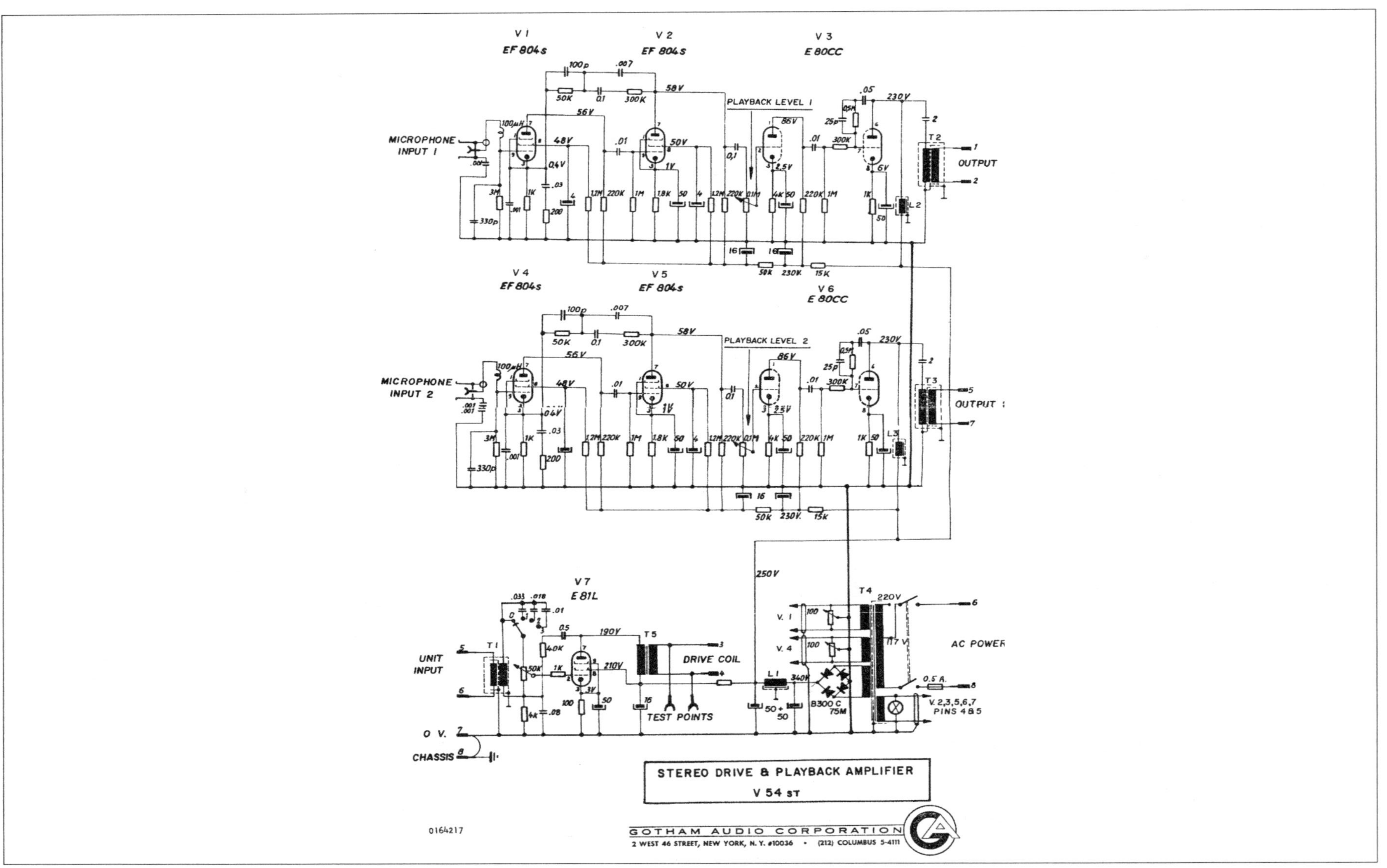
V 1
EF 804s
V 2
EF 804s
V 3
E 80CC
V 4
EF 804s
V 5
EF 804s
V 6
E 80CC
V 7
E 81L
MICROPHONE INPUT 1
MICROPHONE INPUT 2
PLAYBACK LEVEL 1
PLAYBACK LEVEL 2
OUTPUT
UNIT INPUT
DRIVE COIL
TEST POINTS
AC POWER
O V.
CHASSIS
250V
220V
117 V
0.5 A.
V.2,3,5,6,7 PINS 4 & 5
B300 C 75M
STEREO DRIVE & PLAYBACK AMPLIFIER
V 54 ST
GOTHAM AUDIO CORPORATION
2 WEST 46 STREET, NEW YORK, N. Y. #10036 • (212) COLUMBUS 5-4111
0164217

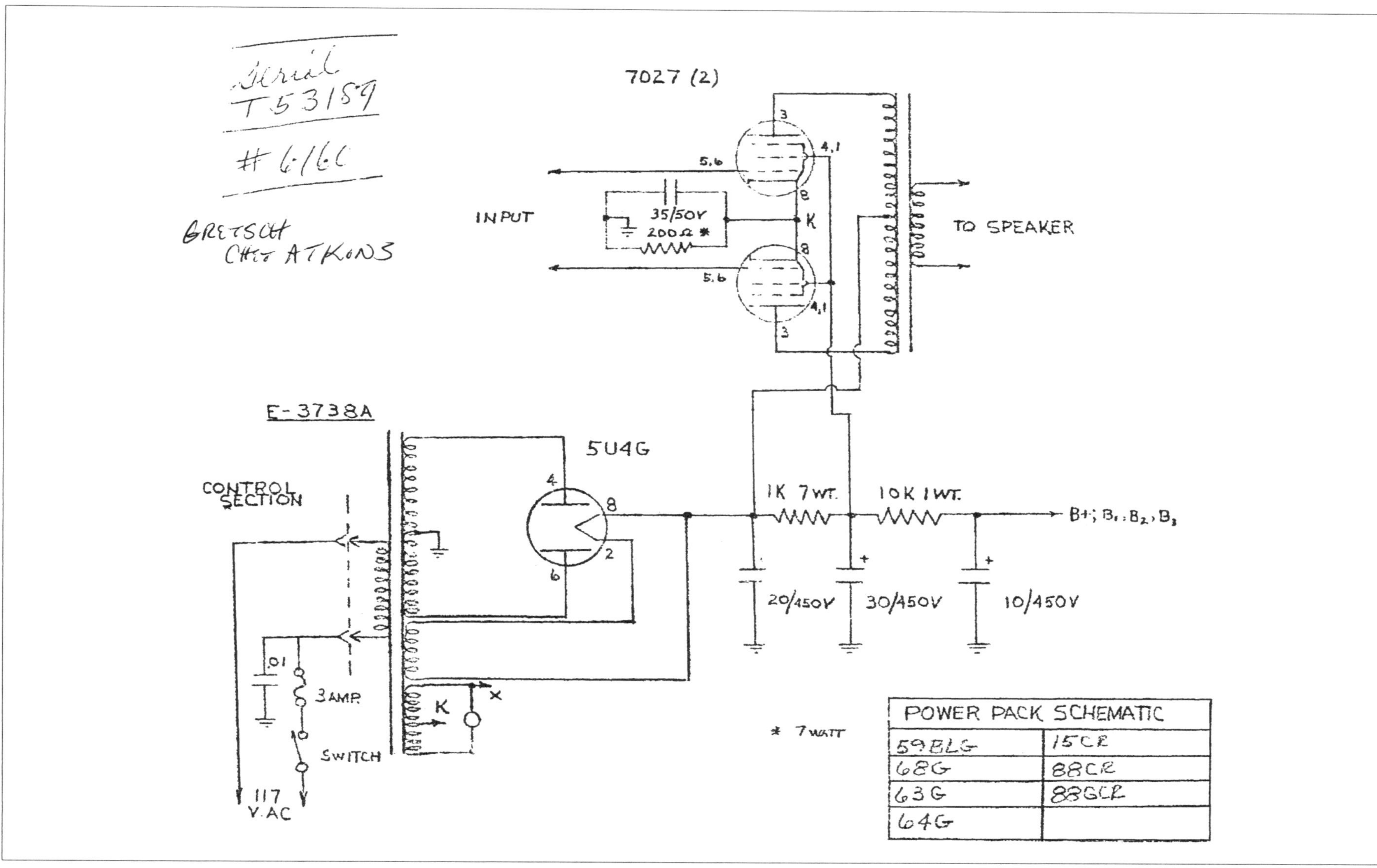
Serial T53189
#6160
GRETSCH CHET ATKINS
7027 (2)
INPUT
35/50V
200Ω *
K
TO SPEAKER
E-3738A
5U4G
CONTROL SECTION
1K 7WT.
10K 1WT.
B+; B1, B2, B3
20/450V
30/450V
10/450V
.01
3AMP.
SWITCH
117 V.AC
* 7WATT
POWER PACK SCHEMATIC
59BLG 15CR
68G 88CR
63G 88GCR
64G

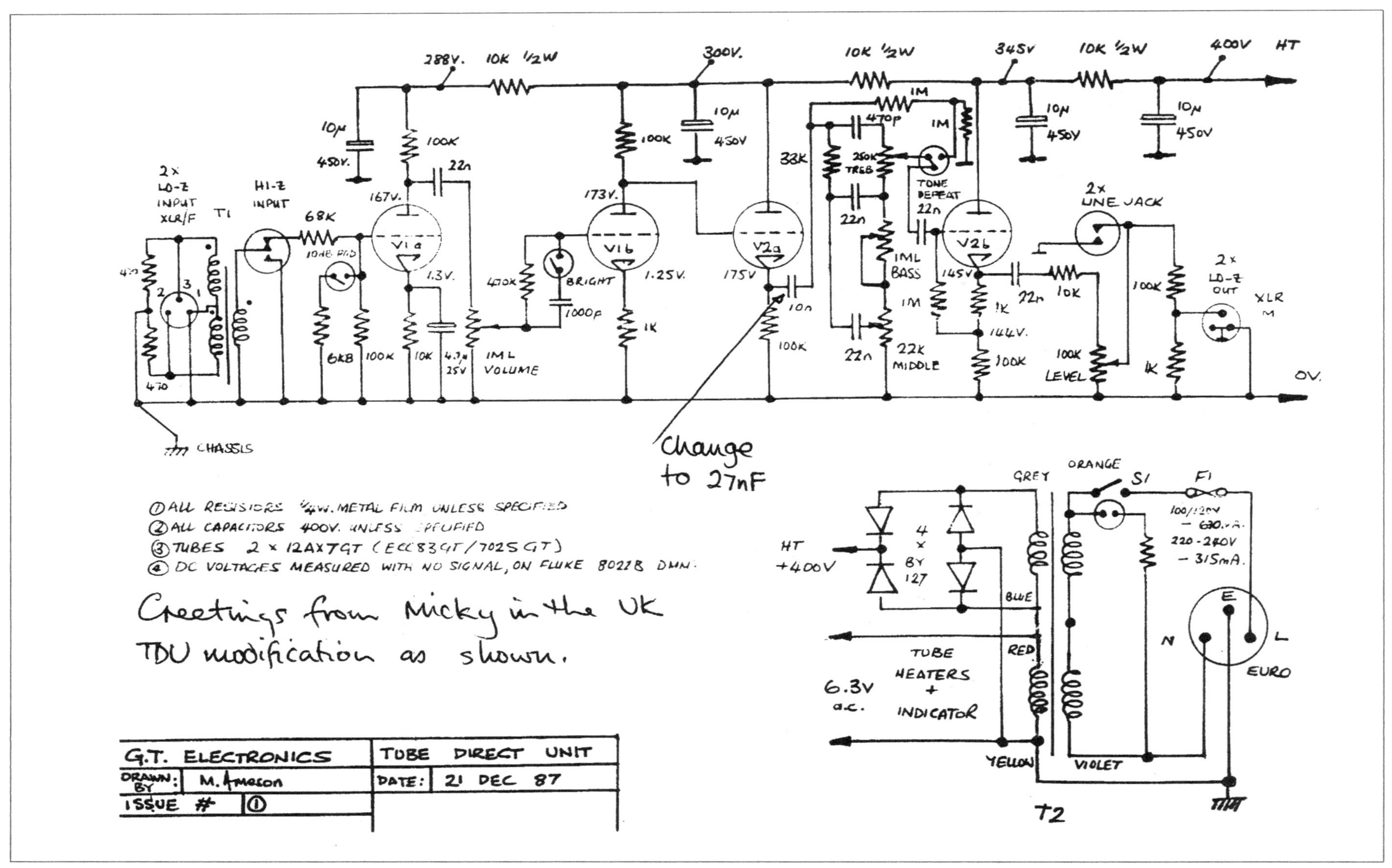
288V.
10K ½W
300V.
10K ½W
345V
10K ½W
400V
HT
10µ 450V.
100K
22n
2x LO-Z INPUT XLR/F
T1
HI-Z INPUT
68K
167V.
V1a
10dB PAD
1.3V.
6K8
100K
10K
4.7µ 25V
IML VOLUME
470K
BRIGHT
1000P
173V.
V1b
1.25V.
1K
100K
10µ 450V
V2a
175V
10n
100K
33K
470P
250K TREB
22n
IML BASS
1M
22n
22K MIDDLE
1M
TONE DEFEAT
22n
V2b
145V
1K
144V.
100K
22n
10K
2x LINE JACK
100K
100K LEVEL
1K
2x LO-Z OUT
XLR M
0V.
CHASSIS
Change to 27nF
① ALL RESISTORS ¼W. METAL FILM UNLESS SPECIFIED
② ALL CAPACITORS 400V. UNLESS SPECIFIED
③ TUBES 2 × 12AX7GT (ECC83GT/7025GT)
④ DC VOLTAGES MEASURED WITH NO SIGNAL, ON FLUKE 8022B DMM.
Greetings from Micky in the UK
TDU modification as shown.
G.T. ELECTRONICS
TUBE DIRECT UNIT
DRAWN BY: M. Ameson
DATE: 21 DEC 87
ISSUE # ①
HT +400V
4 × BY 127
GREY
BLUE
RED
YELLOW
TUBE HEATERS + INDICATOR
6.3V a.c.
ORANGE
S1
F1
100/120V – 630mA.
220-240V – 315mA.
VIOLET
E
N
L
EURO
T2

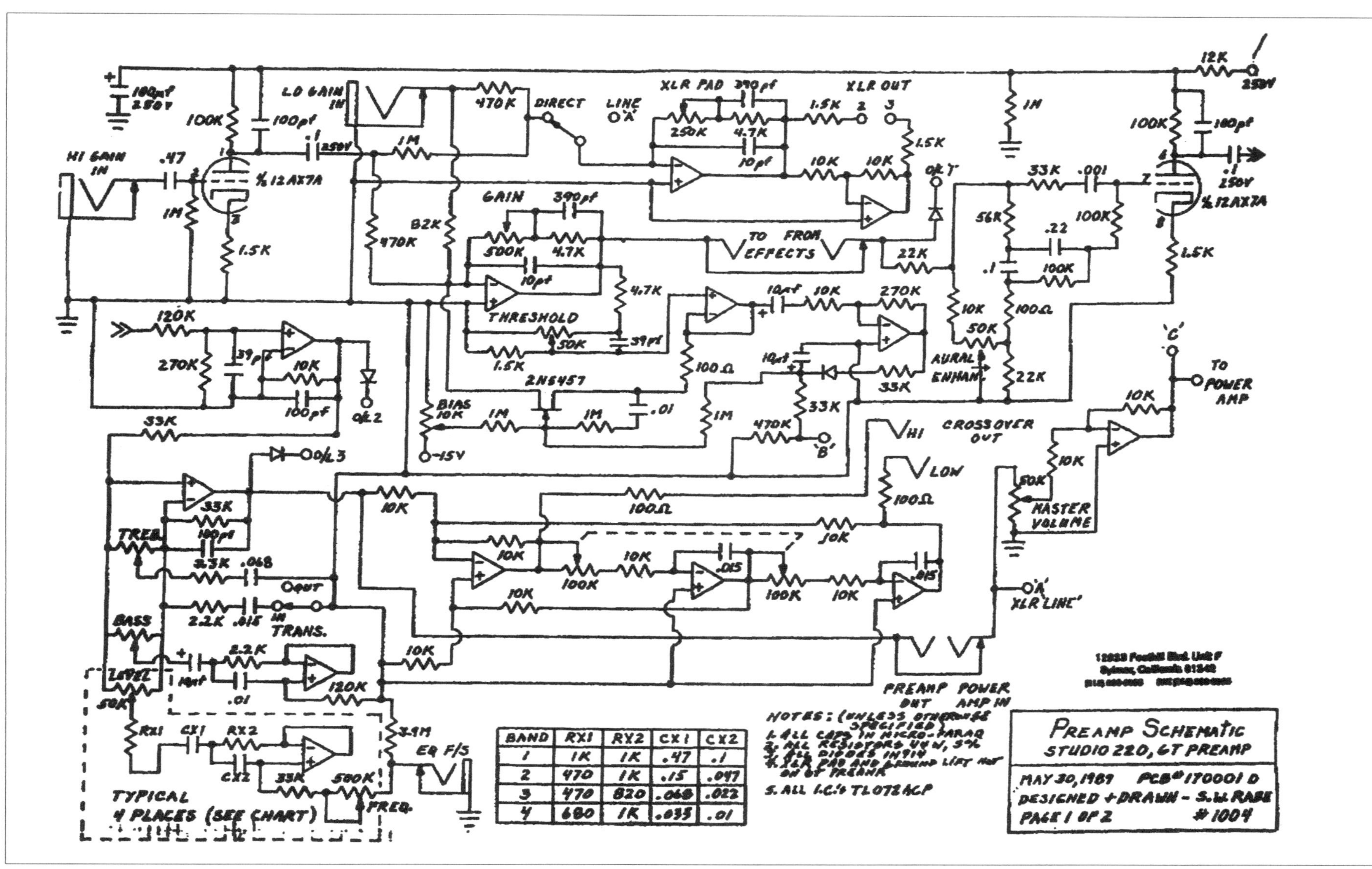

BAND	RX1	RX2	CX1	CX2
1	1K	1K	.47	.1
2	470	1K	.15	.047
3	470	820	.068	.022
4	680	1K	.033	.01

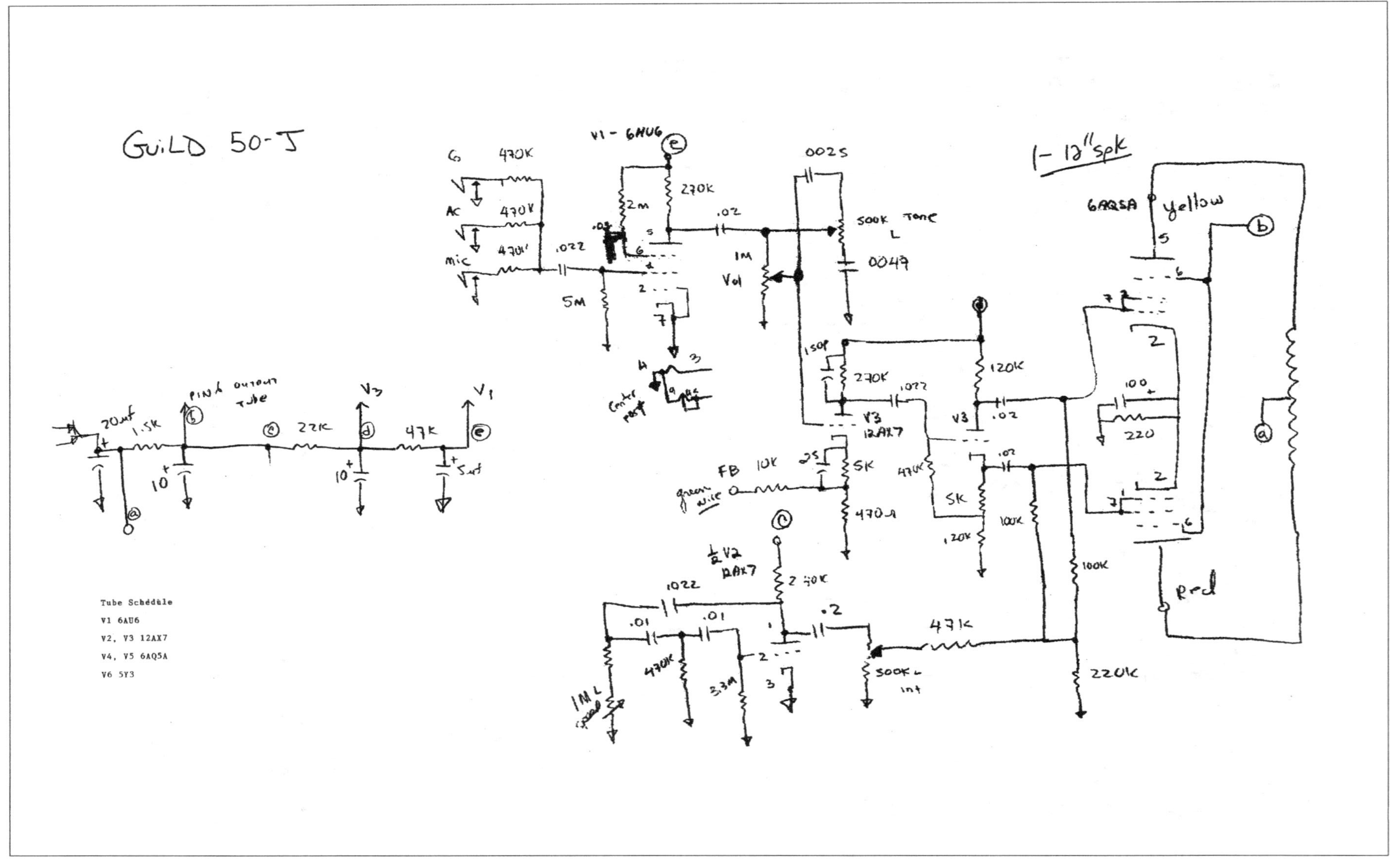
GUILD 50-J
V1 - 6AU6
470K
470K
470K
AC
mic
.022
5M
2m
270K
.02
1M
Vol
.0025
500K Tone
.0047
Center post
150P
270K
.022
V3
12AX7
5K
470Ω
FB
10K
green wire
.25
120K
.02
47K
5K
120K
100K
100K
220K
1- 12" spk
6AQ5A
yellow
Red
100
220
½ V2
12AX7
.022
.01
.01
470K
3.3M
.2
240K
47K
500K
int
1M
20mf
1.5K
10
Pink output tube
22K
V3
10
47K
V1
5mf
Tube Schedule
V1 6AU6
V2, V3 12AX7
V4, V5 6AQ5A
V6 5Y3

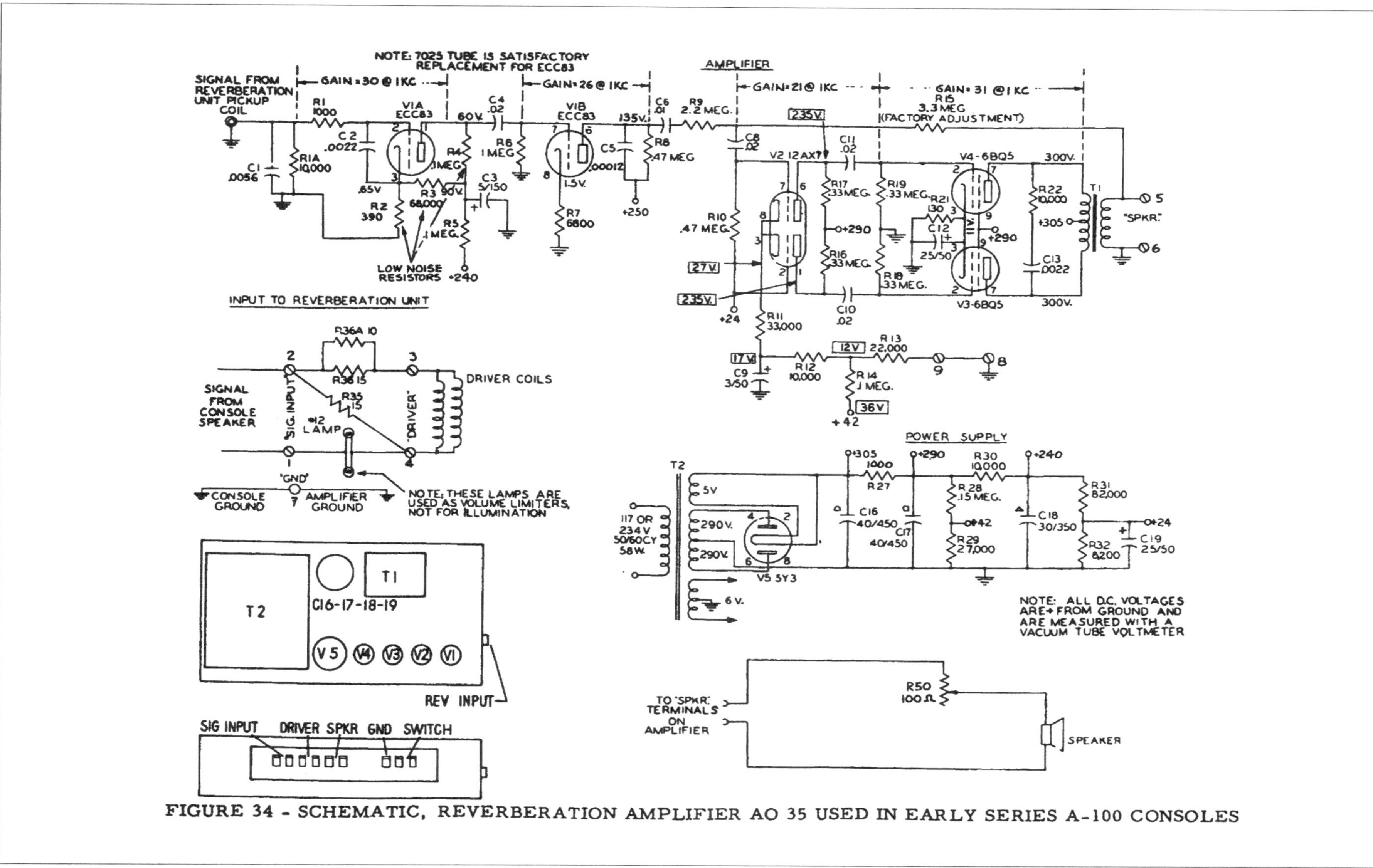

FIGURE 34 - SCHEMATIC, REVERBERATION AMPLIFIER AO 35 USED IN EARLY SERIES A-100 CONSOLES

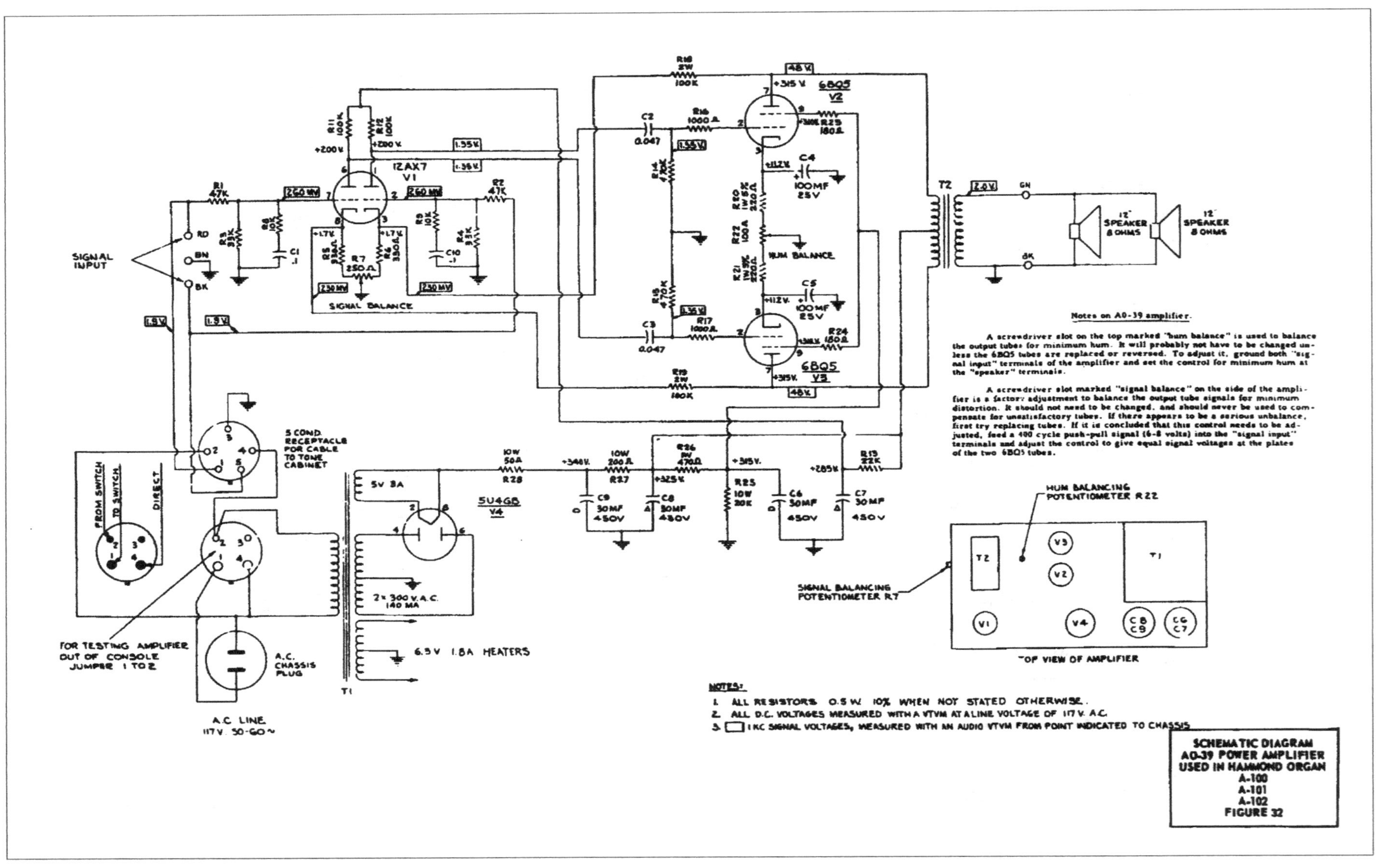
SIGNAL INPUT
RD
BN
BK
R1 47K
R3 33K
R8 10K
C1 .1
12AX7 V1
R11 100K
R12 100K
+200 V.
R2 47K
R9 10K
R4 33K
C10 .1
R5 330Ω
R7 250Ω
R6 330Ω
SIGNAL BALANCE
+1.7V.
C2 0.047
C3 0.047
R16 1000Ω
R17 1000Ω
R14 470K
R15 470K
R18 2W 100K
R19 2W 180K
6BQ5 V2
6BQ5 V3
+315 V.
C4 100MF 25V
C5 100MF 25V
R20 1W 5% 220Ω
R22 100Ω
R21 1W 5% 220Ω
HUM BALANCE
R23 180Ω
R24 180Ω
+112V.
T2
GN
BK
12" SPEAKER 8 OHMS
12" SPEAKER 8 OHMS
5 COND. RECEPTACLE FOR CABLE TO TONE CABINET
FROM SWITCH
TO SWITCH
DIRECT
FOR TESTING AMPLIFIER OUT OF CONSOLE JUMPER 1 TO 2
A.C. CHASSIS PLUG
A.C. LINE 117V. 50-60~
T1
5V 3A
2 x 300 V.A.C. 140 MA
6.5V 1.8A HEATERS
5U4GB V4
10W 50Ω R28
10W 200Ω R27
R26 5W 470Ω
R25 10W 20K
R13 22K
+344V.
+325V.
+315V.
+285V.
C9 30MF 450V
C8 30MF 450V
C6 30MF 450V
C7 30MF 450V
Notes on A0-39 amplifier.
A screwdriver slot on the top marked "hum balance" is used to balance the output tubes for minimum hum. It will probably not have to be changed unless the 6BQ5 tubes are replaced or reversed. To adjust it, ground both "signal input" terminals of the amplifier and set the control for minimum hum at the "speaker" terminals.
A screwdriver slot marked "signal balance" on the side of the amplifier is a factory adjustment to balance the output tube signals for minimum distortion. It should not need to be changed, and should never be used to compensate for unsatisfactory tubes. If there appears to be a serious unbalance, first try replacing tubes. If it is concluded that this control needs to be adjusted, feed a 400 cycle push-pull signal (6-8 volts) into the "signal input" terminals and adjust the control to give equal signal voltages at the plates of the two 6BQ5 tubes.
HUM BALANCING POTENTIOMETER R22
SIGNAL BALANCING POTENTIOMETER R7
T2
T1
V1
V2
V3
V4
C8 C9
C6 C7
TOP VIEW OF AMPLIFIER
NOTES:
1. ALL RESISTORS 0.5 W. 10% WHEN NOT STATED OTHERWISE.
2. ALL D.C. VOLTAGES MEASURED WITH A VTVM AT A LINE VOLTAGE OF 117 V. A.C.
3. [ ] 1 KC SIGNAL VOLTAGES, MEASURED WITH AN AUDIO VTVM FROM POINT INDICATED TO CHASSIS.
SCHEMATIC DIAGRAM
AO-39 POWER AMPLIFIER
USED IN HAMMOND ORGAN
A-100
A-101
A-102
FIGURE 32

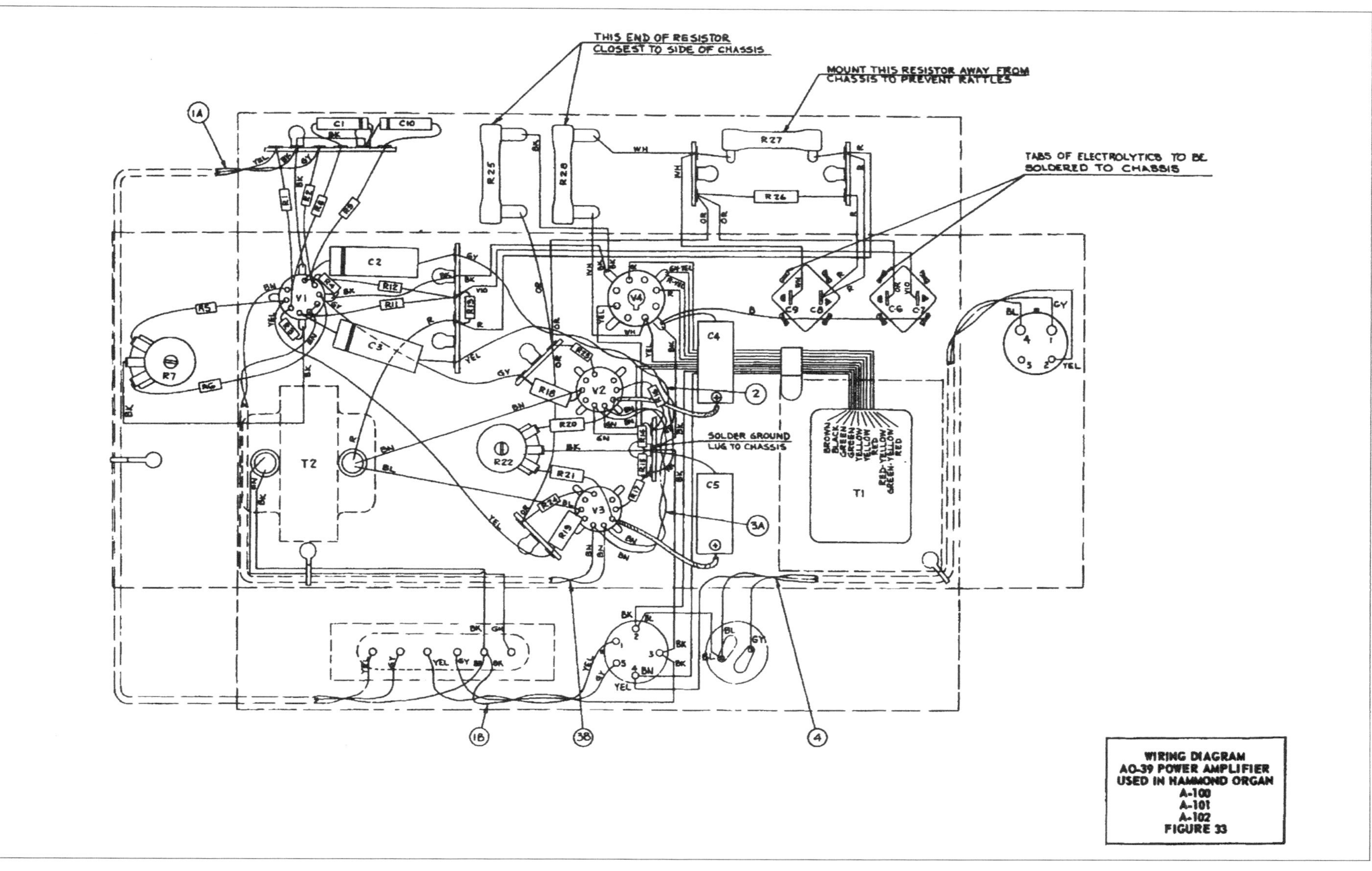

WIRING DIAGRAM
AO-39 POWER AMPLIFIER
USED IN HAMMOND ORGAN
A-100
A-101
A-102
FIGURE 33

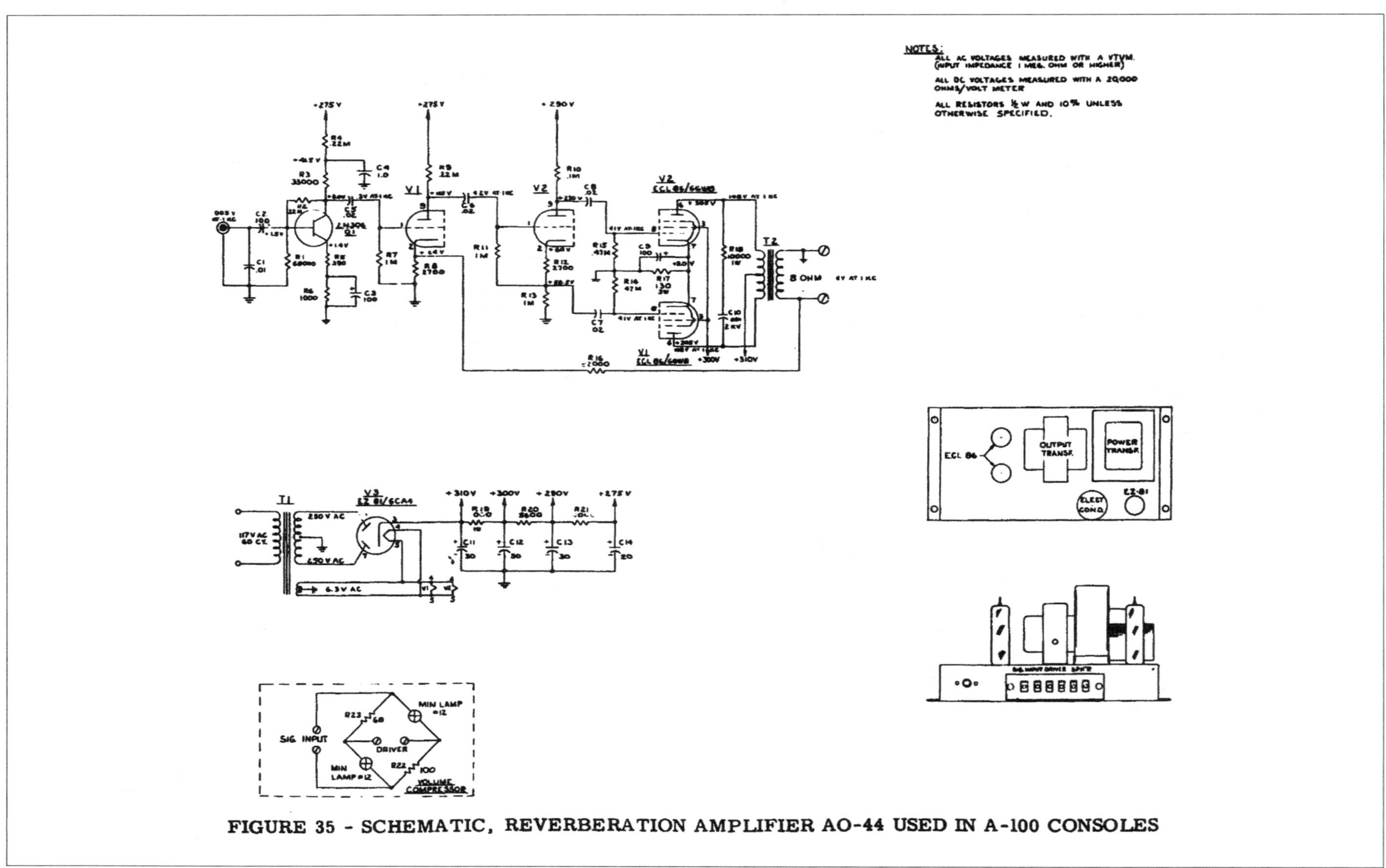

FIGURE 35 - SCHEMATIC, REVERBERATION AMPLIFIER AO-44 USED IN A-100 CONSOLES

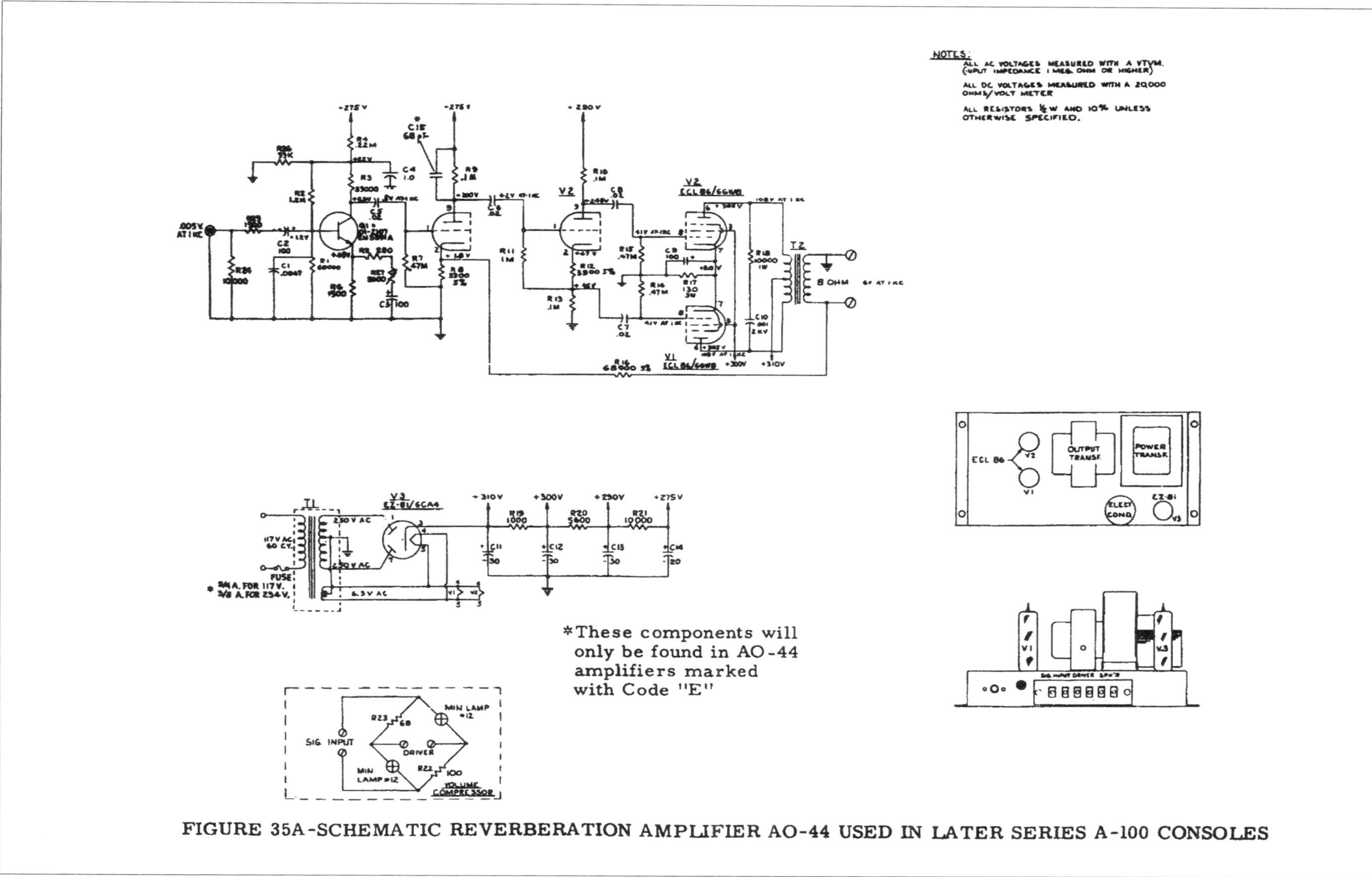

*These components will only be found in AO-44 amplifiers marked with Code "E"

FIGURE 35A-SCHEMATIC REVERBERATION AMPLIFIER AO-44 USED IN LATER SERIES A-100 CONSOLES

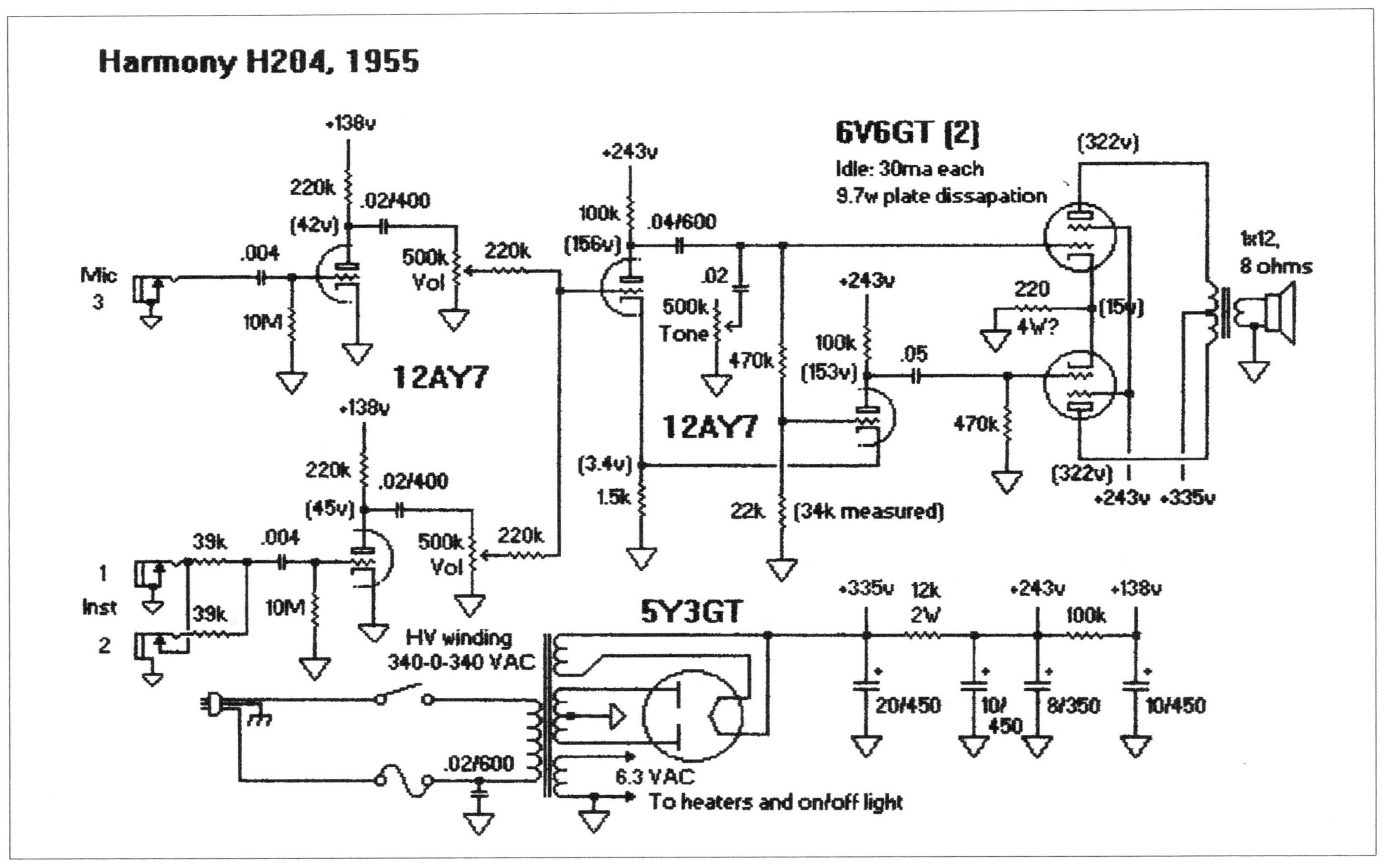
Harmony H204, 1955
+138v
220k
.02/400
(42v)
.004
Mic
3
10M
500k
Vol
220k
12AY7
+138v
220k
.02/400
(45v)
39k
.004
1
Inst
2
39k
10M
500k
Vol
220k
+243v
100k
.04/600
(156v)
.02
500k
Tone
470k
(3.4v)
1.5k
12AY7
22k
(34k measured)
+243v
100k
(153v)
.05
470k
6V6GT (2)
Idle: 30ma each
9.7w plate dissapation
(322v)
220
4W?
(15v)
(322v)
+243v
+335v
1x12,
8 ohms
5Y3GT
HV winding
340-0-340 VAC
.02/600
6.3 VAC
To heaters and on/off light
+335v
12k
2W
+243v
100k
+138v
20/450
10/450
8/350
10/450

High-Power Amplifiers

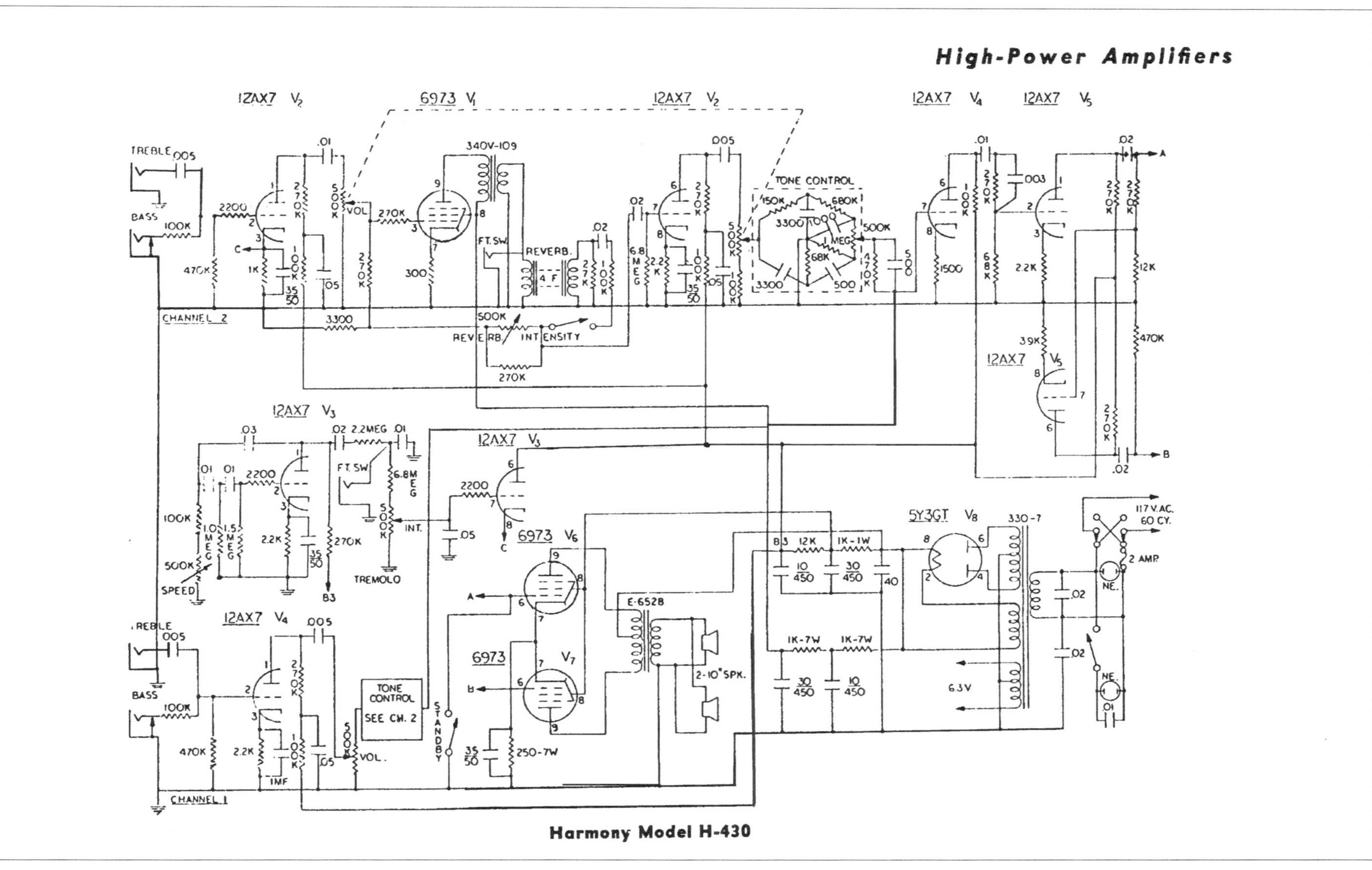

Harmony Model H-430

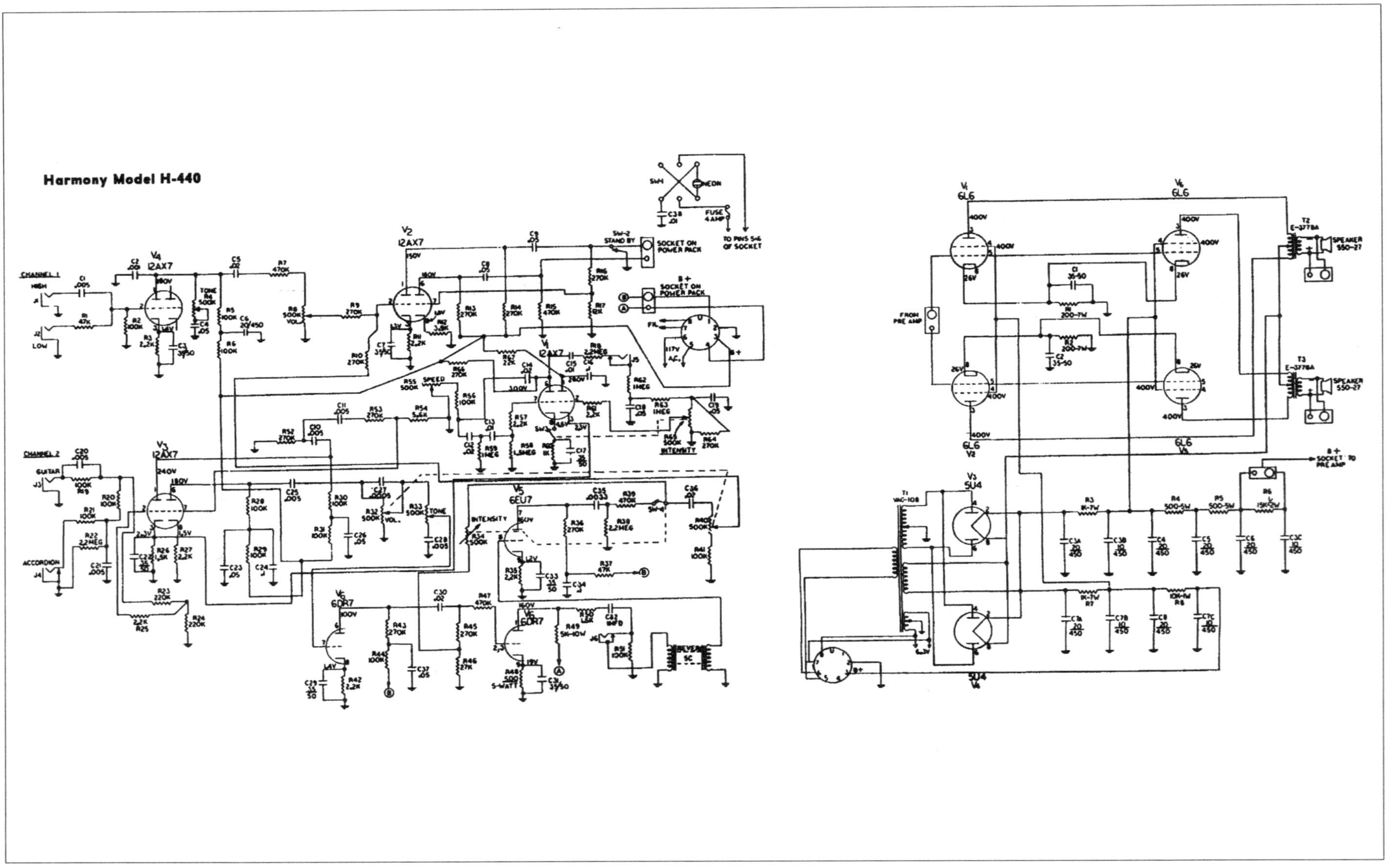
Harmony Model H-440
CHANNEL 1
HIGH
LOW
CHANNEL 2
GUITAR
ACCORDION
V4 12AX7
V2 12AX7
V1 12AX7
V3 12AX7
V5 6EU7
V6 6DR7
INTENSITY
SPEED
TONE
VOL.
STAND BY
SOCKET ON POWER PACK
B+ SOCKET ON POWER PACK
NEON
FUSE 4 AMP
TO PINS 5-6 OF SOCKET
FROM PRE AMP
V1 6L6
V6 6L6
6L6 V2
6L6 V4
V3 5U4
5U4 V4
T1 VAC-108
T2 E-377BA
T3 E-377BA
SPEAKER 550-27
B+ SOCKET TO PRE AMP

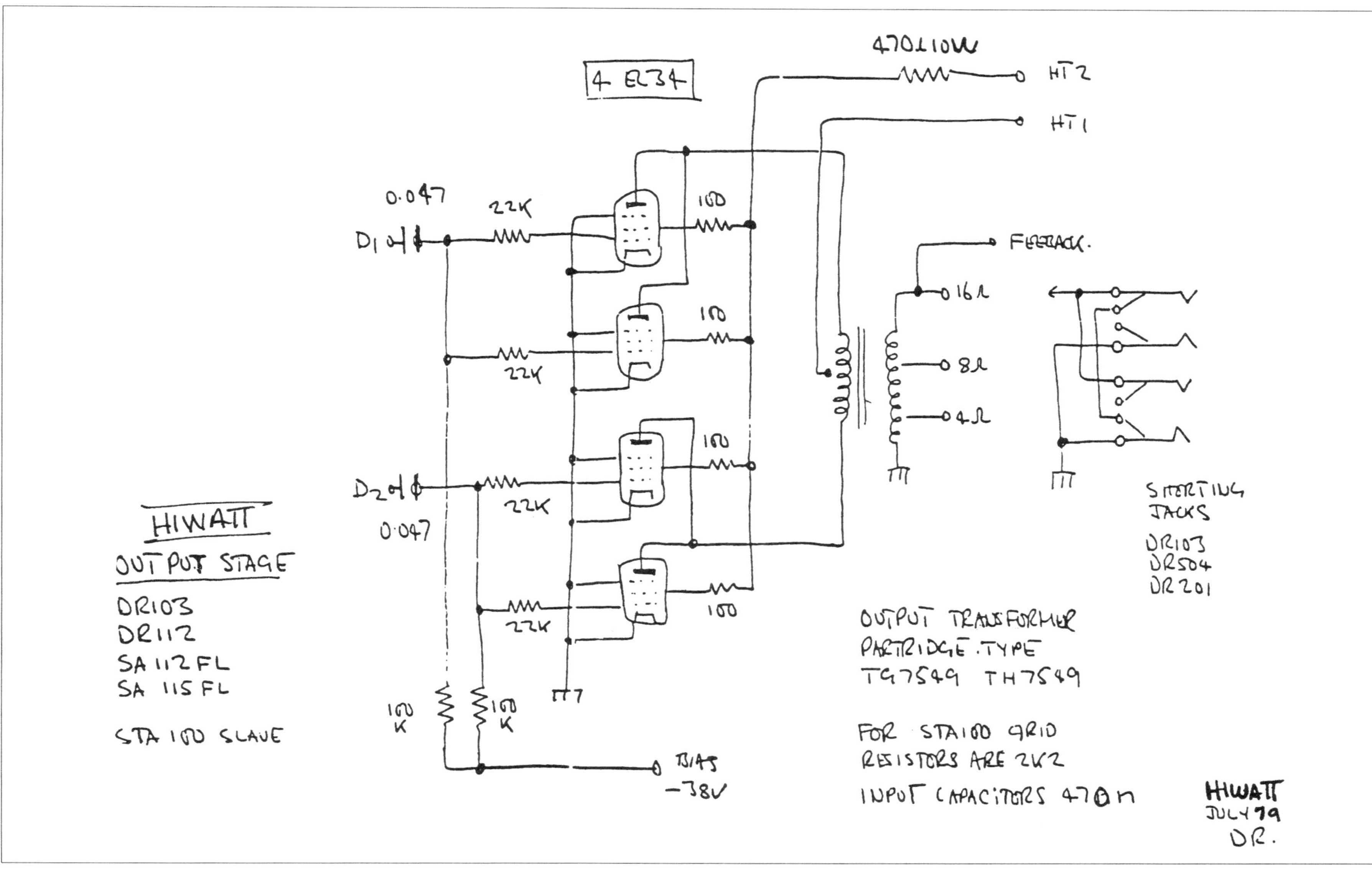
4 EL34
470Ω 10W
HT2
HT1
0.047
D1
22K
100
FEEDBACK.
16Ω
8Ω
4Ω
D2
0.047
HIWATT
OUTPUT STAGE
DR103
DR112
SA 112 FL
SA 115 FL
STA 100 SLAVE
100 K
BIAS
-38V
SHORTING JACKS
DR103
DR504
DR201
OUTPUT TRANSFORMER
PARTRIDGE. TYPE
TG7549 TH7549
FOR STA100 GRID
RESISTORS ARE 2K2
INPUT CAPACITORS 47Ωn
HIWATT
JULY 79
DR.

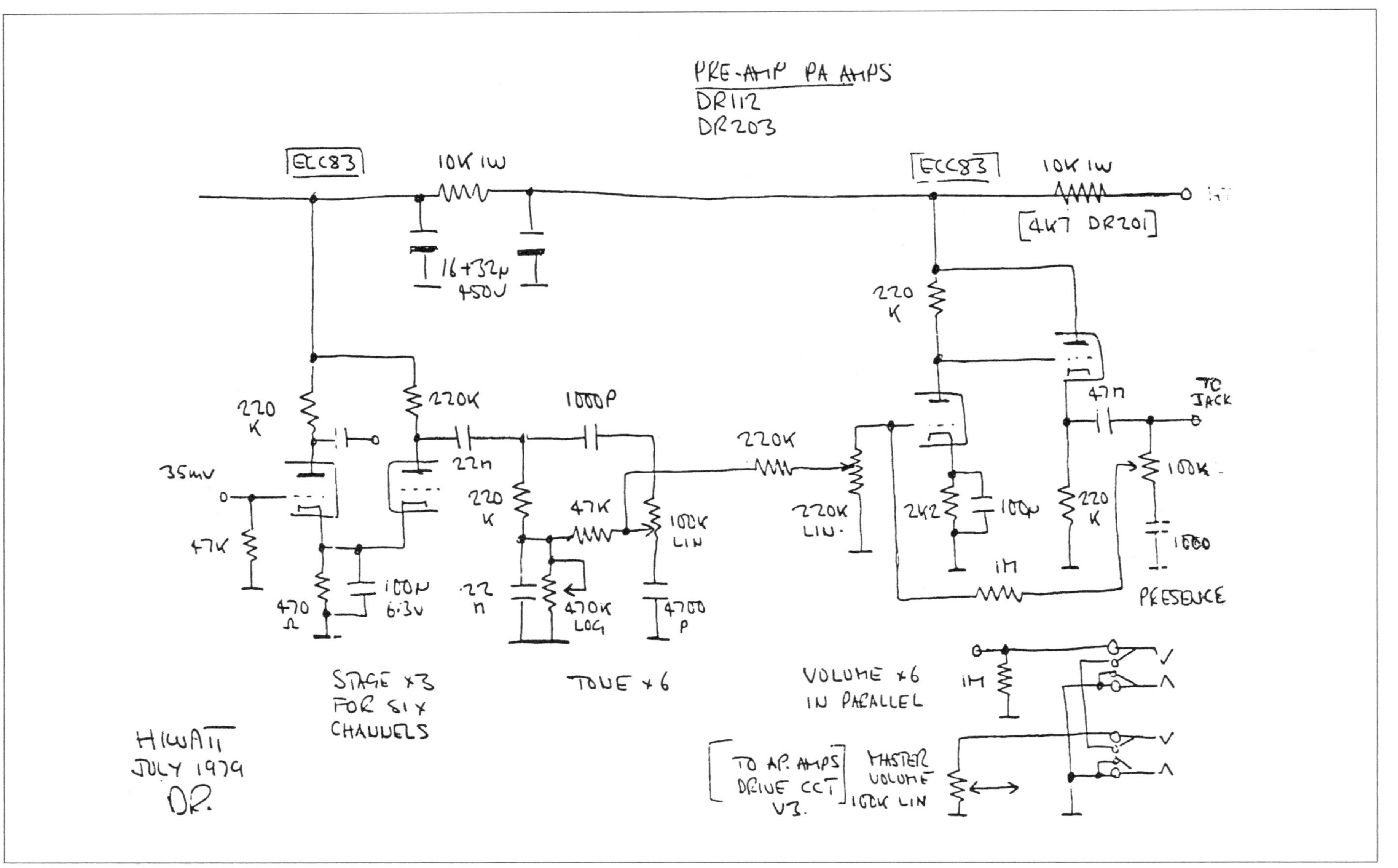
PRE-AMP PA AMPS
DR112
DR203
ECC83
10K 1W
16+32µ
450V
ECC83
10K 1W
[4K7 DR201]
220
K
220
K
220K
1000P
22n
35mV
47K
470
Ω
100µ
6·3V
220
K
47K
·22
n
470K
LOG
100K
LIN
4700
P
220K
220K
LIN·
2K2
100µ
1M
47n
220
K
TO
JACK
100K
1000
PRESENCE
STAGE x3
FOR SIX
CHANNELS
TONE x 6
VOLUME x6
IN PARALLEL
1M
[TO AP. AMPS
DRIVE CCT
V3.]
MASTER
VOLUME
100K LIN
HIWATT
JULY 1979
DR.

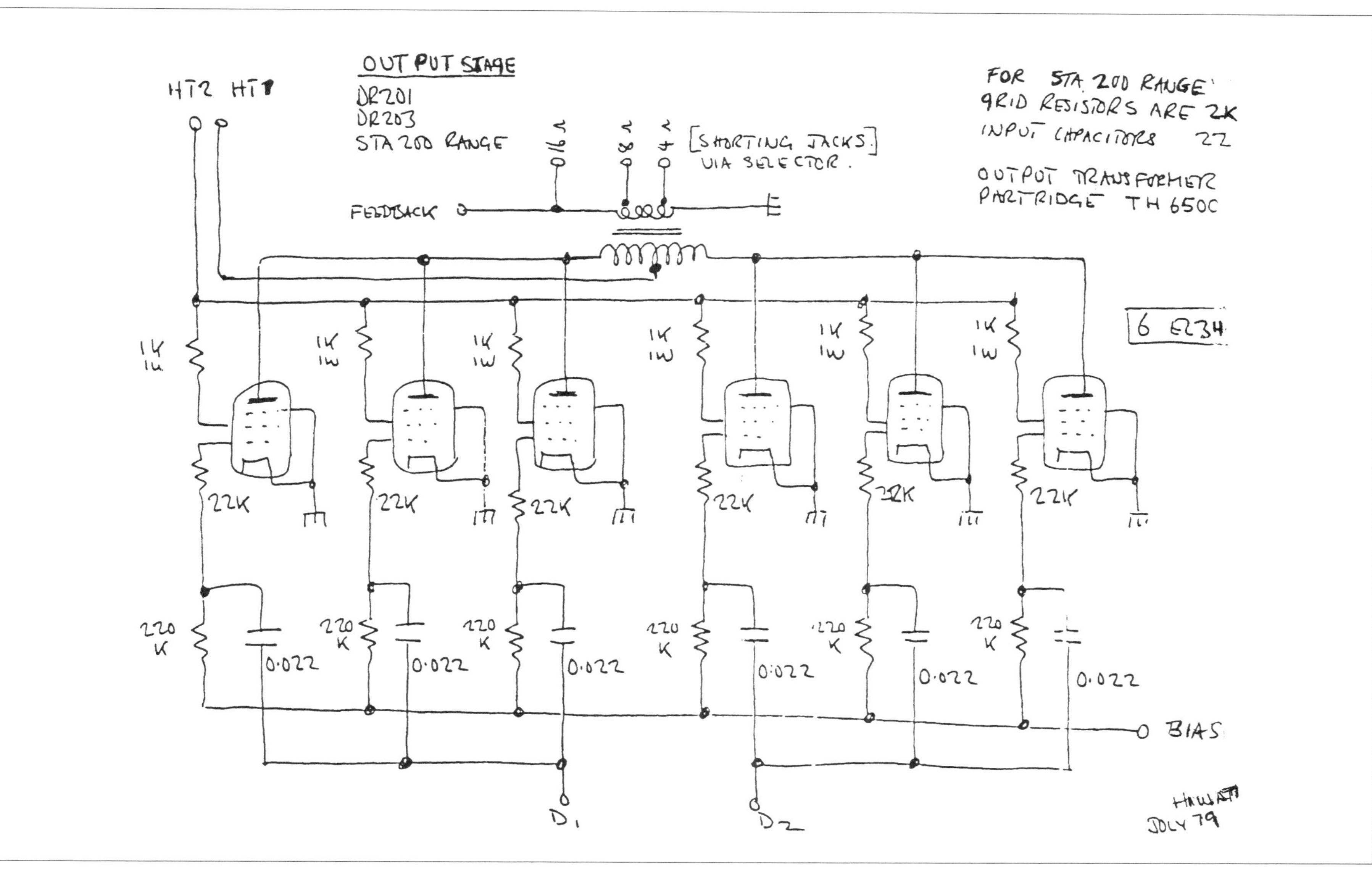
OUTPUT STAGE
DR201
DR203
STA 200 RANGE
HT2 HT1
FEEDBACK
16 Ω
8 Ω
4 Ω
[SHORTING JACKS VIA SELECTOR.]
FOR STA 200 RANGE
GRID RESISTORS ARE 2K
INPUT CAPACITORS 22
OUTPUT TRANSFORMER PARTRIDGE TH 650C
1K 1W
22K
220 K
0.022
6 EL34
BIAS
D1
D2
HIWATT JULY 79

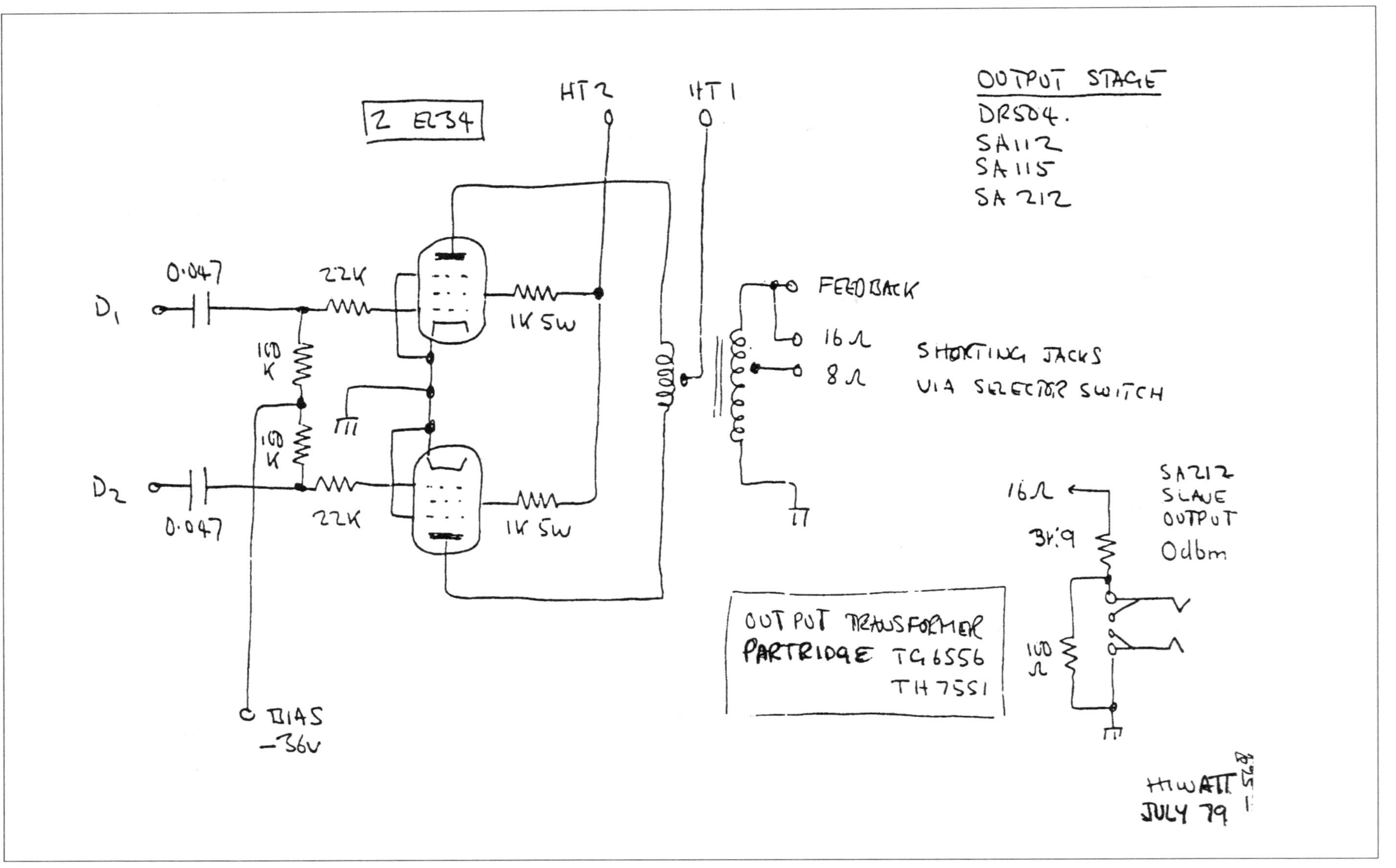
2 EL34
HT2
HT1
OUTPUT STAGE
DR504.
SA112
SA115
SA 212
D1
D2
0.047
0.047
22K
22K
100 K
100 K
1K 5W
1K 5W
FEEDBACK
16 Ω
8 Ω
SHORTING JACKS
VIA SELECTOR SWITCH
BIAS
-36V
OUTPUT TRANSFORMER
PARTRIDGE TG6556
TH7551
16 Ω
SA212
SLAVE
OUTPUT
0dbm
3K9
100 Ω
HIWATT
JULY 79

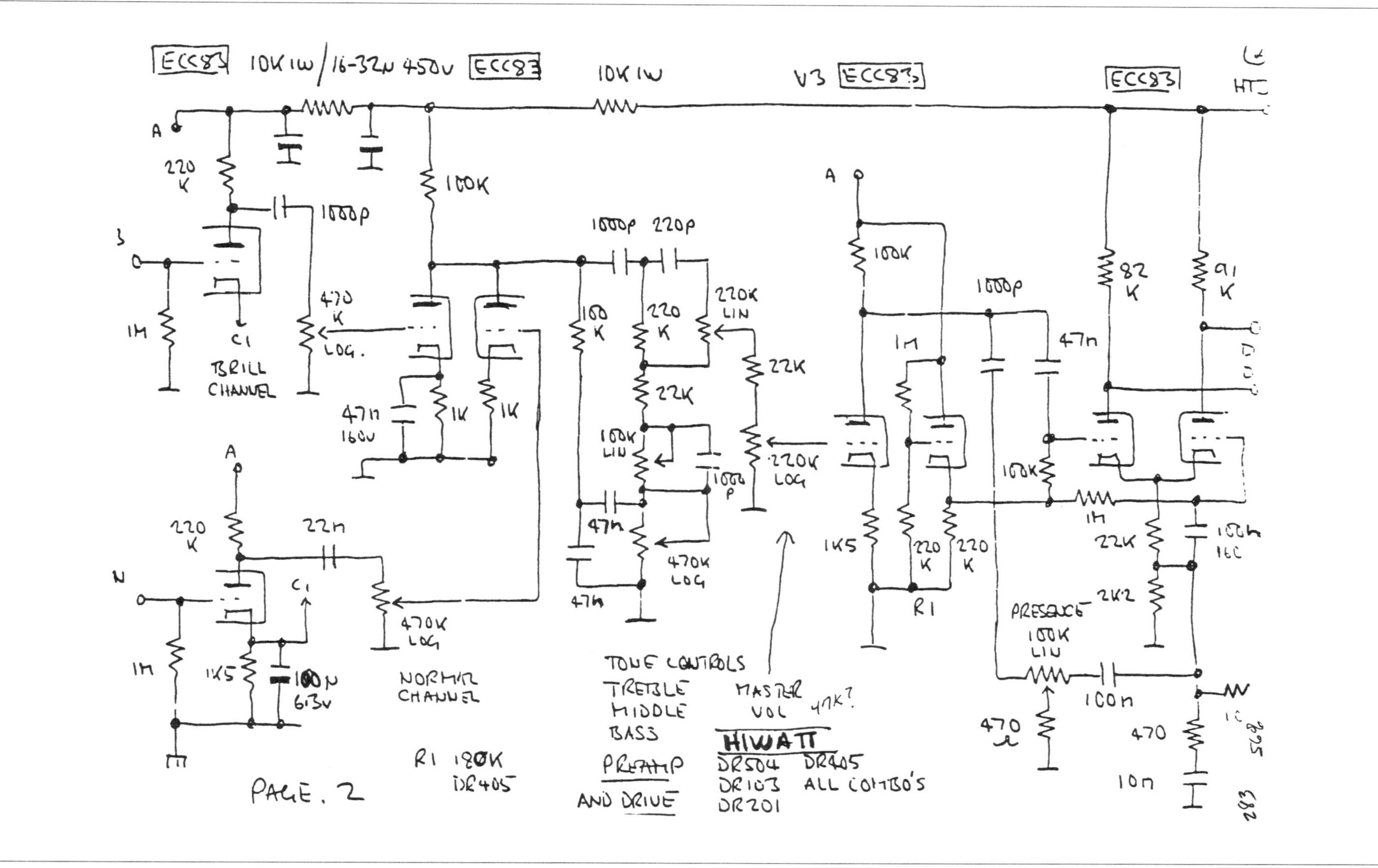
ECC83
10K 1W / 16-32μ 450V
ECC83
10K 1W
V3 ECC83
ECC83
HT
A
220 K
1000p
100K
3
1M
C1
BRILL CHANNEL
470 K LOG.
47n 160V
1K
1K
1000p 220p
100 K
220 K
220K LIN
22K
22K
100K LIN
1000 P
220K LOG
47n
470K LOG
47n
A
100K
1M
1000p
47n
82 K
91 K
100K
1M
22K
100n 16V
1K5
220 K
220 K
R1
2K2
PRESENCE 100K LIN
100n
470
470
10n
A
220 K
22n
N
1M
C1
1K5
100μ 6.3V
470K LOG
NORMAL CHANNEL
TONE CONTROLS
TREBLE
MIDDLE
BASS
PREAMP AND DRIVE
MASTER VOL 47K?
HIWATT
DR504 DR405
DR103 ALL COMBO'S
DR201
R1 180K DR405
PAGE. 2
283

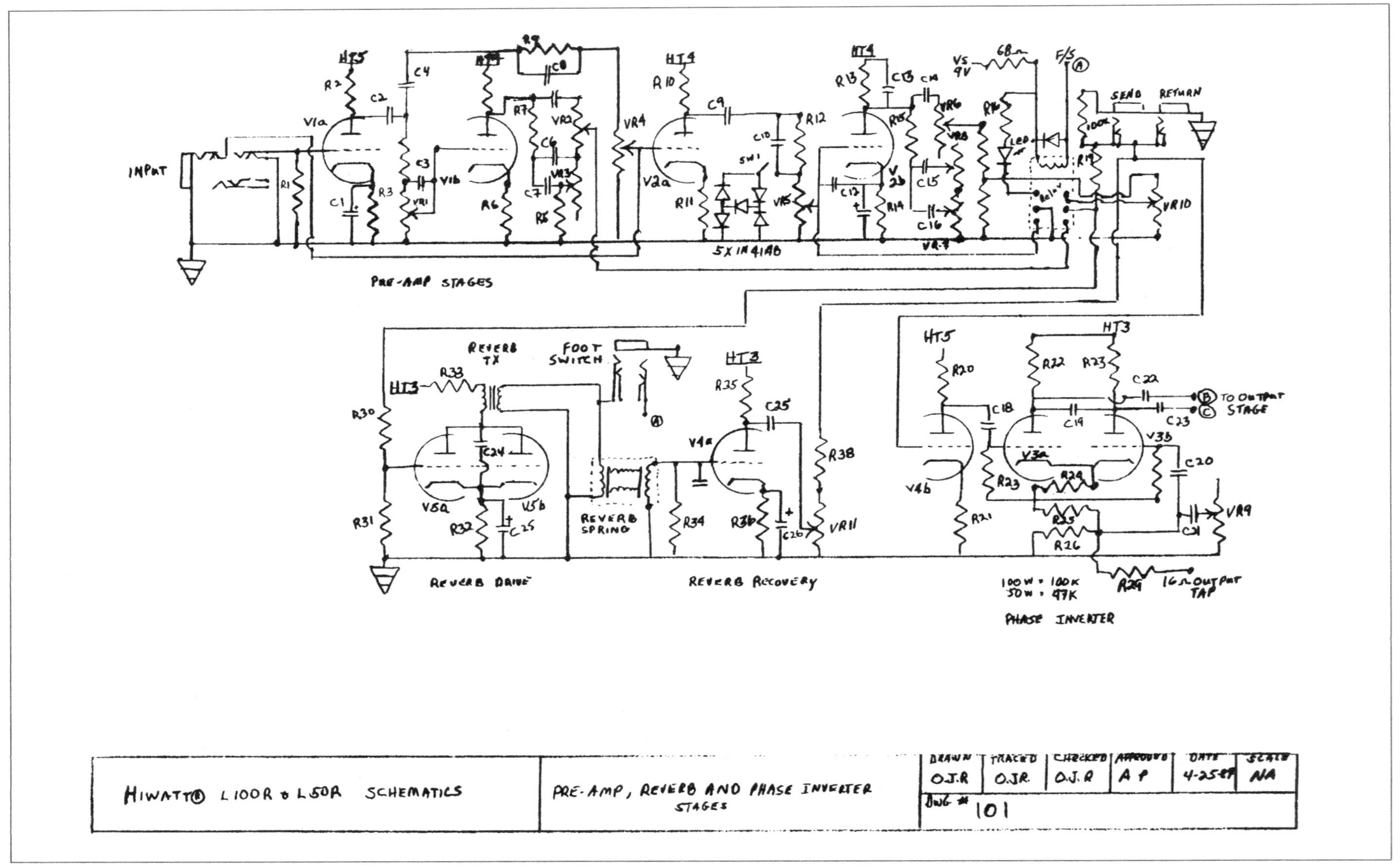
INPUT
PRE-AMP STAGES
5 X IN4148
SEND
RETURN
TO OUTPUT STAGE
REVERB TX
FOOT SWITCH
REVERB SPRING
REVERB DRIVE
REVERB RECOVERY
100W : 100K
50W : 47K
16Ω OUTPUT TAP
PHASE INVERTER
HIWATT® L100R + L50R SCHEMATICS
PRE-AMP, REVERB AND PHASE INVERTER STAGES
DRAWN O.J.R
TRACED O.J.R
CHECKED O.J.R
APPROVED A P
SCALE NA
DWG # 101

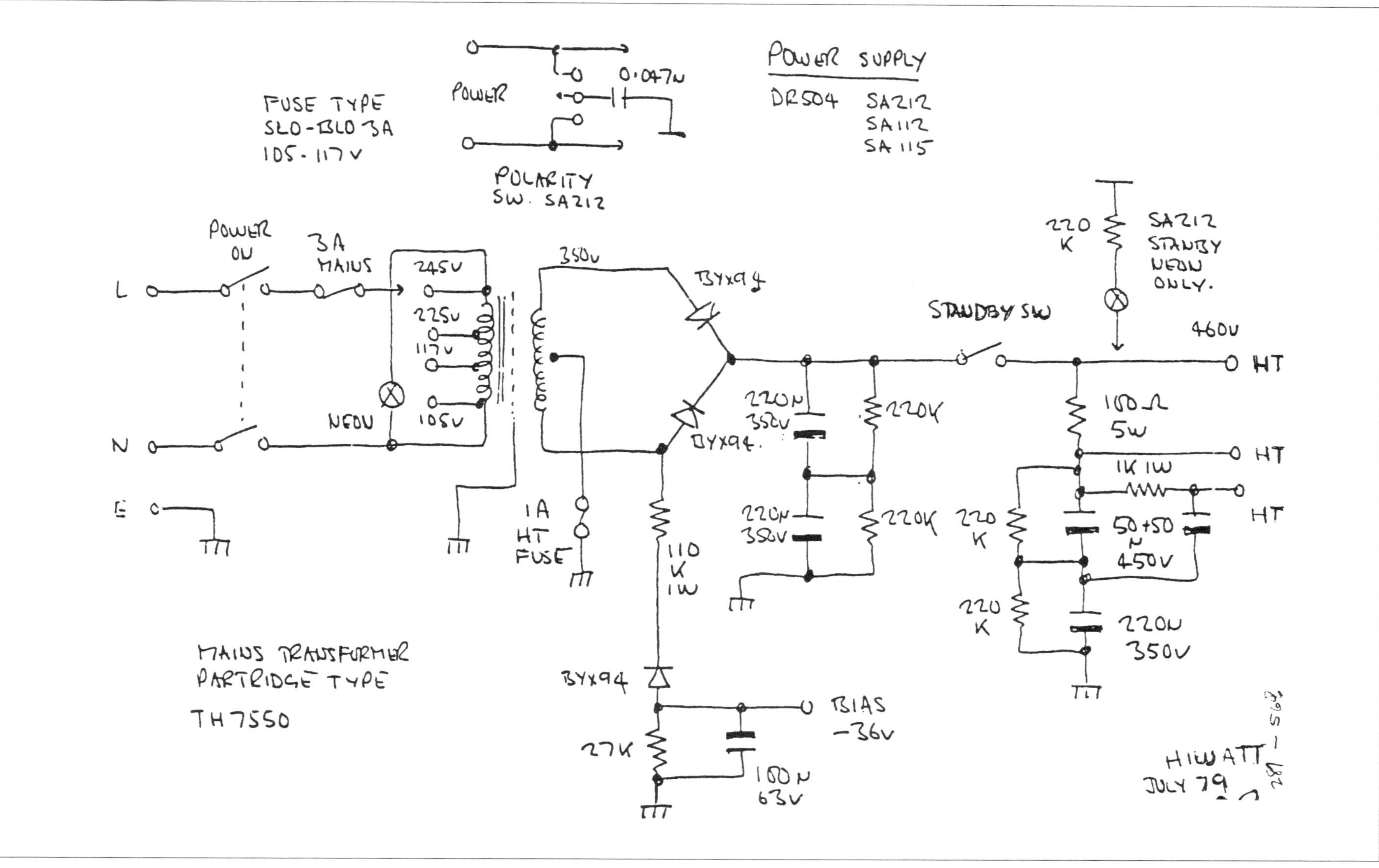

FUSE TYPE
SLO-BLO 3A
105-117V
POWER
0.047u
POLARITY
SW. SA212
POWER SUPPLY
DR504
SA212
SA112
SA115
POWER
ON
3A
MAINS
245V
225V
117V
105V
NEON
350V
BYX94
L
N
E
1A
HT
FUSE
110
K
1W
BYX94
220u
350V
220K
STANDBY SW
220
K
SA212
STANBY
NEON
ONLY.
460V
HT
100Ω
5W
1K 1W
50+50
450V
220u
350V
MAINS TRANSFORMER
PARTRIDGE TYPE
TH7550
27K
BIAS
-36V
100u
63V
HIWATT
JULY 79

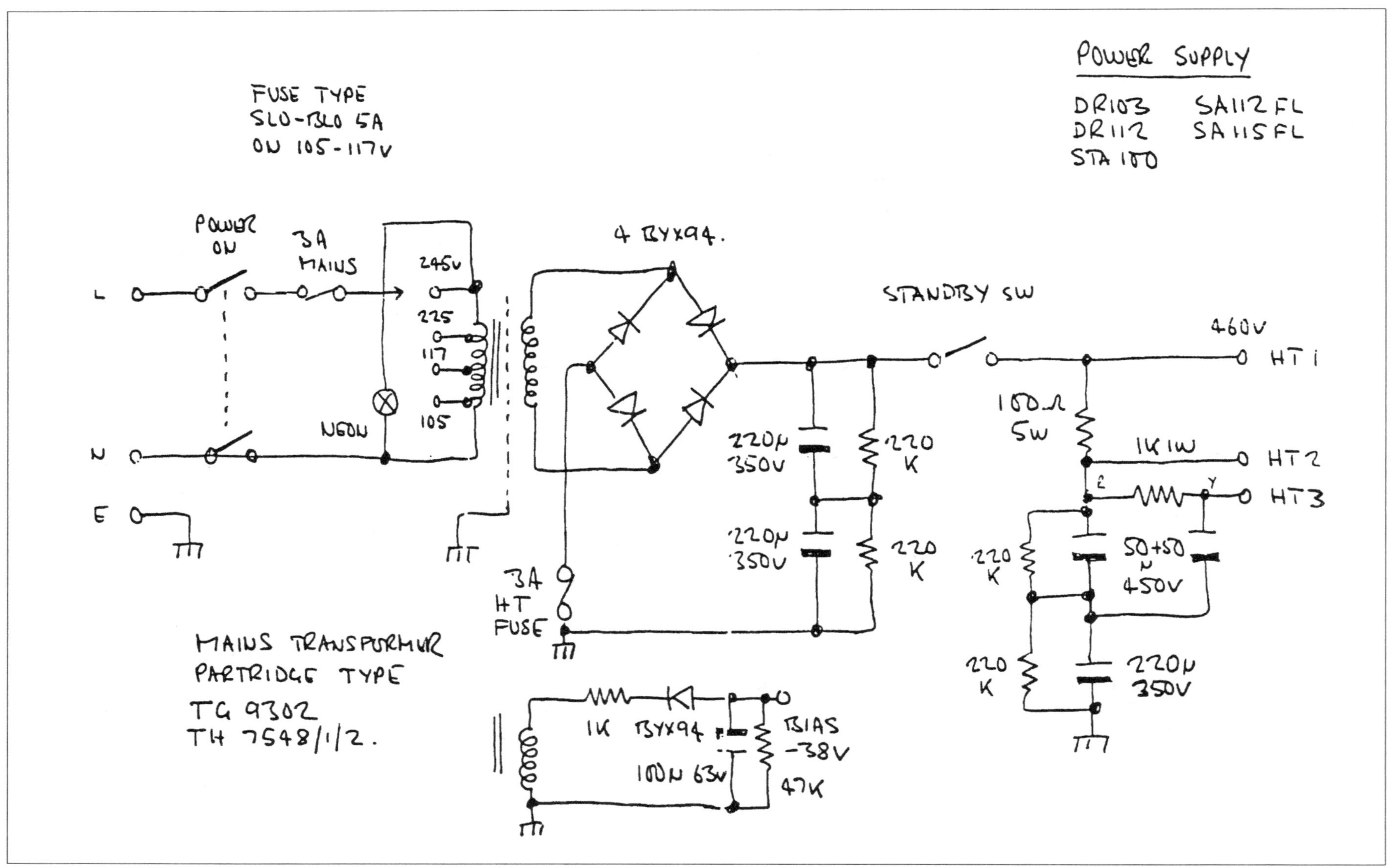
POWER SUPPLY
DR103
DR112
STA 100
SA112FL
SA115FL
FUSE TYPE
SLO-BLO 5A
ON 105-117V
POWER ON
3A MAINS
245v
225
117
105
NEON
L
N
E
4 BYX94.
STANDBY SW
460V
HT 1
HT 2
HT 3
100Ω 5W
1K 1W
220µ 350V
220 K
50+50µ 450V
3A HT FUSE
MAINS TRANSFORMER
PARTRIDGE TYPE
TG 9302
TH 7548/1/2.
1K
BYX94
100µ 63V
BIAS -38V
47K

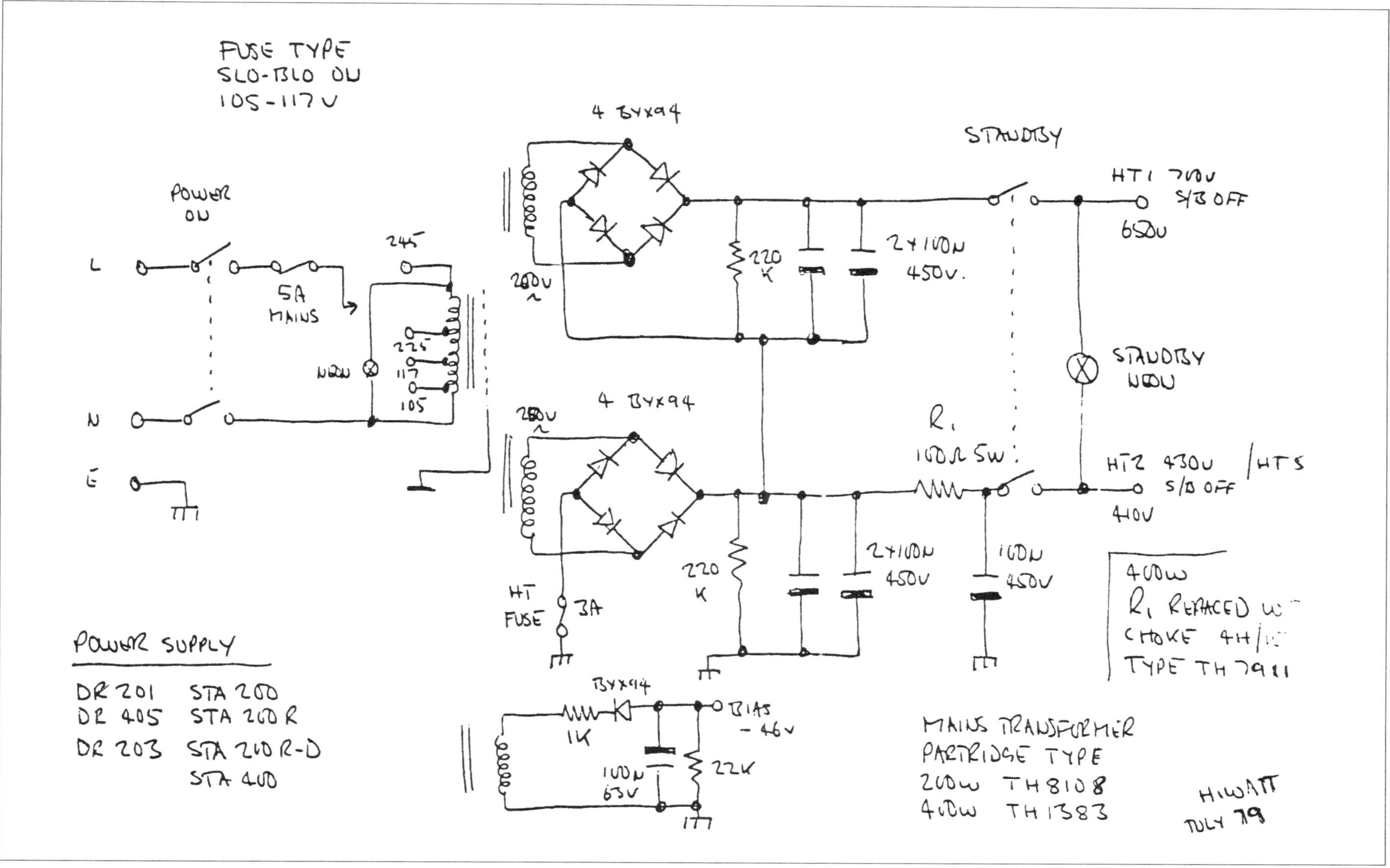
FUSE TYPE
SLO-BLO ON
105-117V
POWER ON
L
N
E
5A
MAINS
NEON
245
225
117
105
4 BYX94
200V
220 K
2×100N
450V.
STANDBY
HT1 700V S/B OFF
650V
STANDBY NEON
4 BYX94
200V
R1
100Ω 5W
HT2 430V S/B OFF
HTS
410V
2×100N
450V
100N
450V
220 K
HT FUSE
3A
400W
R1 REPLACED W
CHOKE 4H/
TYPE TH 7911
POWER SUPPLY
DR 201 STA 200
DR 405 STA 200 R
DR 203 STA 200 R-D
STA 400
BYX94
1K
BIAS
-46V
100N
63V
22K
MAINS TRANSFORMER
PARTRIDGE TYPE
200W TH8108
400W TH1383
HIWATT
JULY 79

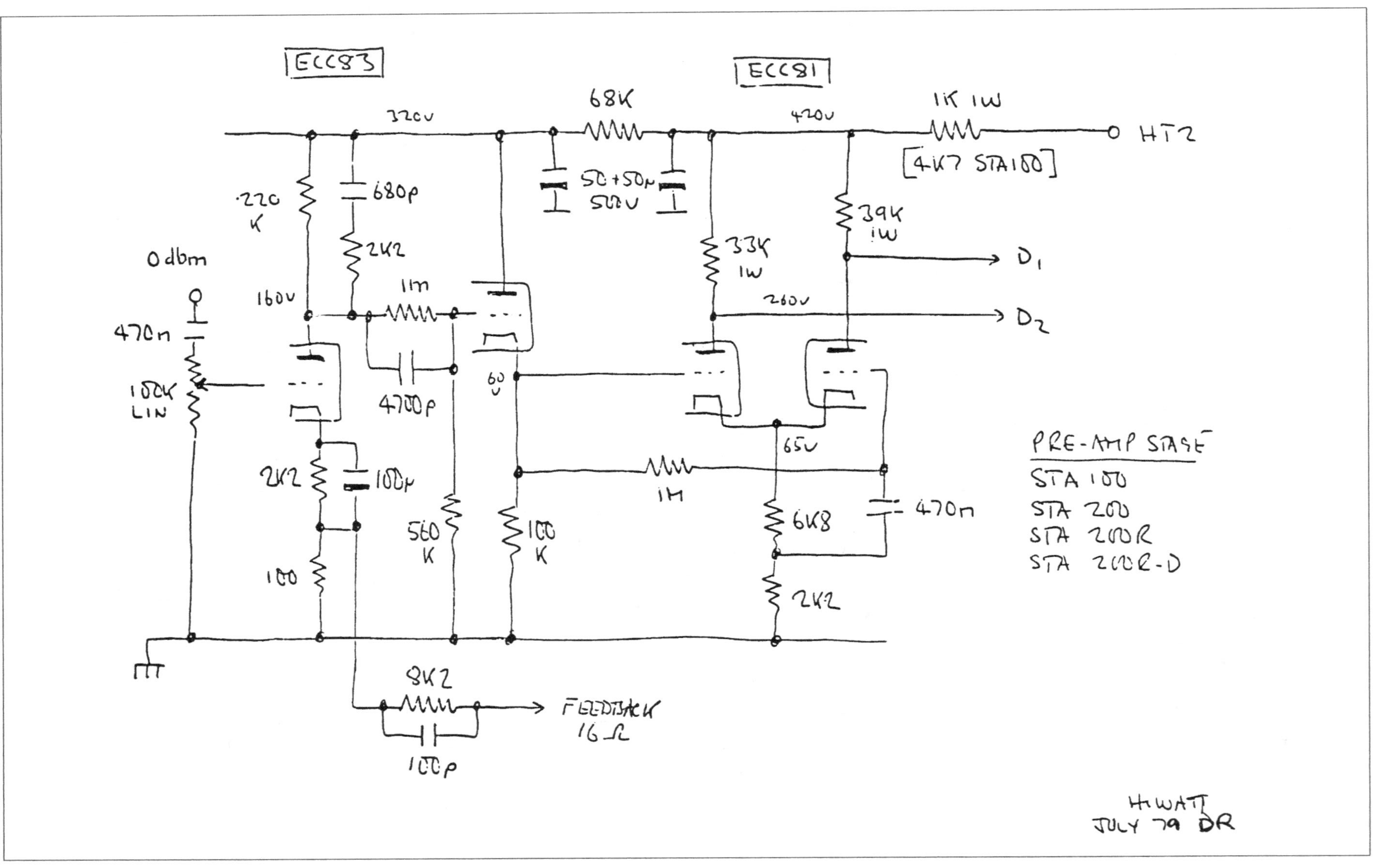
ECC83
ECC81
320v
68K
420v
1K 1W
HT2
[4K7 STA100]
220 K
680p
2K2
50+50μ 500v
39K 1W
33K 1W
0dbm
160v
1m
260v
D1
D2
470n
100K LIN
4700p
60 v
65v
PRE-AMP STAGE
STA 100
STA 200
STA 200R
STA 200R-D
2K2
100μ
1M
560 K
100 K
6K8
470n
100
2K2
8K2
FEEDBACK 16Ω
100p
HIWATT
JULY 79 DR

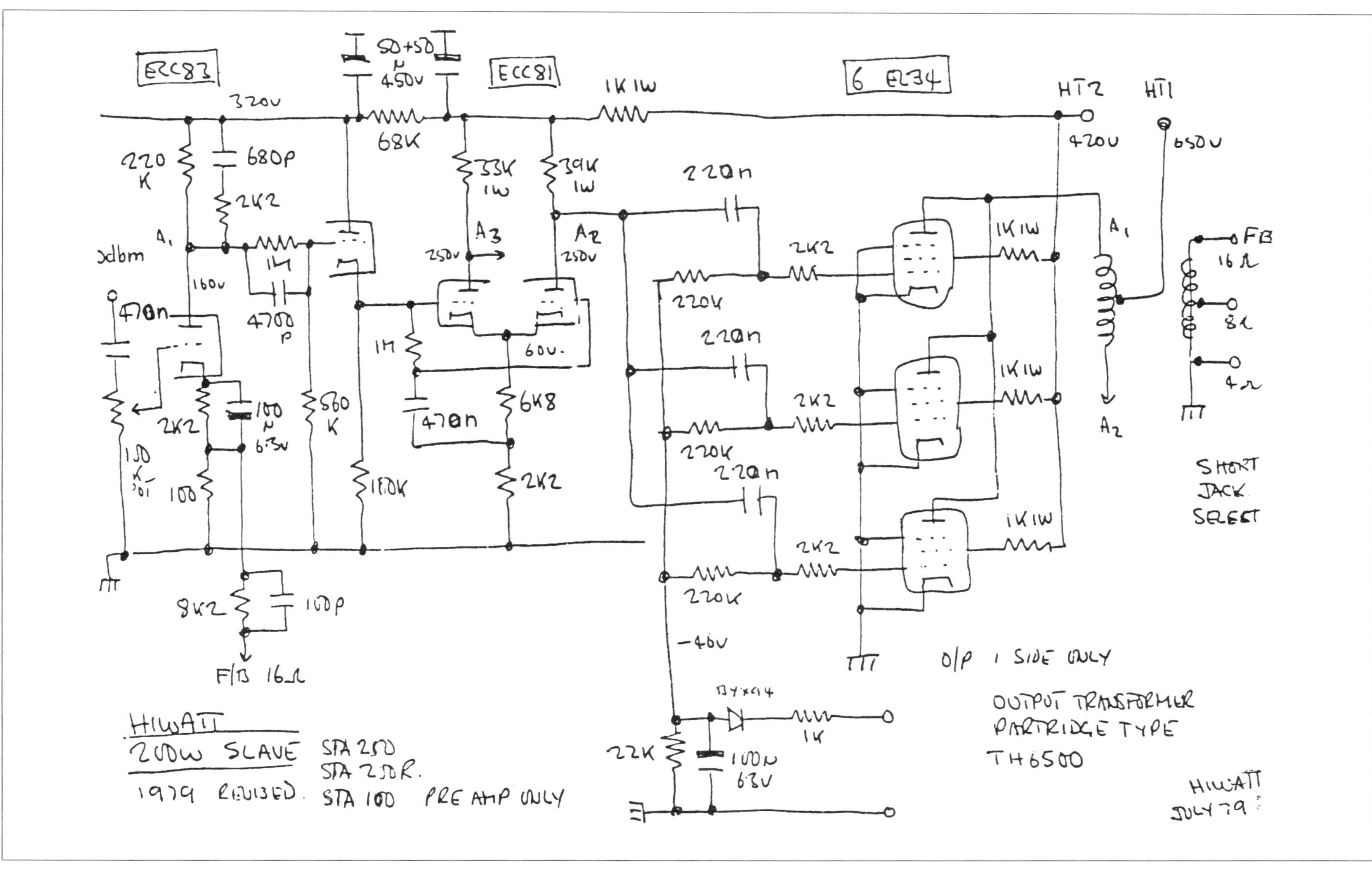
ECC83
ECC81
6 EL34
320v
50+50 µ 450v
68K
1K 1W
HT2
HT1
420v
650v
220 K
680p
2K2
33K 1W
39K 1W
220n
A1
A3
A2
250v
160v
1M
4700 p
470n
0dbm
60v.
100 µ 6.3v
560 K
2K2
100 K
100
180K
6K8
470n
2K2
220K
1K1W
FB 16Ω
8Ω
4Ω
SHORT JACK SELECT
8K2
100p
F/B 16Ω
-46v
BYX14
1K
22K
100µ 63V
O/P 1 SIDE ONLY
OUTPUT TRANSFORMER PARTRIDGE TYPE TH6500
HIWATT
200W SLAVE
STA 250
STA 250R.
1979 REVISED.
STA 100 PRE AMP ONLY
HIWATT JULY 79

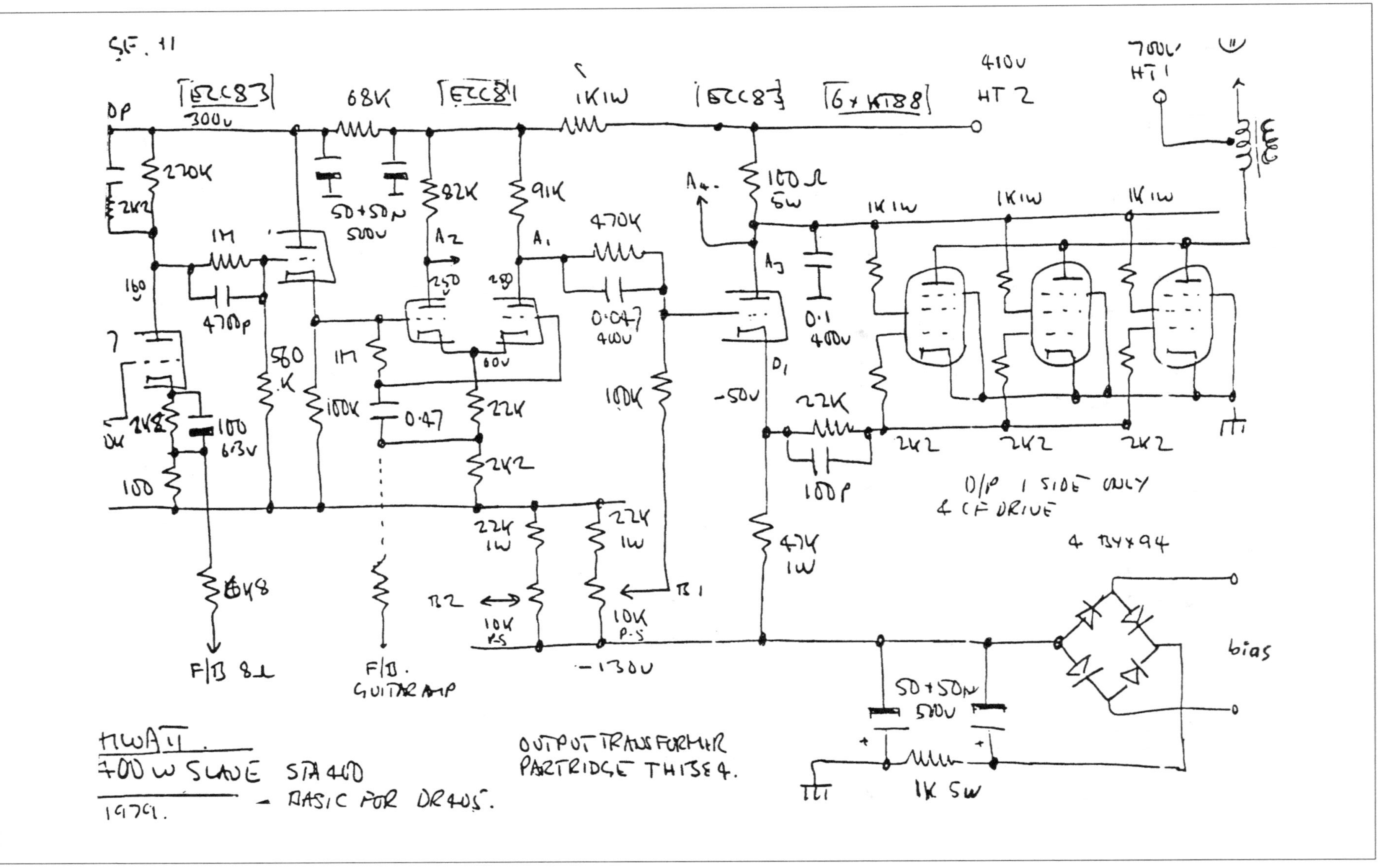
ECC83
300v
68K
ECC81
1K1W
ECC83
6 x KT88
410v
HT 2
700v
HT 1
220K
2K2
1M
470p
560K
100K
50+50µ
500v
82K
91K
A2
A1
470K
0·047
400v
100K
100Ω
5W
A4
A3
0·1
400v
D1
-50v
22K
100p
2K2
1K1W
0/P 1 SIDE ONLY
& CF DRIVE
4 BYX94
47K
1W
bias
1K 5W
22K
1W
10K
-130v
F/B 8Ω
F/B.
GUITAR AMP
HIWATT.
400 W SLAVE STA 400
1979.
– BASIC FOR DR405.
OUTPUT TRANSFORMER
PARTRIDGE TH1384.

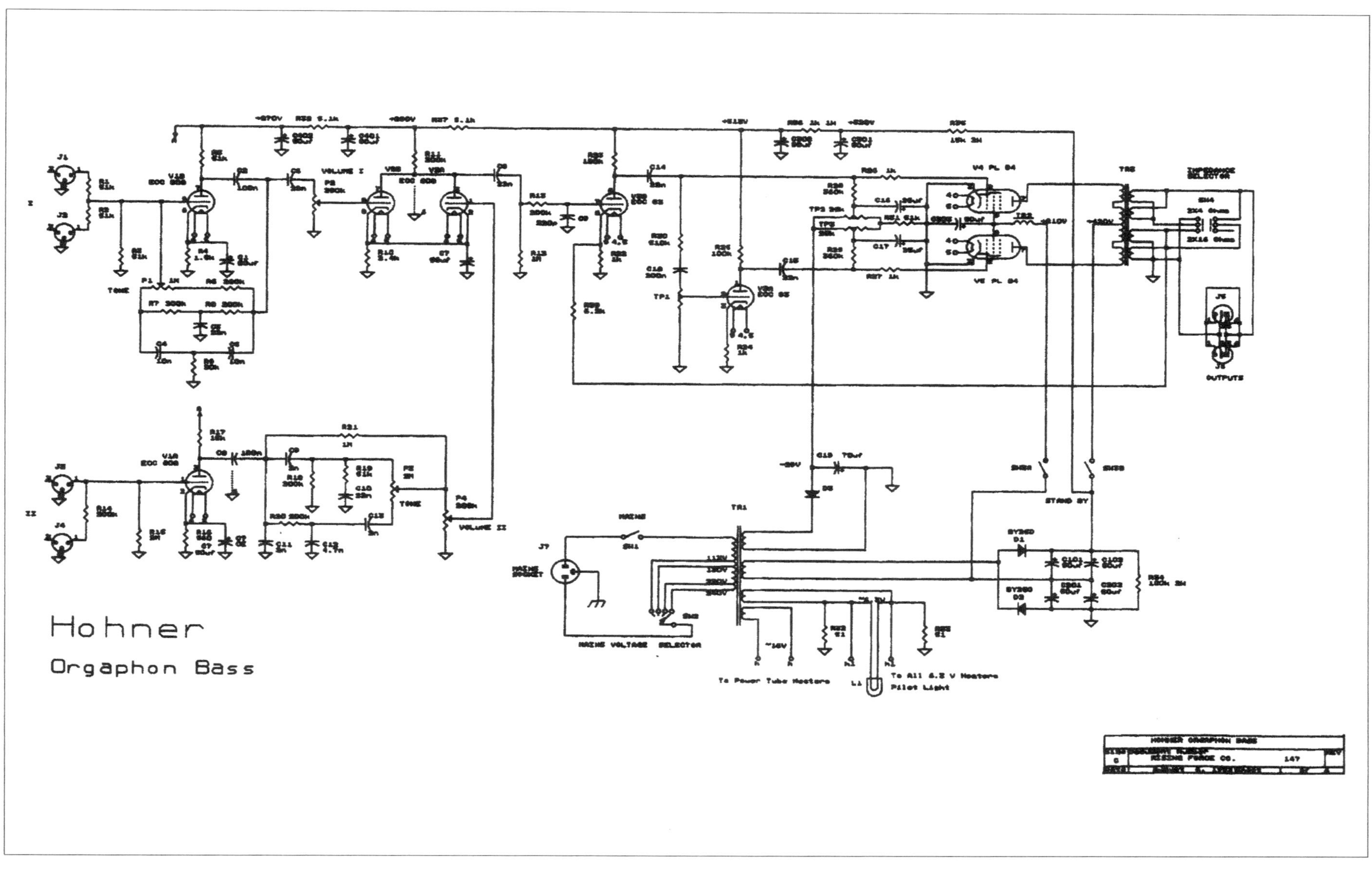
Hohner
Orgaphon Bass

ADVANCED TUBE BASS PRE AMP

Juergen Simon

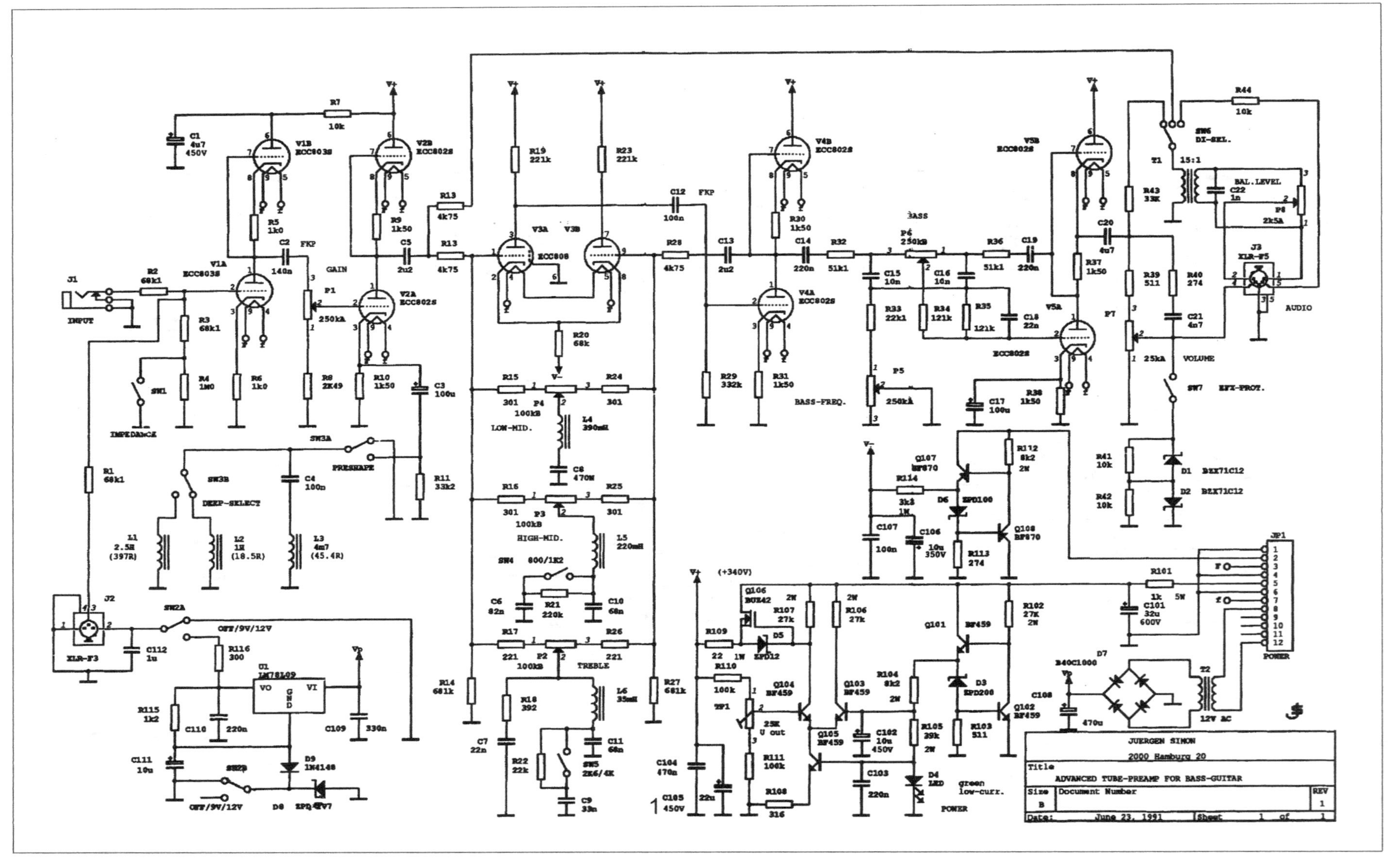

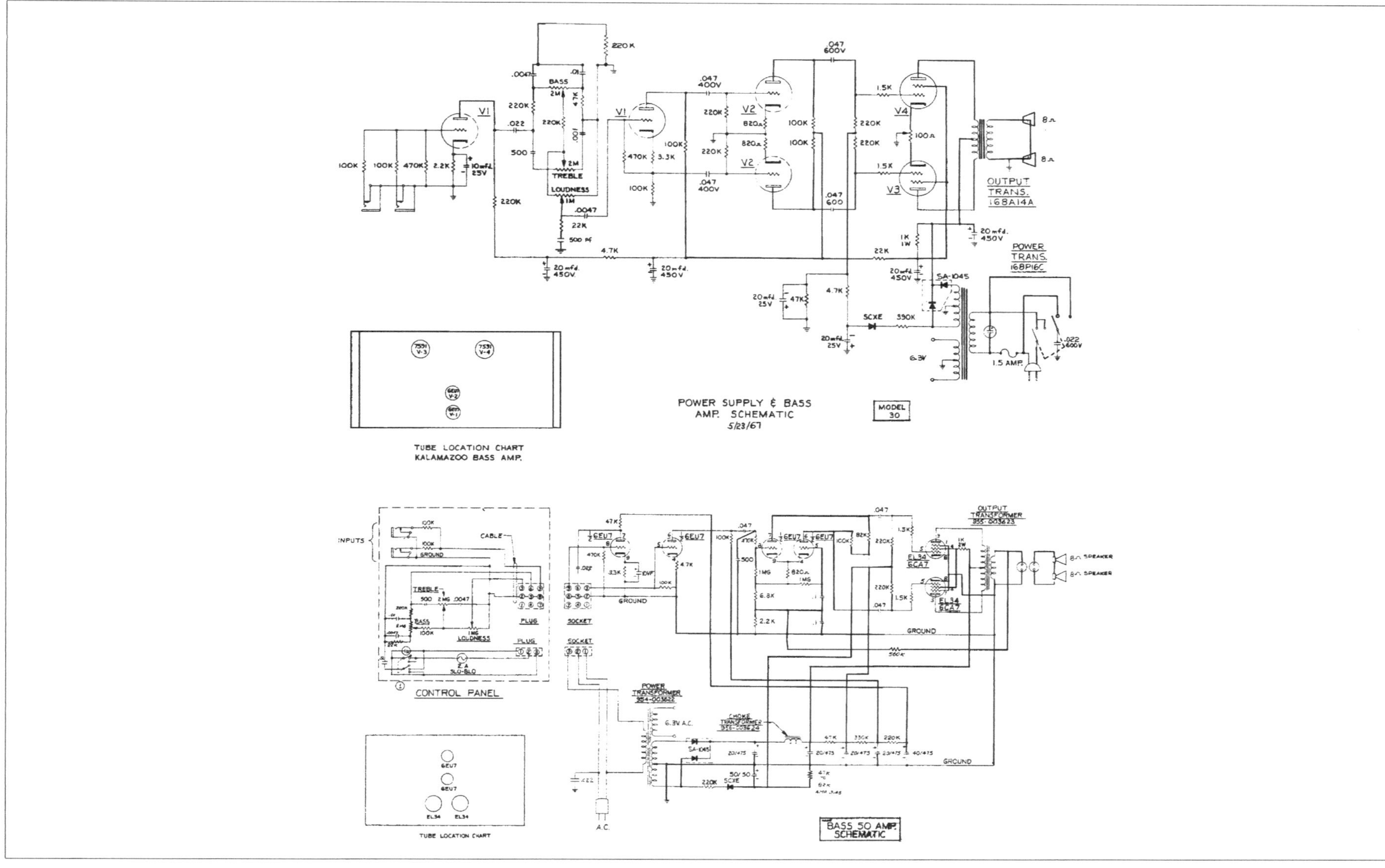
POWER SUPPLY & BASS AMP. SCHEMATIC
5/23/67
MODEL 30
TUBE LOCATION CHART
KALAMAZOO BASS AMP.
OUTPUT TRANS. 168A14A
POWER TRANS. 168P16C
CONTROL PANEL
INPUTS
TUBE LOCATION CHART
BASS 50 AMP. SCHEMATIC
OUTPUT TRANSFORMER
POWER TRANSFORMER 354-003622
GROUND
A.C.

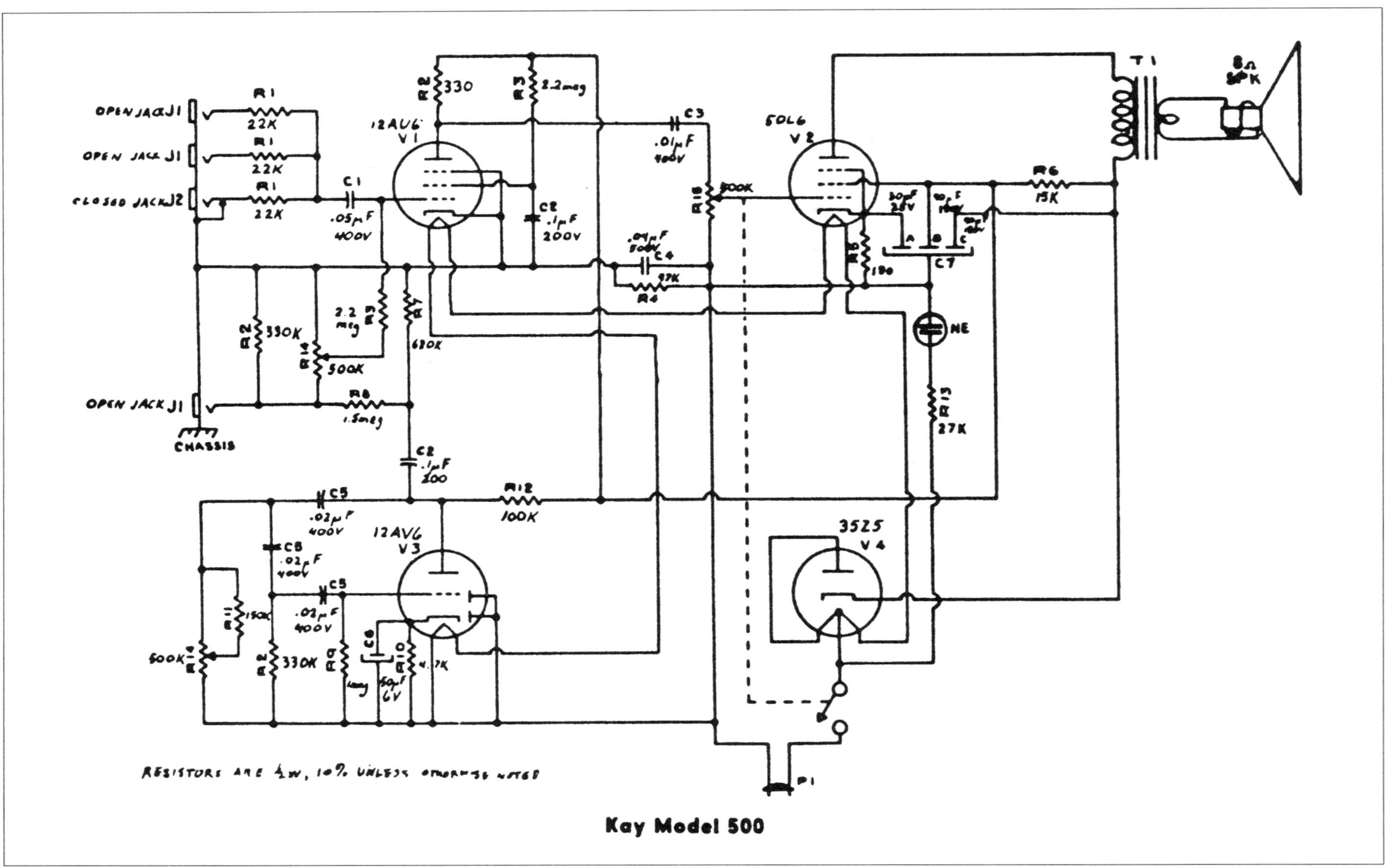

Kay Model 500

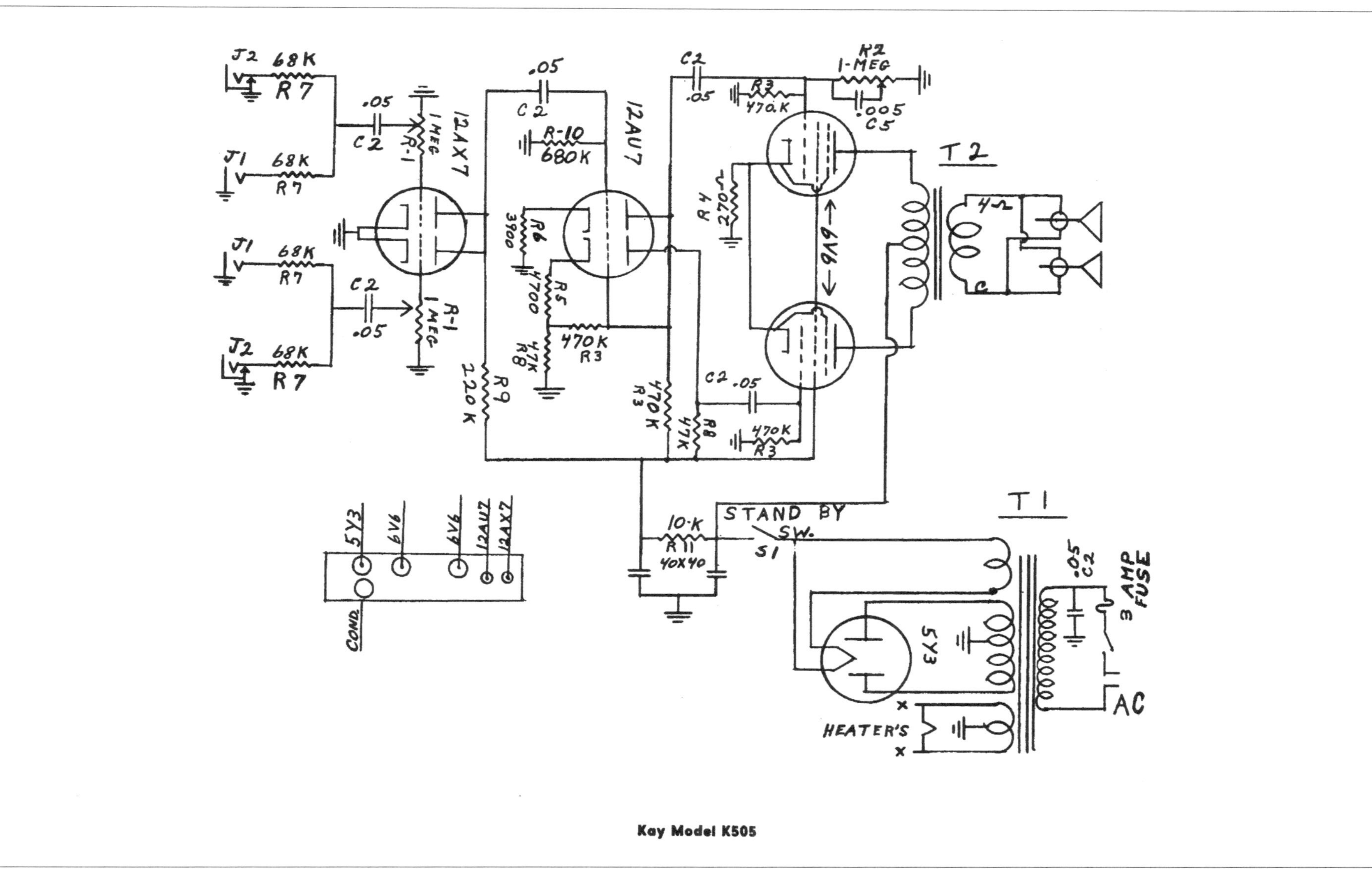

Kay Model K505

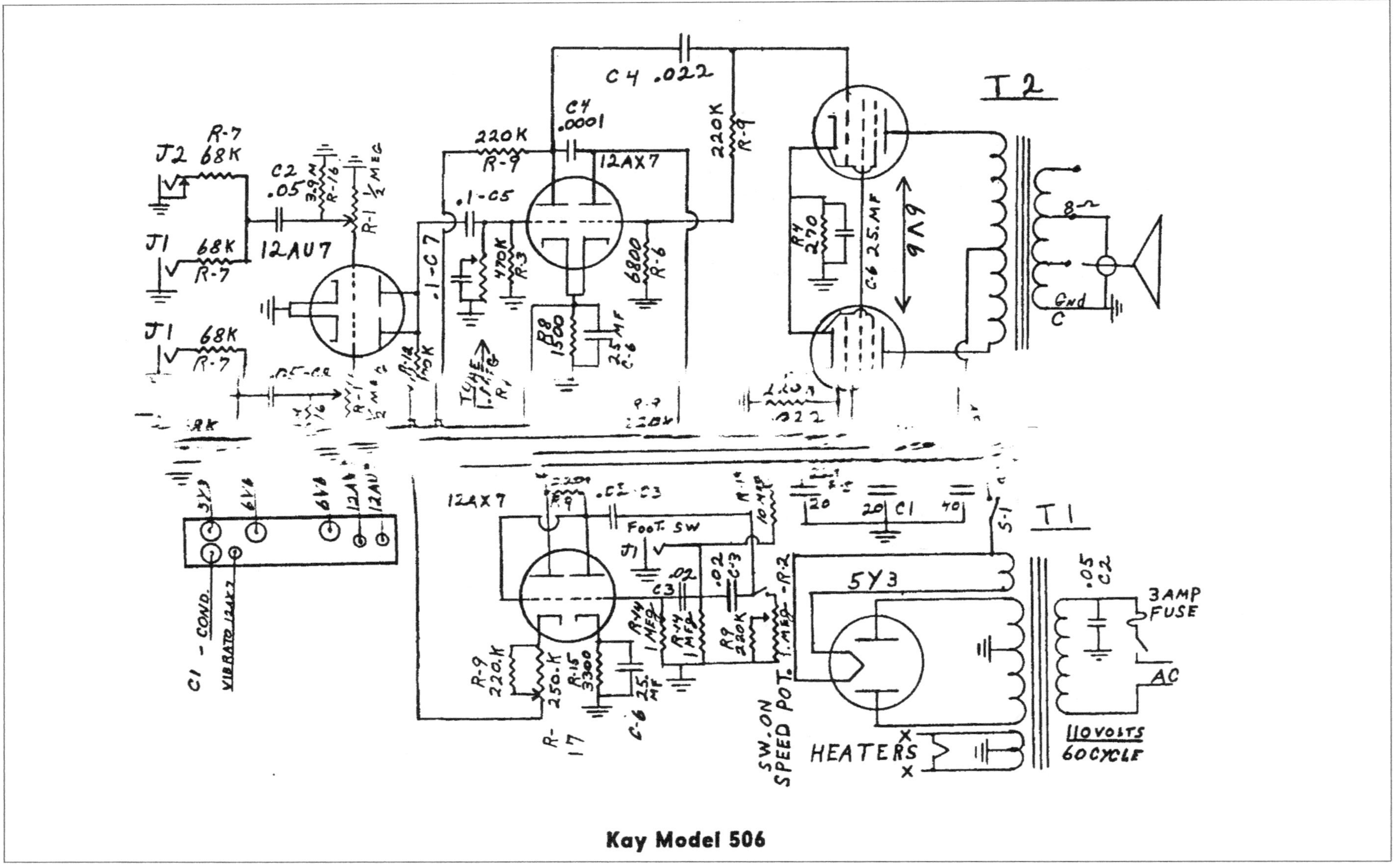

Kay Model 506

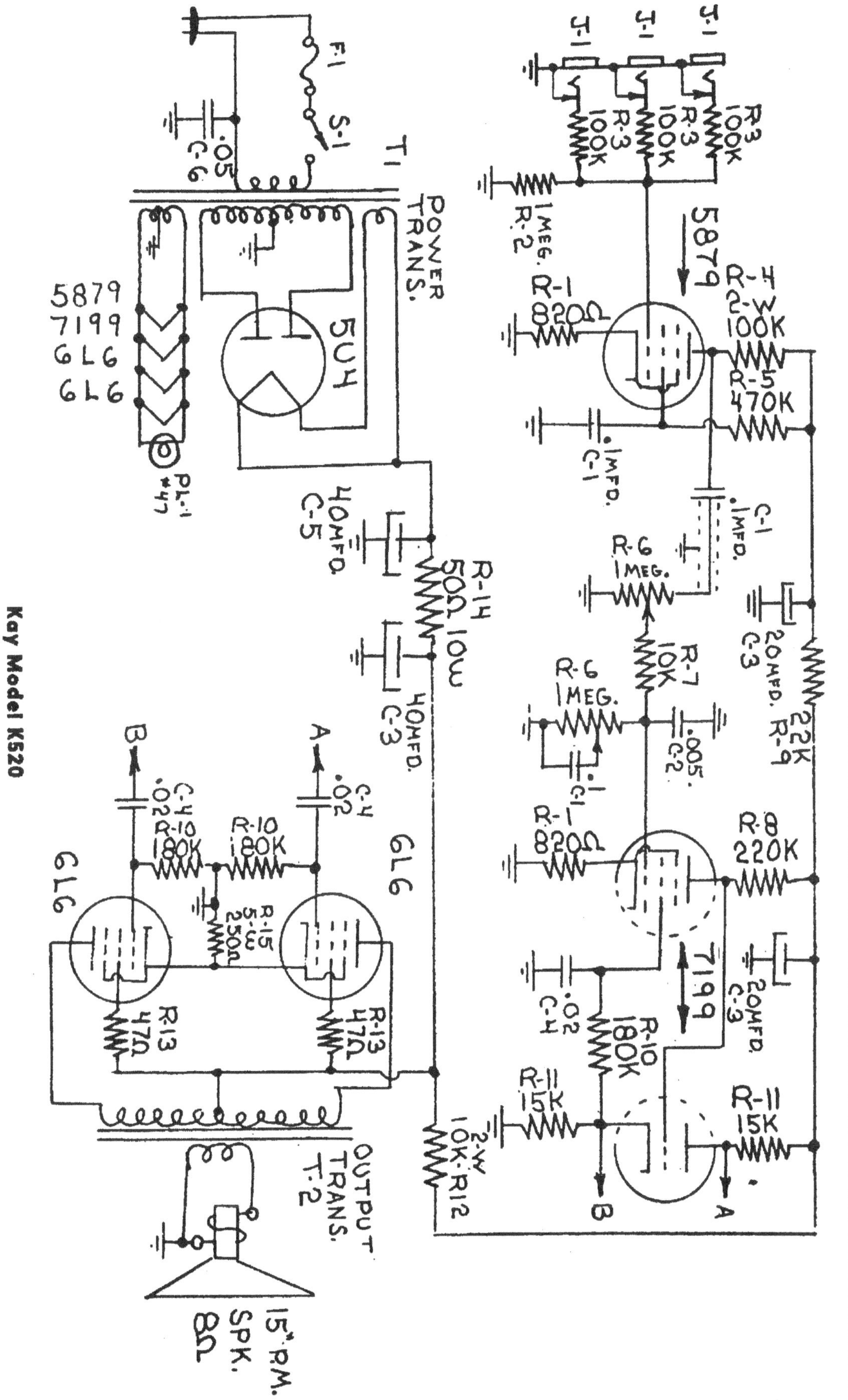

Kay

Kay Model K520

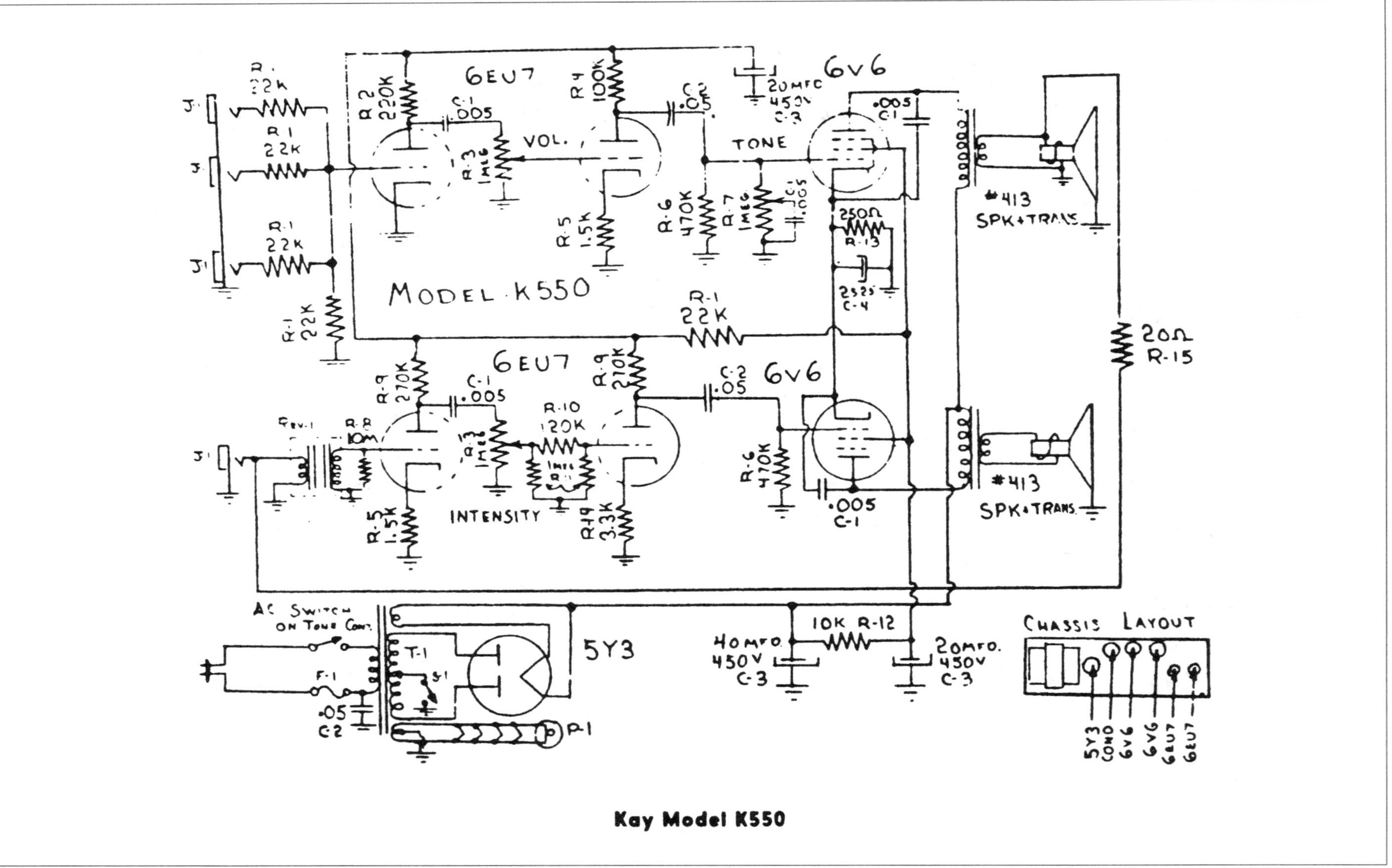

Kay Model K550

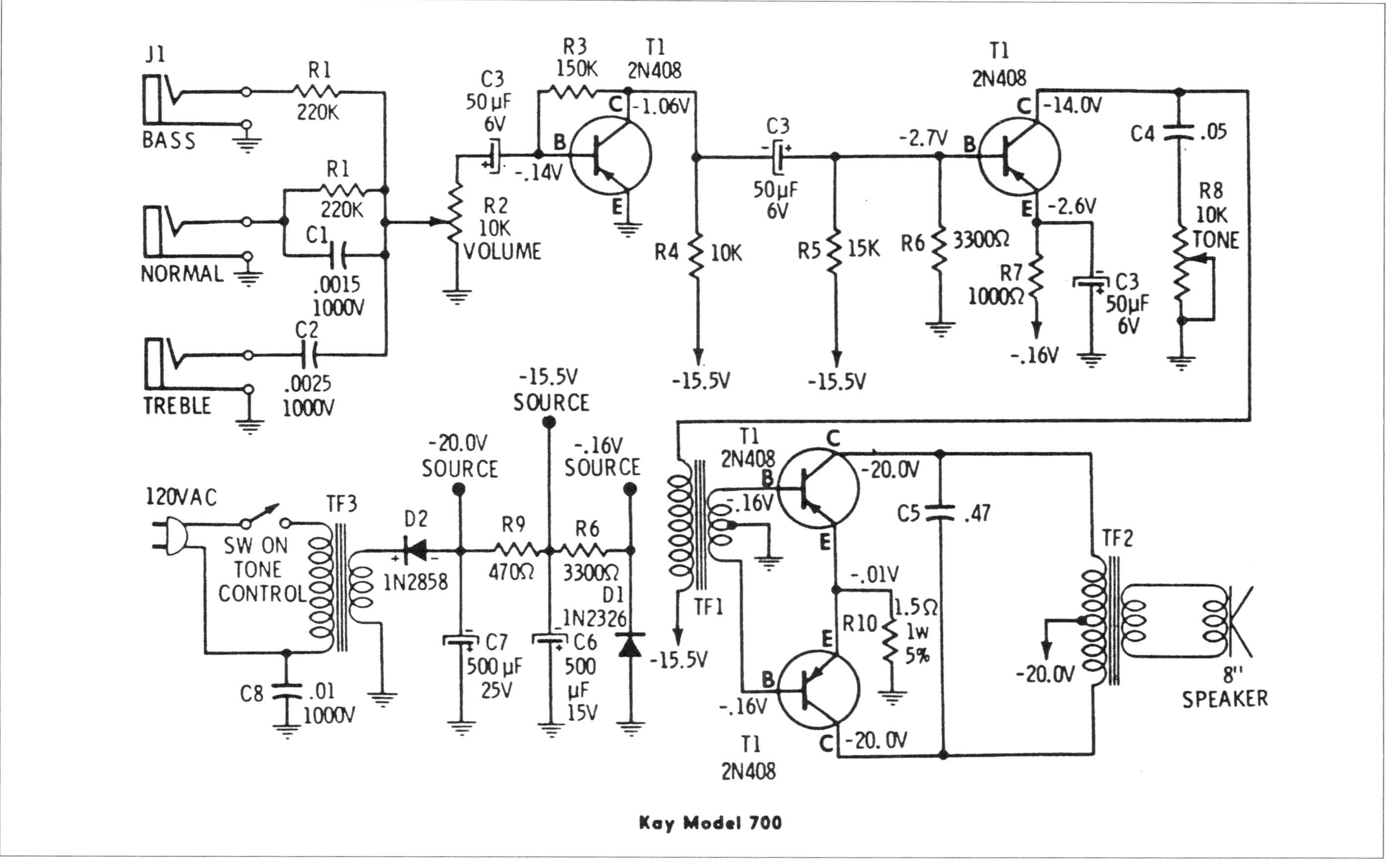

Kay Model 700

Small and Medium Amplifiers

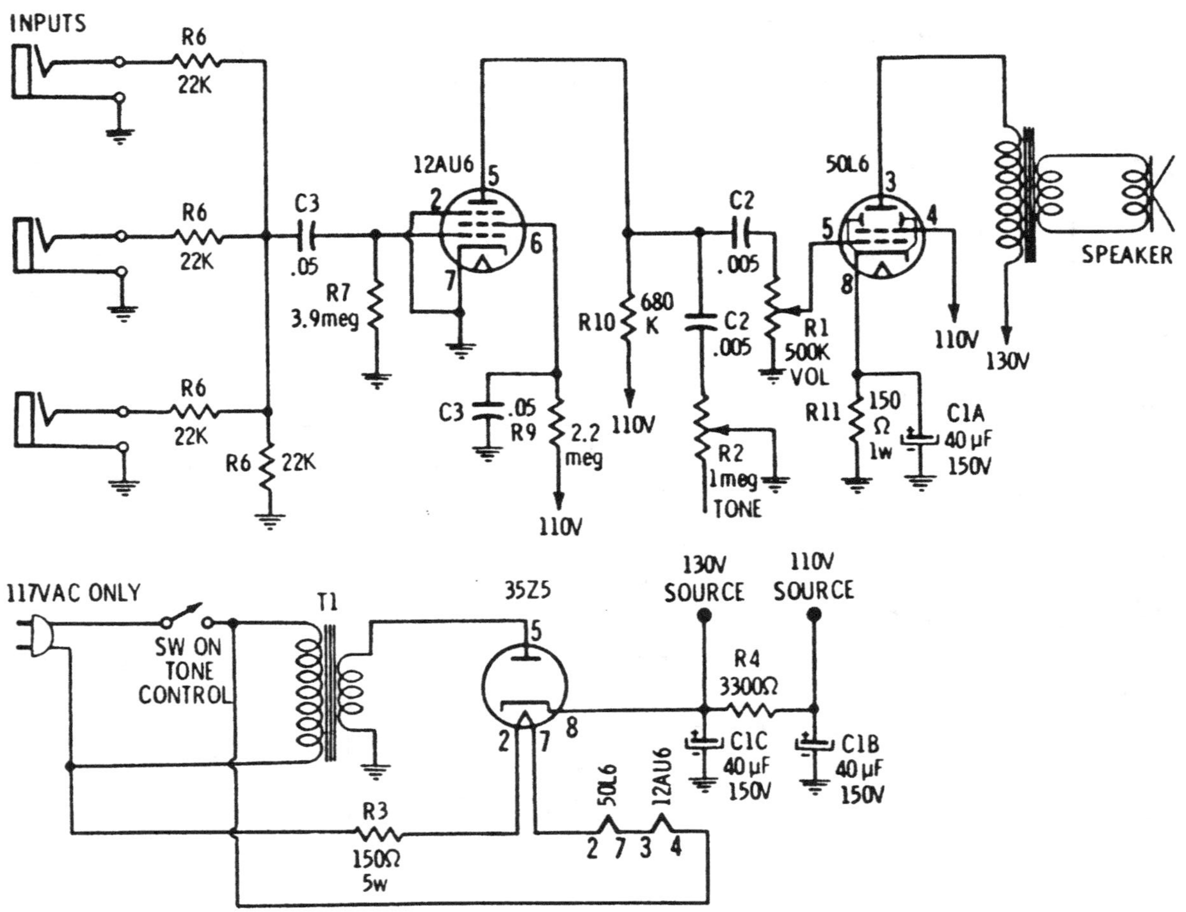

Kay Model 703

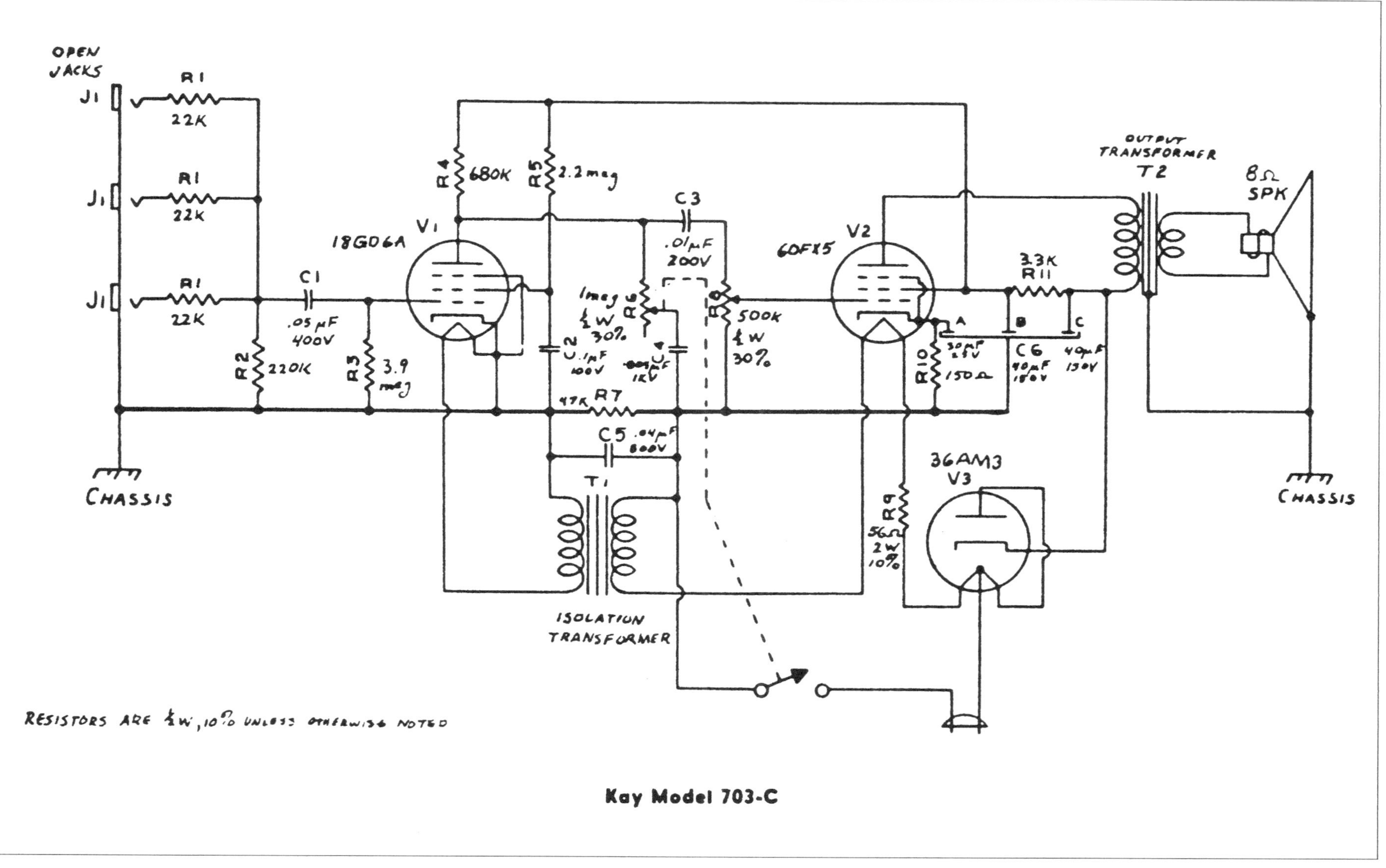

Kay Model 703-C

Small and Medium Amplifiers

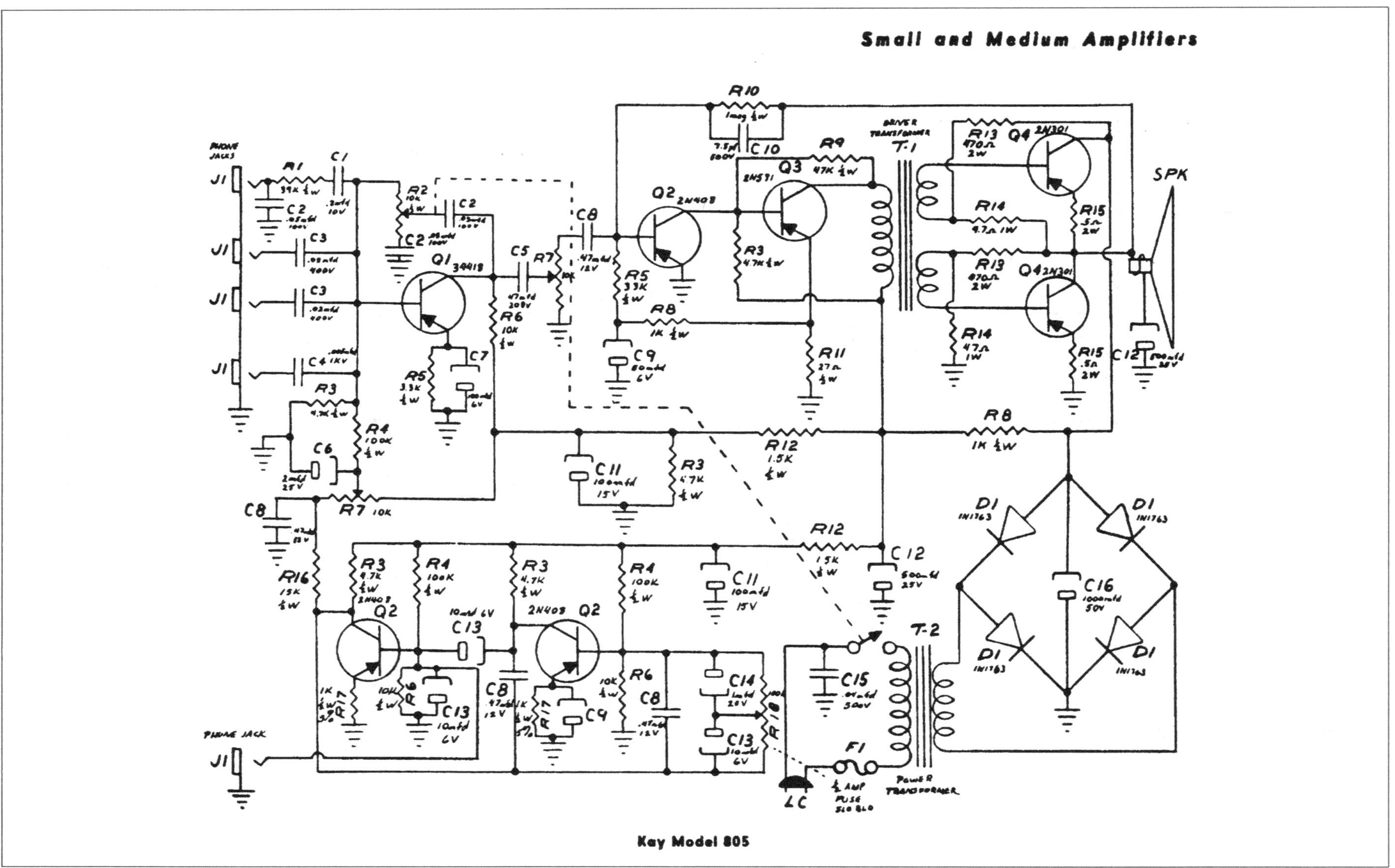

Kay Model 805

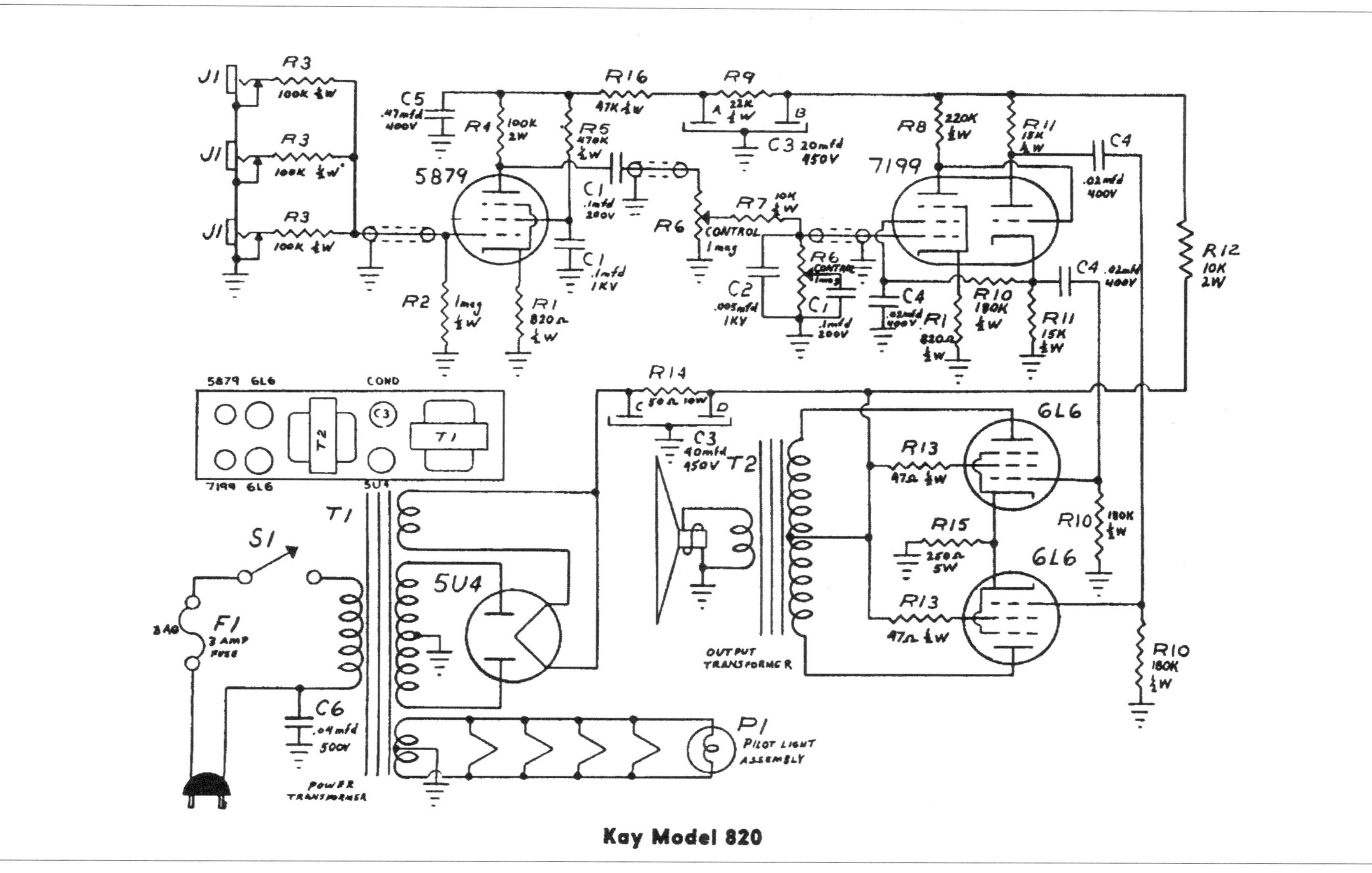

Kay Model 820

High-Power Amplifiers

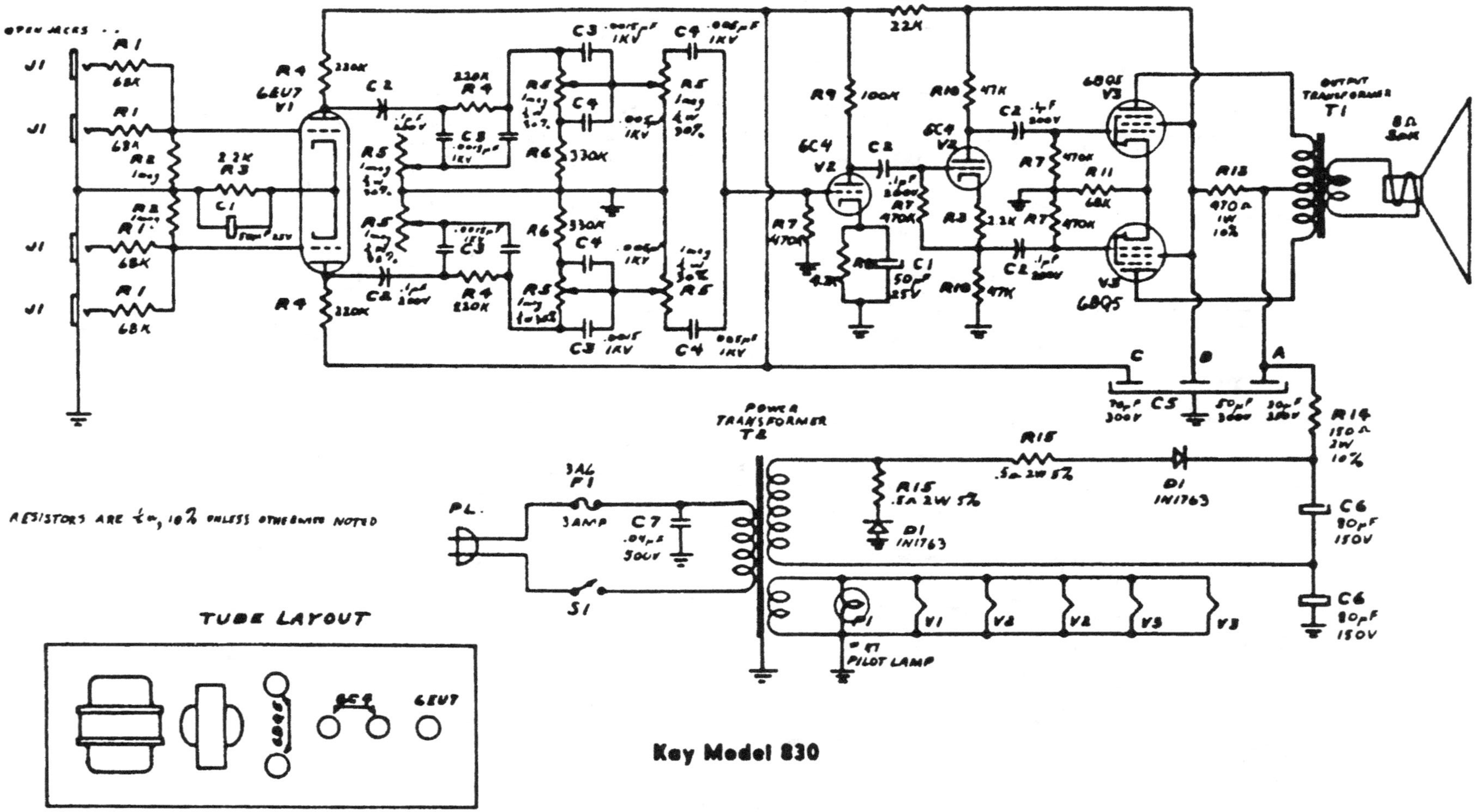

RESISTORS ARE ¼w, 10% UNLESS OTHERWISE NOTED

TUBE LAYOUT

Kay Model 830

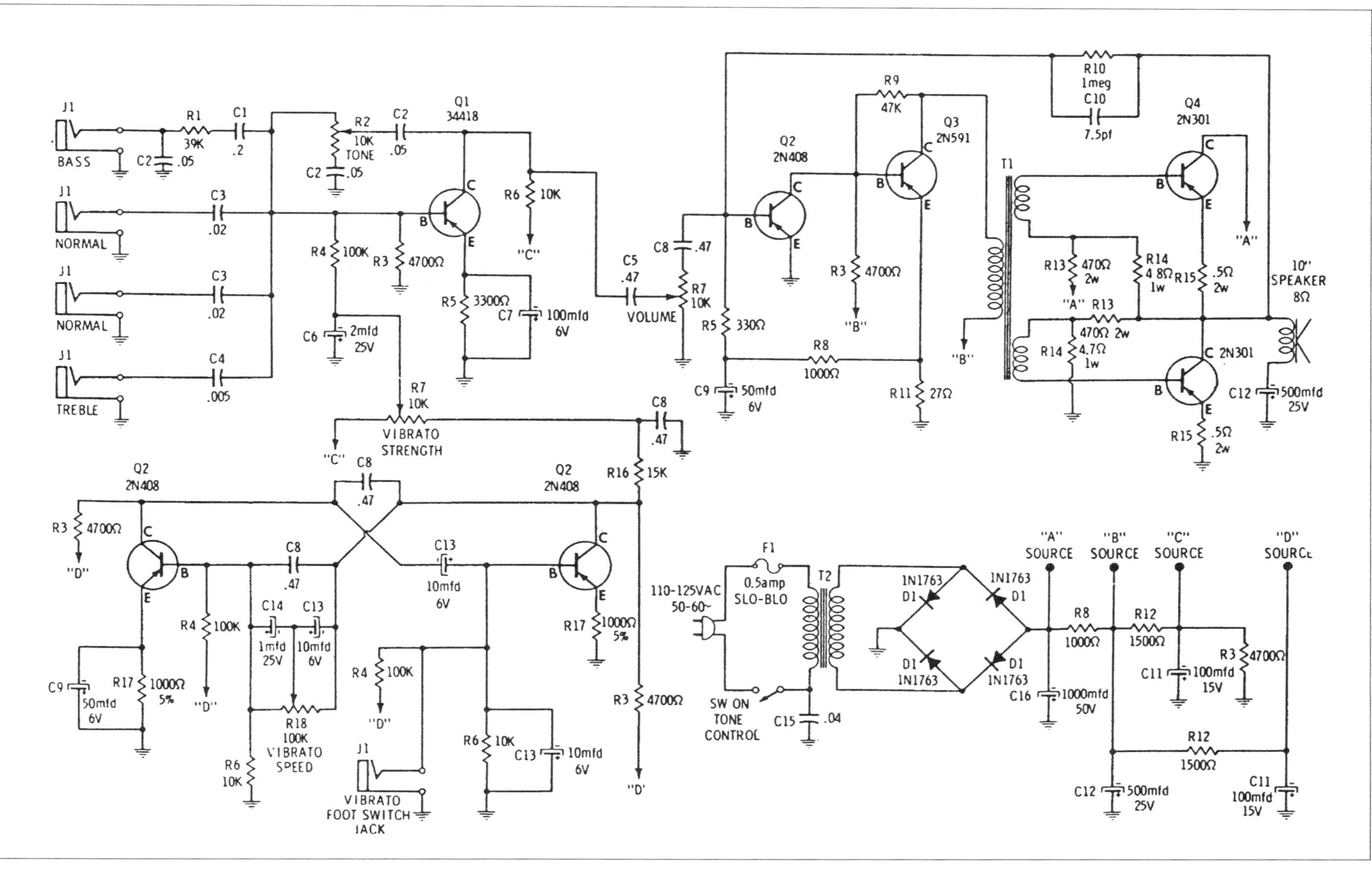
J1
BASS
R1
39K
C1
.2
C2
.05
R2
10K
TONE
Q1
34418
J1
NORMAL
C3
.02
J1
NORMAL
C3
.02
J1
TREBLE
C4
.005
R4
100K
R3
4700Ω
R6
10K
"C"
R5
3300Ω
C7
100mfd
6V
C6
2mfd
25V
C5
.47
VOLUME
R7
10K
C8
.47
Q2
2N408
R5
330Ω
C9
50mfd
6V
R8
1000Ω
R9
47K
Q3
2N591
R3
4700Ω
"B"
R11
27Ω
R10
1meg
C10
7.5pf
T1
Q4
2N301
R13
470Ω
2w
"A"
R14
4.8Ω
1w
R15
.5Ω
2w
10" SPEAKER
8Ω
470Ω 2w
4.7Ω
2N301
C12
500mfd
25V
R7
10K
VIBRATO
STRENGTH
C8
.47
R16
15K
Q2
2N408
Q2
2N408
R3
4700Ω
"D"
C8
.47
C13
10mfd
6V
R4
100K
C14
1mfd
25V
C13
10mfd
6V
R18
100K
VIBRATO
SPEED
R6
10K
C9
50mfd
6V
R17
1000Ω
5%
J1
VIBRATO
FOOT SWITCH
JACK
R6
10K
C13
10mfd
6V
R17
1000Ω
5%
R3
4700Ω
"D"
F1
0.5amp
SLO-BLO
110-125VAC
50-60~
T2
SW ON
TONE
CONTROL
C15
.04
1N1763
D1
C16
1000mfd
50V
"A" SOURCE
"B" SOURCE
"C" SOURCE
"D" SOURCE
R8
1000Ω
R12
1500Ω
C11
100mfd
15V
R3
4700Ω
R12
1500Ω
C12
500mfd
25V
C11
100mfd
15V

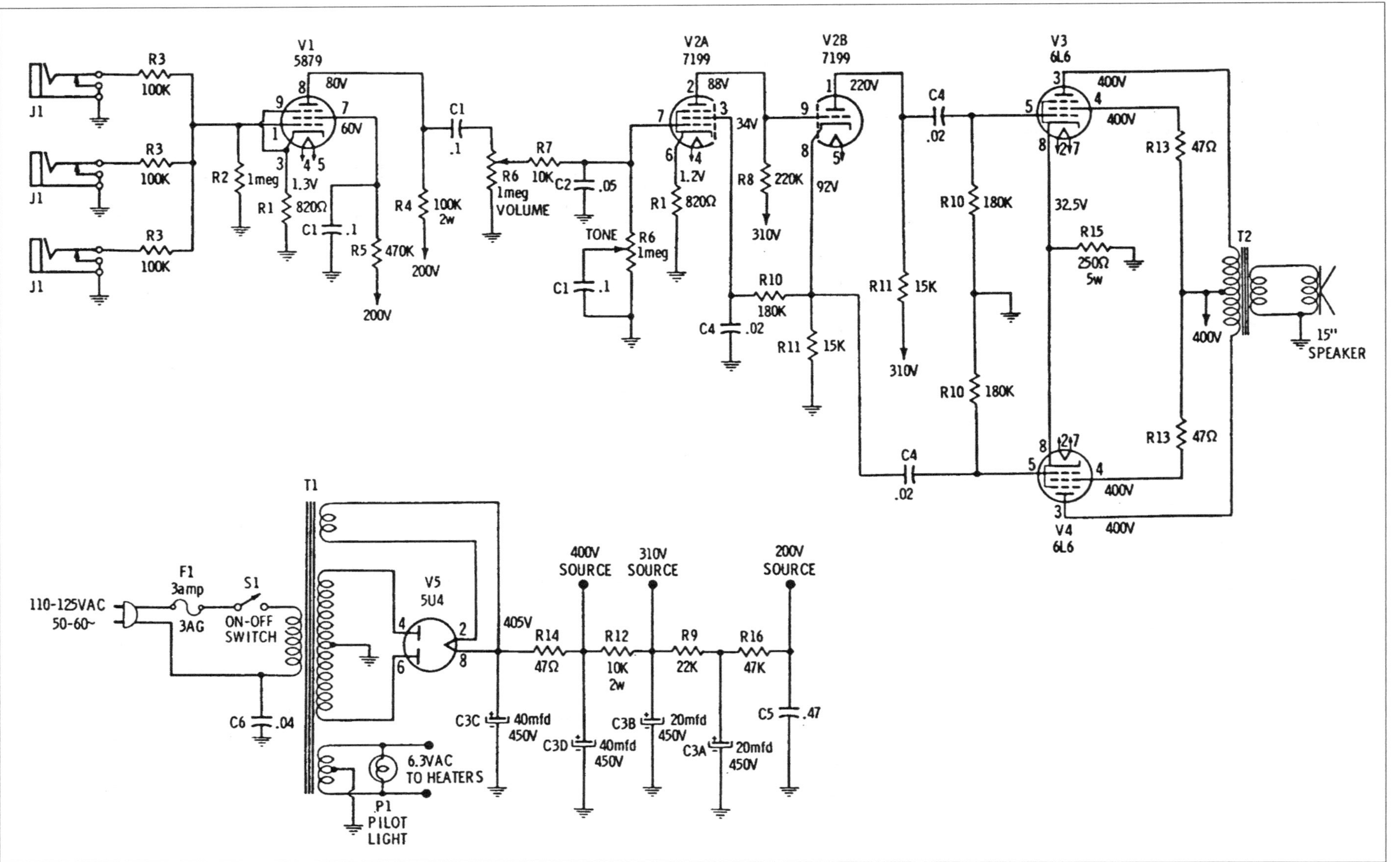
J1
R3
100K
V1
5879
80V
60V
1.3V
R2 1meg
R1 820Ω
C1 .1
R5 470K
200V
R4 100K 2w
C1 .1
R6 1meg VOLUME
R7 10K
C2 .05
TONE
R6 1meg
C1 .1
V2A
7199
88V
34V
1.2V
R1 820Ω
R8 220K
310V
V2B
7199
220V
92V
R10 180K
C4 .02
R11 15K
C4 .02
R10 180K
R11 15K
310V
V3
6L6
400V
400V
32.5V
R15 250Ω 5w
R13 47Ω
V4
6L6
400V
400V
T2
400V
15" SPEAKER
T1
110-125VAC 50-60~
F1 3amp 3AG
S1 ON-OFF SWITCH
C6 .04
V5
5U4
405V
R14 47Ω
R12 10K 2w
R9 22K
R16 47K
400V SOURCE
310V SOURCE
200V SOURCE
C3C 40mfd 450V
C3D 40mfd 450V
C3B 20mfd 450V
C3A 20mfd 450V
C5 .47
6.3VAC TO HEATERS
P1 PILOT LIGHT

A100H

Laney

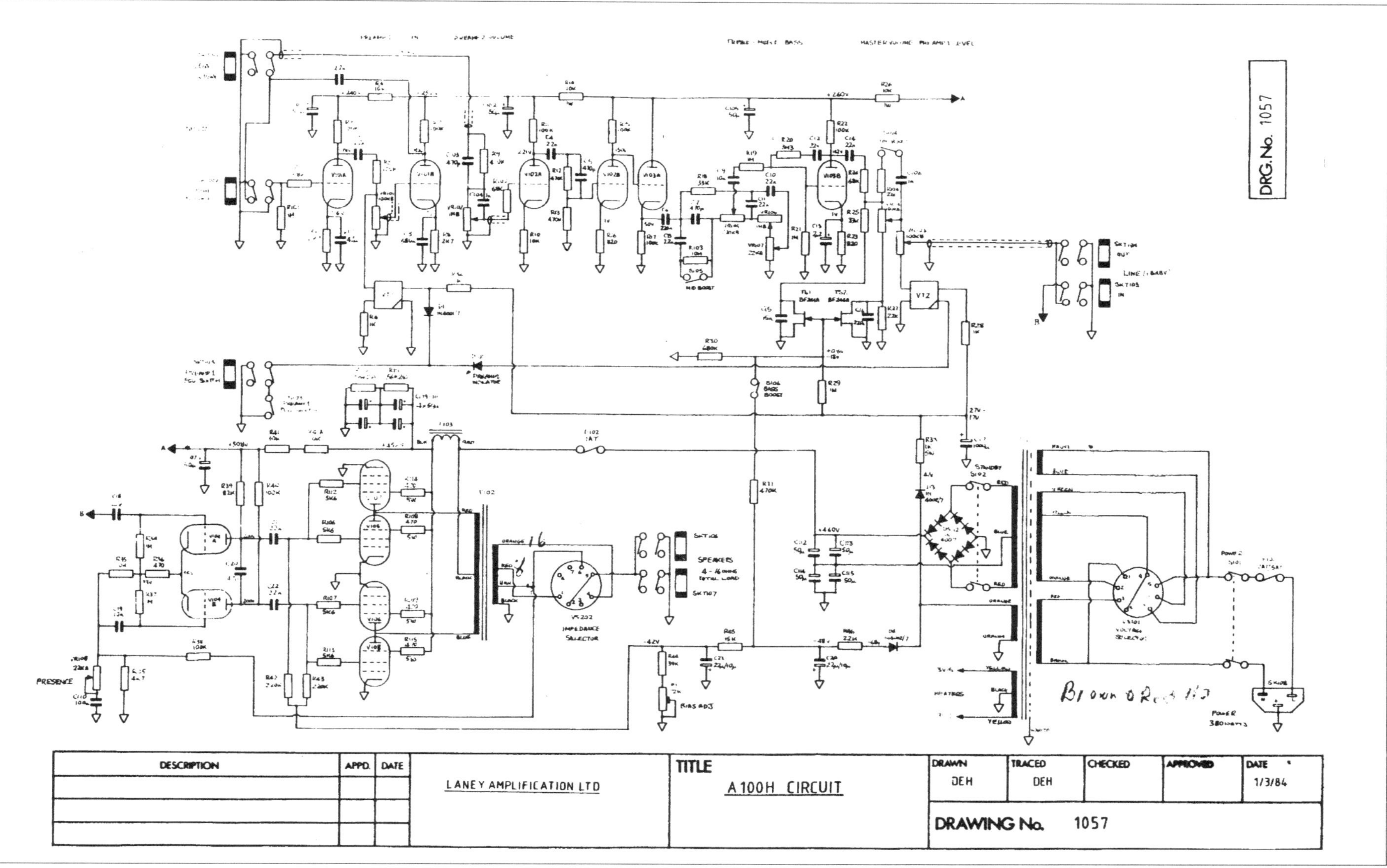

DRG.No. 1057
SPEAKERS
4 - 16 OHMS TOTAL LOAD
IMPEDANCE SELECTOR
PRESENCE
DESCRIPTION
APPD.
DATE
LANEY AMPLIFICATION LTD
TITLE
A100H CIRCUIT
DRAWN
DEH
TRACED
DEH
CHECKED
APPROVED
DATE
1/3/84
DRAWING No. 1057

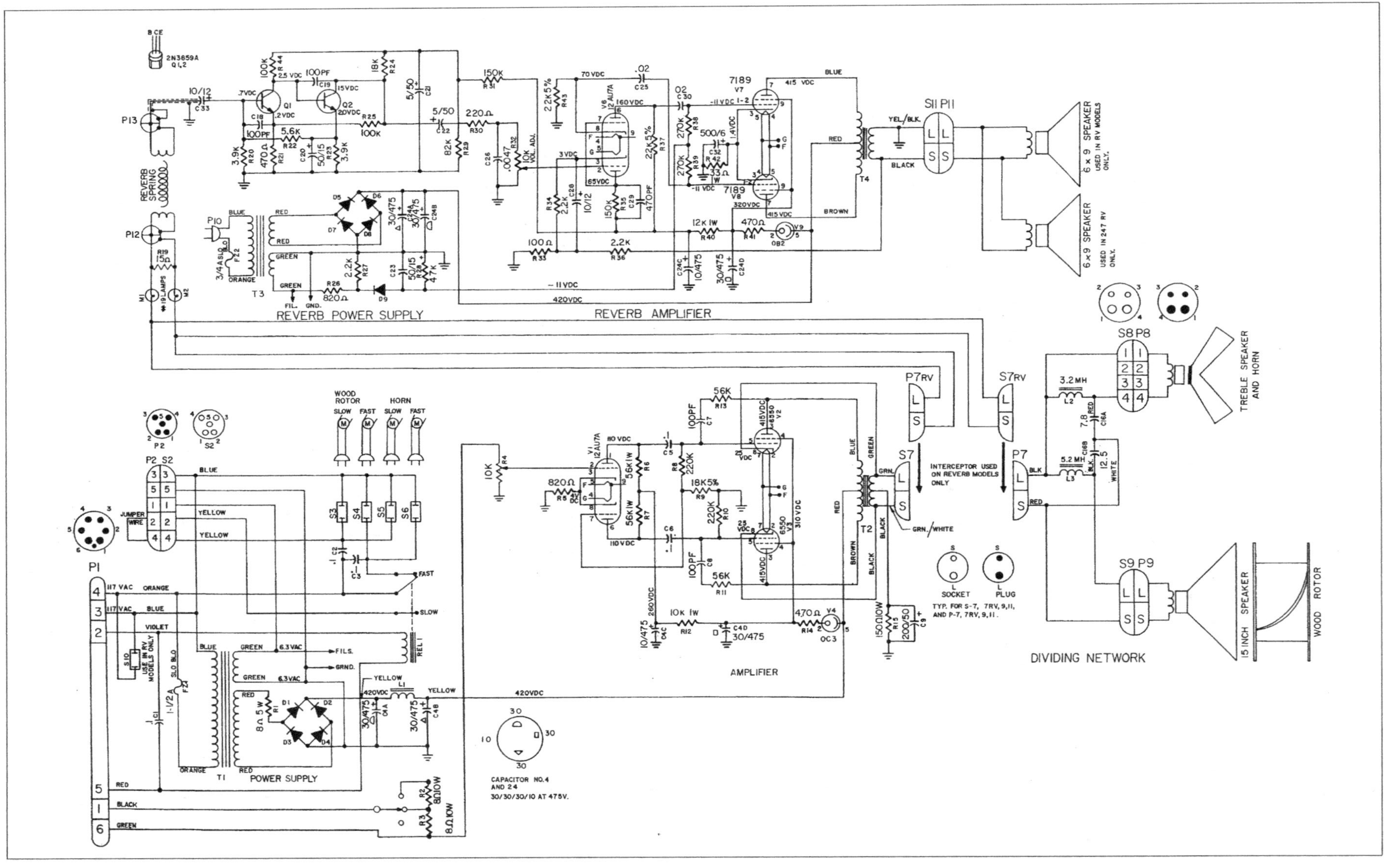
REVERB POWER SUPPLY
REVERB AMPLIFIER
AMPLIFIER
POWER SUPPLY
DIVIDING NETWORK
6 x 9 SPEAKER USED IN RV MODELS ONLY.
6 x 9 SPEAKER USED IN 247 RV ONLY.
TREBLE SPEAKER AND HORN
15 INCH SPEAKER
WOOD ROTOR
INTERCEPTOR USED ON REVERB MODELS ONLY
SOCKET
PLUG
TYP. FOR S-7, 7RV, 9,11, AND P-7, 7RV, 9,11.
CAPACITOR NO.4 AND 24 30/30/30/10 AT 475V.
REVERB SPRING
WOOD ROTOR
HORN
SLOW
FAST
USE IN RV MODELS ONLY

ROTOR **122 AMPLIFIER**

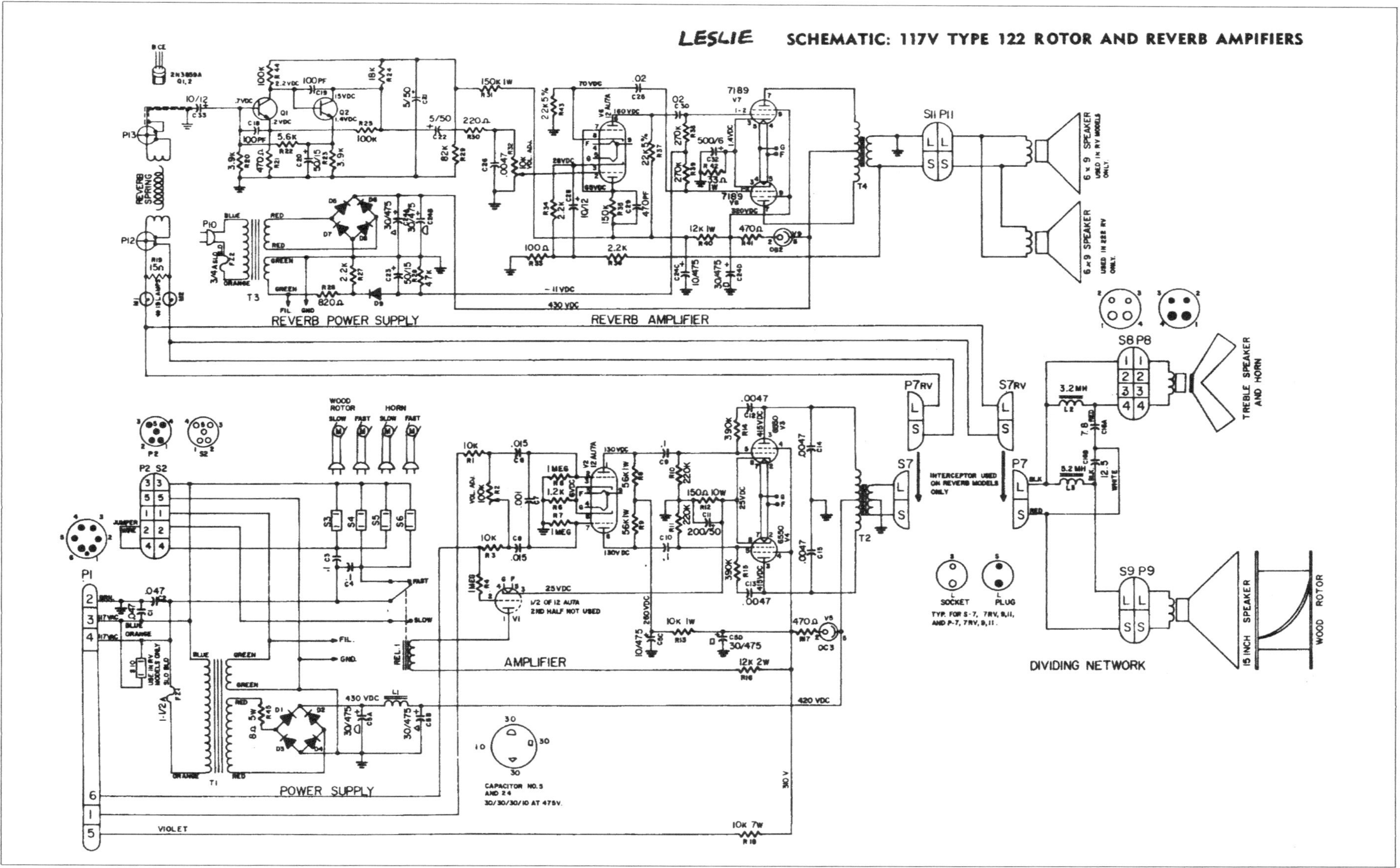

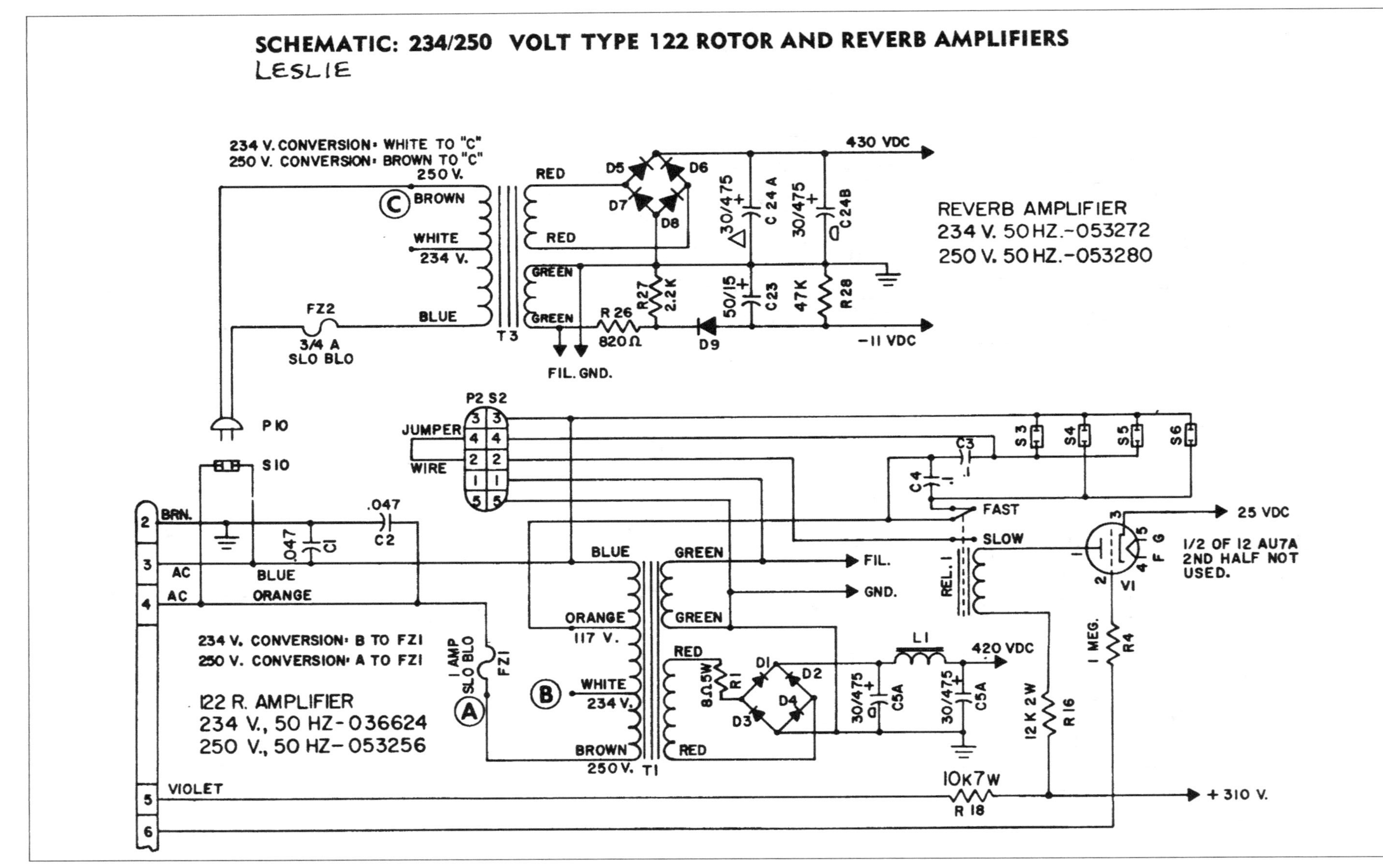
SCHEMATIC: 234/250 VOLT TYPE 122 ROTOR AND REVERB AMPLIFIERS
LESLIE
234 V. CONVERSION: WHITE TO "C"
250 V. CONVERSION: BROWN TO "C"
250 V.
C
BROWN
WHITE
234 V.
BLUE
FZ2
3/4 A
SLO BLO
T3
RED
RED
GREEN
GREEN
R 26
820Ω
FIL. GND.
D5
D6
D7
D8
R27
2.2K
D9
30/475
C 24 A
30/475
C24B
50/15
C23
47K
R28
430 VDC
-11 VDC
REVERB AMPLIFIER
234 V. 50 HZ.-053272
250 V. 50 HZ.-053280
P2 S2
JUMPER
WIRE
P10
S10
BRN.
AC
AC
.047
C1
.047
C2
BLUE
ORANGE
VIOLET
234 V. CONVERSION: B TO FZ1
250 V. CONVERSION: A TO FZ1
1 AMP
SLO BLO
FZ1
A
B
122 R. AMPLIFIER
234 V., 50 HZ-036624
250 V., 50 HZ-053256
BLUE
ORANGE
117 V.
WHITE
234 V.
BROWN
250 V.
T1
GREEN
GREEN
RED
RED
FIL.
GND.
8Ω.5W
R1
D1
D2
D3
D4
30/475
C5A
L1
30/475
C5A
420 VDC
C3
C4
S3
S4
S5
S6
FAST
SLOW
REL. 1
V1
25 VDC
1/2 OF 12 AU7A
2ND HALF NOT
USED.
1 MEG.
R4
12 K 2W
R16
10K7W
R 18
+ 310 V.

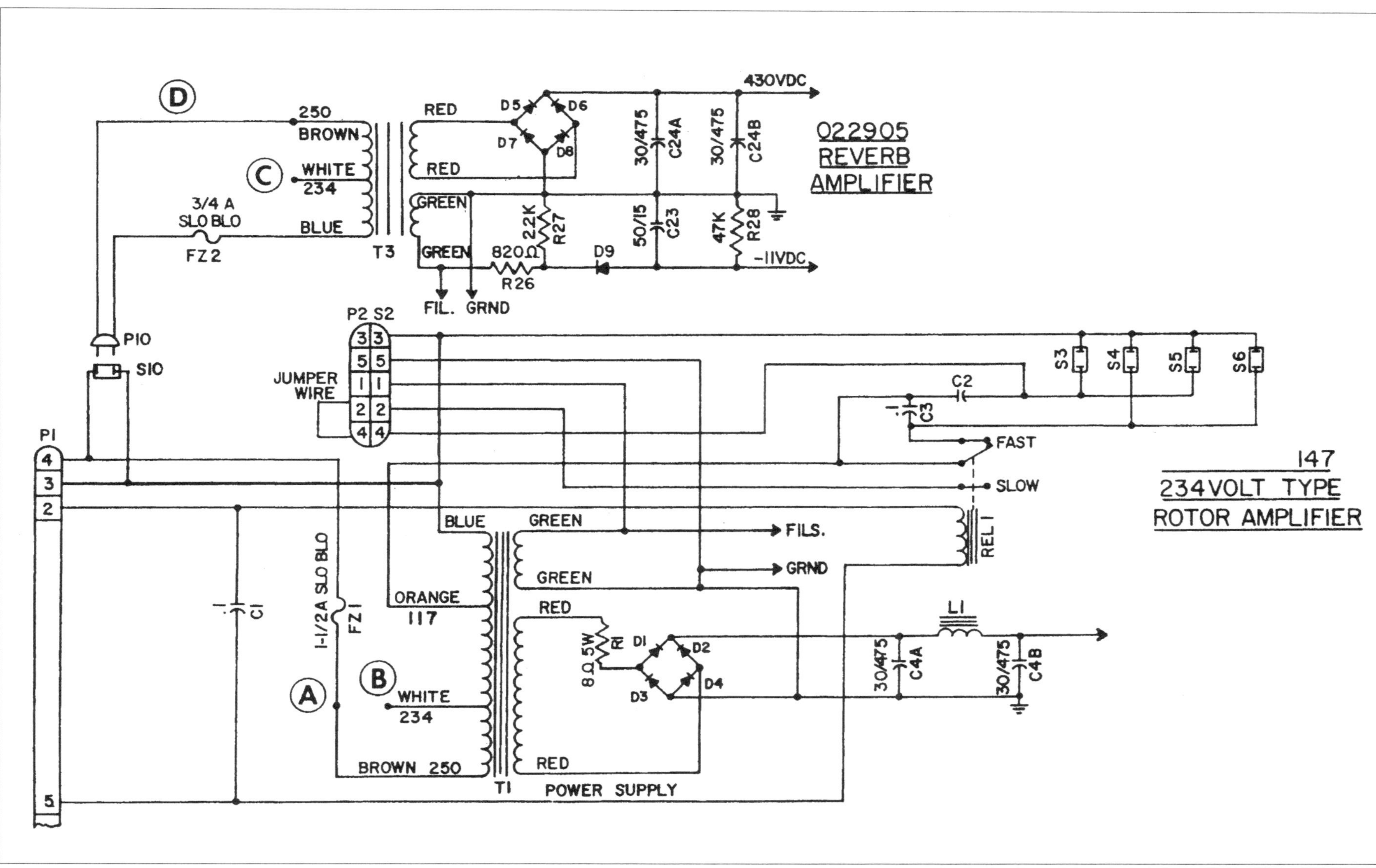
D
250
BROWN
C
WHITE
234
3/4 A
SLO BLO
FZ2
BLUE
T3
RED
RED
GREEN
GREEN
D5
D6
D7
D8
22K
R27
820Ω
R26
D9
FIL.
GRND
430VDC
30/475
C24A
30/475
C24B
50/15
C23
47K
R28
-11VDC
022905
REVERB
AMPLIFIER
P2 S2
P10
S10
JUMPER
WIRE
P1
C2
C3
FAST
SLOW
S3
S4
S5
S6
147
234VOLT TYPE
ROTOR AMPLIFIER
C1
1-1/2A SLO BLO
FZ1
A
B
WHITE
234
BROWN 250
ORANGE
117
BLUE
GREEN
GREEN
RED
RED
T1
POWER SUPPLY
FILS.
GRND
REL 1
8Ω 5W
R1
D1
D2
D3
D4
L1
30/475
C4A
30/475
C4B

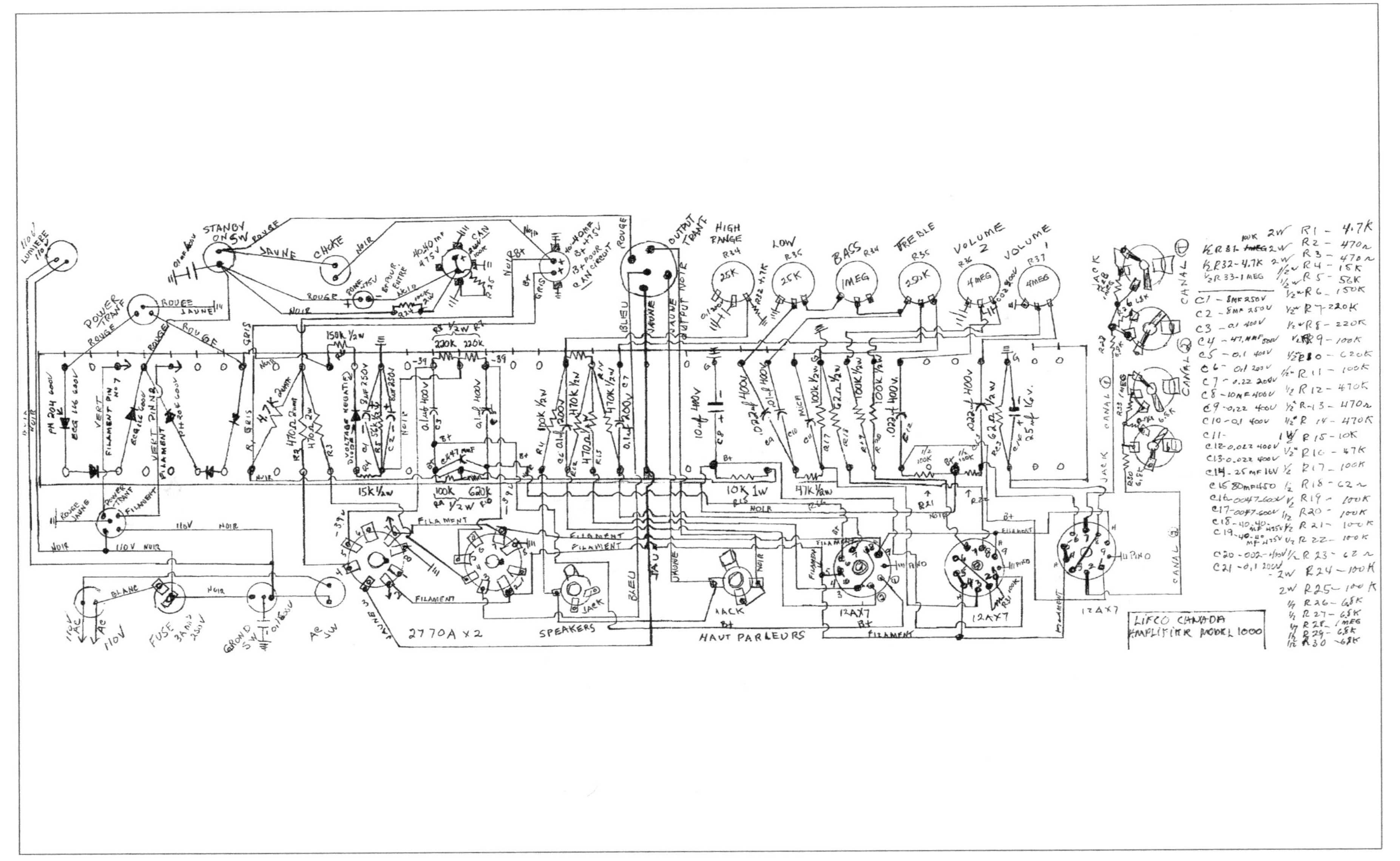
LIFCO CANADA AMPLIFIER MODEL 1000
HIGH RANGE
LOW
BASS
TREBLE
VOLUME 2
VOLUME 1
OUTPUT TRANS
STANDBY ON SW
CHOKE
POWER TRANS
HAUT PARLEURS
SPEAKERS
2770A x 2
12AX7
FUSE
GROND SW
110V
CANAL

VOLTAGE AND RESISTANCE CHART

Tube	Pin No.	DC Volts No. Signal	AC Volts at rated output	Resistance in Ohms Unit Off
12AX7 (V1)	1	132	2.1	150K*
	2	0	.47	22K
	3	1.2	.44	3.3K
	4 & 5	0	0	0
	6	0	0	0
	7	0	0	0
	8	0	0	0
	9	0	6.3	–
12AU7 (V2)	1	265	12	36K
	2	132	2.1	150K
	3	140	.96	18K
	4 & 5	0	0	0
	6	265	12	38K*
	7	105	0	2.4M*
	8	140	.96	18K
	9	0	6.3	0
12BH7 (V3)	1	365	152	12K*
	2	26	12	200K
	3	48	.7	2.7K
	4 & 5	0	0	0
	6	365	152	12K
	7	26	12	200K
	8	48	.7	2.7K
	9	0	6.3	0
12AT7/12AZ7 (V4)	1	235	112	30**
	2	–56	152	1M
	3	–58	149	220K
	4 & 5	0	0	0
	6	235	112	30**
	7	–56	152	1M
	8	–58	149	220K
	9	0	6.3	0
KT88/6550 (V5, V6)	1	–	–	–
	2	0	0 or 6.3	0
	3	460	116	20*
	4	460	112	240*
	5	–58	149	220K
	6	–	–	–
	7	0	6.3 or 0	0
	8	.5	116	14

*This resistance measured with condenser C15A shorted to ground.
**This resistance measured with condenser C13 shorted to ground.
High impedance VTVM is used to measure the operating voltages.

SCHEMATIC NOTES

1. Unless otherwise specified, resistance values are in ohms, 1/2 watt, and 10% tolerance. Resistors marked with an asterisk (*) are 5% tolerance. Capacitance values smaller than 1 are in microfarads (µF); values greater than 1 are in picofarads (pF). Inductors are in microhenries (µH).

2. R18 and R21 are matched to 1%.

3. In units with serial numbers below 470B0: R11 is 27K 1/2W; R12 is 30K 1/2W; C20 is not used; C18 is used; R29 is 10K 2W; R30 is used. R11 & R12 are connected to the power supply as shown by dotted line there is no connection at point marked with "X".

4. Power transformer T2 is connected as shown in detail "A" in early units.

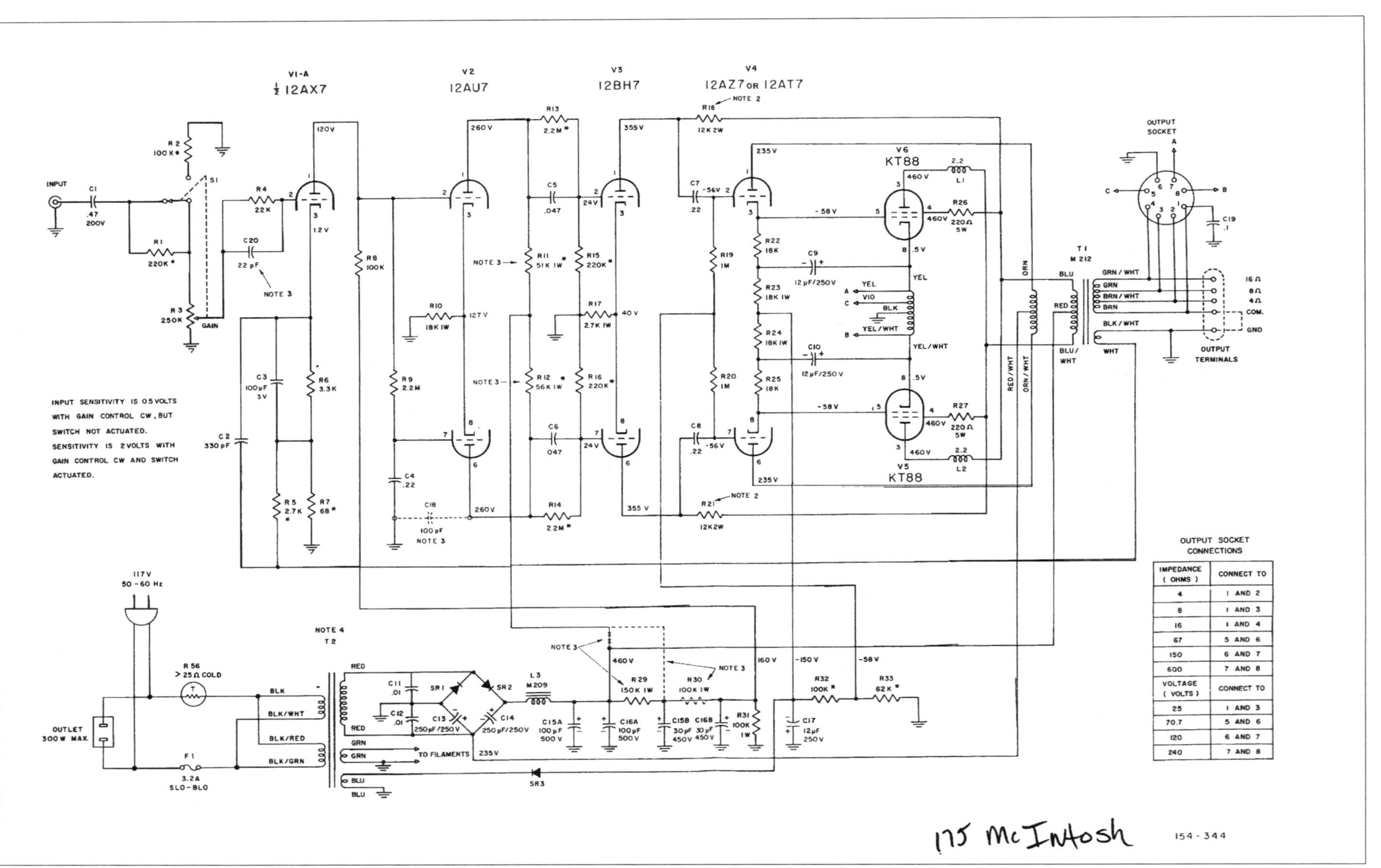
V1-A
½ 12AX7
V2
12AU7
V3
12BH7
V4
12AZ7 OR 12AT7
V6
KT88
V5
KT88
T1
M 212
OUTPUT SOCKET
OUTPUT TERMINALS
16 Ω
8 Ω
4 Ω
COM.
GND
INPUT
GAIN
NOTE 2
NOTE 3
NOTE 4
T2
L3
M209
TO FILAMENTS
117 V
50 - 60 Hz
OUTLET
300 W MAX.
R 56
> 25 Ω COLD
F1
3.2A
SLO-BLO
INPUT SENSITIVITY IS 0.5 VOLTS WITH GAIN CONTROL CW, BUT SWITCH NOT ACTUATED.
SENSITIVITY IS 2 VOLTS WITH GAIN CONTROL CW AND SWITCH ACTUATED.
OUTPUT SOCKET CONNECTIONS
IMPEDANCE (OHMS) | CONNECT TO
4 | 1 AND 2
8 | 1 AND 3
16 | 1 AND 4
67 | 5 AND 6
150 | 6 AND 7
600 | 7 AND 8
VOLTAGE (VOLTS) | CONNECT TO
25 | 1 AND 3
70.7 | 5 AND 6
120 | 6 AND 7
240 | 7 AND 8
175 McIntosh
154 - 344

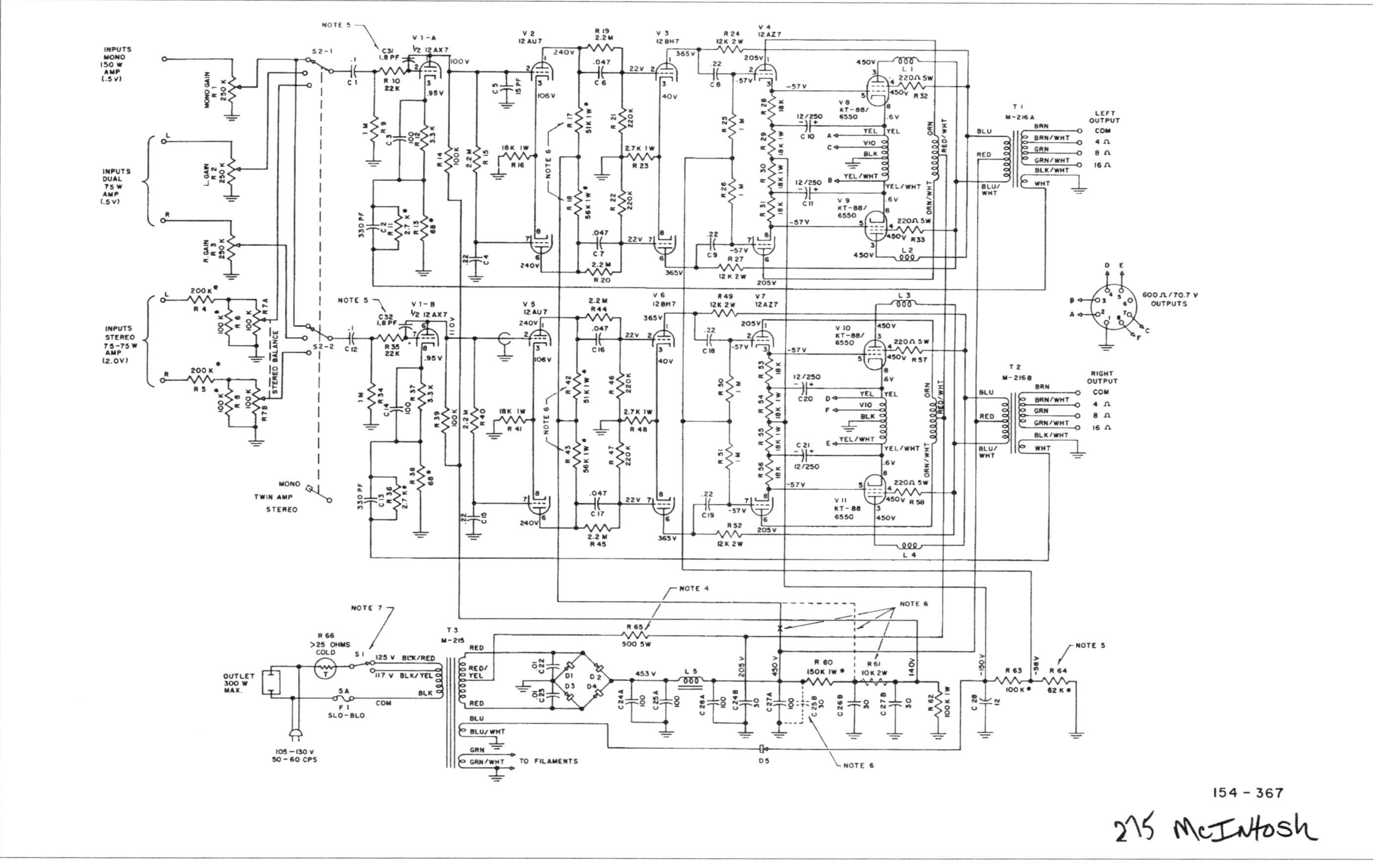
INPUTS MONO 150 W AMP (.5V)
MONO GAIN R 1 250 K
INPUTS DUAL 75 W AMP (.5V)
L. GAIN R 2 250 K
R. GAIN R 3 250 K
INPUTS STEREO 75-75 W AMP (2.0V)
STEREO BALANCE
MONO
TWIN AMP
STEREO
NOTE 5
NOTE 6
NOTE 4
NOTE 7
LEFT OUTPUT
RIGHT OUTPUT
600 Ω/70.7 V OUTPUTS
OUTLET 300 W MAX.
105-130 V 50-60 CPS
TO FILAMENTS
154 - 367
275 McIntosh

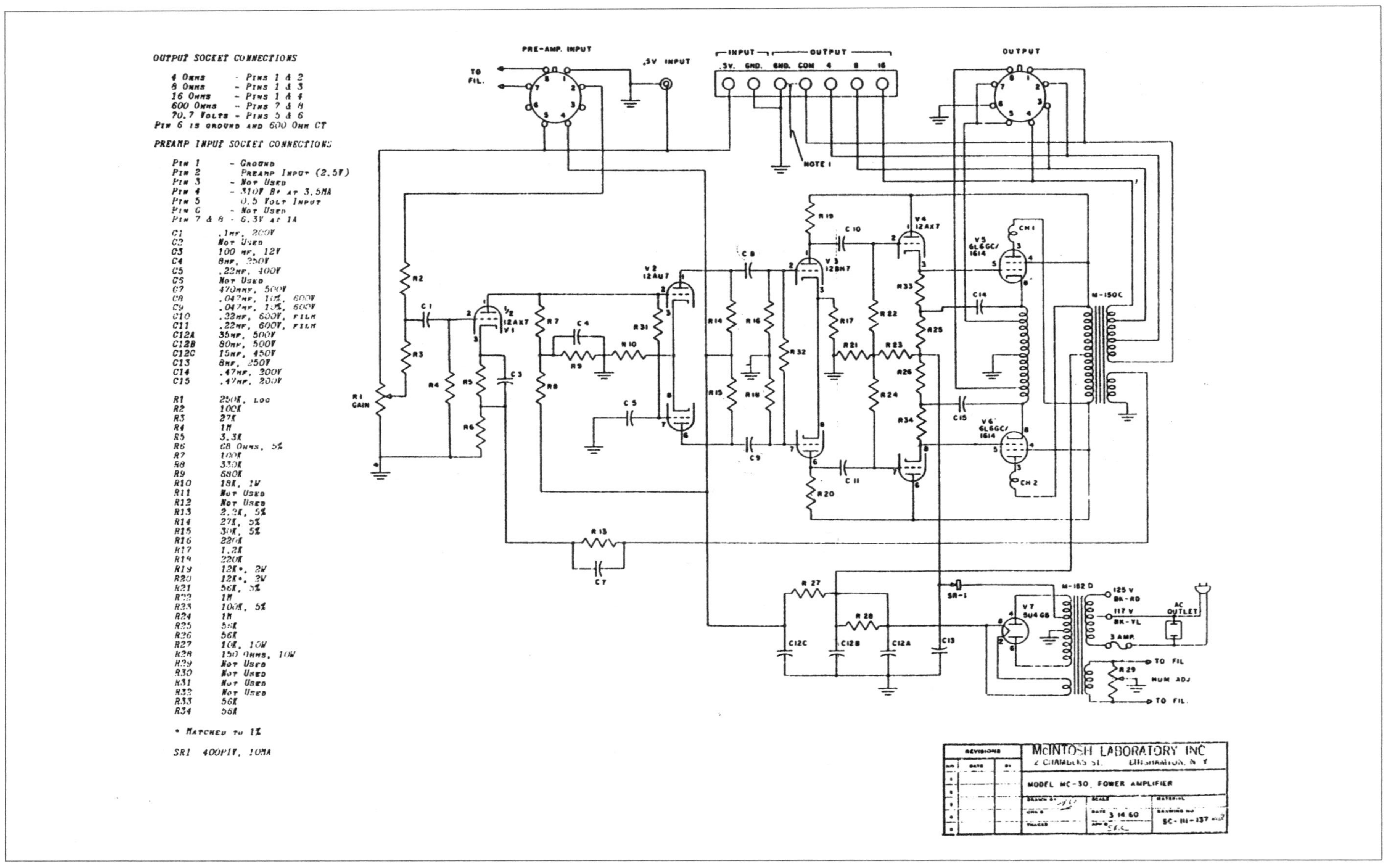
OUTPUT SOCKET CONNECTIONS
4 OHMS - PINS 1 & 2
8 OHMS - PINS 1 & 3
16 OHMS - PINS 1 & 4
600 OHMS - PINS 7 & 8
70.7 VOLTS - PINS 5 & 6
PIN 6 IS GROUND AND 600 OHM CT
PREAMP INPUT SOCKET CONNECTIONS
PIN 1 - GROUND
PIN 2 PREAMP INPUT (2.5V)
PIN 3 - NOT USED
PIN 4 - 310V B+ AT 3.5MA
PIN 5 0.5 VOLT INPUT
PIN 6 - NOT USED
PIN 7 & 8 - 6.3V AT 1A
C1 .1MF, 200V
C2 NOT USED
C3 100 MF, 12V
C4 8MF, 350V
C5 .22MF, 400V
C6 NOT USED
C7 470MMF, 500V
C8 .047MF, 10%, 600V
C9 .047MF, 10%, 600V
C10 .22MF, 600V, FILM
C11 .22MF, 600V, FILM
C12A 35MF, 500V
C12B 80MF, 500V
C12C 15MF, 450V
C13 8MF, 250V
C14 .47MF, 300V
C15 .47MF, 200V
R1 250K, LOG
R2 100K
R3 27K
R4 1M
R5 3.3K
R6 68 OHMS, 5%
R7 100K
R8 330K
R9 680K
R10 18K, 1W
R11 NOT USED
R12 NOT USED
R13 2.2K, 5%
R14 27K, 5%
R15 30K, 5%
R16 220K
R17 1.2K
R18 220K
R19 12K*, 2W
R20 12K*, 2W
R21 56K, 5%
R22 1M
R23 100K, 5%
R24 1M
R25 56K
R26 56K
R27 10K, 10W
R28 150 OHMS, 10W
R29 NOT USED
R30 NOT USED
R31 NOT USED
R32 NOT USED
R33 56K
R34 56K
* MATCHED TO 1%
SR1 400PIV, 10MA
PRE-AMP. INPUT
TO FIL.
.5V INPUT
INPUT
OUTPUT
.5V. GND. GND. COM 4 8 16
NOTE 1
OUTPUT
R1 GAIN
V1 1/2 12AX7
V2 12AU7
V3 12BH7
V4 12AX7
V5 6L6GC/1614
V6 6L6GC/1614
CH 1
CH 2
M-150C
M-182 D
V7 5U4GB
SR-1
125 V BK-RD
117 V BK-YL
AC OUTLET
3 AMP.
TO FIL
HUM ADJ
TO FIL.
REVISIONS
McINTOSH LABORATORY INC
2 CHAMBERS ST. BINGHAMTON, N. Y.
MODEL MC-30 POWER AMPLIFIER
DATE 3 14 60
SC-111-137

MC-30 TYPE A **116b**

McIntosh

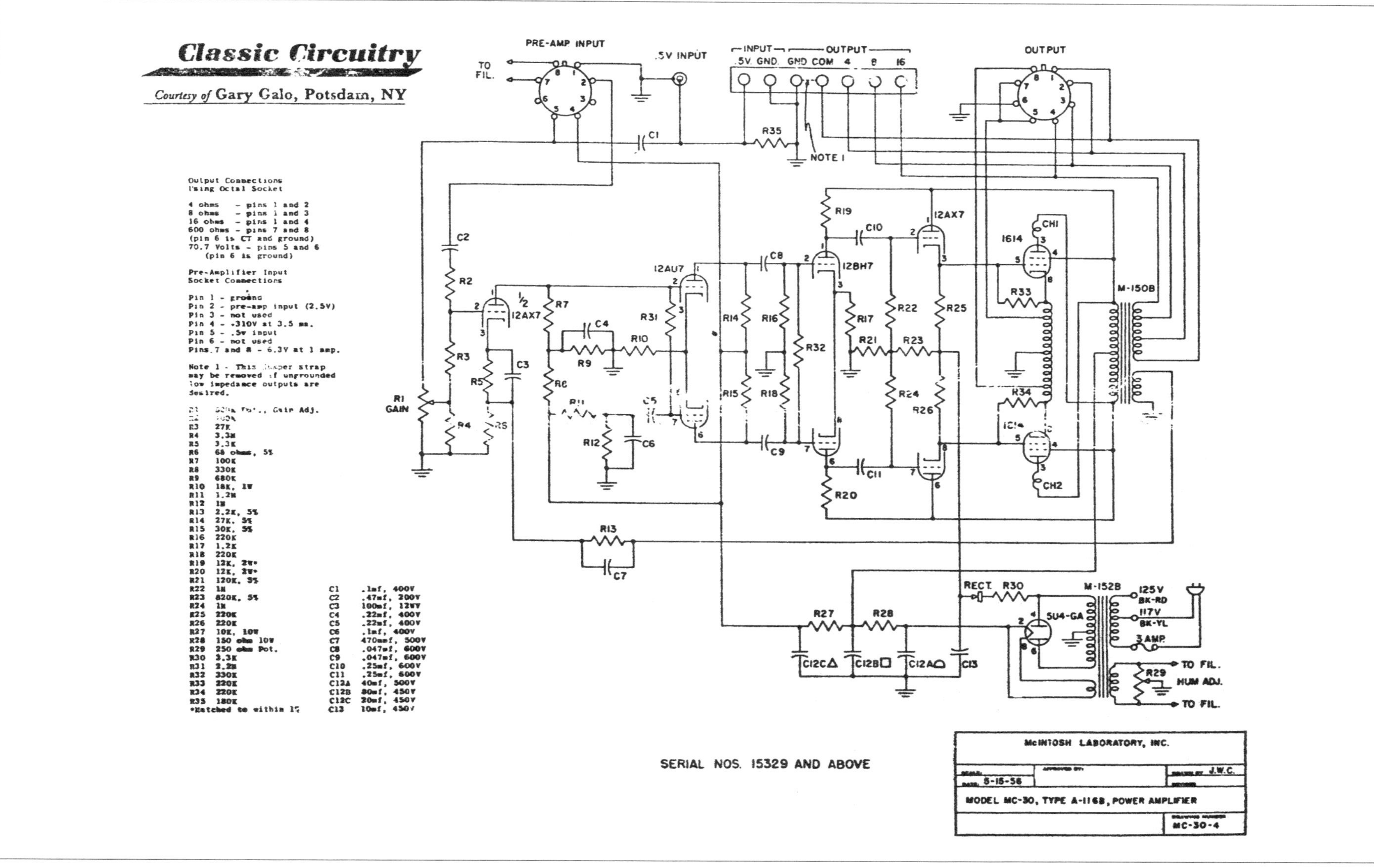

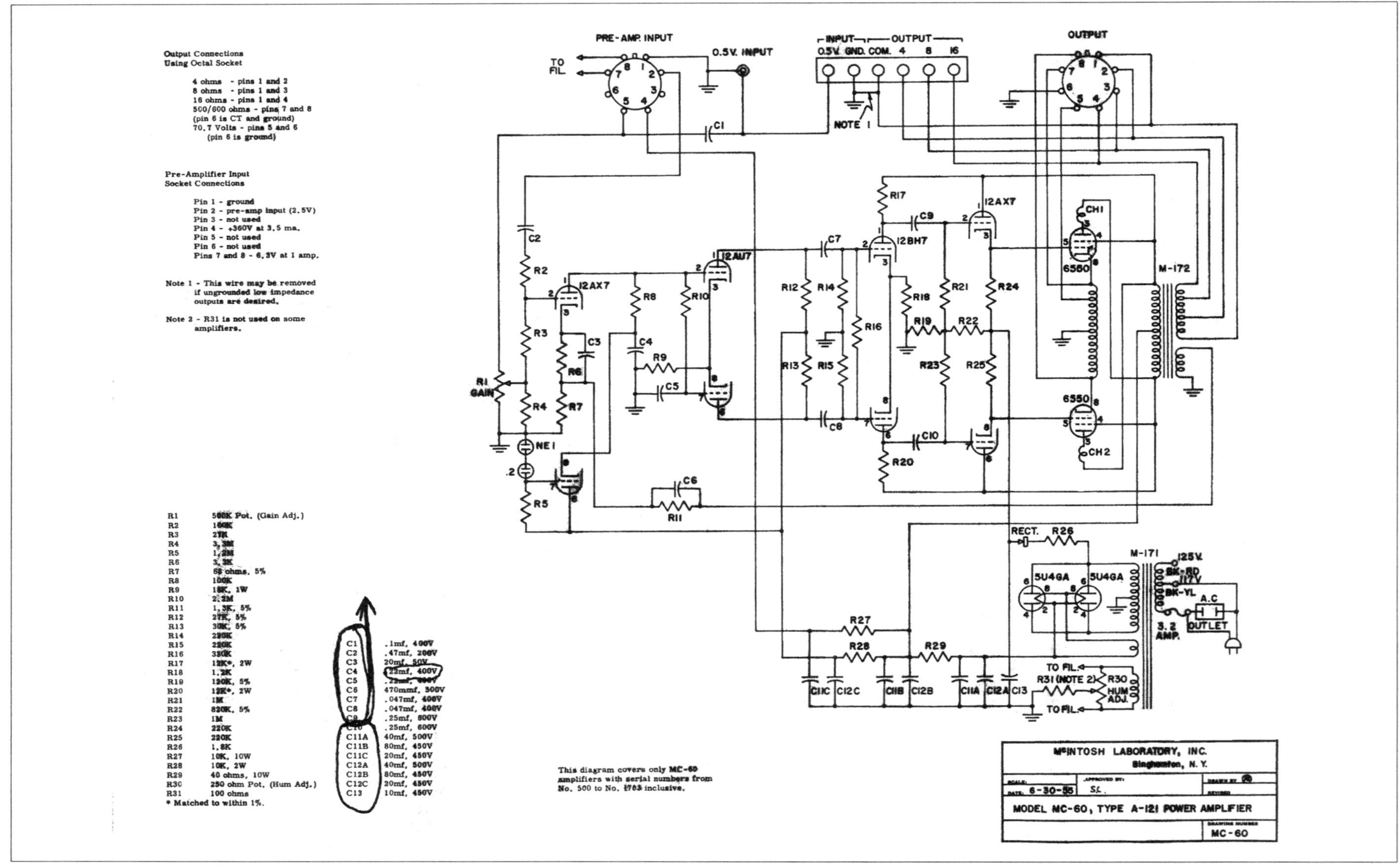
Output Connections
Using Octal Socket
4 ohms - pins 1 and 2
8 ohms - pins 1 and 3
16 ohms - pins 1 and 4
500/600 ohms - pins 7 and 8
(pin 6 is CT and ground)
70.7 Volts - pins 5 and 6
(pin 6 is ground)
Pre-Amplifier Input
Socket Connections
Pin 1 - ground
Pin 2 - pre-amp input (2.5V)
Pin 3 - not used
Pin 4 - +360V at 3.5 ma.
Pin 5 - not used
Pin 6 - not used
Pins 7 and 8 - 6.3V at 1 amp.
Note 1 - This wire may be removed if ungrounded low impedance outputs are desired.
Note 2 - R31 is not used on some amplifiers.
R1 500K Pot. (Gain Adj.)
R2 100K
R3 27K
R4 3.3M
R5 1.2M
R6 3.3K
R7 68 ohms, 5%
R8 100K
R9 18K, 1W
R10 2.2M
R11 1.3K, 5%
R12 27K, 5%
R13 30K, 5%
R14 220K
R15 220K
R16 330K
R17 12K*, 2W
R18 1.2K
R19 120K, 5%
R20 12K*, 2W
R21 1M
R22 820K, 5%
R23 1M
R24 220K
R25 220K
R26 1.8K
R27 10K, 10W
R28 10K, 2W
R29 40 ohms, 10W
R30 250 ohm Pot. (Hum Adj.)
R31 100 ohms
* Matched to within 1%.
C1 .1mf, 400V
C2 .47mf, 200V
C3 20mf, 50V
C4 .22mf, 400V
C5 .22mf, 400V
C6 470mmf, 300V
C7 .047mf, 400V
C8 .047mf, 400V
C9 .25mf, 600V
C10 .25mf, 600V
C11A 40mf, 500V
C11B 80mf, 450V
C11C 20mf, 450V
C12A 40mf, 500V
C12B 80mf, 450V
C12C 20mf, 450V
C13 10mf, 450V
PRE-AMP. INPUT
TO FIL.
0.5V. INPUT
INPUT
OUTPUT
0.5V. GND. COM. 4 8 16
NOTE 1
OUTPUT
12AX7
12AU7
12BH7
6550
CH1
CH2
M-172
M-171
R1 GAIN
NE1
5U4GA
RECT.
125V.
BK-RD
117V
BK-YL
A.C
OUTLET
3.2 AMP.
TO FIL.
R31 (NOTE 2)
HUM ADJ.
This diagram covers only MC-60 amplifiers with serial numbers from No. 500 to No. 1763 inclusive.
McINTOSH LABORATORY, INC.
Binghamton, N. Y.
6-30-55
MODEL MC-60, TYPE A-121 POWER AMPLIFIER
MC-60

MC-60 TYPE A **121 REVISED**

McIntosh

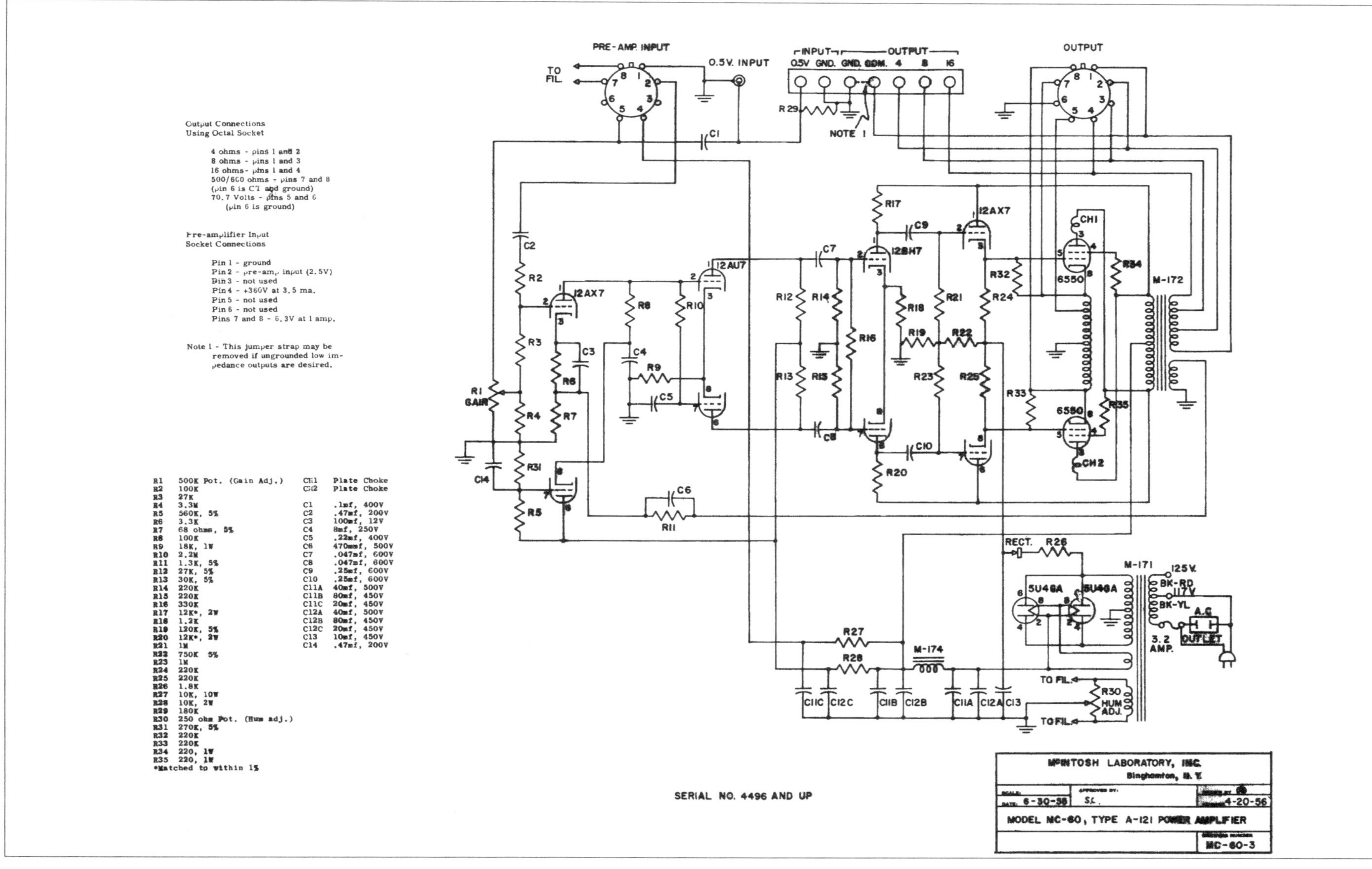

MC-60 TYPE A **125**

McIntosh

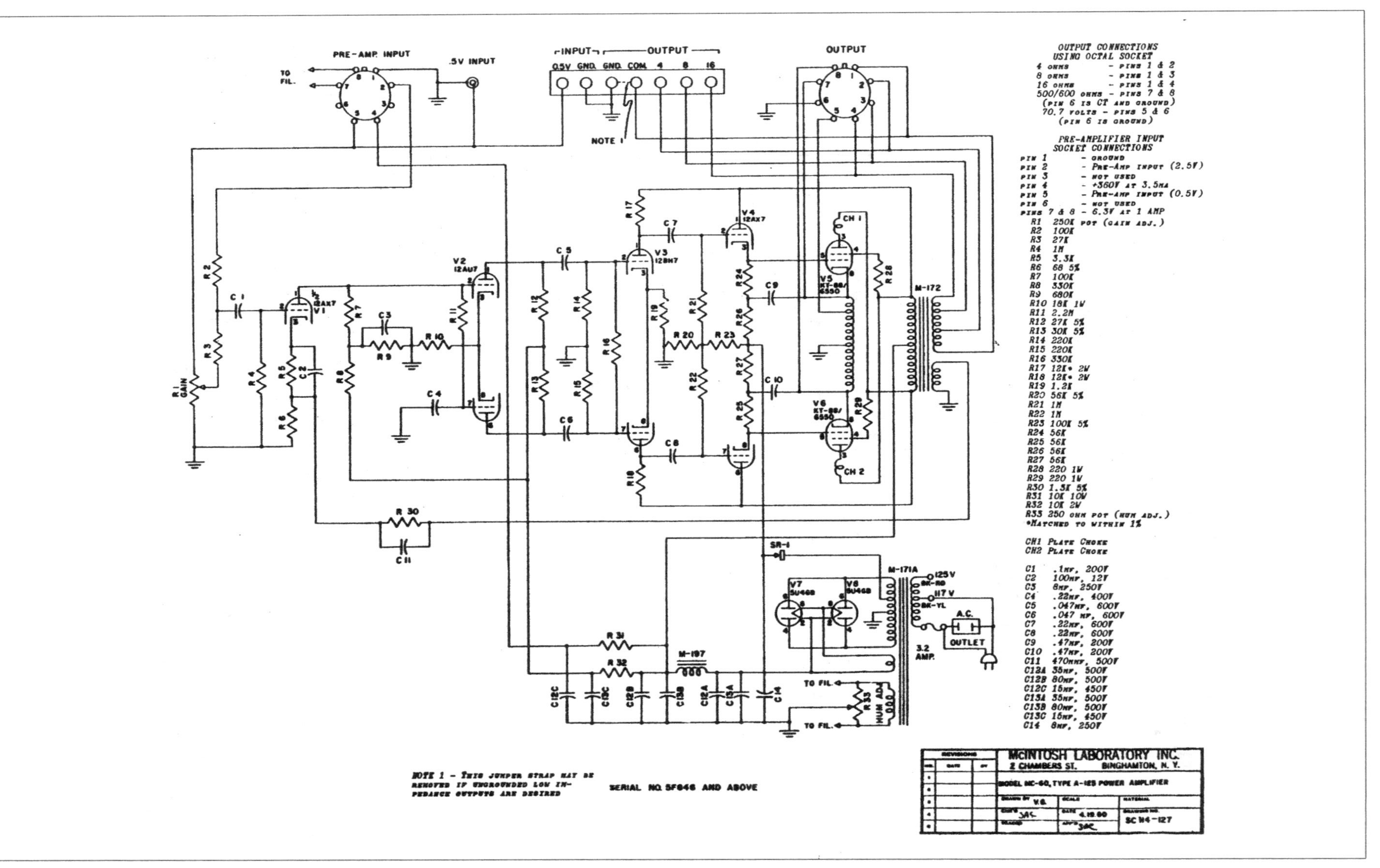

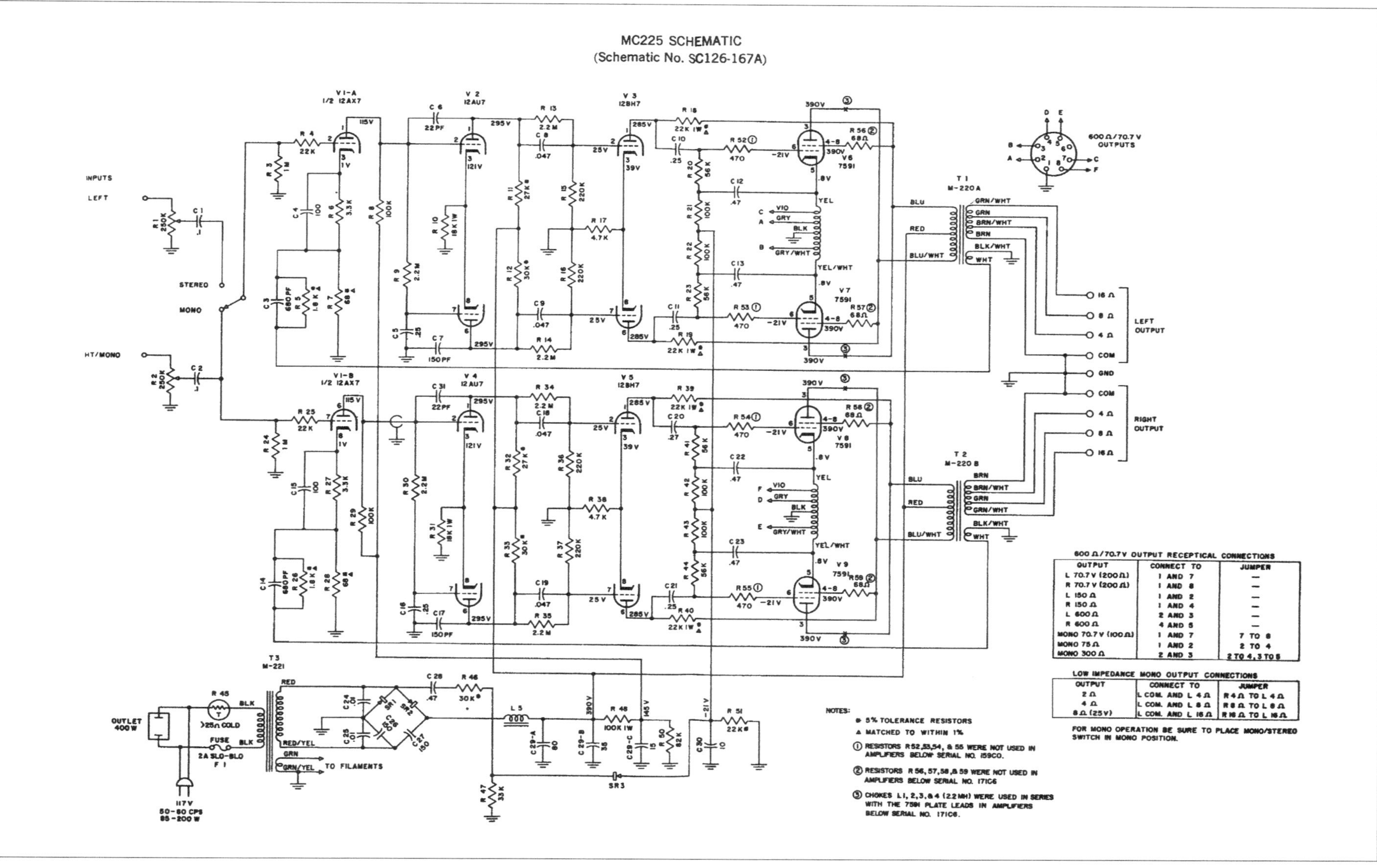
MC225 SCHEMATIC
(Schematic No. SC126-167A)
INPUTS
LEFT
STEREO
MONO
HT/MONO
LEFT OUTPUT
RIGHT OUTPUT
600 Ω/70.7 V OUTPUTS
600 Ω/70.7 V OUTPUT RECEPTICAL CONNECTIONS
OUTPUT | CONNECT TO | JUMPER
L 70.7 V (200 Ω) | 1 AND 7 | —
R 70.7 V (200 Ω) | 1 AND 8 | —
L 150 Ω | 1 AND 2 | —
R 150 Ω | 1 AND 4 | —
L 600 Ω | 2 AND 3 | —
R 600 Ω | 4 AND 5 | —
MONO 70.7 V (100 Ω) | 1 AND 7 | 7 TO 8
MONO 75 Ω | 1 AND 2 | 2 TO 4
MONO 300 Ω | 2 AND 3 | 2 TO 4, 3 TO 5
LOW IMPEDANCE MONO OUTPUT CONNECTIONS
OUTPUT | CONNECT TO | JUMPER
2 Ω | L COM. AND L 4 Ω | R 4 Ω TO L 4 Ω
4 Ω | L COM. AND L 8 Ω | R 8 Ω TO L 8 Ω
8 Ω (25V) | L COM. AND L 16 Ω | R 16 Ω TO L 16 Ω
FOR MONO OPERATION BE SURE TO PLACE MONO/STEREO SWITCH IN MONO POSITION.
NOTES:
● 5% TOLERANCE RESISTORS
▲ MATCHED TO WITHIN 1%
① RESISTORS R52,53,54, & 55 WERE NOT USED IN AMPLIFIERS BELOW SERIAL NO. 159CO.
② RESISTORS R 56, 57, 58, & 59 WERE NOT USED IN AMPLIFIERS BELOW SERIAL NO. 171C6
③ CHOKES L1, 2, 3, & 4 (2.2 MH) WERE USED IN SERIES WITH THE 7591 PLATE LEADS IN AMPLIFIERS BELOW SERIAL NO. 171C6.
OUTLET 400 W
117 V 50-60 CPS 85-200 W
TO FILAMENTS

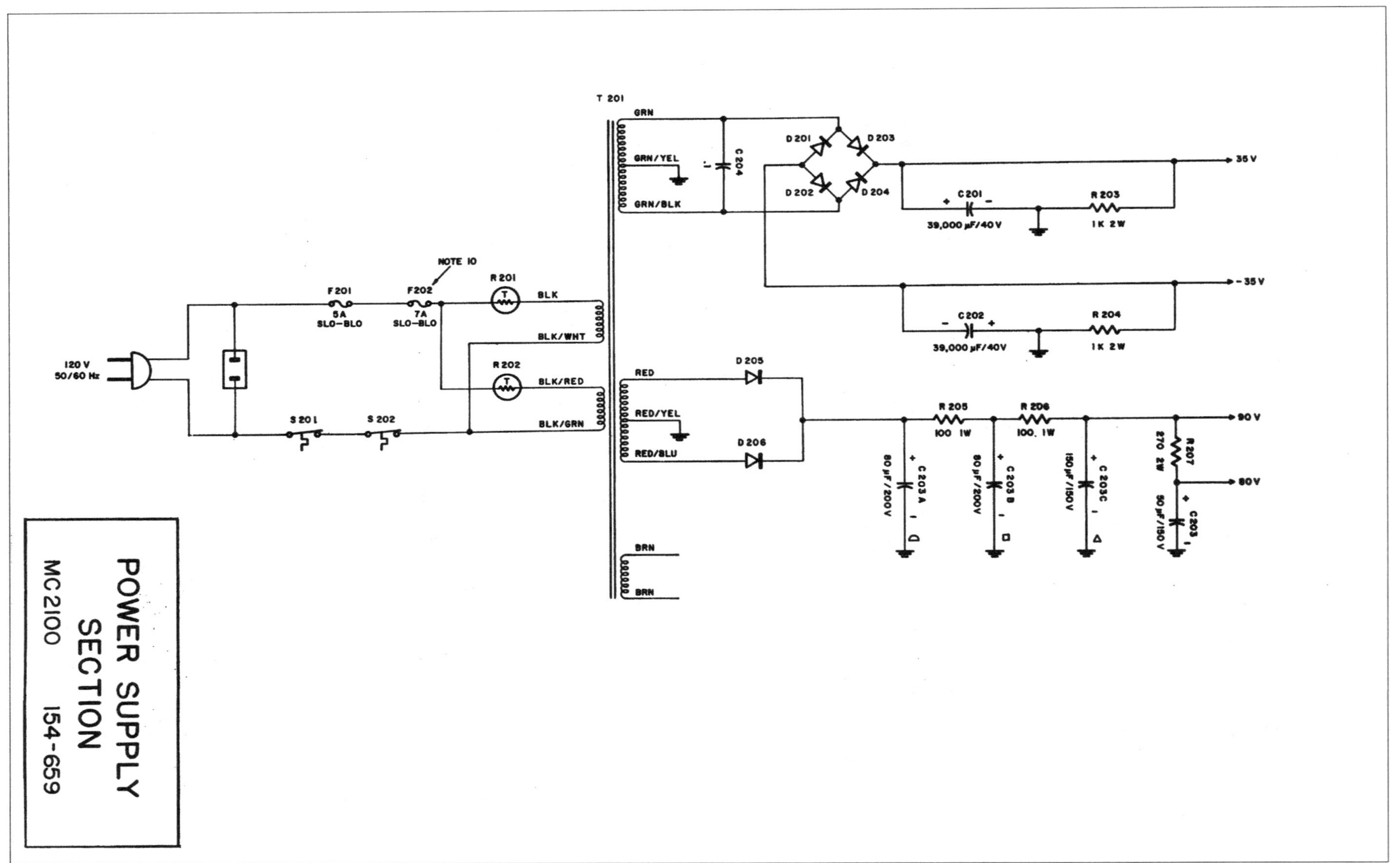
35 V
-35 V
90 V
80 V
R 207
270 2W
C 203
50 µF/150 V
C 203C
150 µF/150 V
R 206
100, 1W
C 203 B
80 µF/200V
R 205
100 1W
C 203 A
80 µF/200V
R 203
1K 2W
R 204
1K 2W
C201
39,000 µF/40V
C202
39,000 µF/40V
D 203
D 204
D 201
D 202
C 204
.1
D 205
D 206
GRN
GRN/YEL
GRN/BLK
RED
RED/YEL
RED/BLU
BRN
BRN
T 201
BLK
BLK/WHT
BLK/RED
BLK/GRN
R201
R202
NOTE 10
F202
7A
SLO-BLO
F201
5A
SLO-BLO
S 202
S 201
120 V
50/60 Hz
POWER SUPPLY
SECTION
MC 2100
154-659

6V6 TREMOLO AMP

Maestro

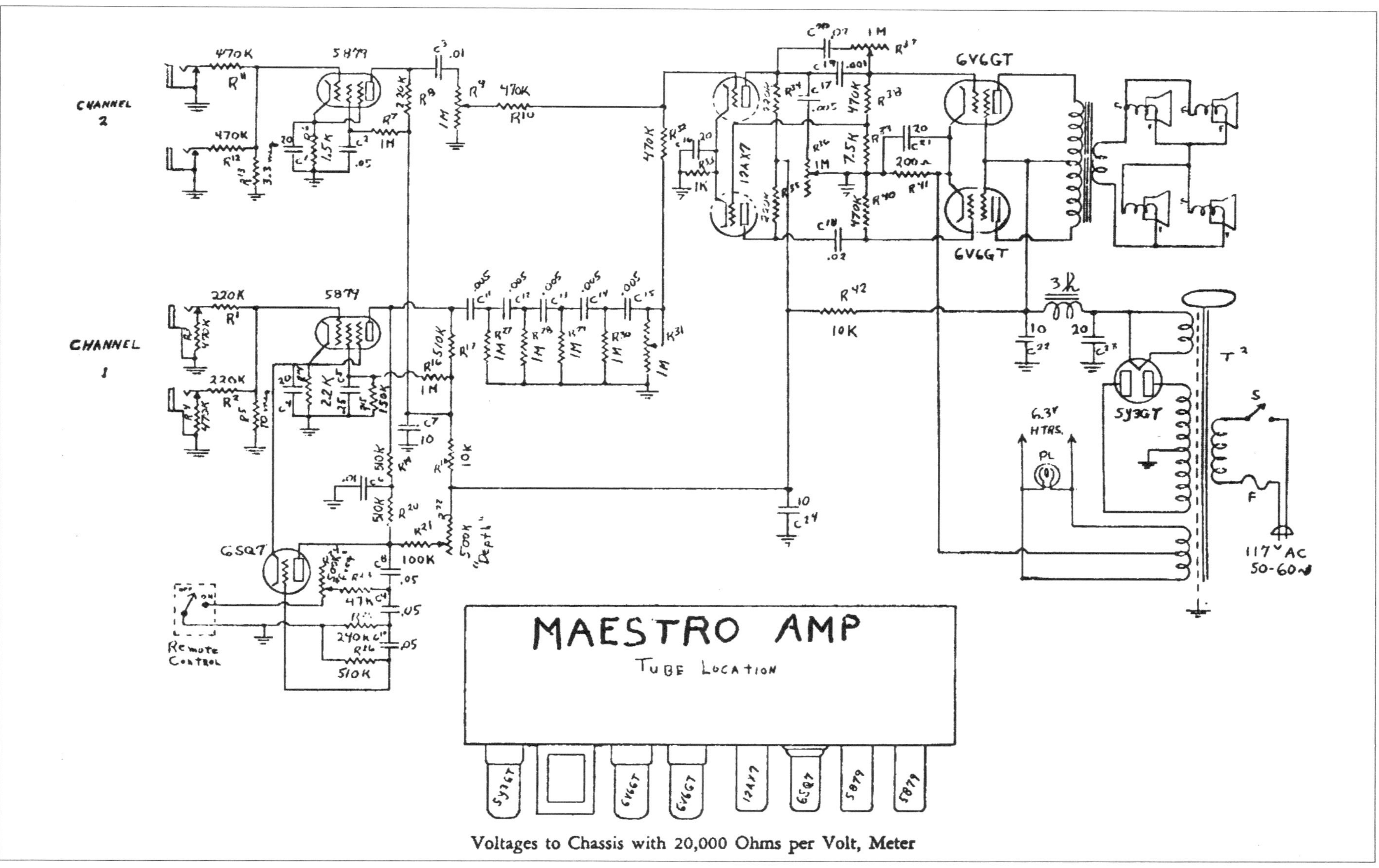

Voltages to Chassis with 20,000 Ohms per Volt, Meter

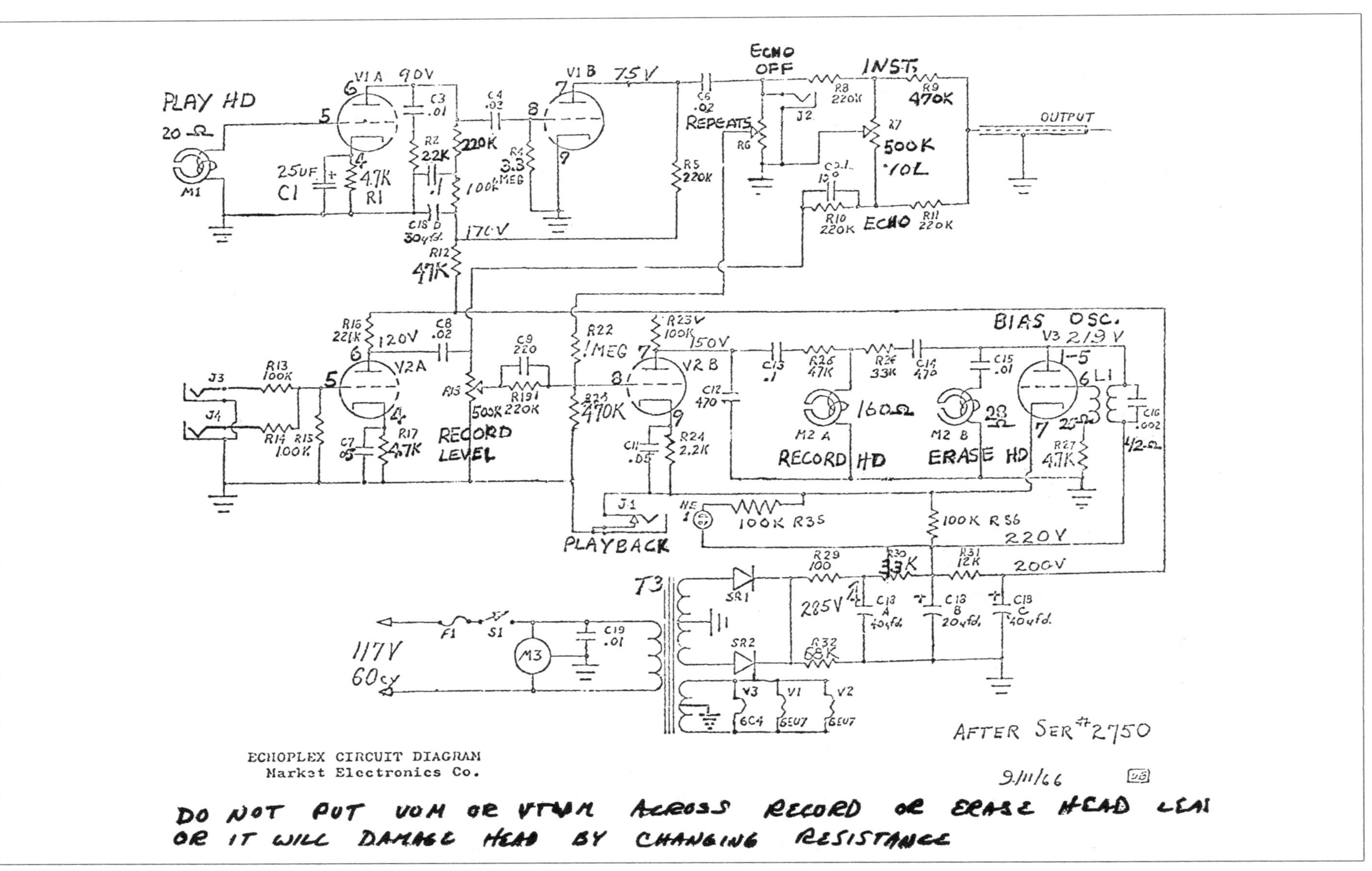

ECHOPLEX CIRCUIT DIAGRAM
Market Electronics Co.

DO NOT PUT VOM OR VTVM ACROSS RECORD OR ERASE HEAD LEA
OR IT WILL DAMAGE HEAD BY CHANGING RESISTANCE

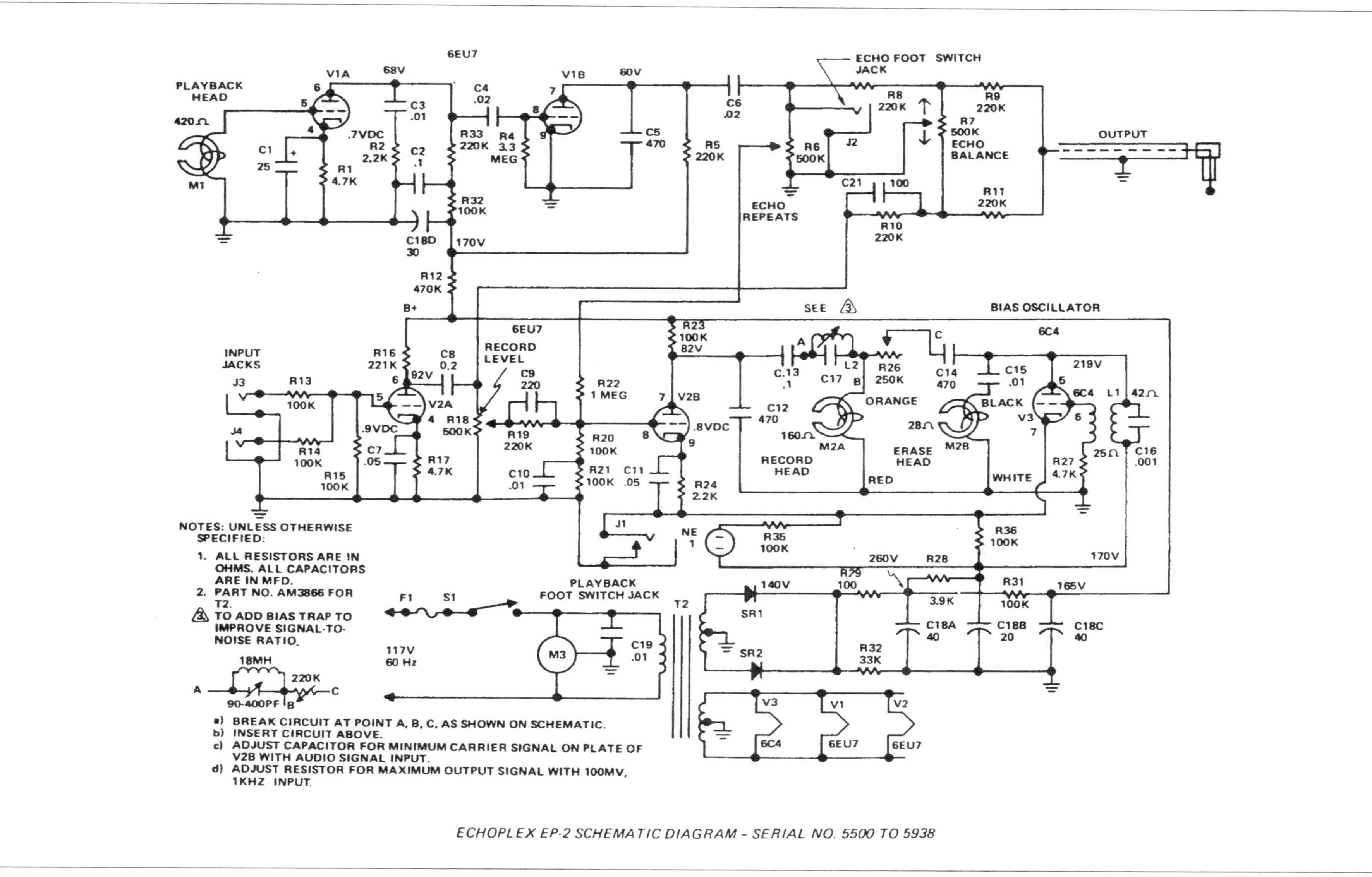

ECHOPLEX EP-2 SCHEMATIC DIAGRAM - SERIAL NO. 5500 TO 5938

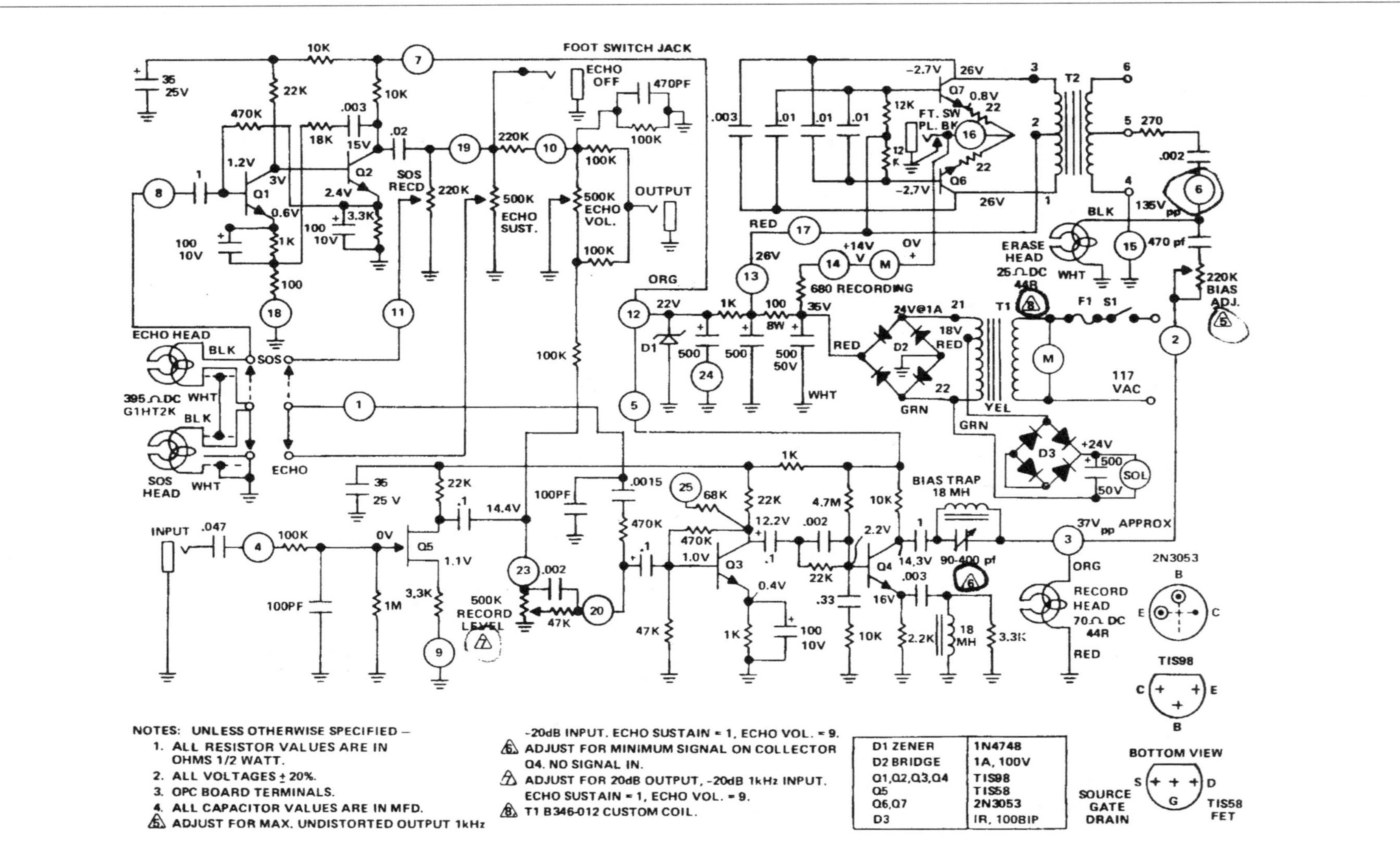

ECHOPLEX EP-3 SCHEMATIC DIAGRAM – SERIAL NO. 12961 TO 28591

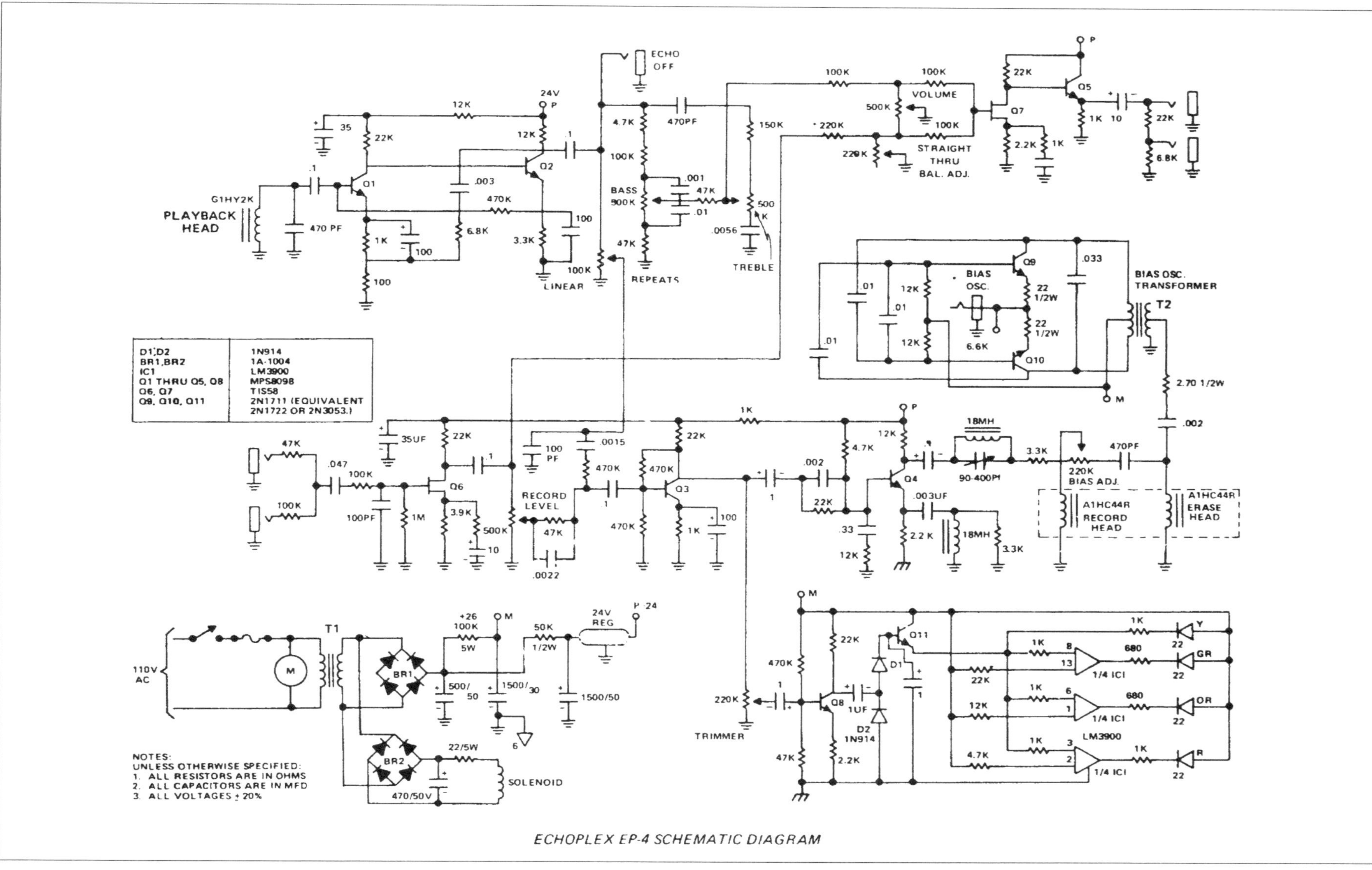

D1,D2	1N914
BR1,BR2	1A-1004
IC1	LM3900
Q1 THRU Q5, Q8	MPS8098
Q6, Q7	TIS58
Q9, Q10, Q11	2N1711 (EQUIVALENT 2N1722 OR 2N3053.)

ECHOPLEX EP-4 SCHEMATIC DIAGRAM

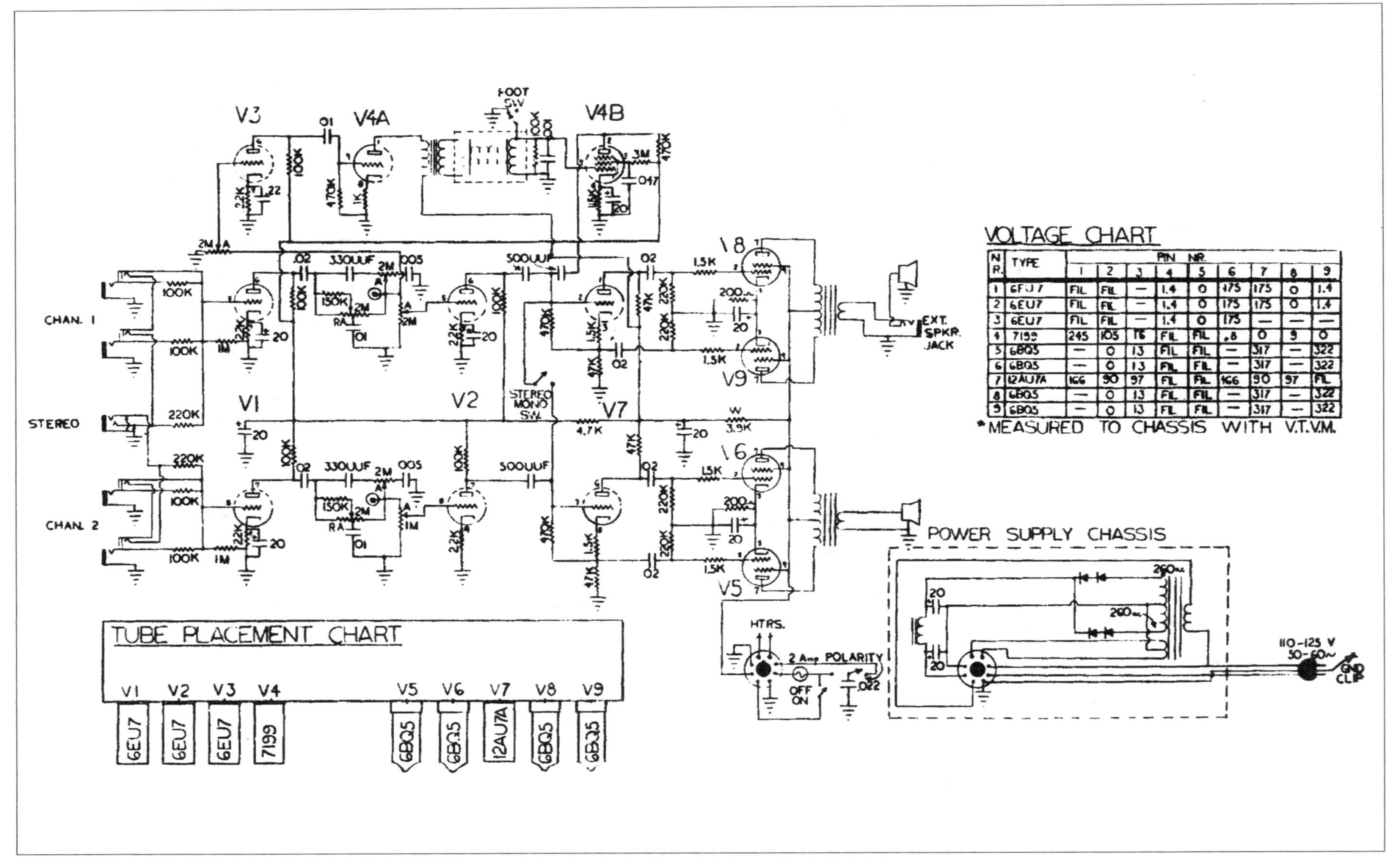

VOLTAGE CHART

NR.	TYPE	PIN NR. 1	2	3	4	5	6	7	8	9
1	6FJ7	FIL	FIL	—	1.4	0	175	175	0	1.4
2	6EU7	FIL	FIL	—	1.4	0	175	175	0	1.4
3	6EU7	FIL	FIL	—	1.4	0	175	—	—	—
4	7199	245	105	16	FIL	FIL	.8	0	9	0
5	6BQ5	—	0	13	FIL	FIL	—	317	—	322
6	6BQ5	—	0	13	FIL	FIL	—	317	—	322
7	12AU7A	166	90	97	FIL	FIL	166	90	97	FIL
8	6BQ5	—	0	13	FIL	FIL	—	317	—	322
9	6BQ5	—	0	13	FIL	FIL	—	317	—	322

*MEASURED TO CHASSIS WITH V.T.V.M.

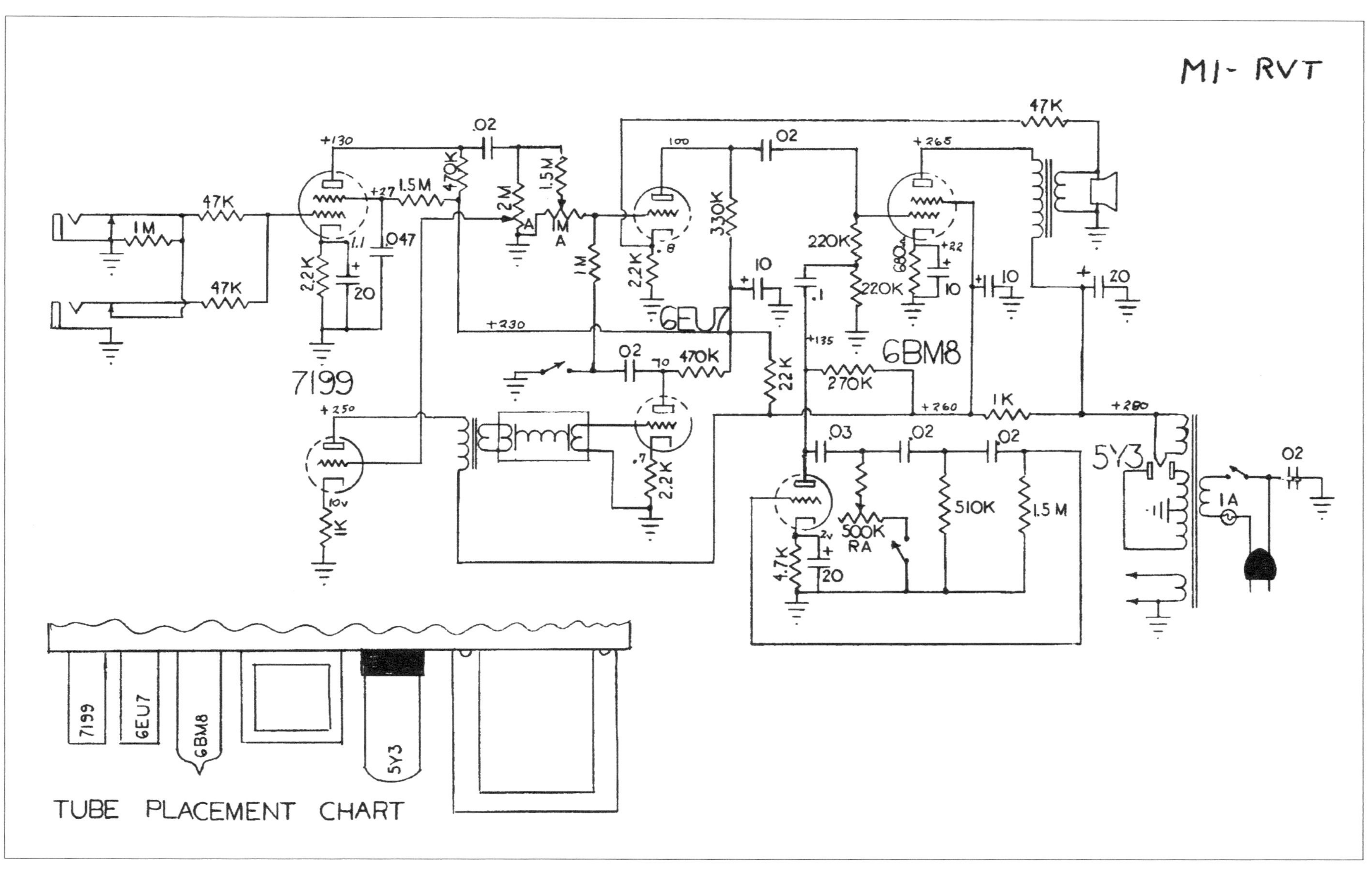
M1- RVT
47K
.02
+130
+27
1.5M
470K
2M
1.5M
A
M
A
1M
47K
47K
2.2K
.1
.047
20
7199
+230
100
02
330K
.8
2.2K
6EU7
02
.70
470K
22K
+265
220K
220K
680Ω
+22
10
10
10
20
.1
+135
6BM8
270K
+260
1K
+280
+250
10v
1K
.7
2.2K
.03
.02
.02
500K
RA
510K
1.5 M
4.7K
2v
20
5Y3
1A
02
TUBE PLACEMENT CHART
7199
6EU7
6BM8
5Y3

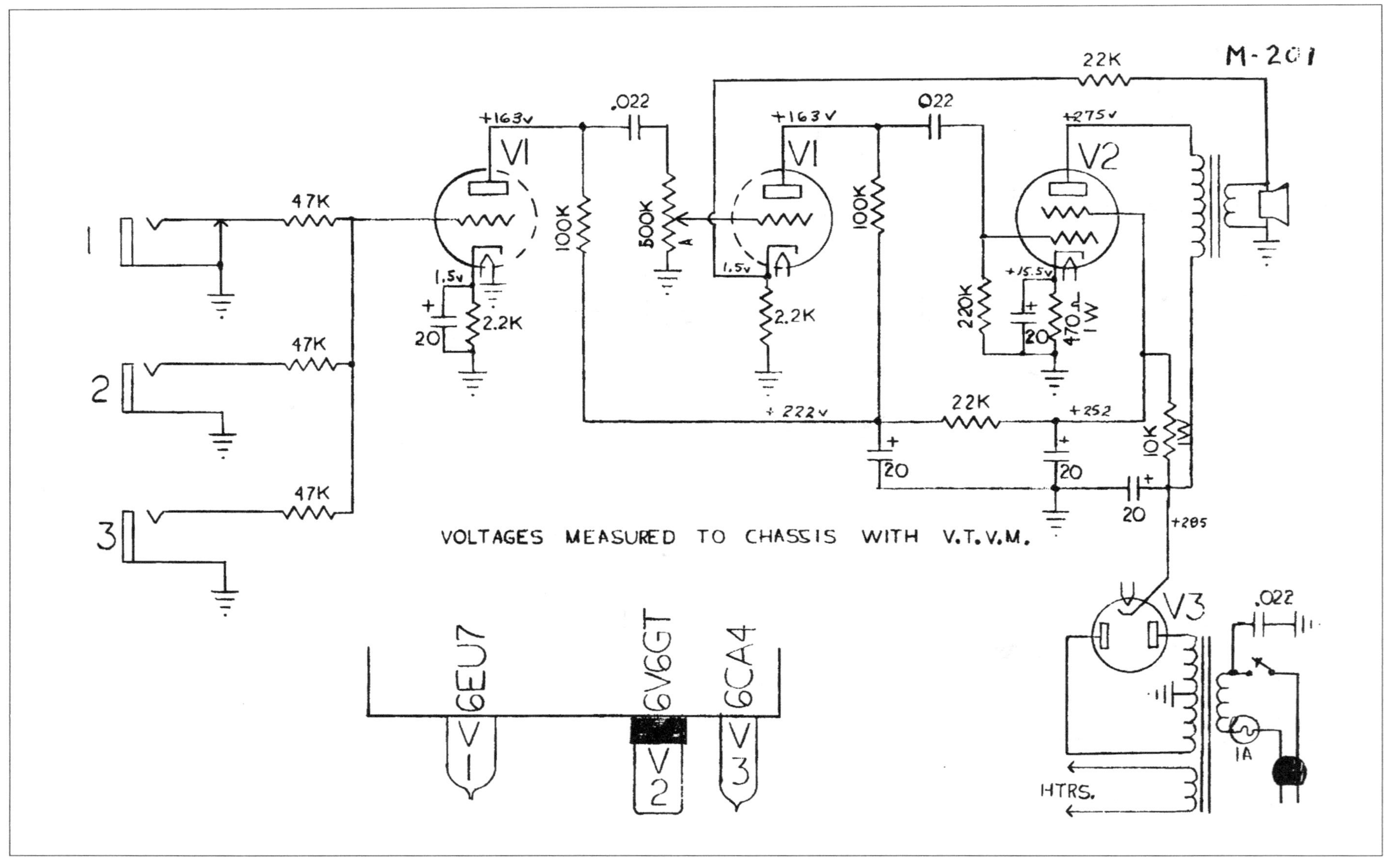
M-201
22K
.022
+163v
V1
.022
+163v
V1
022
+275v
V2
47K
47K
47K
1
2
3
100K
500K
A
1.5v
2.2K
20
1.5v
2.2K
100K
220K
+15.5v
20
470Ω 1W
+222v
22K
+252
20
20
10K 1W
20
+285
VOLTAGES MEASURED TO CHASSIS WITH V.T.V.M.
V3
.022
1A
HTRS.
6EU7
V1
6V6GT
V2
6CA4
V3

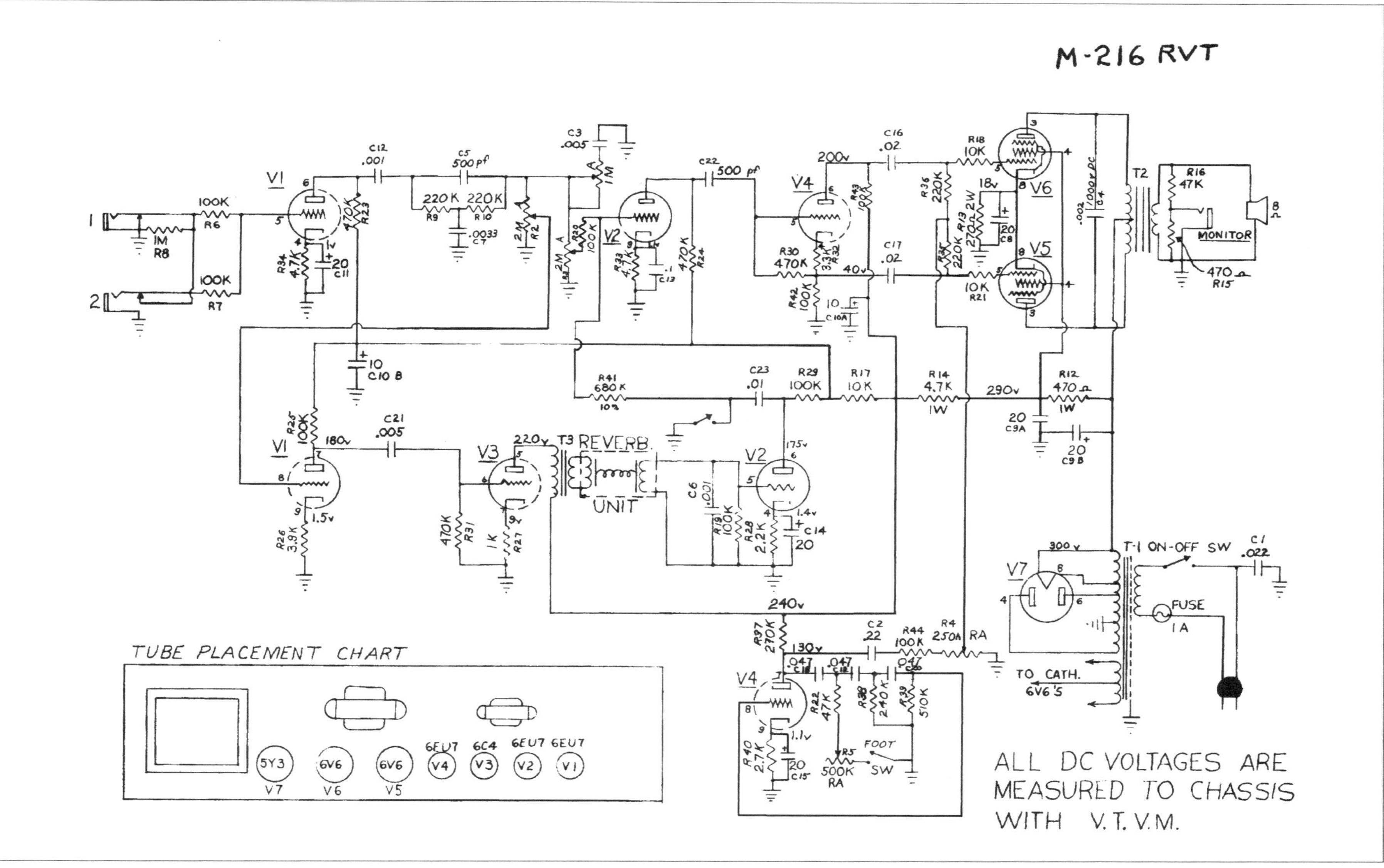
M-216 RVT
TUBE PLACEMENT CHART
5Y3 V7
6V6 V6
6V6 V5
6EU7 V4
6C4 V3
6EU7 V2
6EU7 V1
T3 REVERB. UNIT
T-1 ON-OFF SW
FUSE 1A
TO CATH. 6V6'S
MONITOR
FOOT SW
ALL DC VOLTAGES ARE MEASURED TO CHASSIS WITH V.T.V.M.

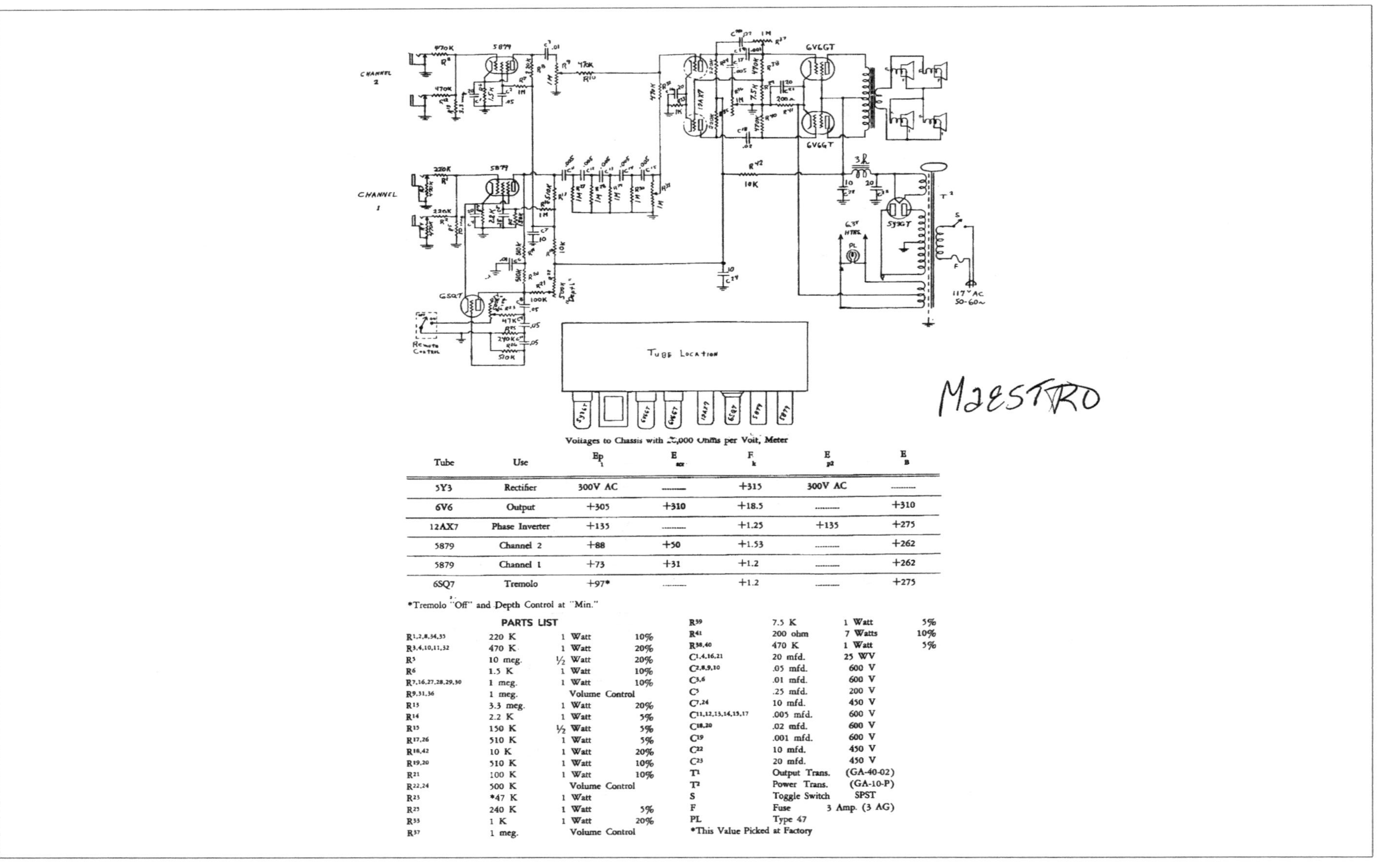

Voltages to Chassis with 20,000 Ohms per Volt Meter

Tube	Use	E_{p1}	E_{sc}	F_{k}	E_{p2}	E_{B}
5Y3	Rectifier	300V AC	———	+315	300V AC	———
6V6	Output	+305	+310	+18.5	———	+310
12AX7	Phase Inverter	+135	———	+1.25	+135	+275
5879	Channel 2	+88	+50	+1.53	———	+262
5879	Channel 1	+73	+31	+1.2	———	+262
6SQ7	Tremolo	+97*	———	+1.2	———	+275

*Tremolo "Off" and Depth Control at "Min."

PARTS LIST

Part	Value	Rating	Tolerance
$R^{1,2,8,34,35}$	220 K	1 Watt	10%
$R^{3,4,10,11,32}$	470 K	1 Watt	20%
R^{5}	10 meg.	½ Watt	20%
R^{6}	1.5 K	1 Watt	10%
$R^{7,16,27,28,29,30}$	1 meg.	1 Watt	10%
$R^{9,31,36}$	1 meg.	Volume Control	
R^{13}	3.3 meg.	1 Watt	20%
R^{14}	2.2 K	1 Watt	5%
R^{15}	150 K	½ Watt	5%
$R^{17,26}$	510 K	1 Watt	5%
$R^{18,42}$	10 K	1 Watt	20%
$R^{19,20}$	510 K	1 Watt	10%
R^{21}	100 K	1 Watt	10%
$R^{22,24}$	500 K	Volume Control	
R^{23}	*47 K	1 Watt	
R^{25}	240 K	1 Watt	5%
R^{33}	1 K	1 Watt	20%
R^{37}	1 meg.	Volume Control	
R^{39}	7.5 K	1 Watt	5%
R^{41}	200 ohm	7 Watts	10%
$R^{38,40}$	470 K	1 Watt	5%
$C^{1,4,16,21}$	20 mfd.	25 WV	
$C^{2,8,9,10}$	.05 mfd.	600 V	
$C^{3,6}$	.01 mfd.	600 V	
C^{5}	.25 mfd.	200 V	
$C^{7,24}$	10 mfd.	450 V	
$C^{11,12,13,14,15,17}$	.005 mfd.	600 V	
$C^{18,20}$	.02 mfd.	600 V	
C^{19}	.001 mfd.	600 V	
C^{22}	10 mfd.	450 V	
C^{23}	20 mfd.	450 V	
T^{1}	Output Trans.	(GA-40-02)	
T^{2}	Power Trans.	(GA-10-P)	
S	Toggle Switch	SPST	
F	Fuse	3 Amp. (3 AG)	
PL	Type 47		

*This Value Picked at Factory

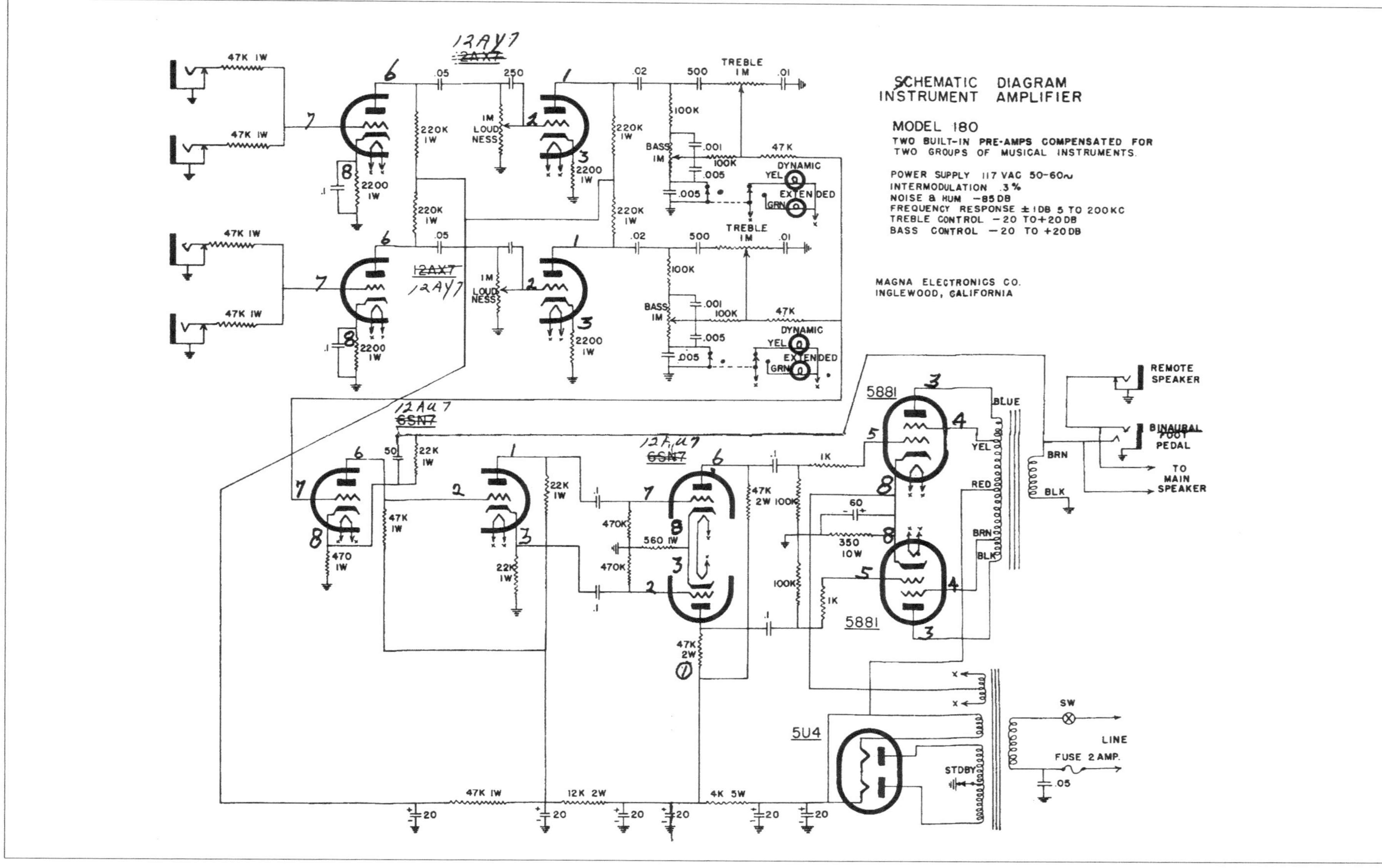
SCHEMATIC DIAGRAM
INSTRUMENT AMPLIFIER
MODEL 180
TWO BUILT-IN PRE-AMPS COMPENSATED FOR
TWO GROUPS OF MUSICAL INSTRUMENTS.
POWER SUPPLY 117 VAC 50-60~
INTERMODULATION .3%
NOISE & HUM -85DB
FREQUENCY RESPONSE ±1DB 5 TO 200KC
TREBLE CONTROL -20 TO +20DB
BASS CONTROL -20 TO +20DB
MAGNA ELECTRONICS CO.
INGLEWOOD, CALIFORNIA
12AY7
12AX7
12AU7
6SN7
5881
5U4
TREBLE
BASS
LOUDNESS
DYNAMIC
EXTENDED
STDBY
REMOTE SPEAKER
BINAURAL FOOT PEDAL
TO MAIN SPEAKER
SW
LINE
FUSE 2 AMP.

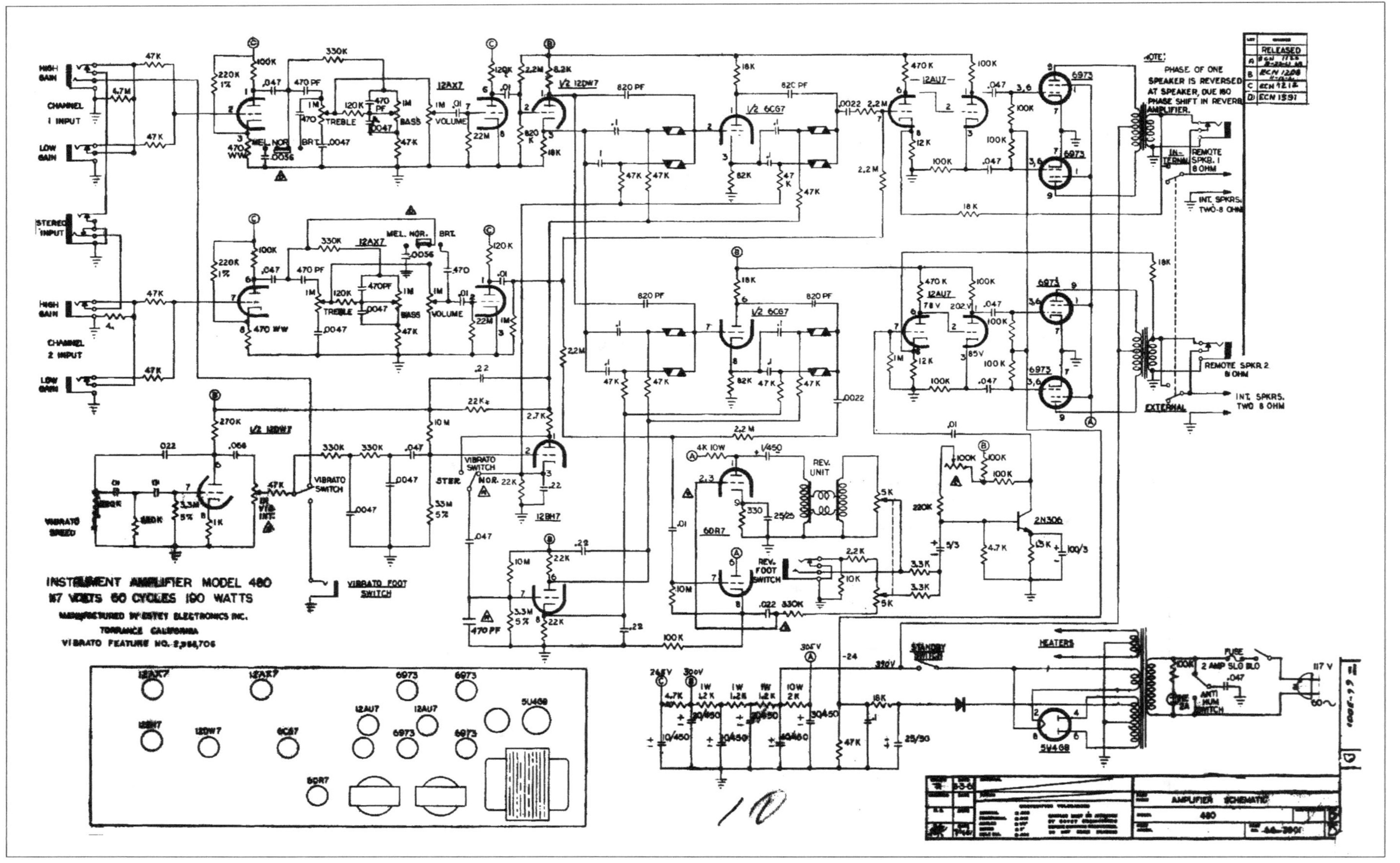
INSTRUMENT AMPLIFIER MODEL 480
117 VOLTS 60 CYCLES 190 WATTS
MANUFACTURED BY ESTEY ELECTRONICS INC.
TORRANCE CALIFORNIA
VIBRATO FEATURE NO. 2,988,706
NOTE:
PHASE OF ONE SPEAKER IS REVERSED AT SPEAKER, DUE 180 PHASE SHIFT IN REVERB AMPLIFIER.
RELEASED
B ECN 1208
C ECN 1212
D ECN 1391
HIGH GAIN
CHANNEL 1 INPUT
LOW GAIN
STEREO INPUT
CHANNEL 2 INPUT
VIBRATO SPEED
VIBRATO SWITCH
VIBRATO FOOT SWITCH
REV. UNIT
REV. FOOT SWITCH
STANDBY SWITCH
HEATERS
FUSE 2 AMP SLO BLO
ANTI HUM SWITCH
117 V
REMOTE SPKR. 1 8 OHM
INT. SPKRS. TWO-8 OHM
REMOTE SPKR 2 8 OHM
INT. SPKRS. TWO 8 OHM
EXTERNAL
AMPLIFIER SCHEMATIC
480

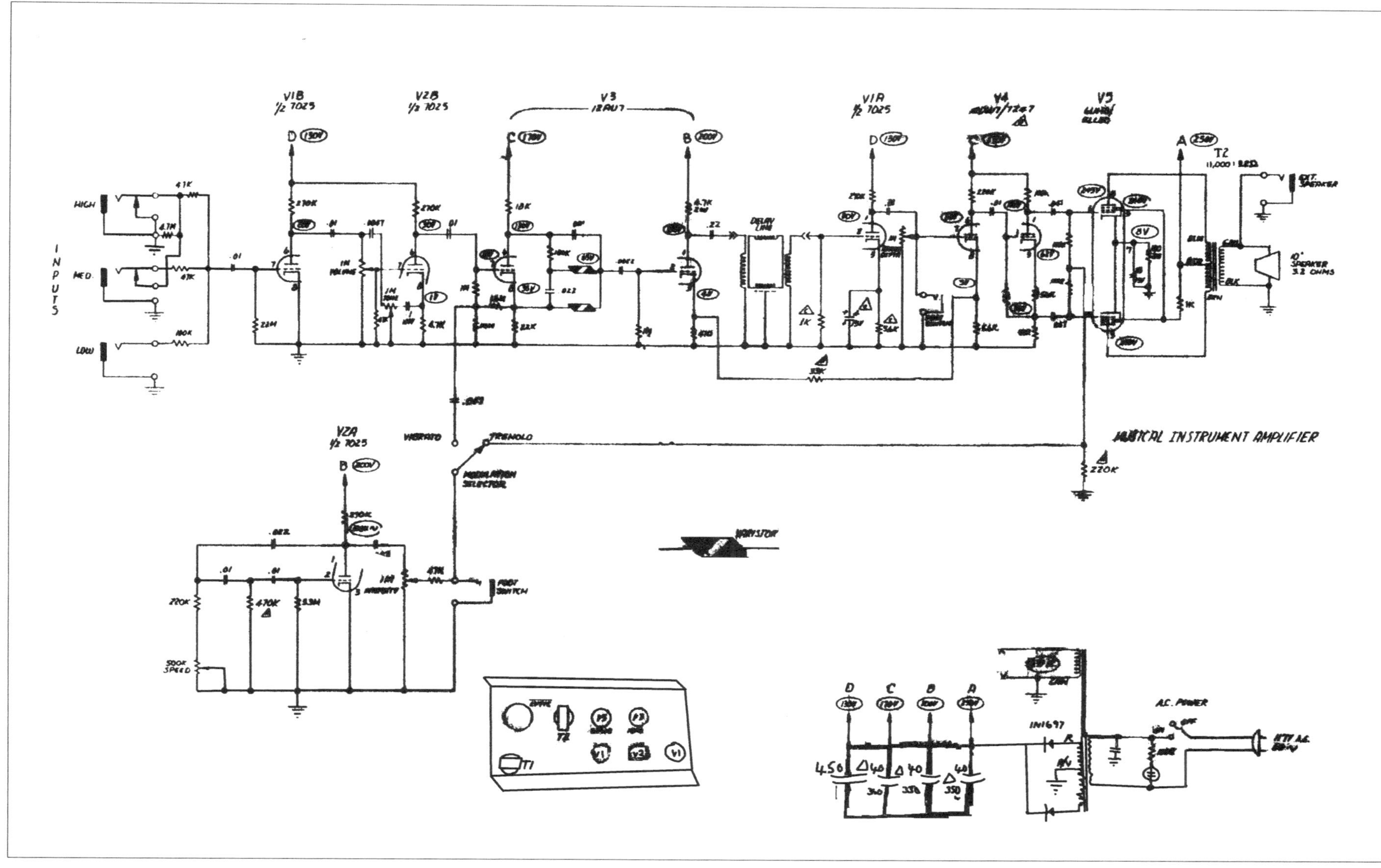
INPUTS
HIGH
MED.
LOW
V1B
1/2 7025
V2B
1/2 7025
V3
12AU7
V1A
1/2 7025
V4
V5
T2
EXT. SPEAKER
10" SPEAKER 3.2 OHMS
DELAY LINE
VIBRATO
TREMOLO
MODULATION SELECTOR
V2A
1/2 7025
FOOT SWITCH
500K SPEED
VARISTOR
MUSICAL INSTRUMENT AMPLIFIER
220K
A.C. POWER
1N1697

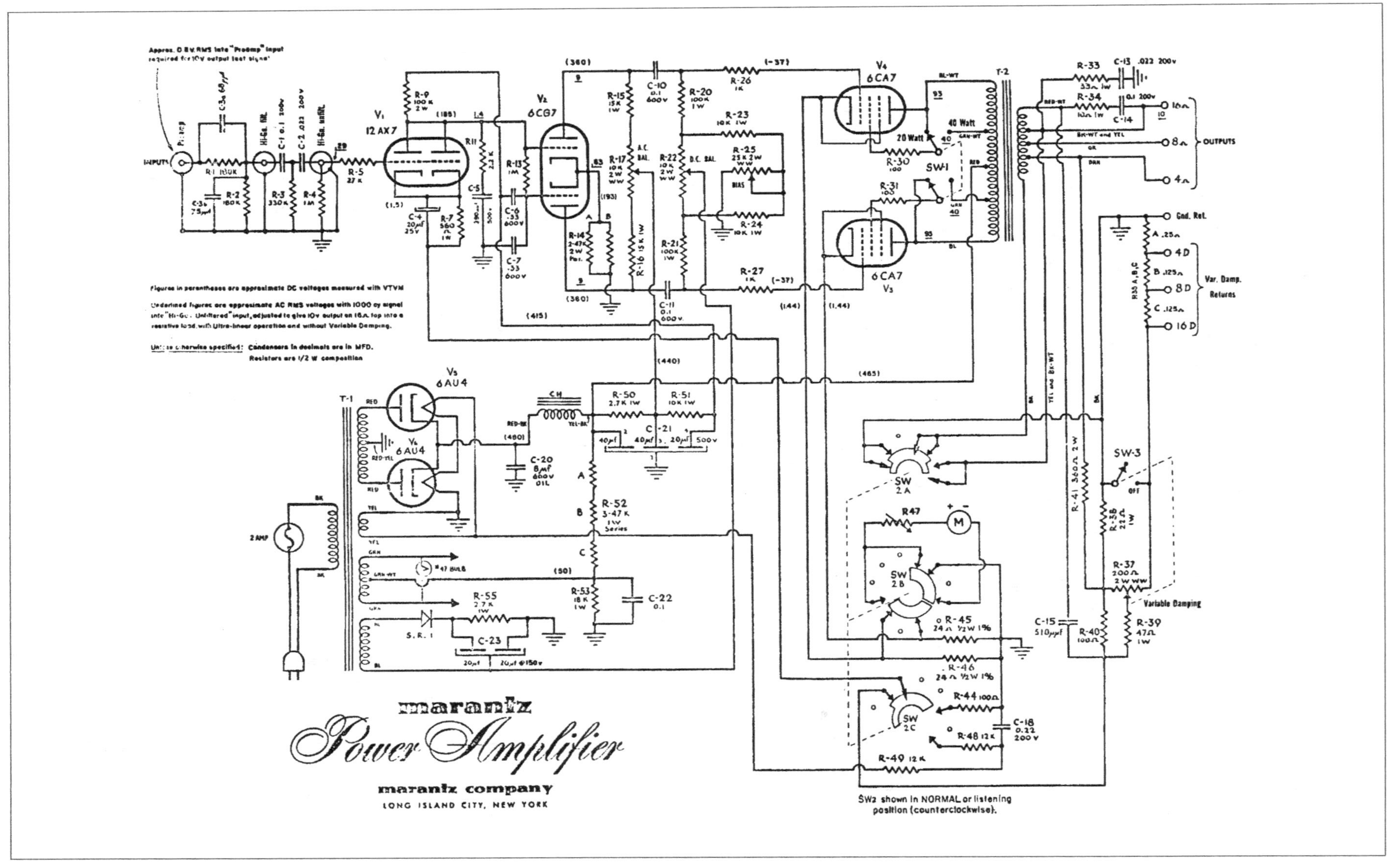

Approx. 0.8V RMS into "Preamp" input required for 10V output test signal
INPUTS
R-1 180K
R-2 180K
R-3 330K
R-4 1M
R-5 27K
V1 12AX7
V2 6CG7
V4 6CA7
V3 6CA7
V5 6AU4
V6 6AU4
T-1
T-2
SW-1
SW-3
20 Watt
40 Watt
OUTPUTS
Gnd. Ret.
Var. Damp. Returns
Variable Damping
Figures in parentheses are approximate DC voltages measured with VTVM
Underlined figures are approximate AC RMS voltages with 1000 cy signal into "Hi-Ga. Unfiltered" input, adjusted to give 10v output on 16Ω tap into a resistive load, with Ultra-linear operation and without Variable Damping.
Unless otherwise specified: Condensers in decimals are in MFD.
Resistors are 1/2 W composition
2 AMP
#47 BULB
S.R. 1
SW2 shown in NORMAL or listening position (counterclockwise).
marantz
Power Amplifier
marantz company
LONG ISLAND CITY, NEW YORK

MARSHALL 50W VOLT. CHART

MK 2

TUBE		PIN NO. 1	2	3	4	5	6	7	8	9
ECC83 12AX7A	V1	100	—	2.9	H	H	205	—	2	H
ECC83 12AX7A	V2	150	—	1	H	H	290	150	150	H
ECC83 12AX7A	V3	210	—	40	H	H	200	—	40	H
6550 (USA)	V4	—	H	395	390	-44	NP	H	0	—
6550 (USA)	V5	—	H	395	390	-44	NP	H	0	—
BIAS	APPROX. -44V. PRI. CURRENT 0.7 TO 0.8 AMPS.									

MARSHALL 200W VOLT CHART

TUBE		PIN NO. 1	2	3	4	5	6	7	8	9
ECC83 12AX7A	V1	190	—	1.75	H	H	235	—	2.4	H
ECC83 12AX7A	V2	210	—	1.6	H	H	300		75	H
ECC82 12AU7	V3	310	—	1.5	H	H	310	—	15	H
KT88	V4	—	H	620	620	-85	—	H	—	—
KT88	V5	—	H	620	620	-85	—	H	—	—
KT88	V6	—	H	620	620	-85	—	H	—	—
KT88	V7	—	H	620	620	-85	—	H	—	—
	V8									

MARSHALL 100W. VOLT. CHART

TUBE		PIN NO. 1	2	3	4	5	6	7	8	9
ECC83 12AX7A	V1	135	—	1.0	H	H	195	—	1.5	H
ECC83 12AX7A	V2	153	—	1.1	H	H	260	153	153	H
ECC83 12AX7A	V3	210	—	18.5	H	H	195	—	18.5	H
6550 (USA)	V4	—	H	453	450	-51*	NP	H	—	—
6550 (USA)	V5	—	H	453	450	-51*	NP	H	—	—
6550 (USA)	V6	—	H	453	450	-51*	NP	H	—	—
6550 (USA)	V7	—	H	453	450	-51*	NP	H	—	—

*(Bias approx. -51vdc; primary current 1.2 - 1.3 amps at 120vac)

MARSHALL 100W. VOLT. CHART

TUBE		PIN NO. 1	2	3	4	5	6	7	8	9
ECC83 12AX7A	V1	165	—	1.0	H	H	165	—	1.5	H
ECC83 12AX7A	V2	175	—	1.1	H	H	305	175	175	H
ECC83 12AX7A	V3	230	—	24	H	H	230	—	24	H
EL34 6CA7	V4	—	H	530	520	-50	NP	H	—	—
EL34 6CA7	V5	—	H	530	520	-50	NP	H	—	—
EL34 6CA7	V6	—	H	530	520	-50	NP	H	—	—
EL34 6CA7	V7	—	H	530	520	-50	NP	H	—	—
	V8			530	520	-50				

MARSHALL 50W. VOLT. CHART

TUBE		PIN NO. 1	2	3	4	5	6	7	8	9
ECC83 12AX7A	V1	180	—	1.6	H	H	160	—	1.6	H
ECC83 12AX7A	V2	140	—	1	H	H	250	140	140	H
ECC83 12AX7A	V3	220	—	40	H	H	210	—	40	H
EL34 6CA7	V4	—	H	380	390	-30	—	H	—	—
EL34 6CA7	V5	—	H	380	390	-30	—	H	—	—
GZ34-5AR4 SOME MODELS	V6	—	H	AC 320	—	AC 320	—	—	400	—
HAVE SOLID STATE REC.	V7									
	V8									

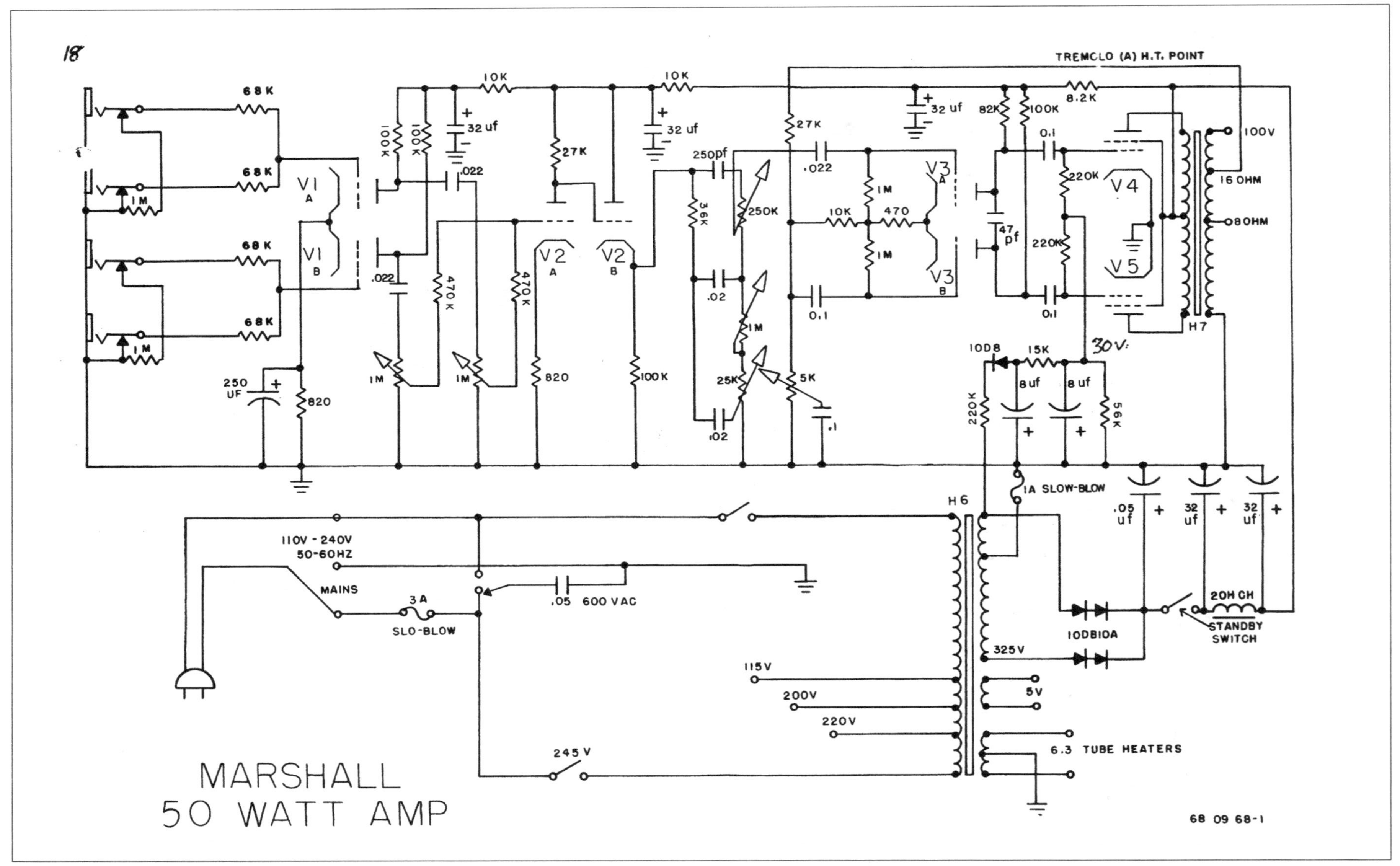
18
68K
68K
68K
68K
1M
1M
V1 A
V1 B
250 UF
820
100K
100K
32 uf
.022
.022
10K
10K
27K
470K
470K
1M
1M
V2 A
V2 B
820
100K
32 uf
250pf
36K
250K
.02
1M
25K
.02
5K
.1
27K
.022
0.1
1M
10K
470
1M
V3 A
V3 B
32 uf
82K
100K
8.2K
0.1
220K
220K
47 pf
0.1
V4
V5
TREMCLO (A) H.T. POINT
100V
16 OHM
8 OHM
H7
30V
10D8
15K
220K
8 uf
8 uf
56K
1A SLOW-BLOW
.05 uf
32 uf
32 uf
20H CH
STANDBY SWITCH
10DB10A
H6
325V
110V - 240V
50-60HZ
MAINS
3A
SLO-BLOW
.05 600 VAC
115V
200V
220V
245V
5V
6.3 TUBE HEATERS
MARSHALL
50 WATT AMP
68 09 68-1

50w MASTER VOL **(MODEL 2204)**

Marshall

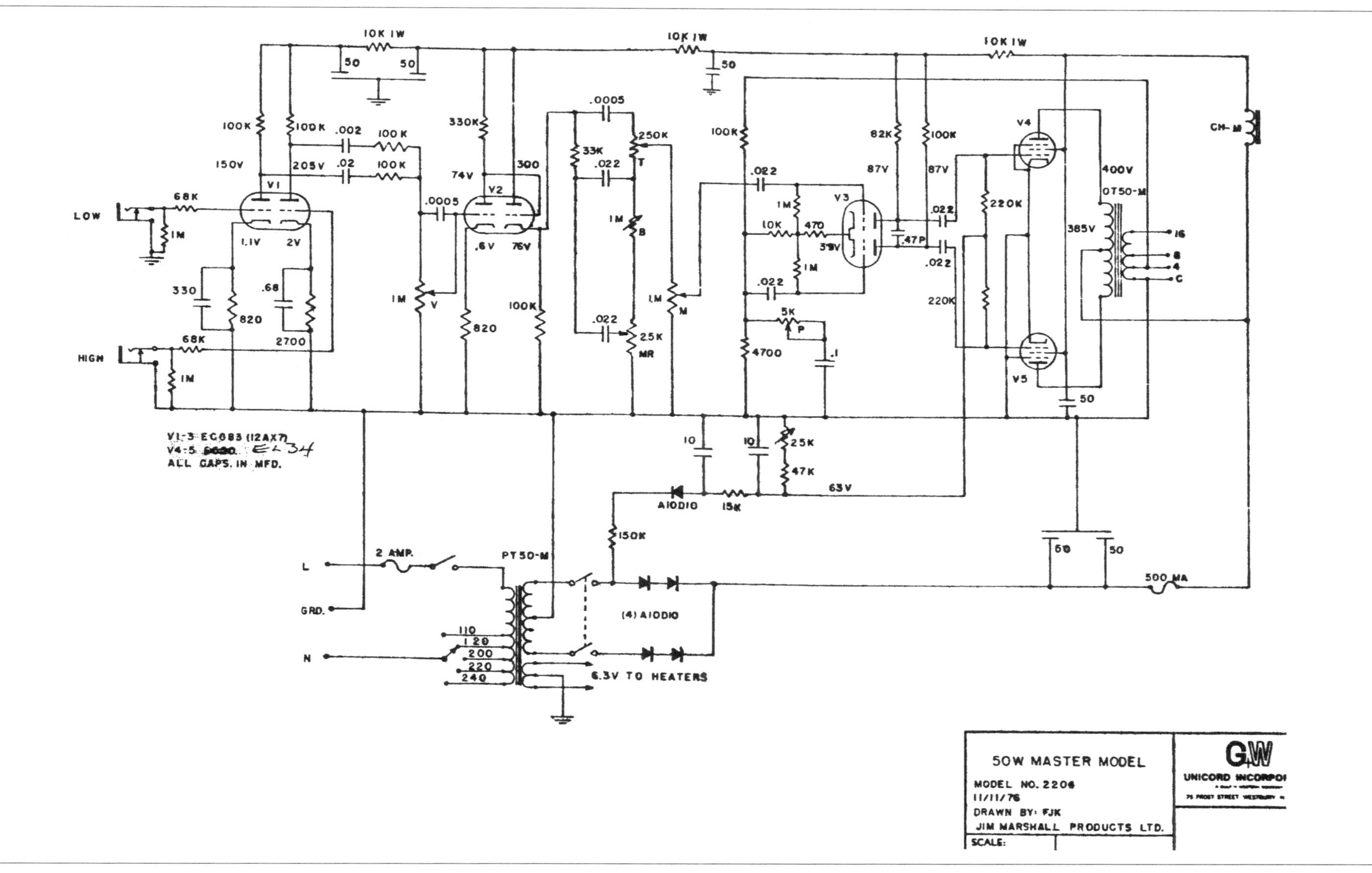

50w MASTER VOL (MODEL 2204) REV

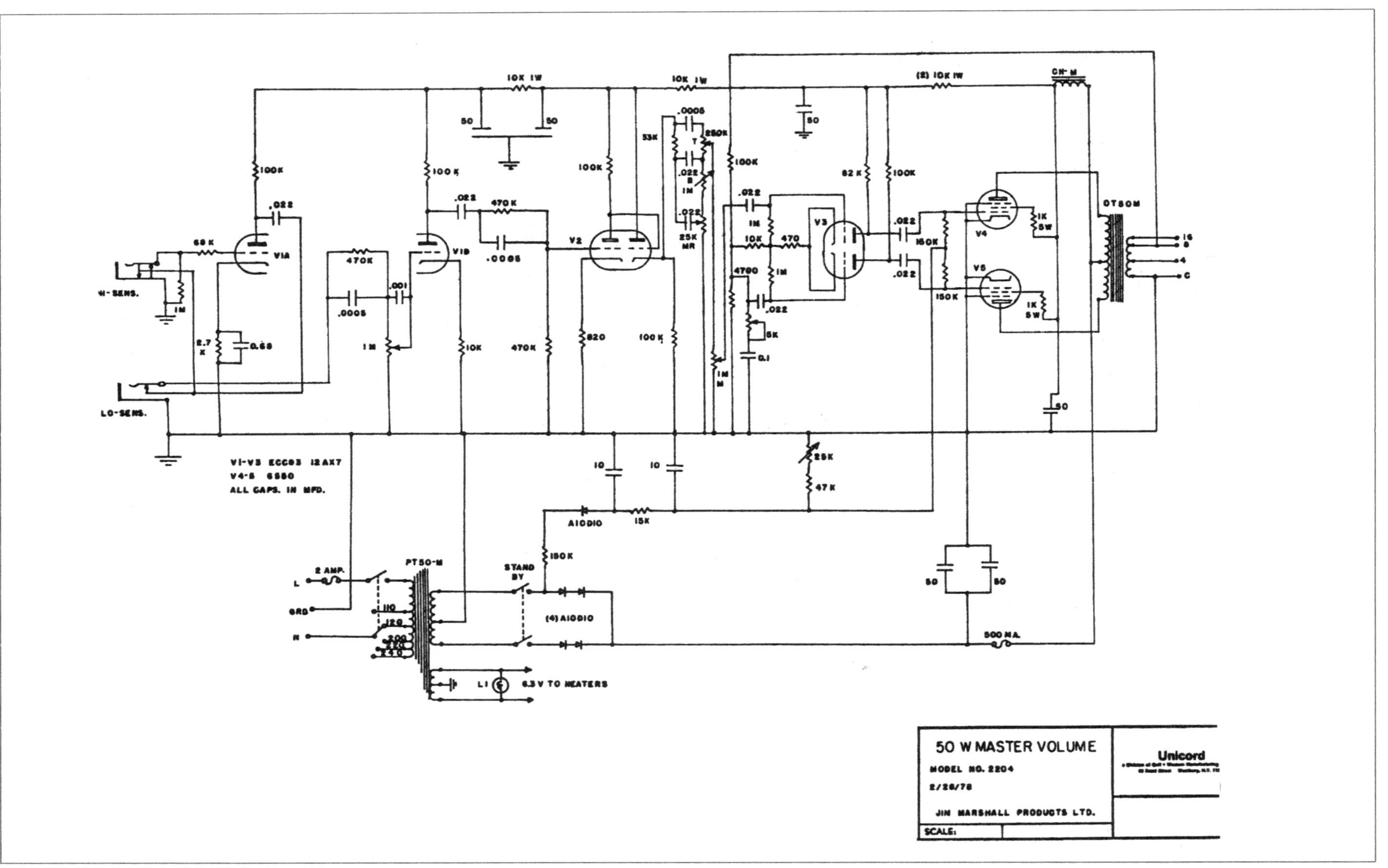

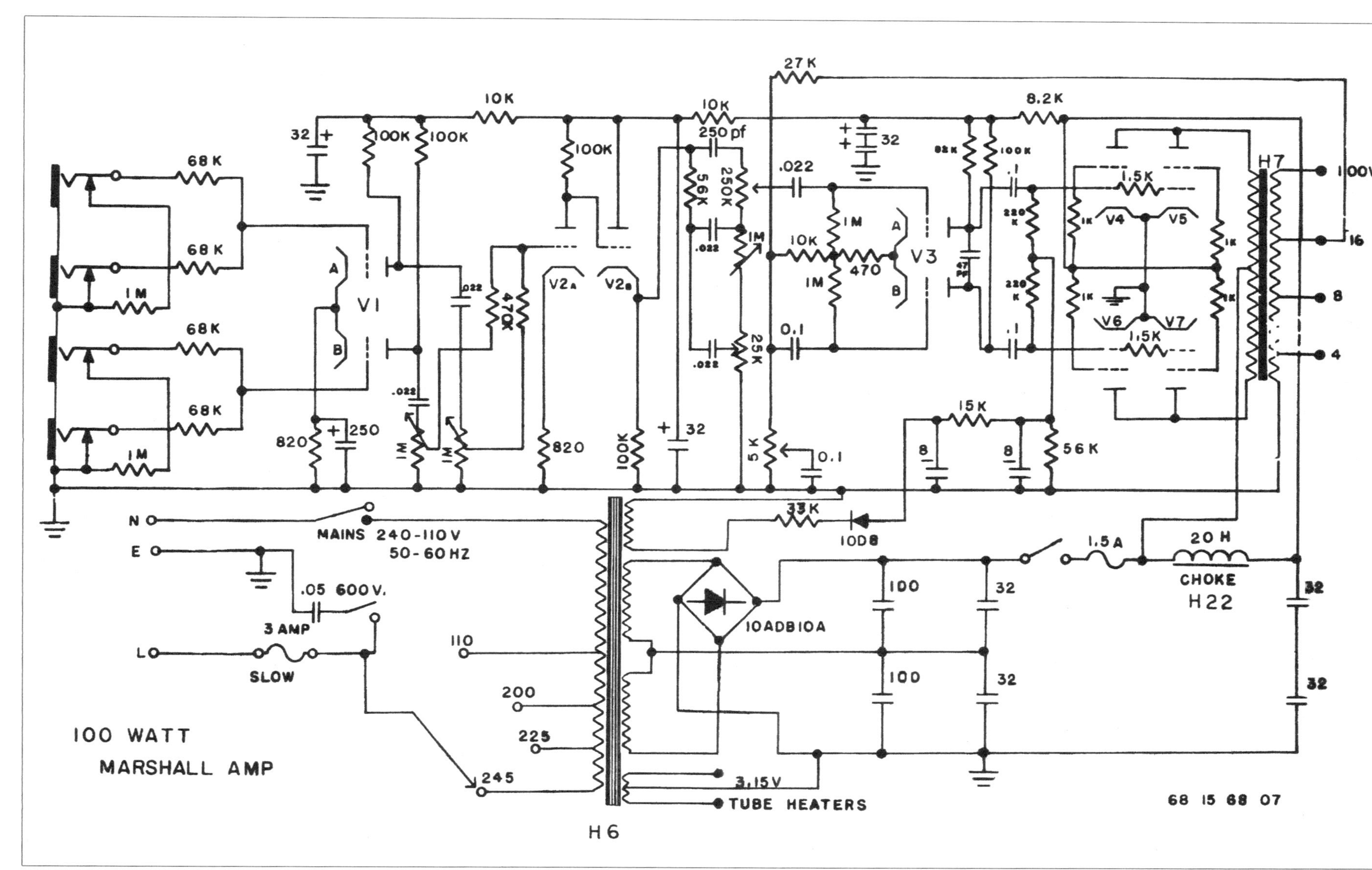
100 WATT
MARSHALL AMP
MAINS 240-110V
50-60HZ
.05 600V.
3 AMP
SLOW
N
E
L
110
200
225
245
H6
3.15V
TUBE HEATERS
10ADB10A
10D8
1.5A
20 H
CHOKE
H22
H7
100V
16
8
4
V1
V2A
V2B
V3
V4
V5
V6
V7
68 15 68 07

100w **PA (MODEL 1968)**

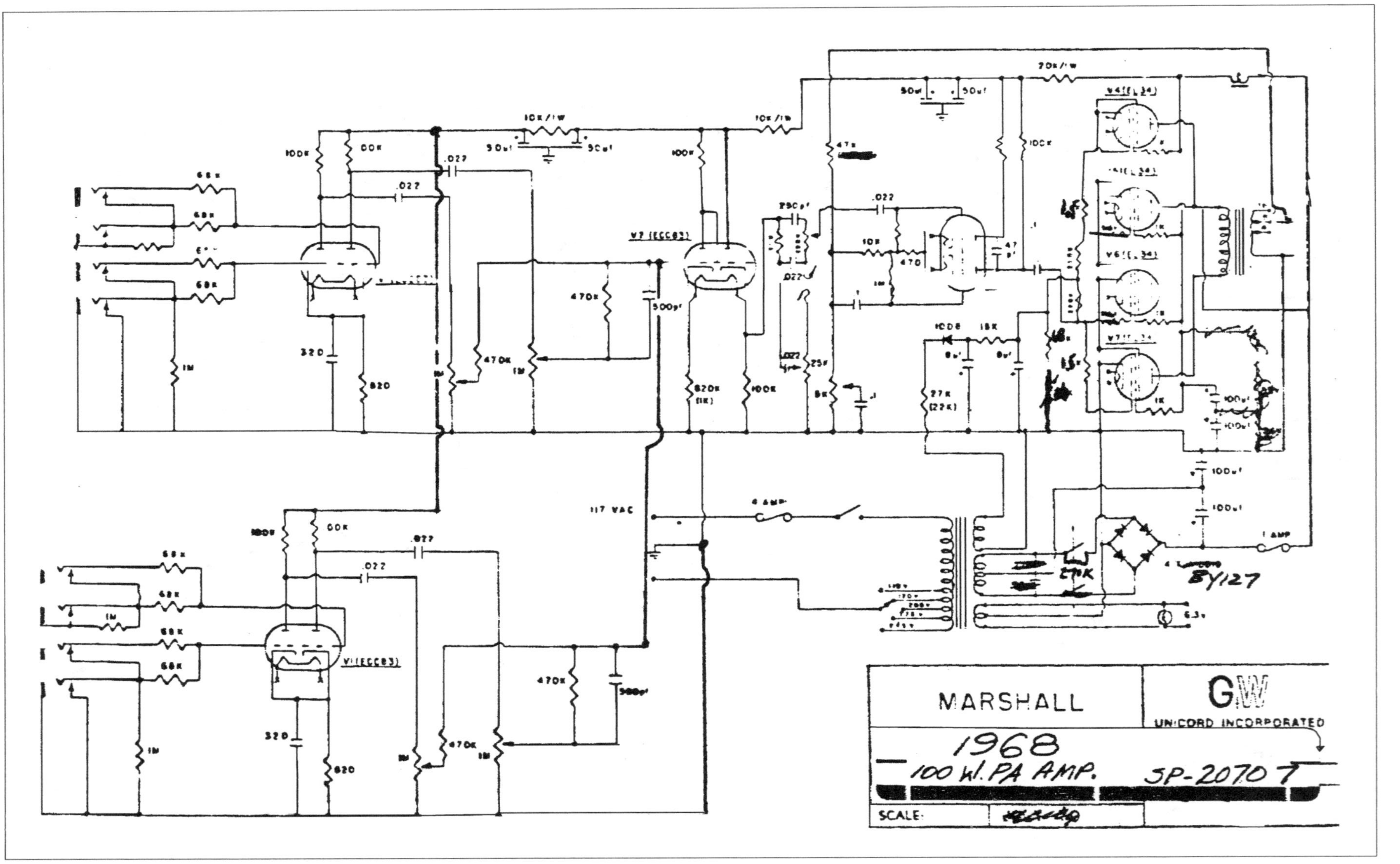

200w **PA (MODEL 1966)**

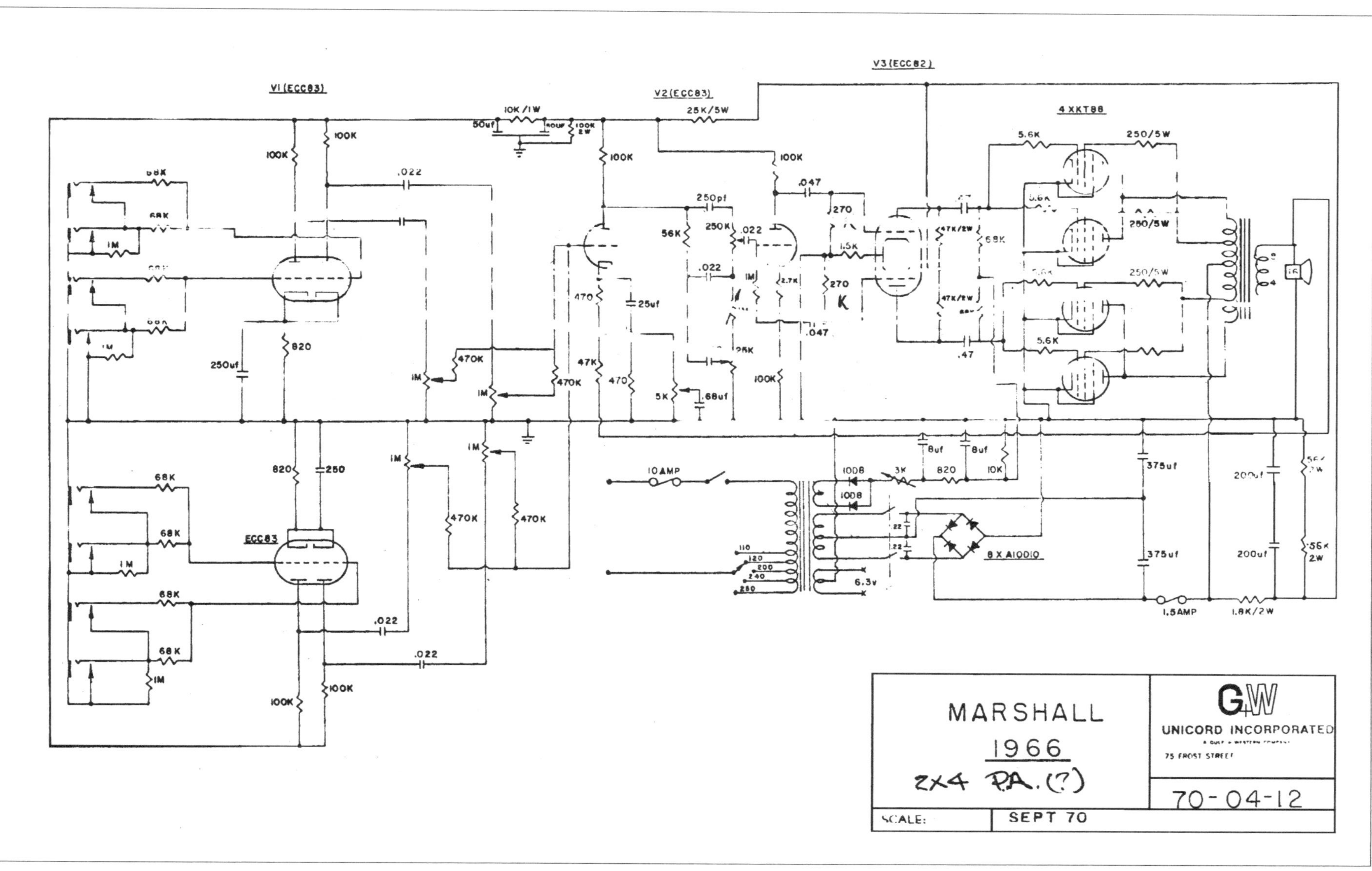

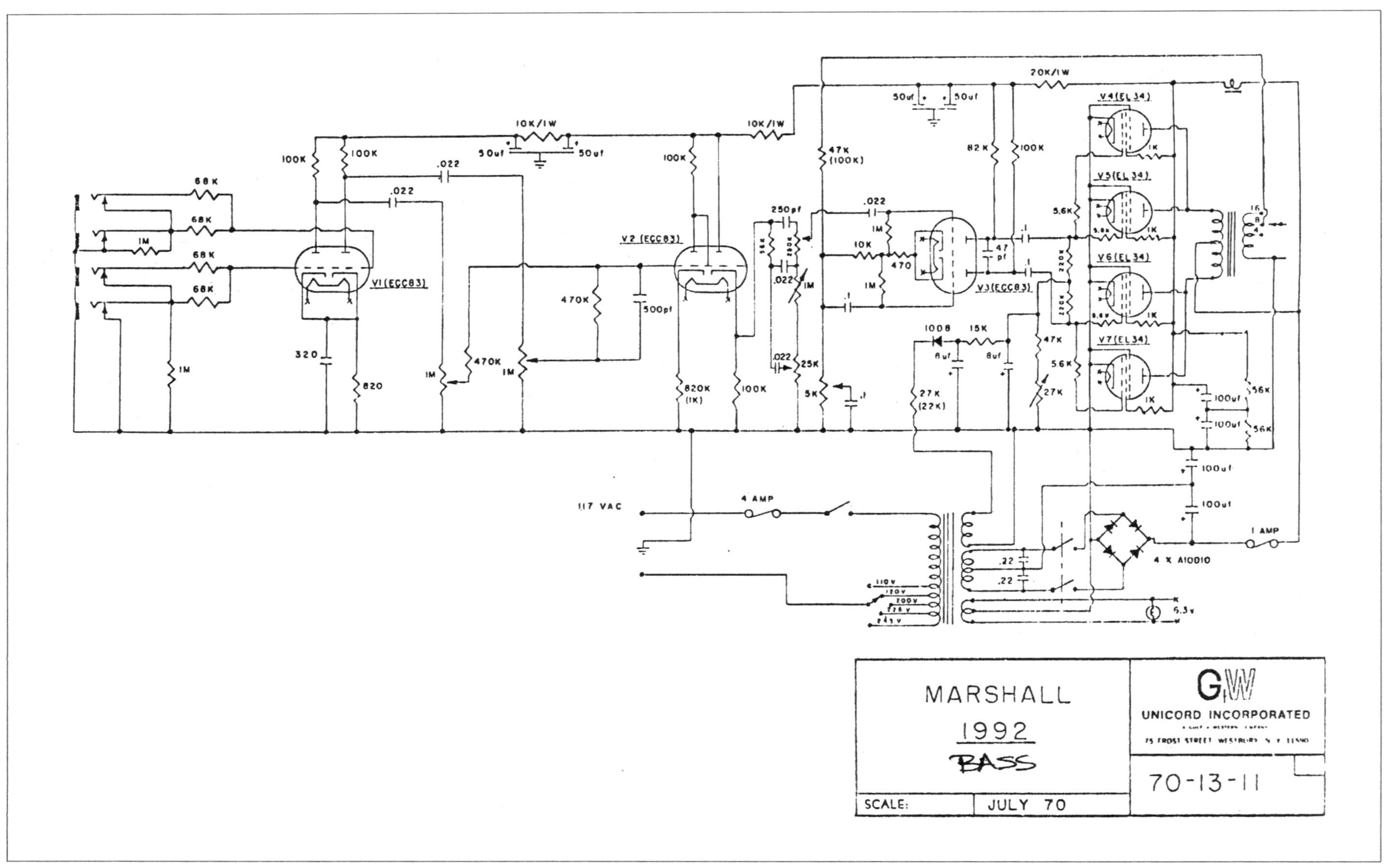
V1 (ECC83)
V2 (ECC83)
V3 (ECC83)
V4 (EL34)
V5 (EL34)
V6 (EL34)
V7 (EL34)
117 VAC
4 AMP
1 AMP
4 X A10010
6.3v
MARSHALL
1992
BASS
SCALE:
JULY 70
GW
UNICORD INCORPORATED
75 FROST STREET WESTBURY N. Y. 11590
70-13-11

BLUESBREAKER REISSUE **(MODEL 1962)**

Marshall

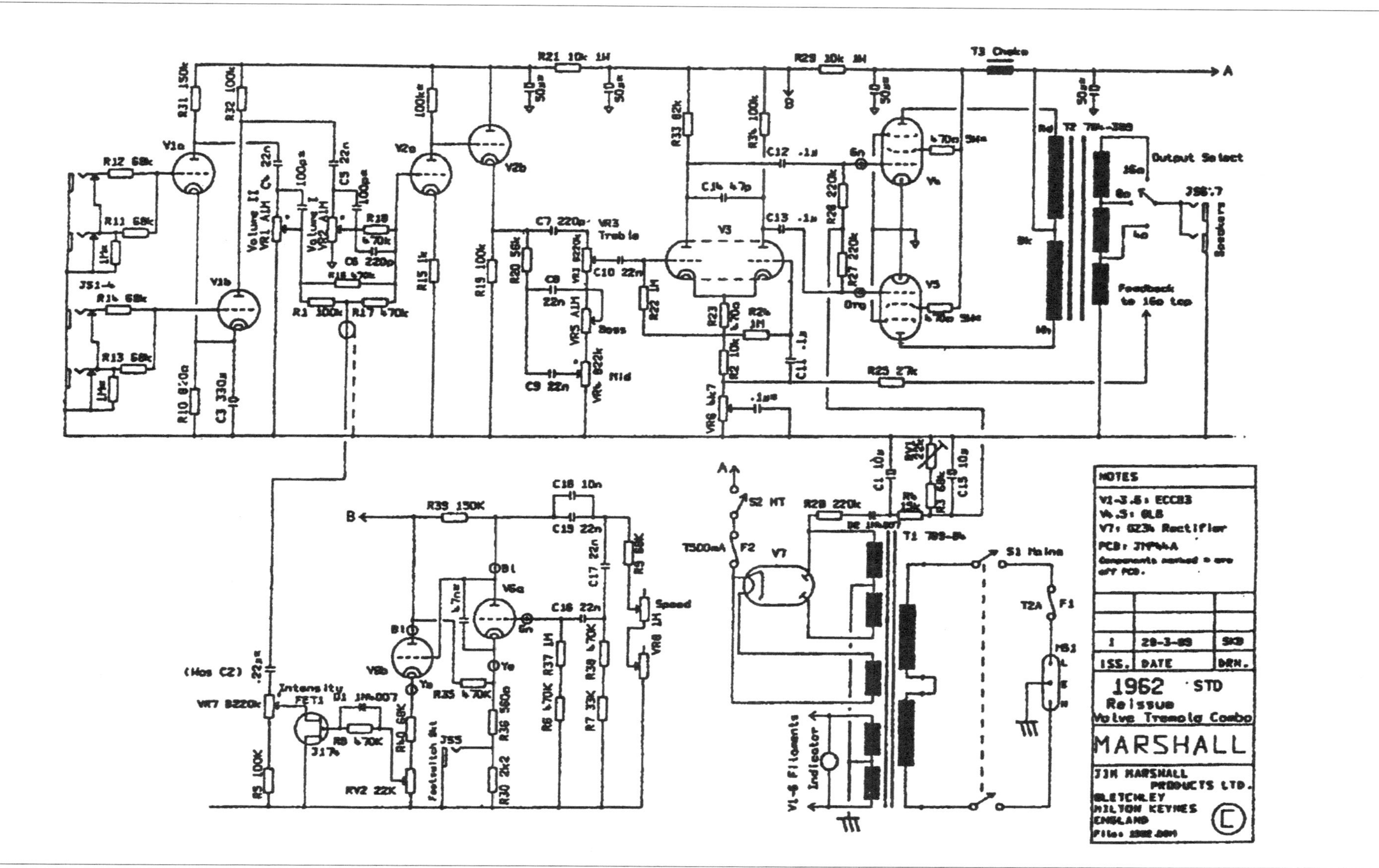

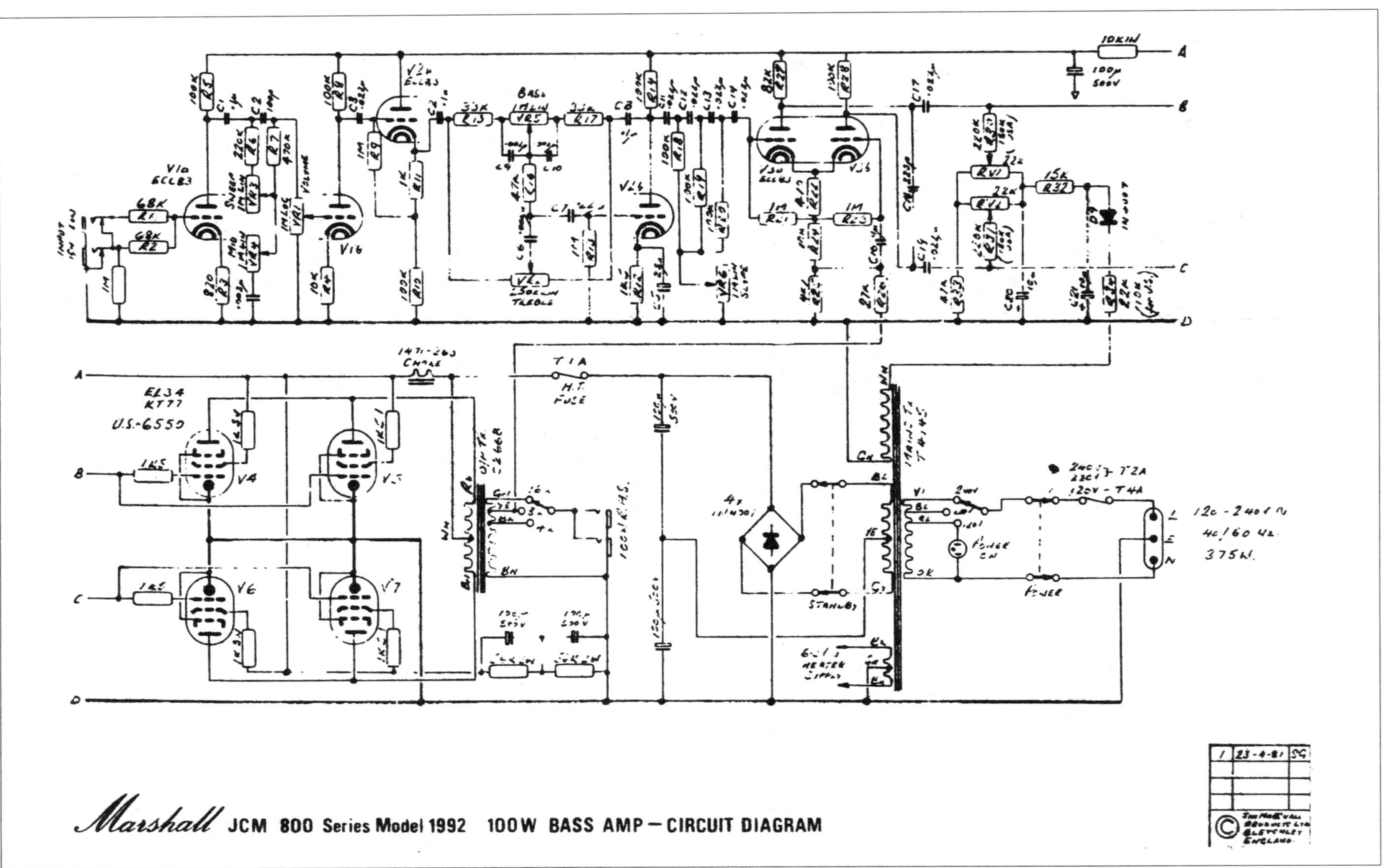

Marshall JCM 800 Series Model 1992 100W BASS AMP – CIRCUIT DIAGRAM

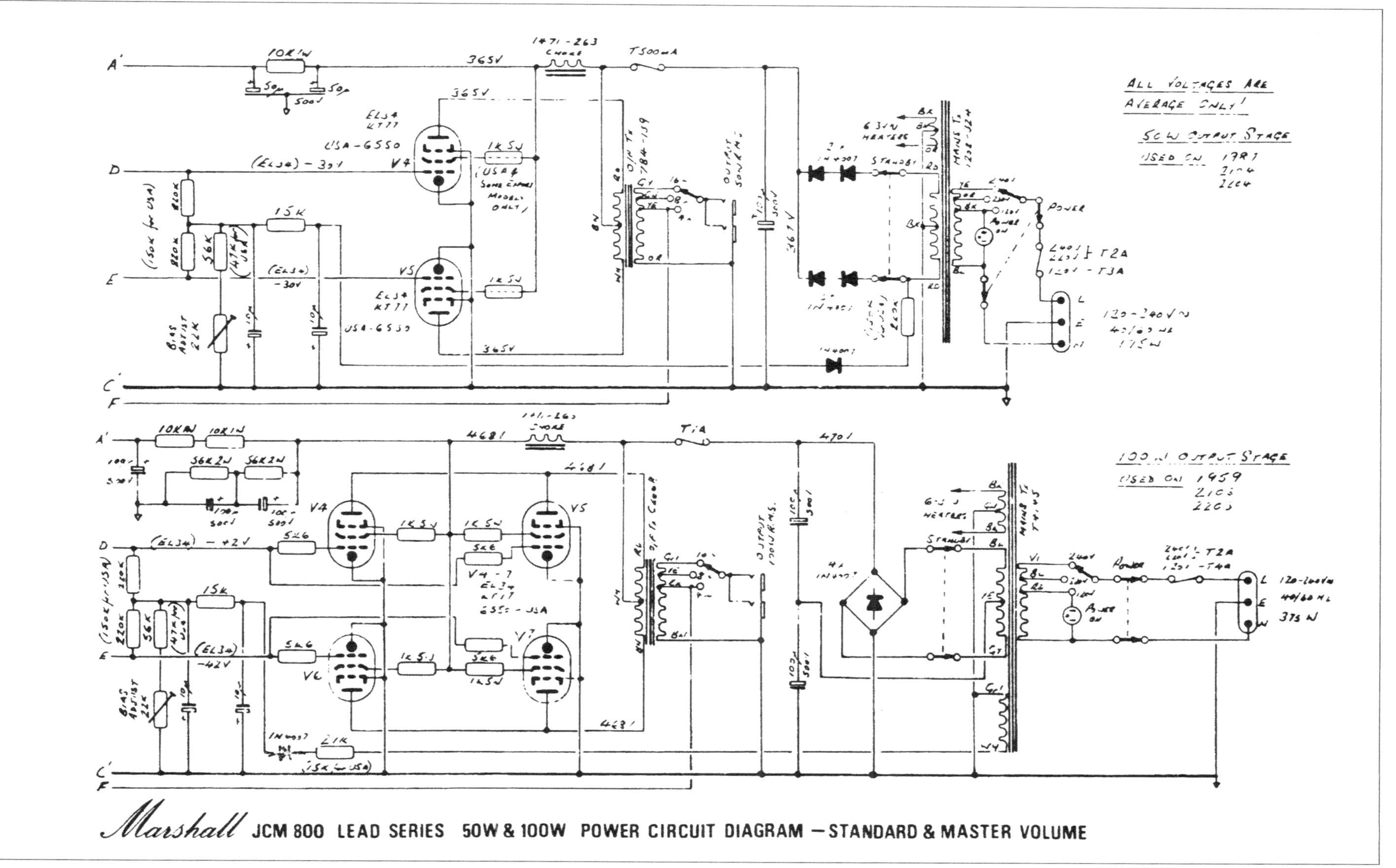

Marshall JCM 800 LEAD SERIES 50W & 100W POWER CIRCUIT DIAGRAM – STANDARD & MASTER VOLUME

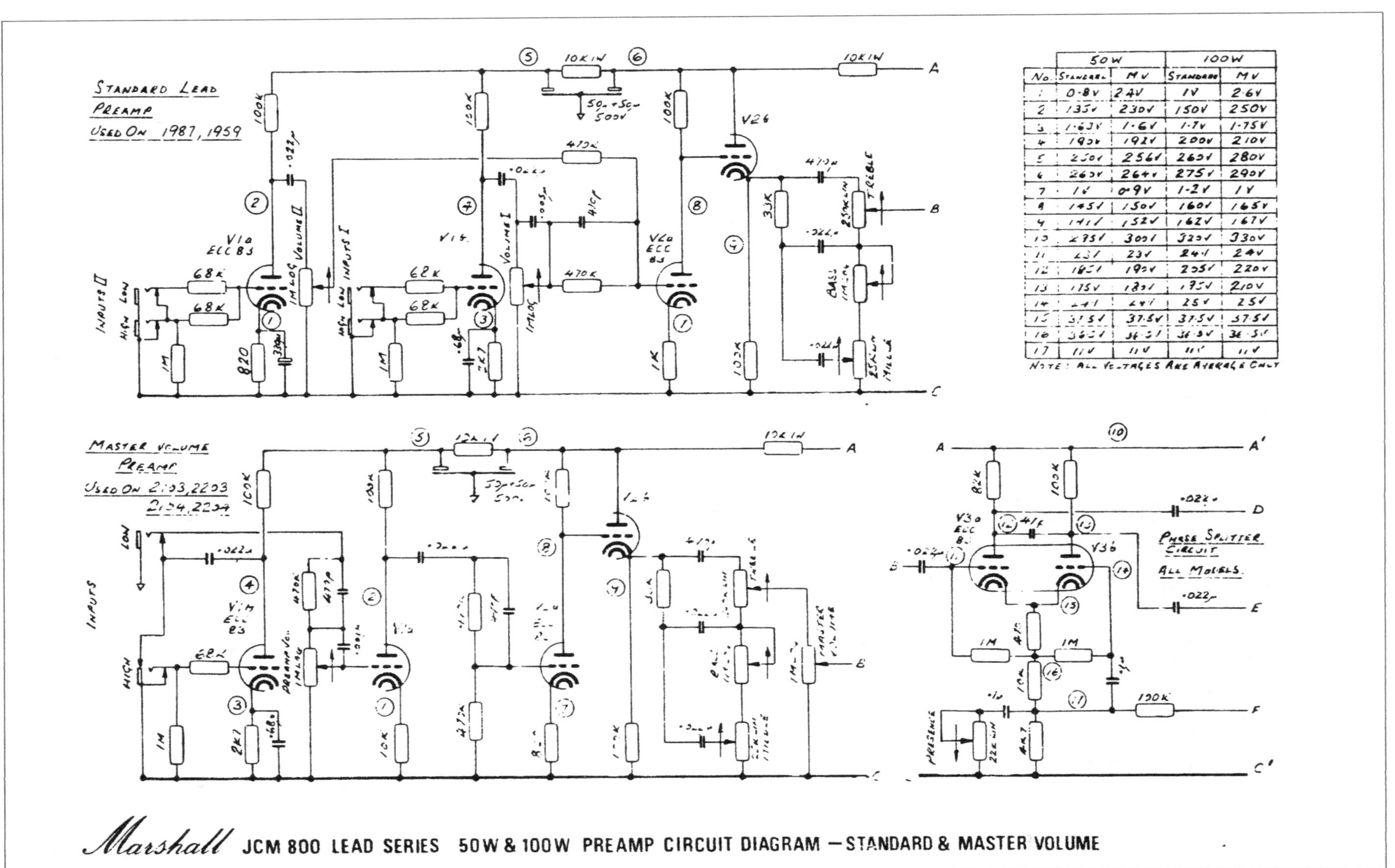

	50W		100W	
No.	STANDARD	MV	STANDARD	MV
1	0.8V	2.4V	1V	2.6V
2	135V	230V	150V	250V
3	1.63V	1.6V	1.7V	1.75V
4	190V	192V	200V	210V
5	250V	256V	265V	280V
6	260V	264V	275V	290V
7	1V	0.9V	1.2V	1V
8	145V	150V	160V	165V
9	141V	152V	162V	167V
10	275V	300V	320V	330V
11	23V	23V	24V	24V
12	185V	190V	205V	220V
13	175V	180V	175V	210V
14	24V	24V	25V	25V
15	37.5V	37.5V	37.5V	37.5V
16	36.5V	36.5V	36.5V	36.5V
17	11V	11V	11V	11V

NOTE: ALL VOLTAGES ARE AVERAGE ONLY

Marshall JCM 800 LEAD SERIES 50W & 100W PREAMP CIRCUIT DIAGRAM — STANDARD & MASTER VOLUME

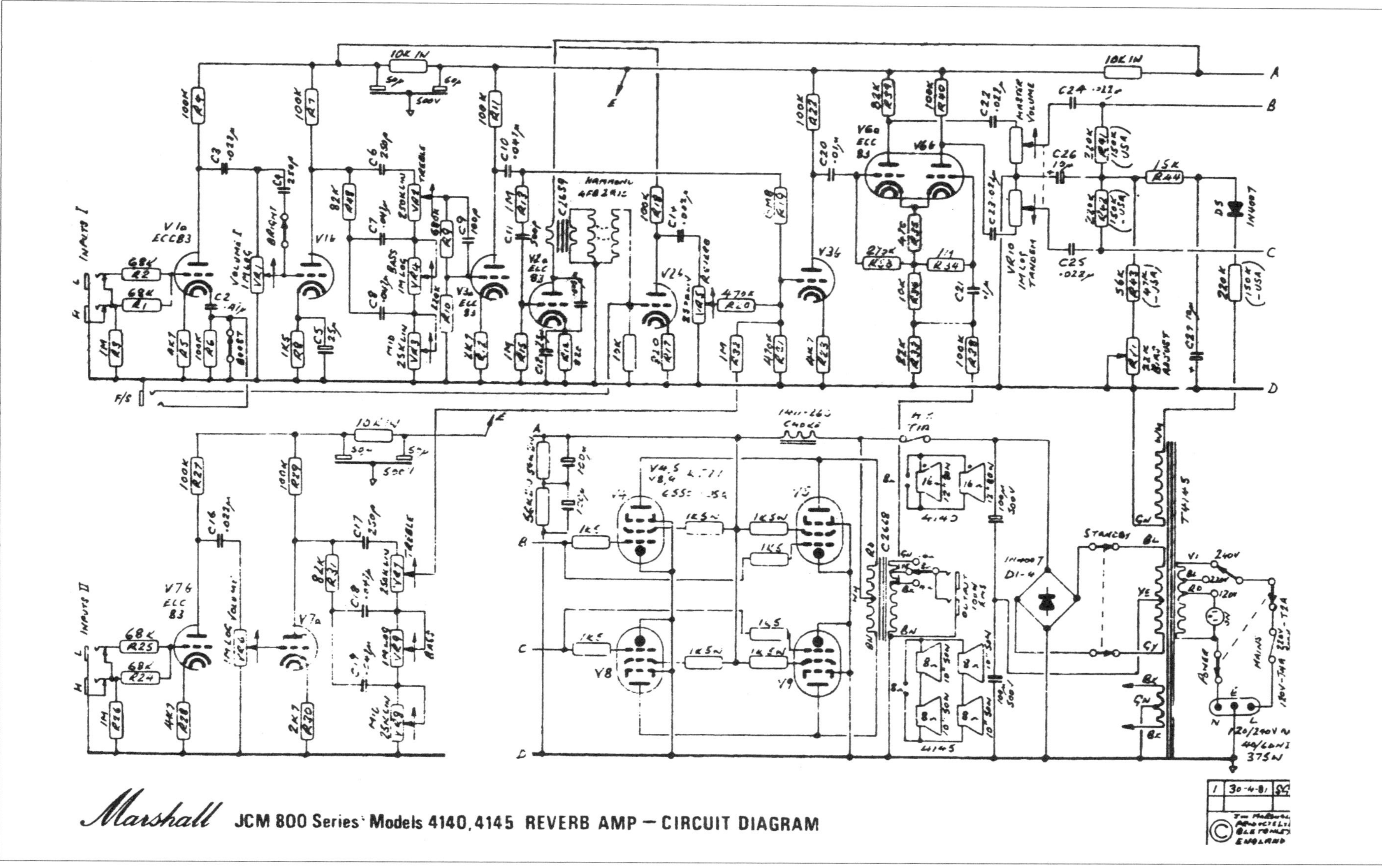
Marshall
JCM 800 Series Models 4140, 4145 REVERB AMP — CIRCUIT DIAGRAM
INPUTS I
INPUTS II
VOLUME I
BRIGHT
BOOST
TREBLE
BASS
MID
REVERB
MASTER VOLUME
VR10 1M LOG TANDEM
BIAS ADJUST
HAMMOND 4FB3R1C
STANDBY
POWER
MAINS
T4145
C2668
C1659
4140
4145
1N4007
ECC83
A
B
C
D

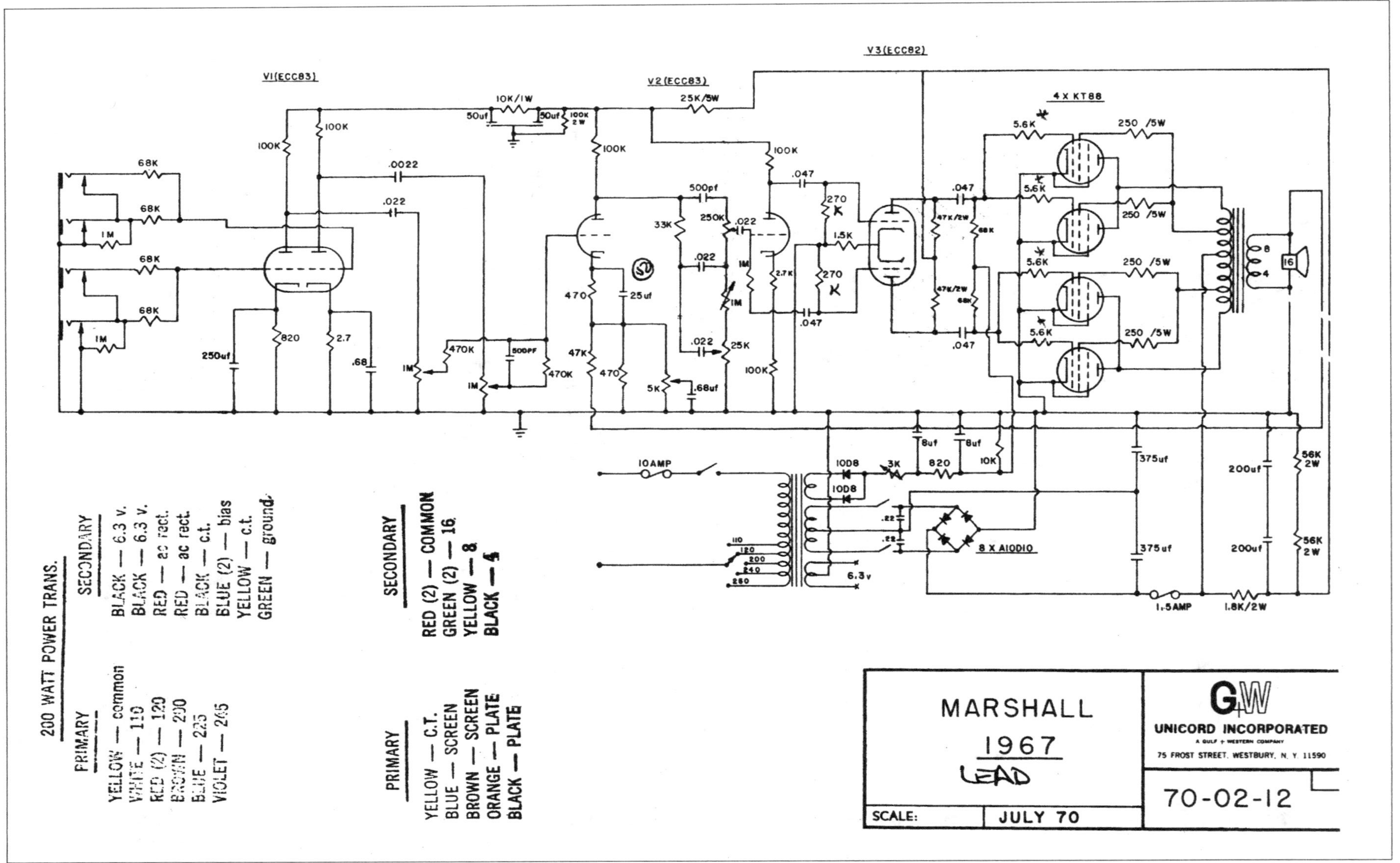
V1(ECC83)
V2(ECC83)
V3(ECC82)
4 X KT88
10K/1W
50uf
100K 2W
25K/5W
100K
68K
1M
.0022
.022
250uf
820
2.7
.68
470K
500PF
47K
470
25uf
5K
.68uf
33K
500pf
250K
25K
2.7K
1.5K
270
.047
47K/2W
68K
5.6K
250 /5W
8
4
16
10AMP
10D8
3K
8uf
10K
375uf
200uf
56K 2W
1.8K/2W
1.5AMP
8 X A10DIO
.22
6.3v
110
120
200
240
260
200 WATT POWER TRANS.
PRIMARY
YELLOW — common
WHITE — 110
RED (2) — 120
BROWN — 200
BLUE — 225
VIOLET — 245
SECONDARY
BLACK — 6.3 v.
BLACK — 6.3 v.
RED — ac rect.
RED — ac rect.
BLACK — c.t.
BLUE (2) — bias
YELLOW — c.t.
GREEN — ground.
PRIMARY
YELLOW — C.T.
BLUE — SCREEN
BROWN — SCREEN
ORANGE — PLATE
BLACK — PLATE
SECONDARY
RED (2) — COMMON
GREEN (2) — 16
YELLOW — 8
BLACK — 4
MARSHALL
1967
LEAD
SCALE:
JULY 70
G+W
UNICORD INCORPORATED
A GULF + WESTERN COMPANY
75 FROST STREET, WESTBURY, N. Y. 11590
70-02-12

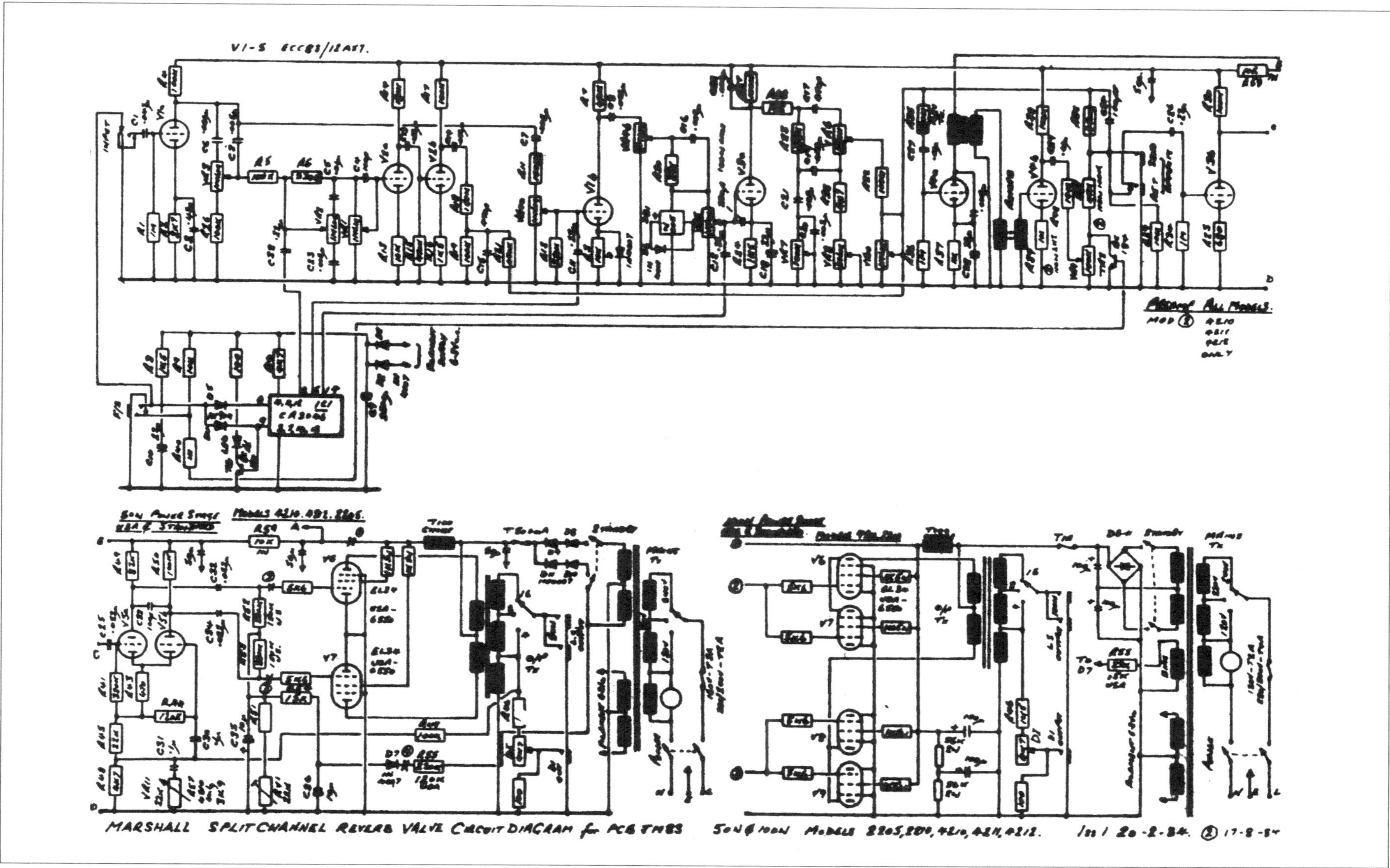
V1-5 ECC83/12AX7.
INPUT
ALTERNATIVE ALL MODELS.
50W POWER STAGE
MODELS 4210, 4211, 2205.
MARSHALL SPLIT CHANNEL REVERB VALVE CIRCUIT DIAGRAM for PCB TM83 50W & 100W MODELS 2205, 2210, 4210, 4211, 4212.
Iss 1 20-2-84. ② 17-8-84

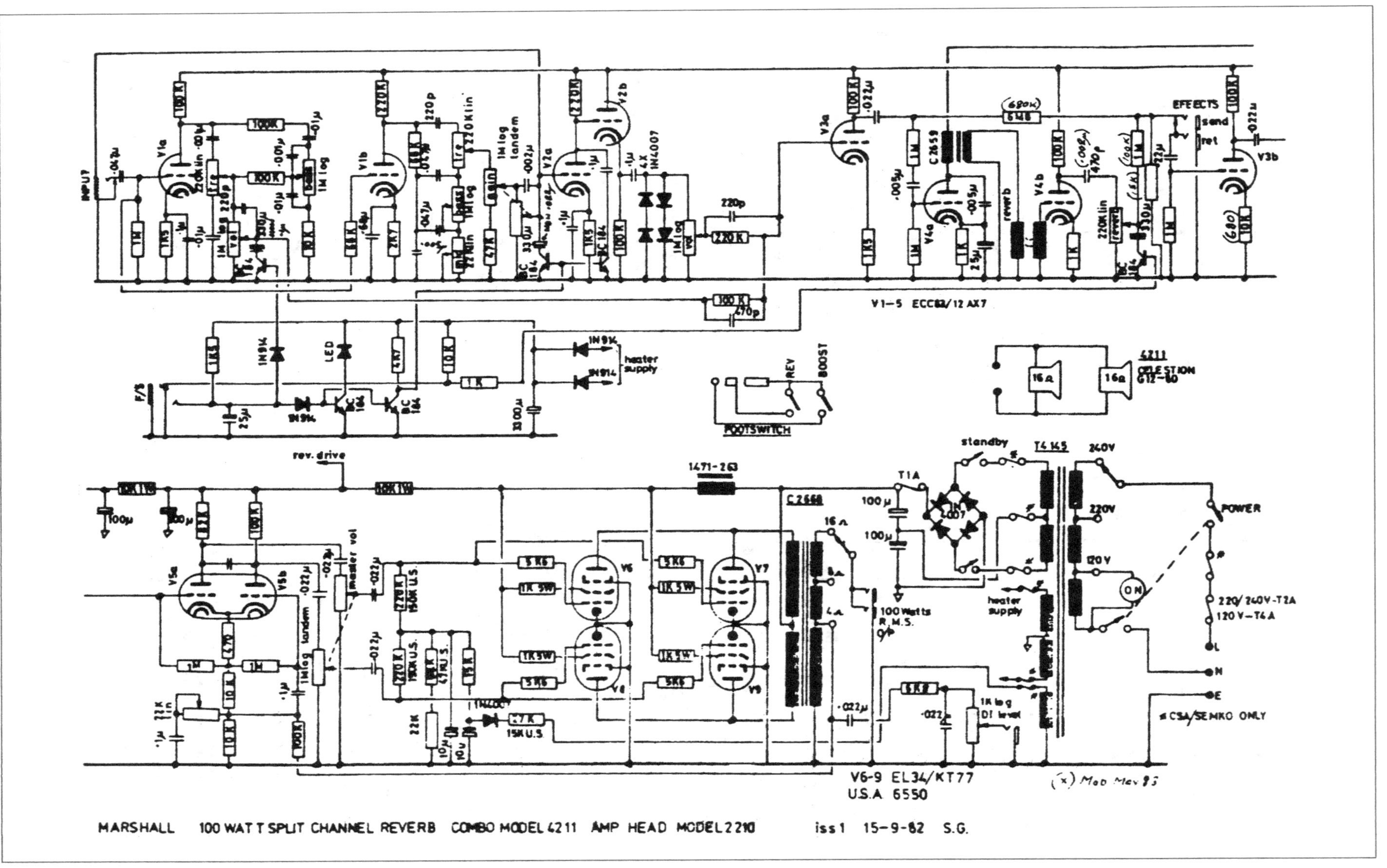

MARSHALL 100 WATT SPLIT CHANNEL REVERB COMBO MODEL 4211 AMP HEAD MODEL 2210 iss1 15-9-82 S.G.

STUDIO 15 (MODEL 4001)

Marshall

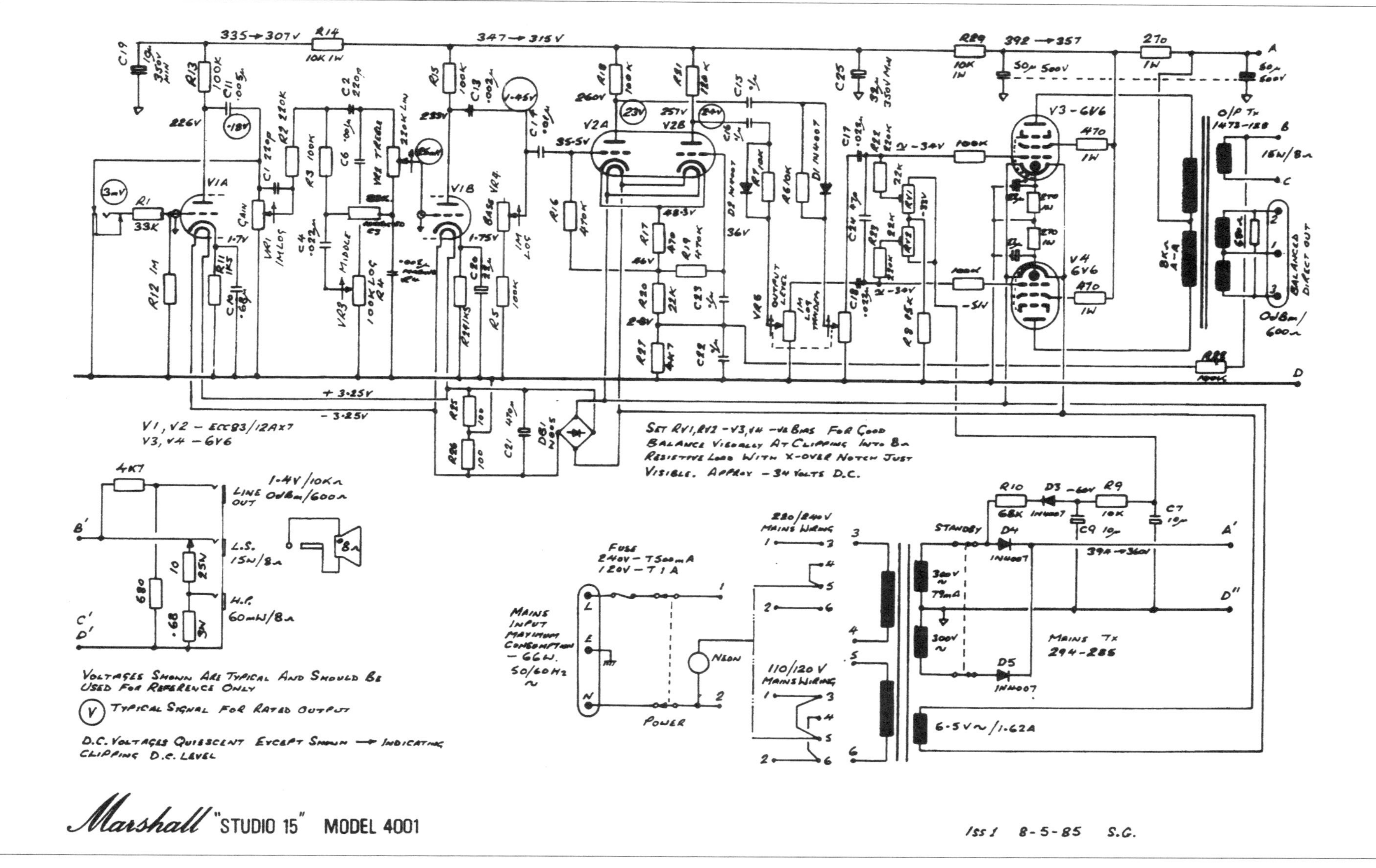

SUPER PA (MODEL 1963)

Marshall

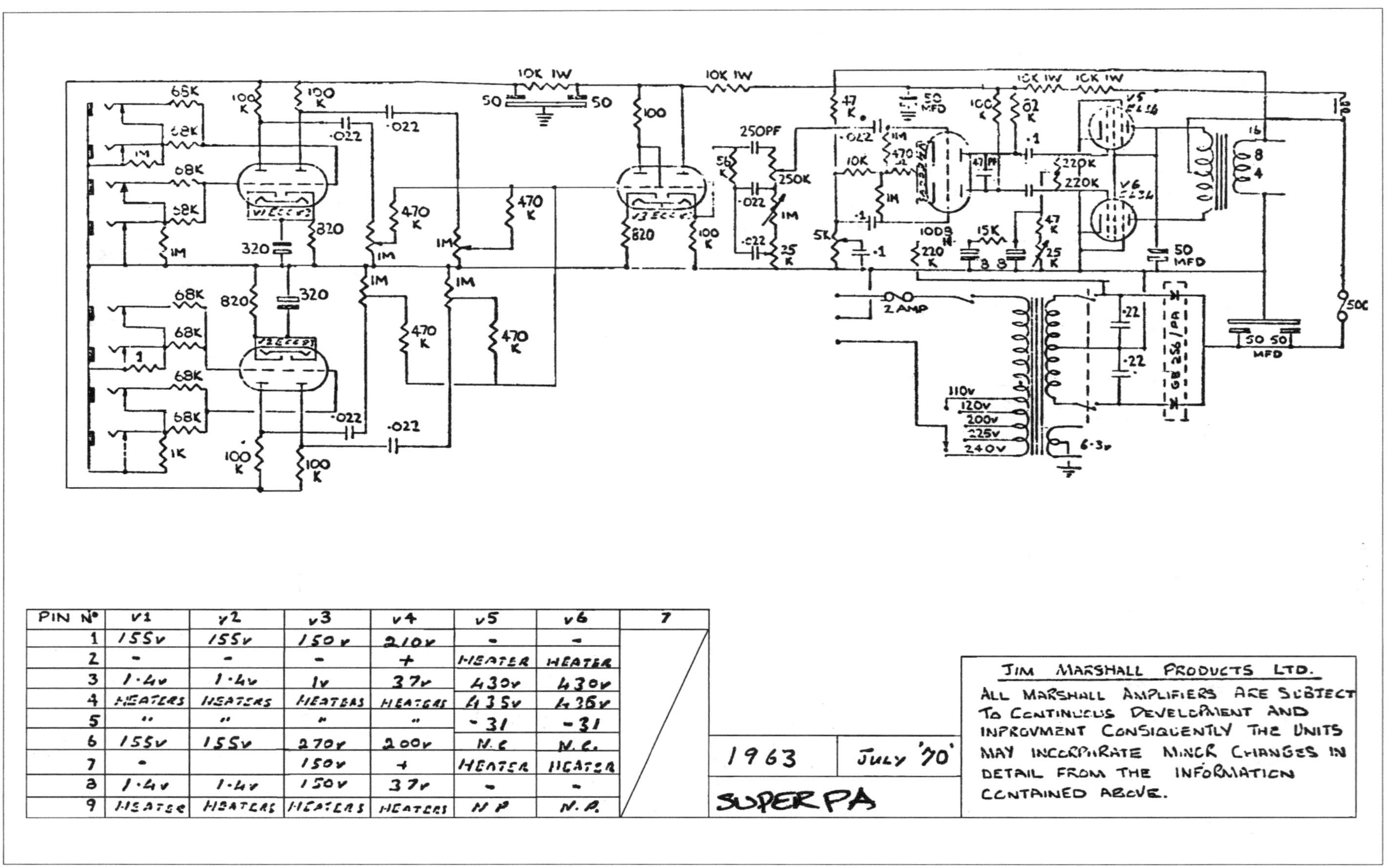

PIN N°	V1	V2	V3	V4	V5	V6	7
1	155v	155v	150v	210v	-	-	
2	-	-	-	+	HEATER	HEATER	
3	1.4v	1.4v	1v	37v	430v	430v	
4	HEATERS	HEATERS	HEATERS	HEATERS	435v	436v	
5	"	"	"	"	-31	-31	
6	155v	155v	270v	200v	N.C	N.C.	
7	-		150v	+	HEATER	HEATER	
8	1.4v	1.4v	150v	37v	-	-	
9	HEATER	HEATERS	HEATERS	HEATERS	N P	N.P.	

1963 | JULY '70
SUPER PA

JIM MARSHALL PRODUCTS LTD.

ALL MARSHALL AMPLIFIERS ARE SUBJECT TO CONTINUOUS DEVELOPMENT AND INPROVMENT CONSIQUENTLY THE UNITS MAY INCORPORATE MINOR CHANGES IN DETAIL FROM THE INFORMATION CONTAINED ABOVE.

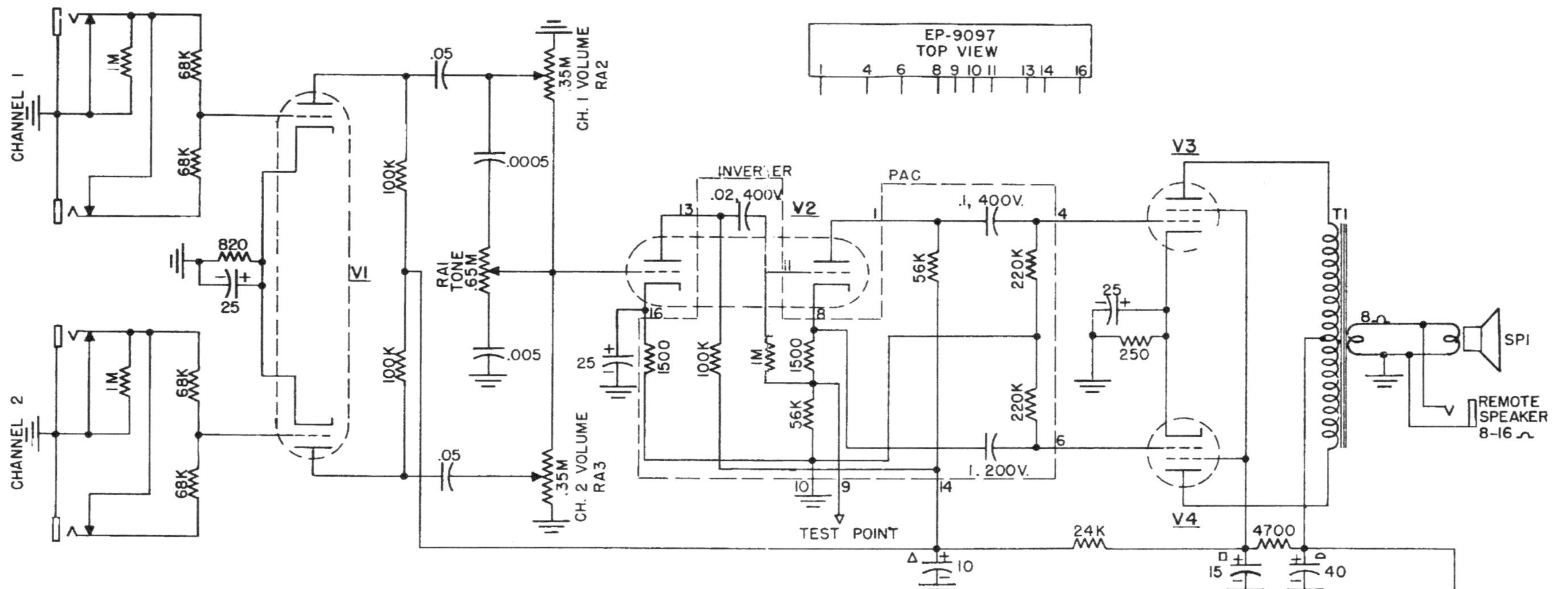

PART LIST

PART	PART NUMBER
INVERTER PAC	EP-9097
TRANSFORMER T1	T-703
TRANSFORMER T2	PT-5201
POWER SWITCH S1	S-1105
SPEAKER SP1	SP-5401
CONTROL RA1	RA-2902
CONTROLS RA2, RA3	RA-2903

* VOLTAGE CHART

PIN NUMBER	V1 12AY7	V2 12AX7	V3 6V6	V4 6V6	V5 5Y3
1	110	140	N.C.	N.C.	N.C.
2	0	0	6.3 A.C.	6.3 A.C.	5 A.C.#
3	1.75	1.15	330	330	N.C.
4	0	0	300	300	300
5	0	0	0	0	N.C.
6	110	175	N.C.	N.C.	500
7	0	0	0	0	N.C.
8	1.75	42	19	19	340
9	6.3 A.C.	6.3 A.C.	—	—	—

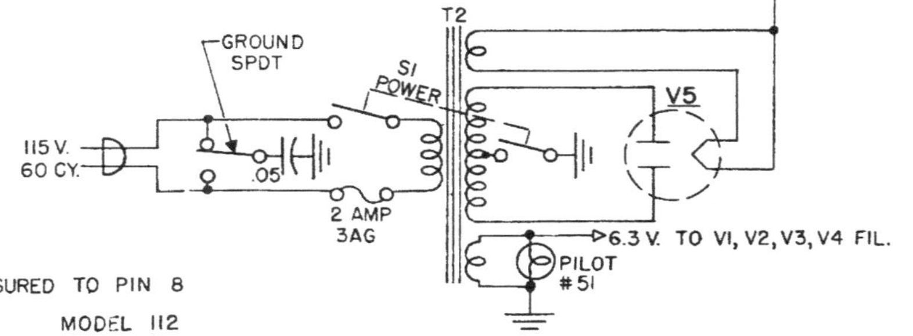

* VOLTS, MEASURED TO GROUND WITH VTVM. # MEASURED TO PIN 8

C.F. MARTIN & CO INC. NAZARETH PENN. MODEL 112

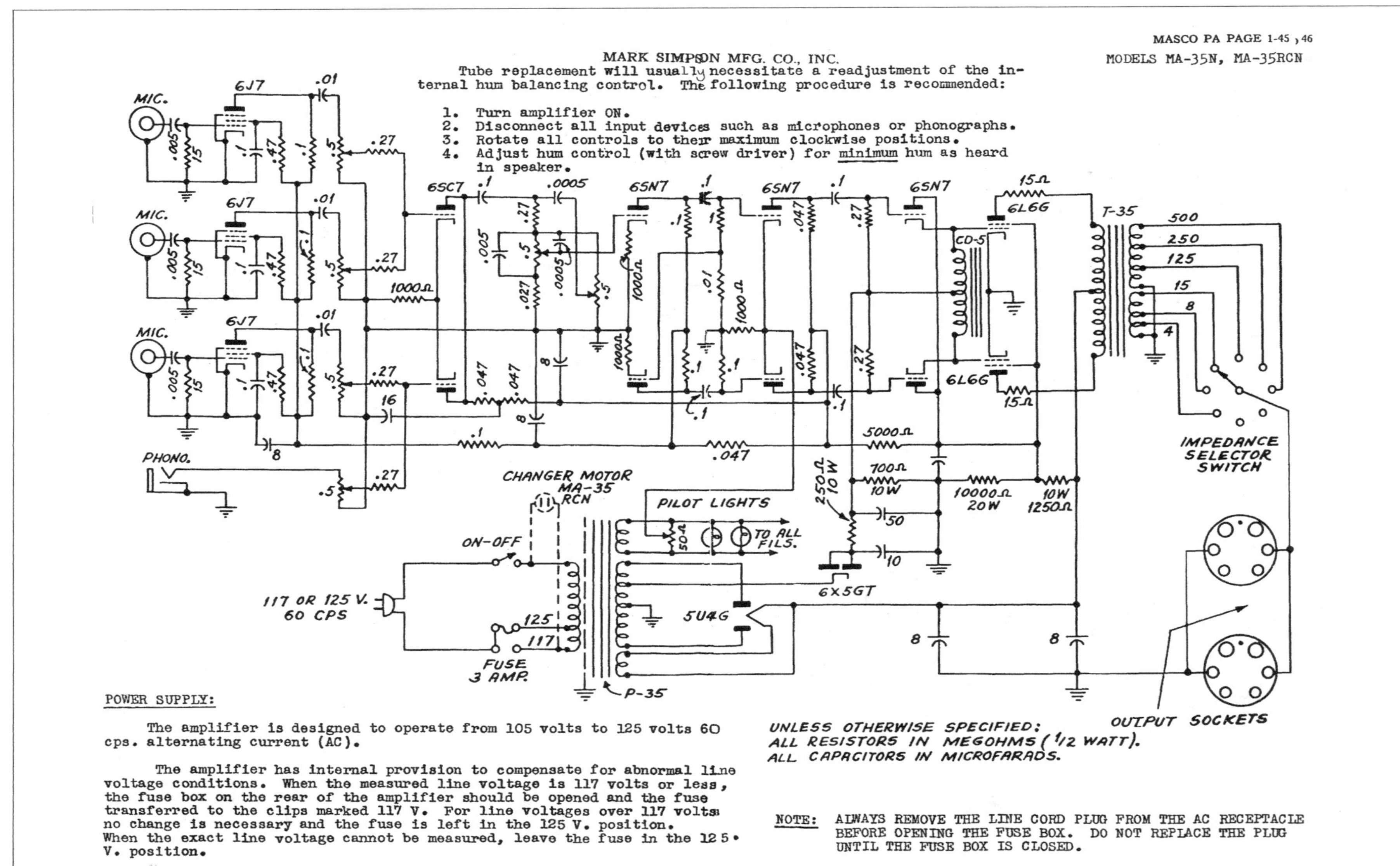

MASCO PA PAGE 1-45, 46
MODELS MA-35N, MA-35RCN

MARK SIMPSON MFG. CO., INC.

Tube replacement will usually necessitate a readjustment of the internal hum balancing control. The following procedure is recommended:

1. Turn amplifier ON.
2. Disconnect all input devices such as microphones or phonographs.
3. Rotate all controls to their maximum clockwise positions.
4. Adjust hum control (with screw driver) for <u>minimum</u> hum as heard in speaker.

POWER SUPPLY:

The amplifier is designed to operate from 105 volts to 125 volts 60 cps. alternating current (AC).

The amplifier has internal provision to compensate for abnormal line voltage conditions. When the measured line voltage is 117 volts or less, the fuse box on the rear of the amplifier should be opened and the fuse transferred to the clips marked 117 V. For line voltages over 117 volts no change is necessary and the fuse is left in the 125 V. position. When the exact line voltage cannot be measured, leave the fuse in the 125 V. position.

NOTE: ALWAYS REMOVE THE LINE CORD PLUG FROM THE AC RECEPTACLE BEFORE OPENING THE FUSE BOX. DO NOT REPLACE THE PLUG UNTIL THE FUSE BOX IS CLOSED.

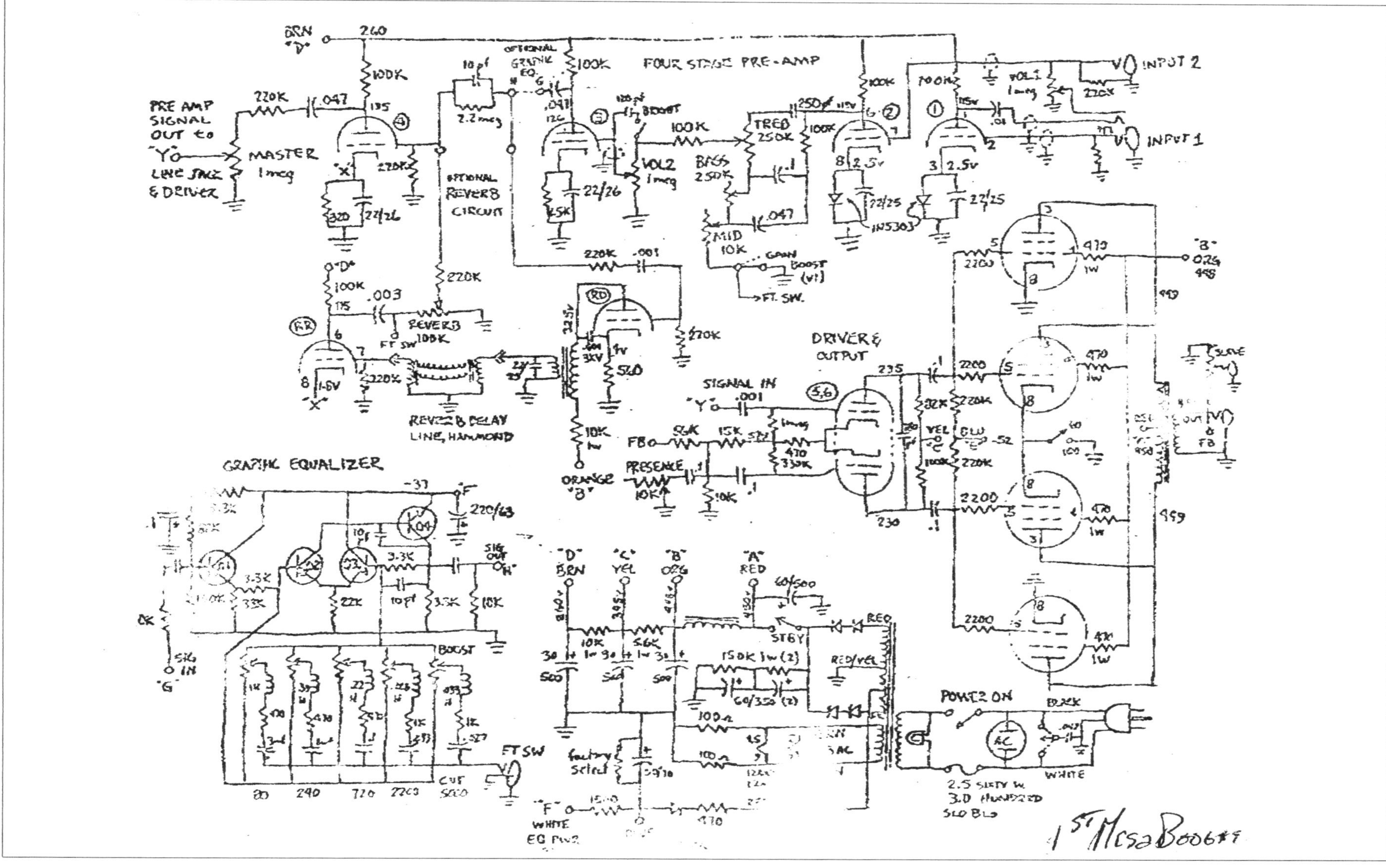
FOUR STAGE PRE-AMP
INPUT 2
INPUT 1
PRE AMP SIGNAL OUT to "Y" LINE JACK & DRIVER
MASTER 1meg
OPTIONAL GRAPHIC EQ.
OPTIONAL REVERB CIRCUIT
REVERB 100K
REVERB DELAY LINE, HAMMOND
VOL2 1meg
VOL1 1meg
TREB 250K
BASS 250K
MID 10K
GAIN BOOST
FT. SW.
DRIVER & OUTPUT
SIGNAL IN
PRESENCE 10K
GRAPHIC EQUALIZER
SIG OUT "H"
SIG IN "G"
BOOST
CUT
FT SW
STBY
RED/YEL
POWER ON
BLACK
WHITE
2.5 SIXTY W. 3.0 HUNDRED SLO BLO
factory Select
1st Mesa Boogie

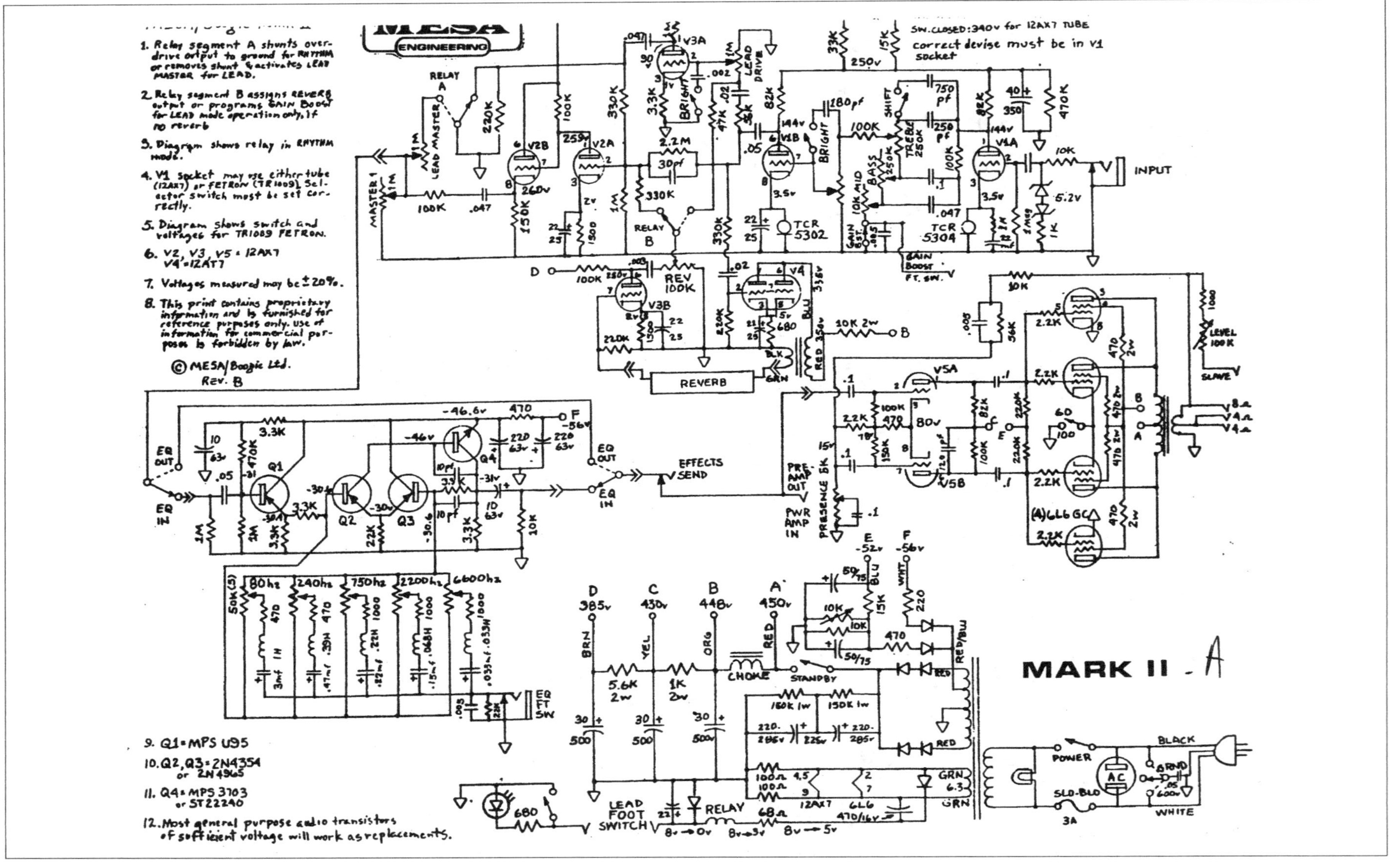
MESA ENGINEERING
1. Relay segment A shunts over-drive output to ground for RHYTHM or removes shunt & activates LEAD MASTER for LEAD.
2. Relay segment B assigns REVERB output or programs GAIN BOOST for LEAD mode operation only, if no reverb
3. Diagram shows relay in RHYTHM mode.
4. V1 socket may use either tube (12AX7) or FETRON (TR1009). Selector switch must be set correctly.
5. Diagram shows switch and voltages for TR1009 FETRON.
6. V2, V3, V5 = 12AX7 V4 = 12AT7
7. Voltages measured may be ± 20%.
8. This print contains proprietary information and is furnished for reference purposes only. Use of information for commercial purposes is forbidden by law.
© MESA/Boogie Ltd. Rev. B
SW. CLOSED: 340v for 12AX7 TUBE correct devise must be in V1 socket
INPUT
RELAY A
LEAD MASTER
MASTER
LEAD DRIVE
BRIGHT
SHIFT
TREBLE 250K
BASS 250K
MID 10K
GAIN BOOST
GAIN BOOST FT. SW.
TCR 5302
TCR 5304
RELAY B
REV 100K
REVERB
EQ OUT
EQ IN
EFFECTS SEND
PRE-AMP OUT
PWR AMP IN
PRESENCE 5K
LEVEL 100K
SLAVE
(4) 6L6 GC
80hz
240hz
750hz
2200hz
6600hz
EQ FT SW
STANDBY
CHOKE
POWER
SLO-BLO 3A
BLACK
WHITE
GRND
LEAD FOOT SWITCH
RELAY
MARK II - A
9. Q1 = MPS U95
10. Q2, Q3 = 2N4354 or 2N4965
11. Q4 = MPS 3703 or ST22240
12. Most general purpose audio transistors of sufficient voltage will work as replacements.

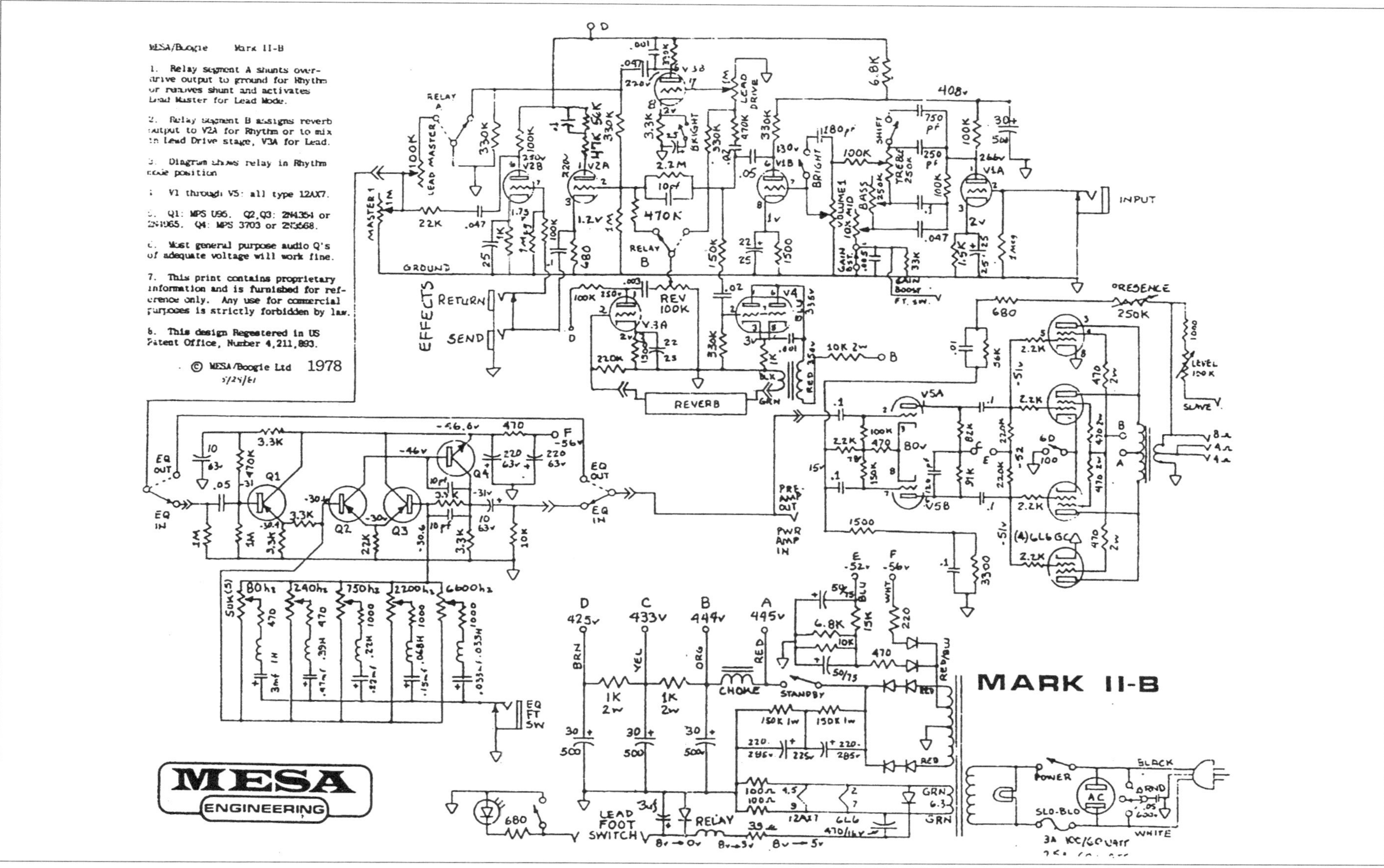
MESA/Boogie Mark II-B
1. Relay segment A shunts overdrive output to ground for Rhythm or removes shunt and activates Lead Master for Lead Mode.
2. Relay segment B assigns reverb output to V2A for Rhythm or to mix in Lead Drive stage, V3A for Lead.
3. Diagram shows relay in Rhythm mode position
4. V1 through V5: all type 12AX7.
5. Q1: MPS U96. Q2,Q3: 2N4354 or 2N3965. Q4: MPS 3703 or 2N3668.
6. Most general purpose audio Q's of adequate voltage will work fine.
7. This print contains proprietary information and is furnished for reference only. Any use for commercial purposes is strictly forbidden by law.
8. This design Regestered in US Patent Office, Number 4,211,893.
© MESA/Boogie Ltd 1978
5/24/81
MASTER 1
LEAD MASTER
RELAY A
RELAY B
GROUND
EFFECTS
RETURN
SEND
LEAD DRIVE
BRIGHT
VOLUME 1
GAIN BST.
GAIN BOOST FT. SW.
BASS
TREBLE
SHIFT
INPUT
REV
REVERB
PRE-AMP OUT
PWR AMP IN
PRESENCE
LEVEL
SLAVE
EQ OUT
EQ IN
EQ FT SW
80hz
240hz
750hz
2200hz
6600hz
CHOKE
STANDBY
MARK II-B
POWER
SLO-BLO
BLACK
WHITE
LEAD FOOT SWITCH
RELAY
MESA ENGINEERING

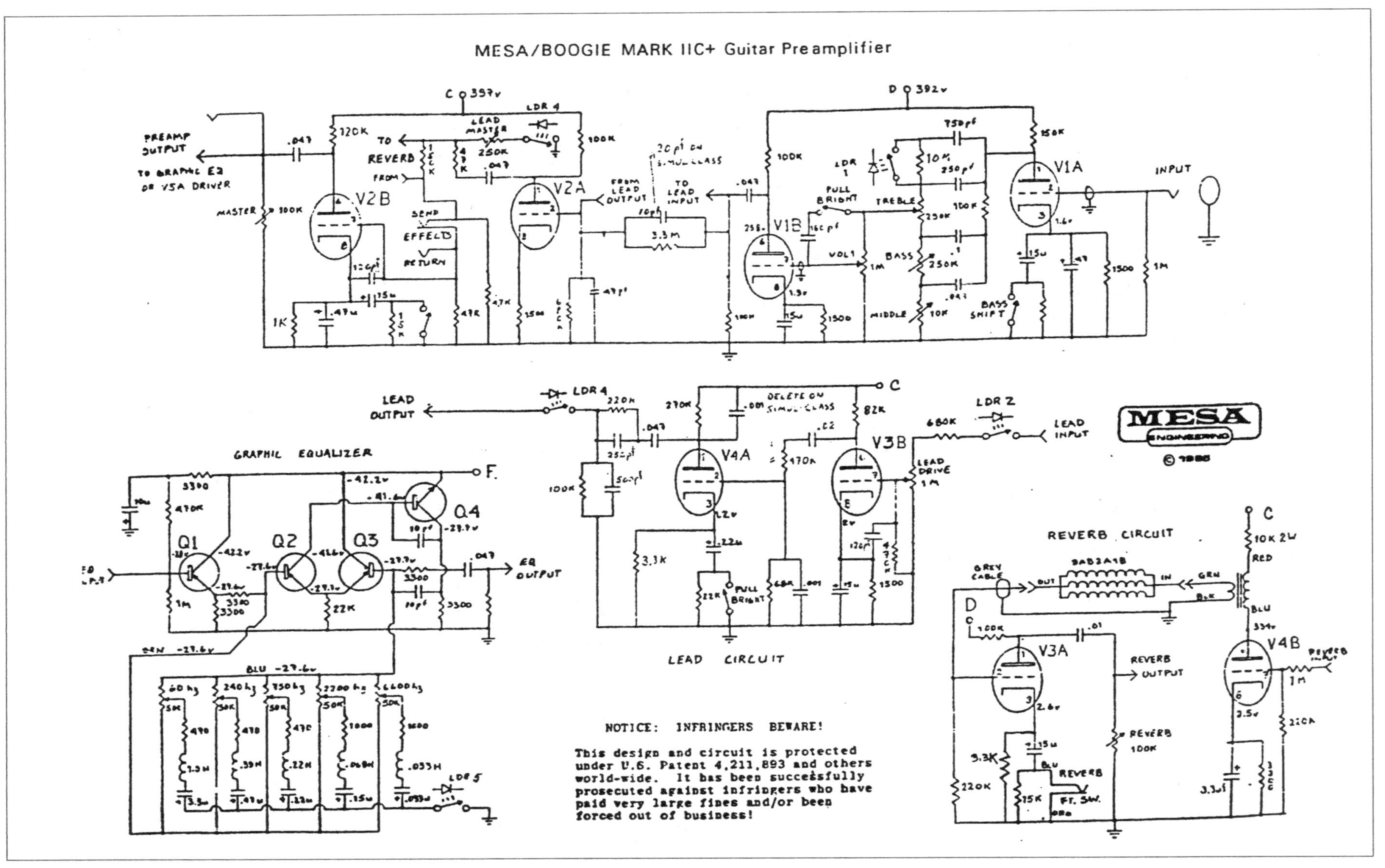
MESA/BOOGIE MARK IIC+ Guitar Preamplifier
PREAMP OUTPUT
TO GRAPHIC EQ OR VSA DRIVER
MASTER
TO REVERB
FROM
LEAD MASTER
LDR 4
SEND
EFFECTS
RETURN
V2B
V2A
FROM LEAD OUTPUT
TO LEAD INPUT
INPUT
V1A
V1B
PULL BRIGHT
TREBLE
BASS
MIDDLE
BASS SHIFT
VOL 1
LEAD OUTPUT
LEAD INPUT
LEAD DRIVE
LDR 2
V4A
V3B
LEAD CIRCUIT
PULL BRIGHT
GRAPHIC EQUALIZER
Q1
Q2
Q3
Q4
EQ OUTPUT
LDR 5
REVERB CIRCUIT
GREY CABLE
RED
BLU
V3A
V4B
REVERB OUTPUT
REVERB
FT. SW.
MESA ENGINEERING
NOTICE: INFRINGERS BEWARE!
This design and circuit is protected under U.S. Patent 4,211,893 and others world-wide. It has been successfully prosecuted against infringers who have paid very large fines and/or been forced out of business!

BOOGIE **MARK III PWR/OUTPUT**

Mesa/Boogie

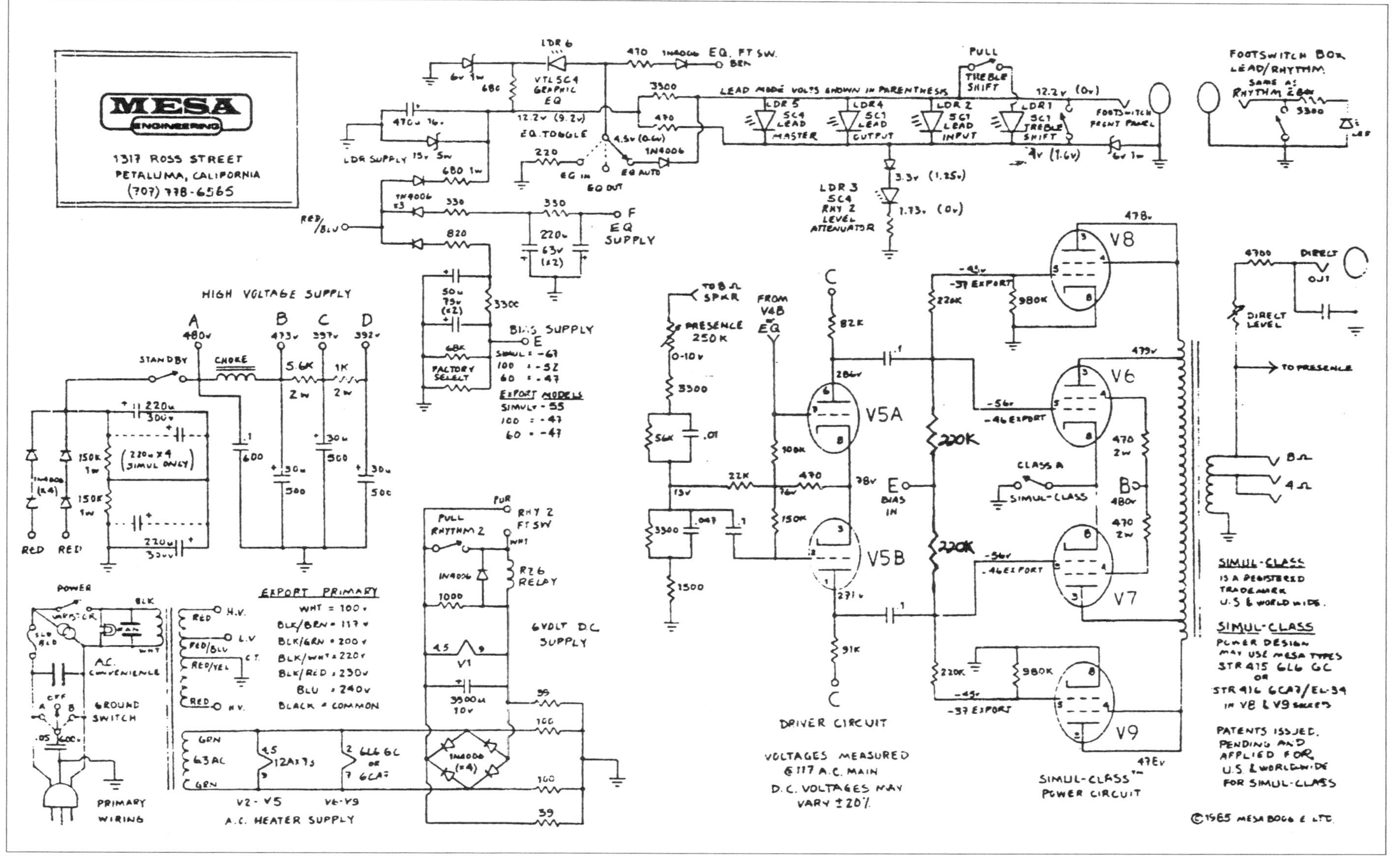

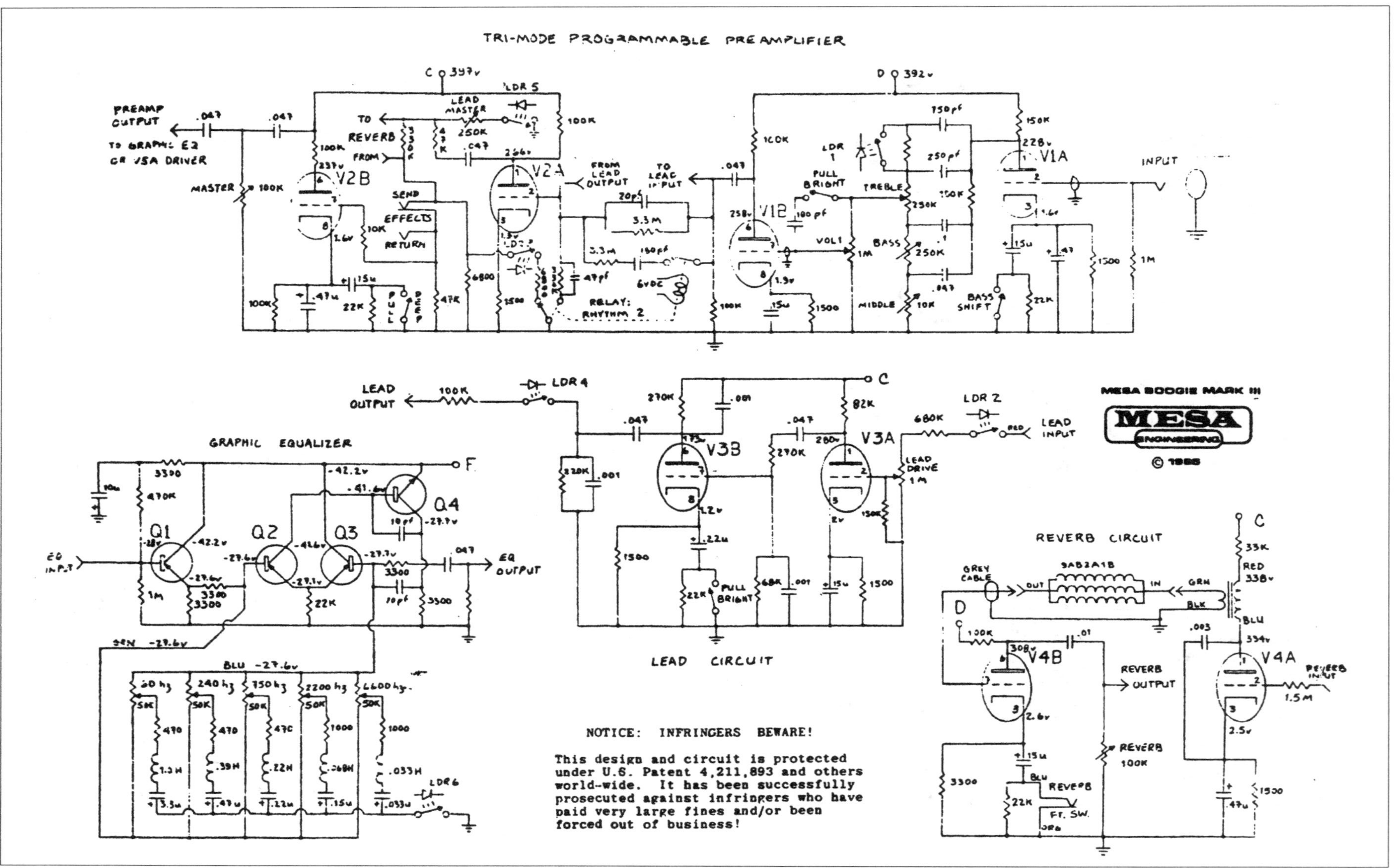
TRI-MODE PROGRAMMABLE PREAMPLIFIER
PREAMP OUTPUT
TO GRAPHIC EQ OR VSA DRIVER
MASTER
TO REVERB
FROM
LEAD MASTER
SEND
EFFECTS
RETURN
FROM LEAD OUTPUT
TO LEAD INPUT
RELAY: RHYTHM 2
PULL BRIGHT
TREBLE
BASS
MIDDLE
VOL 1
BASS SHIFT
INPUT
V1A
V1B
V2A
V2B
GRAPHIC EQUALIZER
EQ INPUT
EQ OUTPUT
Q1
Q2
Q3
Q4
LEAD OUTPUT
LEAD INPUT
LEAD DRIVE
V3A
V3B
LEAD CIRCUIT
MESA BOOGIE MARK III
MESA ENGINEERING
© 1986
REVERB CIRCUIT
GREY CABLE
REVERB OUTPUT
REVERB INPUT
V4A
V4B
NOTICE: INFRINGERS BEWARE!
This design and circuit is protected under U.S. Patent 4,211,893 and others world-wide. It has been successfully prosecuted against infringers who have paid very large fines and/or been forced out of business!

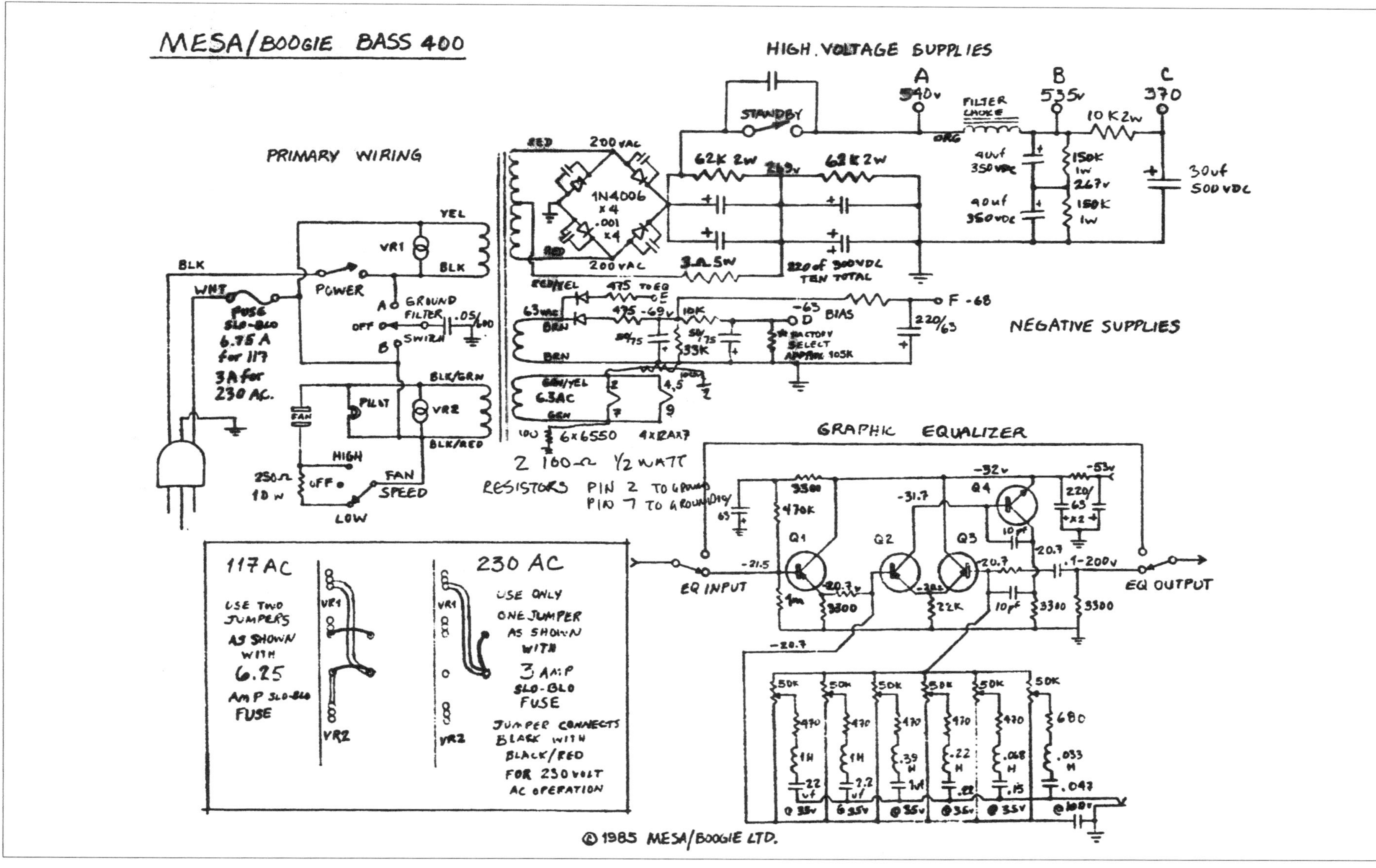
MESA/BOOGIE BASS 400
PRIMARY WIRING
HIGH VOLTAGE SUPPLIES
NEGATIVE SUPPLIES
GRAPHIC EQUALIZER
FUSE SLO-BLO 6.75 A for 117 3A for 230 AC.
POWER
GROUND FILTER SWITCH
STANDBY
FILTER CHOKE
1N4006 x4
.001 x4
200 VAC
62K 2W
220uf 300VDC TEN TOTAL
40uf 350VDC
150K 1W
10K 2W
30uf 500VDC
F -68
BIAS
FACTORY SELECT
220/63
6x6550
4x12AX7
2 100-Ω 1/2 WATT RESISTORS PIN 2 TO GROUND PIN 7 TO GROUND
FAN SPEED
HIGH
LOW
EQ INPUT
EQ OUTPUT
Q1
Q2
Q3
Q4
117 AC
USE TWO JUMPERS AS SHOWN WITH 6.25 AMP SLO-BLO FUSE
230 AC
USE ONLY ONE JUMPER AS SHOWN WITH 3 AMP SLO-BLO FUSE
JUMPER CONNECTS BLACK WITH BLACK/RED FOR 230 VOLT AC OPERATION
© 1985 MESA/BOOGIE LTD.

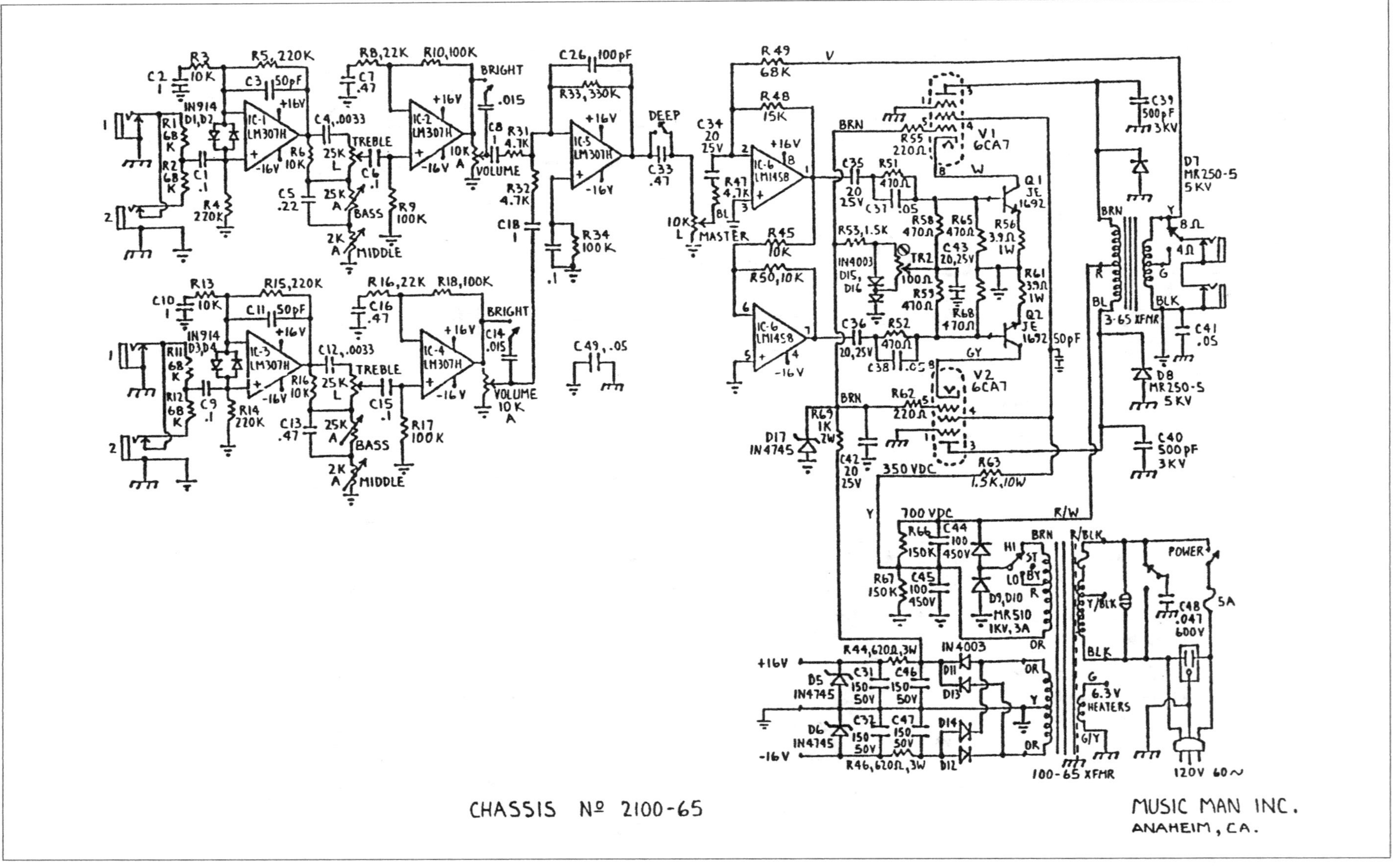
CHASSIS Nº 2100-65
MUSIC MAN INC.
ANAHEIM, CA.

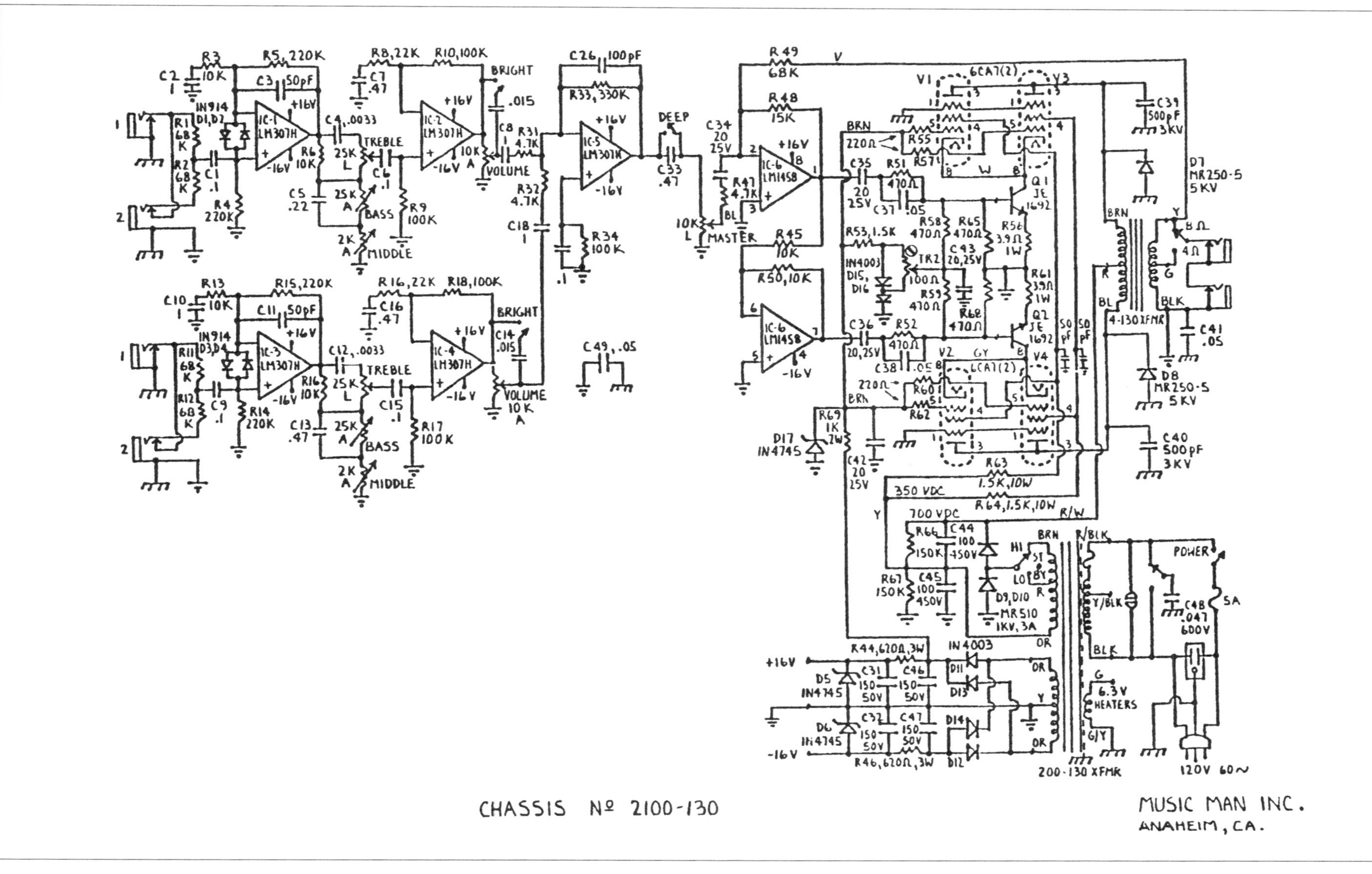
CHASSIS № 2100-130
MUSIC MAN INC.
ANAHEIM, CA.

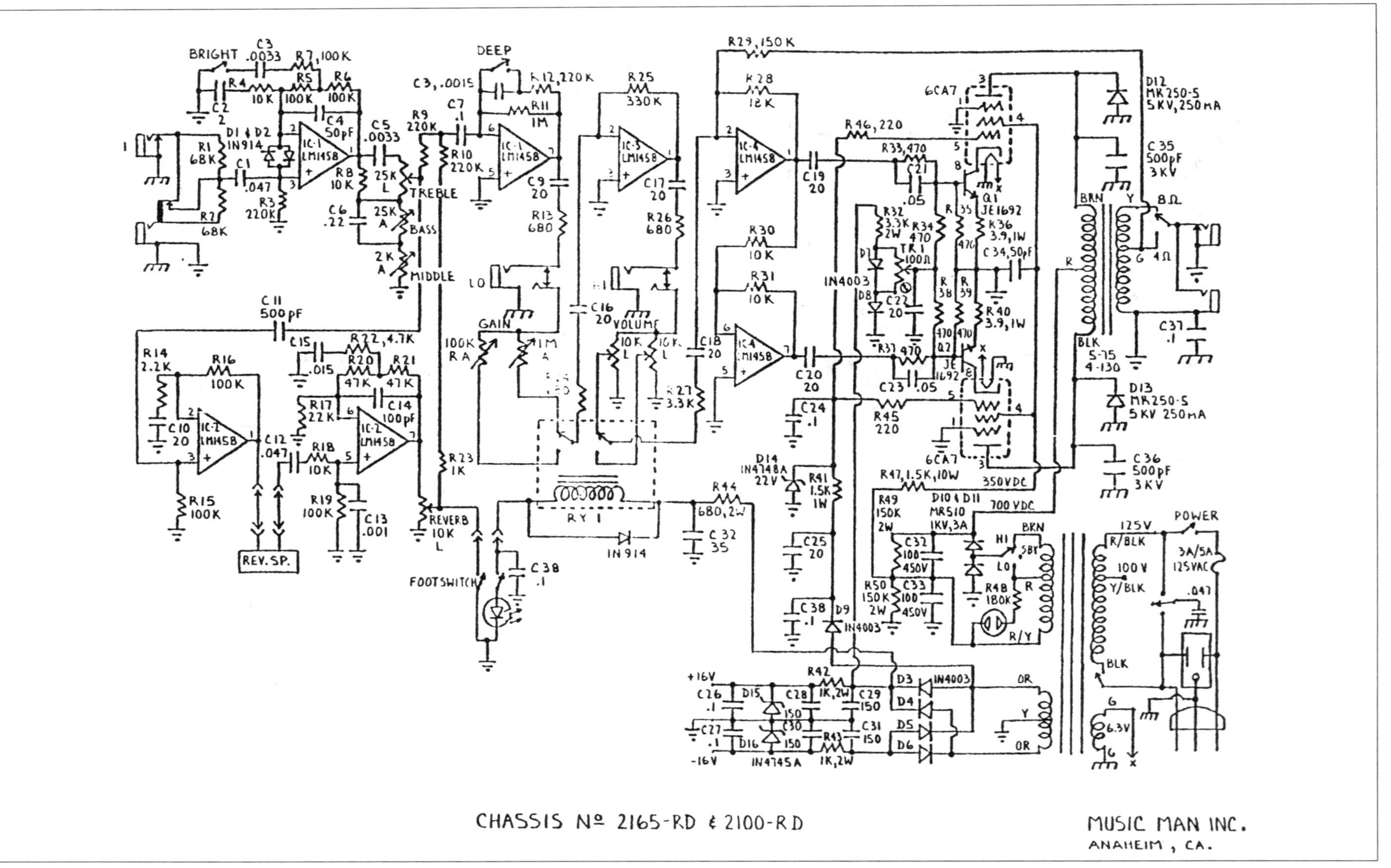
BRIGHT
DEEP
TREBLE
BASS
MIDDLE
GAIN
VOLUME
REVERB
FOOTSWITCH
REV. SP.
POWER
6CA7
LM1458
RY 1
350VDC
700 VDC
CHASSIS No 2165-RD & 2100-RD
MUSIC MAN INC.
ANAHEIM, CA.

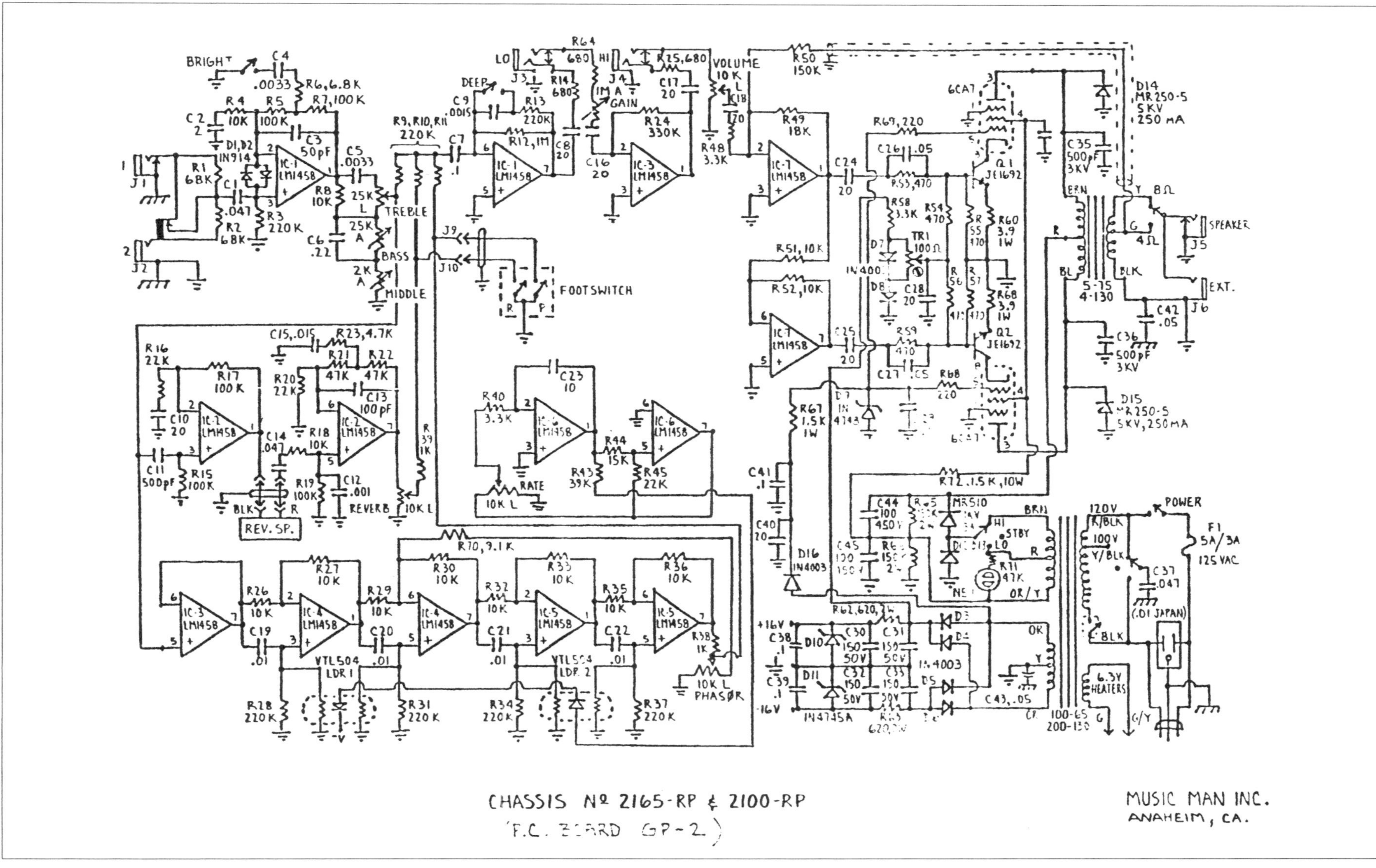
BRIGHT
FOOTSWITCH
TREBLE
BASS
MIDDLE
VOLUME
GAIN
REVERB
RATE
PHASØR
REV. SP.
SPEAKER
EXT.
POWER
6.3V HEATERS
CHASSIS № 2165-RP & 2100-RP
(P.C. BOARD GP-2)
MUSIC MAN INC.
ANAHEIM, CA.

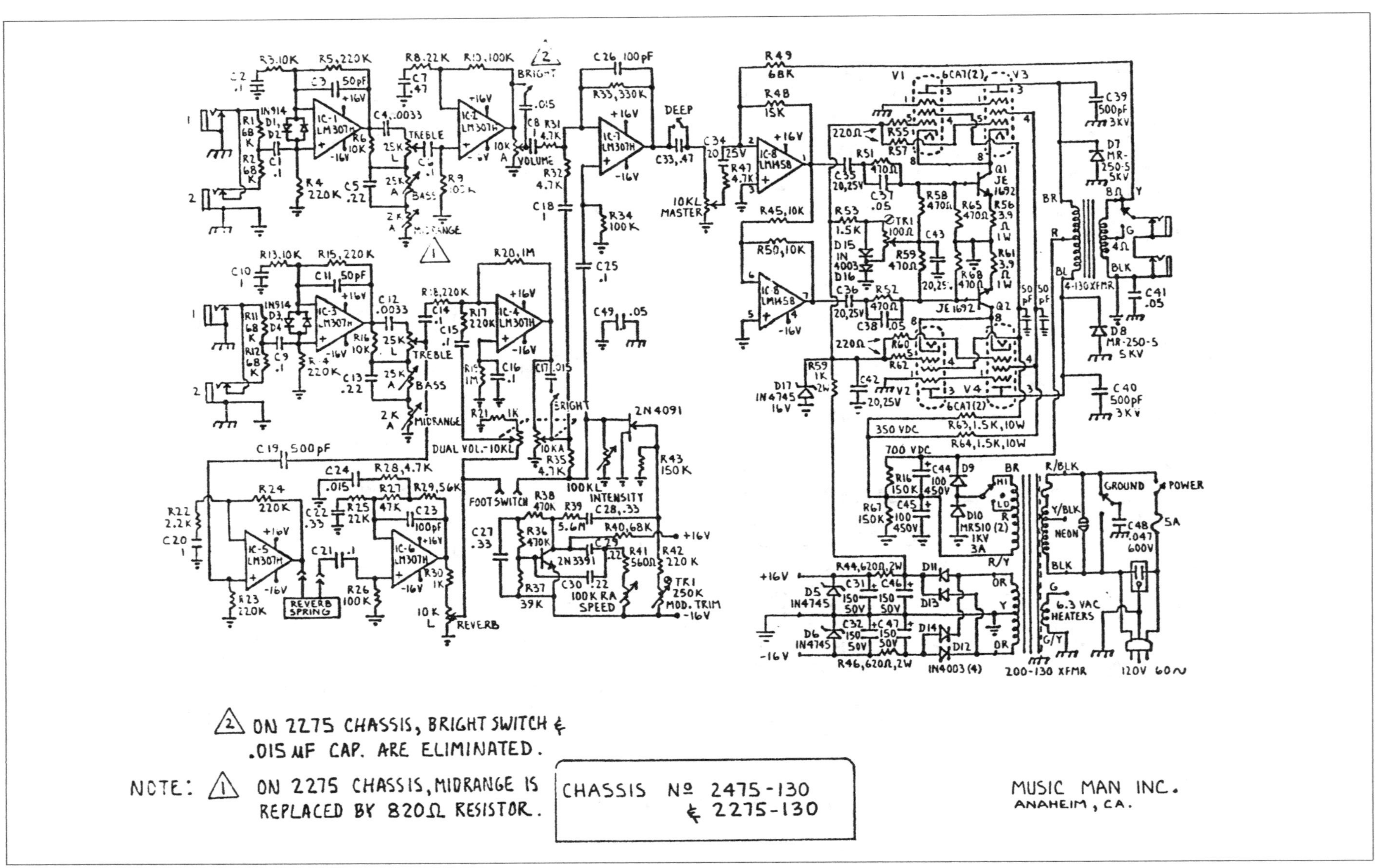
ON 2275 CHASSIS, BRIGHT SWITCH & .015 µF CAP. ARE ELIMINATED.
NOTE: ON 2275 CHASSIS, MIDRANGE IS REPLACED BY 820Ω RESISTOR.
CHASSIS No 2475-130 & 2275-130
MUSIC MAN INC.
ANAHEIM, CA.

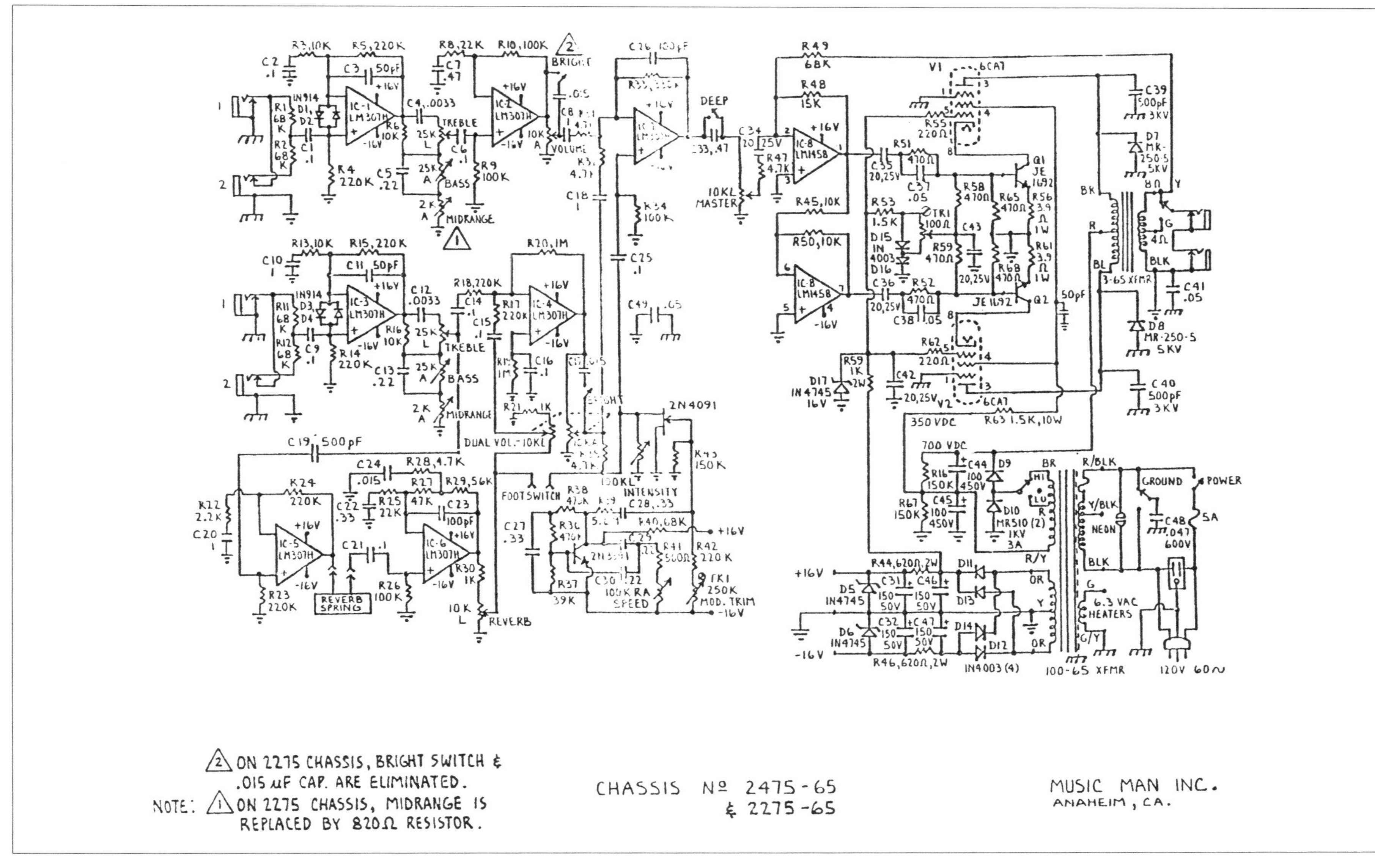

2 ON 2275 CHASSIS, BRIGHT SWITCH & .015 uF CAP. ARE ELIMINATED.
NOTE: 1 ON 2275 CHASSIS, MIDRANGE IS REPLACED BY 820Ω RESISTOR.
CHASSIS № 2475-65 & 2275-65
MUSIC MAN INC.
ANAHEIM, CA.

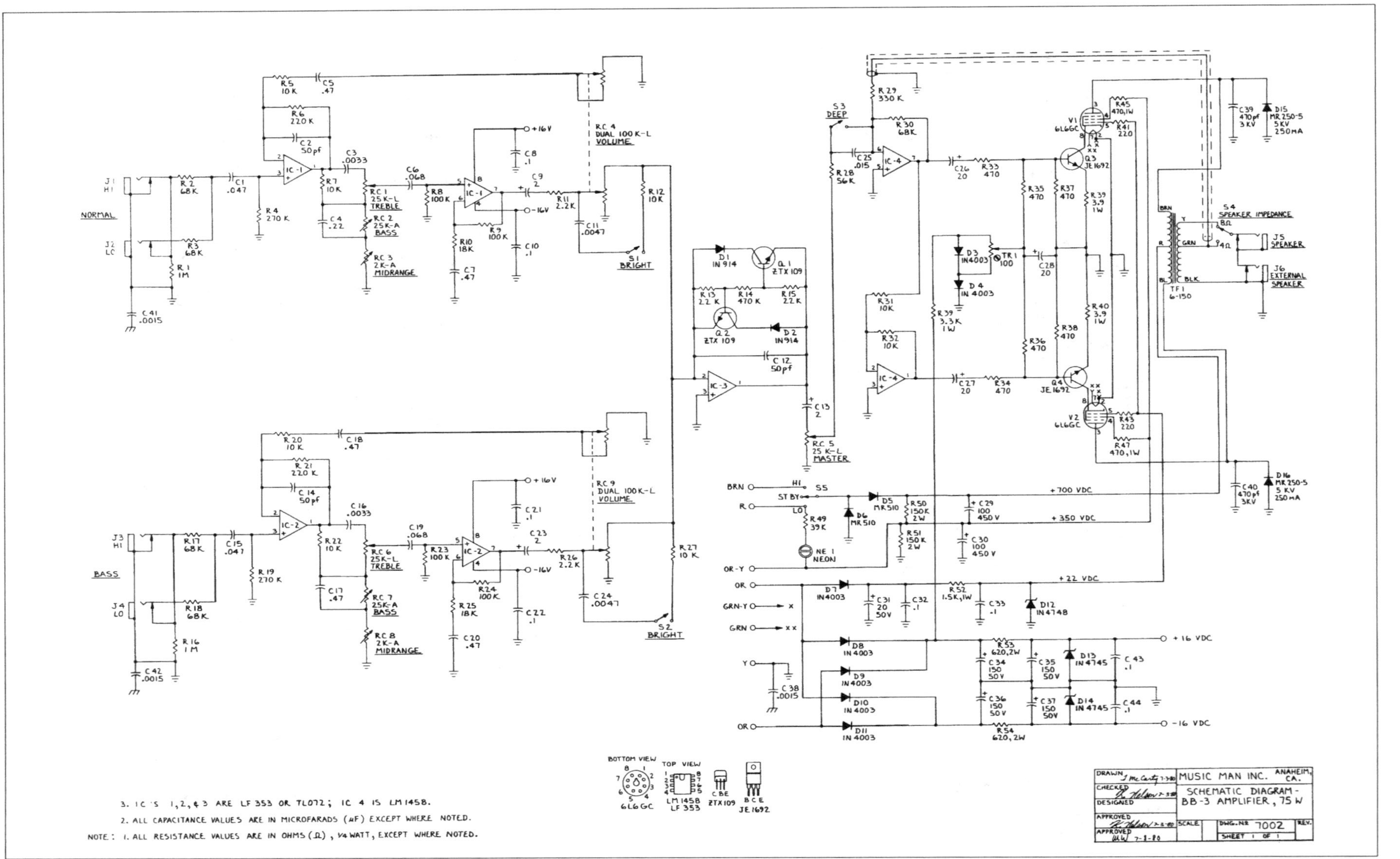
NORMAL
BASS
RC 4 DUAL 100K-L VOLUME
RC 9 DUAL 100K-L VOLUME
RC 1 25K-L TREBLE
RC 2 25K-A BASS
RC 3 2K-A MIDRANGE
RC 6 25K-L TREBLE
RC 7 25K-A BASS
RC 8 2K-A MIDRANGE
S1 BRIGHT
S2 BRIGHT
S3 DEEP
RC 5 25 K-L MASTER
S4 SPEAKER IMPEDANCE
J5 SPEAKER
J6 EXTERNAL SPEAKER
TF1 6-150
+700 VDC
+350 VDC
+22 VDC
+16 VDC
-16 VDC
3. IC'S 1, 2, & 3 ARE LF 353 OR TL072; IC 4 IS LM1458.
2. ALL CAPACITANCE VALUES ARE IN MICROFARADS (μF) EXCEPT WHERE NOTED.
NOTE: 1. ALL RESISTANCE VALUES ARE IN OHMS (Ω), 1/4 WATT, EXCEPT WHERE NOTED.
MUSIC MAN INC. ANAHEIM, CA.
SCHEMATIC DIAGRAM - BB-3 AMPLIFIER, 75 W
DWG. NO. 7002
SHEET 1 OF 1

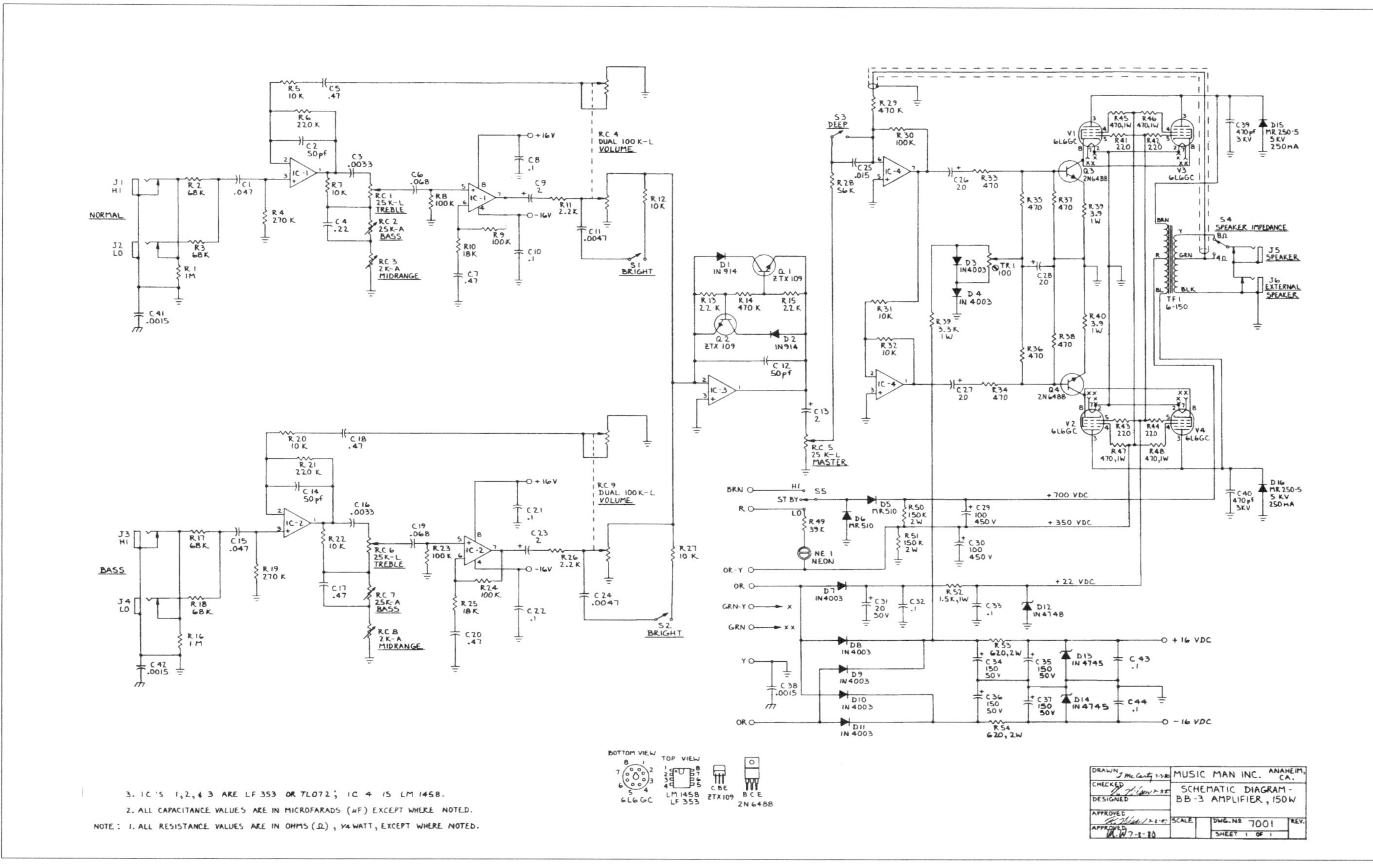
NORMAL
BASS
NOTE: 1. ALL RESISTANCE VALUES ARE IN OHMS (Ω), ¼ WATT, EXCEPT WHERE NOTED.
2. ALL CAPACITANCE VALUES ARE IN MICROFARADS (μF) EXCEPT WHERE NOTED.
3. IC'S 1, 2, & 3 ARE LF 353 OR TL072; IC 4 IS LM 1458.
MUSIC MAN INC. ANAHEIM, CA.
SCHEMATIC DIAGRAM - BB-3 AMPLIFIER, 150W
DWG. NO. 7001
SHEET 1 OF 1

GB-2 (2275 & 2475 CHASSIS)

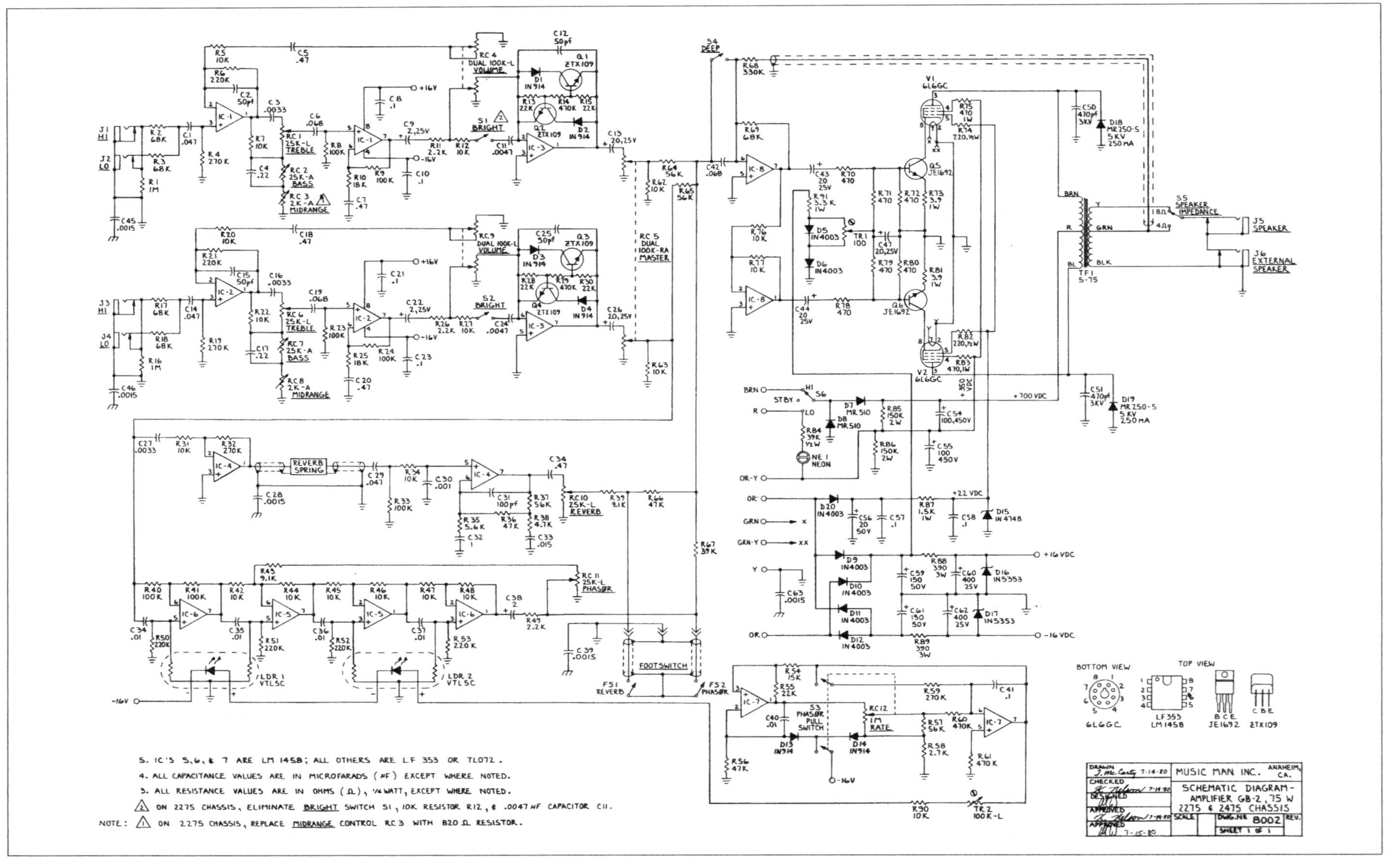

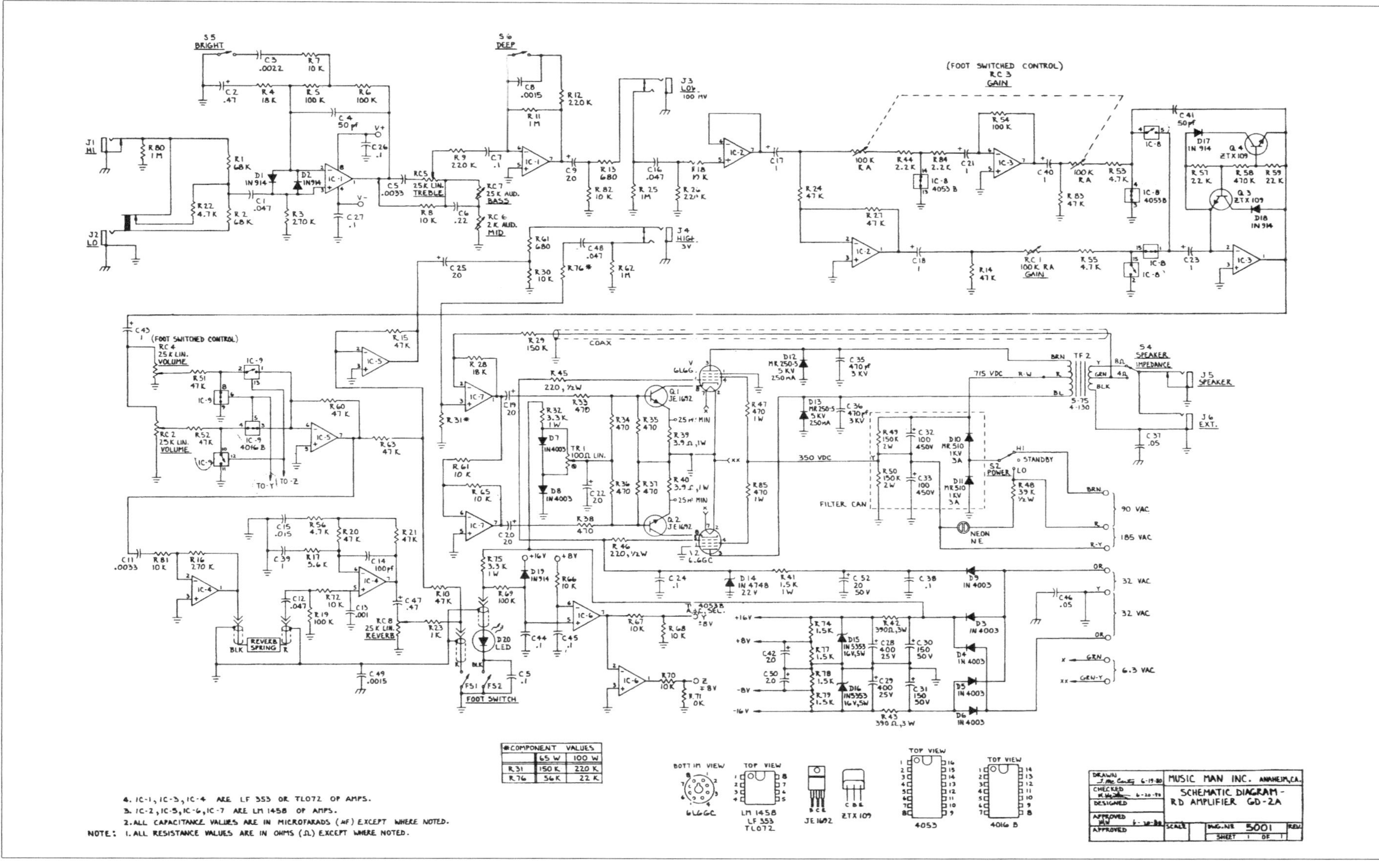
S5 BRIGHT
S6 DEEP
(FOOT SWITCHED CONTROL) RC3 GAIN
J1 HI
J2 LO
J3 LOW 100 MV
J4 HIGH 3V
RC5 25K LIN. TREBLE
RC7 25K AUD. BASS
RC6 2K AUD. MID
(FOOT SWITCHED CONTROL) RC4 25K LIN. VOLUME
RC2 25K LIN. VOLUME
RC8 25K LIN. REVERB
REVERB SPRING
COAX
S4 SPEAKER IMPEDANCE
J5 SPEAKER
J6 EXT.
715 VDC
350 VDC
FILTER CAN
S2 POWER
STANDBY
NEON NE
90 VAC
185 VAC
32 VAC
32 VAC
6.3 VAC
FOOT SWITCH
*COMPONENT VALUES | 65 W | 100 W
R31 | 150 K | 220 K
R76 | 56 K | 22 K
4. IC-1, IC-3, IC-4 ARE LF 353 OR TL072 OP AMPS.
3. IC-2, IC-5, IC-6, IC-7 ARE LM 1458 OP AMPS.
2. ALL CAPACITANCE VALUES ARE IN MICROFARADS (μf) EXCEPT WHERE NOTED.
NOTE: 1. ALL RESISTANCE VALUES ARE IN OHMS (Ω) EXCEPT WHERE NOTED.
BOTTOM VIEW 6L6GC
TOP VIEW LM 1458 LF 353 TL072
JE 1692
ZTX 109
TOP VIEW 4053
TOP VIEW 4016 B
MUSIC MAN INC. ANAHEIM, CA.
SCHEMATIC DIAGRAM - RD AMPLIFIER GD-2A
DWG. NR 5001
SHEET 1 OF 1

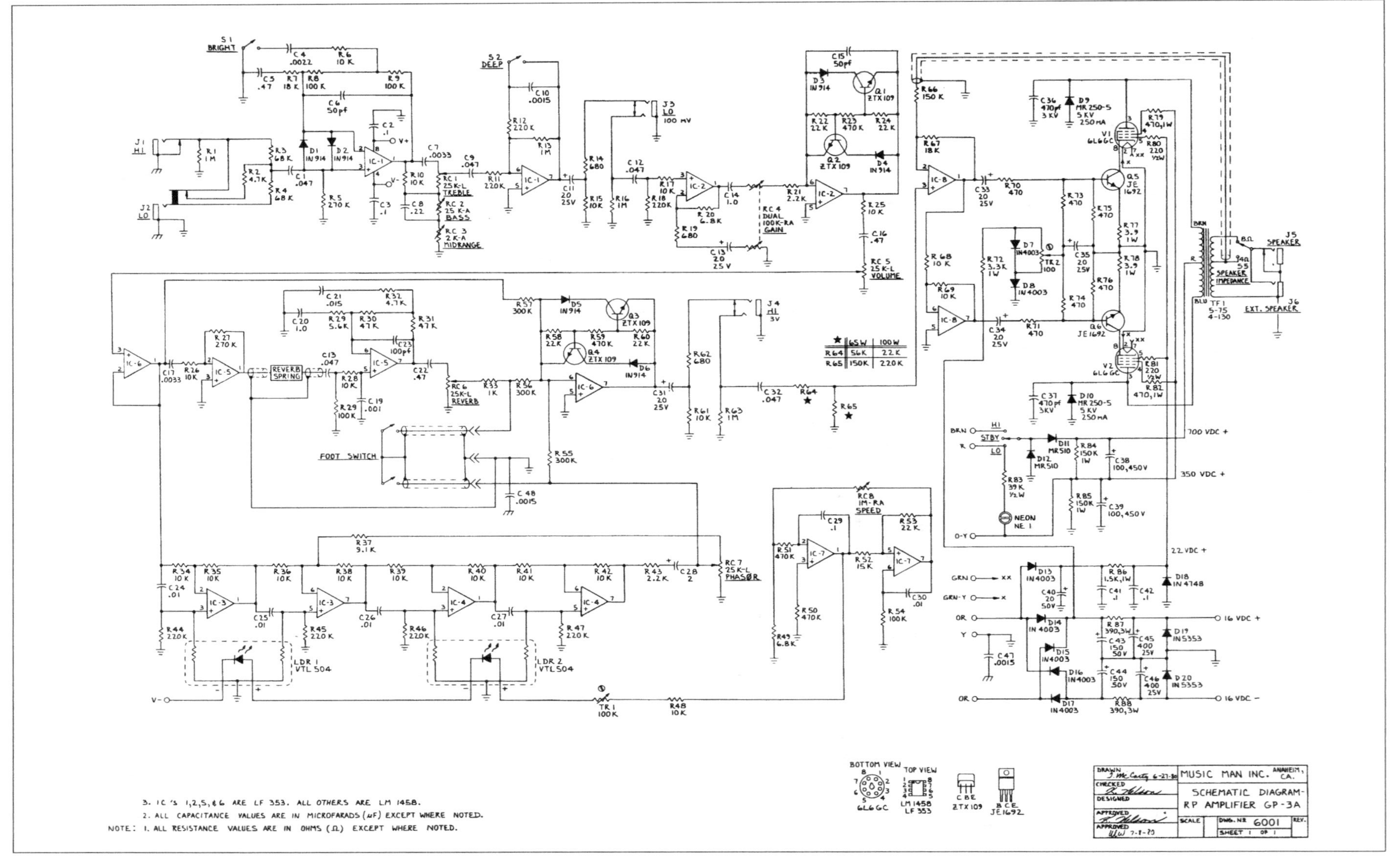
MUSIC MAN INC. ANAHEIM, CA.
SCHEMATIC DIAGRAM-
RP AMPLIFIER GP-3A
DWG. NO 6001
SHEET 1 OF 1
NOTE: 1. ALL RESISTANCE VALUES ARE IN OHMS (Ω) EXCEPT WHERE NOTED.
2. ALL CAPACITANCE VALUES ARE IN MICROFARADS (μF) EXCEPT WHERE NOTED.
3. IC'S 1,2,5,&6 ARE LF 353. ALL OTHERS ARE LM 1458.
BOTTOM VIEW
6L6GC
TOP VIEW
LM1458
LF353
ZTX109
JE1692
S1 BRIGHT
S2 DEEP
J1 HI
J2 LO
J3 LO 100 mV
J4 HI 3V
J5 SPEAKER
J6 EXT. SPEAKER
SPEAKER IMPEDANCE
RC1 25K-L TREBLE
RC2 25K-A BASS
RC3 2K-A MIDRANGE
RC4 DUAL 100K-RA GAIN
RC5 25K-L VOLUME
RC6 25K-L REVERB
RC7 25K-L PHASØR
RC8 1M-RA SPEED
REVERB SPRING
FOOT SWITCH
LDR 1 VTL 504
LDR 2 VTL 504
NEON NE 1
700 VDC +
350 VDC +
22 VDC +
16 VDC +
16 VDC −
★ 65W 100W
R64 56K 22K
R65 150K 220K

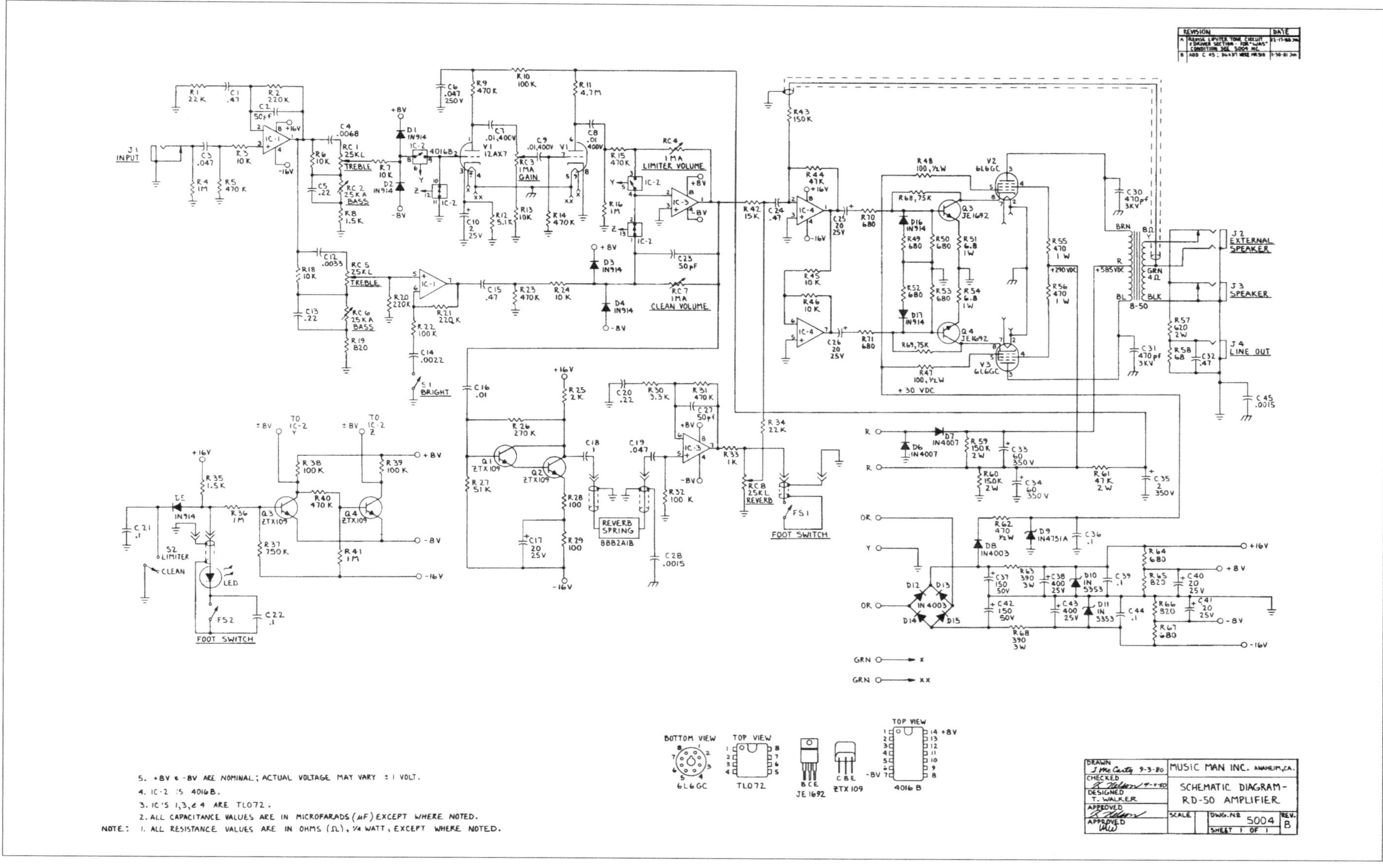
J1 INPUT
J2 EXTERNAL SPEAKER
J3 SPEAKER
J4 LINE OUT
LIMITER VOLUME
CLEAN VOLUME
GAIN
TREBLE
BASS
BRIGHT
REVERB
REVERB SPRING 8BB2A1B
FOOT SWITCH
S2 LIMITER
CLEAN
LED
BOTTOM VIEW 6L6GC
TOP VIEW TL072
JE1692
ZTX109
TOP VIEW 4016B
NOTE: 1. ALL RESISTANCE VALUES ARE IN OHMS (Ω), ¼ WATT, EXCEPT WHERE NOTED.
2. ALL CAPACITANCE VALUES ARE IN MICROFARADS (μF) EXCEPT WHERE NOTED.
3. IC'S 1,3,& 4 ARE TL072.
4. IC-2 IS 4016B.
5. +8V & -8V ARE NOMINAL; ACTUAL VOLTAGE MAY VARY ± 1 VOLT.
MUSIC MAN INC. ANAHEIM, CA.
SCHEMATIC DIAGRAM - RD-50 AMPLIFIER
DESIGNED T. WALKER
DWG. NO 5004
REV. B
SHEET 1 OF 1

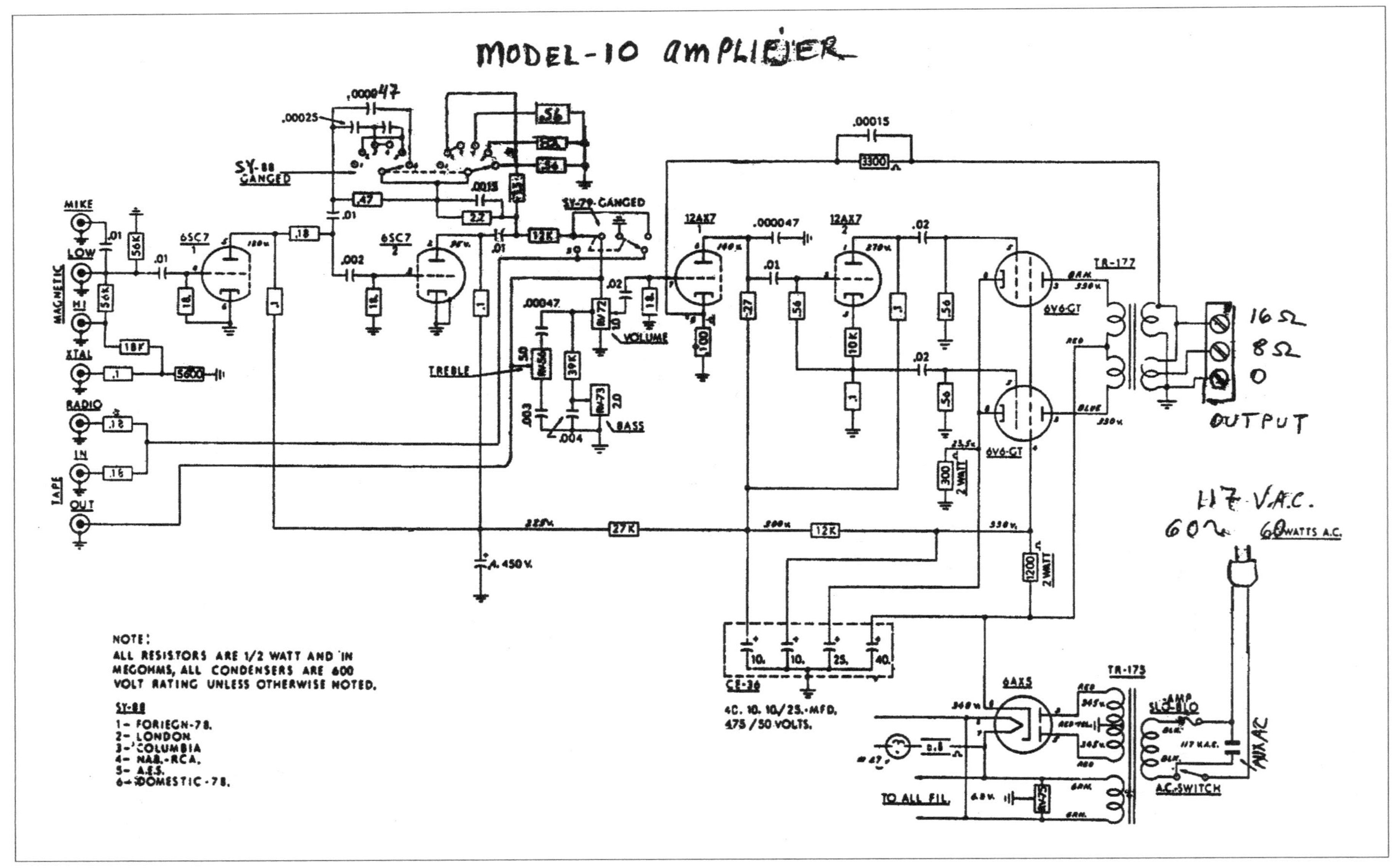
MODEL-10 AMPLIFIER
MIKE
LOW
MAGNETIC
HI
XTAL
RADIO
IN
OUT
TAPE
6SC7 1
6SC7 2
SY-88 GANGED
SY-79 GANGED
TREBLE
VOLUME
BASS
12AX7 1
12AX7 2
6V6-GT
TR-177
OUTPUT
16Ω
8Ω
0
117 V.A.C.
60~ 60 WATTS A.C.
TR-175
6AX5
CE-36
4C. 10. 10/25.-MFD.
475/50 VOLTS.
TO ALL FIL.
AMP SLO-BLO
A.C. SWITCH
AUX AC
A. 450 V.
NOTE:
ALL RESISTORS ARE 1/2 WATT AND IN MEGOHMS, ALL CONDENSERS ARE 600 VOLT RATING UNLESS OTHERWISE NOTED.
SY-88
1- FORIEGN-78.
2- LONDON.
3- COLUMBIA
4- NAB.-RCA.
5- A.E.S.
6- DOMESTIC-78.

120w GRAPHIC MK II (EARLY MODEL)

Orange

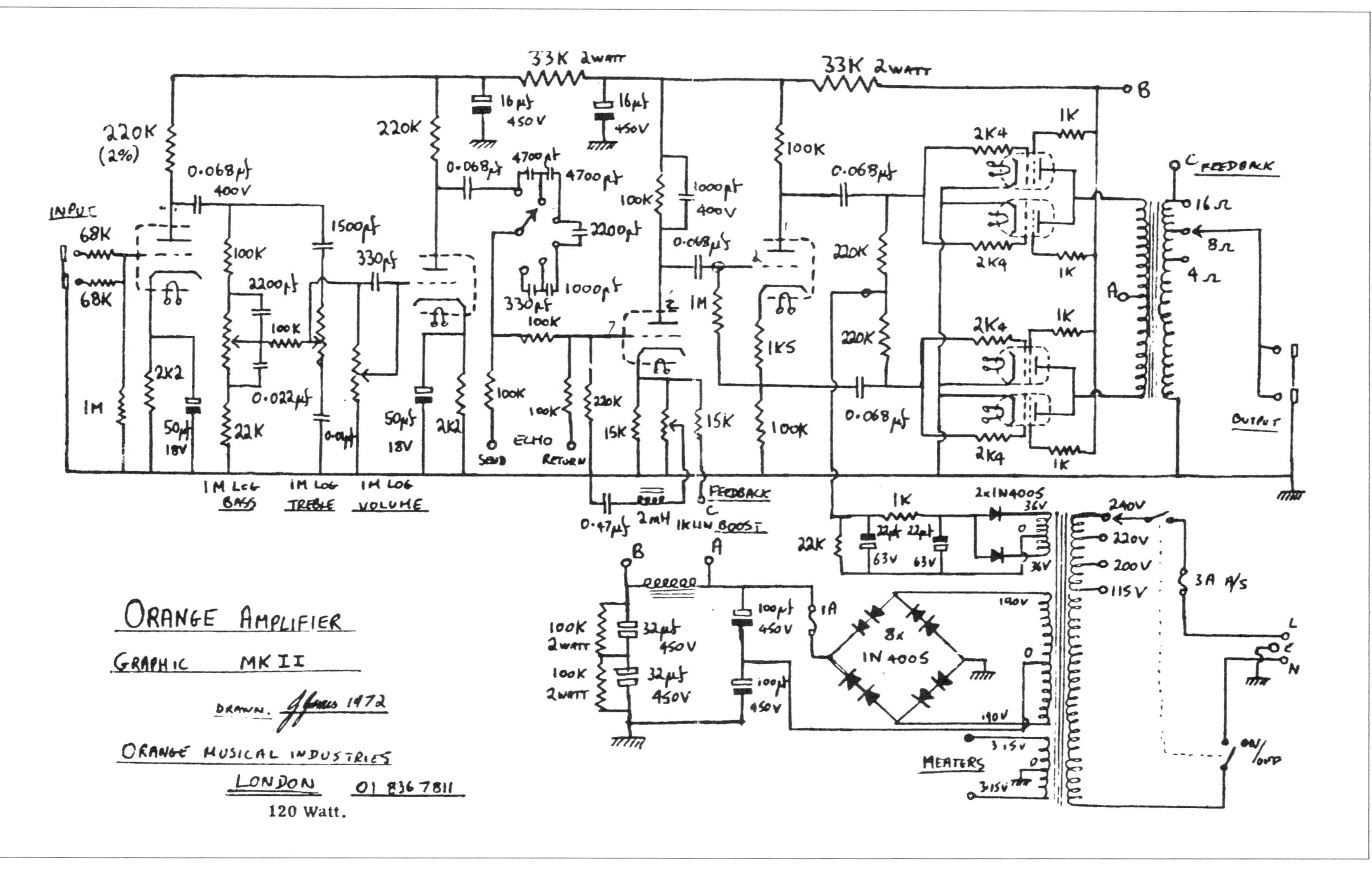

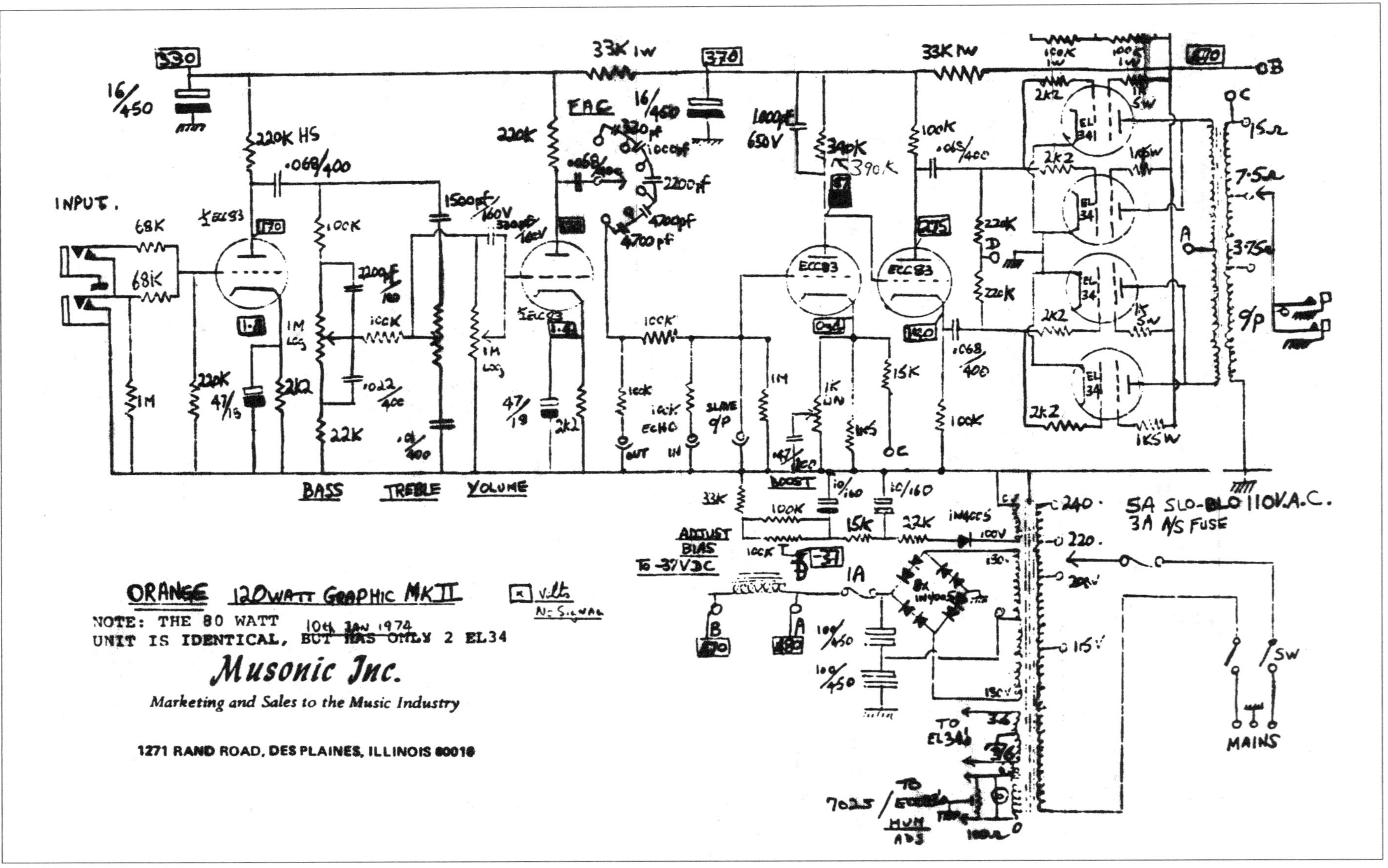
INPUT.
68K
68K
1M
ECC83
220K HS
.068/400
100K
22K
BASS
TREBLE
VOLUME
1M LOG
1500pf/160V
220K
EAC
16/450
16/460
330
370
470
33K 1W
33K 1W
1000pF 650V
390K
ECC83
ECC83
EL34
EL34
EL34
EL34
2K2
2K2
2K2
2K2
1K 5W
15K
100K
.068/400
220K
220K
A
O/P
15Ω
7.5Ω
3.75Ω
OC
OB
1M
1K LIN
SLAVE O/P
ECHO
OUT
IN
BOOST
33K
100K
15K
22K
10/160
10/160
ADJUST BIAS TO -37VDC
1A
100/450
100/450
240
220
200
115
5A SLO-BLO 110V.A.C.
3A A/S FUSE
SW
MAINS
TO EL34
HUM ADJ
ORANGE 120WATT GRAPHIC MK II
NOTE: THE 80 WATT
UNIT IS IDENTICAL, BUT HAS ONLY 2 EL34
10th JAN 1974
Volts
No Signal
Musonic Inc.
Marketing and Sales to the Music Industry
1271 RAND ROAD, DES PLAINES, ILLINOIS 60018

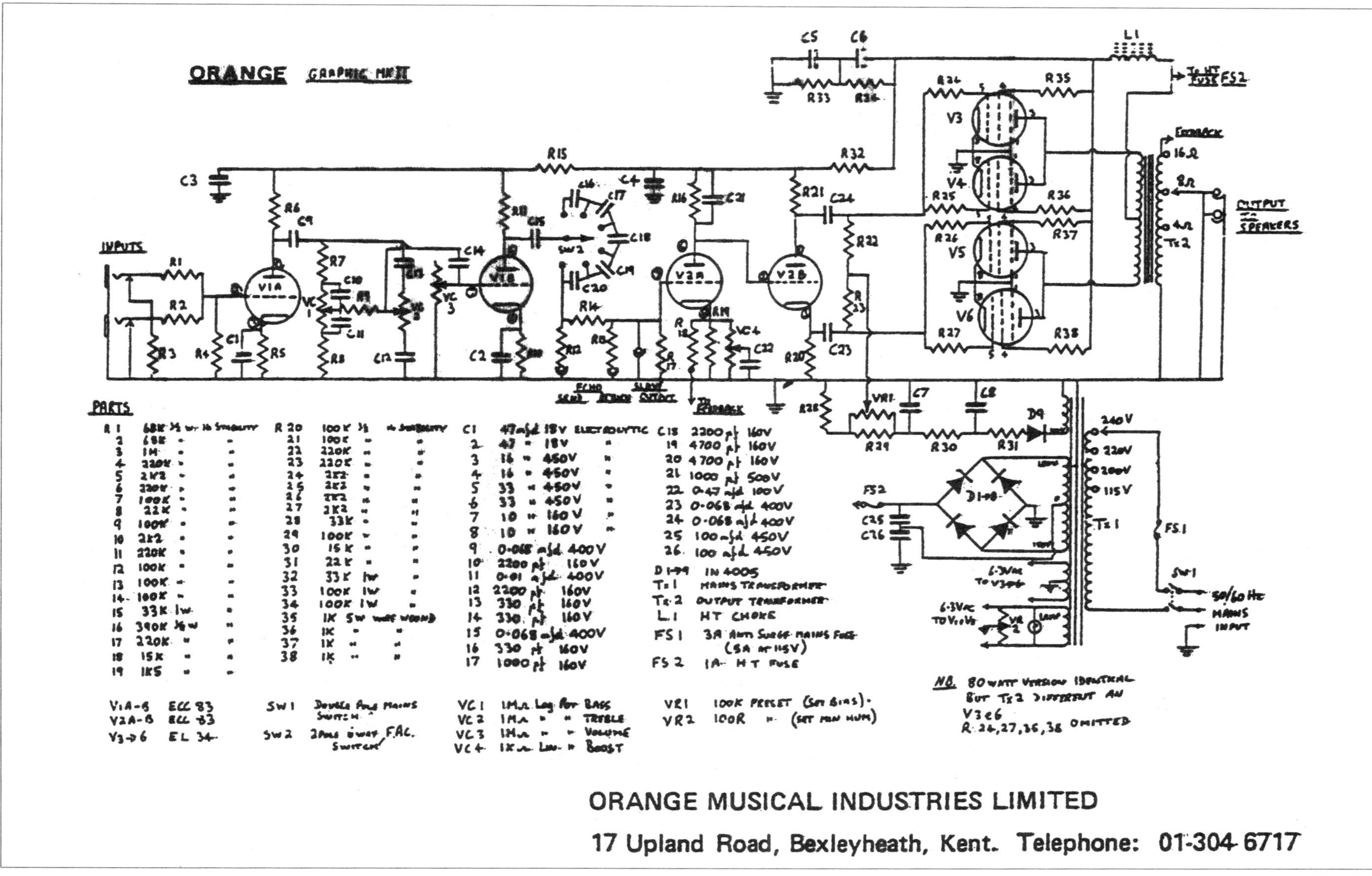
ORANGE GRAPHIC MK II
INPUTS
PARTS
OUTPUT TO SPEAKERS
ORANGE MUSICAL INDUSTRIES LIMITED
17 Upland Road, Bexleyheath, Kent. Telephone: 01-304 6717

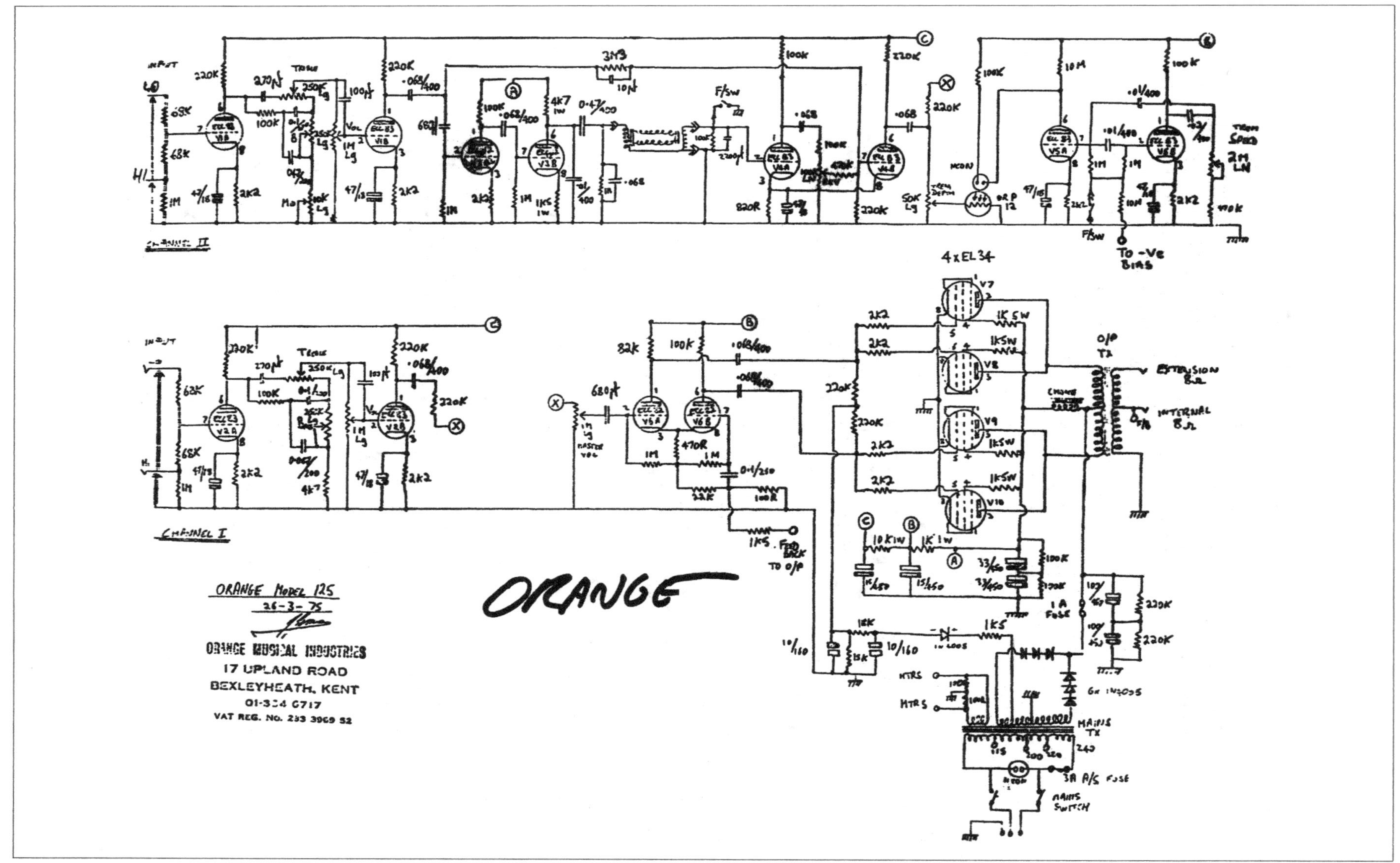

CHANNEL II
CHANNEL I
INPUT
4 x EL34
O/P TX
EXTENSION 8Ω
INTERNAL 8Ω
To -Ve BIAS
MASTER VOL
FEED BACK TO O/P
MAINS TX
3A A/S FUSE
MAINS SWITCH
1A FUSE
ORANGE
ORANGE Model 12S
26-3-75
ORANGE MUSICAL INDUSTRIES
17 UPLAND ROAD
BEXLEYHEATH, KENT
01-304 6717
VAT REG. No. 233 3969 52

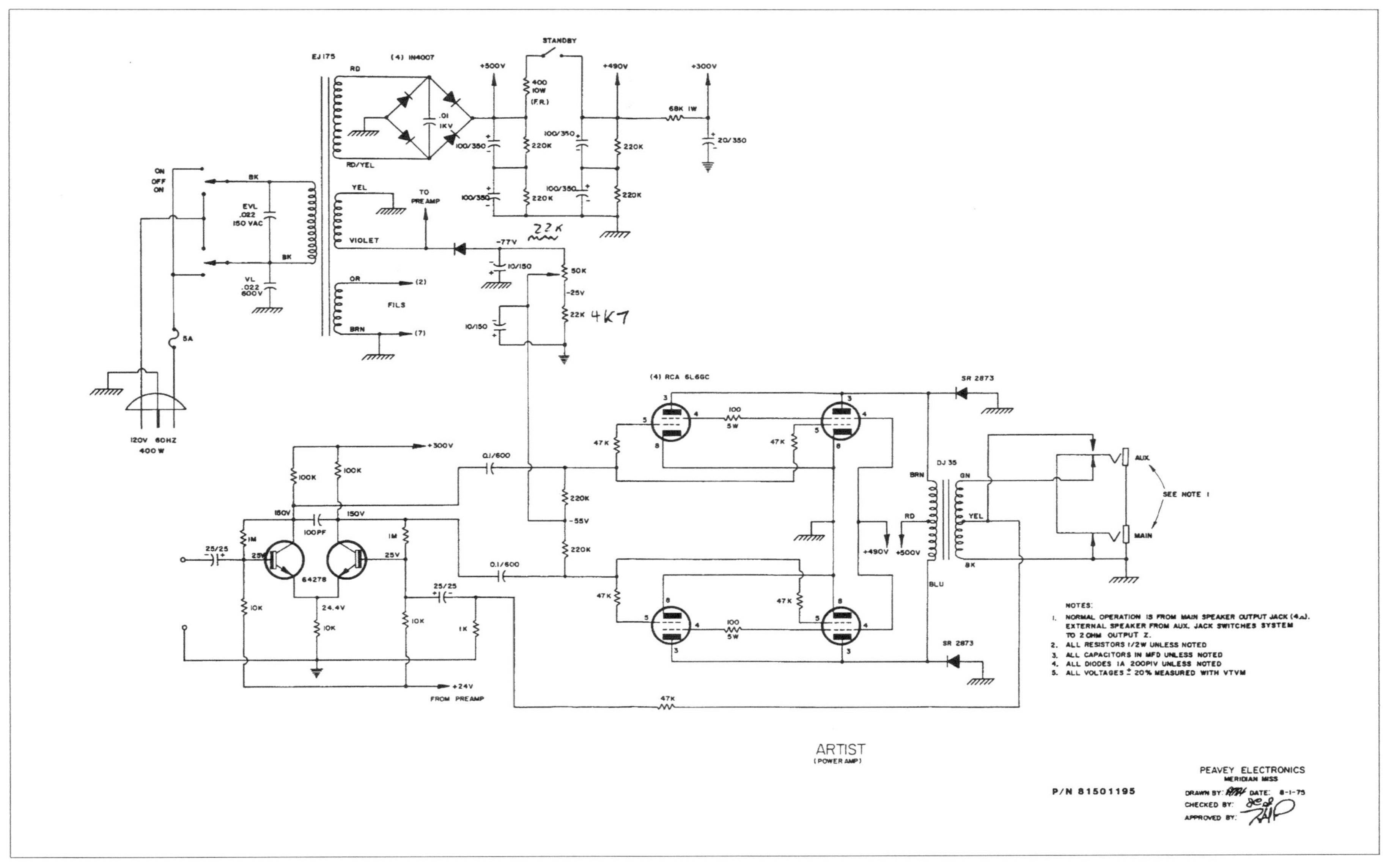
EJ175
(4) IN4007
RD
RD/YEL
YEL
VIOLET
OR
BRN
FILS
TO PREAMP
STANDBY
+500V
+490V
+300V
400 10W (F.R.)
68K 1W
20/350
.01 1KV
100/350
220K
ON
OFF
ON
BK
EVL .022 150 VAC
VL .022 600V
5A
120V 60HZ 400 W
-77V
22K
10/150
50K
-25V
22K 4K7
(4) RCA 6L6GC
SR 2873
100 5W
47K
DJ 35
GN
YEL
BK
BRN
RD
BLU
+490V
+500V
AUX
MAIN
SEE NOTE 1
+300V
0.1/600
100K
220K
-55V
150V
100PF
1M
25/25
25V
64278
24.4V
10K
1K
+24V
FROM PREAMP
NOTES:
1. NORMAL OPERATION IS FROM MAIN SPEAKER OUTPUT JACK (4Ω). EXTERNAL SPEAKER FROM AUX. JACK SWITCHES SYSTEM TO 2 OHM OUTPUT Z.
2. ALL RESISTORS 1/2W UNLESS NOTED
3. ALL CAPACITORS IN MFD UNLESS NOTED
4. ALL DIODES 1A 200PIV UNLESS NOTED
5. ALL VOLTAGES ± 20% MEASURED WITH VTVM
ARTIST
(POWER AMP)
P/N 81501195
PEAVEY ELECTRONICS
MERIDIAN MISS
DRAWN BY:
DATE: 8-1-75
CHECKED BY:
APPROVED BY:

ARTIST VT PRE-AMP

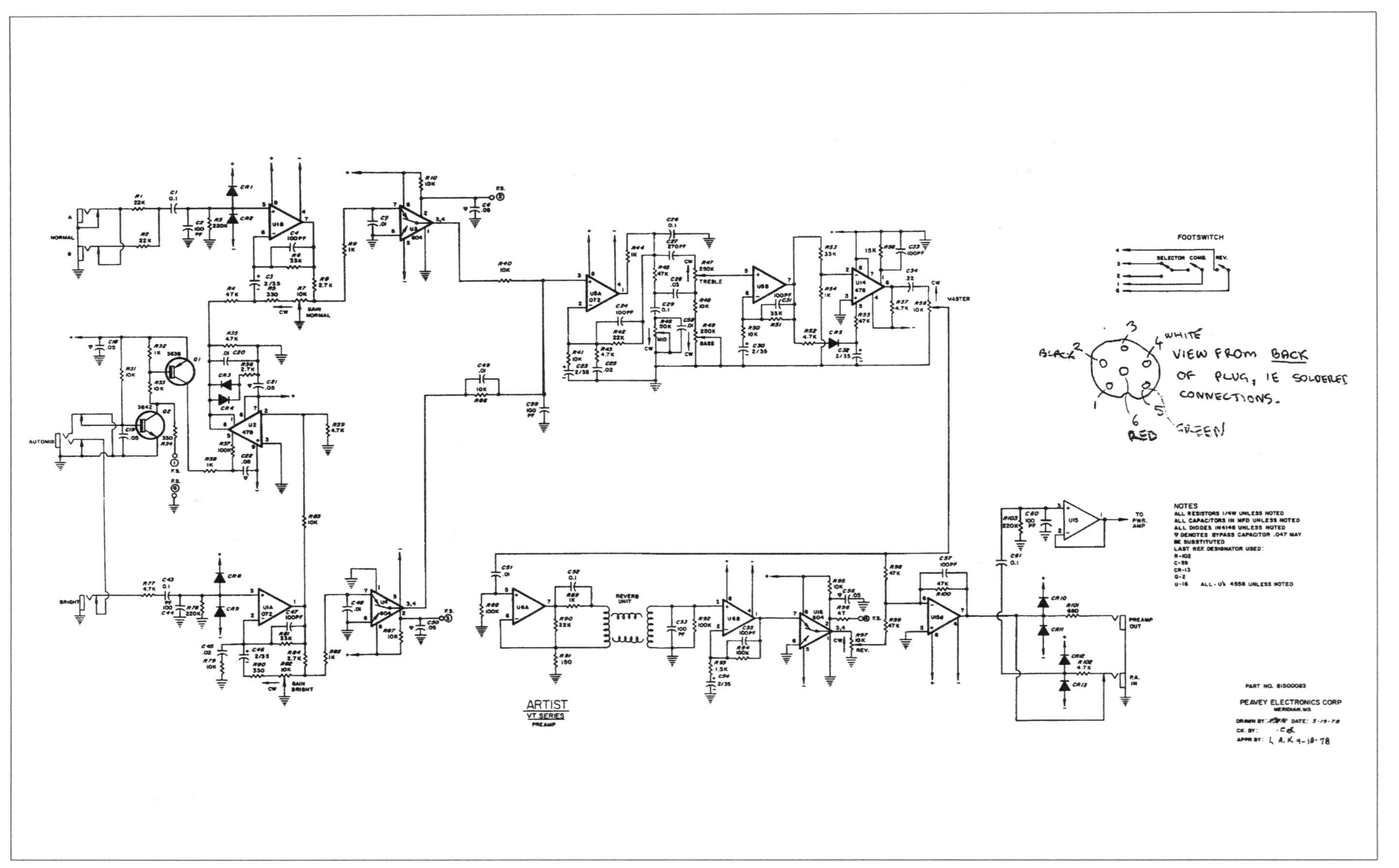

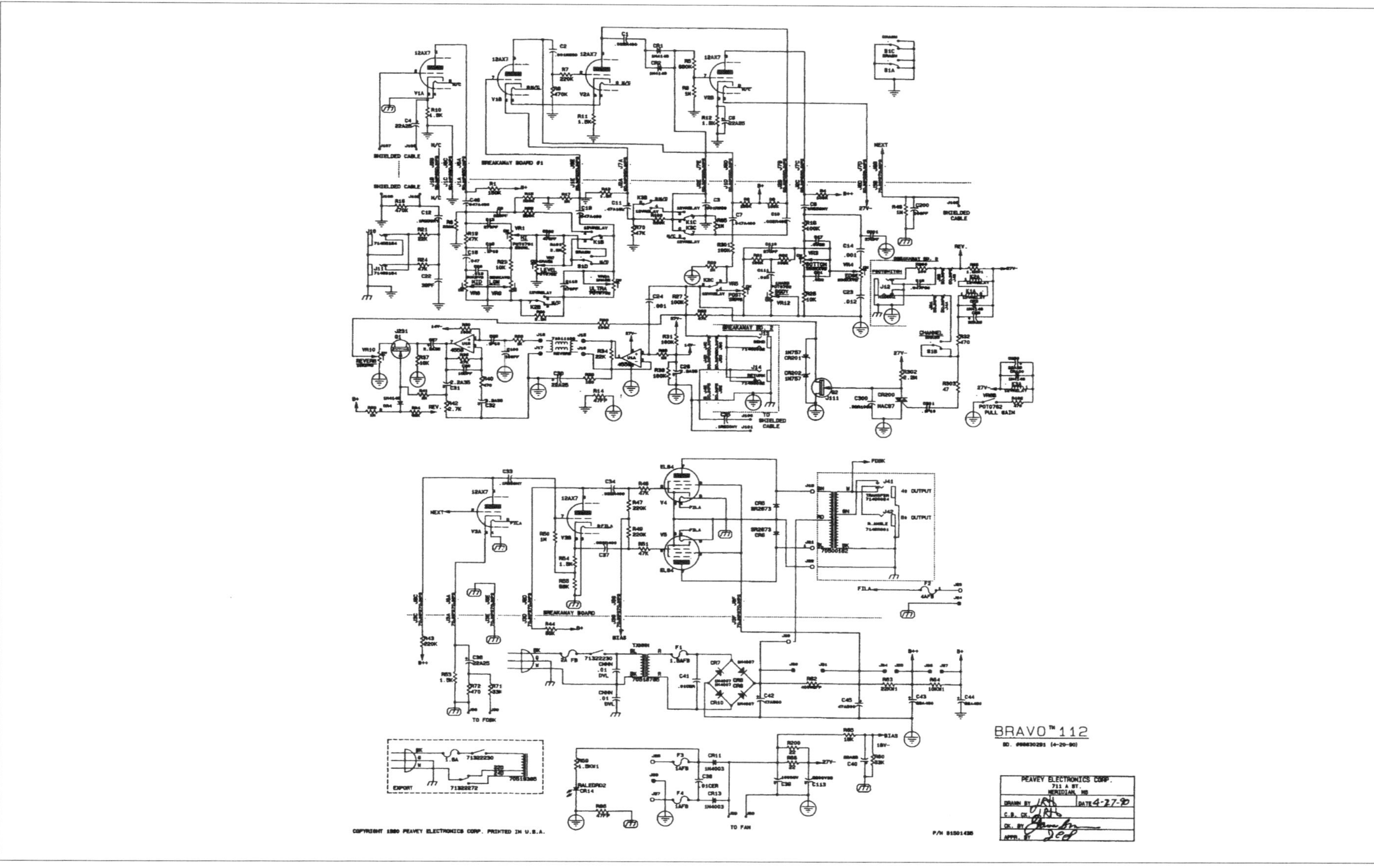
BRAVO™ 112
PEAVEY ELECTRONICS CORP.
12AX7
EL84
BREAKAWAY BOARD
SHIELDED CABLE
NEXT
TO FDBK
TO FAN
EXPORT

CLASSIC 30

Peavey

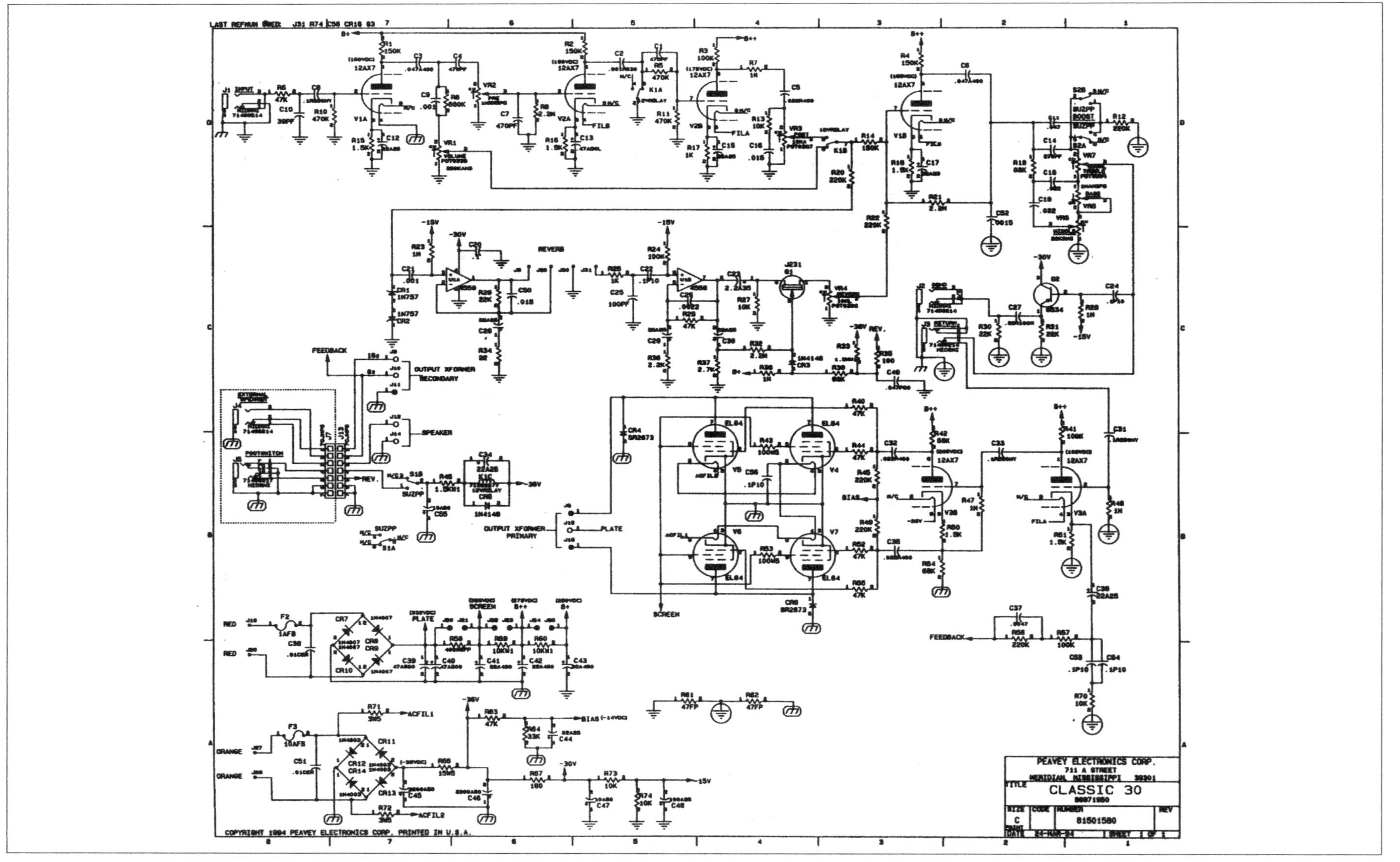

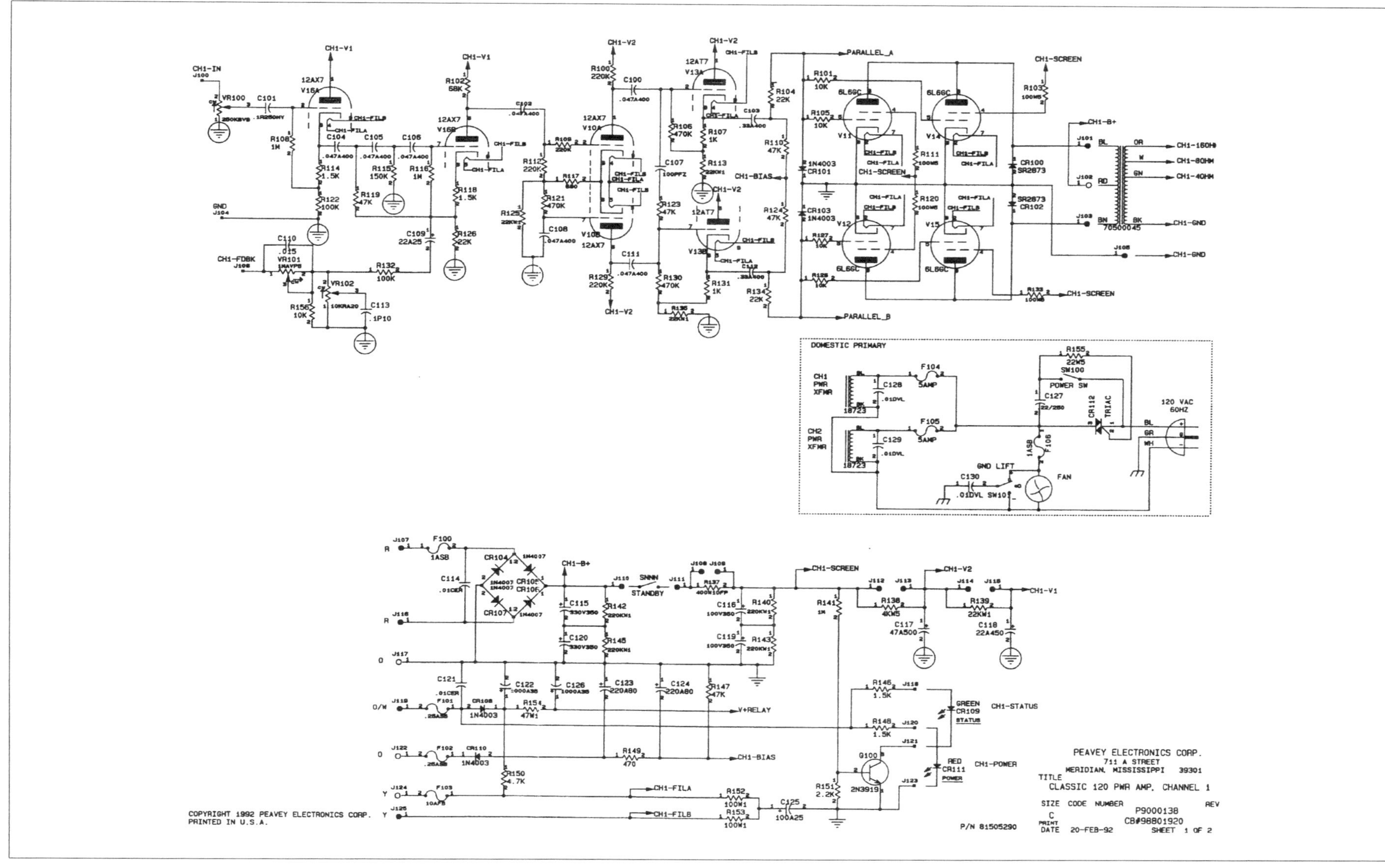
DOMESTIC PRIMARY
COPYRIGHT 1992 PEAVEY ELECTRONICS CORP.
PRINTED IN U.S.A.
PEAVEY ELECTRONICS CORP.
711 A STREET
MERIDIAN, MISSISSIPPI 39301
TITLE
CLASSIC 120 PWR AMP. CHANNEL 1
SIZE CODE NUMBER REV
C P9000138
CB#98801920
P/N 81505290
DATE 20-FEB-92 SHEET 1 OF 2

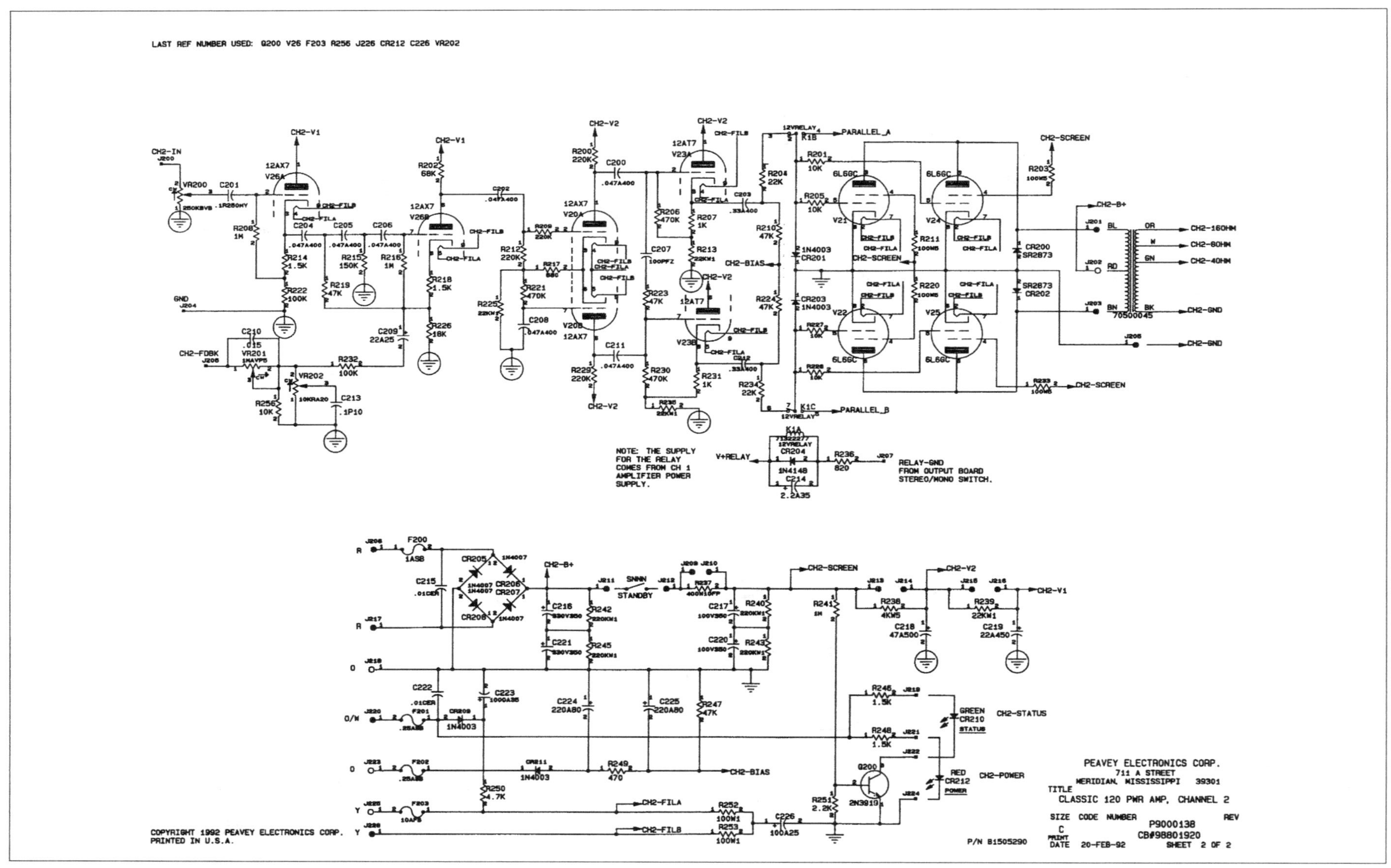
LAST REF NUMBER USED: Q200 V26 F203 R256 J226 CR212 C226 VR202
NOTE: THE SUPPLY FOR THE RELAY COMES FROM CH 1 AMPLIFIER POWER SUPPLY.
RELAY-GND FROM OUTPUT BOARD STEREO/MONO SWITCH.
PEAVEY ELECTRONICS CORP.
711 A STREET
MERIDIAN, MISSISSIPPI 39301
TITLE CLASSIC 120 PWR AMP, CHANNEL 2
SIZE C
CODE NUMBER P9000138
REV
CB#98801920
PRINT DATE 20-FEB-92
SHEET 2 OF 2
P/N 81505290
COPYRIGHT 1992 PEAVEY ELECTRONICS CORP.
PRINTED IN U.S.A.

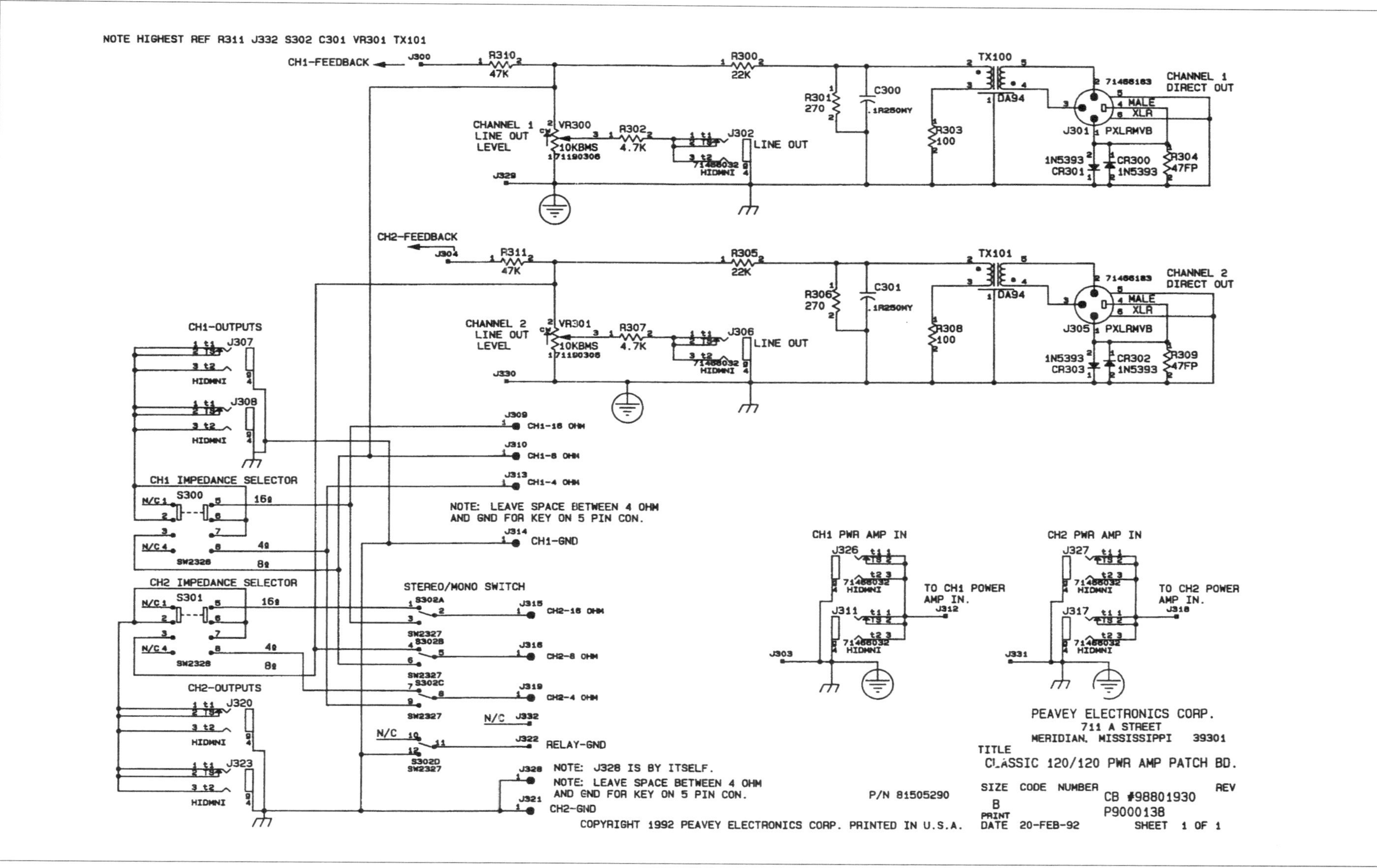
NOTE HIGHEST REF R311 J332 S302 C301 VR301 TX101
CH1-FEEDBACK
J300
R310
47K
R300
22K
TX100
CHANNEL 1
DIRECT OUT
71466183
MALE
XLR
J301
PXLRMVB
R301
270
C300
.1R250MY
DA94
R303
100
CHANNEL 1
LINE OUT
LEVEL
VR300
10KBMS
171190306
R302
4.7K
J302
LINE OUT
71466032
HIDMNI
J329
1N5393
CR301
CR300
1N5393
R304
47FP
CH2-FEEDBACK
J304
R311
47K
R305
22K
TX101
CHANNEL 2
DIRECT OUT
71466183
MALE
XLR
J305
PXLRMVB
R306
270
C301
.1R250MY
DA94
R308
100
CHANNEL 2
LINE OUT
LEVEL
VR301
10KBMS
171190306
R307
4.7K
J306
LINE OUT
J330
1N5393
CR303
CR302
1N5393
R309
47FP
CH1-OUTPUTS
J307
J308
HIDMNI
J309
CH1-16 OHM
J310
CH1-8 OHM
J313
CH1-4 OHM
CH1 IMPEDANCE SELECTOR
S300
N/C
SW2328
16Ω
4Ω
8Ω
NOTE: LEAVE SPACE BETWEEN 4 OHM
AND GND FOR KEY ON 5 PIN CON.
J314
CH1-GND
CH2 IMPEDANCE SELECTOR
S301
STEREO/MONO SWITCH
S302A
S302B
S302C
S302D
SW2327
J315
CH2-16 OHM
J316
CH2-8 OHM
J319
CH2-4 OHM
N/C
J332
J322
RELAY-GND
CH2-OUTPUTS
J320
J323
J328
NOTE: J328 IS BY ITSELF.
NOTE: LEAVE SPACE BETWEEN 4 OHM
AND GND FOR KEY ON 5 PIN CON.
J321
CH2-GND
CH1 PWR AMP IN
J326
J311
J303
TO CH1 POWER
AMP IN.
J312
CH2 PWR AMP IN
J327
J317
J331
TO CH2 POWER
AMP IN.
J318
PEAVEY ELECTRONICS CORP.
711 A STREET
MERIDIAN, MISSISSIPPI 39301
TITLE
CLASSIC 120/120 PWR AMP PATCH BD.
P/N 81505290
SIZE CODE NUMBER
B
CB #98801930
P9000138
REV
PRINT DATE 20-FEB-92
SHEET 1 OF 1
COPYRIGHT 1992 PEAVEY ELECTRONICS CORP. PRINTED IN U.S.A.

MACE DEUCE **PREAMP**

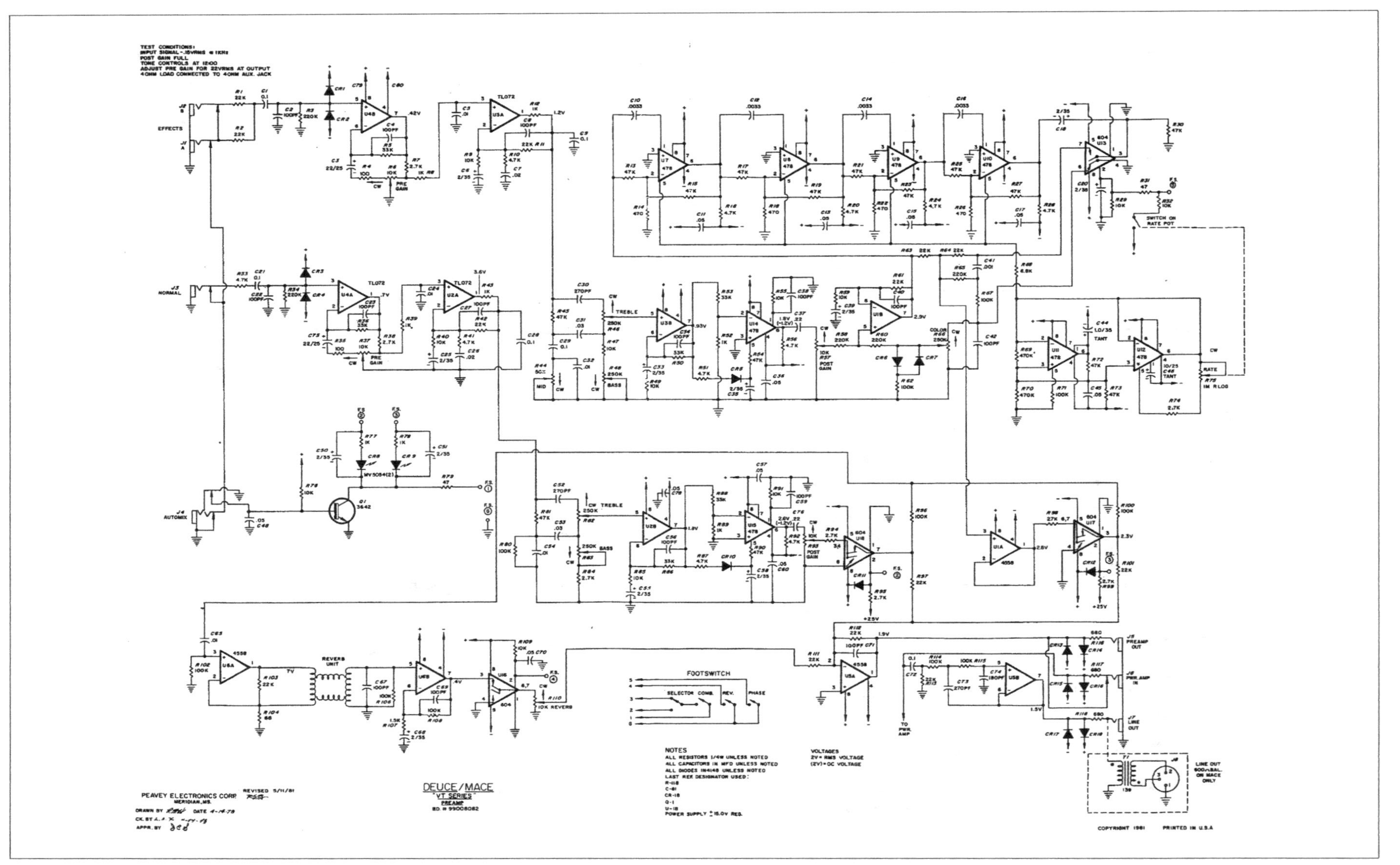

MACE DEUCE **PWR/OUTPUT**

Peavey

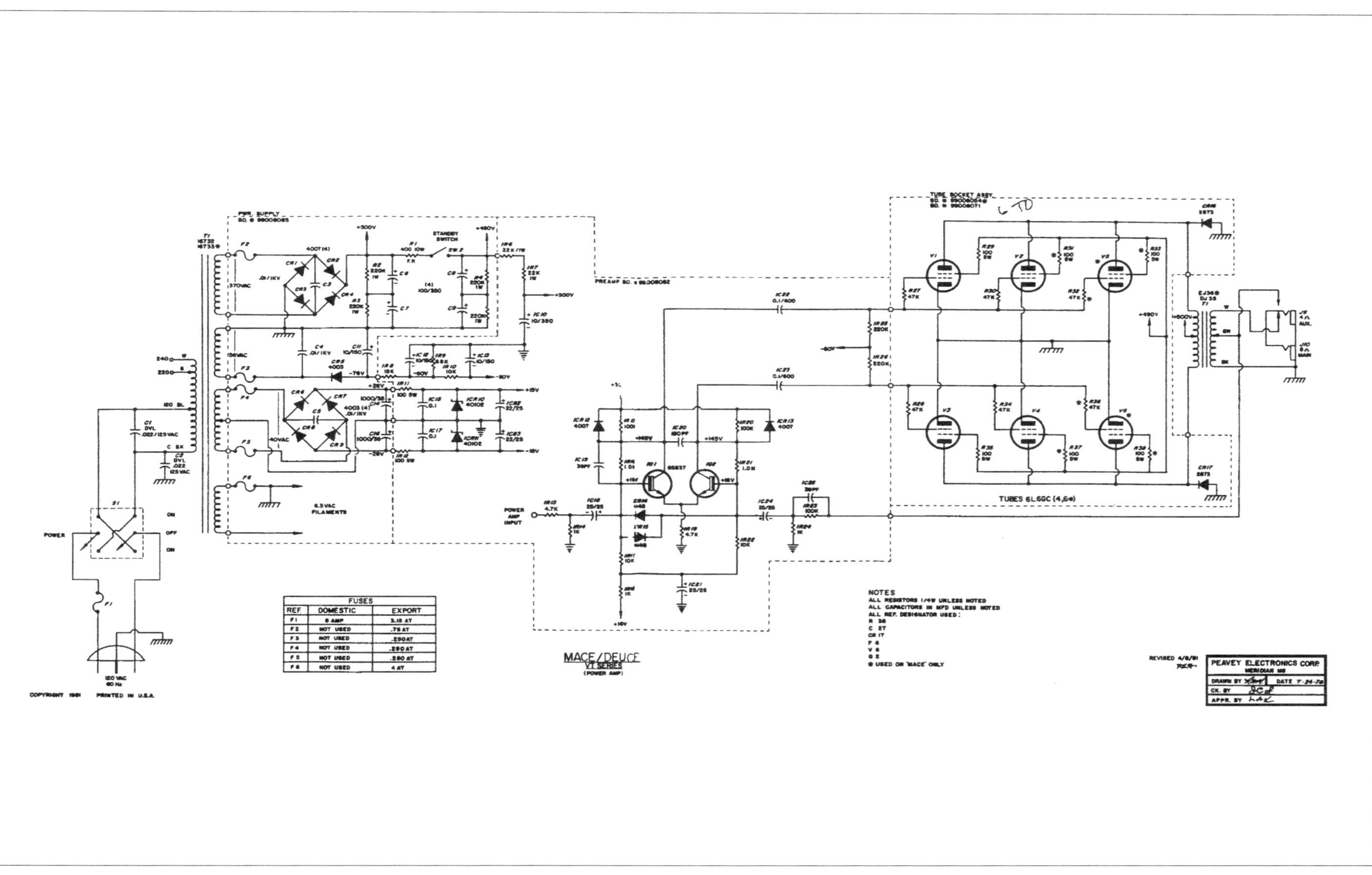

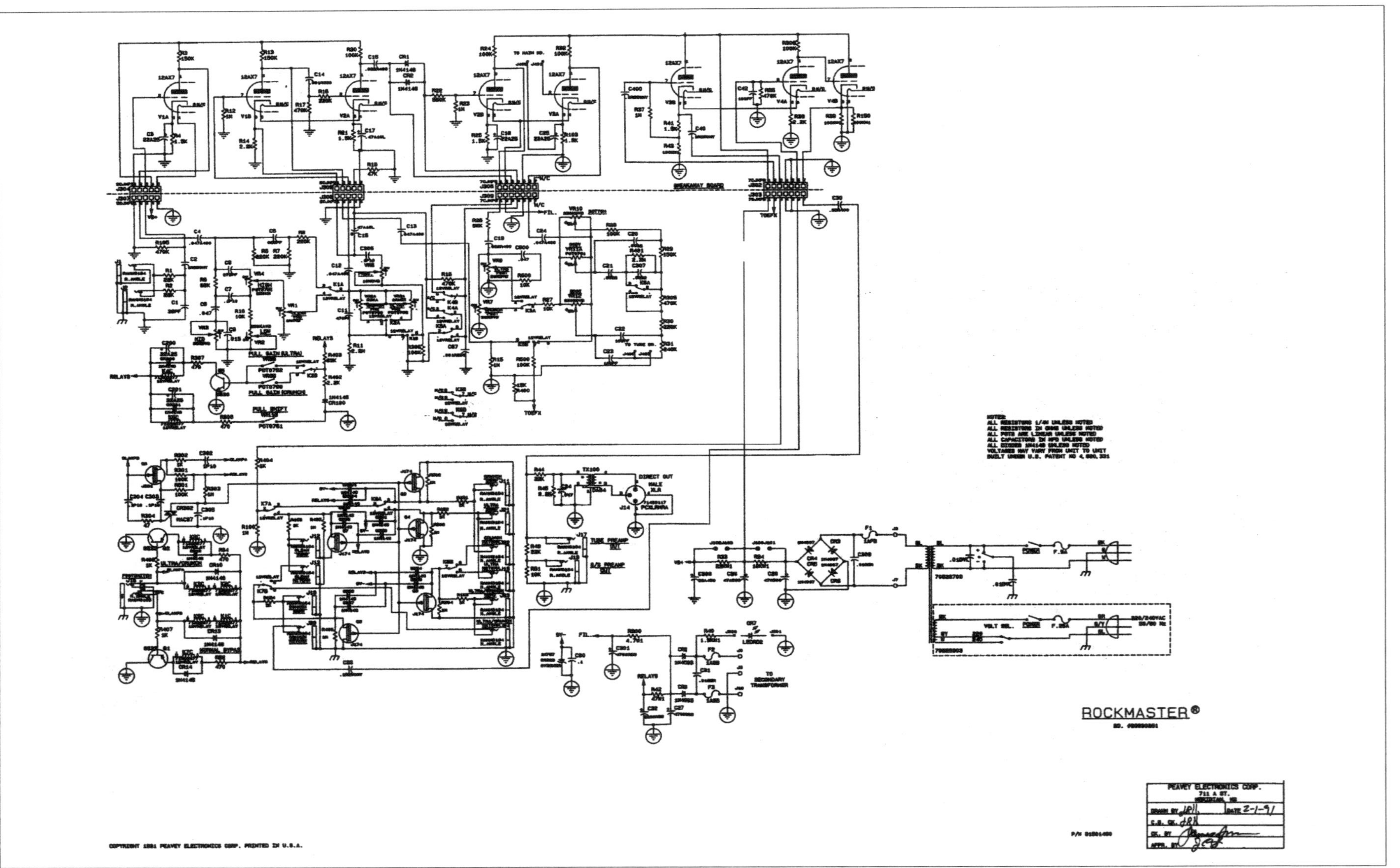
ROCKMASTER®

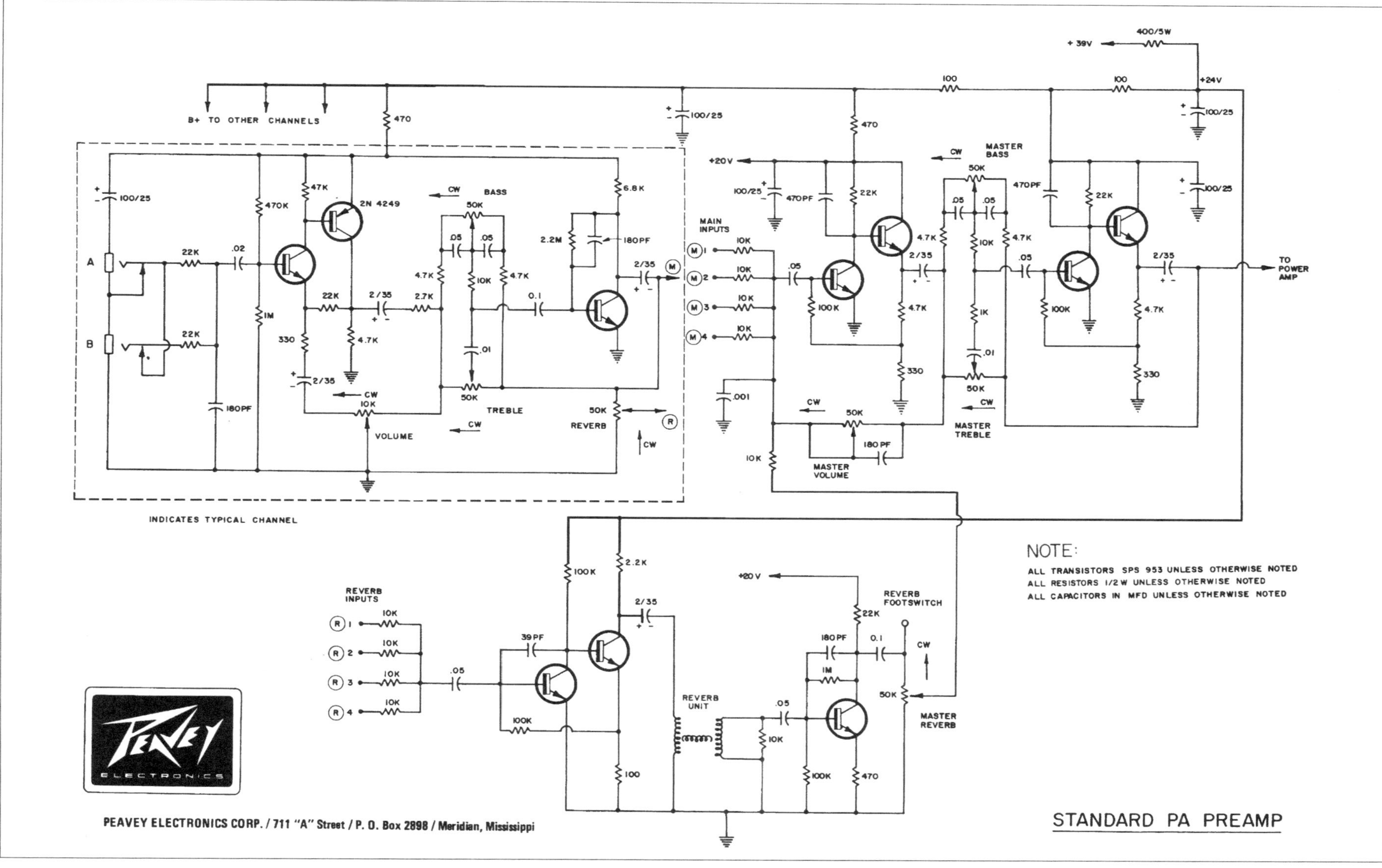

PEAVEY ELECTRONICS CORP. / 711 "A" Street / P. O. Box 2898 / Meridian, Mississippi

NOTE:
ALL TRANSISTORS SPS 953 UNLESS OTHERWISE NOTED
ALL RESISTORS 1/2 W UNLESS OTHERWISE NOTED
ALL CAPACITORS IN MFD UNLESS OTHERWISE NOTED

STANDARD PA PREAMP

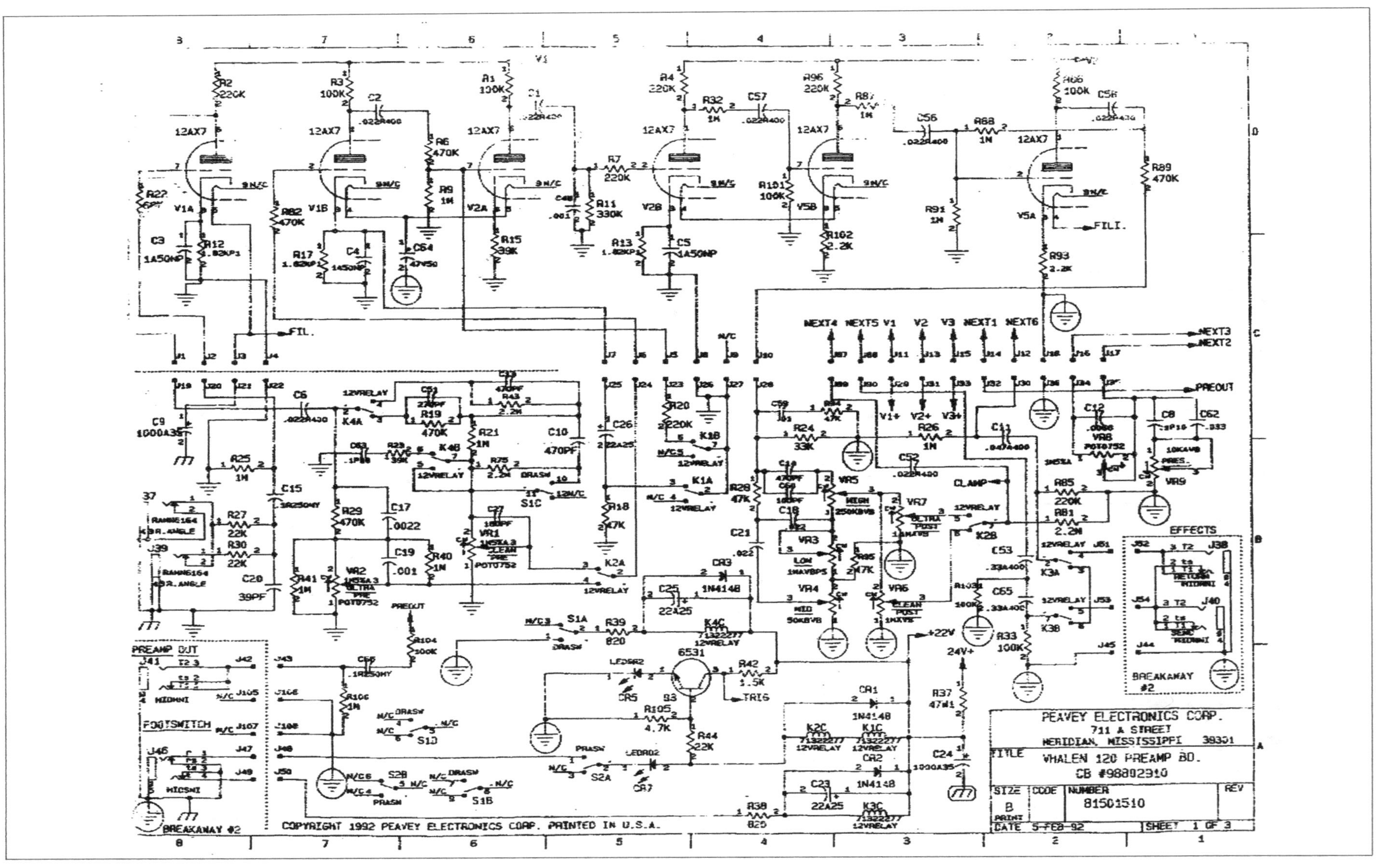
PEAVEY ELECTRONICS CORP.
711 A STREET
MERIDIAN, MISSISSIPPI 39301
TITLE
VHALEN 120 PREAMP BD.
CB #98802910
SIZE
B
CODE
NUMBER
81501510
REV
DATE 5-FEB-92
SHEET 1 OF 3
COPYRIGHT 1992 PEAVEY ELECTRONICS CORP. PRINTED IN U.S.A.
EFFECTS
BREAKAWAY #2
PREAMP OUT
FOOTSWITCH
PREOUT
CLAMP
FIL.
FILI.
TRIG
+22V
24V+
12AX7
V1A
V1B
V2A
V2B
V5A
V5B
NEXT1
NEXT2
NEXT3
NEXT4
NEXT5
NEXT6

VAN HALEN 120 PREAMP **(DIAGRAM 2)**

Peavey

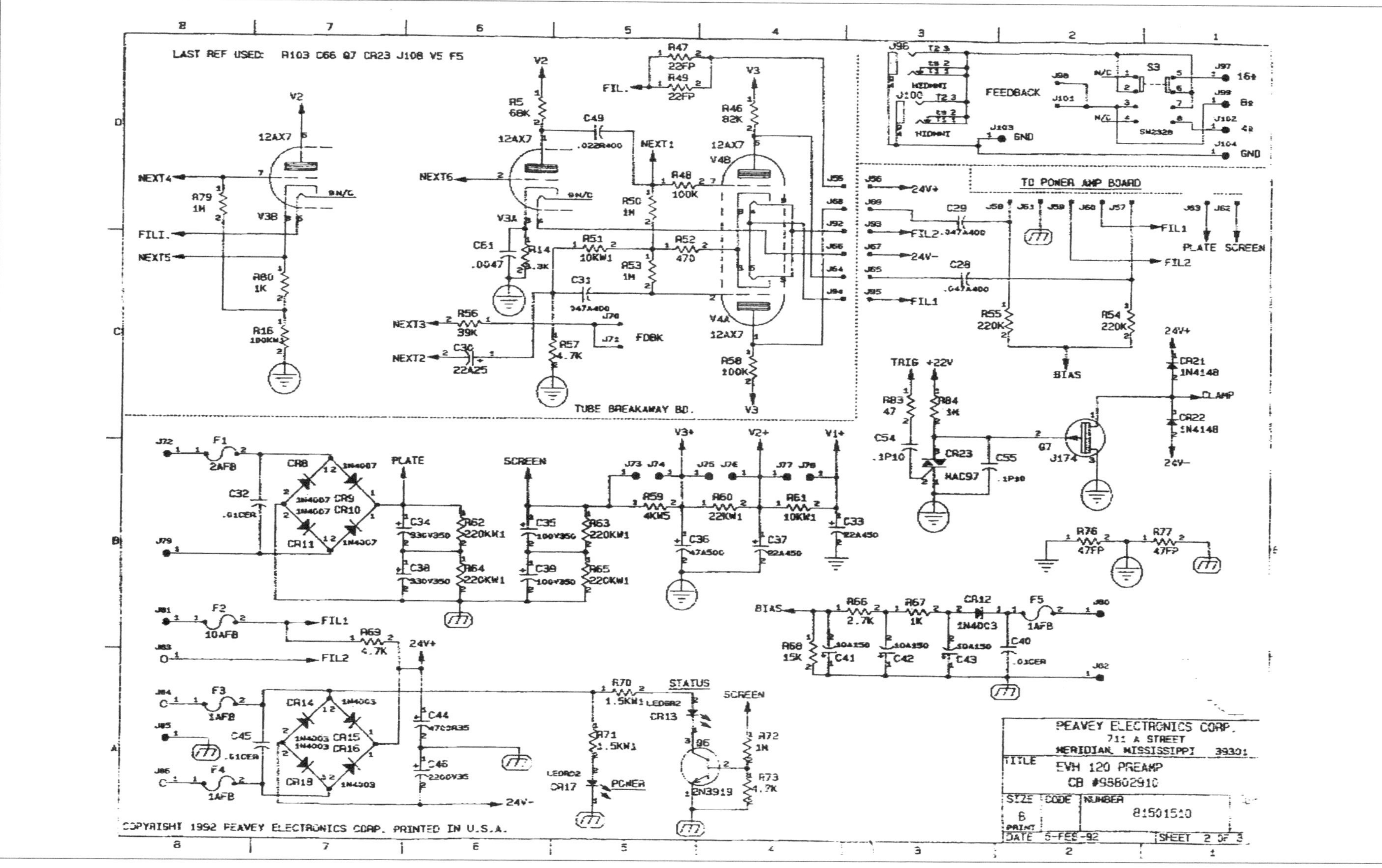

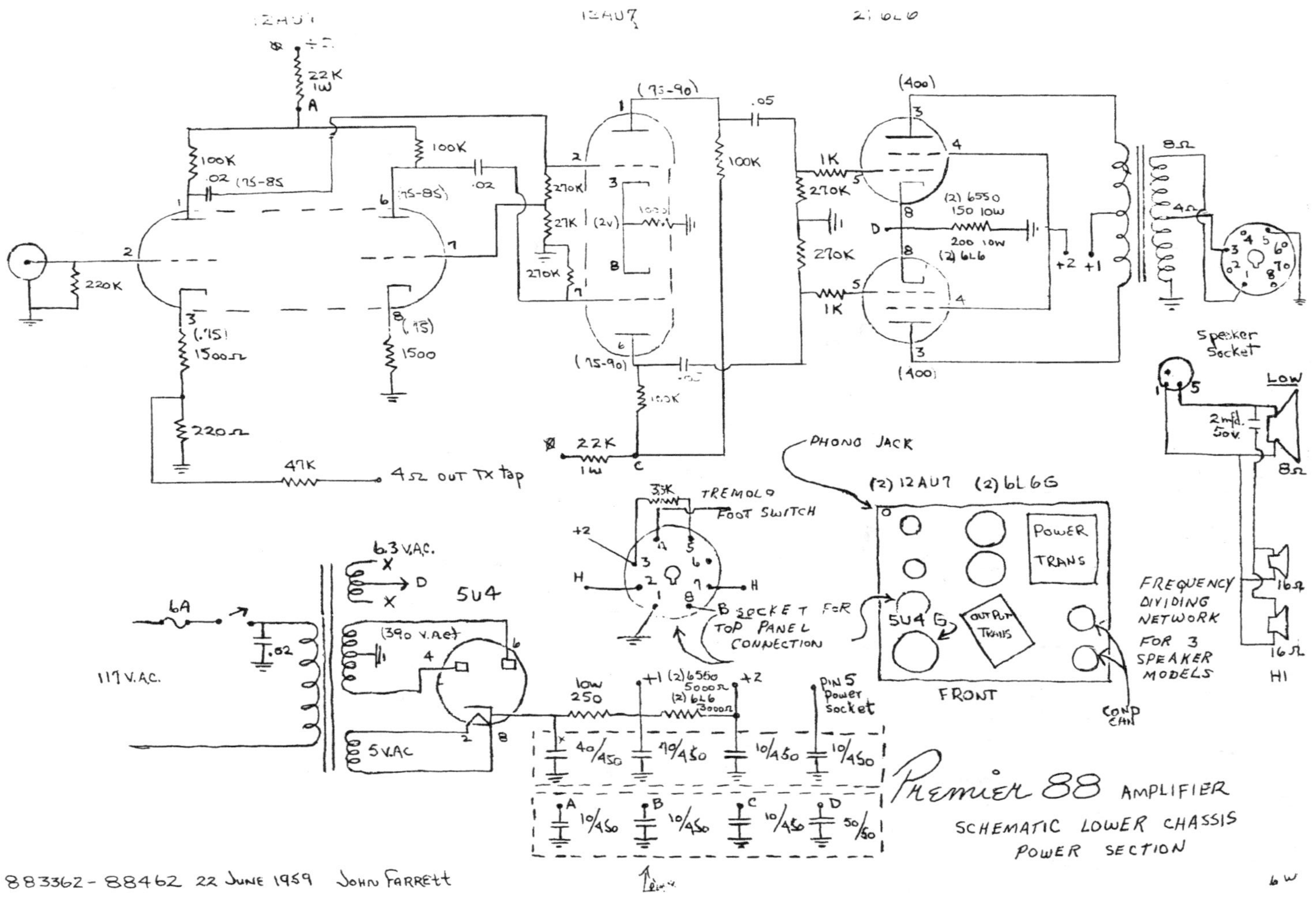

12AU7

12AU7

2) 6L6

PHONO JACK

(2) 12AU7 (2) 6L6G

POWER TRANS

OUTPUT TRANS

5U4 G

FRONT

COND CAN

TREMOLO FOOT SWITCH

SOCKET FOR TOP PANEL CONNECTION

4Ω OUT TX tap

Speaker Socket

LOW

HI

FREQUENCY DIVIDING NETWORK FOR 3 SPEAKER MODELS

117 V.A.C.

6.3 V.A.C.

5U4

5 V.AC

PIN 5 Power Socket

Premier 88 AMPLIFIER

SCHEMATIC LOWER CHASSIS POWER SECTION

883362 - 88462 22 June 1959 John Farrett

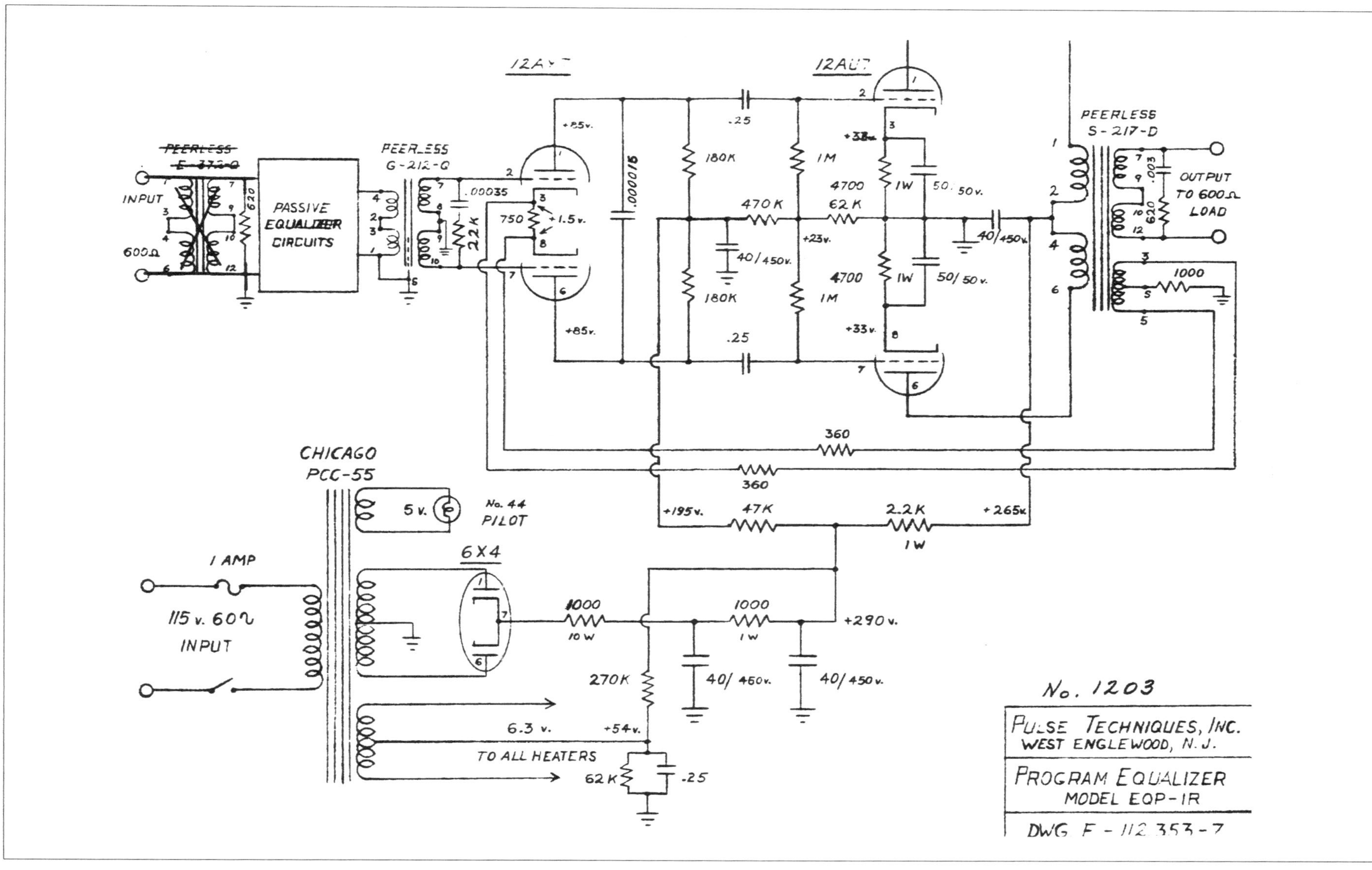
PEERLESS
E-372-Q
INPUT
600Ω
620
PASSIVE
EQUALIZER
CIRCUITS
PEERLESS
G-212-Q
.00035
22K
750
+1.5v.
12AY7
+85v.
.000015
180K
.25
1M
470K
62K
4700
1W
50/50v.
+23v.
40/450v.
12AU7
+33v.
360
360
+195v.
47K
2.2K
1W
+265v.
PEERLESS
S-217-D
.003
620
OUTPUT
TO 600Ω
LOAD
1000
CHICAGO
PCC-55
5v.
No. 44
PILOT
6X4
1 AMP
115 v. 60~
INPUT
1000
10W
1000
1W
+290v.
270K
40/450v.
40/450v.
6.3 v.
+54v.
TO ALL HEATERS
62K
.25
No. 1203
PULSE TECHNIQUES, INC.
WEST ENGLEWOOD, N. J.
PROGRAM EQUALIZER
MODEL EQP-1R
DWG F-112 353-7

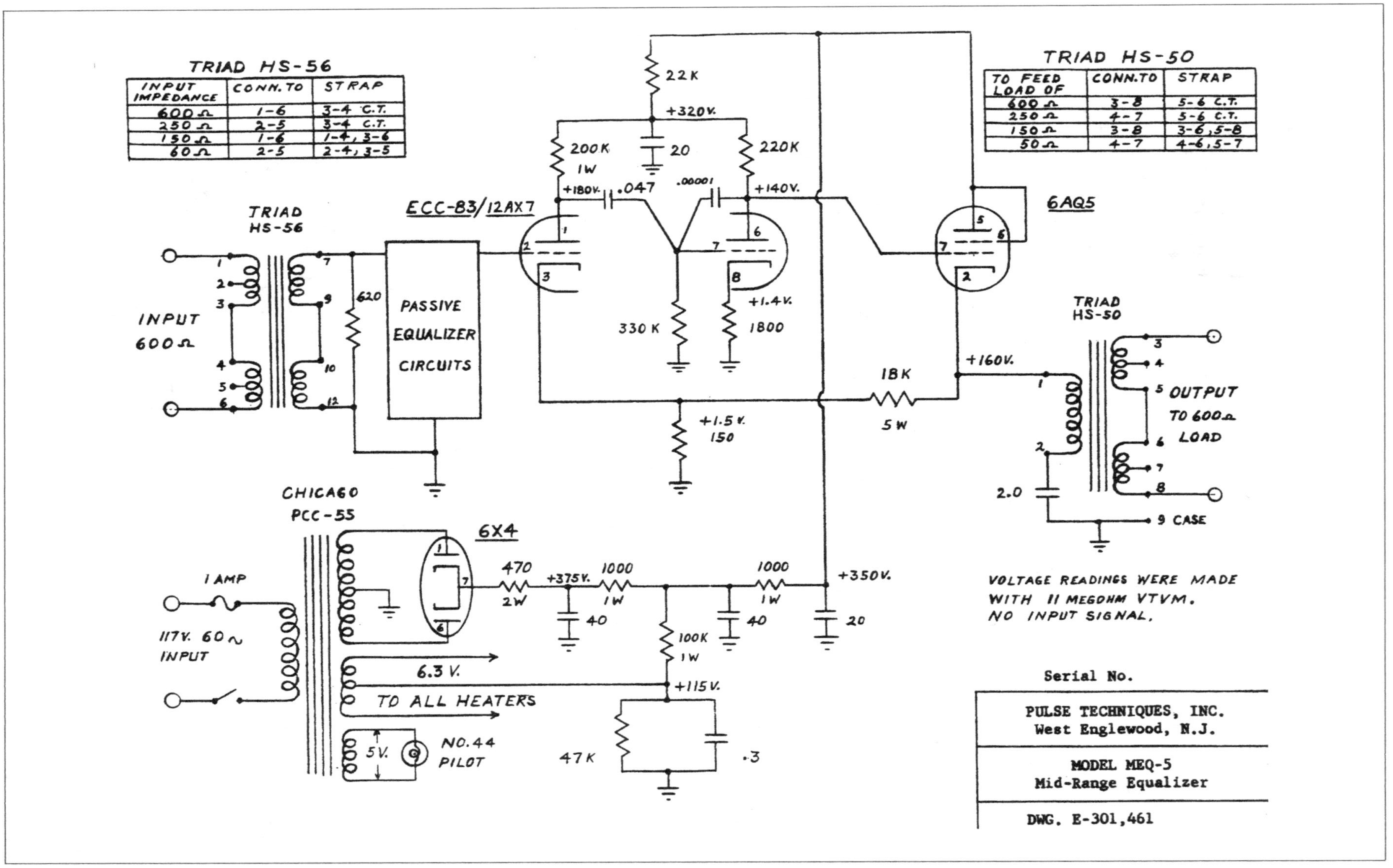
TRIAD HS-56
INPUT IMPEDANCE | CONN. TO | STRAP
600 Ω | 1-6 | 3-4 C.T.
250 Ω | 2-5 | 3-4 C.T.
150 Ω | 1-6 | 1-4, 3-6
60 Ω | 2-5 | 2-4, 3-5
TRIAD HS-50
TO FEED LOAD OF | CONN. TO | STRAP
600 Ω | 3-8 | 5-6 C.T.
250 Ω | 4-7 | 5-6 C.T.
150 Ω | 3-8 | 3-6, 5-8
50 Ω | 4-7 | 4-6, 5-7
TRIAD HS-56
INPUT 600 Ω
620
PASSIVE EQUALIZER CIRCUITS
ECC-83/12AX7
22K
+320V.
20
200K 1W
220K
+180V.
.047
.00001
+140V.
330K
+1.4V.
1800
+1.5V.
150
6AQ5
18K
5W
+160V.
TRIAD HS-50
OUTPUT TO 600 Ω LOAD
9 CASE
2.0
CHICAGO PCC-55
6X4
1 AMP
117V. 60 ~ INPUT
470 2W
+375V.
1000 1W
40
1000 1W
+350V.
20
100K 1W
+115V.
6.3 V.
TO ALL HEATERS
5V.
NO. 44 PILOT
47K
.3
VOLTAGE READINGS WERE MADE WITH 11 MEGOHM VTVM. NO INPUT SIGNAL.
Serial No.
PULSE TECHNIQUES, INC.
West Englewood, N.J.
MODEL MEQ-5
Mid-Range Equalizer
DWG. E-301,461

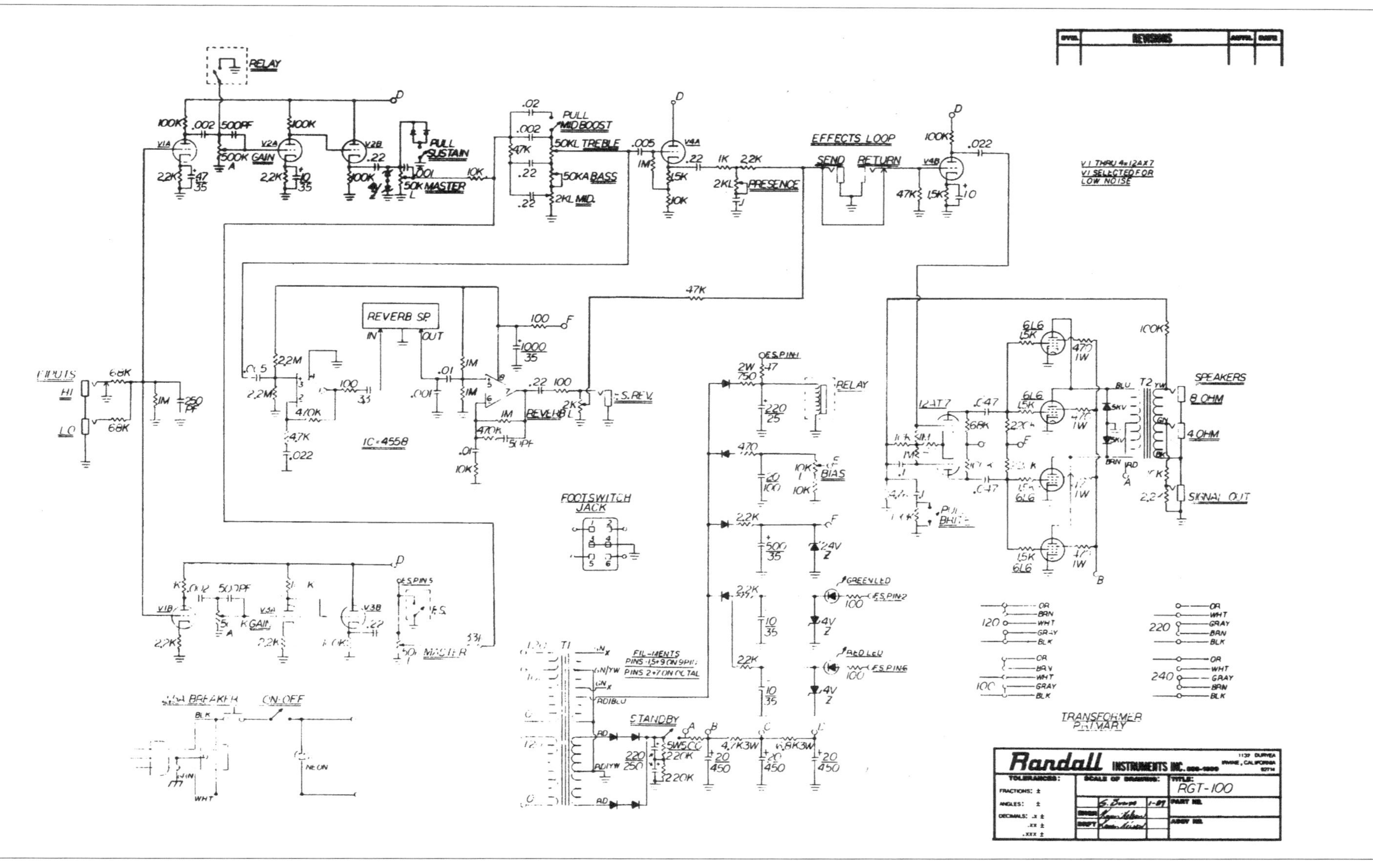
RELAY
PULL MIDBOOST
50KL TREBLE
50KA BASS
2KL MID
PULL SUSTAIN
50K MASTER
500K GAIN
EFFECTS LOOP
SEND
RETURN
PRESENCE
V1 THRU 4=12AX7
V1 SELECTED FOR LOW NOISE
REVERB SP
IN
OUT
IC-4558
INPUTS
HI
LO
FOOTSWITCH JACK
BIAS
GREEN LED
RED LED
SPEAKERS
8 OHM
4 OHM
SIGNAL OUT
12AT7
6L6
STANDBY
FILAMENTS
PINS 1-5+9 ON 9PIN
PINS 2+7 ON OCTAL
ON-OFF
NEON
TRANSFORMER PRIMARY
Randall INSTRUMENTS INC.
TITLE: RGT-100
PART NO.
ASSY NO.
REVISIONS

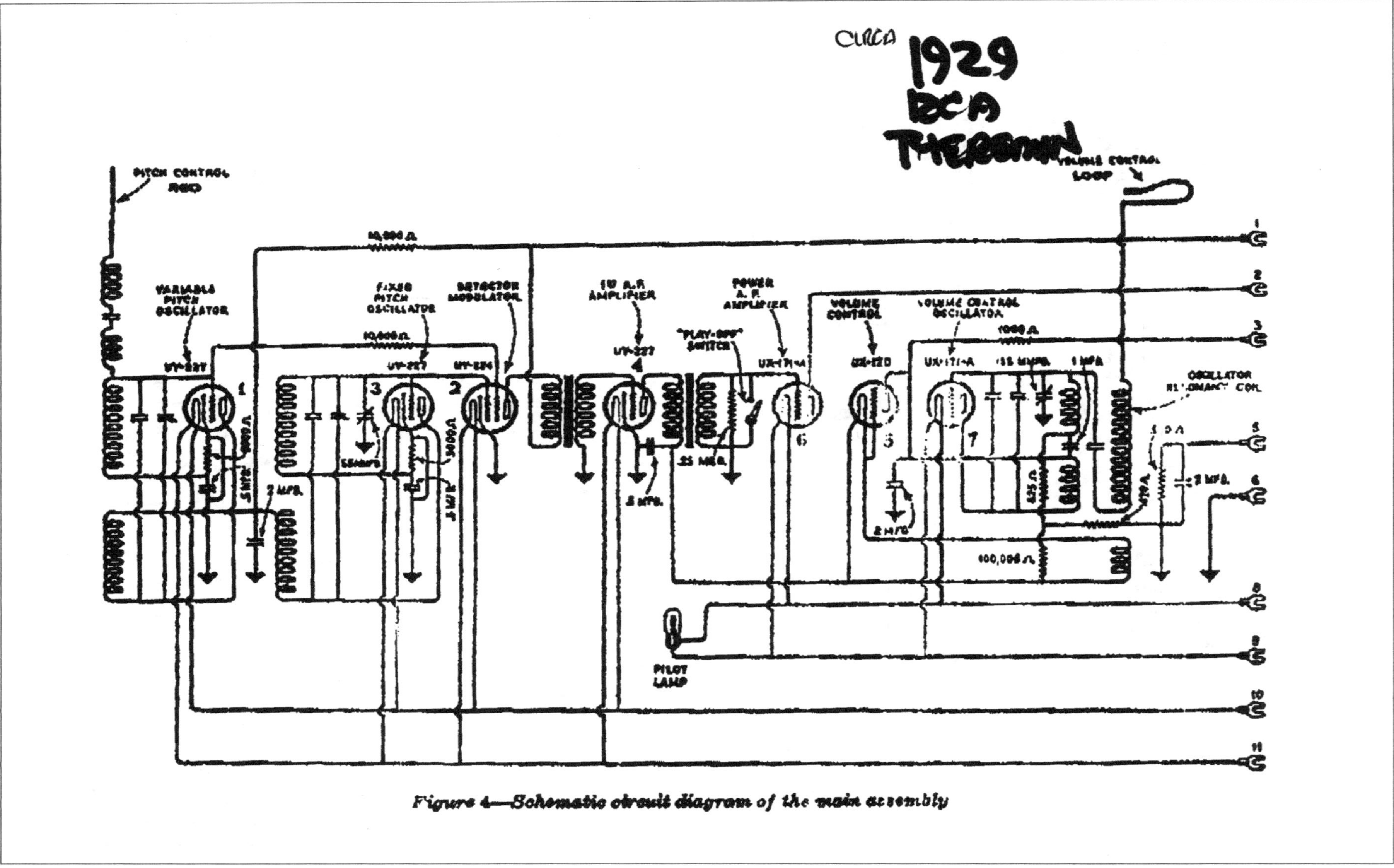

Figure 4—Schematic circuit diagram of the main assembly

Revox Stereo Tape Recorder

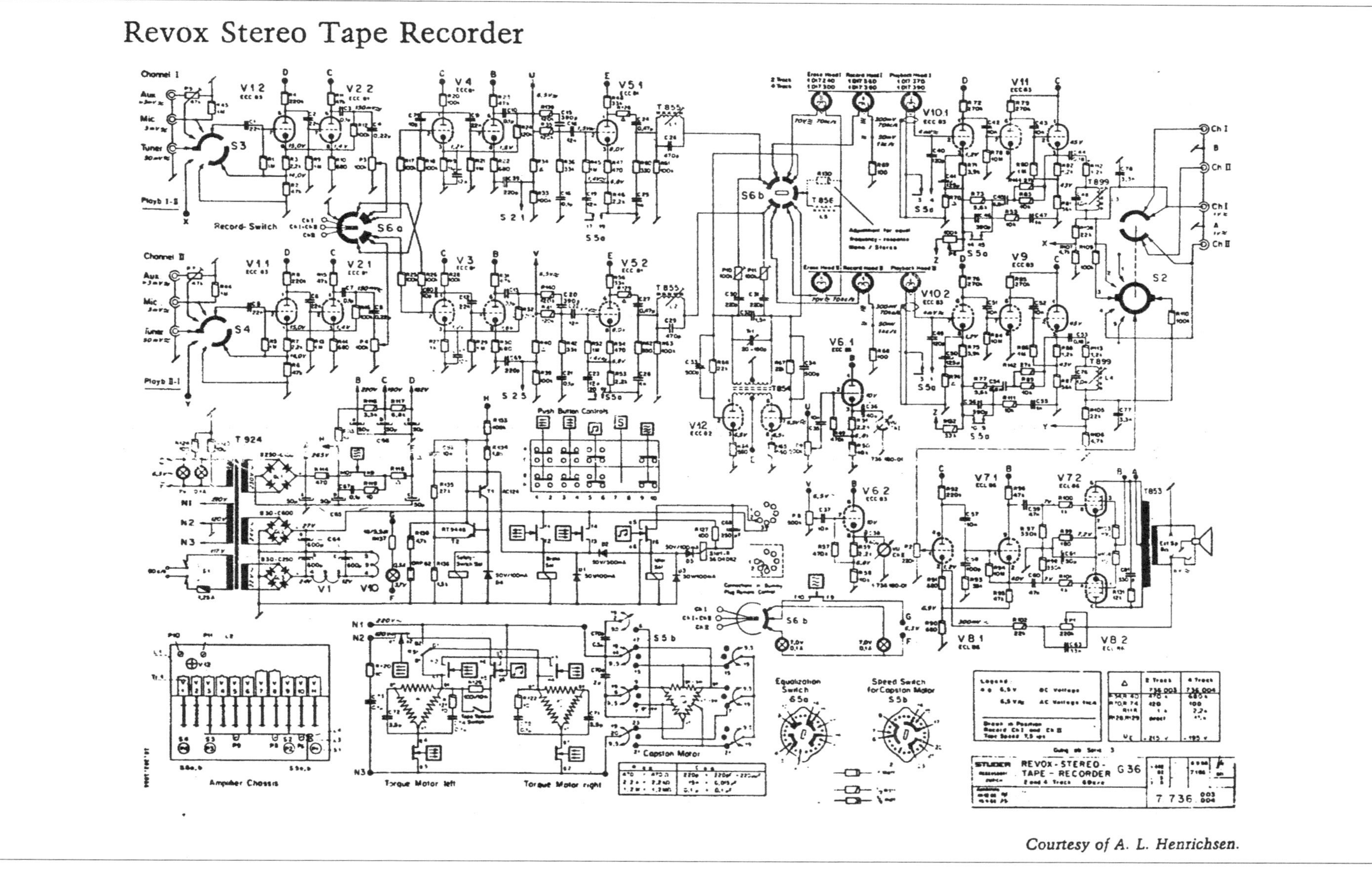

Courtesy of A. L. Henrichsen.

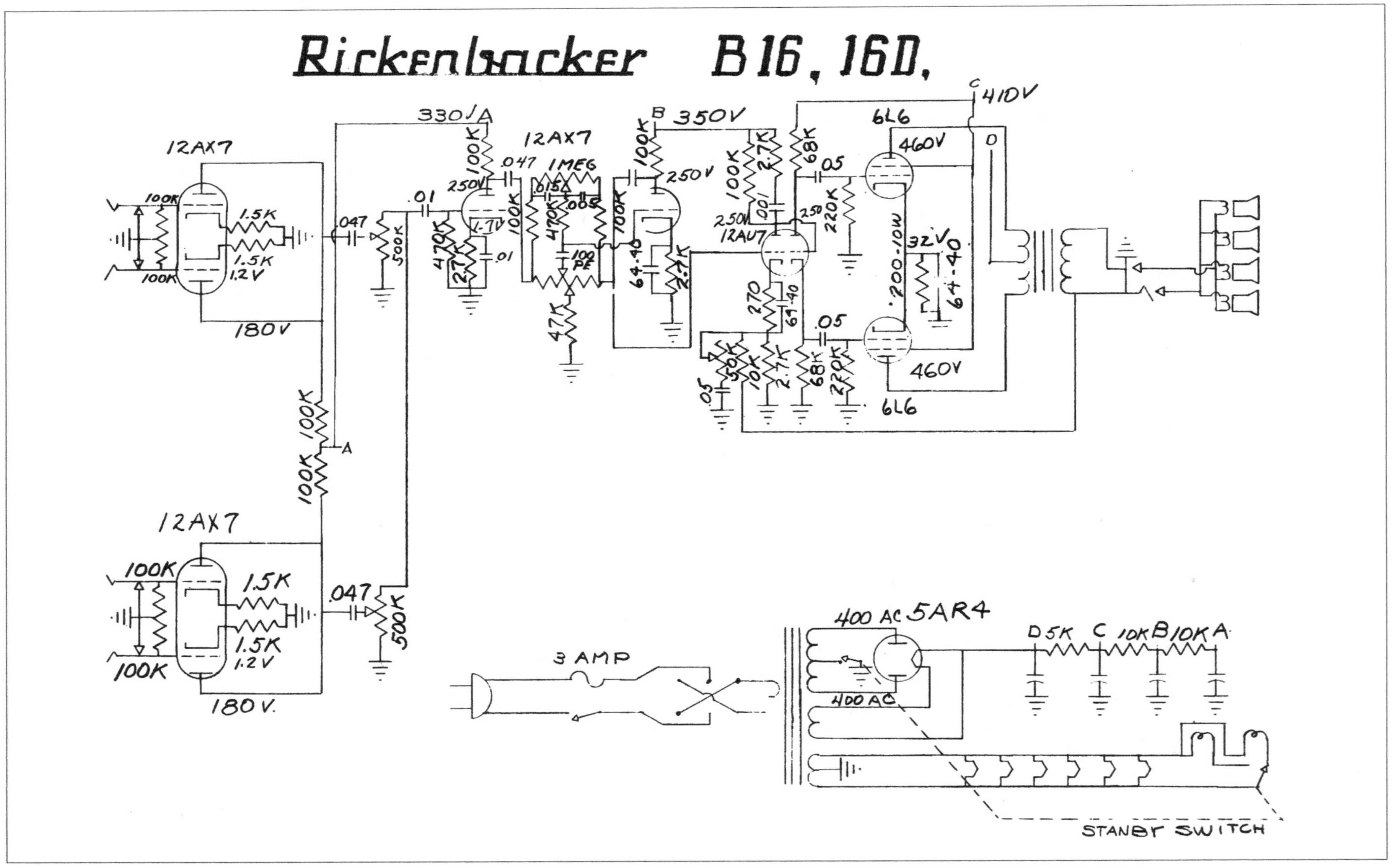
Rickenbacker B16, 16D.
12AX7
100K
1.5K
1.5K
1.2V
100K
180V
.047
500K
100K
100K
A
12AX7
100K
1.5K
1.5K
1.2V
100K
180V.
.047
500K
330/A
.01
100K
250V
.047
470K
2.7K
1.7V
.01
12AX7
1MEG
.015
470K
.005
100K
470K
100K
100 PF
47K
B 350V
100K
250V
64.40
2.7K
100K
2.7K
.001
250V
12AU7
250
68K
.05
220K
270
64.40
.05
50K
10K
2.7K
68K
220K
.05
6L6
460V
200-10W
32V
64-40
460V
6L6
C 410V
D
3 AMP
400 AC 5AR4
400 AC
D 5K C 10K B 10K A
STANBY SWITCH

B SERIES GUITAR

Rickenbacker

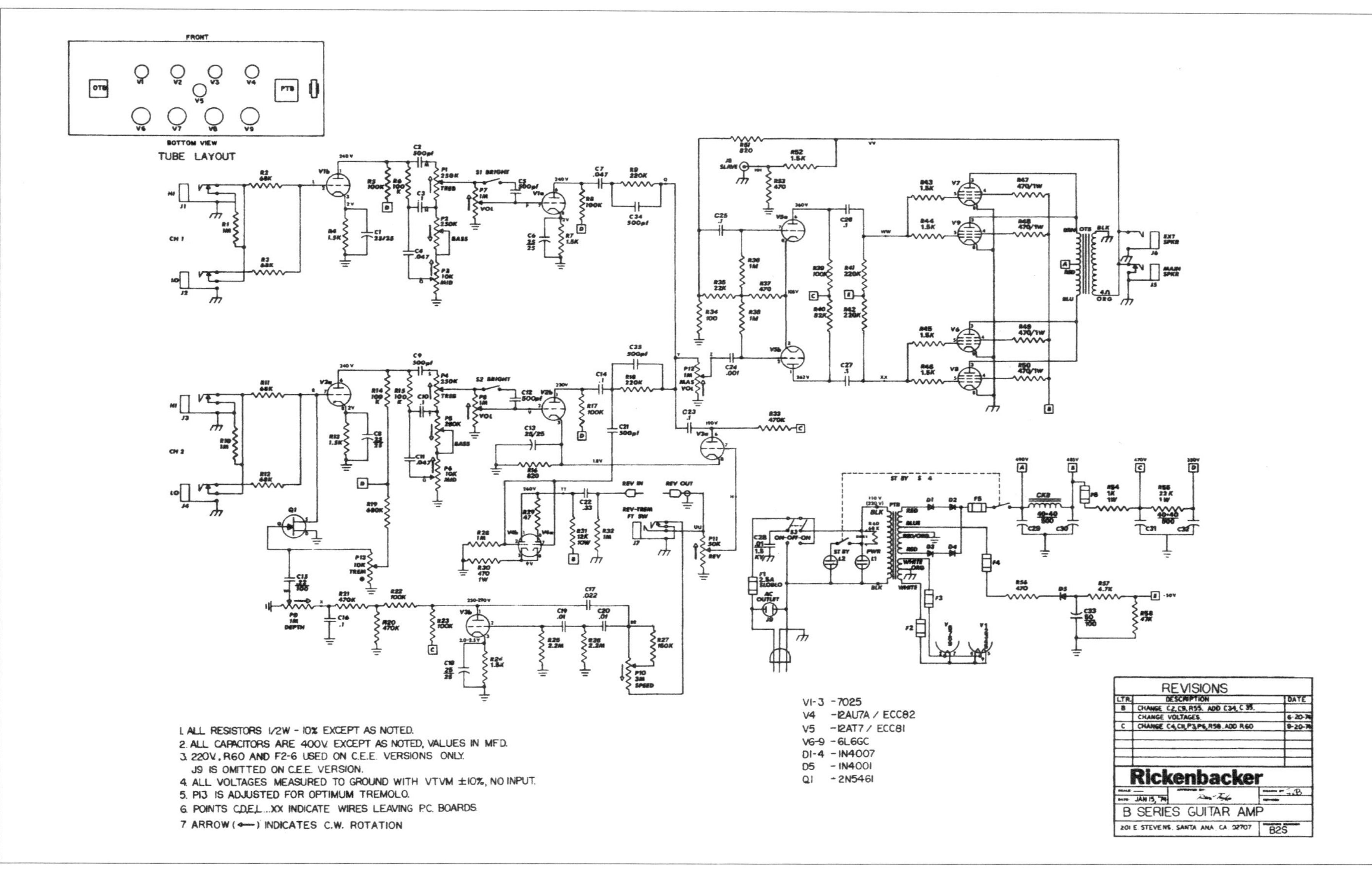

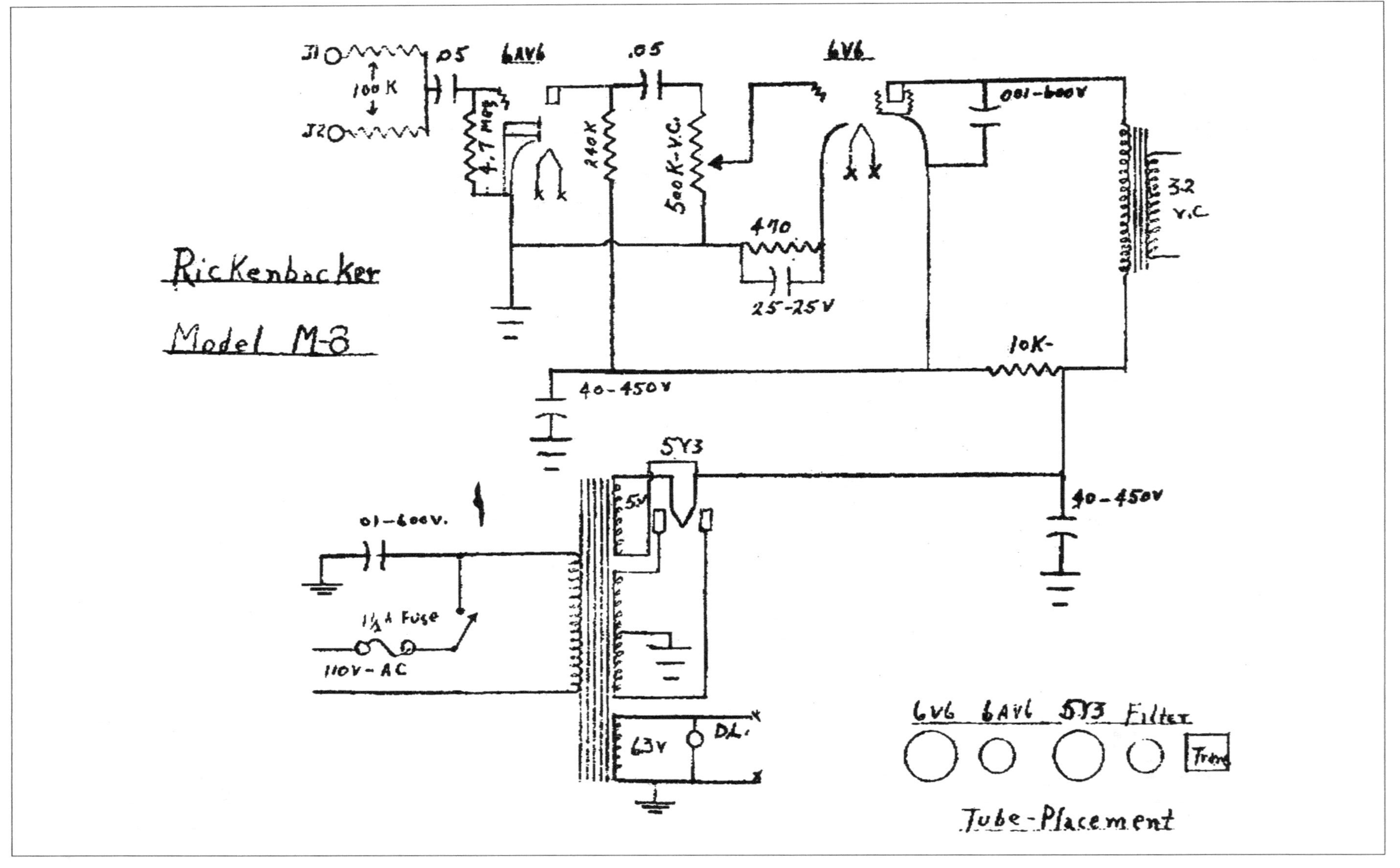
J1
J2
100 K
.05
6AV6
4.7 Meg
240K
.05
500K-V.C.
470
25-25V
6V6
.001-600V
3.2
V.C
10K-
40-450V
5Y3
5V
40-450V
.01-600V.
1½A Fuse
110V-AC
6.3V
DL.
Rickenbacker
Model M-8
6V6 6AV6 5Y3 Filter
Trans
Tube-Placement

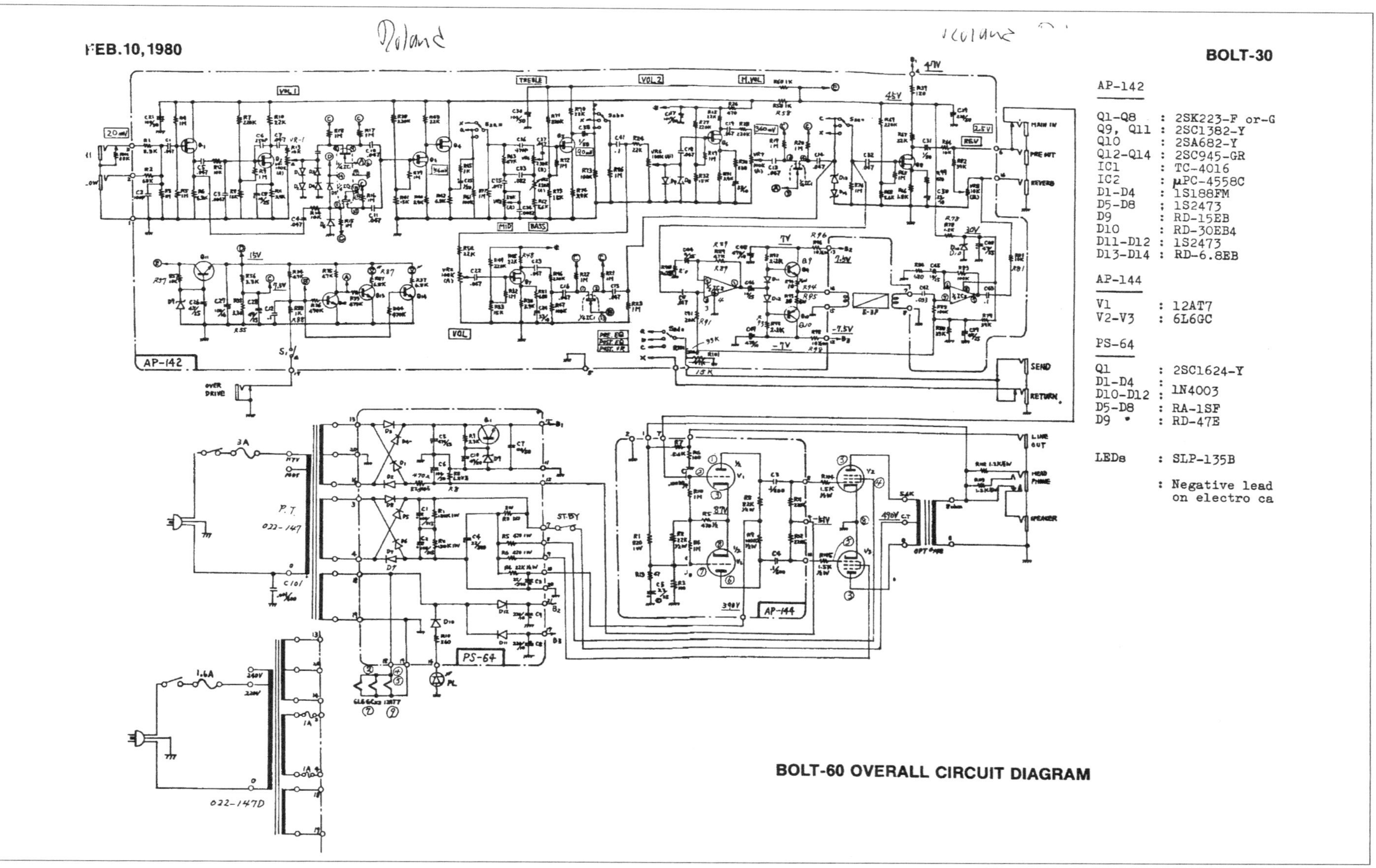

BOLT-60 OVERALL CIRCUIT DIAGRAM

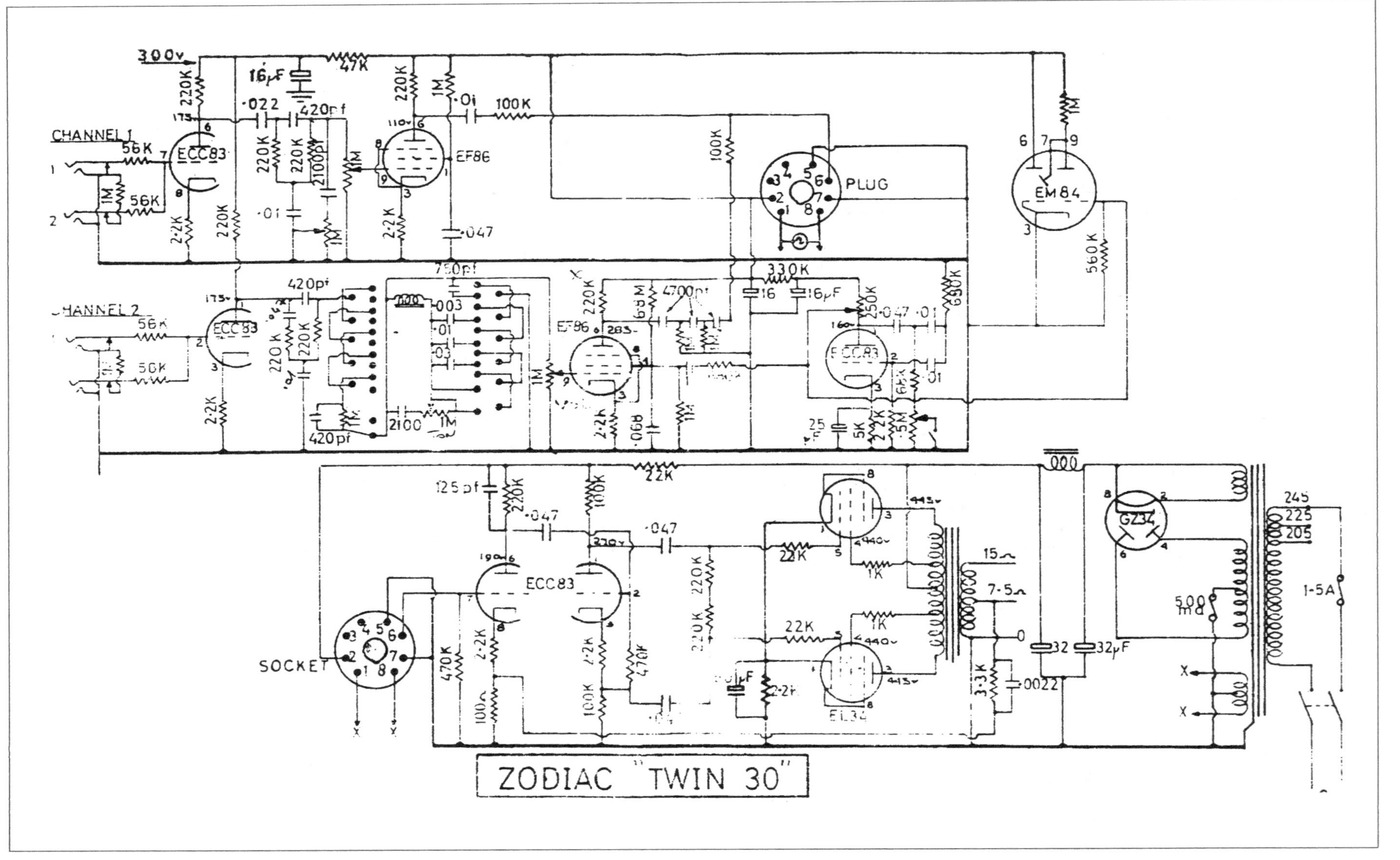
300v
CHANNEL 1
CHANNEL 2
ECC83
EF86
EM84
PLUG
SOCKET
EL34
GZ34
ZODIAC "TWIN 30"
245
225
205
1-5A
500 mA
15Ω
7·5Ω
0
32
32μF
·0022
3·3K
22K
220K
100K
470K
2·2K
56K
1M
47K
16μF
·022
420pf
2100pf
·01
·047
750pf
330K
16
250K
690K
560K
4700pf
6·8M
125pf

ZODIAC **TWIN 50**

Selmer

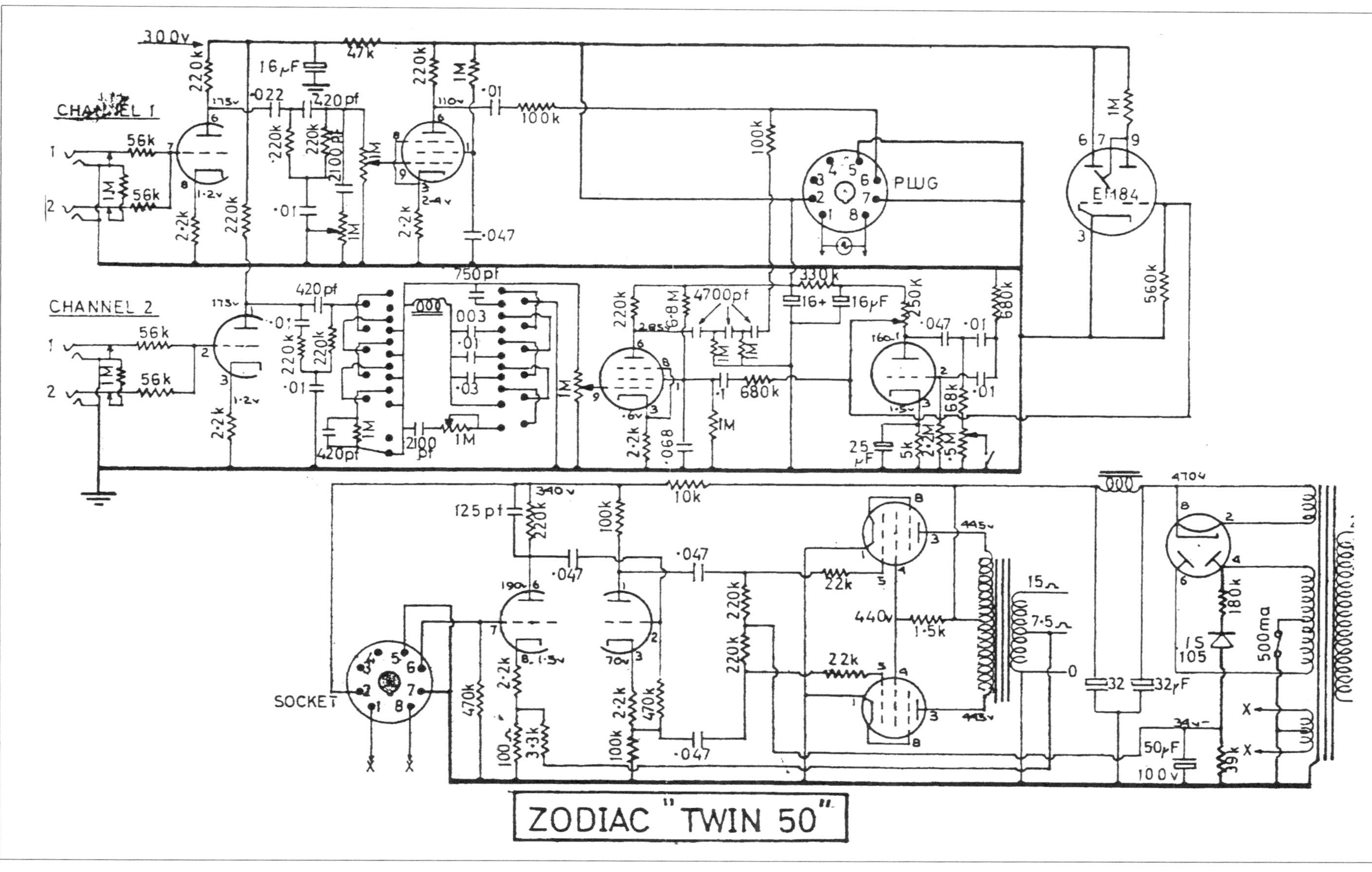

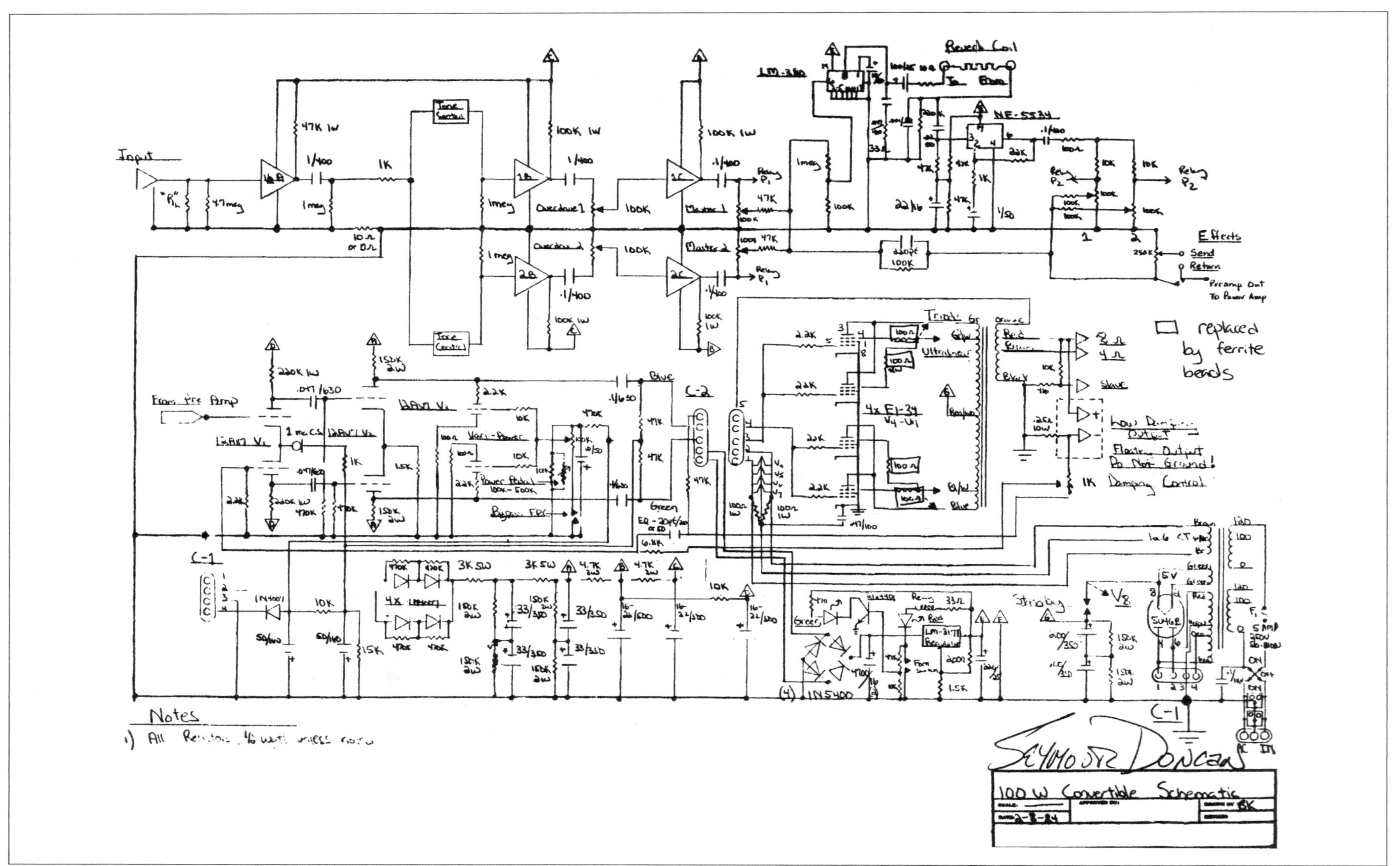
Input
47K 1W
1/400
1K
1meg
Tone Control
100K 1W
Overdrive 1
Overdrive 2
100K
Master 1
Master 2
Reverb Coil
LM-380
NE-5534
Effects
Send
Return
Preamp Out To Power Amp
replaced by ferrite beads
From Pre Amp
Blue
Green
C-2
C-1
Vari-Power
Power Pedal 100K-500K
Bypass FRC
Triode 6L
Ultralinear
Pentode
4x EL-34 V4-V1
8 Ω
4 Ω
Slave
Low Damping Output
Floating Output Do Not Ground!
1K Damping Control
Standby
5U4GB
LM-317T Regulator
1N5400
1N4001
5 Amp
Notes
1) All Resistors ½ watt unless noted
Seymour Duncan
100 W Convertible Schematic

2X6L6 TREMOLO AMP **(OLD)**

Silvertone

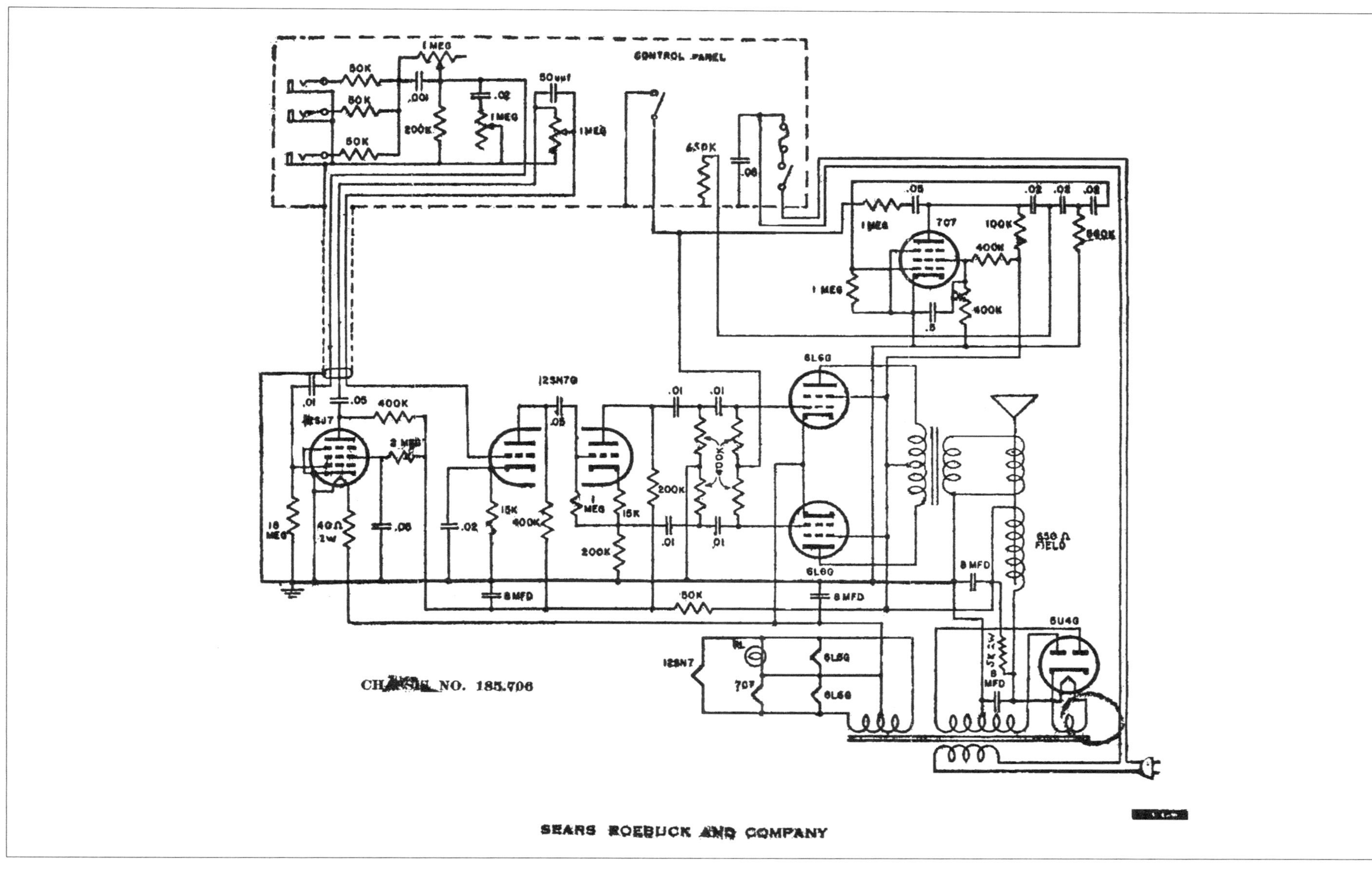

2X6V6 AMP

Silvertone

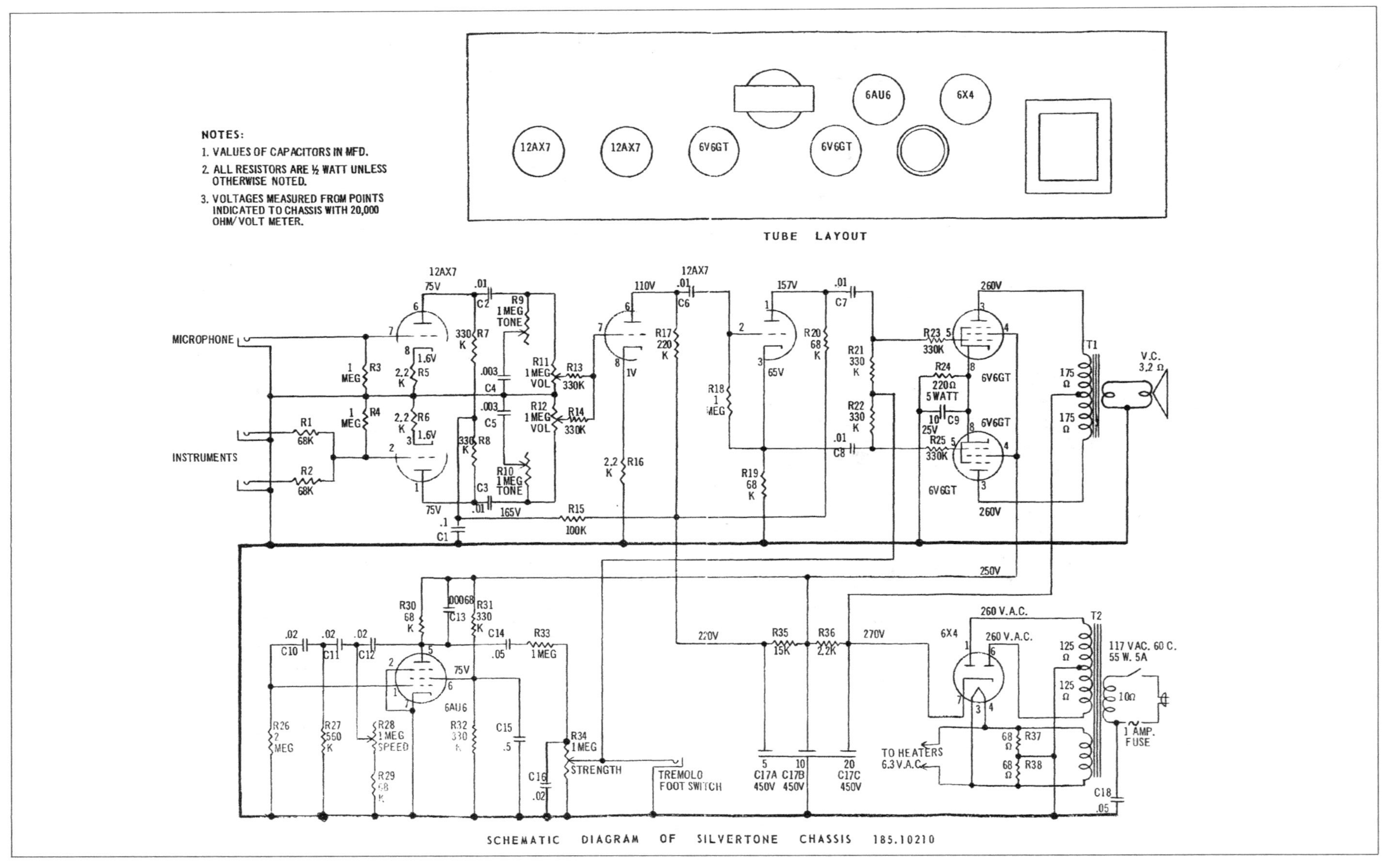

SCHEMATIC DIAGRAM OF SILVERTONE CHASSIS 185.10210

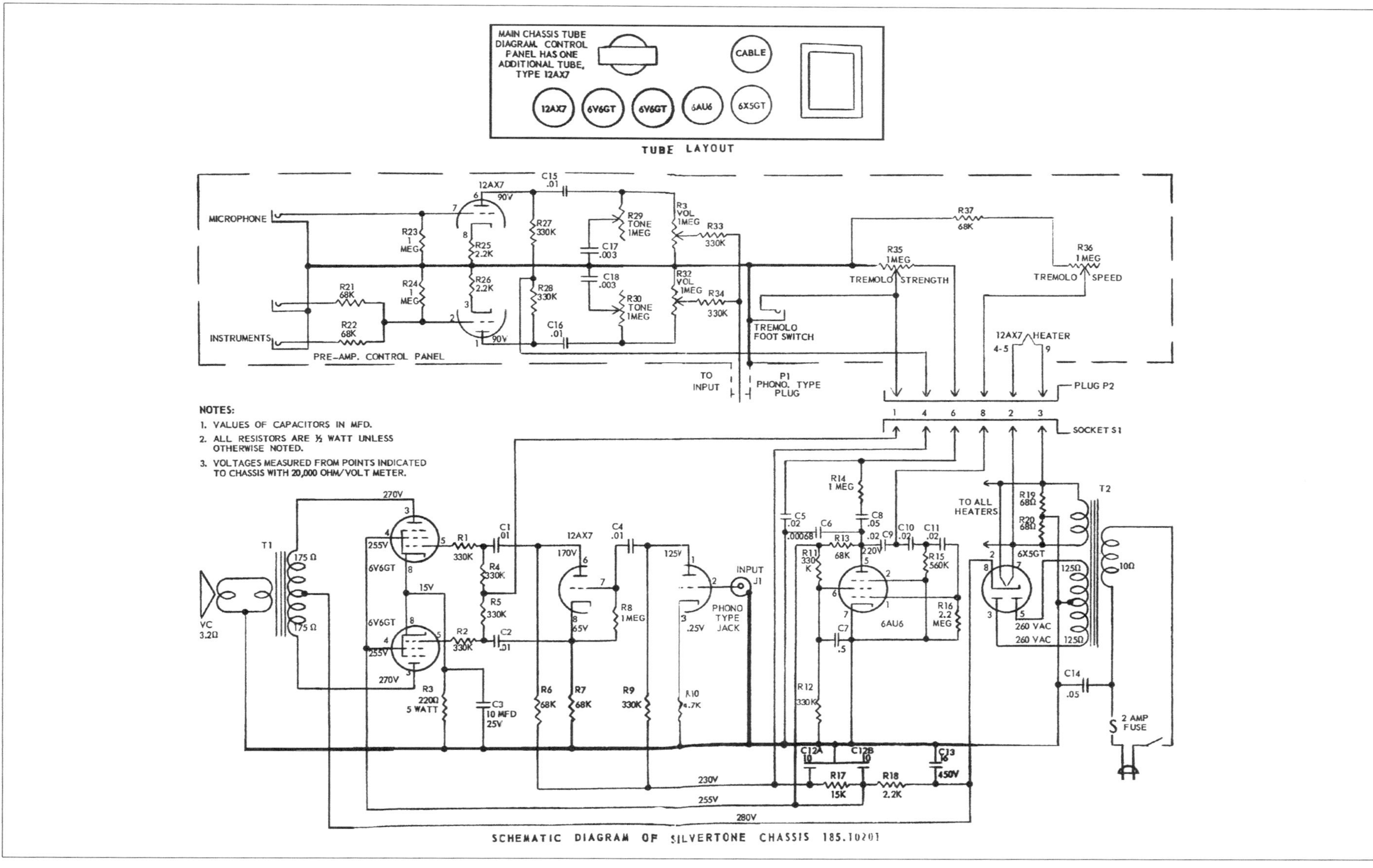
MAIN CHASSIS TUBE DIAGRAM. CONTROL PANEL HAS ONE ADDITIONAL TUBE, TYPE 12AX7
CABLE
12AX7
6V6GT
6V6GT
6AU6
6X5GT
TUBE LAYOUT
MICROPHONE
INSTRUMENTS
PRE-AMP. CONTROL PANEL
TREMOLO FOOT SWITCH
TO INPUT
P1 PHONO. TYPE PLUG
TREMOLO STRENGTH
TREMOLO SPEED
12AX7 HEATER
PLUG P2
SOCKET S1
NOTES:
1. VALUES OF CAPACITORS IN MFD.
2. ALL RESISTORS ARE ½ WATT UNLESS OTHERWISE NOTED.
3. VOLTAGES MEASURED FROM POINTS INDICATED TO CHASSIS WITH 20,000 OHM/VOLT METER.
INPUT J1
PHONO TYPE JACK
TO ALL HEATERS
2 AMP FUSE
SCHEMATIC DIAGRAM OF SILVERTONE CHASSIS 185.10201

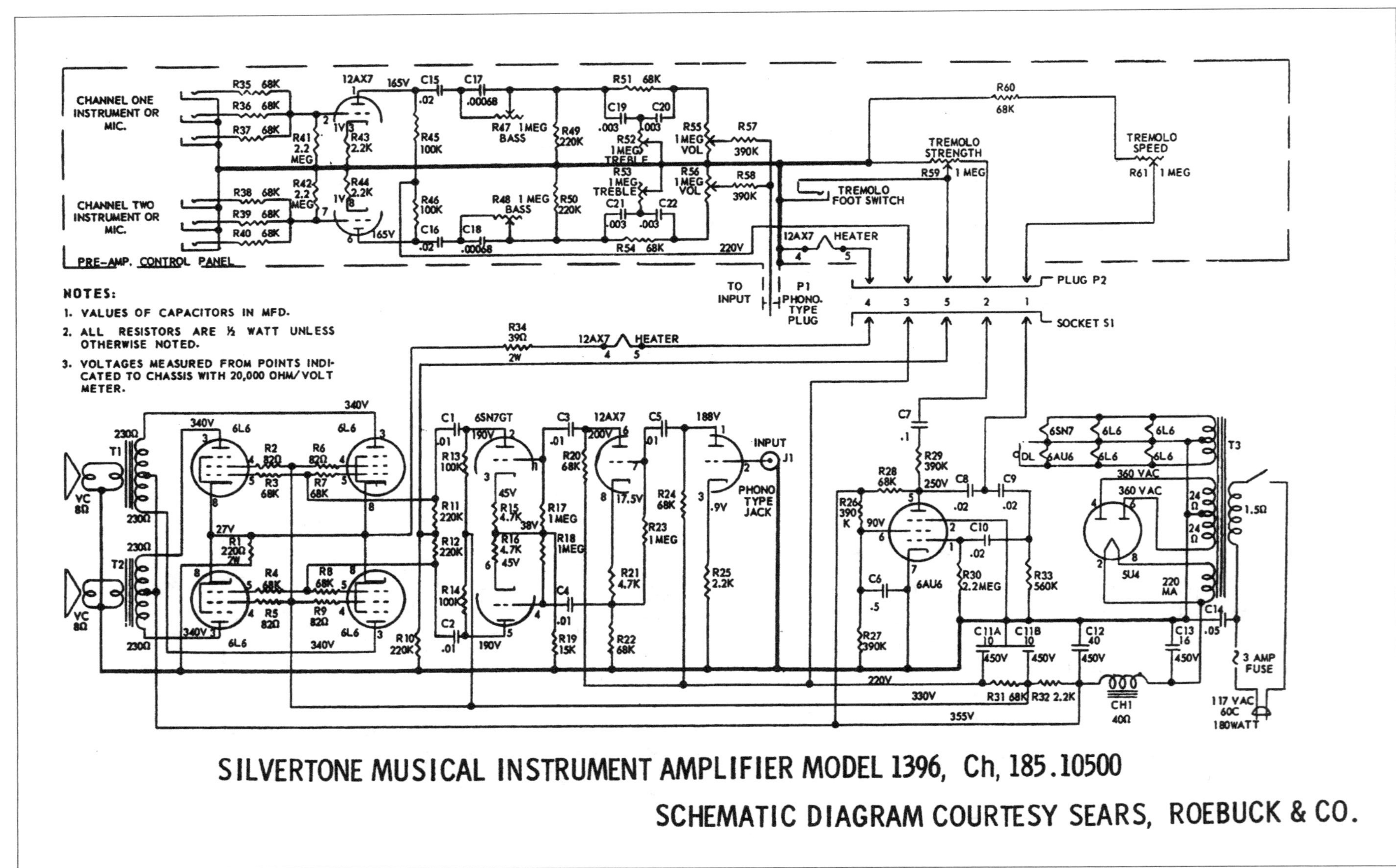
CHANNEL ONE INSTRUMENT OR MIC.
CHANNEL TWO INSTRUMENT OR MIC.
PRE-AMP. CONTROL PANEL
TREMOLO STRENGTH
TREMOLO SPEED
TREMOLO FOOT SWITCH
PLUG P2
SOCKET S1
TO INPUT
P1 PHONO. TYPE PLUG
INPUT J1
PHONO TYPE JACK
12AX7 HEATER
3 AMP FUSE
117 VAC 60C 180WATT
NOTES:
1. VALUES OF CAPACITORS IN MFD.
2. ALL RESISTORS ARE ½ WATT UNLESS OTHERWISE NOTED.
3. VOLTAGES MEASURED FROM POINTS INDICATED TO CHASSIS WITH 20,000 OHM/VOLT METER.
SILVERTONE MUSICAL INSTRUMENT AMPLIFIER MODEL 1396, Ch, 185.10500
SCHEMATIC DIAGRAM COURTESY SEARS, ROEBUCK & CO.

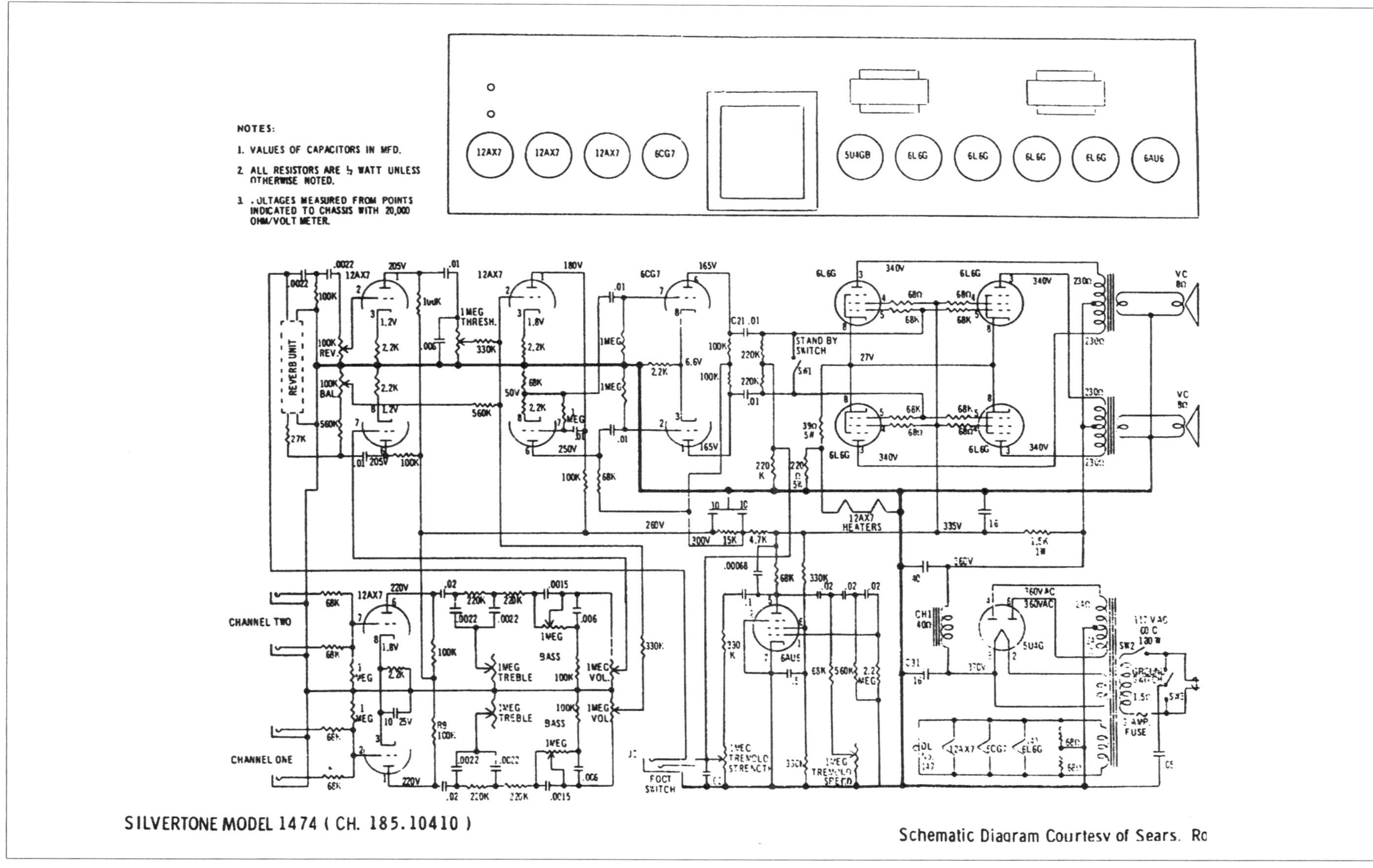
NOTES:
1. VALUES OF CAPACITORS IN MFD.
2. ALL RESISTORS ARE ½ WATT UNLESS OTHERWISE NOTED.
3. .ULTAGES MEASURED FROM POINTS INDICATED TO CHASSIS WITH 20,000 OHM/VOLT METER.
12AX7
12AX7
12AX7
6CG7
5U4GB
6L6G
6L6G
6L6G
6L6G
6AU6
REVERB UNIT
CHANNEL TWO
CHANNEL ONE
FOOT SWITCH
STAND BY SWITCH
12AX7 HEATERS
SILVERTONE MODEL 1474 (CH. 185.10410)
Schematic Diagram Courtesy of Sears, Ro

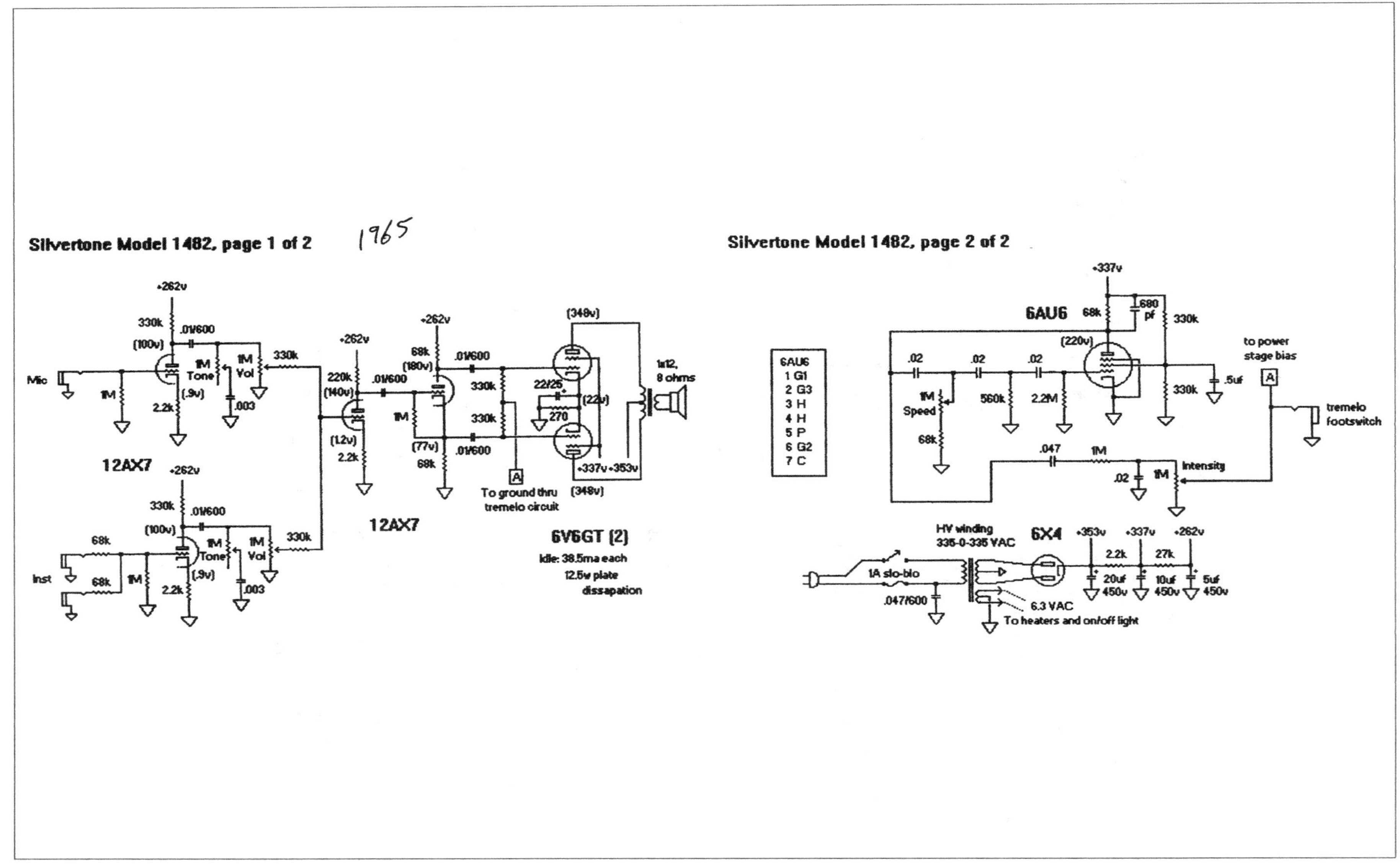
Silvertone Model 1482, page 1 of 2
1965
+262v
330k
.01/600
(100v)
Mic
1M
1M Tone
1M Vol
330k
(.9v)
2.2k
.003
12AX7
+262v
220k
(140v)
.01/600
(1.2v)
2.2k
+262v
68k
(180v)
.01/600
1M
330k
330k
(77v)
.01/600
68k
To ground thru tremelo circuit
(348v)
22/25
(22v)
270
+337v +353v
(348v)
1x12, 8 ohms
6V6GT (2)
Idle: 38.5ma each
12.5w plate dissapation
12AX7
Inst
68k
68k
Silvertone Model 1482, page 2 of 2
6AU6
1 G1
2 G3
3 H
4 H
5 P
6 G2
7 C
+337v
6AU6
68k
680 pf
330k
(220v)
.02
.02
.02
1M Speed
560k
2.2M
68k
330k
.5uf
to power stage bias
A
tremelo footswitch
.047
1M
.02
1M
Intensity
HV winding 335-0-335 VAC
6X4
+353v
+337v
+262v
2.2k
27k
1A slo-blo
20uf 450v
10uf 450v
5uf 450v
.047/600
6.3 VAC
To heaters and on/off light

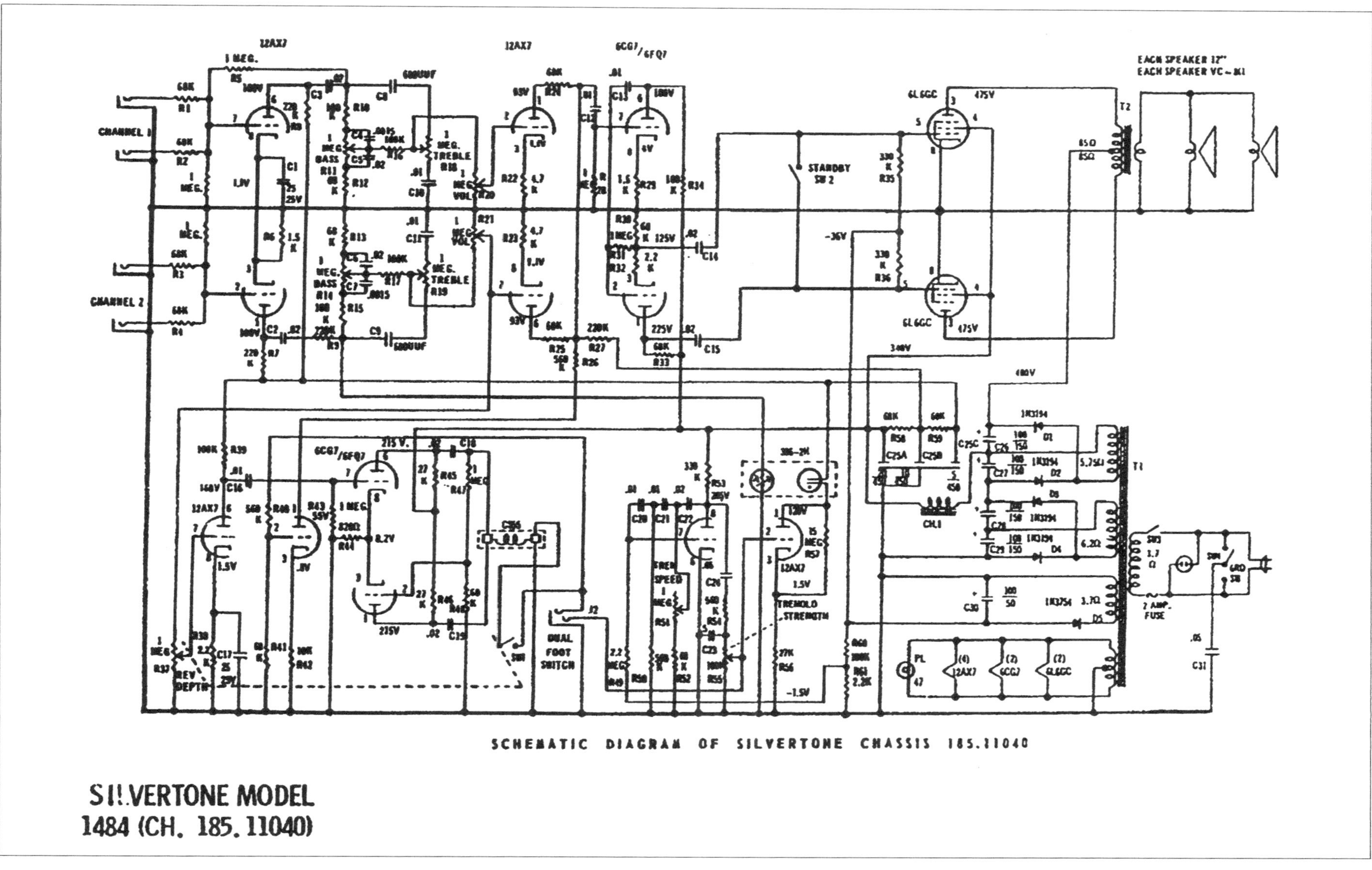

SI!.VERTONE MODEL
1484 (CH. 185. 11040)

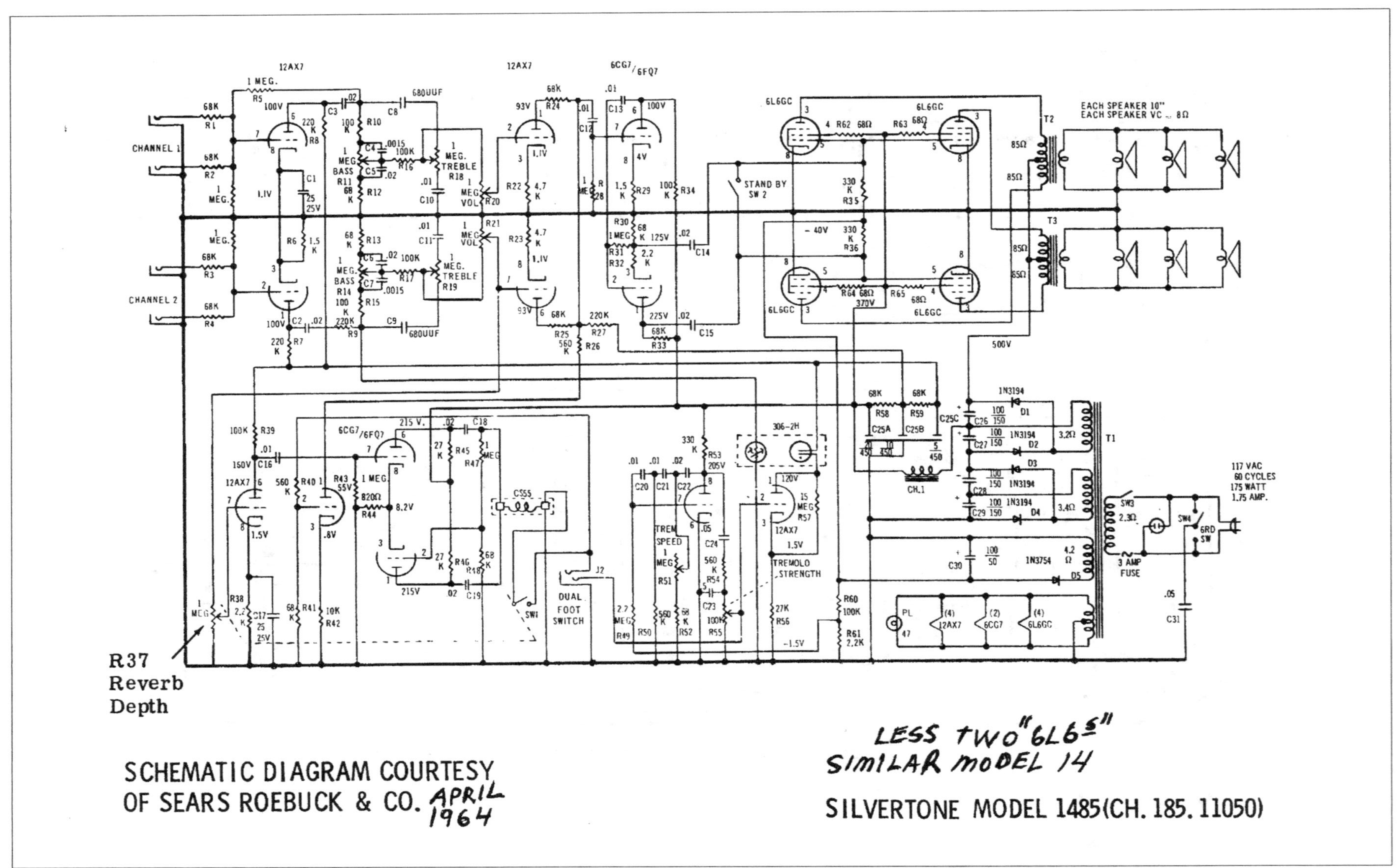

SCHEMATIC DIAGRAM COURTESY OF SEARS ROEBUCK & CO. APRIL 1964

SILVERTONE MODEL 1485(CH. 185. 11050)

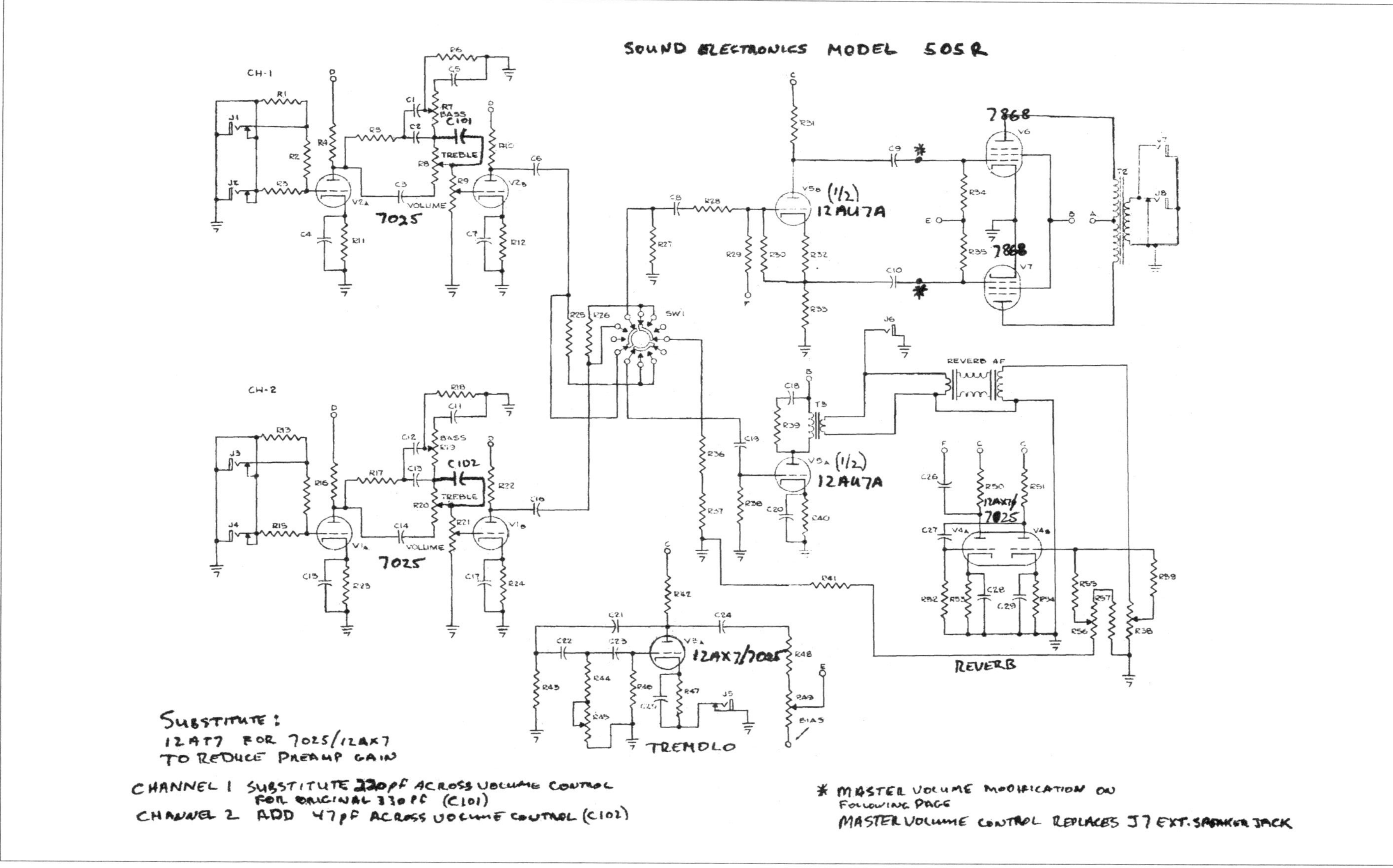
SOUND ELECTRONICS MODEL 505R
CH-1
CH-2
7025
7025
7868
7868
(1/2) 12AU7A
(1/2) 12AU7A
12AX7/7025
12AX7/7025
TREMOLO
REVERB
REVERB AF
BIAS
SUBSTITUTE:
12AT7 FOR 7025/12AX7
TO REDUCE PREAMP GAIN
CHANNEL 1 SUBSTITUTE 220pF ACROSS VOLUME CONTROL FOR ORIGINAL 330pF (C101)
CHANNEL 2 ADD 47pF ACROSS VOLUME CONTROL (C102)
* MASTER VOLUME MODIFICATION ON FOLLOWING PAGE
MASTER VOLUME CONTROL REPLACES J7 EXT. SPEAKER JACK

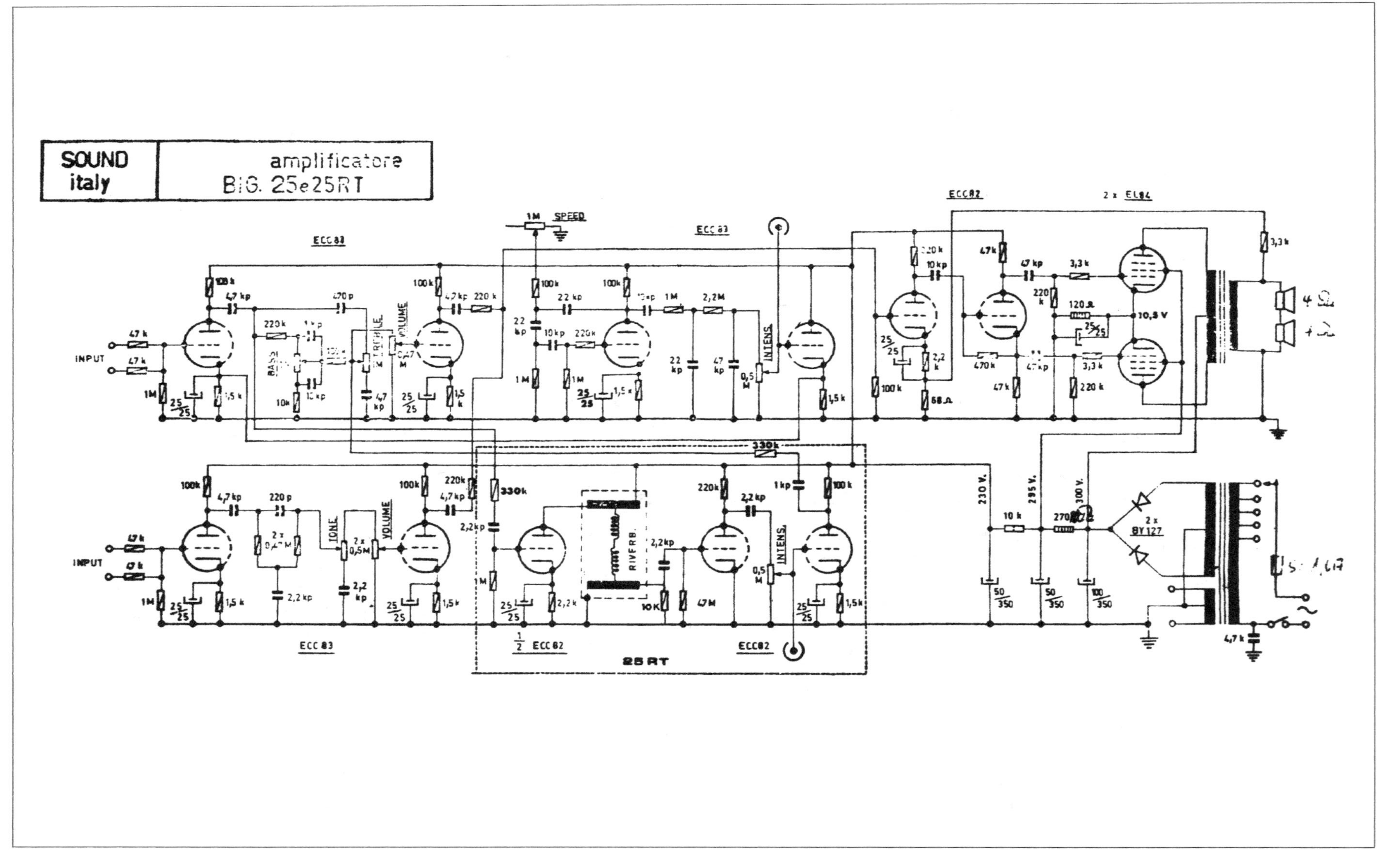
SOUND
italy
amplificatore
BIG. 25e25RT
INPUT
ECC 83
ECC 82
2 x EL 84
SPEED
VOLUME
TREBLE
BASS
TONE
INTENS.
RIVERB.
25 RT
1/2 ECC 82
230 V.
295 V.
300 V.
10,5 V
2 x
BY 127
4 Ω

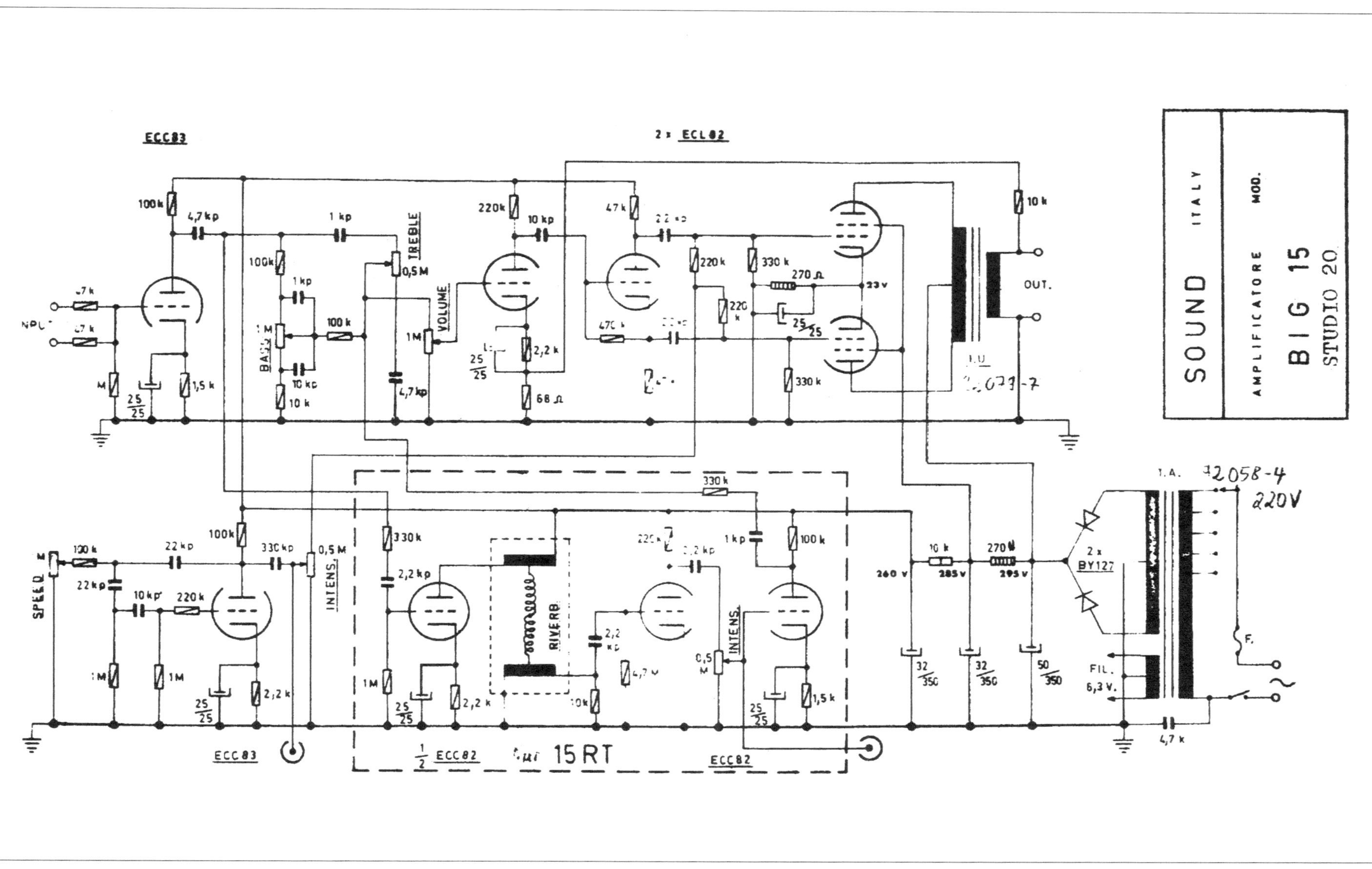

ECC83
2 x ECL82
TREBLE
VOLUME
BASS
OUT.
SPEED
INTENS.
RIVERB
ECC83
1/2 ECC82
15 RT
ECC82
2 x BY127
FIL. 6,3 V.
T.A.
220V
SOUND ITALY
AMPLIFICATORE MOD.
BIG 15
STUDIO 20

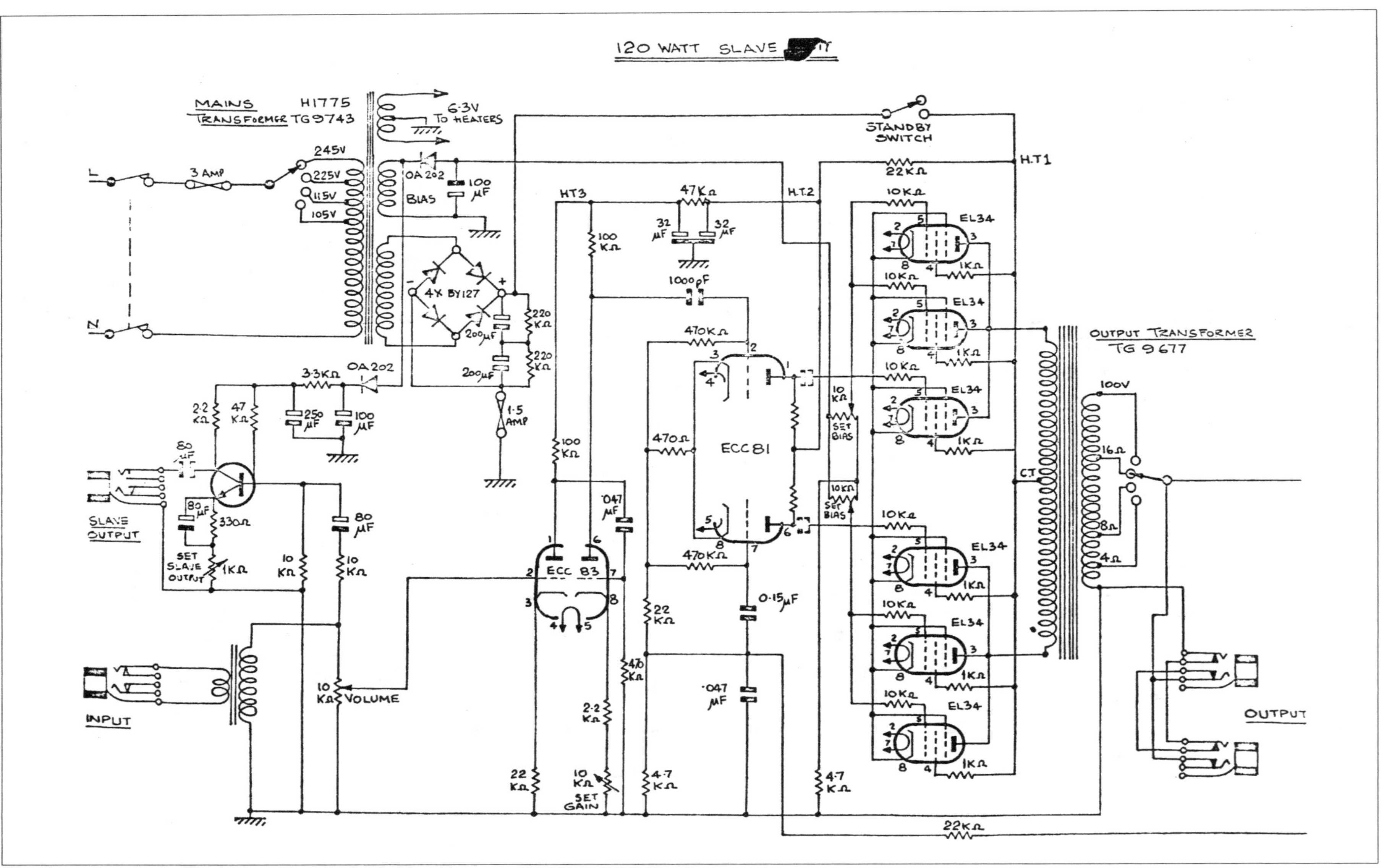
120 WATT SLAVE
MAINS TRANSFORMER H1775 TG9743
6·3V TO HEATERS
245V
225V
115V
105V
3 AMP
L
N
OA 202
BIAS
100 µF
4X BY127
200µF
220 KΩ
1·5 AMP
3·3KΩ
250 µF
2·2 KΩ
47 KΩ
80 µF
330Ω
SET SLAVE OUTPUT
1KΩ
SLAVE OUTPUT
INPUT
10 KΩ
VOLUME
HT3
H.T.2
H.T1
47KΩ
32 µF
1000pF
470KΩ
470Ω
ECC81
ECC 83
·047 µF
0·15µF
22 KΩ
4·7 KΩ
2·2 KΩ
SET GAIN
STANDBY SWITCH
EL34
10KΩ
SET BIAS
C.T
OUTPUT TRANSFORMER TG 9677
100V
16Ω
8Ω
4Ω
OUTPUT

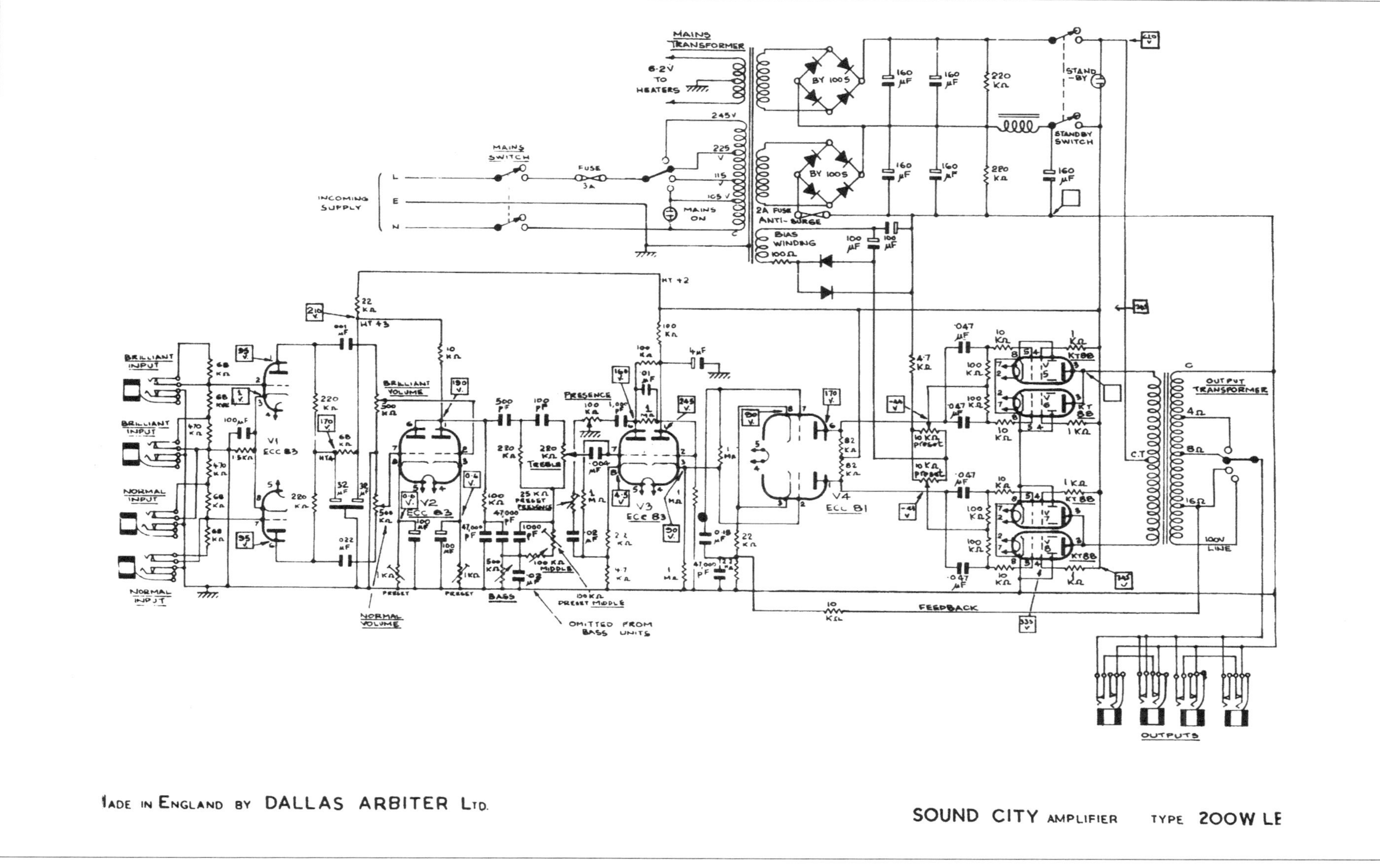

Made in England by DALLAS ARBITER Ltd.

SOUND CITY AMPLIFIER TYPE 200W LE

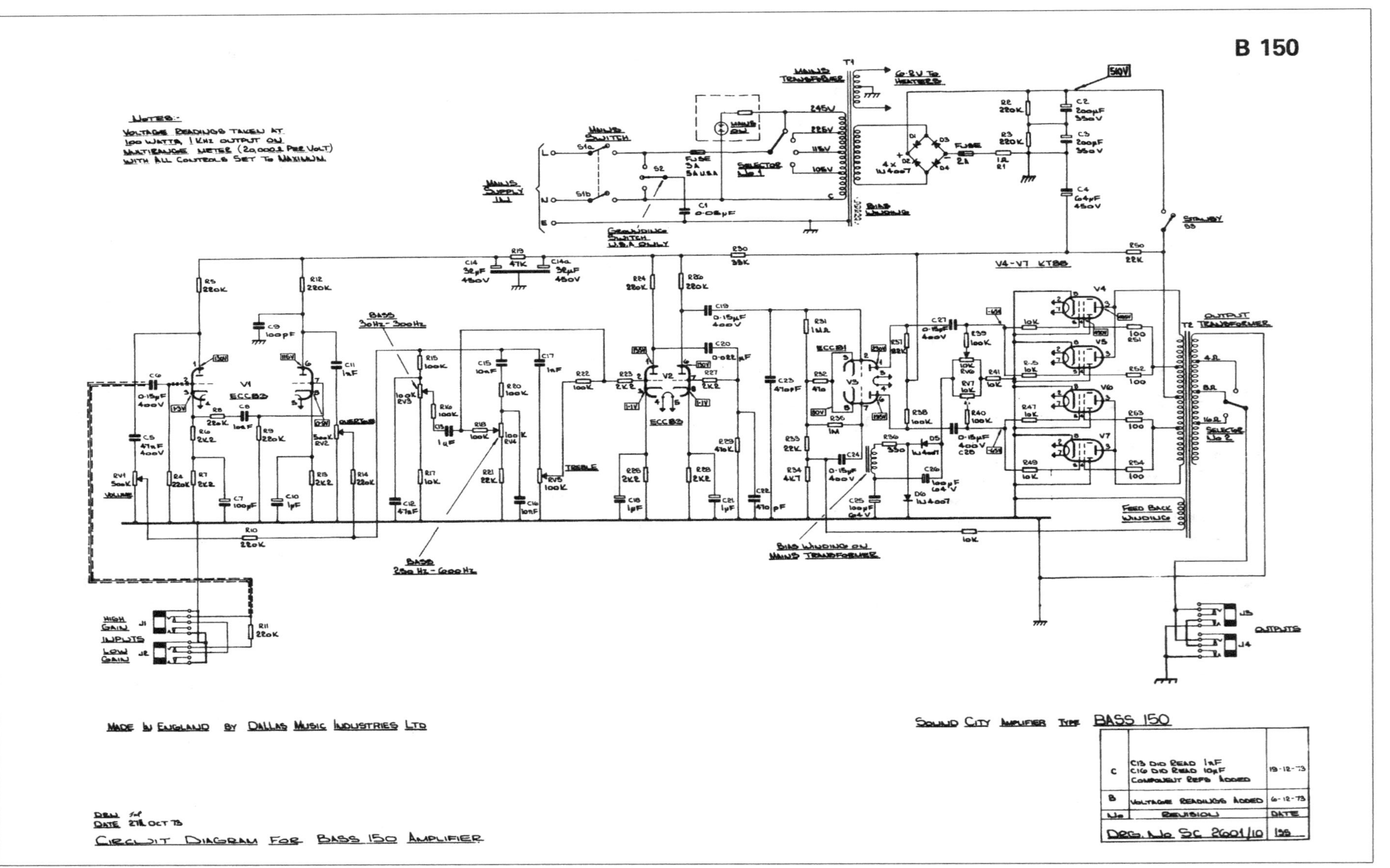
B 150
NOTES:-
VOLTAGE READINGS TAKEN AT
100 WATTS, 1 KHZ OUTPUT ON
MULTIRANGE METER (20,000Ω PER VOLT)
WITH ALL CONTROLS SET TO MAXIMUM.
MAINS SUPPLY
MAINS SWITCH
GROUNDING SWITCH U.S.A. ONLY
MAINS TRANSFORMER
SELECTOR No 1
BIAS WINDING
6·3V TO HEATERS
STANDBY S3
V4-V7 KT88
OUTPUT TRANSFORMER
SELECTOR No 2
FEED BACK WINDING
BIAS WINDING ON MAINS TRANSFORMER
OUTPUTS
HIGH GAIN
INPUTS
LOW GAIN
VOLUME
OVERDRIVE
BASS 30Hz - 300Hz
BASS 250 Hz - 6000 Hz
TREBLE
ECC83
ECC81
MADE IN ENGLAND BY DALLAS MUSIC INDUSTRIES LTD
SOUND CITY AMPLIFIER TYPE BASS 150
C | C13 DID READ 1nF C16 DID READ 10µF COMPONENT REFS ADDED | 19-12-73
B | VOLTAGE READINGS ADDED | 6-12-73
No | REVISION | DATE
DRG. No SC 2601/10 | ISS
DATE 27th OCT 73
CIRCUIT DIAGRAM FOR BASS 150 AMPLIFIER

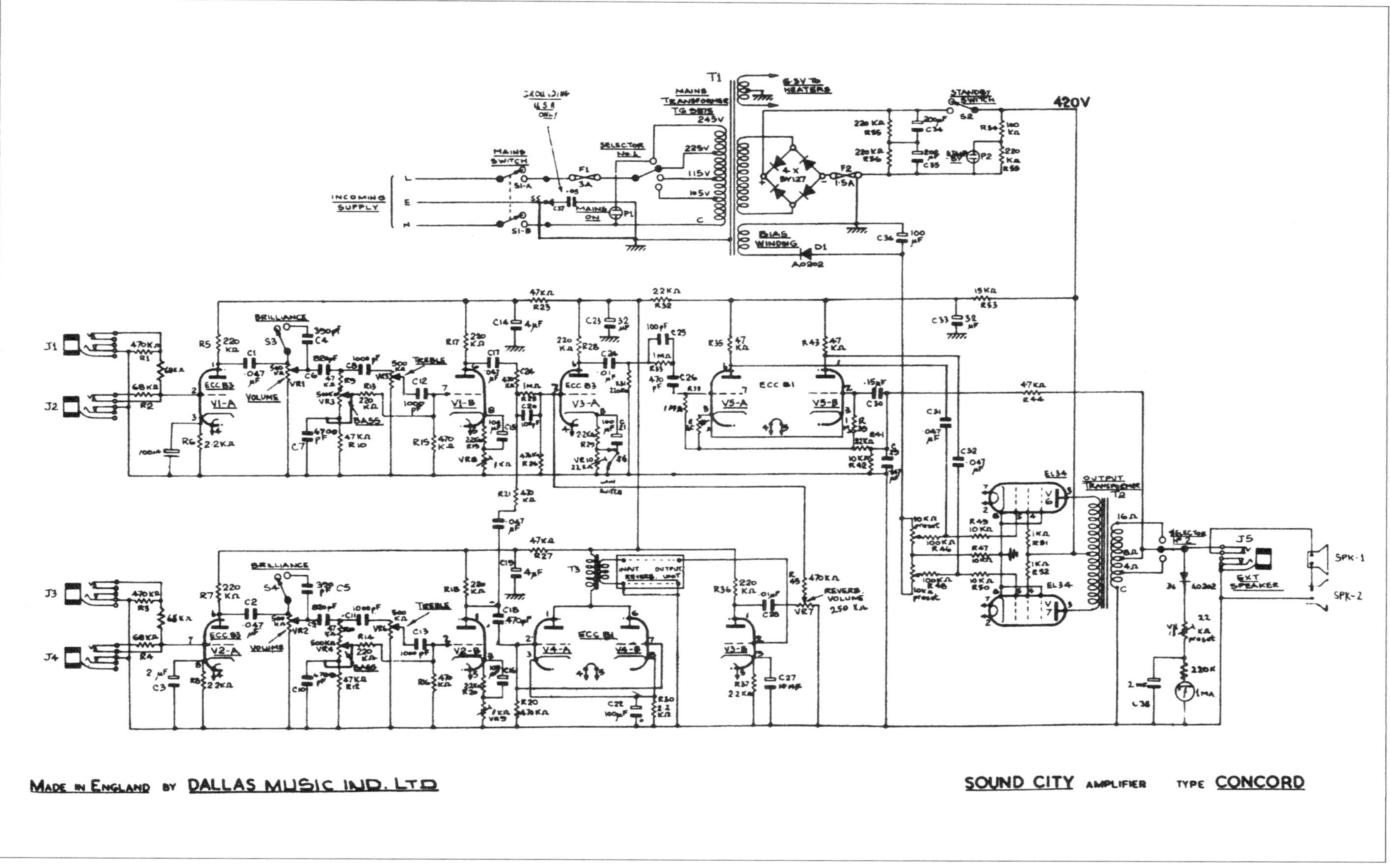
MADE IN ENGLAND BY DALLAS MUSIC IND. LTD
SOUND CITY AMPLIFIER TYPE CONCORD
INCOMING SUPPLY
MAINS SWITCH
SELECTOR No.1
MAINS TRANSFORMER
6.3V TO HEATERS
STANDBY SWITCH
420V
BIAS WINDING
OUTPUT TRANSFORMER
EXT SPEAKER
SPK-1
SPK-2
BRILLIANCE
VOLUME
BASS
TREBLE
REVERB. VOLUME
J1
J2
J3
J4
J5

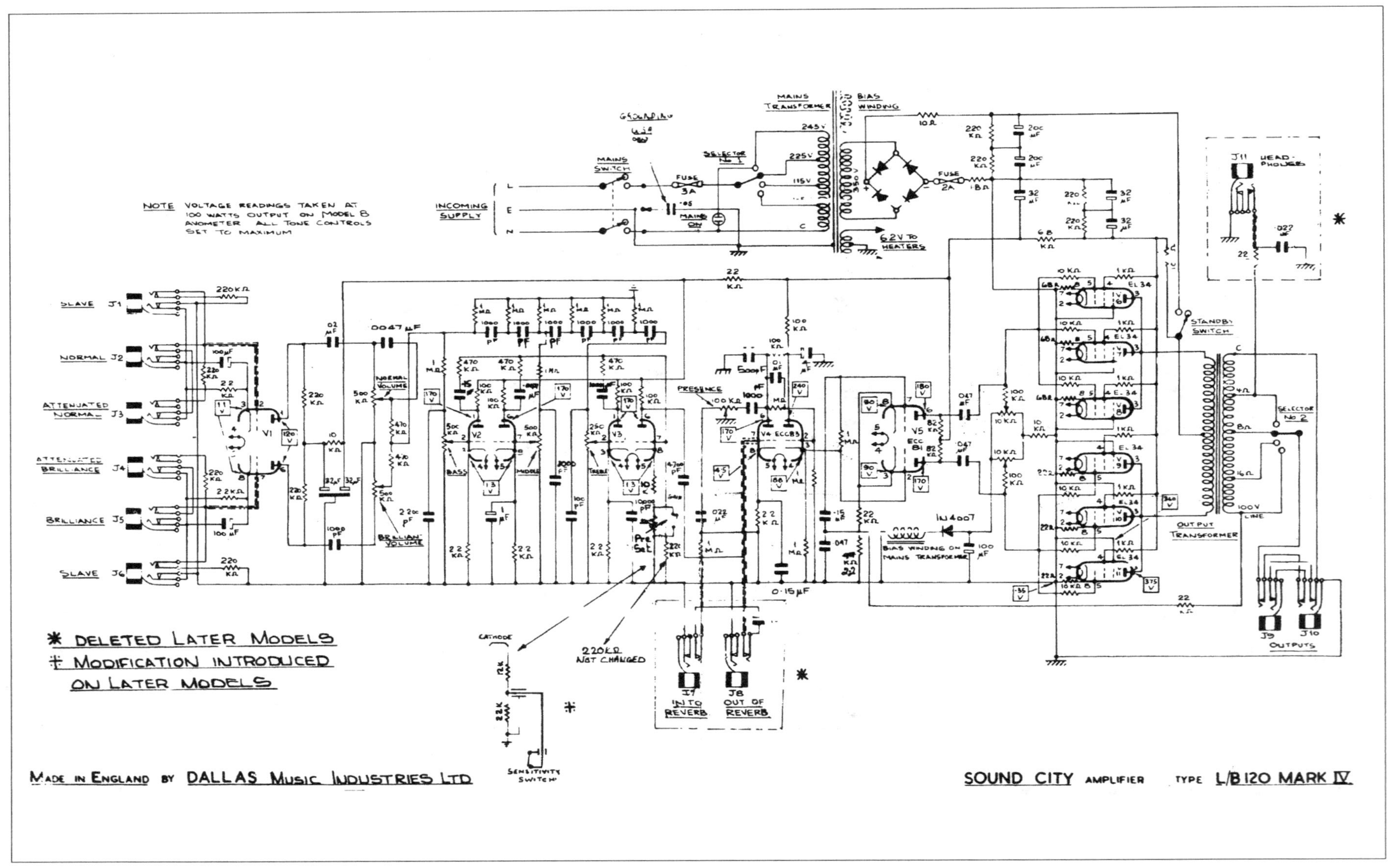
NOTE VOLTAGE READINGS TAKEN AT 100 WATTS OUTPUT ON MODEL 8 AVOMETER ALL TONE CONTROLS SET TO MAXIMUM
INCOMING SUPPLY
MAINS SWITCH
FUSE 3A
SELECTOR No 1
MAINS TRANSFORMER
BIAS WINDING
FUSE 2A
6.2V TO HEATERS
J11 HEAD-PHONES
SLAVE J1
NORMAL J2
ATTENUATED NORMAL J3
ATTENUATED BRILLIANCE J4
BRILLIANCE J5
SLAVE J6
NORMAL VOLUME
BRILLIANCE VOLUME
BASS
MIDDLE
TREBLE
PRESENCE
STANDBY SWITCH
SELECTOR No 2
OUTPUT TRANSFORMER
100V LINE
BIAS WINDING ON MAINS TRANSFORMER
IN4007
220KΩ NOT CHANGED
CATHODE
SENSITIVITY SWITCH
J7 INTO REVERB
J8 OUT OF REVERB
J9 J10 OUTPUTS
* DELETED LATER MODELS
‡ MODIFICATION INTRODUCED ON LATER MODELS
MADE IN ENGLAND BY DALLAS MUSIC INDUSTRIES LTD
SOUND CITY AMPLIFIER TYPE L/B120 MARK IV

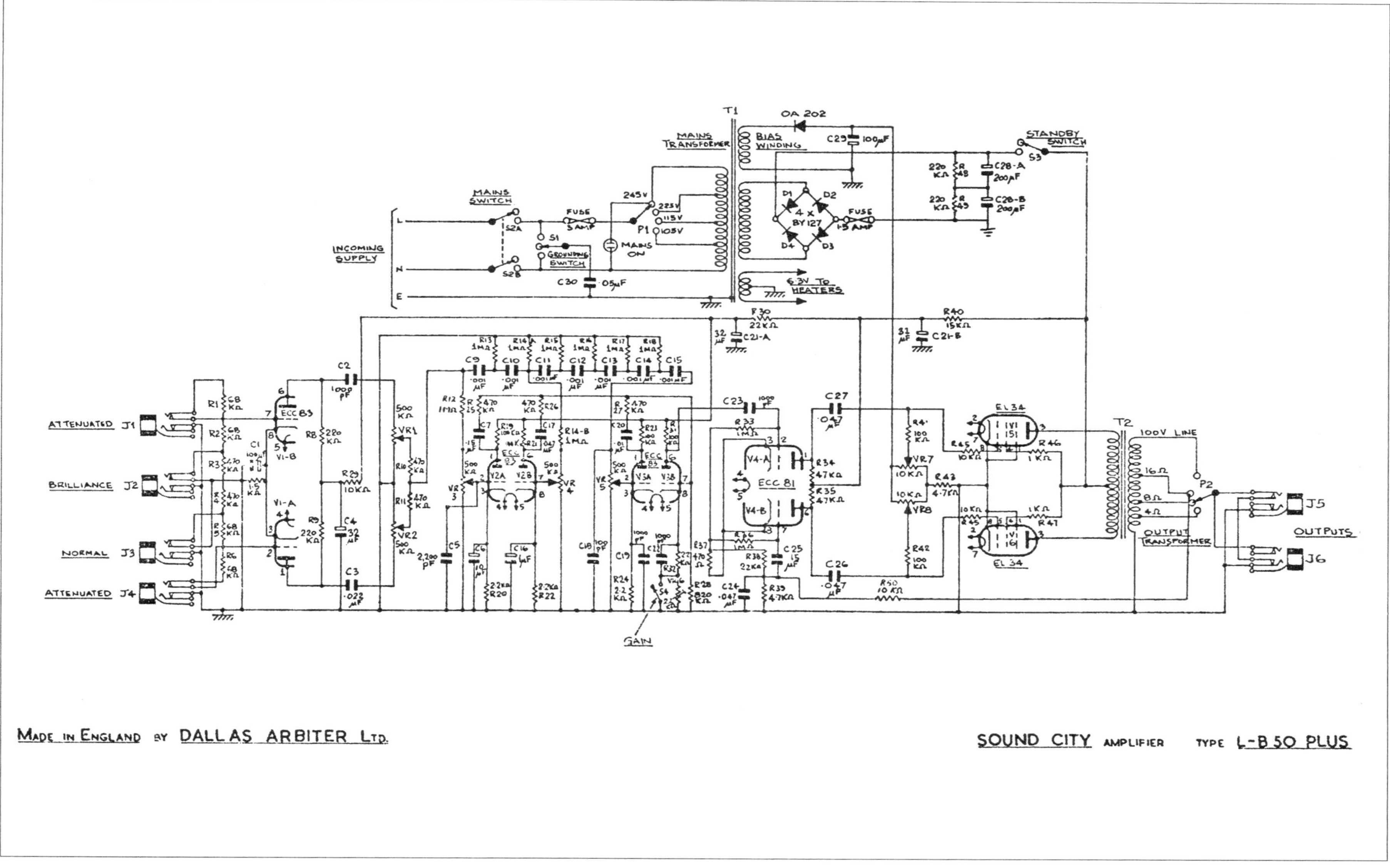
ATTENUATED J1
BRILLIANCE J2
NORMAL J3
ATTENUATED J4
INCOMING SUPPLY
MAINS SWITCH
FUSE 3 AMP
GROUNDING SWITCH
MAINS ON
MAINS TRANSFORMER
BIAS WINDING
OA 202
4 x BY 127
FUSE 1·5 AMP
STANDBY SWITCH
6·3V TO HEATERS
ECC 83
ECC 81
EL 34
GAIN
100V LINE
OUTPUT TRANSFORMER
OUTPUTS
J5
J6
MADE IN ENGLAND BY DALLAS ARBITER LTD.
SOUND CITY AMPLIFIER TYPE L-B 50 PLUS

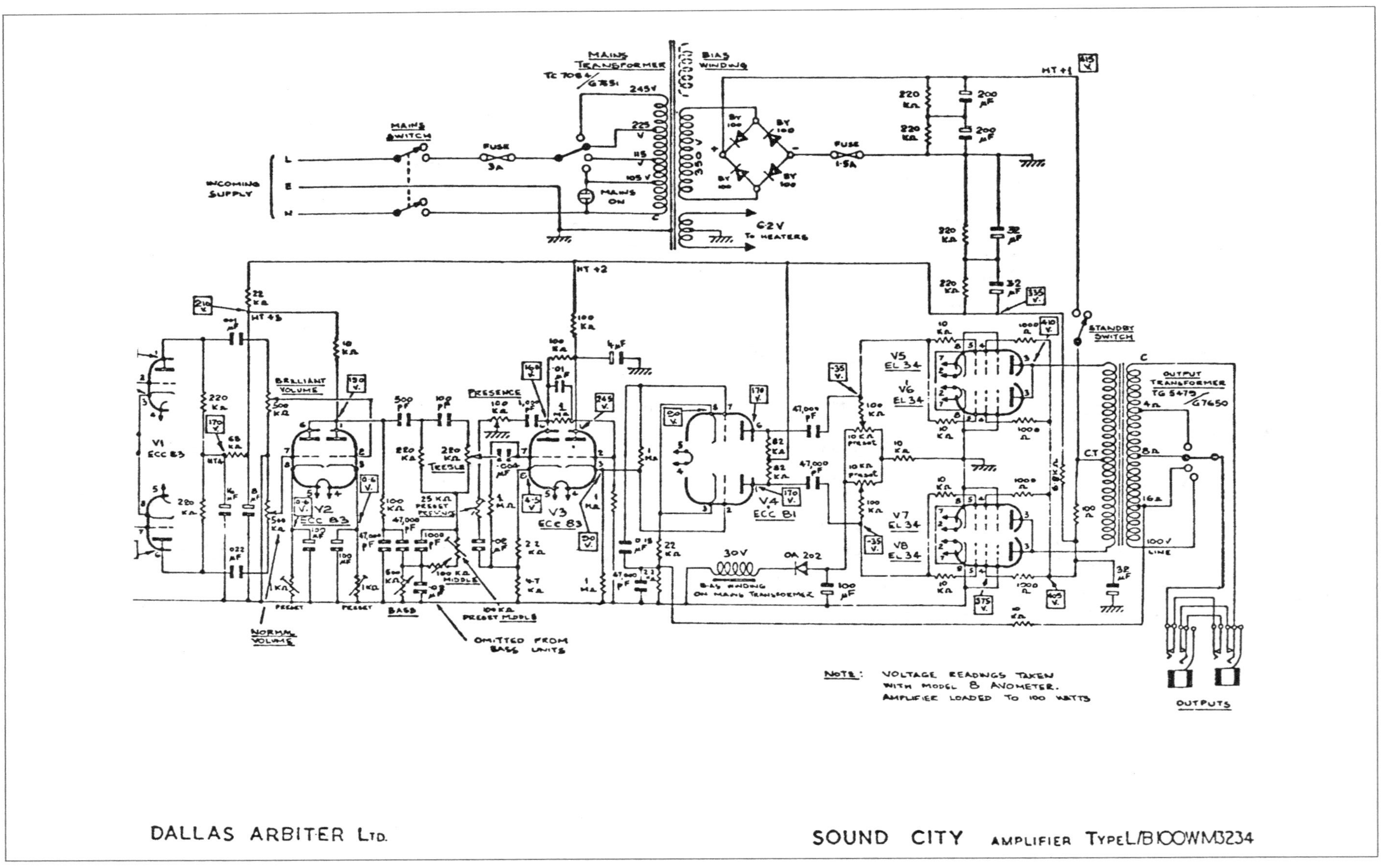

DALLAS ARBITER LTD.

SOUND CITY AMPLIFIER TYPE L/B100W M3234

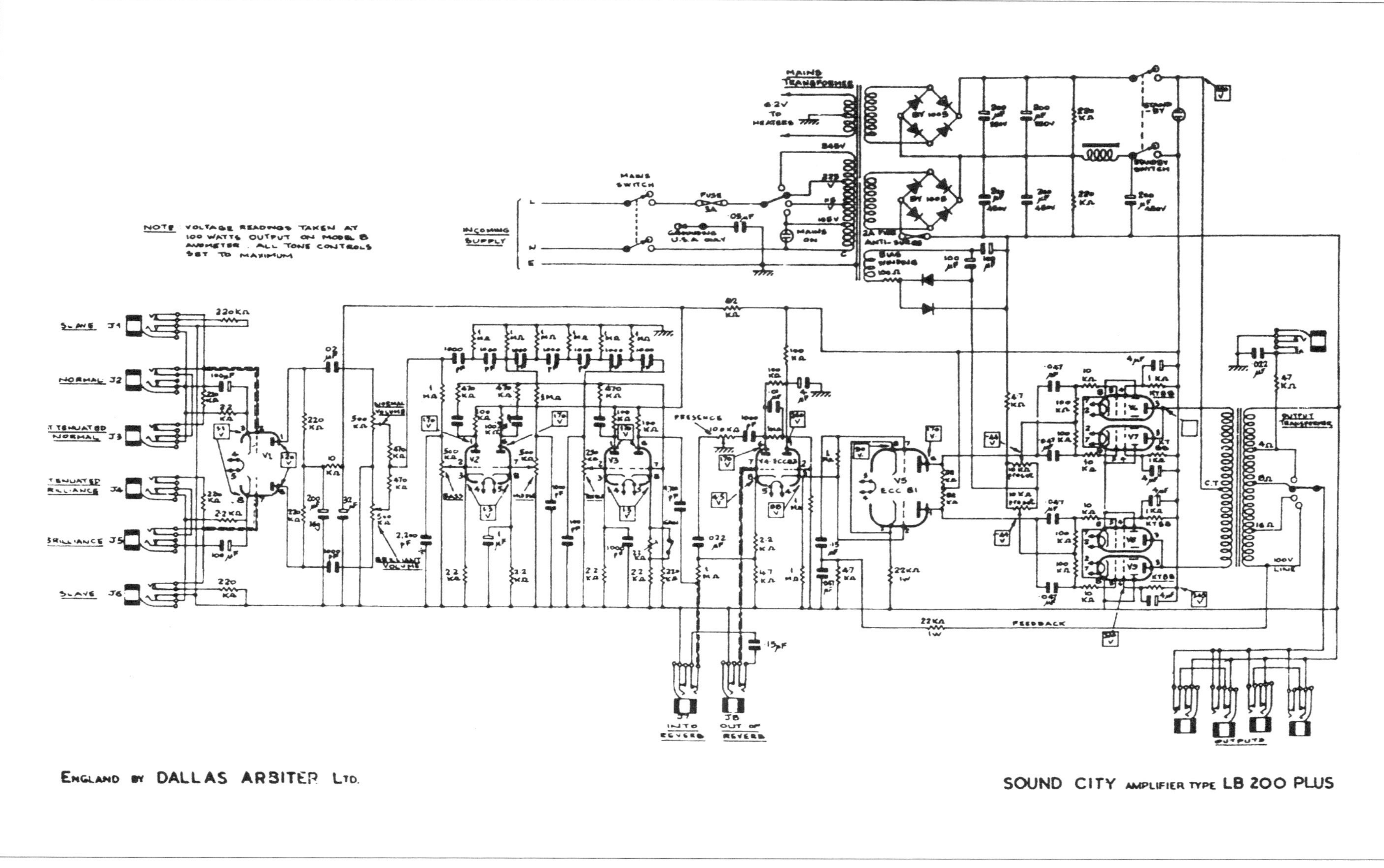
NOTE VOLTAGE READINGS TAKEN AT 100 WATTS OUTPUT ON MODEL 8 AVOMETER. ALL TONE CONTROLS SET TO MAXIMUM
MAINS TRANSFORMER
6.2V TO HEATERS
MAINS SWITCH
FUSE 3A
INCOMING SUPPLY
MAINS ON
GROUNDING U.S.A. ONLY
2A FUSE ANTI-SURGE
BIAS WINDING
STANDBY SWITCH
SLAVE J1
NORMAL J2
ATTENUATED NORMAL J3
ATTENUATED BRILLIANCE J4
BRILLIANCE J5
SLAVE J6
NORMAL VOLUME
BRILLIANT VOLUME
BASS
PRESENCE
V1
V2
V3
V4 ECC83
V5 ECC 81
V6
V7
V8
V9
KT88
J7 INTO REVERB
J8 OUT OF REVERB
FEEDBACK
OUTPUT TRANSFORMER
C.T.
4Ω
8Ω
16Ω
100V LINE
OUTPUTS
ENGLAND BY DALLAS ARBITER LTD.
SOUND CITY AMPLIFIER TYPE LB 200 PLUS

MIG 50 MASTER VOLUME

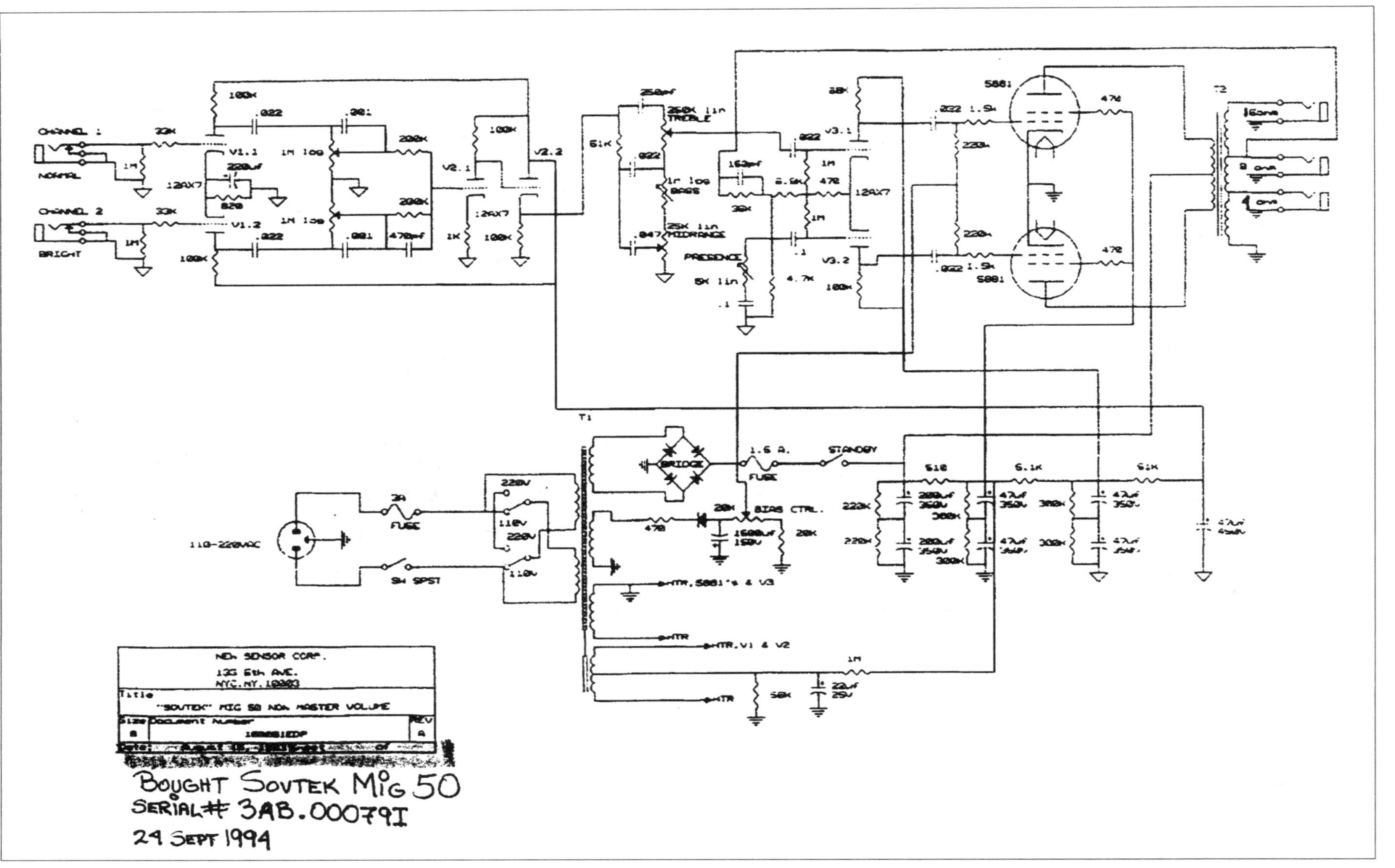

MIG 60 LEAD

Sovtek

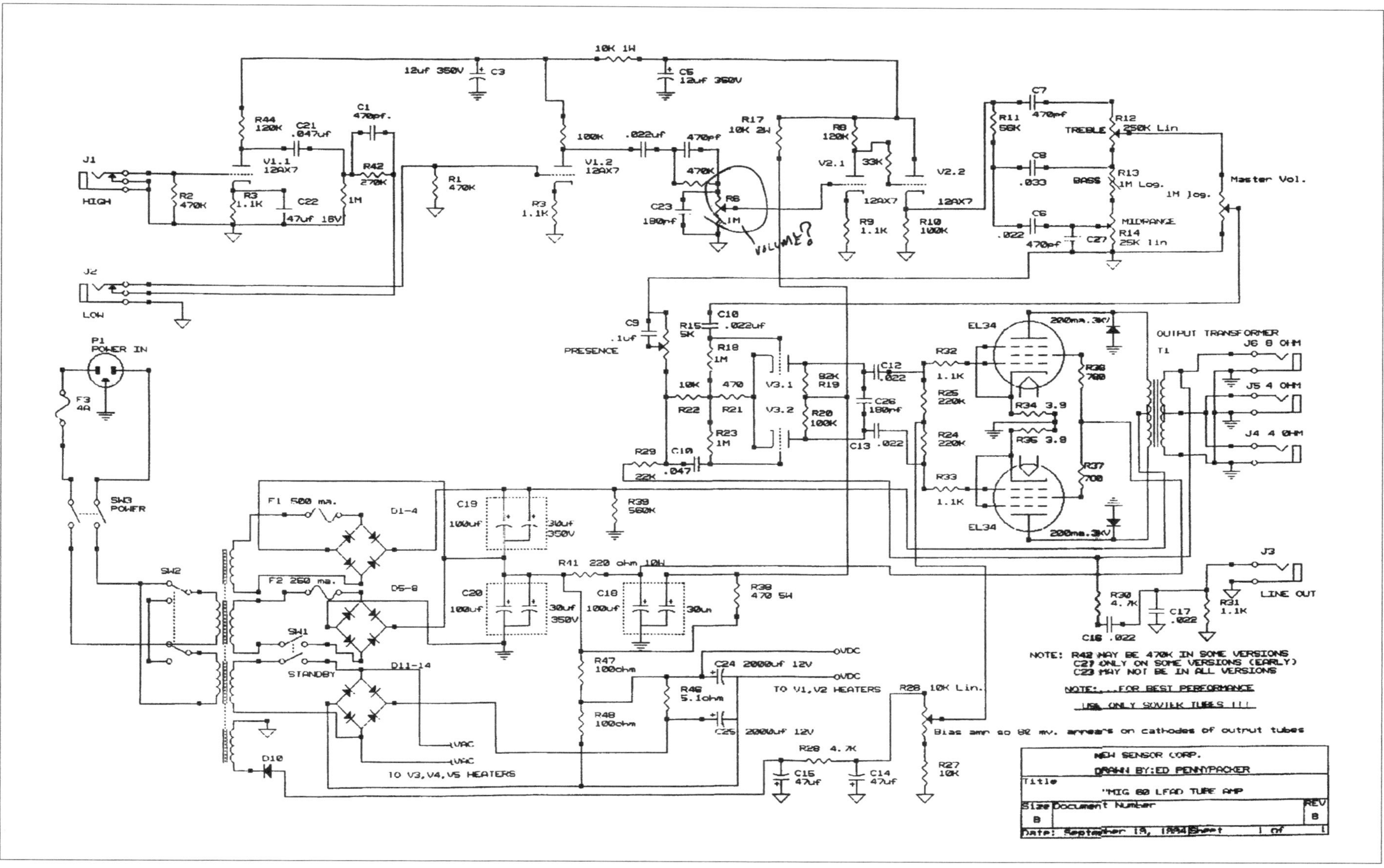

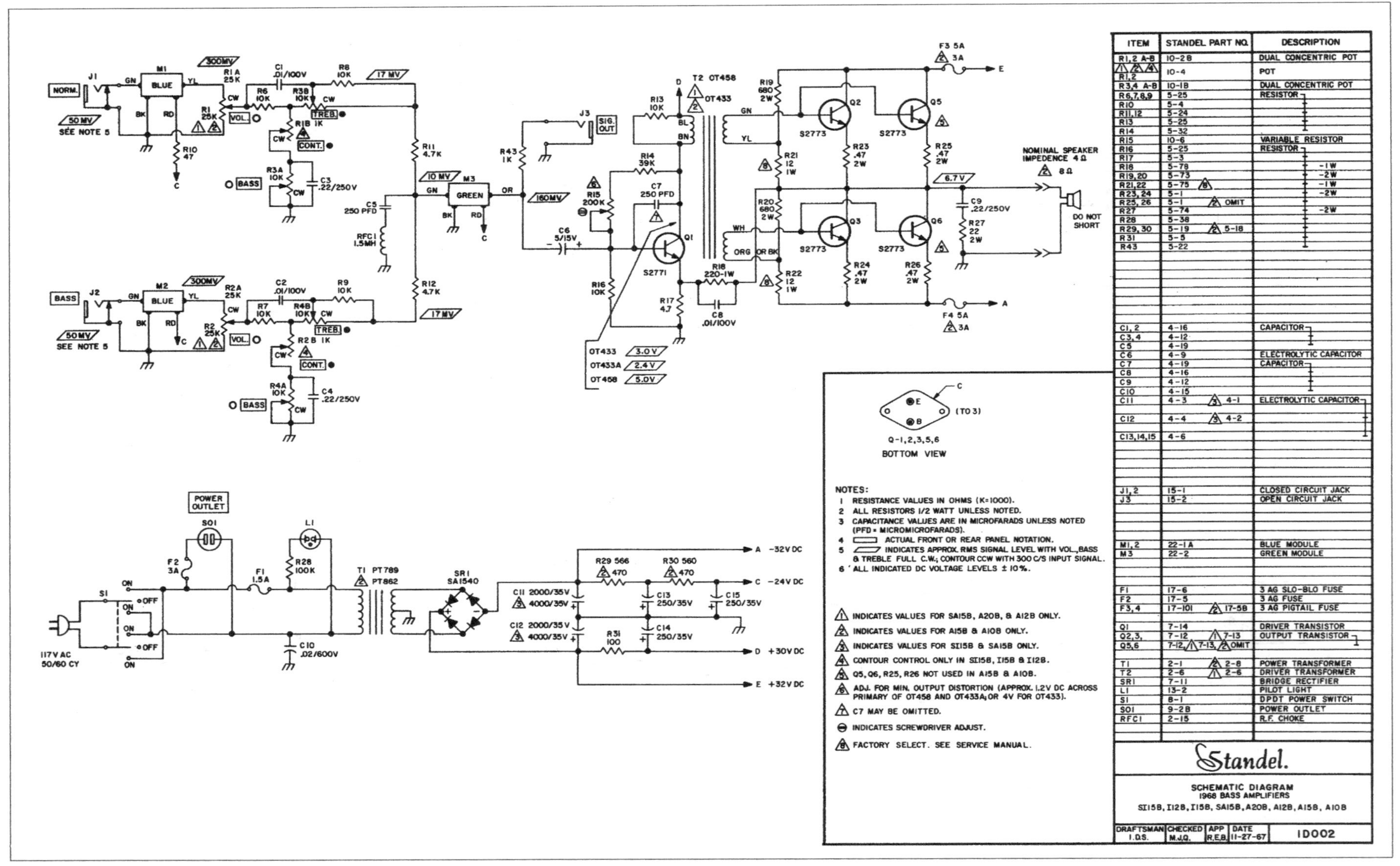
NOTES:
1 RESISTANCE VALUES IN OHMS (K=1000).
2 ALL RESISTORS 1/2 WATT UNLESS NOTED.
3 CAPACITANCE VALUES ARE IN MICROFARADS UNLESS NOTED (PFD = MICROMICROFARADS).
4 ACTUAL FRONT OR REAR PANEL NOTATION.
5 INDICATES APPROX. RMS SIGNAL LEVEL WITH VOL., BASS & TREBLE FULL C.W.; CONTOUR CCW WITH 300 C/S INPUT SIGNAL.
6 ALL INDICATED DC VOLTAGE LEVELS ± 10%.
1 INDICATES VALUES FOR SA15B, A20B, & A12B ONLY.
2 INDICATES VALUES FOR A15B & A10B ONLY.
3 INDICATES VALUES FOR SI15B & SA15B ONLY.
4 CONTOUR CONTROL ONLY IN SI15B, I15B & I12B.
5 Q5, Q6, R25, R26 NOT USED IN A15B & A10B.
6 ADJ. FOR MIN. OUTPUT DISTORTION (APPROX. 1.2V DC ACROSS PRIMARY OF OT458 AND OT433A; OR 4V FOR OT433).
7 C7 MAY BE OMITTED.
INDICATES SCREWDRIVER ADJUST.
8 FACTORY SELECT. SEE SERVICE MANUAL.
Q-1,2,3,5,6
BOTTOM VIEW
(TO 3)
NOMINAL SPEAKER IMPEDENCE 4 Ω
8 Ω
DO NOT SHORT
117V AC 50/60 CY
POWER OUTLET
NORM.
BASS
SEE NOTE 5
SIG. OUT
Standel.
SCHEMATIC DIAGRAM
1968 BASS AMPLIFIERS
SI15B, I12B, I15B, SA15B, A20B, A12B, A15B, A10B
DRAFTSMAN I.D.S.
CHECKED M.J.Q.
APP R.E.B.
DATE 11-27-67
1D002

ITEM	STANDEL PART NO.	DESCRIPTION
R1,2 A-B	10-28	DUAL CONCENTRIC POT
1 2 4 R1,2	10-4	POT
R3,4 A-B	10-18	DUAL CONCENTRIC POT
R6,7,8,9	5-25	RESISTOR
R10	5-4	
R11,12	5-24	
R13	5-25	
R14	5-32	
R15	10-6	VARIABLE RESISTOR
R16	5-25	RESISTOR
R17	5-3	
R18	5-78	-1W
R19,20	5-73	-2W
R21,22	5-75 8	-1W
R23,24	5-1	-2W
R25,26	5-1 2 OMIT	
R27	5-74	-2W
R28	5-38	
R29,30	5-19 2 5-18	
R31	5-5	
R43	5-22	
C1,2	4-16	CAPACITOR
C3,4	4-12	
C5	4-19	
C6	4-9	ELECTROLYTIC CAPACITOR
C7	4-19	CAPACITOR
C8	4-16	
C9	4-12	
C10	4-15	
C11	4-3 3 4-1	ELECTROLYTIC CAPACITOR
C12	4-4 3 4-2	
C13,14,15	4-6	
J1,2	15-1	CLOSED CIRCUIT JACK
J3	15-2	OPEN CIRCUIT JACK
M1,2	22-1A	BLUE MODULE
M3	22-2	GREEN MODULE
F1	17-6	3 AG SLO-BLO FUSE
F2	17-5	3 AG FUSE
F3,4	17-101 2 17-5B	3 AG PIGTAIL FUSE
Q1	7-14	DRIVER TRANSISTOR
Q2,3,	7-12 1 7-13	OUTPUT TRANSISTOR
Q5,6	7-12, 1 7-13, 2 OMIT	
T1	2-1 2 2-8	POWER TRANSFORMER
T2	2-6 1 2-6	DRIVER TRANSFORMER
SR1	7-11	BRIDGE RECTIFIER
L1	13-2	PILOT LIGHT
S1	8-1	DPDT POWER SWITCH
SO1	9-2B	POWER OUTLET
RFC1	2-15	R.F. CHOKE

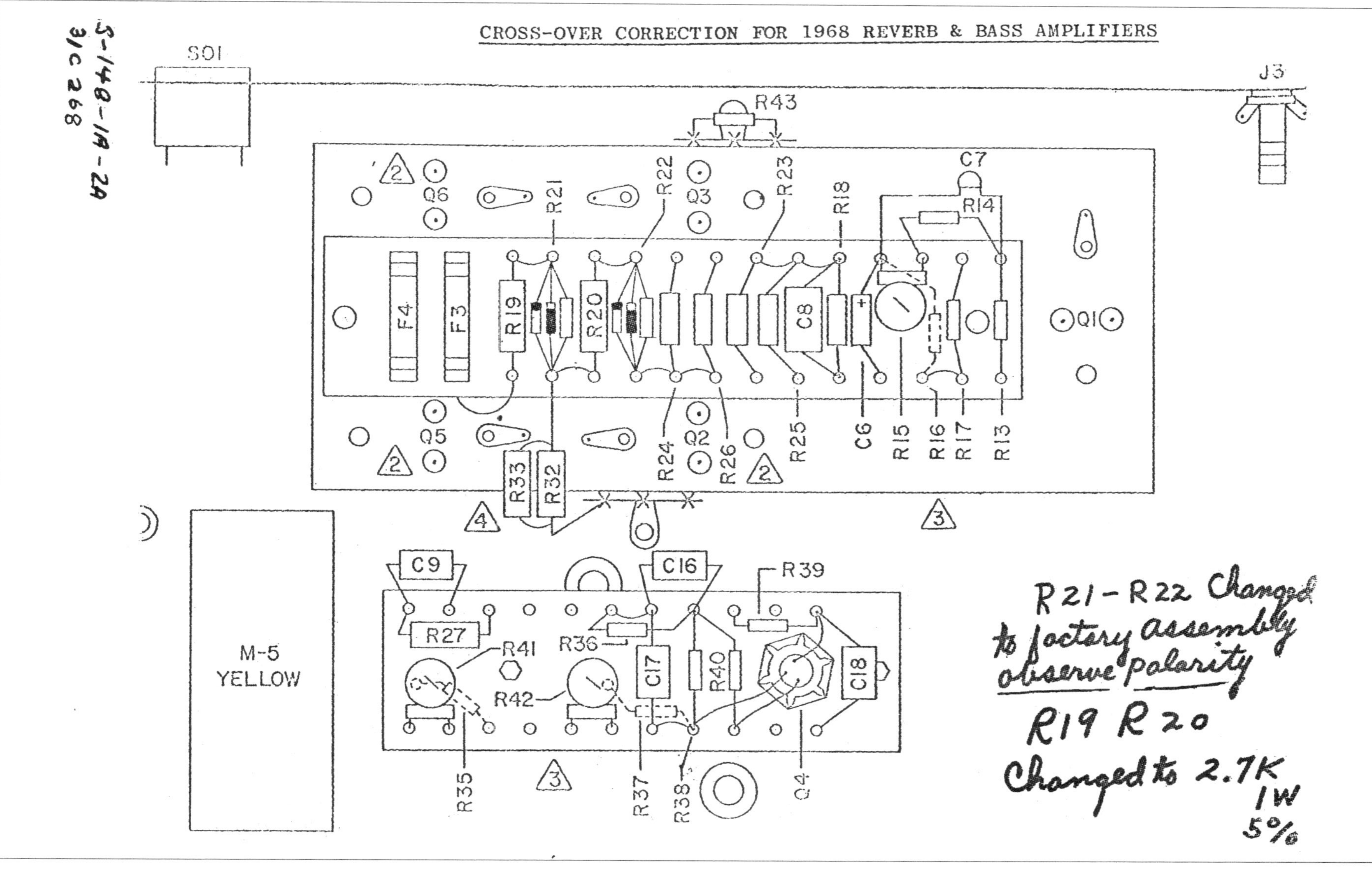
CROSS-OVER CORRECTION FOR 1968 REVERB & BASS AMPLIFIERS
S-148-1A-2A
3/C 268
SO1
J3
R43
C7
R14
Q6
R21
R22
Q3
R23
R18
F4
F3
R19
R20
C8
Q1
Q5
Q2
R24
R26
R25
C6
R15
R16
R17
R13
R33
R32
M-5
YELLOW
C9
C16
R39
R27
R41
R36
C17
R40
C18
R42
R35
R37
R38
Q4
R21 - R22 Changed
to factory assembly
observe polarity
R19 R20
Changed to 2.7K
1W
5%

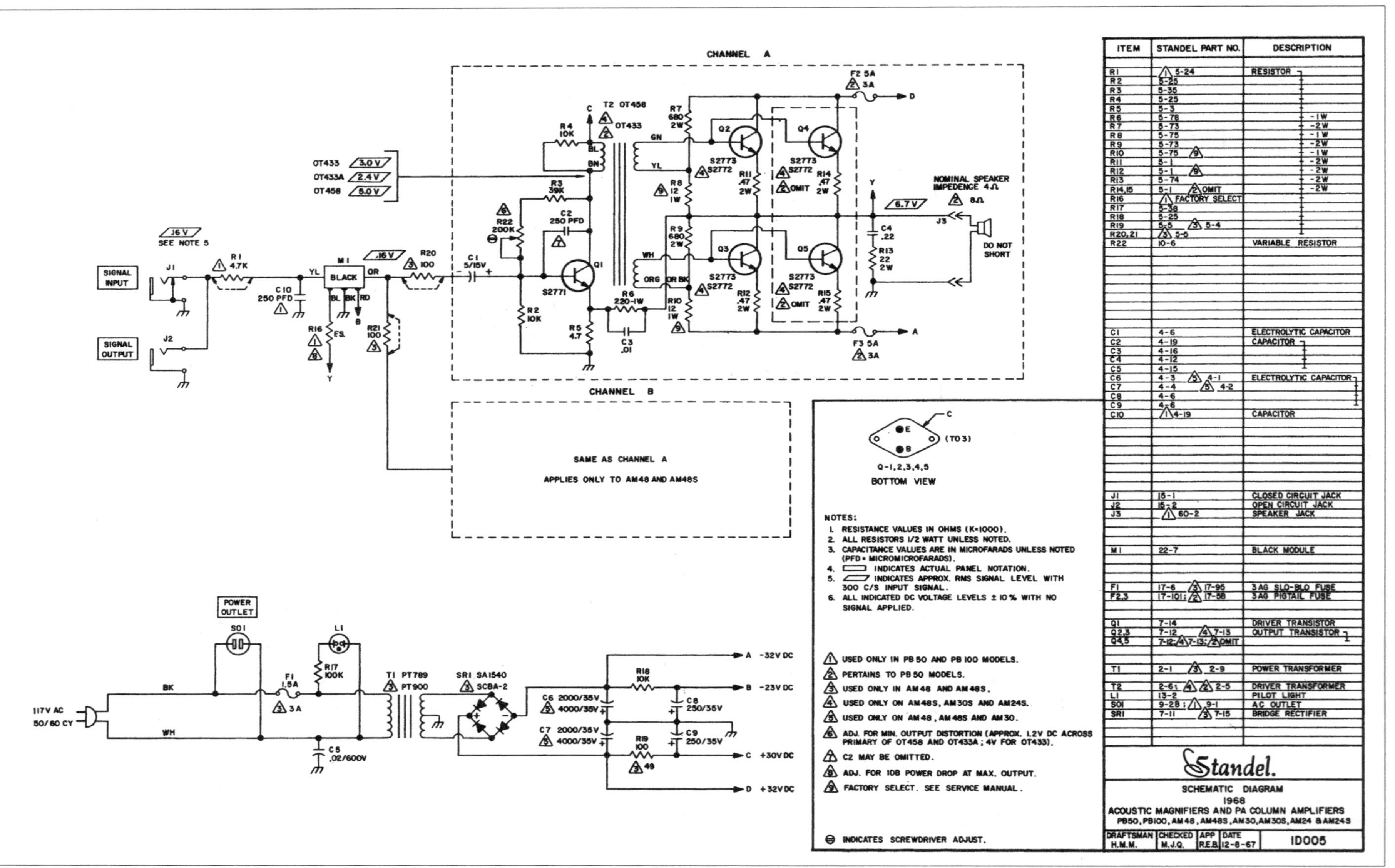

ITEM	STANDEL PART NO.	DESCRIPTION
R1	(1) 5-24	RESISTOR
R2	5-25	
R3	5-35	
R4	5-25	
R5	5-3	
R6	5-78	-1W
R7	5-73	-2W
R8	5-75	-1W
R9	5-73	-2W
R10	5-75 (9)	-1W
R11	5-1	-2W
R12	5-1 (9)	-2W
R13	5-74	-2W
R14,15	5-1 (2) OMIT	-2W
R16	(1) FACTORY SELECT	
R17	5-38	
R18	5-25	
R19	5-5 (3) 5-4	
R20,21	(3) 5-5	
R22	10-6	VARIABLE RESISTOR
C1	4-6	ELECTROLYTIC CAPACITOR
C2	4-19	CAPACITOR
C3	4-16	
C4	4-12	
C5	4-15	
C6	4-3 (5) 4-1	ELECTROLYTIC CAPACITOR
C7	4-4 (5) 4-2	
C8	4-6	
C9	4-6	
C10	(1) 4-19	CAPACITOR
J1	15-1	CLOSED CIRCUIT JACK
J2	15-2	OPEN CIRCUIT JACK
J3	(1) 60-2	SPEAKER JACK
M1	22-7	BLACK MODULE
F1	17-6 (3) 17-95	3AG SLO-BLO FUSE
F2,3	17-101; (2) 17-58	3AG PIGTAIL FUSE
Q1	7-14	DRIVER TRANSISTOR
Q2,3	7-12 (4) 7-13	OUTPUT TRANSISTOR
Q4,5	7-12; (4) 7-13; (2) OMIT	
T1	2-1 (3) 2-9	POWER TRANSFORMER
T2	2-6; (4) (2) 2-5	DRIVER TRANSFORMER
L1	13-2	PILOT LIGHT
SO1	9-28; (1) 9-1	AC OUTLET
SR1	7-11 (3) 7-15	BRIDGE RECTIFIER

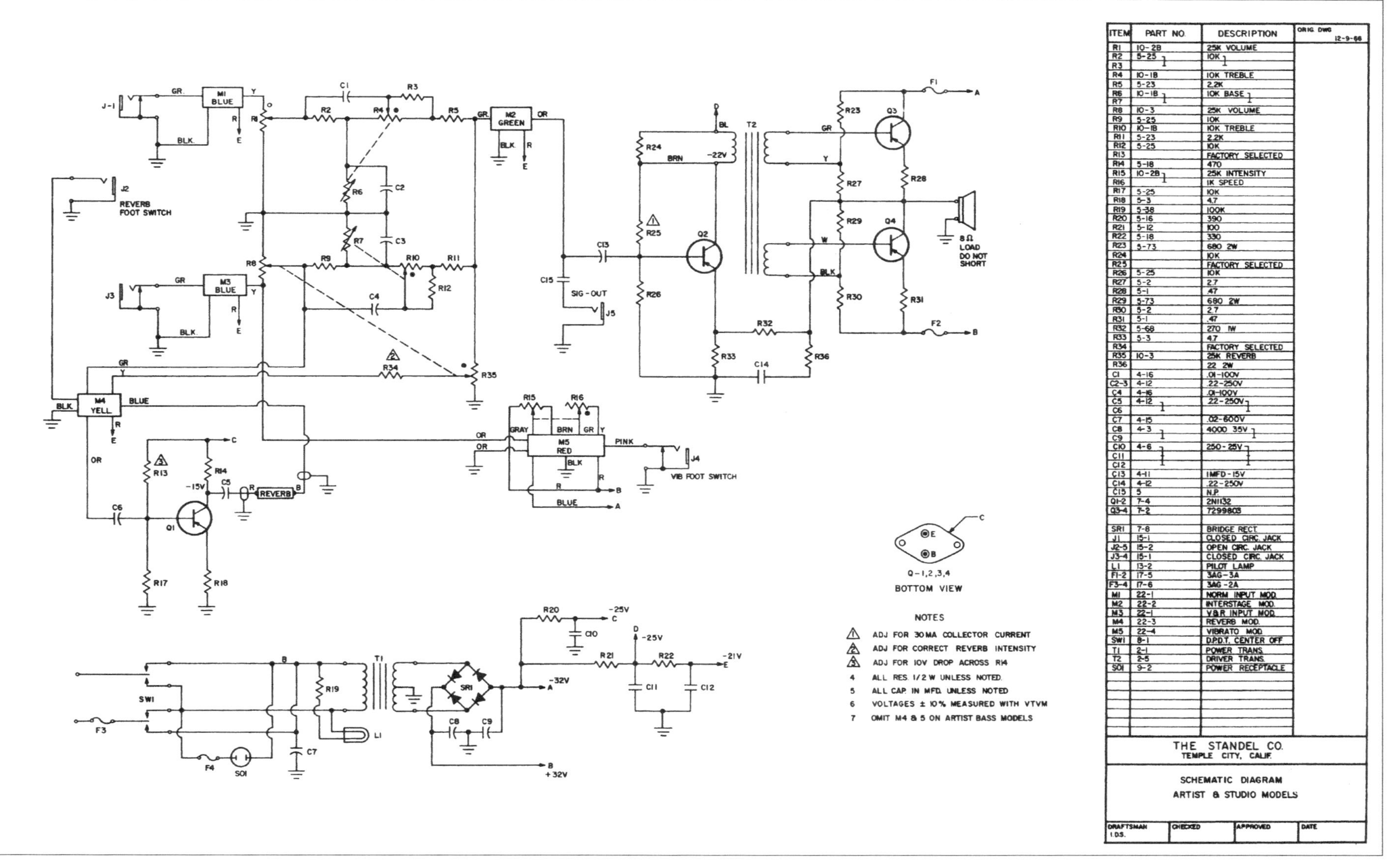
Q-1,2,3,4
BOTTOM VIEW
NOTES
1 ADJ FOR 30MA COLLECTOR CURRENT
2 ADJ FOR CORRECT REVERB INTENSITY
3 ADJ FOR 10V DROP ACROSS R14
4 ALL RES. 1/2 W UNLESS NOTED.
5 ALL CAP. IN MFD. UNLESS NOTED
6 VOLTAGES ± 10% MEASURED WITH VTVM
7 OMIT M4 & 5 ON ARTIST BASS MODELS
ORIG. DWG 12-9-66
THE STANDEL CO.
TEMPLE CITY, CALIF.
SCHEMATIC DIAGRAM
ARTIST & STUDIO MODELS
DRAFTSMAN I.D.S.
CHECKED
APPROVED
DATE

ITEM	PART NO.	DESCRIPTION
R1	10-2B	25K VOLUME
R2	5-25	10K
R3		
R4	10-1B	10K TREBLE
R5	5-23	2.2K
R6	10-1B	10K BASE
R7		
R8	10-3	25K VOLUME
R9	5-25	10K
R10	10-1B	10K TREBLE
R11	5-23	2.2K
R12	5-25	10K
R13		FACTORY SELECTED
R14	5-18	470
R15	10-2B	25K INTENSITY
R16		1K SPEED
R17	5-25	10K
R18	5-3	4.7
R19	5-38	100K
R20	5-16	390
R21	5-12	100
R22	5-18	330
R23	5-73	680 2W
R24		10K
R25		FACTORY SELECTED
R26	5-25	10K
R27	5-2	2.7
R28	5-1	.47
R29	5-73	680 2W
R30	5-2	2.7
R31	5-1	.47
R32	5-68	270 1W
R33	5-3	4.7
R34		FACTORY SELECTED
R35	10-3	25K REVERB
R36		22 2W
C1	4-16	.01-100V
C2-3	4-12	.22-250V
C4	4-16	.01-100V
C5	4-12	.22-250V
C6		
C7	4-15	.02-600V
C8	4-3	4000 35V
C9		
C10	4-6	250-25V
C11		
C12		
C13	4-11	1MFD-15V
C14	4-12	.22-250V
C15	5	N.P.
Q1-2	7-4	2N1132
Q3-4	7-2	7299803
SR1	7-8	BRIDGE RECT.
J1	15-1	CLOSED CIRC. JACK
J2-5	15-2	OPEN CIRC. JACK
J3-4	15-1	CLOSED CIRC. JACK
L1	13-2	PILOT LAMP
F1-2	17-5	3AG-3A
F3-4	17-6	3AG-2A
M1	22-1	NORM INPUT MOD.
M2	22-2	INTERSTAGE MOD.
M3	22-1	V&R INPUT MOD.
M4	22-3	REVERB MOD.
M5	22-4	VIBRATO MOD.
SW1	8-1	D.P.D.T. CENTER OFF
T1	2-1	POWER TRANS.
T2	2-5	DRIVER TRANS.
SO1	9-2	POWER RECEPTACLE

ARTIST MODELS **SUFFIX "A"**

Standel

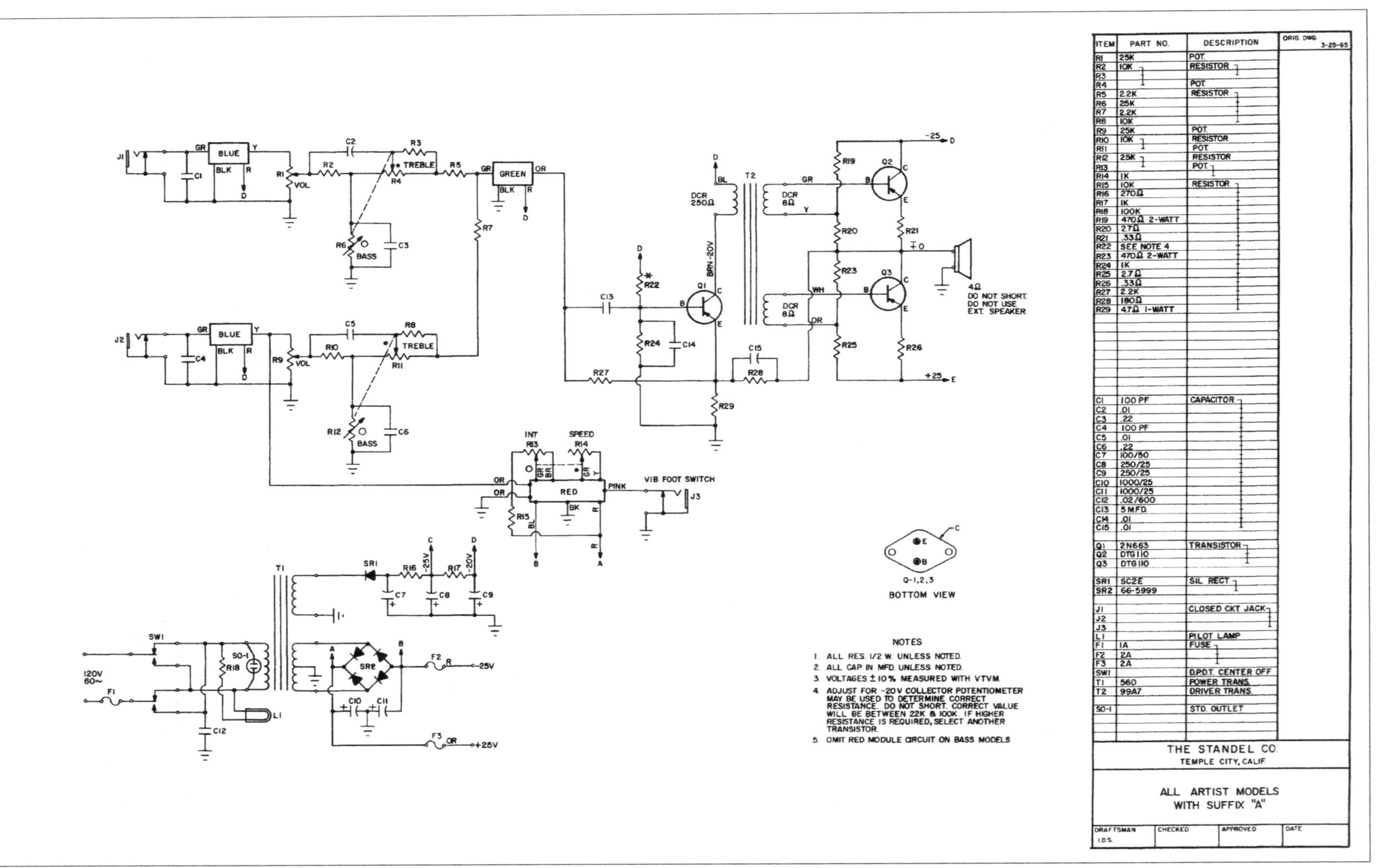

ITEM | PART NO. | DESCRIPTION
ORIG. DWG. 3-25-65
R1 25K POT.
R2 10K RESISTOR
R3
R4 POT.
R5 2.2K RESISTOR
R6 25K
R7 2.2K
R8 10K
R9 25K POT.
R10 10K RESISTOR
R11 POT.
R12 25K RESISTOR
R13 POT.
R14 1K RESISTOR
R15 10K
R16 270Ω
R17 1K
R18 100K
R19 470Ω 2-WATT
R20 27Ω
R21 .33Ω
R22 SEE NOTE 4
R23 470Ω 2-WATT
R24 1K
R25 27Ω
R26 .33Ω
R27 2.2K
R28 180Ω
R29 47Ω 1-WATT
C1 100 PF CAPACITOR
C2 .01
C3 .22
C4 100 PF
C5 .01
C6 .22
C7 100/50
C8 250/25
C9 250/25
C10 1000/25
C11 1000/25
C12 .02/600
C13 5 MFD
C14 .01
C15 .01
Q1 2N663 TRANSISTOR
Q2 DTG110
Q3 DTG110
SR1 SC2E SIL. RECT.
SR2 66-5999
J1 CLOSED CKT JACK
J2
J3
L1 PILOT LAMP
F1 1A FUSE
F2 2A
F3 2A
SW1 D.P.D.T. CENTER OFF
T1 560 POWER TRANS.
T2 99A7 DRIVER TRANS.
SO-1 STD. OUTLET
THE STANDEL CO.
TEMPLE CITY, CALIF.
ALL ARTIST MODELS
WITH SUFFIX "A"
DRAFTSMAN I.D.S.
CHECKED
APPROVED
DATE
NOTES
1. ALL RES. 1/2 W. UNLESS NOTED.
2. ALL CAP IN MFD UNLESS NOTED.
3. VOLTAGES ± 10% MEASURED WITH VTVM.
4. ADJUST FOR -20V COLLECTOR POTENTIOMETER MAY BE USED TO DETERMINE CORRECT RESISTANCE. DO NOT SHORT. CORRECT VALUE WILL BE BETWEEN 22K & 100K. IF HIGHER RESISTANCE IS REQUIRED, SELECT ANOTHER TRANSISTOR.
5. OMIT RED MODULE CIRCUIT ON BASS MODELS
Q-1,2,3
BOTTOM VIEW
4Ω
DO NOT SHORT.
DO NOT USE
EXT. SPEAKER
VIB FOOT SWITCH
INT
SPEED
BLUE
GREEN
RED
TREBLE
BASS
VOL
DCR 250Ω
DCR 8Ω
BRN-20V
120V 60~
-25V
+25V

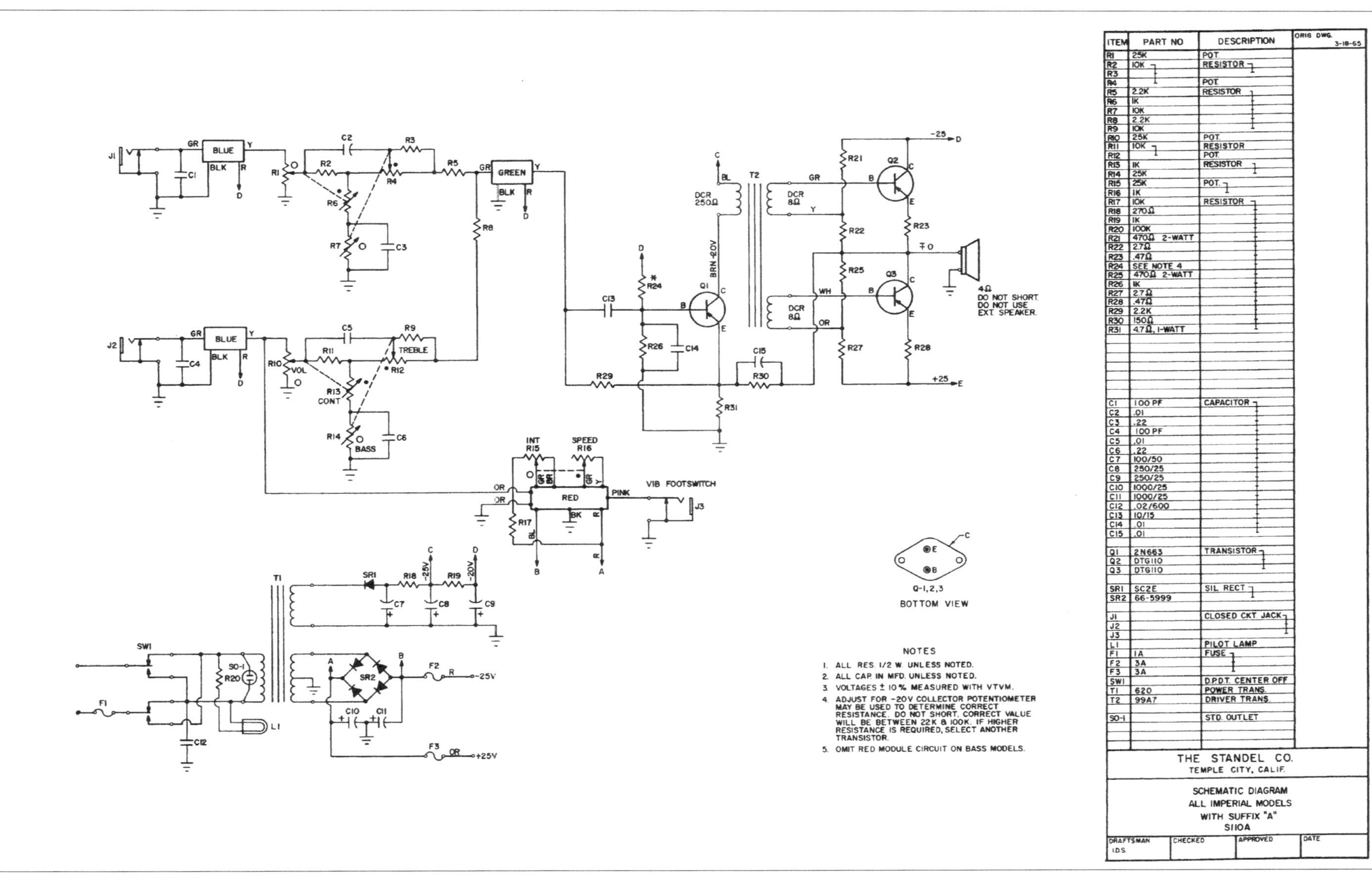
Q-1,2,3
BOTTOM VIEW
4Ω
DO NOT SHORT
DO NOT USE
EXT. SPEAKER.
NOTES
1. ALL RES. 1/2 W. UNLESS NOTED.
2. ALL CAP. IN MFD. UNLESS NOTED.
3. VOLTAGES ± 10% MEASURED WITH VTVM.
4. ADJUST FOR -20V COLLECTOR POTENTIOMETER MAY BE USED TO DETERMINE CORRECT RESISTANCE. DO NOT SHORT. CORRECT VALUE WILL BE BETWEEN 22K & 100K. IF HIGHER RESISTANCE IS REQUIRED, SELECT ANOTHER TRANSISTOR.
5. OMIT RED MODULE CIRCUIT ON BASS MODELS.

ITEM	PART NO	DESCRIPTION
R1	25K	POT.
R2	10K	RESISTOR
R3		
R4		POT.
R5	2.2K	RESISTOR
R6	1K	
R7	10K	
R8	2.2K	
R9	10K	
R10	25K	POT.
R11	10K	RESISTOR
R12		POT.
R13	1K	RESISTOR
R14	25K	
R15	25K	POT.
R16	1K	
R17	10K	RESISTOR
R18	270Ω	
R19	1K	
R20	100K	
R21	470Ω 2-WATT	
R22	2.7Ω	
R23	.47Ω	
R24	SEE NOTE 4	
R25	470Ω 2-WATT	
R26	1K	
R27	2.7Ω	
R28	.47Ω	
R29	2.2K	
R30	150Ω	
R31	4.7Ω, 1-WATT	
C1	100 PF	CAPACITOR
C2	.01	
C3	.22	
C4	100 PF	
C5	.01	
C6	.22	
C7	100/50	
C8	250/25	
C9	250/25	
C10	1000/25	
C11	1000/25	
C12	.02/600	
C13	10/15	
C14	.01	
C15	.01	
Q1	2N663	TRANSISTOR
Q2	DTG110	
Q3	DTG110	
SR1	SC2E	SIL. RECT.
SR2	66-5999	
J1		CLOSED CKT. JACK
J2		
J3		
L1		PILOT LAMP
F1	1A	FUSE
F2	3A	
F3	3A	
SW1		D.P.D.T. CENTER OFF
T1	620	POWER TRANS.
T2	99A7	DRIVER TRANS.
SO-1		STD. OUTLET

ORIG. DWG. 3-18-65

THE STANDEL CO.
TEMPLE CITY, CALIF.

SCHEMATIC DIAGRAM
ALL IMPERIAL MODELS
WITH SUFFIX "A"
S110A

DRAFTSMAN	CHECKED	APPROVED	DATE
I.D.S.			

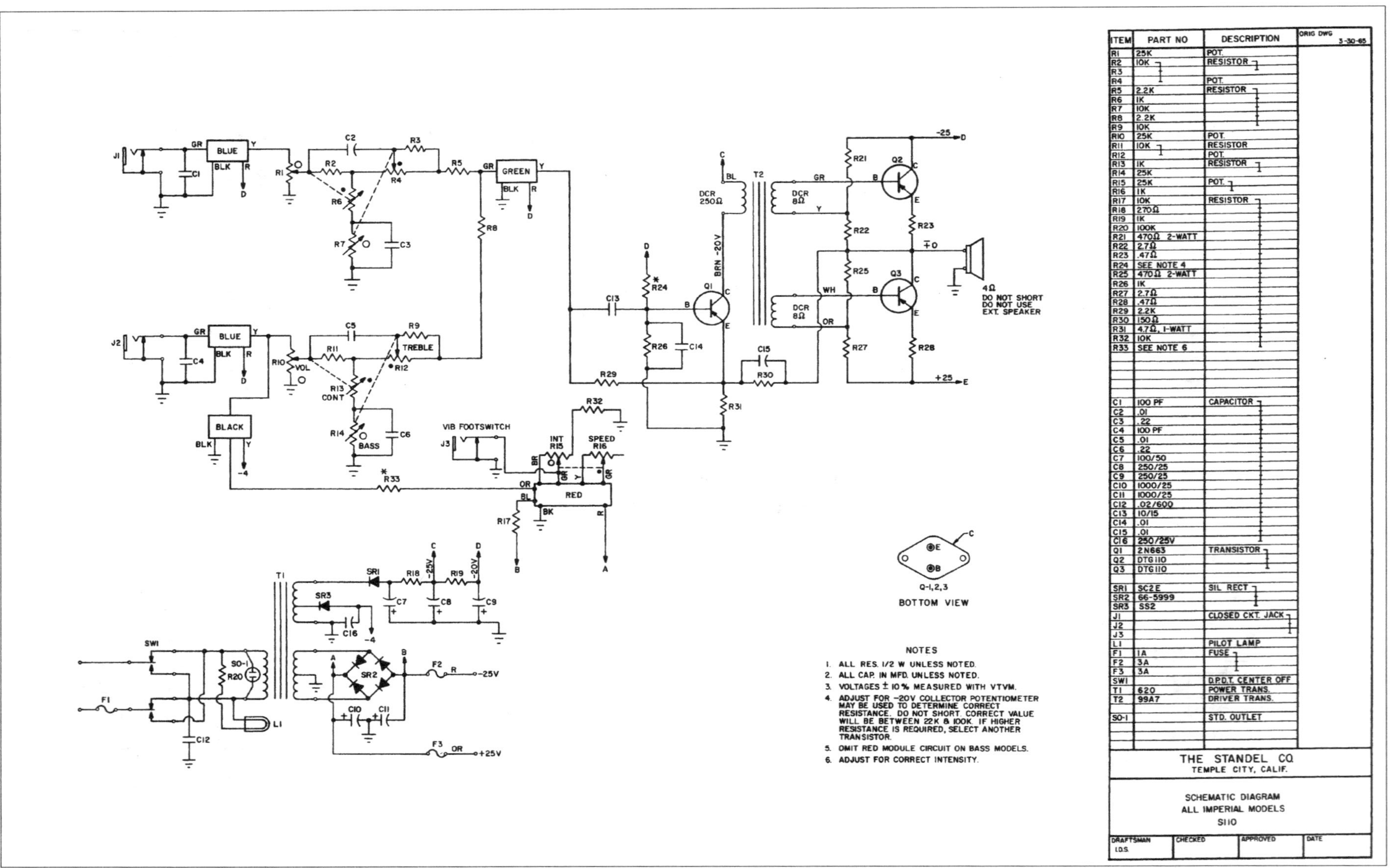
NOTES
1. ALL RES. 1/2 W UNLESS NOTED.
2. ALL CAP. IN MFD. UNLESS NOTED.
3. VOLTAGES ± 10% MEASURED WITH VTVM.
4. ADJUST FOR -20V COLLECTOR POTENTIOMETER MAY BE USED TO DETERMINE CORRECT RESISTANCE. DO NOT SHORT. CORRECT VALUE WILL BE BETWEEN 22K & 100K. IF HIGHER RESISTANCE IS REQUIRED, SELECT ANOTHER TRANSISTOR.
5. OMIT RED MODULE CIRCUIT ON BASS MODELS.
6. ADJUST FOR CORRECT INTENSITY.
Q-1,2,3
BOTTOM VIEW
4Ω
DO NOT SHORT
DO NOT USE
EXT. SPEAKER
VIB FOOTSWITCH
THE STANDEL CO.
TEMPLE CITY, CALIF.
SCHEMATIC DIAGRAM
ALL IMPERIAL MODELS
S110
ORIG DWG 3-30-65
DRAFTSMAN I.D.S.
CHECKED
APPROVED
DATE

ITEM	PART NO	DESCRIPTION
R1	25K	POT.
R2	10K	RESISTOR
R3		
R4		POT.
R5	2.2K	RESISTOR
R6	1K	
R7	10K	
R8	2.2K	
R9	10K	
R10	25K	POT.
R11	10K	RESISTOR
R12		POT.
R13	1K	RESISTOR
R14	25K	
R15	25K	POT.
R16	1K	
R17	10K	RESISTOR
R18	270Ω	
R19	1K	
R20	100K	
R21	470Ω 2-WATT	
R22	2.7Ω	
R23	.47Ω	
R24	SEE NOTE 4	
R25	470Ω 2-WATT	
R26	1K	
R27	2.7Ω	
R28	.47Ω	
R29	2.2K	
R30	150Ω	
R31	4.7Ω, 1-WATT	
R32	10K	
R33	SEE NOTE 6	
C1	100 PF	CAPACITOR
C2	.01	
C3	.22	
C4	100 PF	
C5	.01	
C6	.22	
C7	100/50	
C8	250/25	
C9	250/25	
C10	1000/25	
C11	1000/25	
C12	.02/600	
C13	10/15	
C14	.01	
C15	.01	
C16	250/25V	
Q1	2N663	TRANSISTOR
Q2	DTG110	
Q3	DTG110	
SR1	SC2E	SIL. RECT.
SR2	66-5999	
SR3	SS2	
J1		CLOSED CKT. JACK
J2		
J3		
L1		PILOT LAMP
F1	1A	FUSE
F2	3A	
F3	3A	
SW1		D.P.D.T. CENTER OFF
T1	620	POWER TRANS.
T2	99A7	DRIVER TRANS.
SO-1		STD. OUTLET

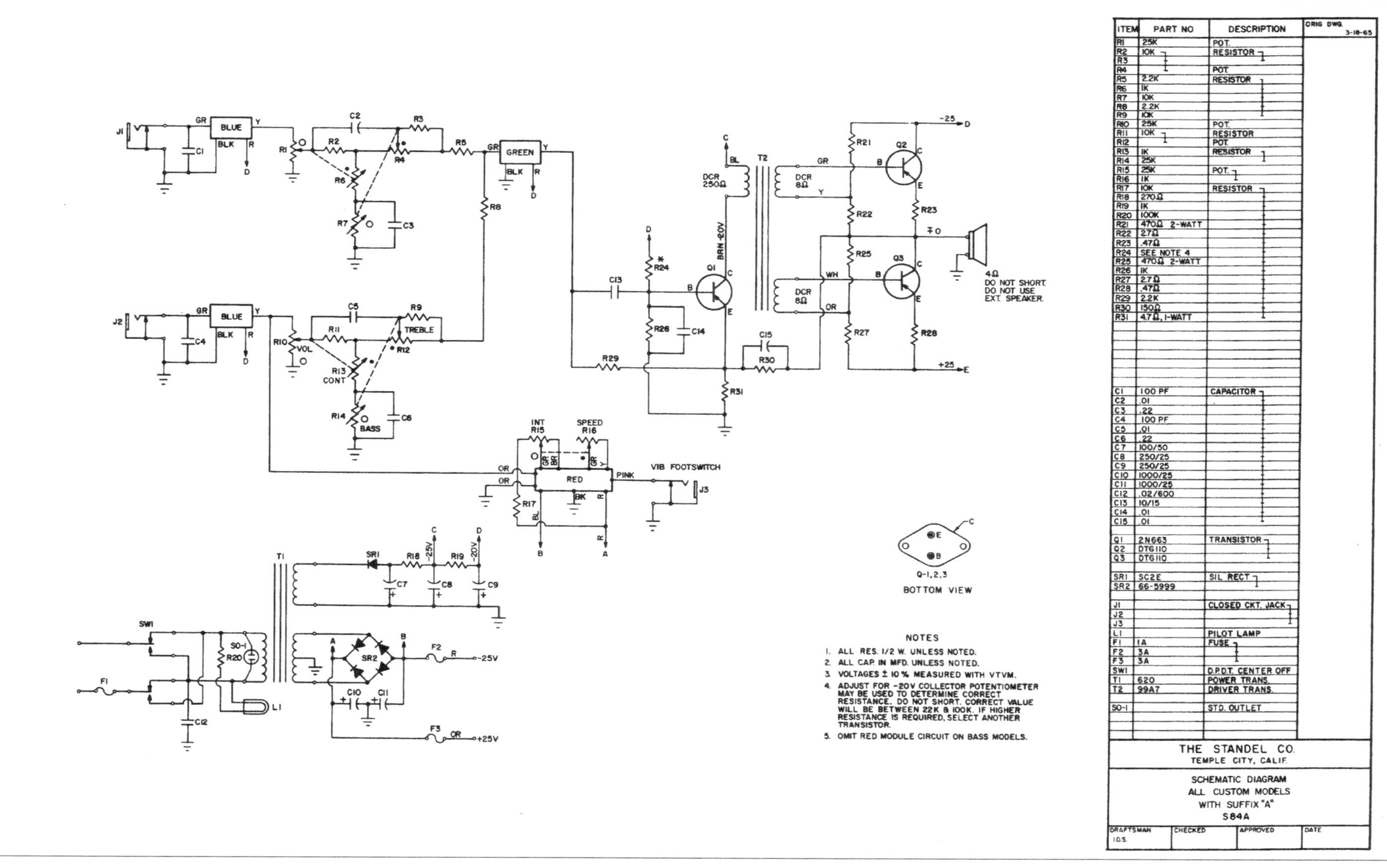
NOTES
1. ALL RES. 1/2 W. UNLESS NOTED.
2. ALL CAP. IN MFD. UNLESS NOTED.
3. VOLTAGES ± 10 % MEASURED WITH VTVM.
4. ADJUST FOR -20V COLLECTOR POTENTIOMETER MAY BE USED TO DETERMINE CORRECT RESISTANCE. DO NOT SHORT. CORRECT VALUE WILL BE BETWEEN 22K & 100K. IF HIGHER RESISTANCE IS REQUIRED, SELECT ANOTHER TRANSISTOR.
5. OMIT RED MODULE CIRCUIT ON BASS MODELS.
Q-1,2,3
BOTTOM VIEW
4Ω
DO NOT SHORT
DO NOT USE
EXT. SPEAKER.
VIB FOOTSWITCH
THE STANDEL CO.
TEMPLE CITY, CALIF.
SCHEMATIC DIAGRAM
ALL CUSTOM MODELS
WITH SUFFIX "A"
S84A
DRAFTSMAN I.D.S.
CHECKED
APPROVED
DATE
ORIG DWG 3-18-65

ITEM	PART NO	DESCRIPTION
R1	25K	POT.
R2	10K	RESISTOR
R3		
R4		POT.
R5	2.2K	RESISTOR
R6	1K	
R7	10K	
R8	2.2K	
R9	10K	
R10	25K	POT.
R11	10K	RESISTOR
R12		POT.
R13	1K	RESISTOR
R14	25K	
R15	25K	POT.
R16	1K	
R17	10K	RESISTOR
R18	270Ω	
R19	1K	
R20	100K	
R21	470Ω 2-WATT	
R22	27Ω	
R23	.47Ω	
R24	SEE NOTE 4	
R25	470Ω 2-WATT	
R26	1K	
R27	27Ω	
R28	.47Ω	
R29	2.2K	
R30	150Ω	
R31	47Ω, 1-WATT	
C1	100 PF	CAPACITOR
C2	.01	
C3	.22	
C4	100 PF	
C5	.01	
C6	.22	
C7	100/50	
C8	250/25	
C9	250/25	
C10	1000/25	
C11	1000/25	
C12	.02/600	
C13	10/15	
C14	.01	
C15	.01	
Q1	2N663	TRANSISTOR
Q2	DTG110	
Q3	DTG110	
SR1	SC2E	SIL. RECT.
SR2	66-5999	
J1		CLOSED CKT. JACK
J2		
J3		
L1		PILOT LAMP
F1	1A	FUSE
F2	3A	
F3	3A	
SW1		D.P.D.T. CENTER OFF
T1	620	POWER TRANS.
T2	99A7	DRIVER TRANS.
SO-1		STD. OUTLET

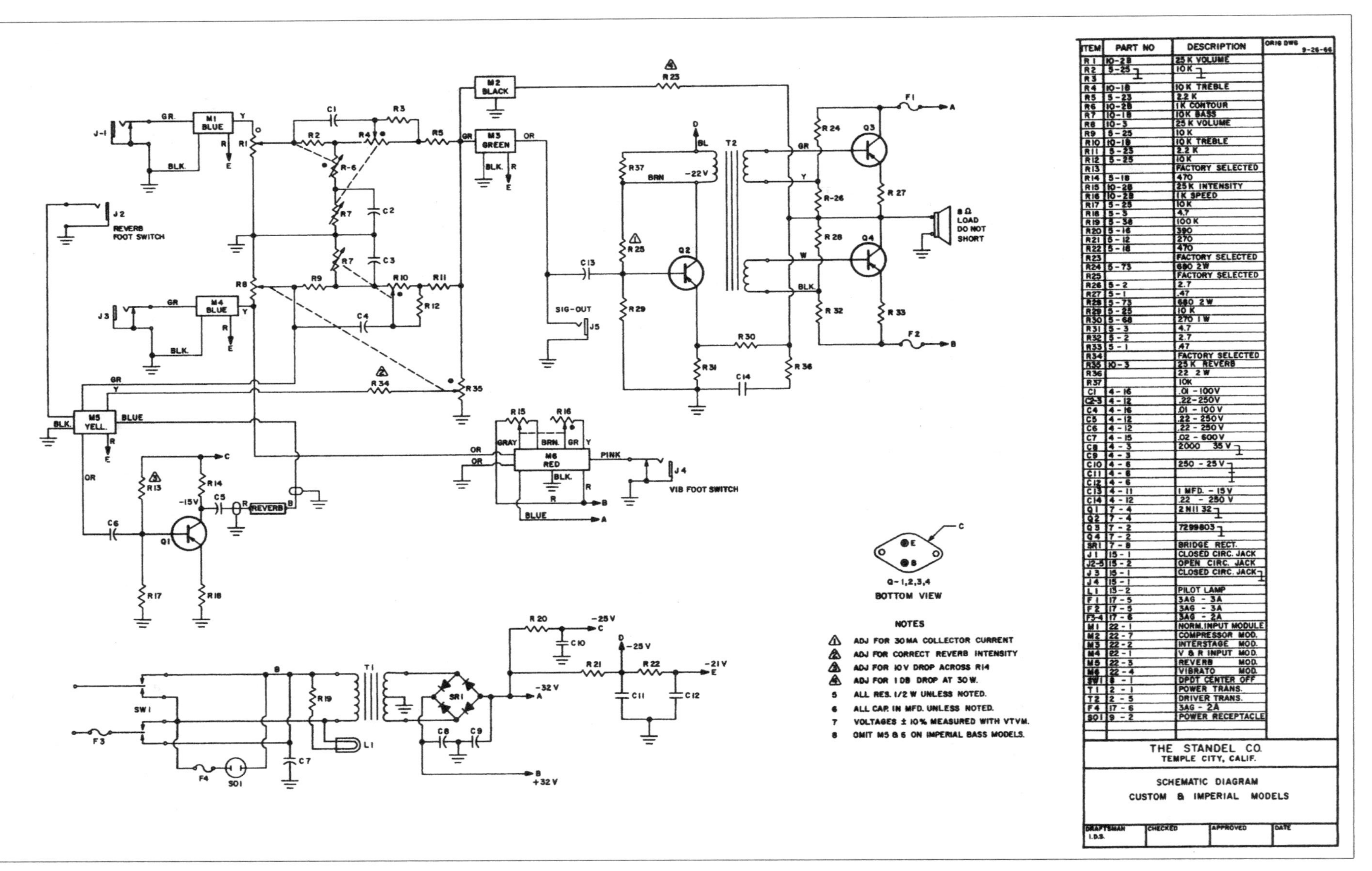
REVERB FOOT SWITCH
VIB FOOT SWITCH
SIG-OUT
8 Ω LOAD DO NOT SHORT
Q-1,2,3,4 BOTTOM VIEW
NOTES
1 ADJ FOR 30 MA COLLECTOR CURRENT
2 ADJ FOR CORRECT REVERB INTENSITY
3 ADJ FOR 10 V DROP ACROSS R14
4 ADJ FOR 1 DB DROP AT 30 W.
5 ALL RES. 1/2 W UNLESS NOTED.
6 ALL CAP. IN MFD. UNLESS NOTED.
7 VOLTAGES ± 10% MEASURED WITH VTVM.
8 OMIT M5 & 6 ON IMPERIAL BASS MODELS.
THE STANDEL CO.
TEMPLE CITY, CALIF.
SCHEMATIC DIAGRAM
CUSTOM & IMPERIAL MODELS
ORIG DWG 9-26-66
DRAFTSMAN I.D.S.
CHECKED
APPROVED
DATE

ITEM	PART NO	DESCRIPTION
R 1	10-28	25 K VOLUME
R 2	5-25	10 K
R 3		
R 4	10-18	10 K TREBLE
R 5	5-23	2.2 K
R 6	10-28	1 K CONTOUR
R 7	10-18	10 K BASS
R 8	10-3	25 K VOLUME
R 9	5-25	10 K
R 10	10-18	10 K TREBLE
R 11	5-23	2.2 K
R 12	5-25	10 K
R 13		FACTORY SELECTED
R 14	5-18	470
R 15	10-28	25 K INTENSITY
R 16	10-28	1 K SPEED
R 17	5-25	10 K
R 18	5-3	4.7
R 19	5-38	100 K
R 20	5-16	390
R 21	5-12	270
R 22	5-18	470
R 23		FACTORY SELECTED
R 24	5-73	680 2 W
R 25		FACTORY SELECTED
R 26	5-2	2.7
R 27	5-1	.47
R 28	5-73	680 2 W
R 29	5-25	10 K
R 30	5-68	270 1 W
R 31	5-3	4.7
R 32	5-2	2.7
R 33	5-1	.47
R 34		FACTORY SELECTED
R 35	10-3	25 K REVERB
R 36		22 2 W
R 37		10K
C 1	4-16	.01 - 100V
C2-3	4-12	.22 - 250V
C 4	4-16	.01 - 100 V
C 5	4-12	.22 - 250 V
C 6	4-12	.22 - 250 V
C 7	4-15	.02 - 600 V
C 8	4-3	2000 35 V
C 9	4-3	
C 10	4-6	250 - 25 V
C 11	4-6	
C 12	4-6	
C 13	4-11	1 MFD. - 15 V
C 14	4-12	.22 - 250 V
Q 1	7-4	2N1132
Q 2	7-4	
Q 3	7-2	7299803
Q 4	7-2	
SR1	7-8	BRIDGE RECT.
J 1	15-1	CLOSED CIRC. JACK
J2-5	15-2	OPEN CIRC. JACK
J 3	15-1	CLOSED CIRC. JACK
J 4	15-1	
L 1	13-2	PILOT LAMP
F 1	17-5	3AG - 3A
F 2	17-5	3AG - 3A
F3-4	17-6	3AG - 2A
M 1	22-1	NORM. INPUT MODULE
M 2	22-7	COMPRESSOR MOD.
M 3	22-2	INTERSTAGE MOD.
M 4	22-1	V & R INPUT MOD.
M 5	22-3	REVERB MOD.
M 6	22-4	VIBRATO MOD.
SW1	8-1	DPDT CENTER OFF
T 1	2-1	POWER TRANS.
T 2	2-5	DRIVER TRANS.
F 4	17-6	3AG - 2A
SO1	9-2	POWER RECEPTACLE

CUSTOM REVERB

Standel

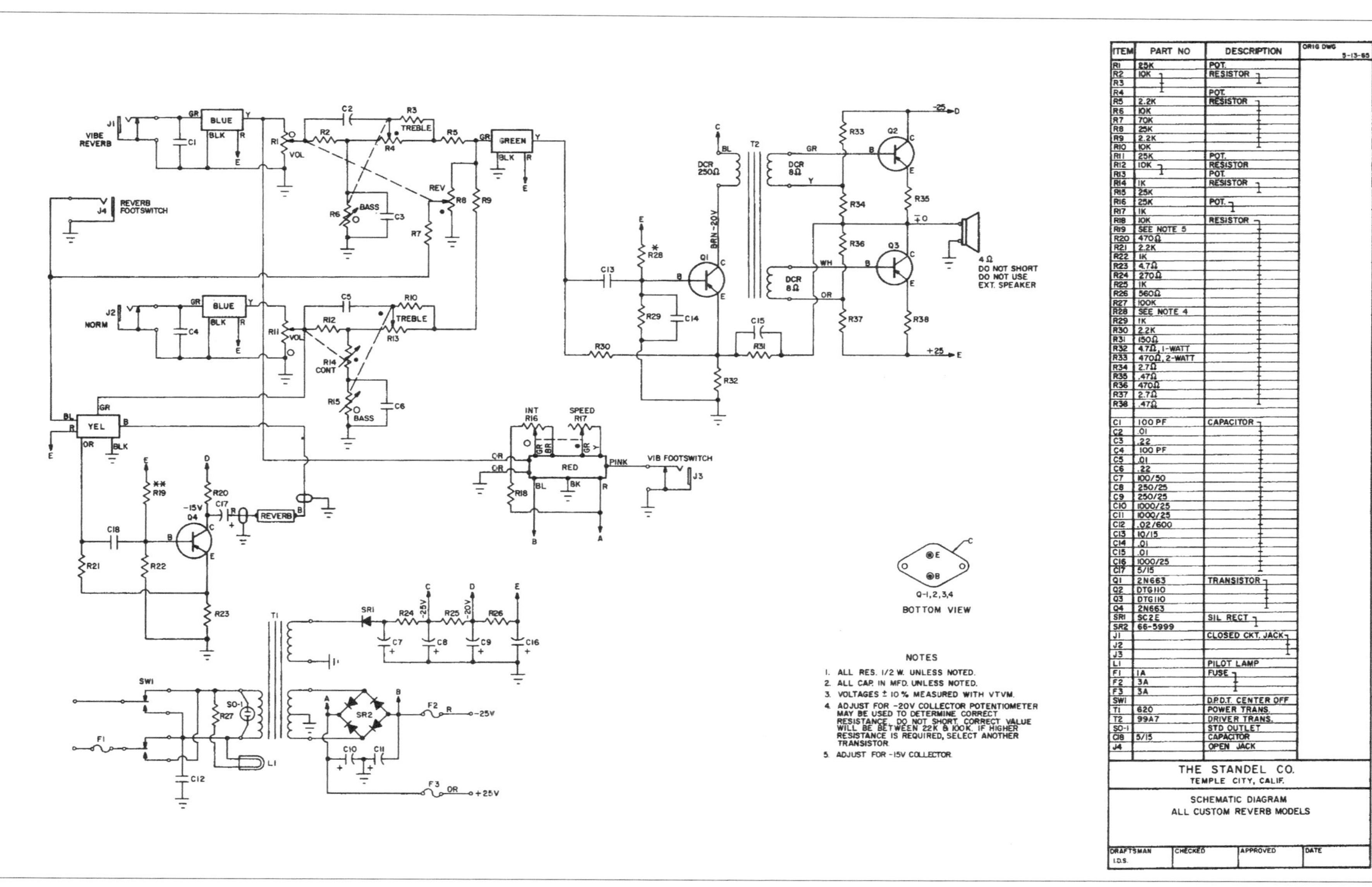

VIBE REVERB
NORM
REVERB FOOTSWITCH
VIB FOOTSWITCH
BLUE
GREEN
YEL
RED
REVERB
VOL
TREBLE
BASS
REV
CONT
INT
SPEED
DCR 250Ω
BRN -20V
DCR 8Ω
4Ω DO NOT SHORT DO NOT USE EXT. SPEAKER
Q-1,2,3,4 BOTTOM VIEW
NOTES
1. ALL RES. 1/2 W. UNLESS NOTED.
2. ALL CAP. IN MFD. UNLESS NOTED.
3. VOLTAGES ± 10 % MEASURED WITH VTVM.
4. ADJUST FOR -20V COLLECTOR POTENTIOMETER MAY BE USED TO DETERMINE CORRECT RESISTANCE. DO NOT SHORT. CORRECT VALUE WILL BE BETWEEN 22K & 100K. IF HIGHER RESISTANCE IS REQUIRED, SELECT ANOTHER TRANSISTOR.
5. ADJUST FOR -15V COLLECTOR.

ITEM	PART NO	DESCRIPTION	ORIG DWG 5-13-65
R1	25K	POT.	
R2	10K	RESISTOR	
R3			
R4		POT.	
R5	2.2K	RESISTOR	
R6	10K		
R7	70K		
R8	25K		
R9	2.2K		
R10	10K		
R11	25K	POT.	
R12	10K	RESISTOR	
R13		POT.	
R14	1K	RESISTOR	
R15	25K		
R16	25K	POT.	
R17	1K		
R18	10K	RESISTOR	
R19	SEE NOTE 5		
R20	470Ω		
R21	2.2K		
R22	1K		
R23	4.7Ω		
R24	270Ω		
R25	1K		
R26	560Ω		
R27	100K		
R28	SEE NOTE 4		
R29	1K		
R30	2.2K		
R31	150Ω		
R32	4.7Ω, 1-WATT		
R33	470Ω, 2-WATT		
R34	2.7Ω		
R35	.47Ω		
R36	470Ω		
R37	2.7Ω		
R38	.47Ω		
C1	100 PF	CAPACITOR	
C2	.01		
C3	.22		
C4	100 PF		
C5	.01		
C6	.22		
C7	100/50		
C8	250/25		
C9	250/25		
C10	1000/25		
C11	1000/25		
C12	.02/600		
C13	10/15		
C14	.01		
C15	.01		
C16	1000/25		
C17	5/15		
Q1	2N663	TRANSISTOR	
Q2	DTG110		
Q3	DTG110		
Q4	2N663		
SR1	SC2E	SIL RECT	
SR2	66-5999		
J1		CLOSED CKT. JACK	
J2			
J3			
L1		PILOT LAMP	
F1	1A	FUSE	
F2	3A		
F3	3A		
SW1		D.P.D.T. CENTER OFF	
T1	620	POWER TRANS.	
T2	99A7	DRIVER TRANS.	
SO-1		STD OUTLET	
C18	5/15	CAPACITOR	
J4		OPEN JACK	

THE STANDEL CO.
TEMPLE CITY, CALIF.

SCHEMATIC DIAGRAM
ALL CUSTOM REVERB MODELS

DRAFTSMAN	CHECKED	APPROVED	DATE
I.D.S.			

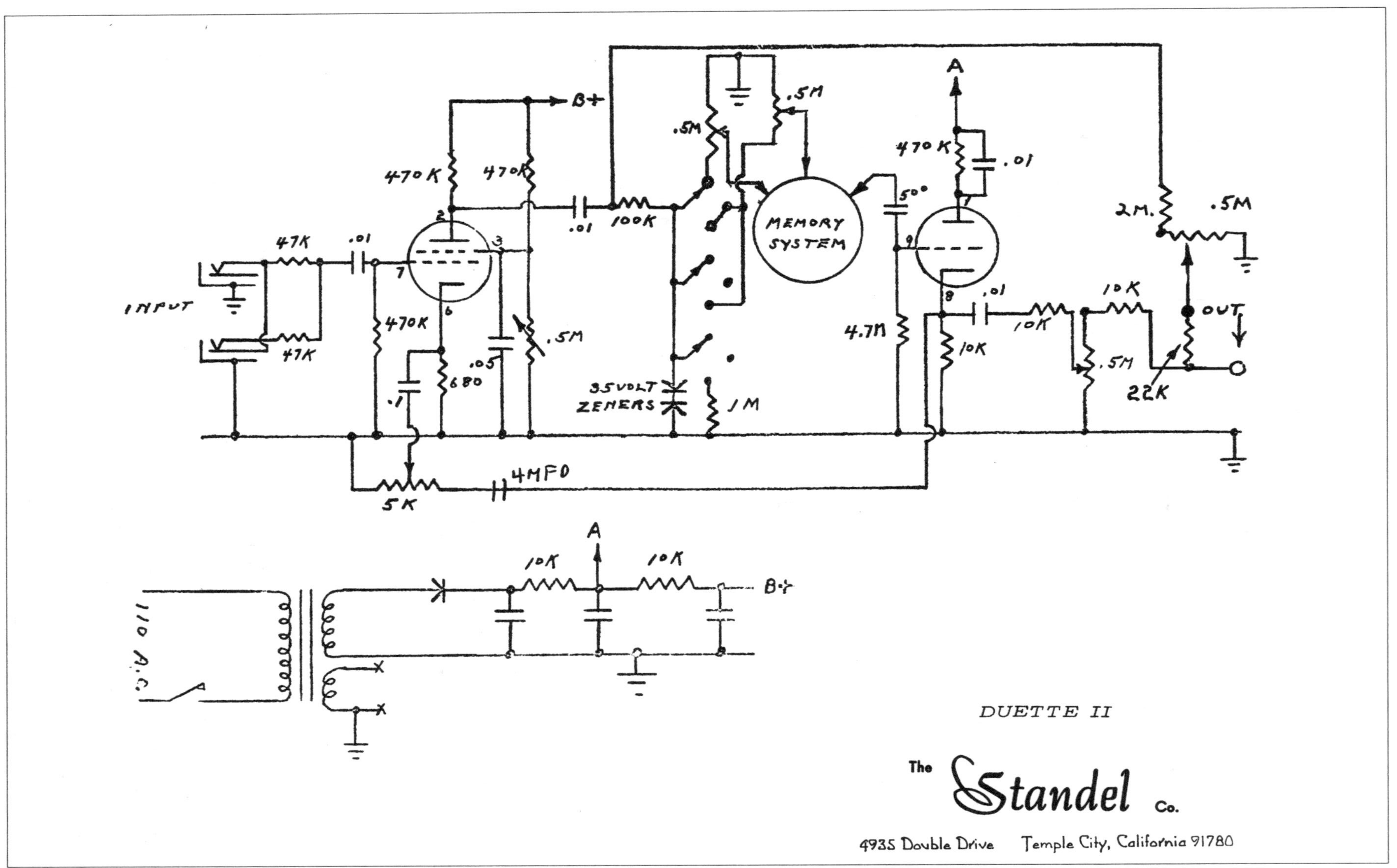

B+
470K
470K
.01
100K
.5M
.5M
A
470K
.01
MEMORY SYSTEM
50°
2M.
.5M
47K
.01
INPUT
47K
470K
.05
.5M
.1
680
35 VOLT ZENERS
1M
4.7M
10K
10K
10K
.5M
OUT
22K
4MFD
5K
A
10K
10K
B+
110 A.C.
DUETTE II
The Standel Co.
4935 Double Drive Temple City, California 91780

Standel

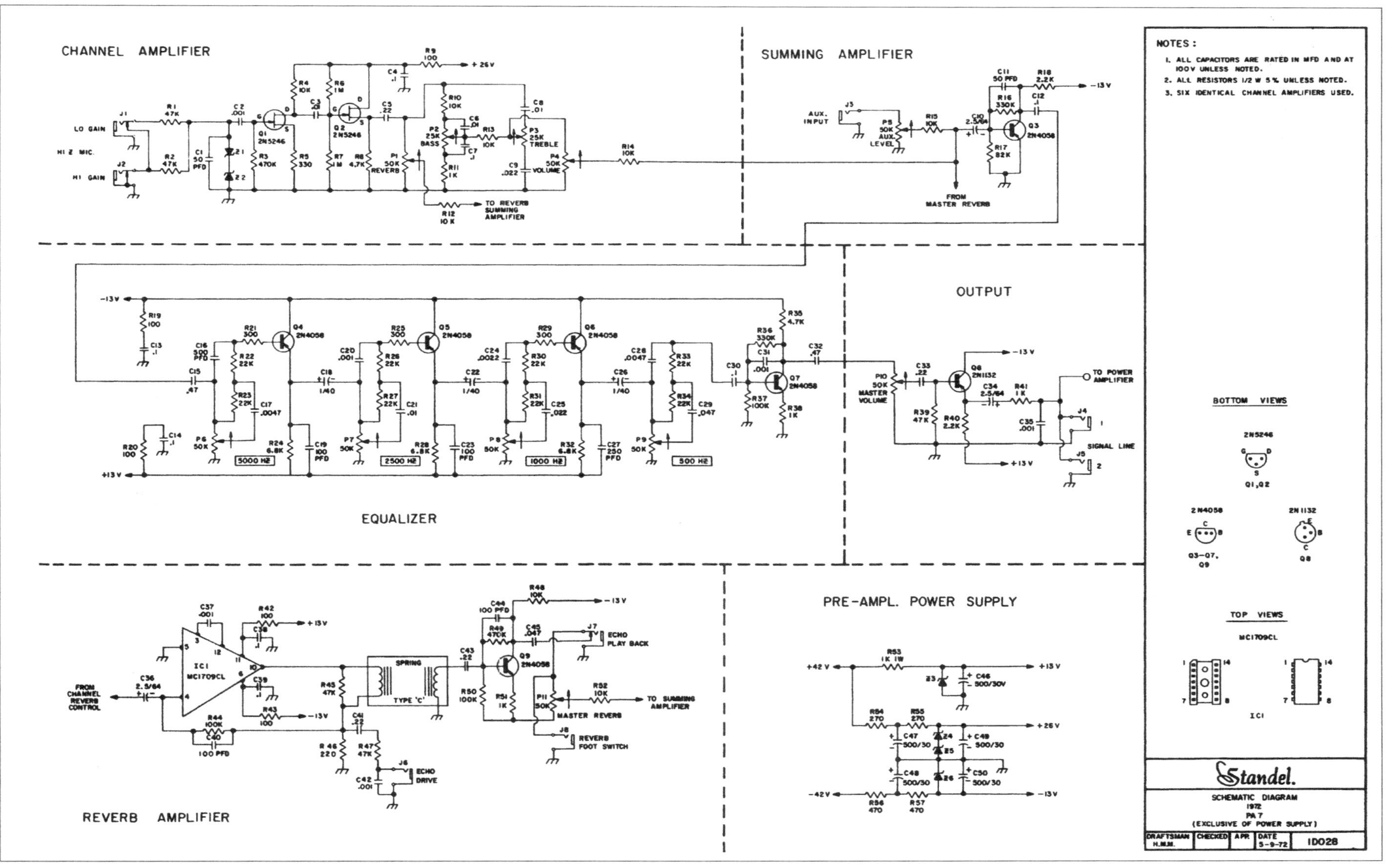
CHANNEL AMPLIFIER
SUMMING AMPLIFIER
EQUALIZER
OUTPUT
REVERB AMPLIFIER
PRE-AMPL. POWER SUPPLY
LO GAIN
HI Z MIC.
HI GAIN
AUX. INPUT
FROM MASTER REVERB
TO REVERB SUMMING AMPLIFIER
TO POWER AMPLIFIER
SIGNAL LINE
FROM CHANNEL REVERB CONTROL
ECHO PLAY BACK
ECHO DRIVE
REVERB FOOT SWITCH
MASTER REVERB
TO SUMMING AMPLIFIER
SPRING
TYPE 'C'
5000 HZ
2500 HZ
1000 HZ
500 HZ
NOTES:
1. ALL CAPACITORS ARE RATED IN MFD AND AT 100V UNLESS NOTED.
2. ALL RESISTORS 1/2 W 5% UNLESS NOTED.
3. SIX IDENTICAL CHANNEL AMPLIFIERS USED.
BOTTOM VIEWS
2N5246
Q1,Q2
2N4058
Q3–Q7, Q9
2N1132
Q8
TOP VIEWS
MC1709CL
IC1
Standel.
SCHEMATIC DIAGRAM
1972
PA 7
(EXCLUSIVE OF POWER SUPPLY)
DRAFTSMAN H.M.M.
CHECKED
APR.
DATE 5-9-72
1D028

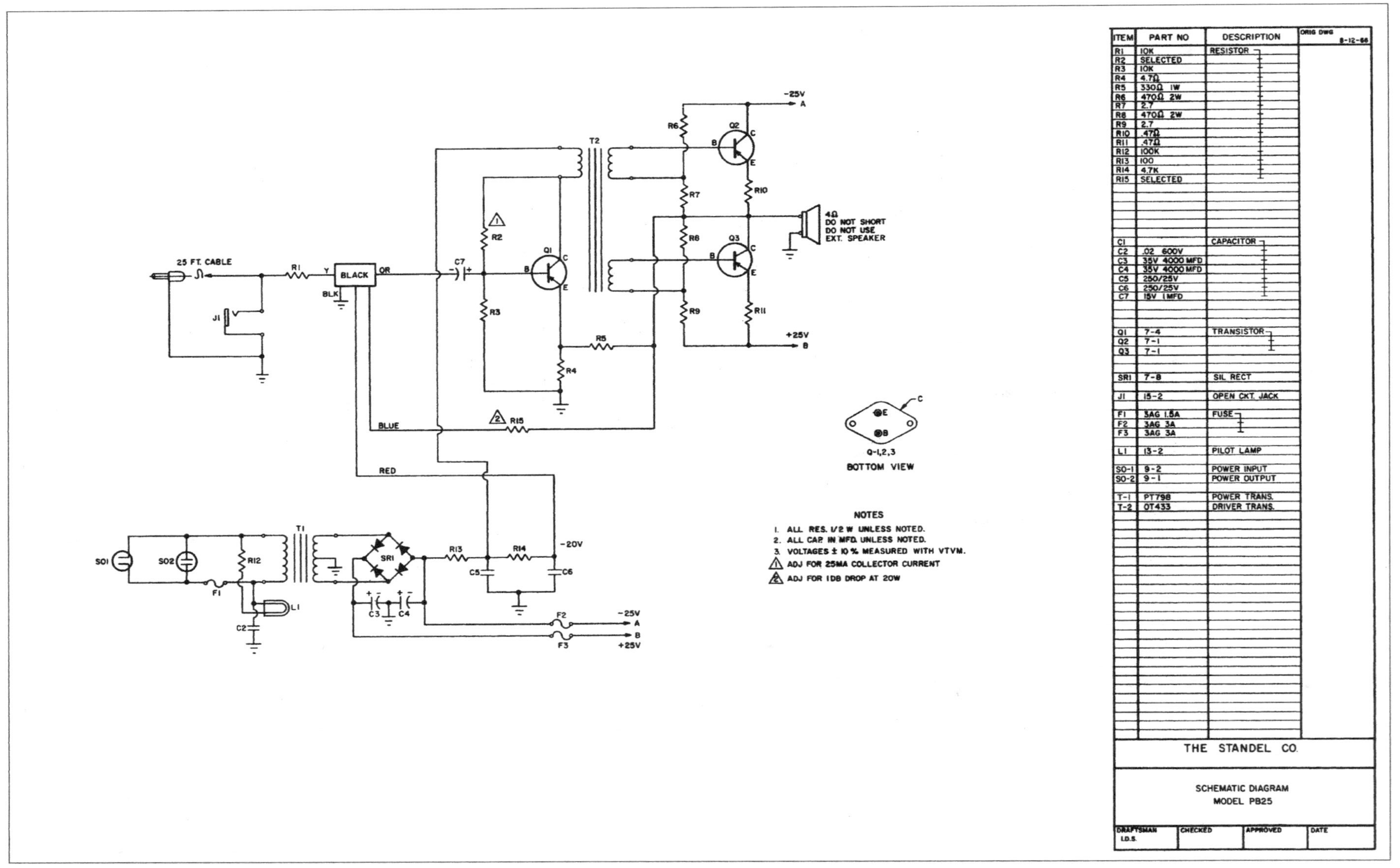

ITEM	PART NO	DESCRIPTION
R1	10K	RESISTOR
R2	SELECTED	
R3	10K	
R4	4.7Ω	
R5	330Ω 1W	
R6	470Ω 2W	
R7	2.7	
R8	470Ω 2W	
R9	2.7	
R10	.47Ω	
R11	.47Ω	
R12	100K	
R13	100	
R14	4.7K	
R15	SELECTED	
C1		CAPACITOR
C2	.02 600V	
C3	35V 4000 MFD	
C4	35V 4000 MFD	
C5	250/25V	
C6	250/25V	
C7	15V 1 MFD	
Q1	7-4	TRANSISTOR
Q2	7-1	
Q3	7-1	
SR1	7-8	SIL. RECT
J1	15-2	OPEN CKT. JACK
F1	3AG 1.5A	FUSE
F2	3AG 3A	
F3	3AG 3A	
L1	13-2	PILOT LAMP
SO-1	9-2	POWER INPUT
SO-2	9-1	POWER OUTPUT
T-1	PT798	POWER TRANS.
T-2	OT433	DRIVER TRANS.

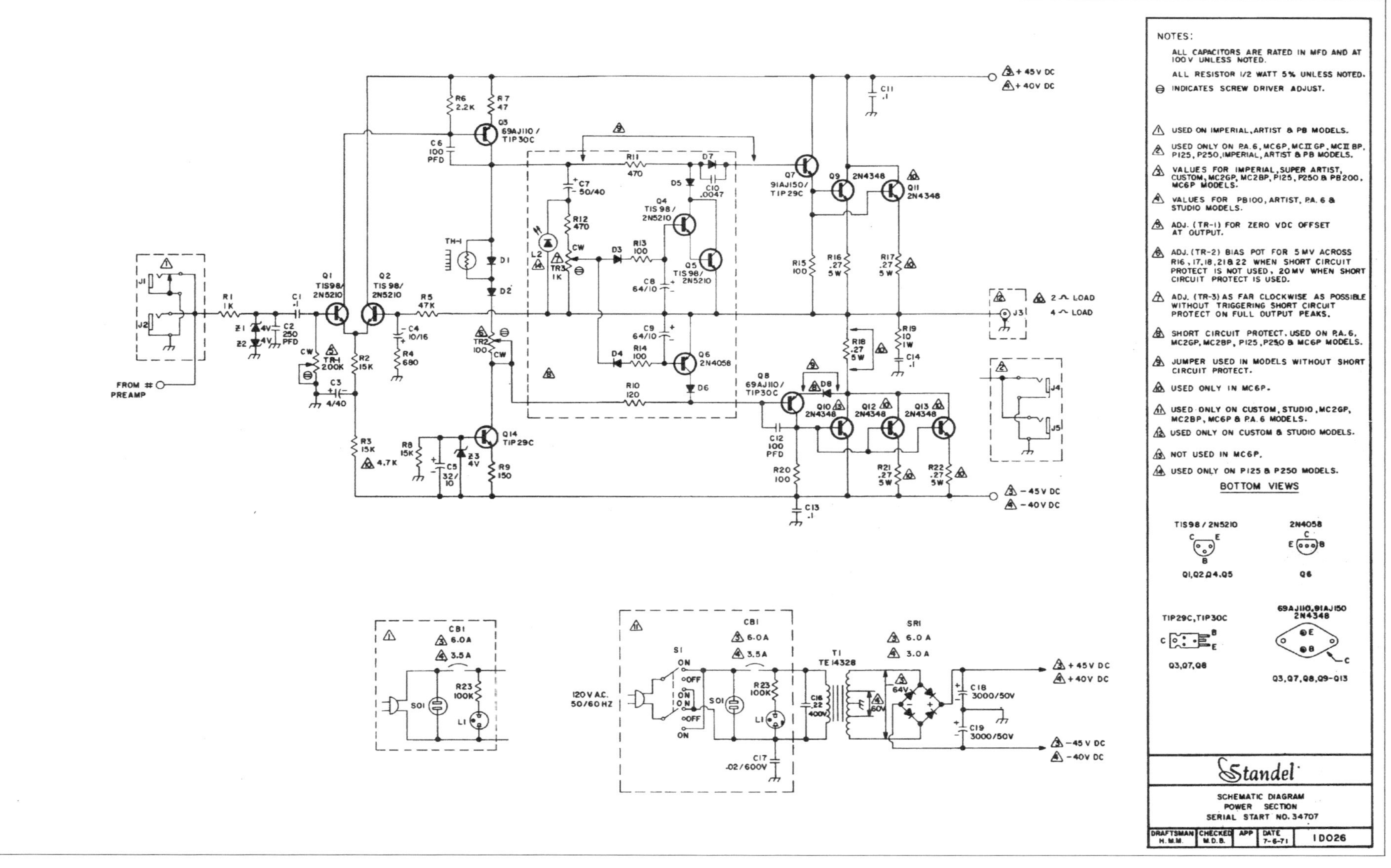
NOTES:
ALL CAPACITORS ARE RATED IN MFD AND AT 100 V UNLESS NOTED.
ALL RESISTOR 1/2 WATT 5% UNLESS NOTED.
⊖ INDICATES SCREW DRIVER ADJUST.
1 USED ON IMPERIAL, ARTIST & PB MODELS.
2 USED ONLY ON P.A. 6, MC6P, MCII GP, MCII BP, P125, P250, IMPERIAL, ARTIST & PB MODELS.
3 VALUES FOR IMPERIAL, SUPER ARTIST, CUSTOM, MC2GP, MC2BP, P125, P250 & PB200, MC6P MODELS.
4 VALUES FOR PB100, ARTIST, P.A. 6 & STUDIO MODELS.
5 ADJ. (TR-1) FOR ZERO VDC OFFSET AT OUTPUT.
6 ADJ. (TR-2) BIAS POT FOR 5 MV ACROSS R16, 17, 18, 21 & 22 WHEN SHORT CIRCUIT PROTECT IS NOT USED, 20 MV WHEN SHORT CIRCUIT PROTECT IS USED.
7 ADJ. (TR-3) AS FAR CLOCKWISE AS POSSIBLE WITHOUT TRIGGERING SHORT CIRCUIT PROTECT ON FULL OUTPUT PEAKS.
8 SHORT CIRCUIT PROTECT. USED ON P.A. 6, MC2GP, MC2BP, P125, P250 & MC6P MODELS.
9 JUMPER USED IN MODELS WITHOUT SHORT CIRCUIT PROTECT.
10 USED ONLY IN MC6P.
11 USED ONLY ON CUSTOM, STUDIO, MC2GP, MC2BP, MC6P & P.A. 6 MODELS.
12 USED ONLY ON CUSTOM & STUDIO MODELS.
13 NOT USED IN MC6P.
14 USED ONLY ON P125 & P250 MODELS.
BOTTOM VIEWS
TIS98 / 2N5210
Q1, Q2, Q4, Q5
2N4058
Q6
TIP29C, TIP30C
Q3, Q7, Q8
69AJ110, 91AJ150
2N4348
Q3, Q7, Q8, Q9-Q13
Standel
SCHEMATIC DIAGRAM
POWER SECTION
SERIAL START NO. 34707
DRAFTSMAN H.M.M.
CHECKED M.D.B.
APP
DATE 7-6-71
1D026
FROM # PREAMP
+45 V DC
+40 V DC
-45 V DC
-40 V DC
2 Ω LOAD
4 Ω LOAD
120 V A.C. 50/60 HZ

PR5B & PR5R **MASTER CONTROL**

Standel

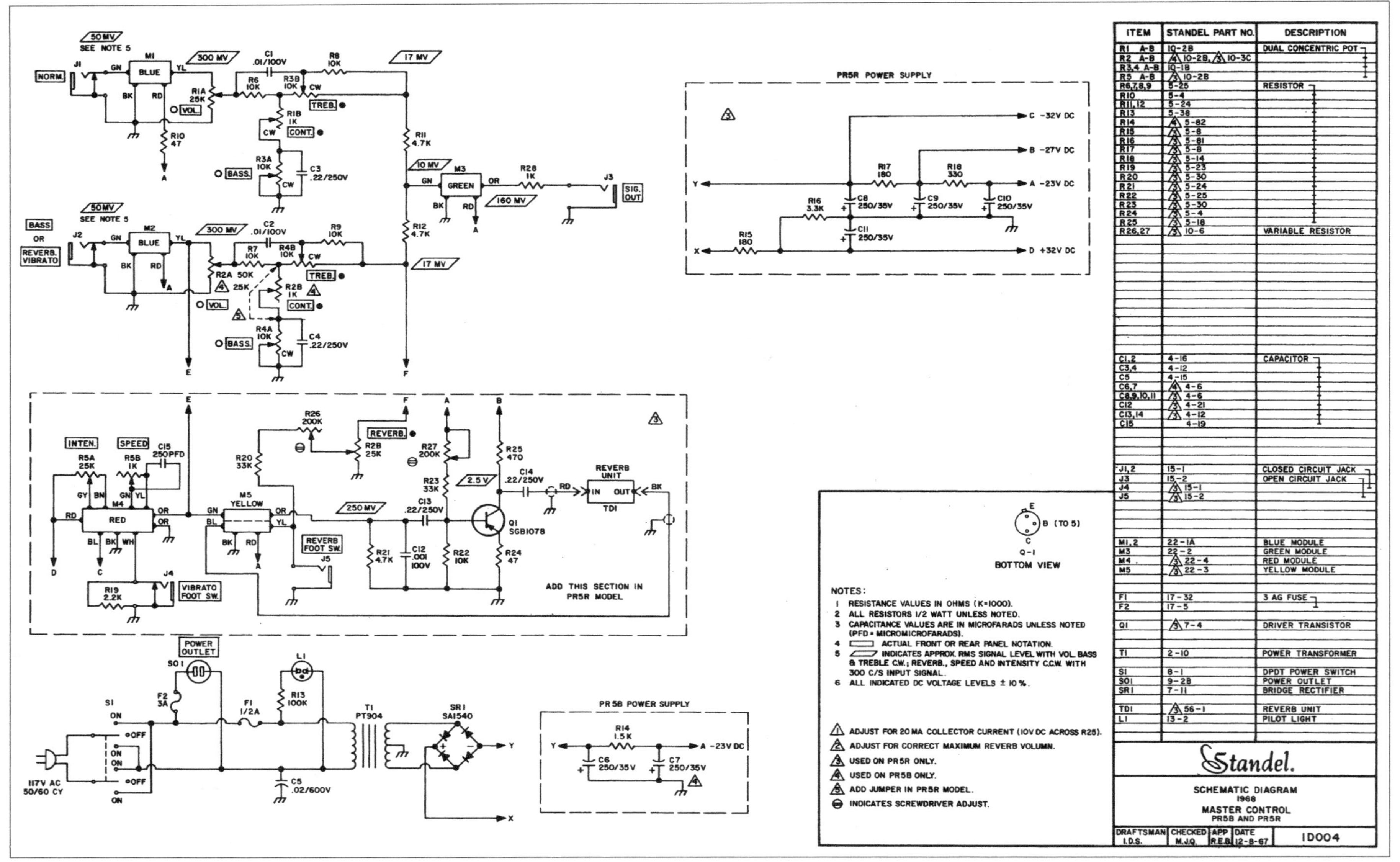

ITEM	STANDEL PART NO.	DESCRIPTION
R1 A-B	10-2B	DUAL CONCENTRIC POT
R2 A-B	(4) 10-2B, (3) 10-3C	
R3,4 A-B	10-1B	
R5 A-B	(3) 10-2B	
R6,7,8,9	5-25	RESISTOR
R10	5-4	
R11,12	5-24	
R13	5-38	
R14	(4) 5-82	
R15	(3) 5-8	
R16	(3) 5-81	
R17	(3) 5-8	
R18	(3) 5-14	
R19	(3) 5-23	
R20	(3) 5-30	
R21	(3) 5-24	
R22	(3) 5-25	
R23	(3) 5-30	
R24	(3) 5-4	
R25	(3) 5-18	
R26,27	(3) 10-6	VARIABLE RESISTOR
C1,2	4-16	CAPACITOR
C3,4	4-12	
C5	4-15	
C6,7	(4) 4-6	
C8,9,10,11	(3) 4-6	
C12	(3) 4-21	
C13,14	(3) 4-12	
C15	4-19	
J1,2	15-1	CLOSED CIRCUIT JACK
J3	15-2	OPEN CIRCUIT JACK
J4	(3) 15-1	
J5	(3) 15-2	
M1,2	22-1A	BLUE MODULE
M3	22-2	GREEN MODULE
M4	(3) 22-4	RED MODULE
M5	(3) 22-3	YELLOW MODULE
F1	17-32	3 AG FUSE
F2	17-5	
Q1	(3) 7-4	DRIVER TRANSISTOR
T1	2-10	POWER TRANSFORMER
S1	8-1	DPDT POWER SWITCH
SO1	9-2B	POWER OUTLET
SR1	7-11	BRIDGE RECTIFIER
TD1	(3) 56-1	REVERB UNIT
L1	13-2	PILOT LIGHT

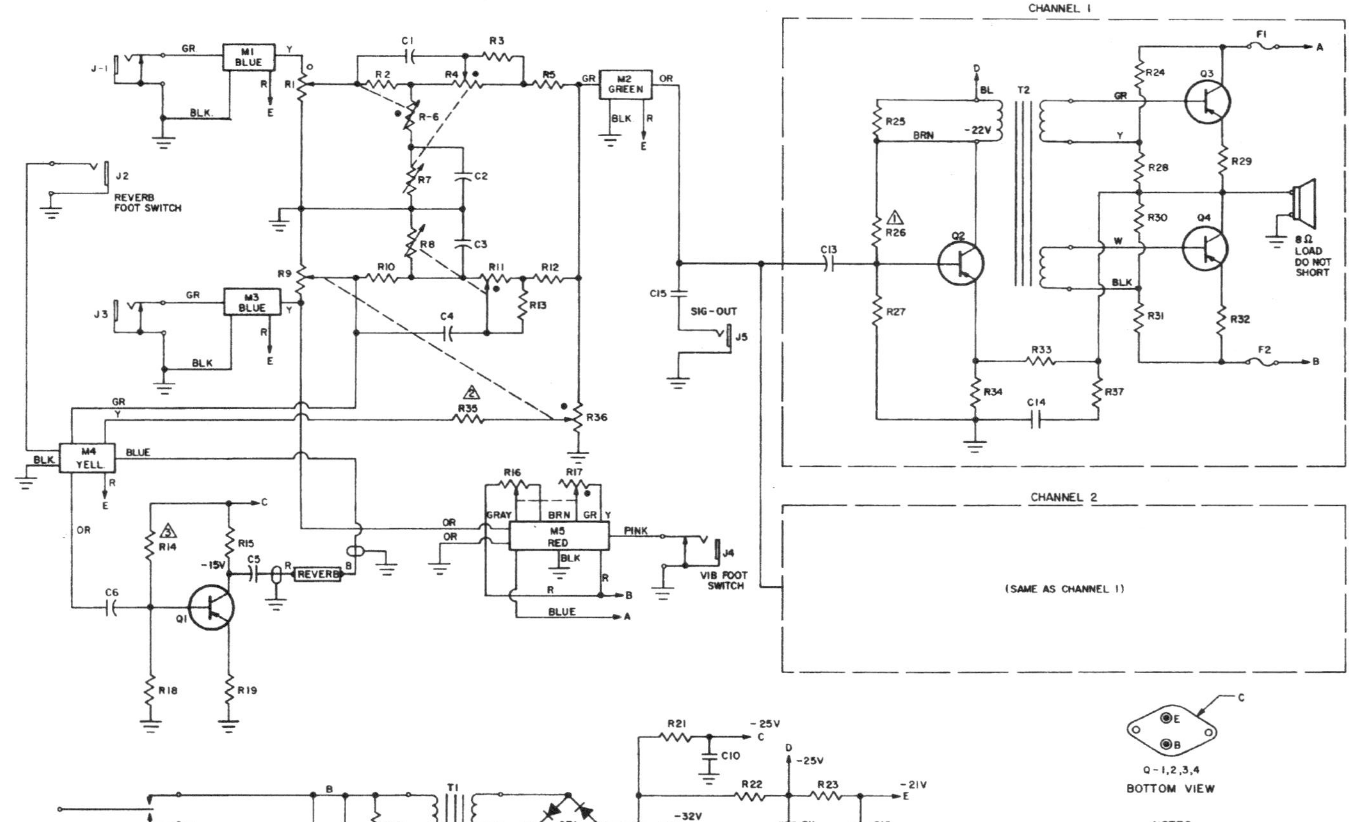

NOTES

1. ADJ FOR 30MA COLLECTOR CURRENT
2. ADJ FOR CORRECT REVERB INTENSITY
3. ADJ FOR 10V DROP ACROSS R14
4. OMIT M4 & 5 ON BASS MODELS
5. ALL RES. 1/2 W UNLESS NOTED.
6. ALL CAP. IN MFD. UNLESS NOTED.
7. VOLTAGES ± 10% MEASURED WITH VTVM
8. R6 OMITTED ON ARTIST MODELS

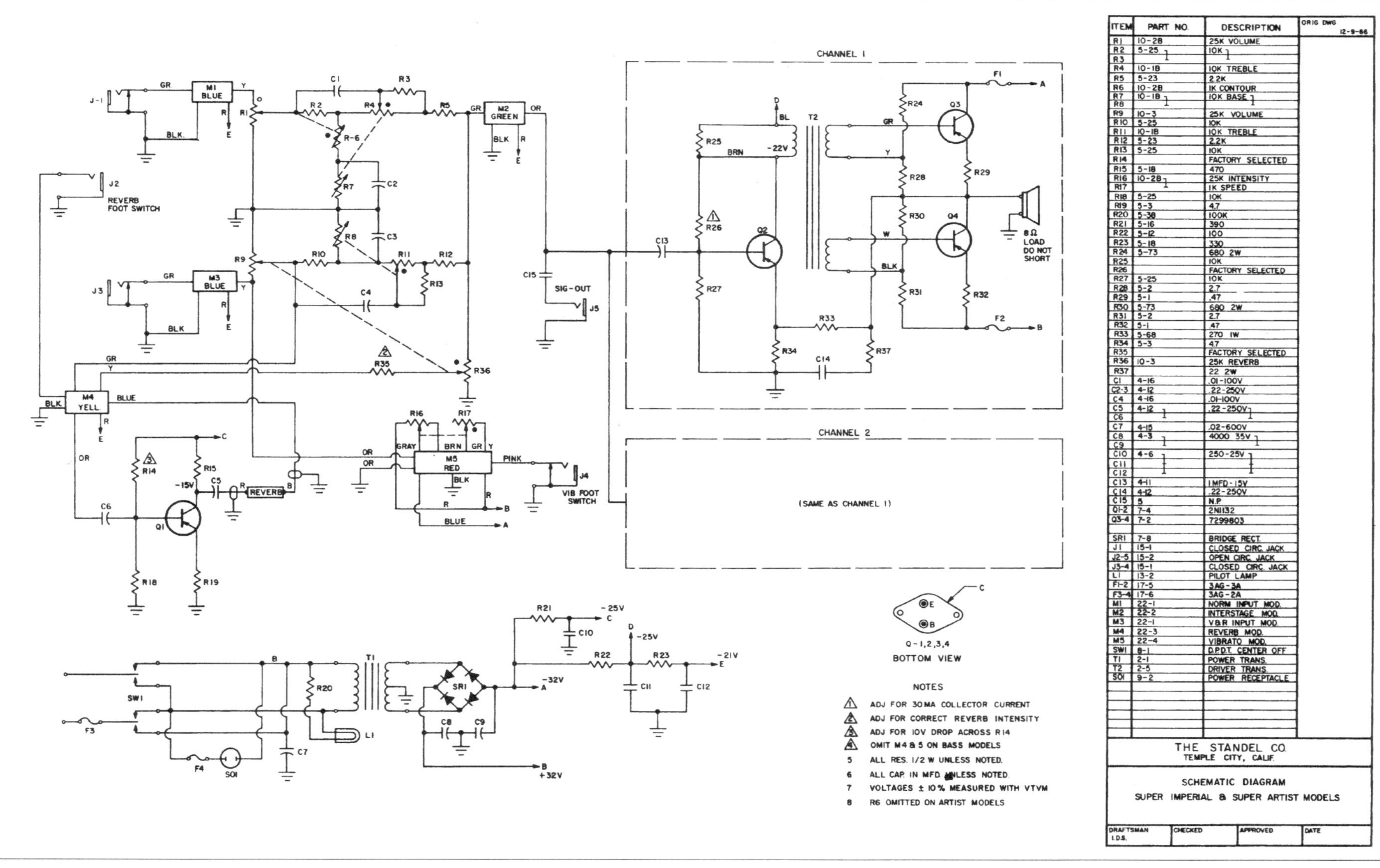

ITEM	PART NO.	DESCRIPTION
R1	10-28	25K VOLUME
R2	5-25	10K
R3		
R4	10-18	10K TREBLE
R5	5-23	2.2K
R6	10-28	1K CONTOUR
R7	10-18	10K BASE
R8		
R9	10-3	25K VOLUME
R10	5-25	10K
R11	10-18	10K TREBLE
R12	5-23	2.2K
R13	5-25	10K
R14		FACTORY SELECTED
R15	5-18	470
R16	10-28	25K INTENSITY
R17		1K SPEED
R18	5-25	10K
R19	5-3	4.7
R20	5-38	100K
R21	5-16	390
R22	5-12	100
R23	5-18	330
R24	5-73	680 2W
R25		10K
R26		FACTORY SELECTED
R27	5-25	10K
R28	5-2	2.7
R29	5-1	.47
R30	5-73	680 2W
R31	5-2	2.7
R32	5-1	.47
R33	5-68	270 1W
R34	5-3	4.7
R35		FACTORY SELECTED
R36	10-3	25K REVERB
R37		22 2W
C1	4-16	.01-100V
C2-3	4-12	.22-250V
C4	4-16	.01-100V
C5	4-12	.22-250V
C6		
C7	4-15	.02-600V
C8	4-3	4000 35V
C9		
C10	4-6	250-25V
C11		
C12		
C13	4-11	1MFD-15V
C14	4-12	.22-250V
C15	5	N.P.
Q1-2	7-4	2N1132
Q3-4	7-2	7299803
SR1	7-8	BRIDGE RECT.
J1	15-1	CLOSED CIRC. JACK
J2-5	15-2	OPEN CIRC. JACK
J3-4	15-1	CLOSED CIRC. JACK
L1	13-2	PILOT LAMP
F1-2	17-5	3AG-3A
F3-4	17-6	3AG-2A
M1	22-1	NORM INPUT MOD.
M2	22-2	INTERSTAGE MOD.
M3	22-1	V&R INPUT MOD.
M4	22-3	REVERB MOD.
M5	22-4	VIBRATO MOD.
SW1	8-1	D.P.D.T. CENTER OFF
T1	2-1	POWER TRANS.
T2	2-5	DRIVER TRANS.
SO1	9-2	POWER RECEPTACLE

ORIG DWG 12-9-66

THE STANDEL CO.
TEMPLE CITY, CALIF.

SCHEMATIC DIAGRAM
SUPER IMPERIAL & SUPER ARTIST MODELS

DRAFTSMAN	CHECKED	APPROVED	DATE
I.D.S.			

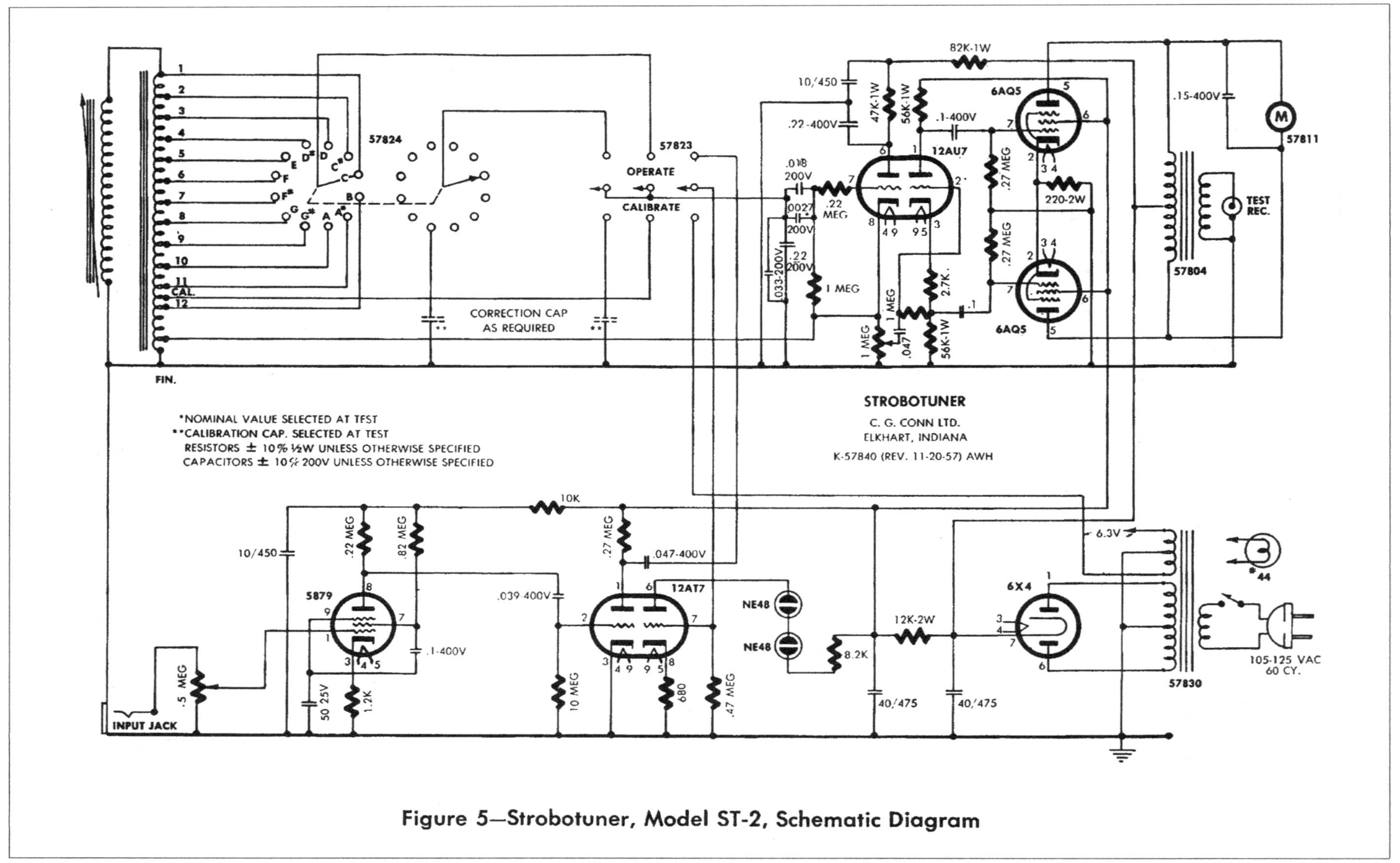

Figure 5—Strobotuner, Model ST-2, Schematic Diagram

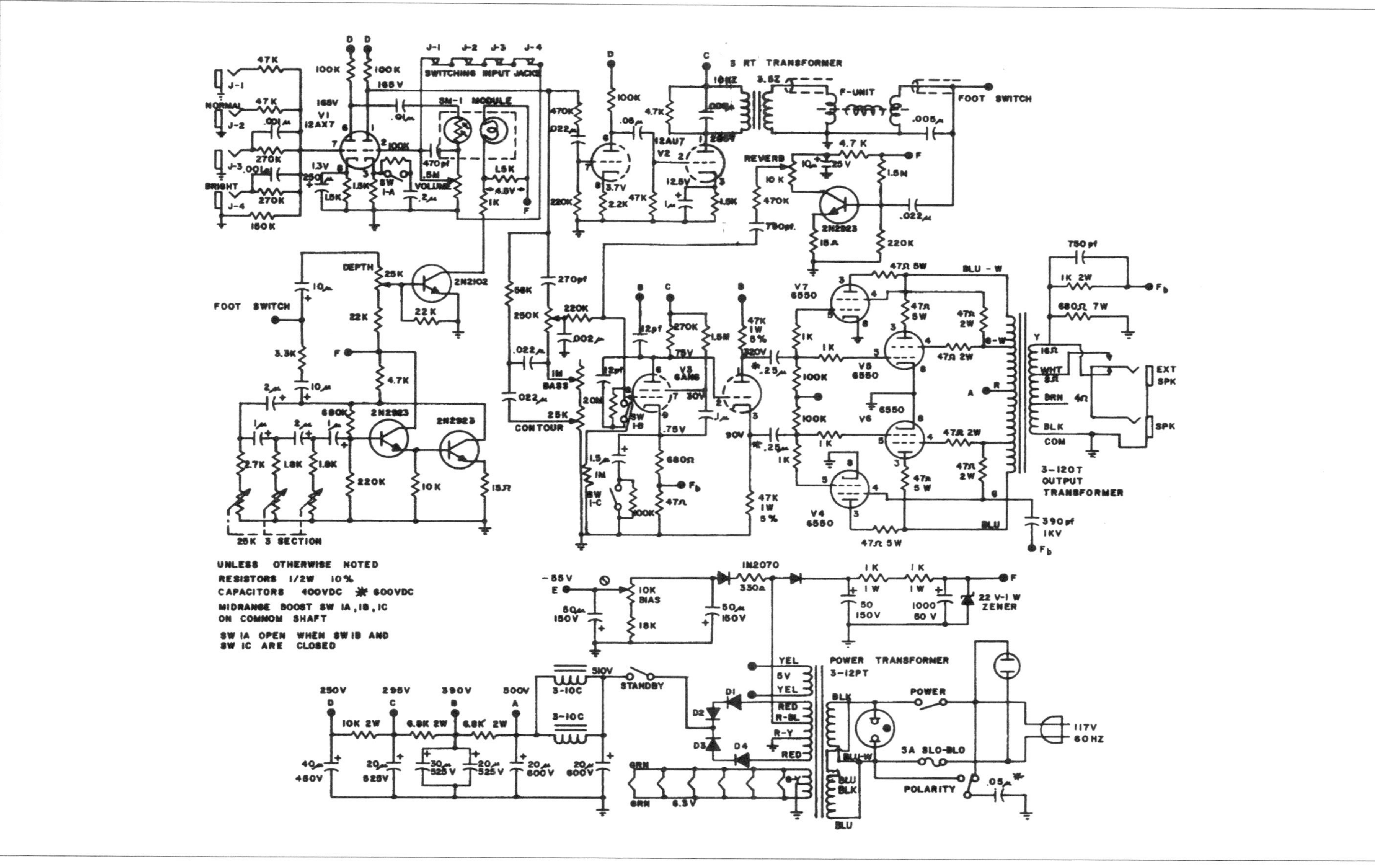
SWITCHING INPUT JACKS
SM-1 MODULE
3 RT TRANSFORMER
F-UNIT
FOOT SWITCH
REVERB
DEPTH
FOOT SWITCH
BASS
CONTOUR
25K 3 SECTION
UNLESS OTHERWISE NOTED
RESISTORS 1/2W 10%
CAPACITORS 400VDC * 600VDC
MIDRANGE BOOST SW IA, IB, IC
ON COMMON SHAFT
SW IA OPEN WHEN SW IB AND
SW IC ARE CLOSED
3-120T
OUTPUT
TRANSFORMER
EXT SPK
SPK
BIAS
22 V-1 W ZENER
POWER TRANSFORMER
3-12PT
STANDBY
POWER
5A SLO-BLO
POLARITY
117V 60HZ

Sunn

sunn®

1200S

schematic

(6) 12" transducers
option – (6) JBL D123F

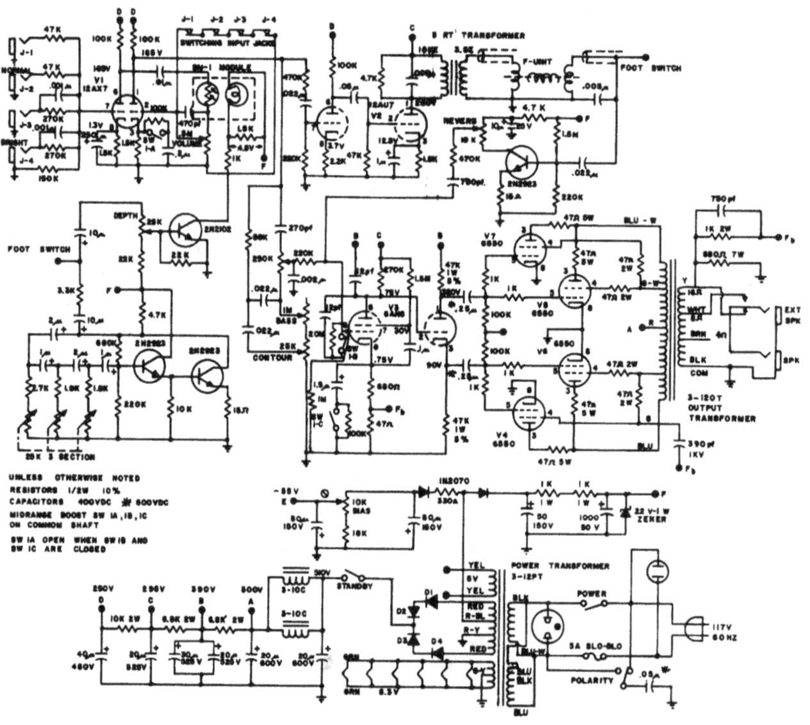

Sunn Musical Equipment Company amburn industrial park, tualatin, oregon 97062 2/18/71

1200s

1211 CROSSOVER

Sunn

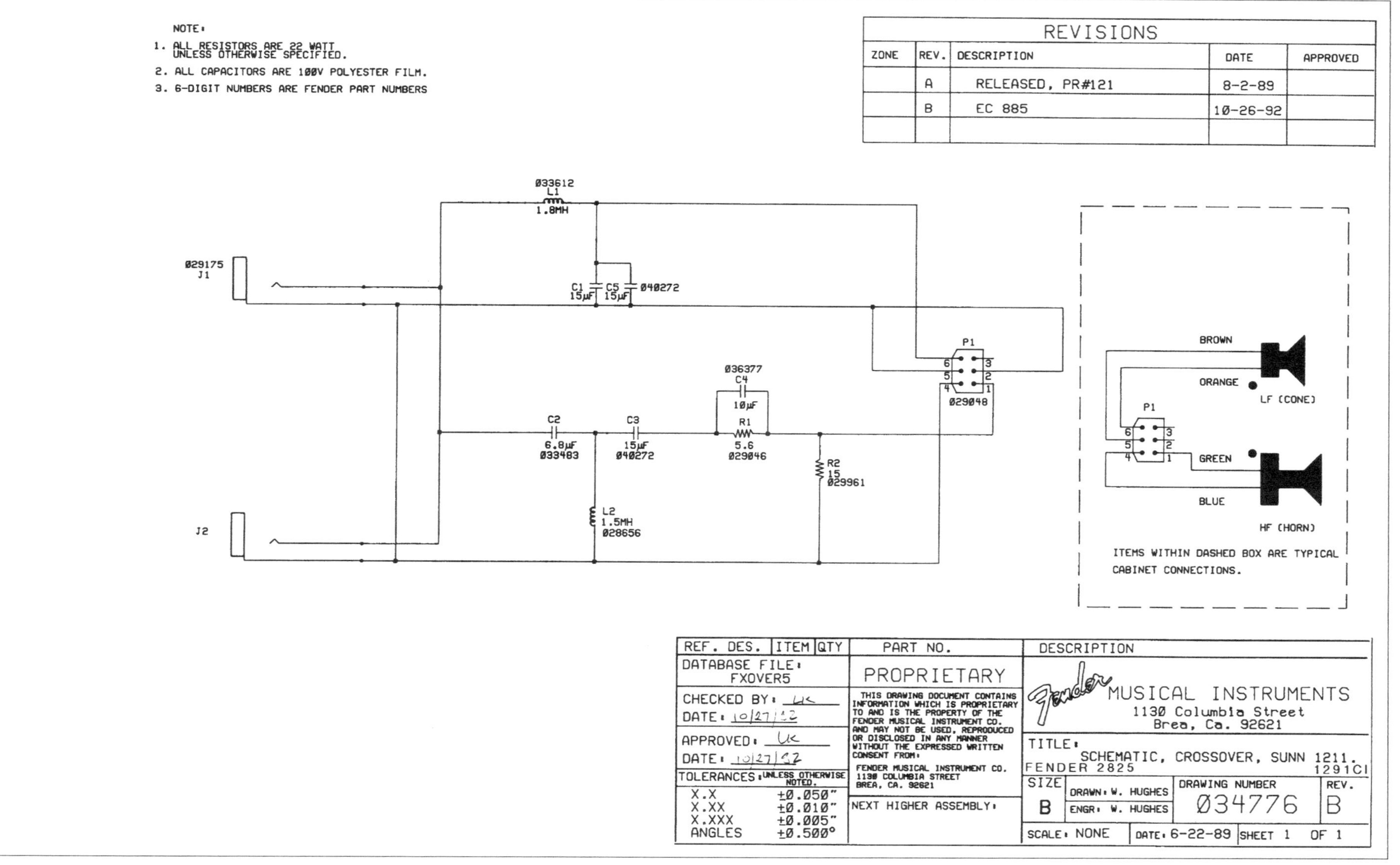

NOTE:
1. ALL RESISTORS ARE 22 WATT UNLESS OTHERWISE SPECIFIED.
2. ALL CAPACITORS ARE 100V POLYESTER FILM.
3. 6-DIGIT NUMBERS ARE FENDER PART NUMBERS
REVISIONS
ZONE | REV. | DESCRIPTION | DATE | APPROVED
A | RELEASED, PR#121 | 8-2-89
B | EC 885 | 10-26-92
033612
L1
1.8MH
029175
J1
C1
15µF
C5
15µF
040272
J2
P1
029048
036377
C4
10µF
R1
5.6
029046
C2
6.8µF
033483
C3
15µF
040272
R2
15
029961
L2
1.5MH
028656
BROWN
ORANGE
LF (CONE)
GREEN
BLUE
HF (HORN)
ITEMS WITHIN DASHED BOX ARE TYPICAL CABINET CONNECTIONS.
REF. DES. | ITEM | QTY | PART NO. | DESCRIPTION
DATABASE FILE: FXOVER5
CHECKED BY: UK
DATE: 10/27/92
APPROVED: UK
DATE: 10/27/92
TOLERANCES: UNLESS OTHERWISE NOTED.
X.X ±0.050"
X.XX ±0.010"
X.XXX ±0.005"
ANGLES ±0.500°
PROPRIETARY
THIS DRAWING DOCUMENT CONTAINS INFORMATION WHICH IS PROPRIETARY TO AND IS THE PROPERTY OF THE FENDER MUSICAL INSTRUMENT CO. AND MAY NOT BE USED, REPRODUCED OR DISCLOSED IN ANY MANNER WITHOUT THE EXPRESSED WRITTEN CONSENT FROM:
FENDER MUSICAL INSTRUMENT CO.
1130 COLUMBIA STREET
BREA, CA. 92621
NEXT HIGHER ASSEMBLY:
Fender MUSICAL INSTRUMENTS
1130 Columbia Street
Brea, Ca. 92621
TITLE:
SCHEMATIC, CROSSOVER, SUNN 1211, 1291CI
FENDER 2825
SIZE B
DRAWN: W. HUGHES
ENGR: W. HUGHES
DRAWING NUMBER 034776
REV. B
SCALE: NONE
DATE: 6-22-89
SHEET 1 OF 1

1225 CROSSOVER

Sunn

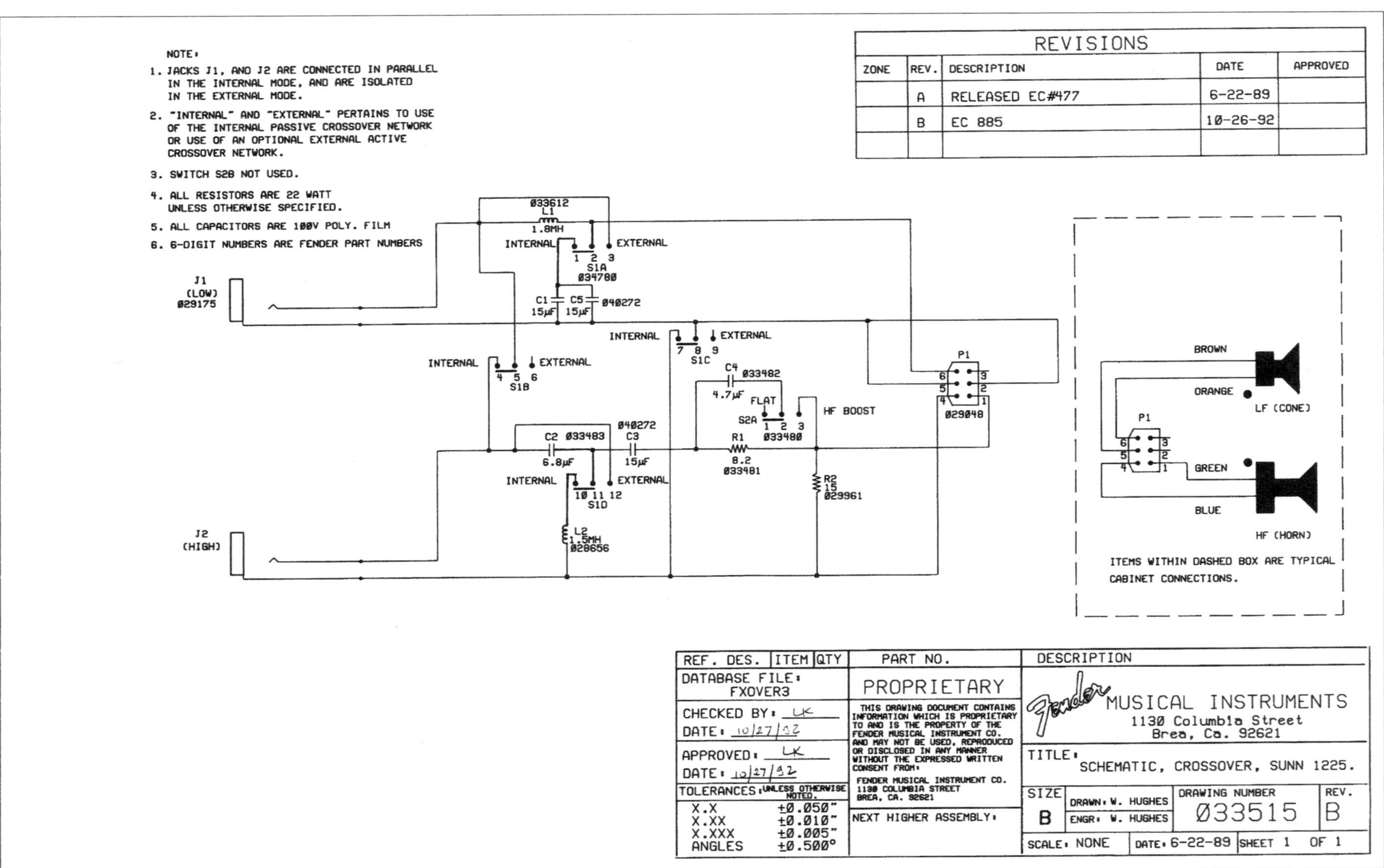

NOTE:
1. JACKS J1, AND J2 ARE CONNECTED IN PARALLEL IN THE INTERNAL MODE, AND ARE ISOLATED IN THE EXTERNAL MODE.
2. "INTERNAL" AND "EXTERNAL" PERTAINS TO USE OF THE INTERNAL PASSIVE CROSSOVER NETWORK OR USE OF AN OPTIONAL EXTERNAL ACTIVE CROSSOVER NETWORK.
3. SWITCH S2B NOT USED.
4. ALL RESISTORS ARE 22 WATT UNLESS OTHERWISE SPECIFIED.
5. ALL CAPACITORS ARE 100V POLY. FILM
6. 6-DIGIT NUMBERS ARE FENDER PART NUMBERS
REVISIONS
ZONE
REV.
DESCRIPTION
DATE
APPROVED
A
RELEASED EC#477
6-22-89
B
EC 885
10-26-92
J1 (LOW) 029175
J2 (HIGH)
033612 L1 1.8MH
INTERNAL
EXTERNAL
1 2 3 S1A 034780
C1 15µF
C5 15µF 040272
4 5 6 S1B
7 8 9 S1C
C4 033482 4.7µF
FLAT
S2A 1 2 3 033480
HF BOOST
R1 8.2 033481
R2 15 029961
C2 033483 6.8µF
040272 C3 15µF
10 11 12 S1D
L2 1.5MH 028656
P1 029048
BROWN
ORANGE
LF (CONE)
GREEN
BLUE
HF (HORN)
ITEMS WITHIN DASHED BOX ARE TYPICAL CABINET CONNECTIONS.
REF. DES.
ITEM
QTY
PART NO.
DESCRIPTION
DATABASE FILE: FXOVER3
CHECKED BY: LK
DATE: 10/27/92
APPROVED: LK
DATE: 10/27/92
TOLERANCES: UNLESS OTHERWISE NOTED.
X.X ±0.050"
X.XX ±0.010"
X.XXX ±0.005"
ANGLES ±0.500°
PROPRIETARY
THIS DRAWING DOCUMENT CONTAINS INFORMATION WHICH IS PROPRIETARY TO AND IS THE PROPERTY OF THE FENDER MUSICAL INSTRUMENT CO. AND MAY NOT BE USED, REPRODUCED OR DISCLOSED IN ANY MANNER WITHOUT THE EXPRESSED WRITTEN CONSENT FROM:
FENDER MUSICAL INSTRUMENT CO. 1130 COLUMBIA STREET BREA, CA. 92621
NEXT HIGHER ASSEMBLY:
Fender MUSICAL INSTRUMENTS
1130 Columbia Street
Brea, Ca. 92621
TITLE: SCHEMATIC, CROSSOVER, SUNN 1225.
SIZE B
DRAWN: W. HUGHES
ENGR: W. HUGHES
DRAWING NUMBER 033515
REV. B
SCALE: NONE
DATE: 6-22-89
SHEET 1 OF 1

1226 CROSSOVER

Sunn

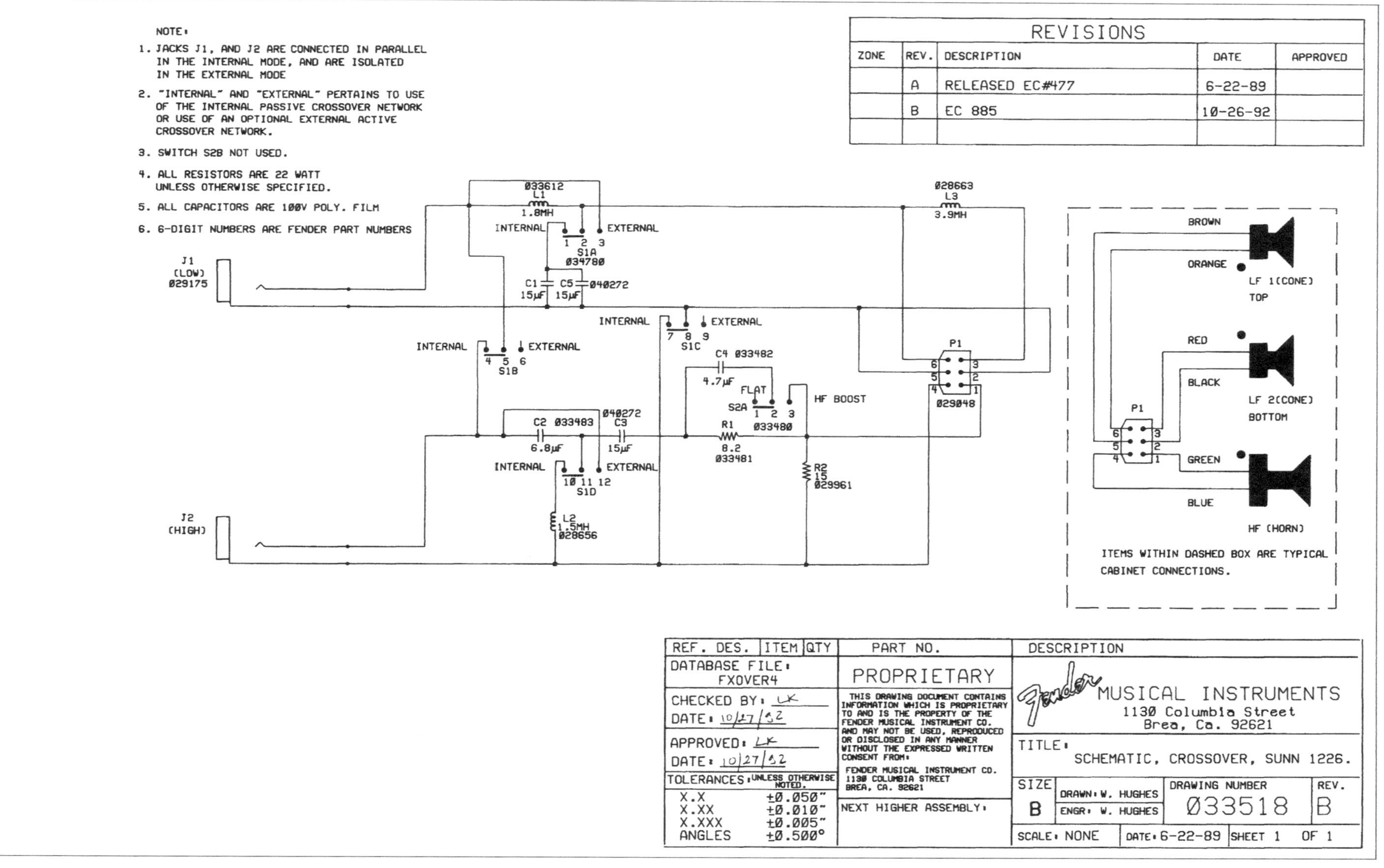

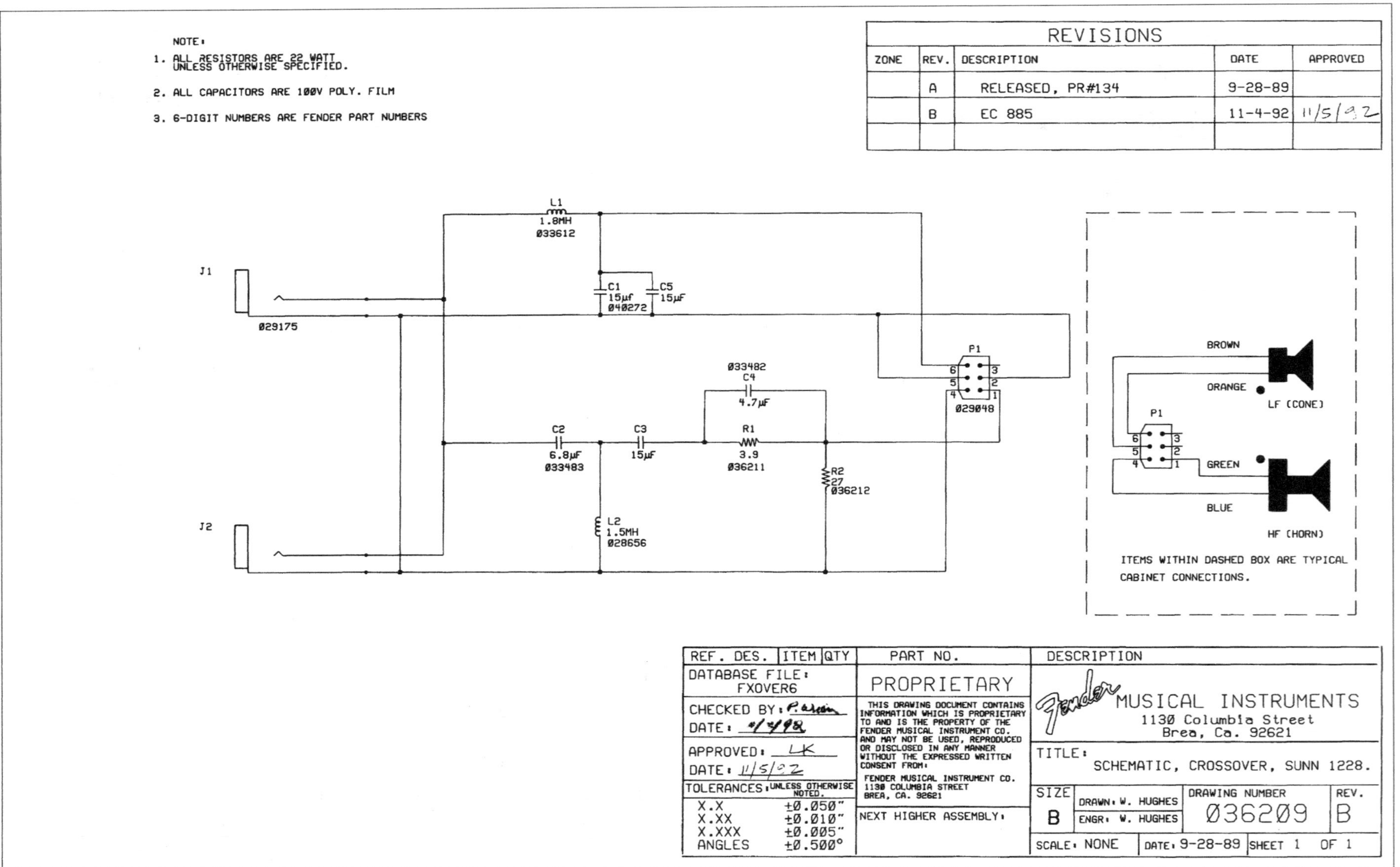

NOTE:
1. ALL RESISTORS ARE 22 WATT UNLESS OTHERWISE SPECIFIED.
2. ALL CAPACITORS ARE 100V POLY. FILM
3. 6-DIGIT NUMBERS ARE FENDER PART NUMBERS
REVISIONS
ZONE | REV. | DESCRIPTION | DATE | APPROVED
A | RELEASED, PR#134 | 9-28-89
B | EC 885 | 11-4-92 | 11/5/92
J1
J2
029175
L1 1.8MH 033612
C1 15µf 040272
C5 15µF
C2 6.8µF 033483
C3 15µF
C4 4.7µF 033482
R1 3.9 036211
R2 27 036212
L2 1.5MH 028656
P1 029048
BROWN
ORANGE
LF (CONE)
GREEN
BLUE
HF (HORN)
ITEMS WITHIN DASHED BOX ARE TYPICAL CABINET CONNECTIONS.
REF. DES. | ITEM | QTY | PART NO. | DESCRIPTION
DATABASE FILE: FXOVER6
CHECKED BY:
APPROVED: LK
DATE: 11/5/92
TOLERANCES: UNLESS OTHERWISE NOTED.
X.X ±0.050"
X.XX ±0.010"
X.XXX ±0.005"
ANGLES ±0.500°
PROPRIETARY
THIS DRAWING DOCUMENT CONTAINS INFORMATION WHICH IS PROPRIETARY TO AND IS THE PROPERTY OF THE FENDER MUSICAL INSTRUMENT CO. AND MAY NOT BE USED, REPRODUCED OR DISCLOSED IN ANY MANNER WITHOUT THE EXPRESSED WRITTEN CONSENT FROM:
FENDER MUSICAL INSTRUMENT CO. 1130 COLUMBIA STREET BREA, CA. 92621
NEXT HIGHER ASSEMBLY:
Fender MUSICAL INSTRUMENTS
1130 Columbia Street
Brea, Ca. 92621
TITLE: SCHEMATIC, CROSSOVER, SUNN 1228.
SIZE B
DRAWN: W. HUGHES
ENGR: W. HUGHES
DRAWING NUMBER 036209
REV. B
SCALE: NONE
DATE: 9-28-89
SHEET 1 OF 1

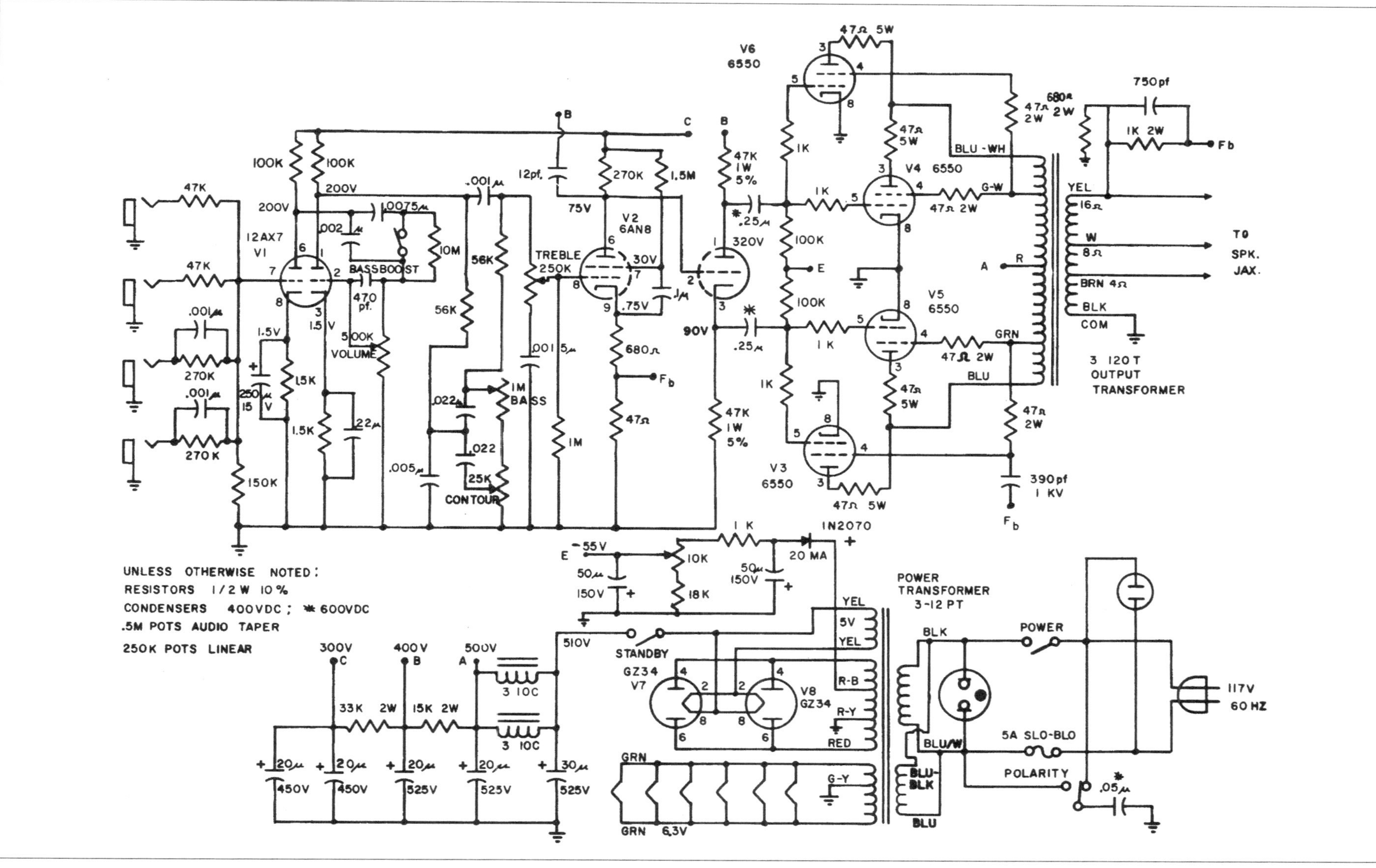
UNLESS OTHERWISE NOTED:
RESISTORS 1/2 W 10%
CONDENSERS 400VDC; ✱ 600VDC
.5M POTS AUDIO TAPER
250K POTS LINEAR
12AX7
V1
V2
6AN8
V3
6550
V4 6550
V5
6550
V6
6550
GZ34
V7
V8
GZ34
BASSBOOST
VOLUME
TREBLE
250K
BASS
CONTOUR
STANDBY
POWER TRANSFORMER
3-12 PT
3 120 T
OUTPUT
TRANSFORMER
TO
SPK.
JAX.
POWER
5A SLO-BLO
POLARITY
117V
60 HZ
1N2070
20 MA
300V
400V
500V
510V
GRN 6.3V

sunn 2000S*

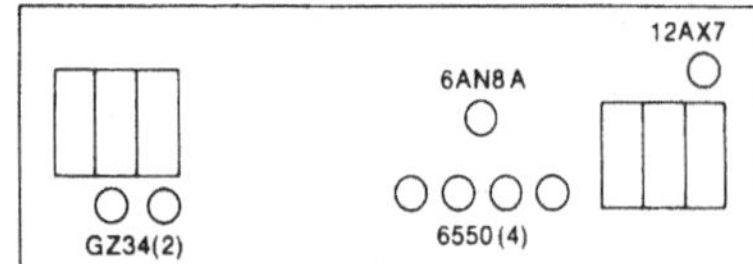

*2000S: With bassboost. After 5-10-69.
Two J.B.L. D140F (D140F-6 after 3-1-69) in rear-loading folded horn enclosure.

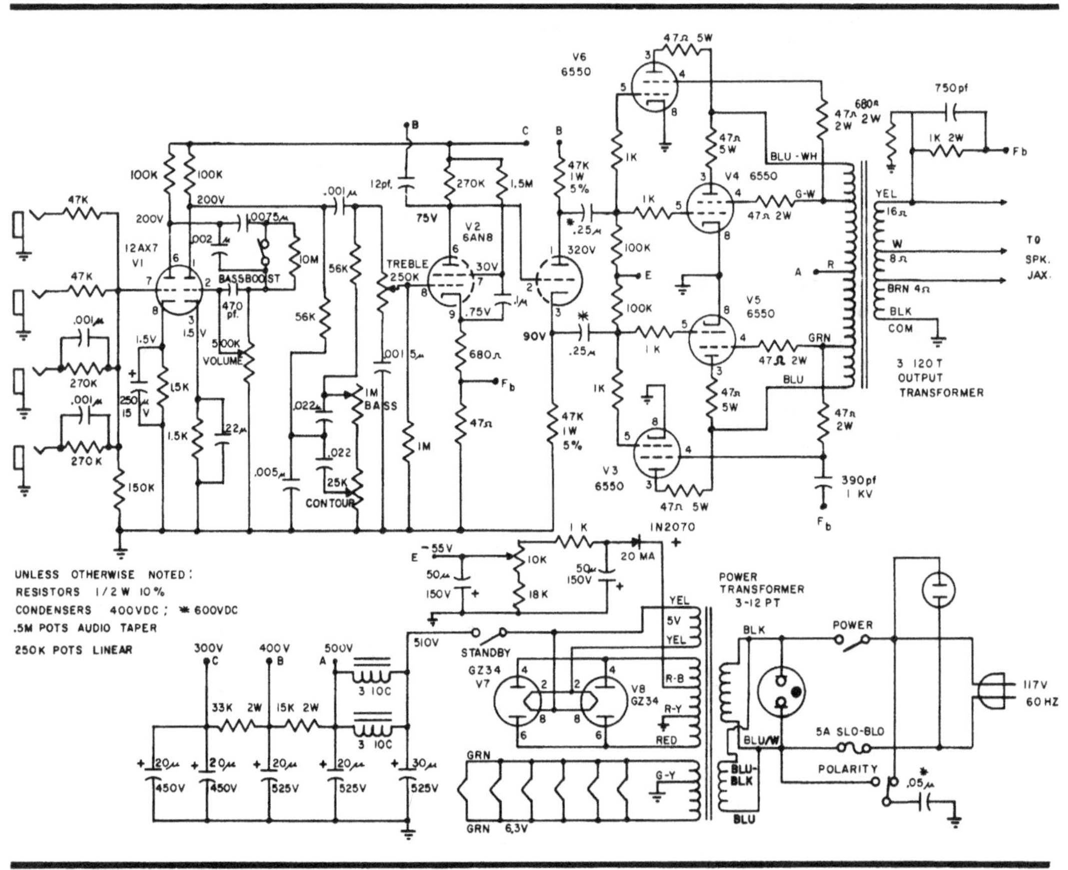

MODEL A

Sunn

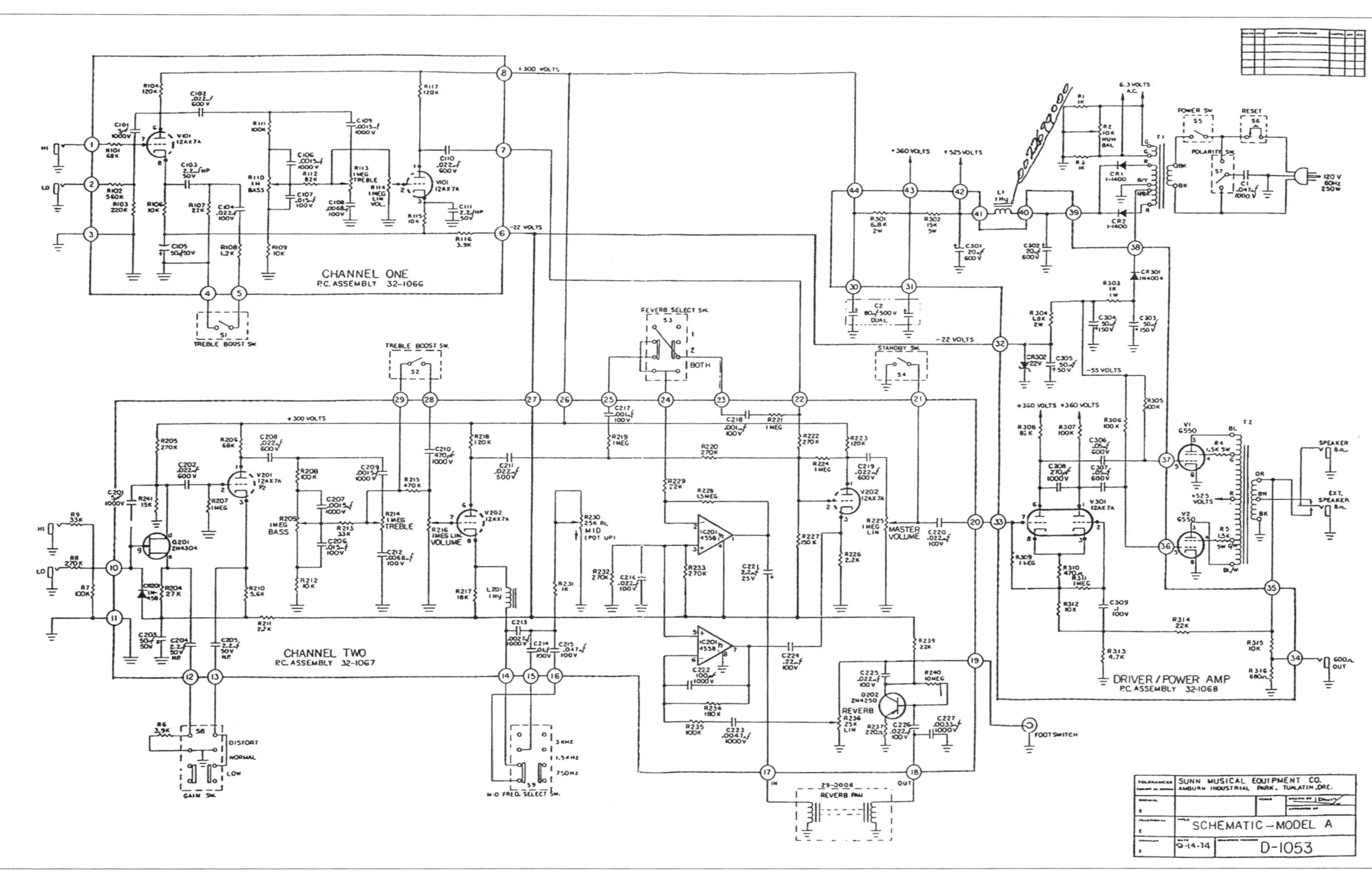
CHANNEL ONE
P.C. ASSEMBLY 32-1066
CHANNEL TWO
P.C. ASSEMBLY 32-1067
DRIVER / POWER AMP
P.C. ASSEMBLY 32-1068
TREBLE BOOST SW.
REVERB SELECT SW.
STANDBY SW.
POWER SW
RESET
POLARITY SW
GAIN SW.
MID FREQ. SELECT SW.
FOOTSWITCH
REVERB PAN
SPEAKER 8Ω
EXT. SPEAKER 8Ω
600Ω OUT
120 V 60Hz 250W
SUNN MUSICAL EQUIPMENT CO.
AMBURN INDUSTRIAL PARK, TUALATIN, ORE.
SCHEMATIC-MODEL A
D-1053

MODEL T SUPER

Sunn

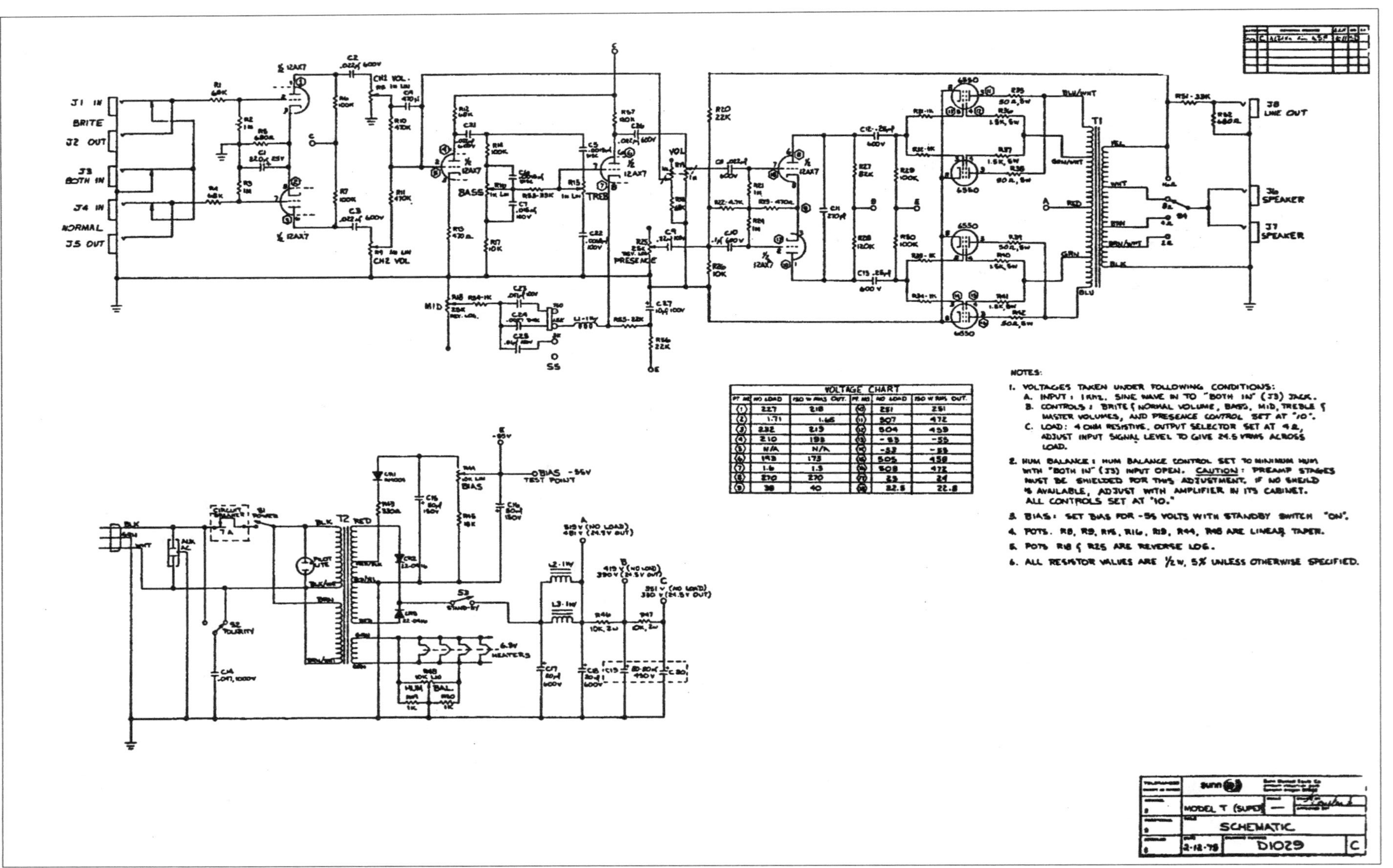

PT NO	NO LOAD	150 W RMS OUT.	PT NO	NO LOAD	150 W RMS OUT.
1	227	218	10	251	251
2	1.71	1.65	11	507	472
3	232	219	12	504	459
4	210	193	13	-53	-55
5	N/A	N/A	14	-53	-55
6	193	173	15	505	458
7	1.6	1.5	16	508	472
8	270	270	17	23	24
9	38	40	18	22.5	22.8

NOTES:

1. VOLTAGES TAKEN UNDER FOLLOWING CONDITIONS:
   A. INPUT: 1KHZ. SINE WAVE IN TO "BOTH IN" (J3) JACK.
   B. CONTROLS: BRITE & NORMAL VOLUME, BASS, MID, TREBLE & MASTER VOLUMES, AND PRESENCE CONTROL SET AT "10".
   C. LOAD: 4 OHM RESISTIVE. OUTPUT SELECTOR SET AT 4Ω, ADJUST INPUT SIGNAL LEVEL TO GIVE 24.5 VRMS ACROSS LOAD.
2. HUM BALANCE: HUM BALANCE CONTROL SET TO MINIMUM HUM WITH "BOTH IN" (J3) INPUT OPEN. CAUTION: PREAMP STAGES MUST BE SHIELDED FOR THIS ADJUSTMENT. IF NO SHEILD IS AVAILABLE, ADJUST WITH AMPLIFIER IN ITS CABINET. ALL CONTROLS SET AT "10."
3. BIAS: SET BIAS FOR -56 VOLTS WITH STANDBY SWITCH "ON".
4. POTS. R8, R9, R15, R16, R19, R44, R48 ARE LINEAR TAPER.
5. POTS R18 & R25 ARE REVERSE LOG.
6. ALL RESISTOR VALUES ARE ½W, 5% UNLESS OTHERWISE SPECIFIED.

100S* Sceptre Sentura II

sunn

*100S: With reverb/vibrato. After 10-1-67. See 100S and Spectrum II page for units without reverb/-vibrato. One J.B.L. D130F, one LE100S driver with H5040 horn, in rear-loading folded horn enclosure. N1200S or N1200 crossover network.

Sceptre: Four 12-inch transducers (Appendix 1) in sealed (infinite) baffle.

Sentura II: Two J.B.L. D130F (D15S after 3-1-69) in rear-loading folded horn enclosure.

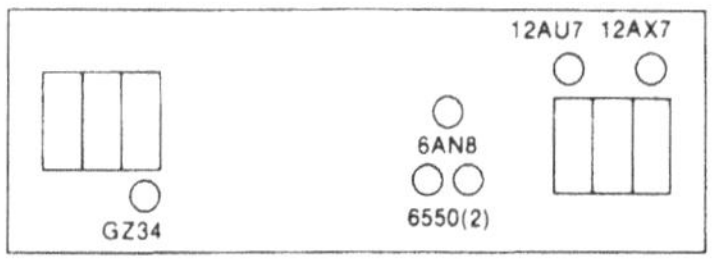

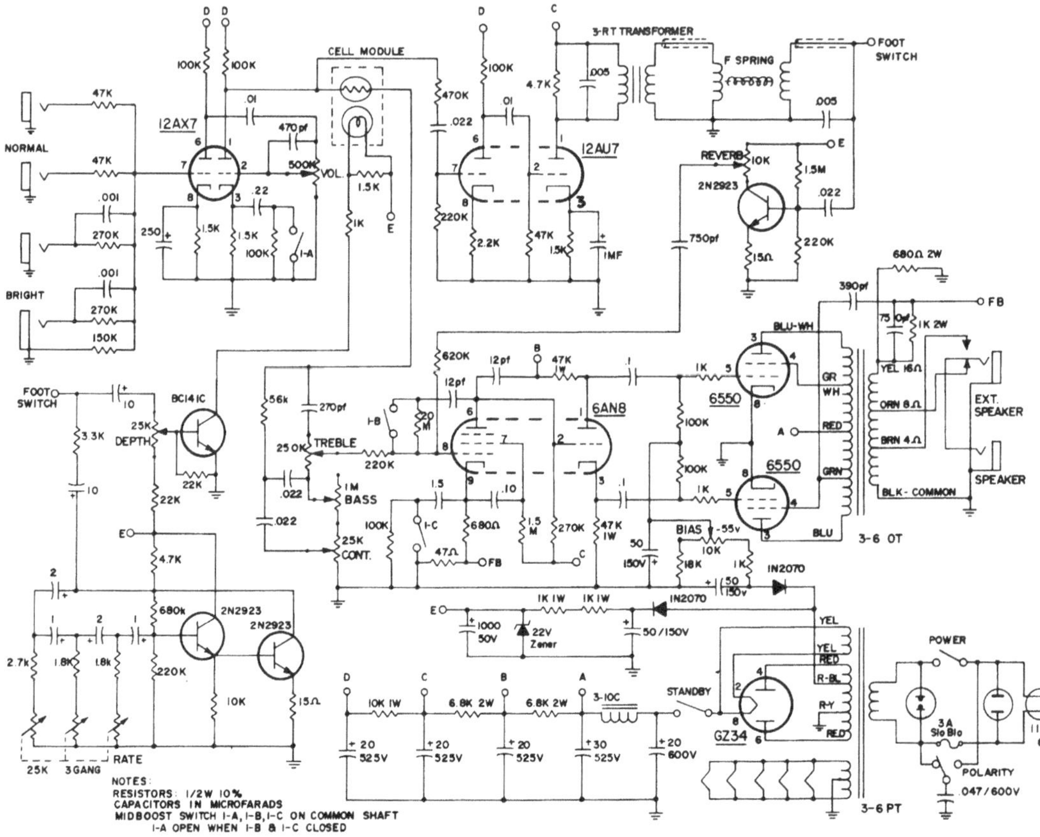

NOTES:
RESISTORS: 1/2W 10%
CAPACITORS IN MICROFARADS
MIDBOOST SWITCH I-A, I-B, I-C ON COMMON SHAFT
I-A OPEN WHEN I-B & I-C CLOSED

design/production: kinetic design

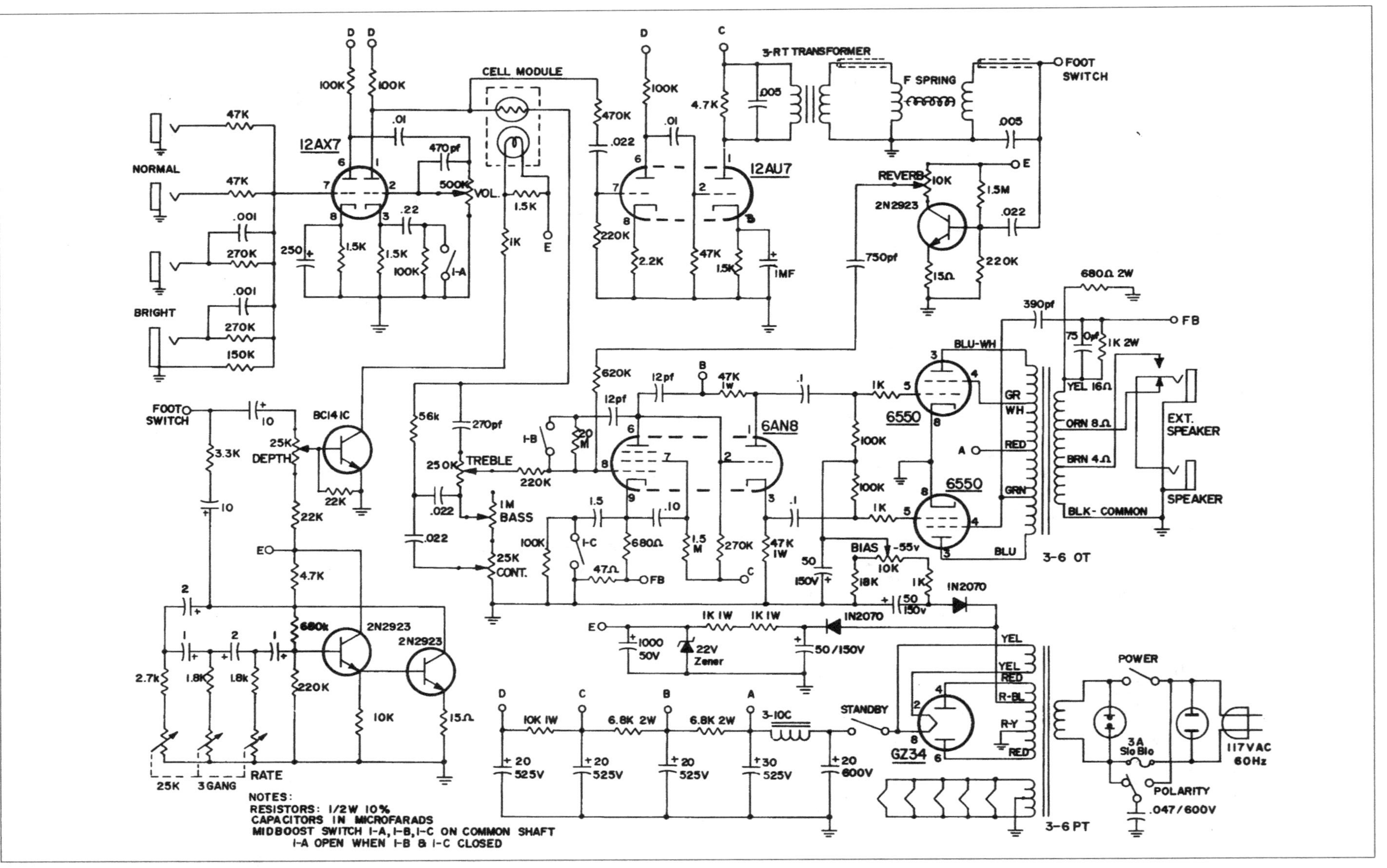
12AX7
12AU7
6AN8
6550
GZ34
2N2923
BC141C
NORMAL
BRIGHT
CELL MODULE
VOL.
500K
TREBLE
BASS
25K CONT.
DEPTH
FOOT SWITCH
REVERB
3-RT TRANSFORMER
F SPRING
BIAS -55V
IN2070
22V Zener
STANDBY
POWER
POLARITY
3A Slo Blo
117VAC 60Hz
.047/600V
3-6 OT
3-6 PT
3-10C
EXT. SPEAKER
SPEAKER
YEL 16Ω
ORN 8Ω
BRN 4Ω
BLK - COMMON
RATE
25K 3GANG
NOTES:
RESISTORS: 1/2W 10%
CAPACITORS IN MICROFARADS
MIDBOOST SWITCH I-A, I-B, I-C ON COMMON SHAFT
I-A OPEN WHEN I-B & I-C CLOSED

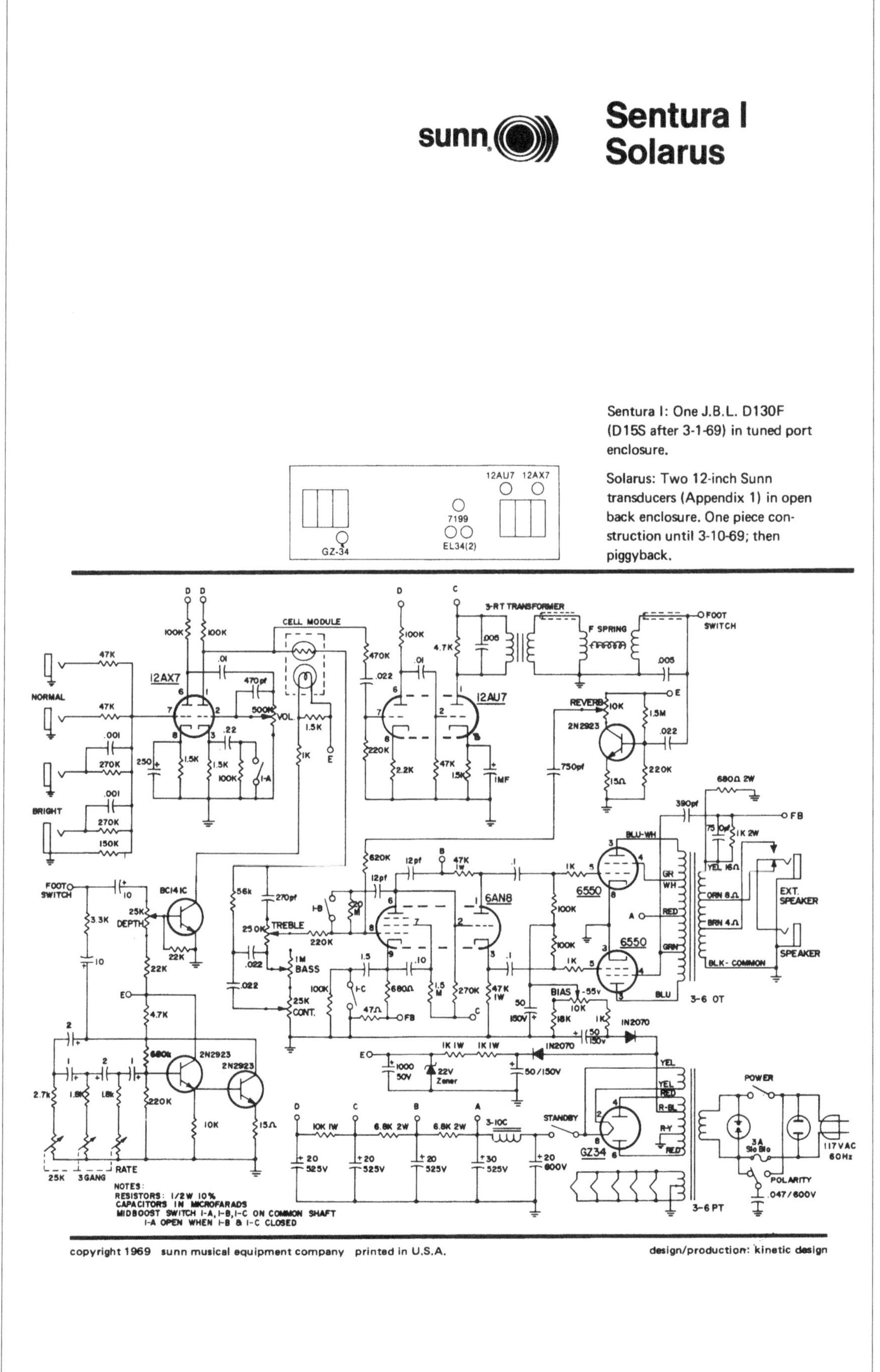

sunn

Sentura I
Solarus

Sentura I: One J.B.L. D130F (D15S after 3-1-69) in tuned port enclosure.

Solarus: Two 12-inch Sunn transducers (Appendix 1) in open back enclosure. One piece construction until 3-10-69; then piggyback.

design/production: kinetic design

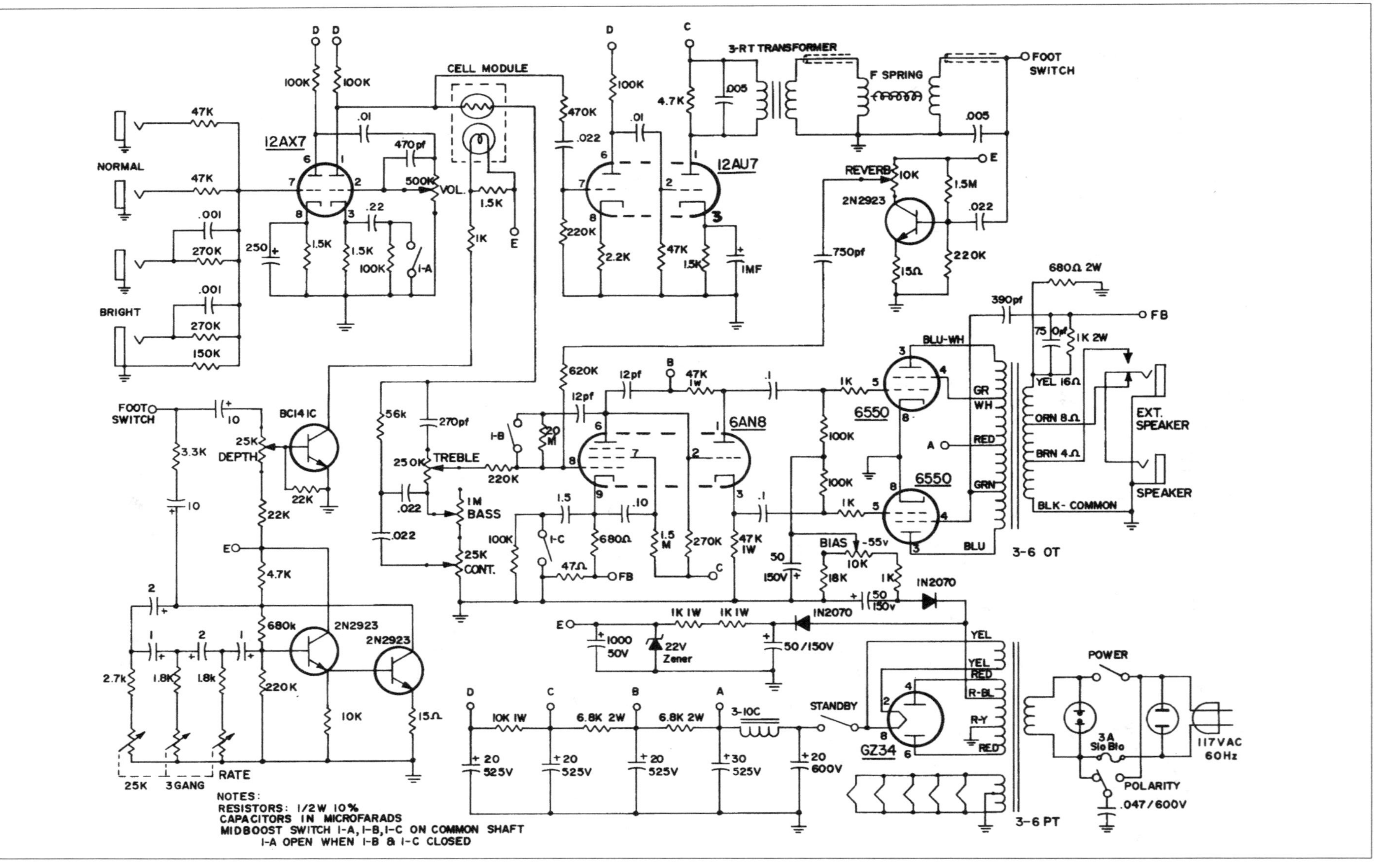
12AX7
12AU7
6AN8
6550
6550
GZ34
2N2923
BC141C
NORMAL
BRIGHT
CELL MODULE
VOL.
3-RT TRANSFORMER
F SPRING
FOOT SWITCH
REVERB
TREBLE
BASS
CONT.
DEPTH
RATE
25K 3GANG
BIAS -55v
IN2070
22V Zener
STANDBY
POWER
POLARITY
3A Slo Blo
117VAC 60Hz
.047/600V
3-10C
3-6 OT
3-6 PT
BLU-WH
GR
WH
RED
GRN
BLU
YEL 16Ω
ORN 8Ω
BRN 4Ω
BLK- COMMON
EXT. SPEAKER
SPEAKER
680Ω 2W
1K 2W
FB
NOTES:
RESISTORS: 1/2W 10%
CAPACITORS IN MICROFARADS
MIDBOOST SWITCH 1-A, 1-B, 1-C ON COMMON SHAFT
1-A OPEN WHEN 1-B & 1-C CLOSED

SONIC II/1-40/200s/SORADO

Sunn

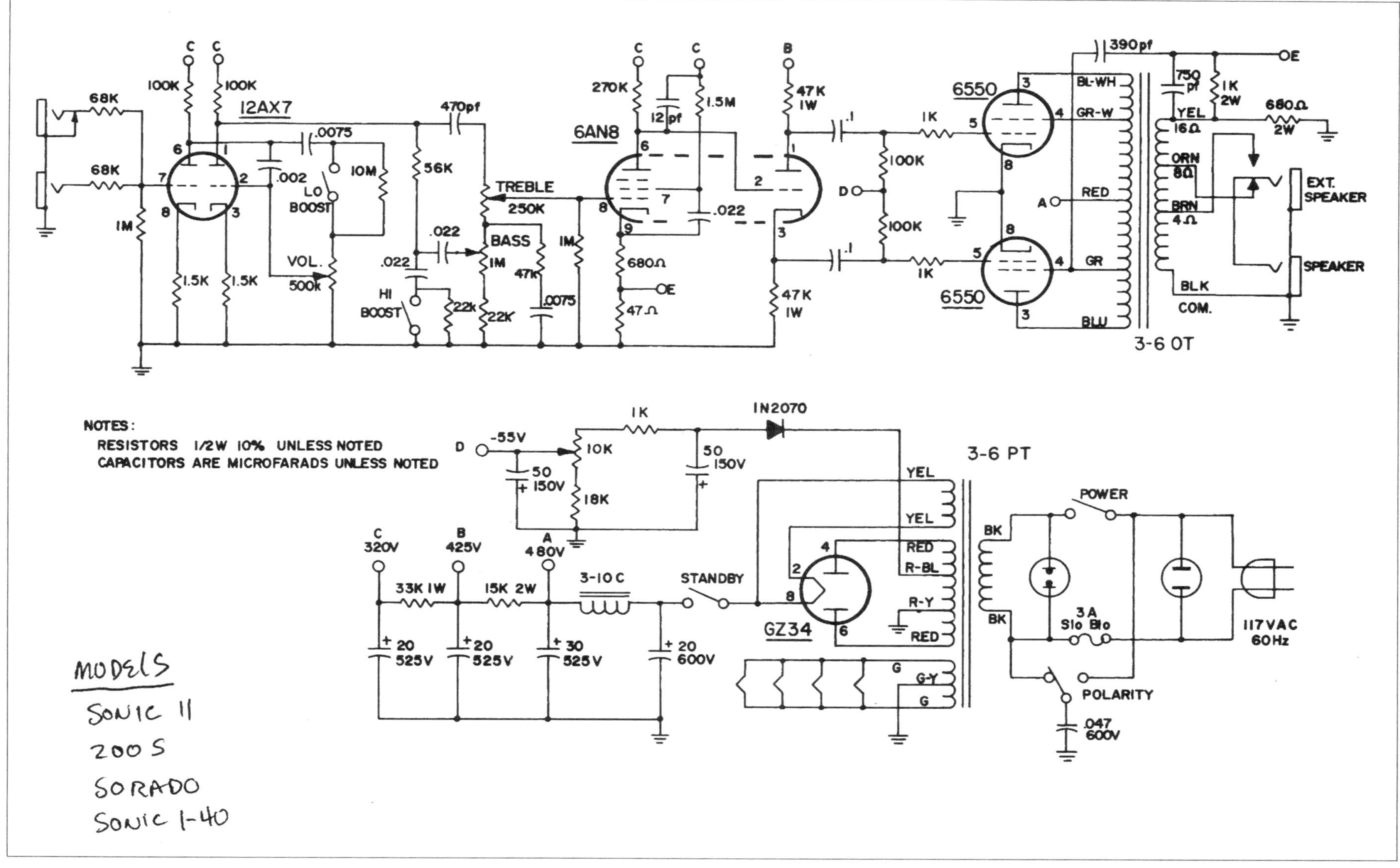

Sonic II*
200S
Sorado
Sonic I-40**

sunn

*Sonic II: Discontinued 4-10-68. Two J.B.L. D130F in rear-loading, folded horn enclosure.

200S: Two J.B.L. D140F (D140F-6 after 3-1-69) in rear-loading, folded horn enclosure.

Sorado: Two 15-inch Sunn transducers (Appendix 1) in tuned port bass reflex enclosure.

**Sonic I-40: 60 watt chassis produced after 5-10-69. One J.B.L. D140F (D140F-6 after 3-1-69) in rear-loading, folded horn enclosure.

MODELS
SONIC II
200S
SORADO
SONIC I-40

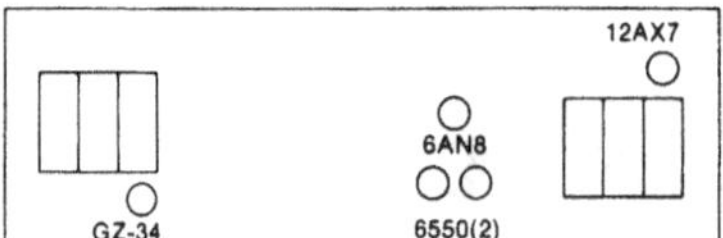

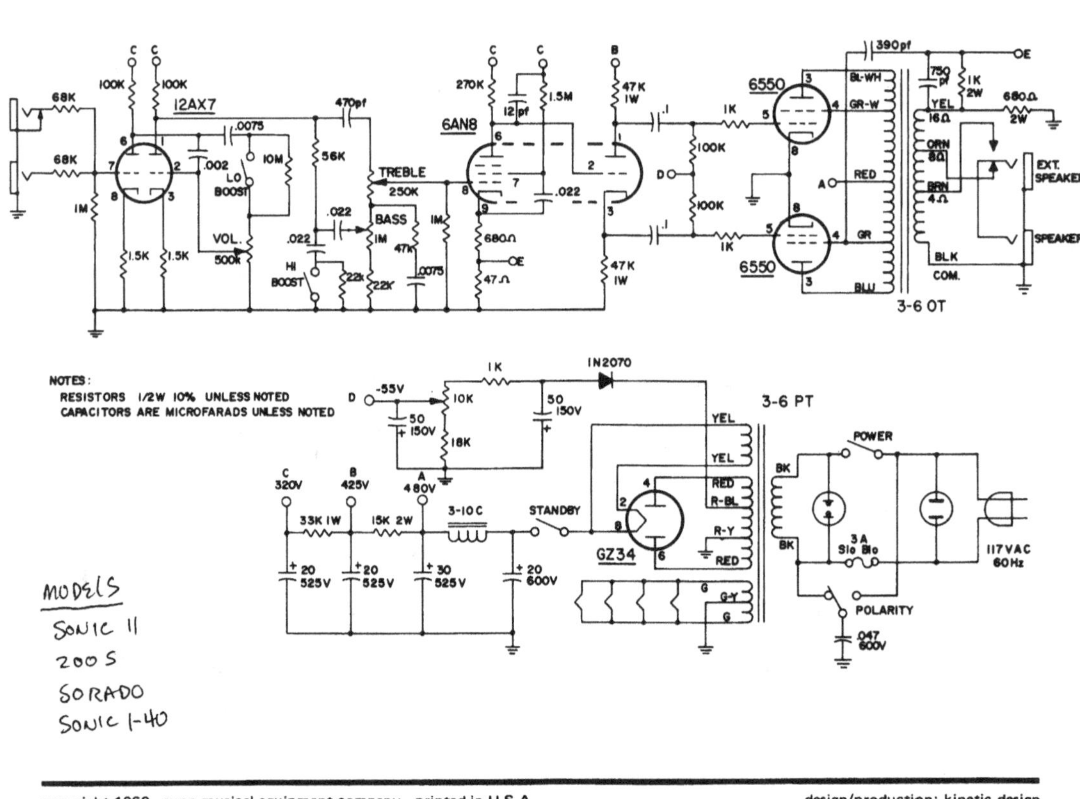

 design/production: kinetic design

100S+ Spectrum II**

+100S: Without reverb/vibrato. Before 10-1-67. One J.B.L. D130F, one LE100S driver with H5040 horn, in rear-loading, folded horn enclosure. N1200S or N1200 crossover network.

**Spectrum II: Discontinued 5-10-69. Two J.B.L. D130F (J.B.L. D15S after 3-1-69) in rear-loading, folded horn enclosure.

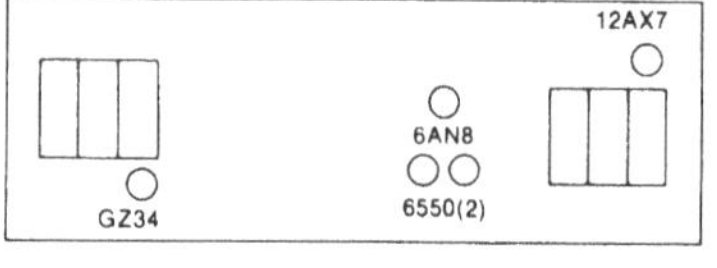

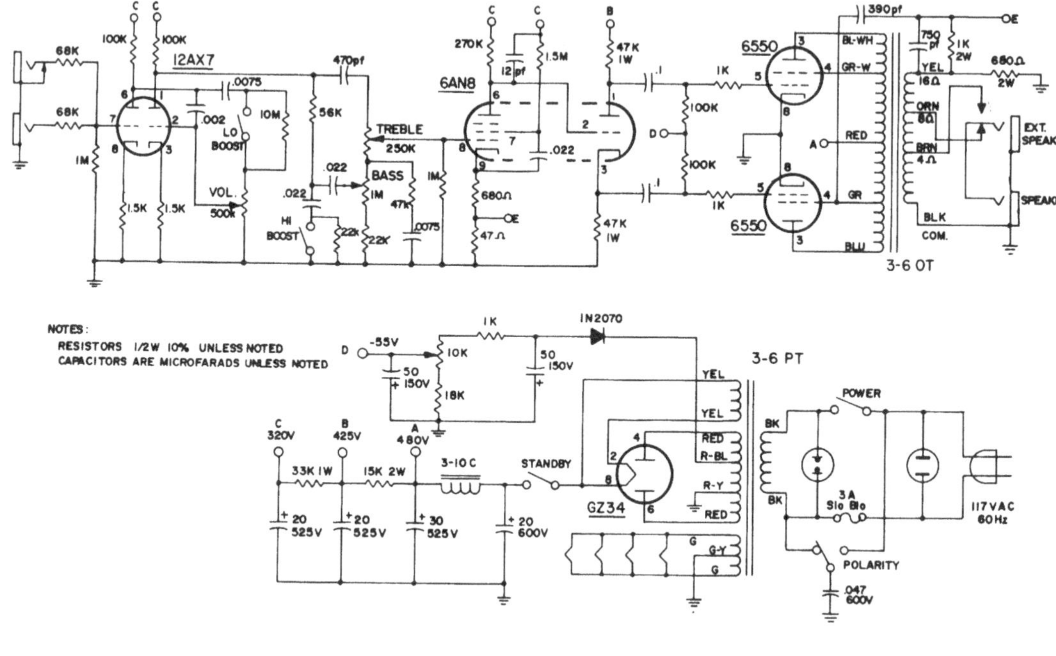

design/production: kinetic design

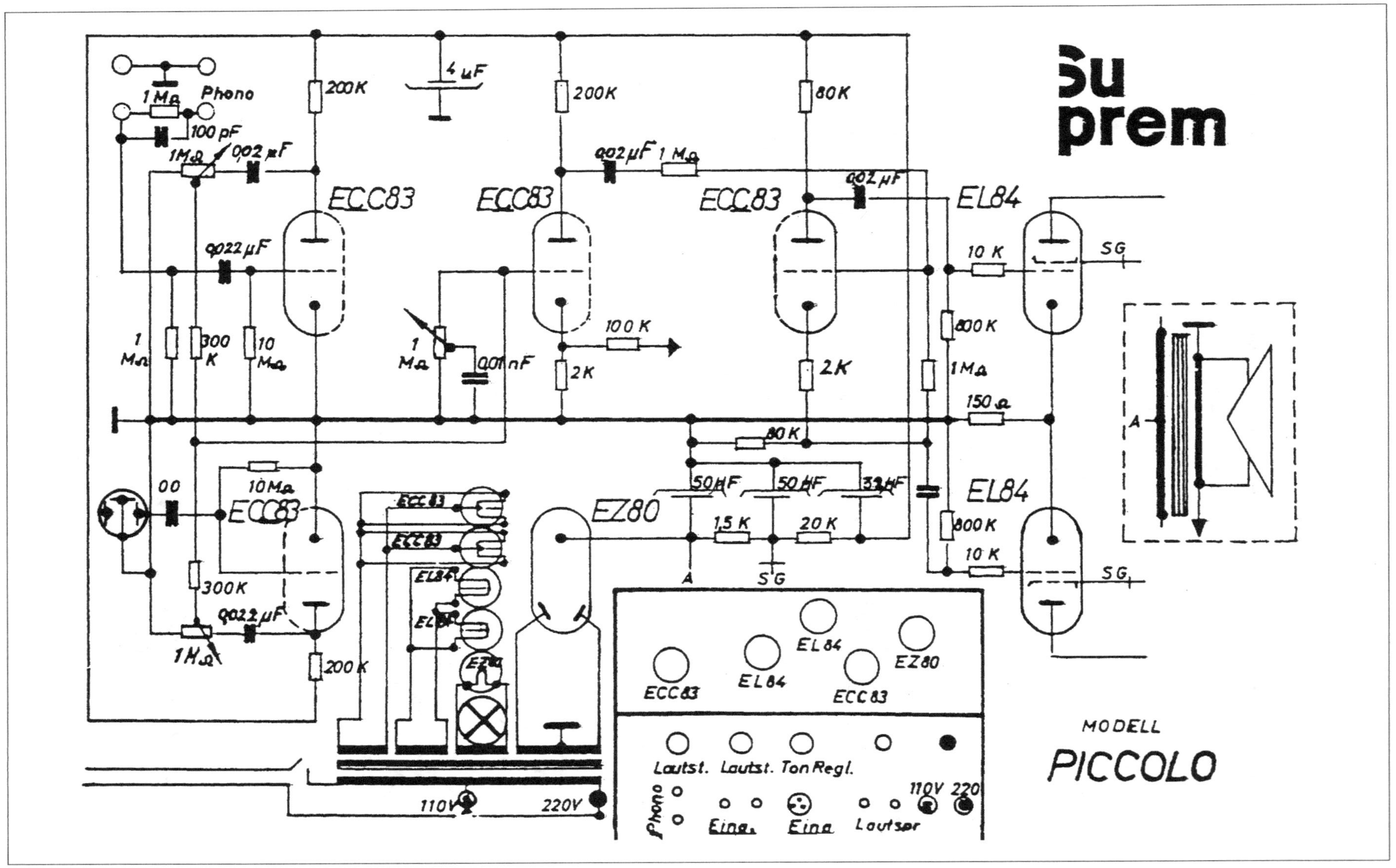
Suprem
Phono
1 MΩ
100 pF
1 MΩ
0,02 µF
200K
4 µF
200 K
80K
0,02 µF
1 MΩ
0,02 µF
ECC83
ECC83
ECC83
EL84
0,022 µF
10 K
SG
1 MΩ
300 K
10 MΩ
1 MΩ
0,01 nF
100 K
2K
2K
800 K
1 MΩ
150 Ω
80 K
A
00
10 MΩ
ECC83
50 µF
50 µF
32 µF
EL84
EZ80
1,5 K
20 K
800 K
10 K
A
SG
SG
300K
0,022 µF
1 MΩ
200 K
ECC83
ECC83
EL84
EL84
EZ80
110V
220V
ECC83
EL84
EL84
ECC83
EZ80
Lautst.
Lautst.
Ton Regl.
110V 220
Phono
Eing.
Eing
Lautspr
MODELL
PICCOLO

2 X E34 AMP

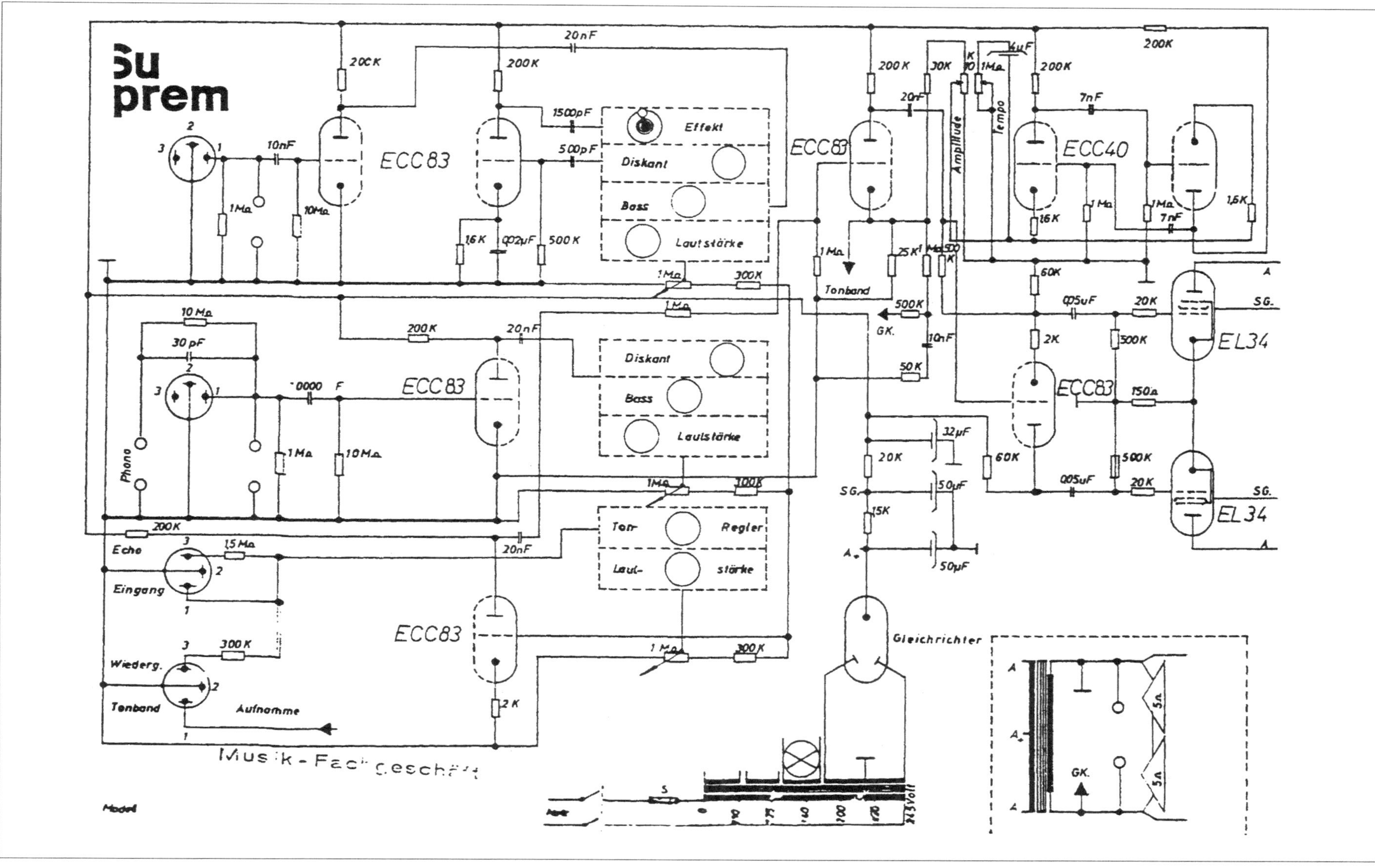
Suprem
ECC83
ECC83
ECC83
ECC83
ECC83
ECC40
EL34
EL34
Effekt
Diskant
Bass
Lautstärke
Ton Regler
Laut- stärke
Phono
Echo
Eingang
Wiederg.
Tonband
Aufnahme
Amplitude
Tempo
Gleichrichter
GK.
SG.
Musik-Fachgeschäft
Model
200K
20nF
1500pF
500pF
10nF
1Ma
10Ma
16K
002µF
500K
300K
30pF
10000 F
15Ma
2K
30K
4µF
7nF
1,6K
60K
005µF
20K
150Ω
50K
25K
32µF
50µF
15K
243 Volt
5Ω

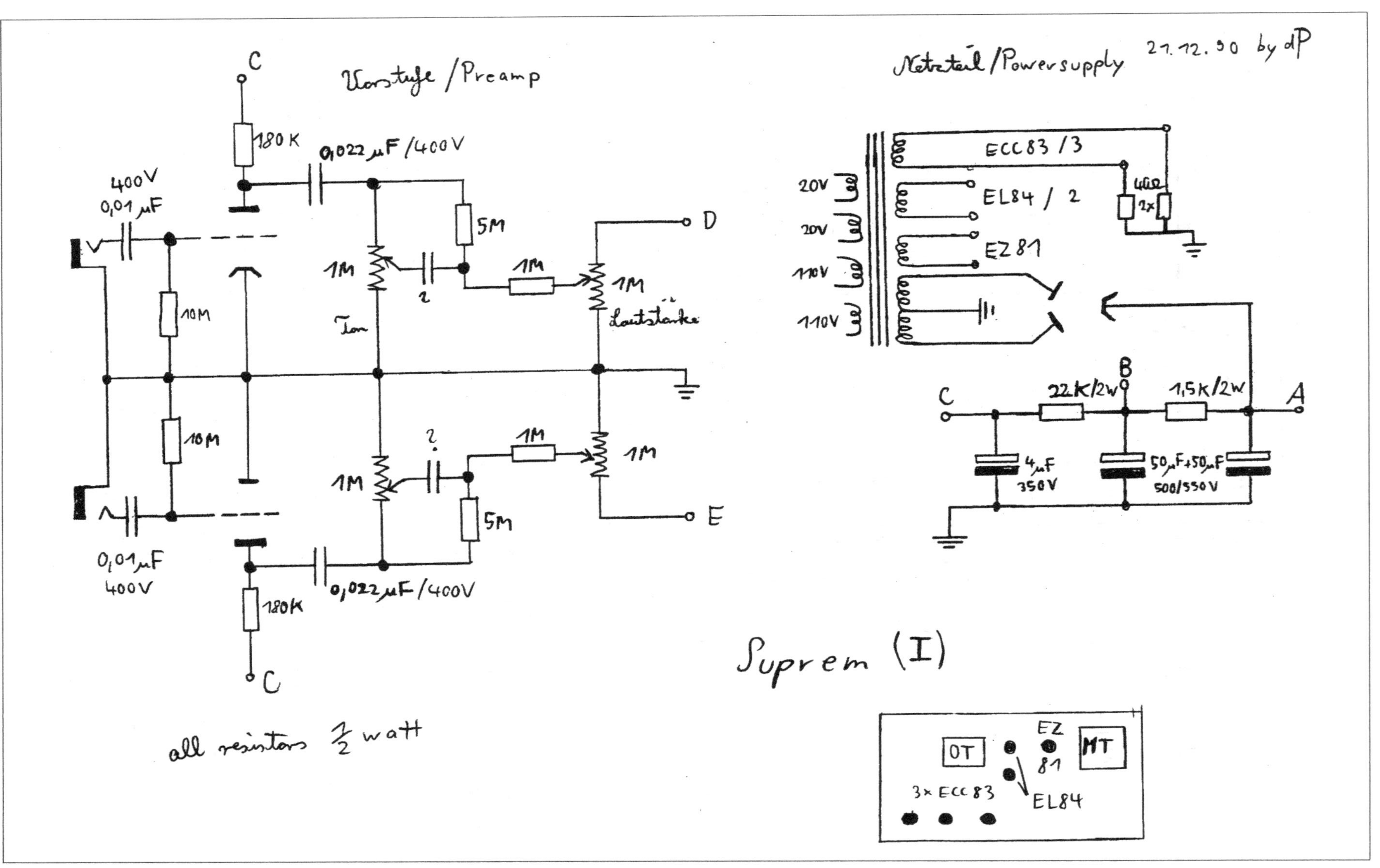
Vorstufe / Preamp
C
180 K
0,022 µF / 400V
400V
0,01 µF
5M
1M
1M
10M
Ton
1M
Lautstärke
D
10M
1M
1M
1M
5M
E
0,01 µF
400V
0,022 µF / 400V
180K
C
all resistors 1/2 watt
Netzteil / Powersupply
21.12.90 by dP
ECC83 / 3
EL84 / 2
EZ81
20V
20V
110V
110V
C
22K/2W
B
1,5K/2W
A
4 µF
350V
50 µF+50 µF
500/550V
Suprem (I)
OT
EZ
81
MT
3x ECC83
EL84

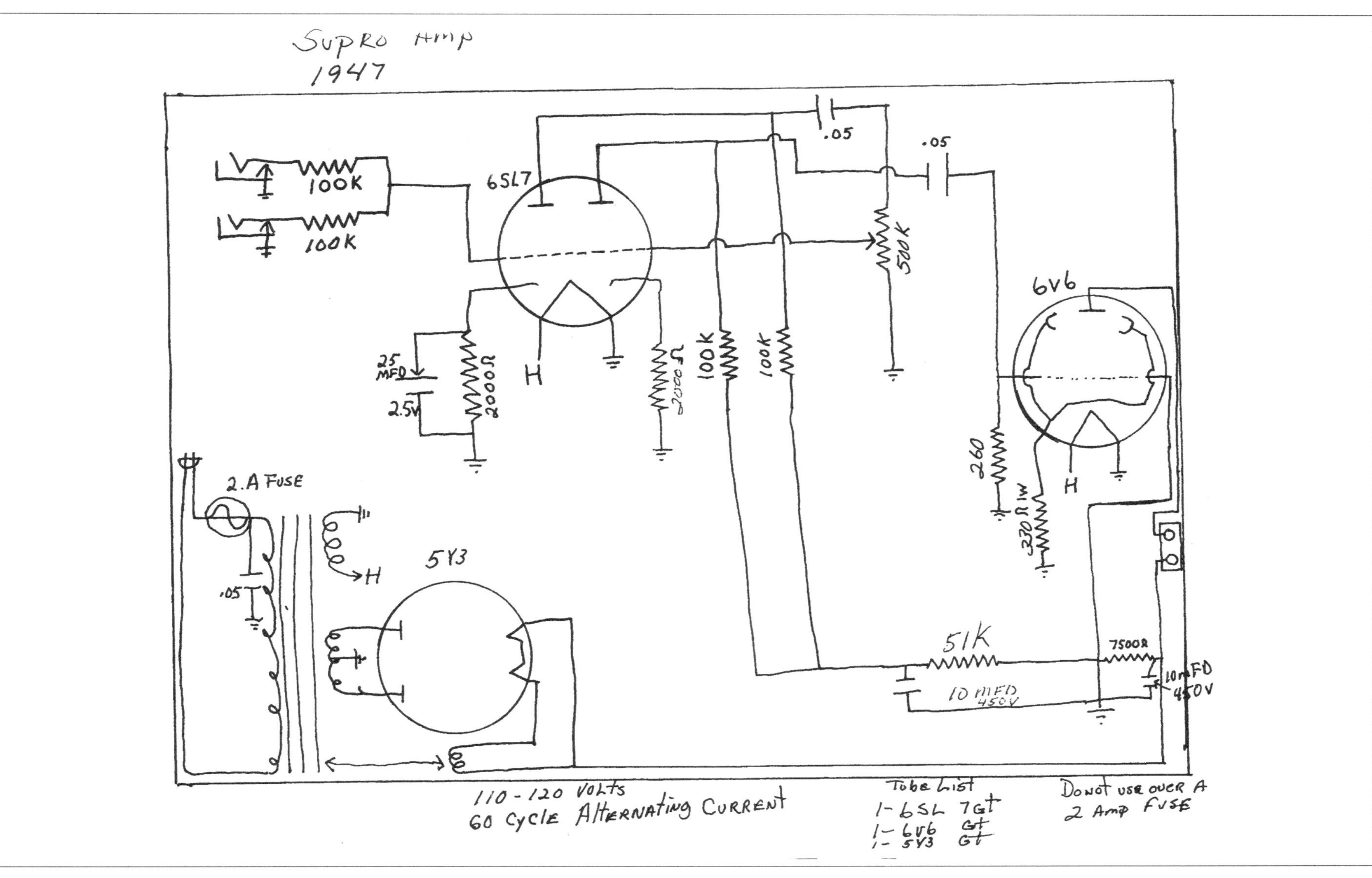
Supro Amp
1947
100K
100K
6SL7
.05
.05
500K
25 MFD
2.5V
2000Ω
H
2000Ω
100K
100K
6V6
260
330 Ω 1W
H
2.A Fuse
H
5Y3
.05
51K
750Ω
10 MFD 450V
10 MFD 450V
110 - 120 Volts
60 Cycle Alternating Current
Tube List
1- 6SL 7Gt
1- 6V6 Gt
1- 5Y3 Gt
Do not use over A 2 Amp Fuse

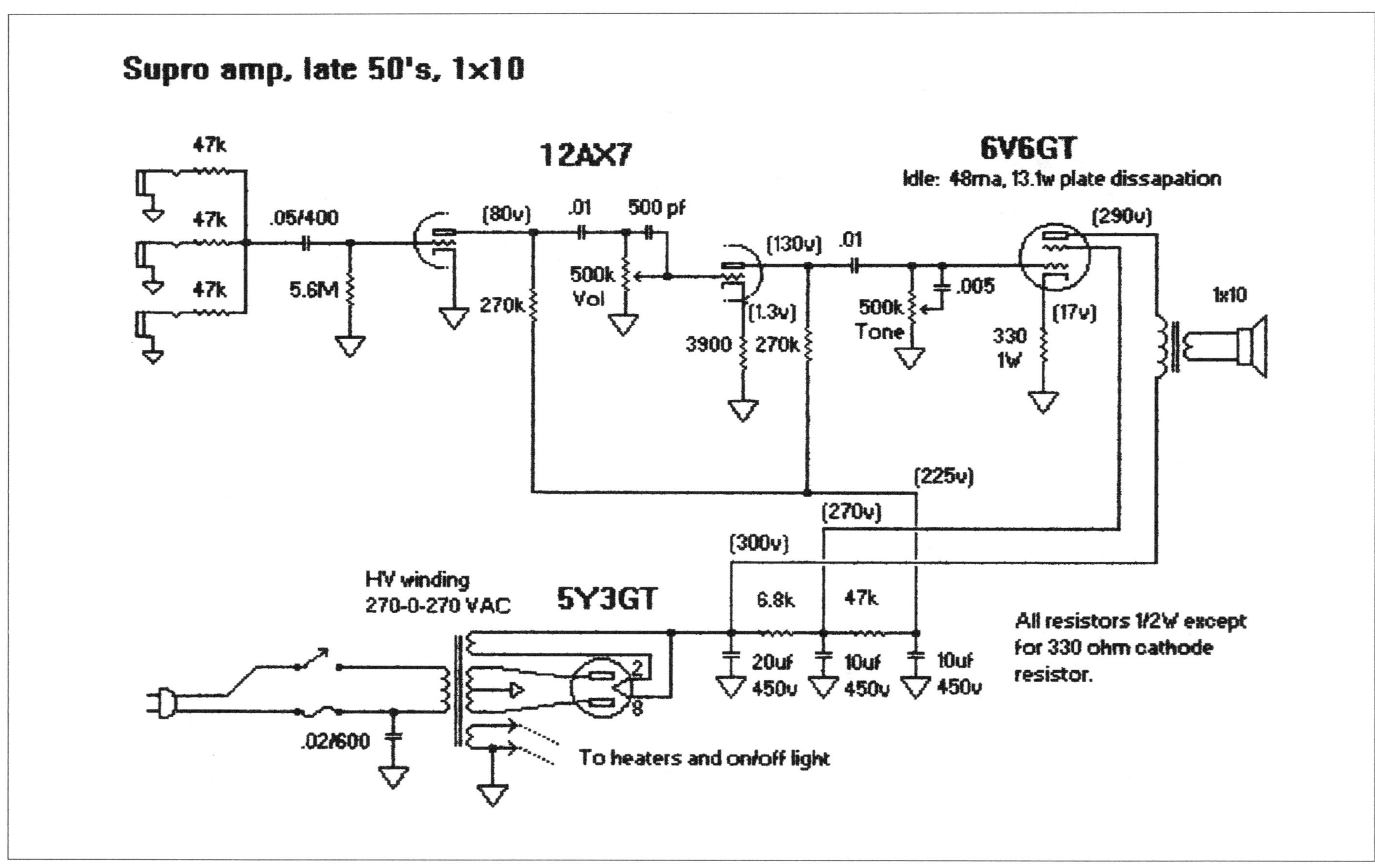
Supro amp, late 50's, 1x10
47k
47k
47k
.05/400
5.6M
12AX7
(80v)
270k
.01
500 pf
500k
Vol
(130v)
.01
(1.3v)
3900
270k
500k
Tone
.005
6V6GT
Idle: 48ma, 13.1w plate dissapation
(290v)
(17v)
330
1w
1x10
(225v)
(270v)
(300v)
HV winding
270-0-270 VAC
5Y3GT
2
8
6.8k
47k
20uf
450v
10uf
450v
10uf
450v
All resistors 1/2W except
for 330 ohm cathode
resistor.
.02/600
To heaters and on/off light

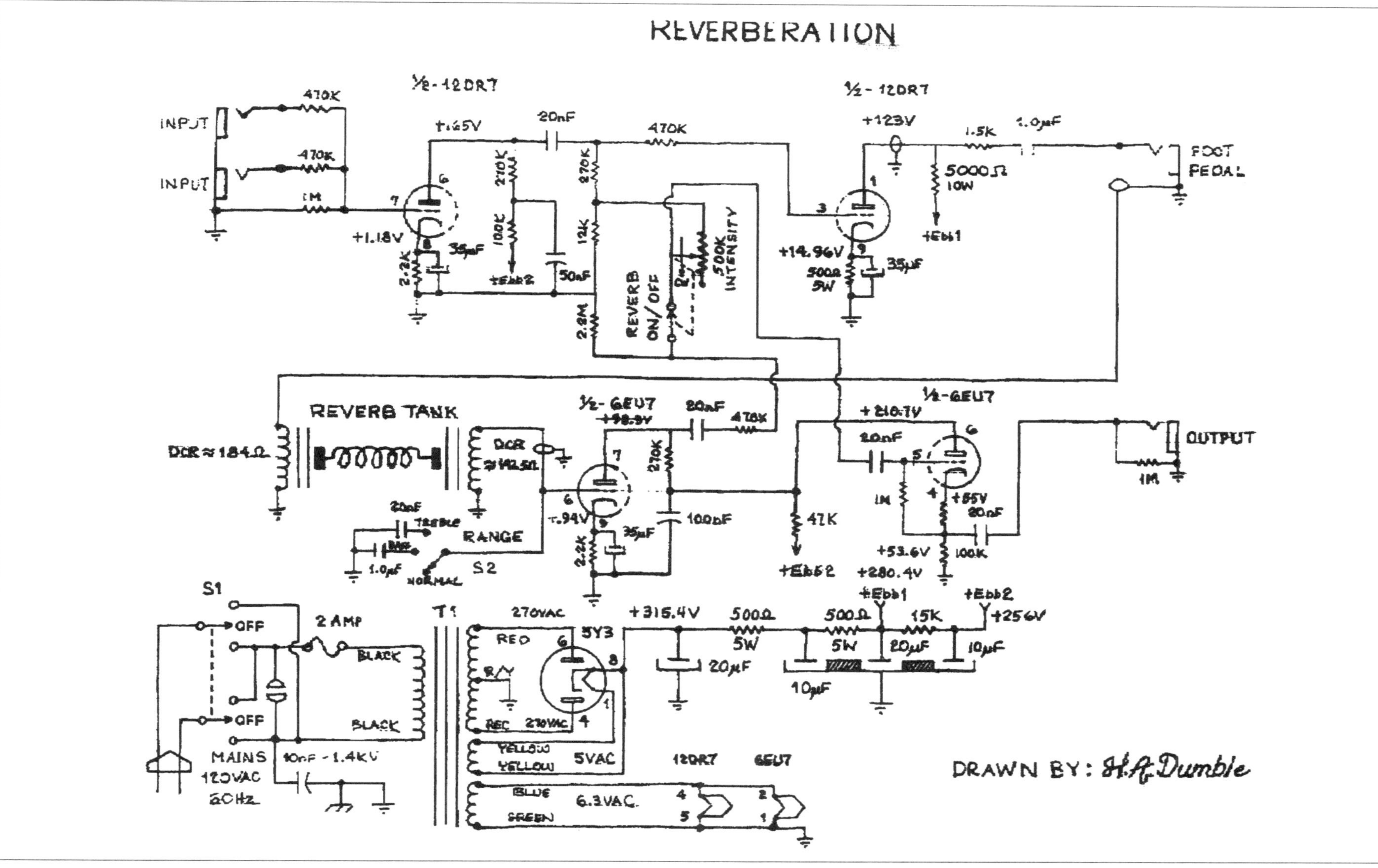
REVERBERATION
½-12DR7
INPUT
INPUT
470K
470K
1M
+1.65V
+1.18V
2.2K
35µF
20nF
270K
100K
+Ebb2
50nF
470K
270K
12K
2.2M
REVERB ON/OFF
500K INTENSITY
½-12DR7
+123V
+14.96V
500Ω 5W
35µF
1.5K
1.0µF
5000Ω 10W
+Ebb1
FOOT PEDAL
REVERB TANK
DCR≈184Ω
½-6EU7
+98.9V
20nF
470K
270K
100nF
35µF
2.2K
20nF
TREBLE
RANGE
1.0µF
NORMAL
S2
47K
+Ebb2
½-6EU7
+210.1V
20nF
1M
+55V
20nF
+53.6V
100K
+280.4V
OUTPUT
1M
S1
OFF
OFF
2 AMP
BLACK
BLACK
MAINS 120VAC 60Hz
10nF-1.4KV
T1
270VAC
RED
5Y3
REC 270VAC
YELLOW
YELLOW
5VAC
BLUE
GREEN
6.3VAC.
+315.4V
500Ω 5W
20µF
500Ω 5W
+Ebb1
15K
20µF
+Ebb2
+256V
10µF
10µF
12DR7
6EU7
DRAWN BY: H.A. Dumble

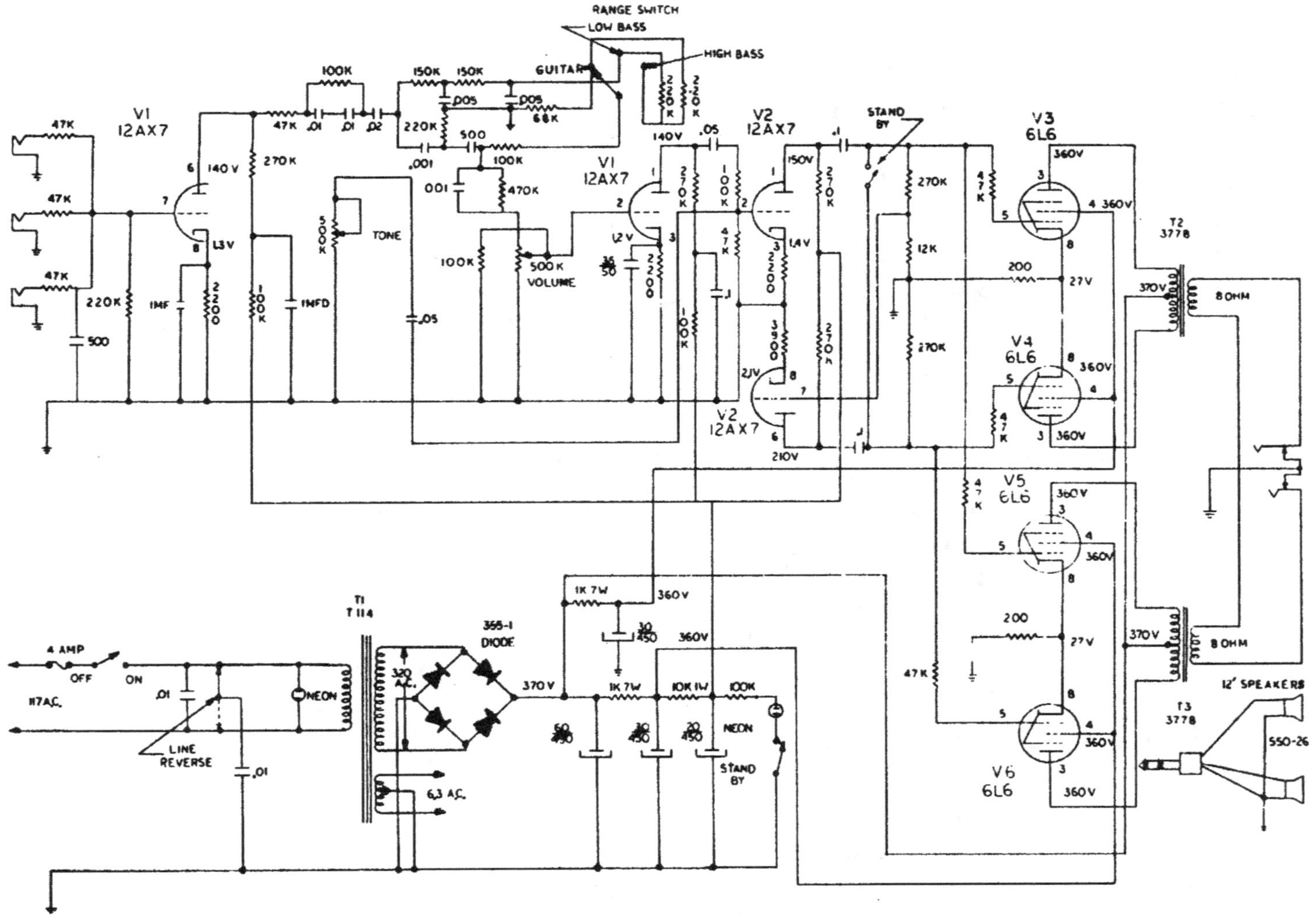

Supro Model S6625

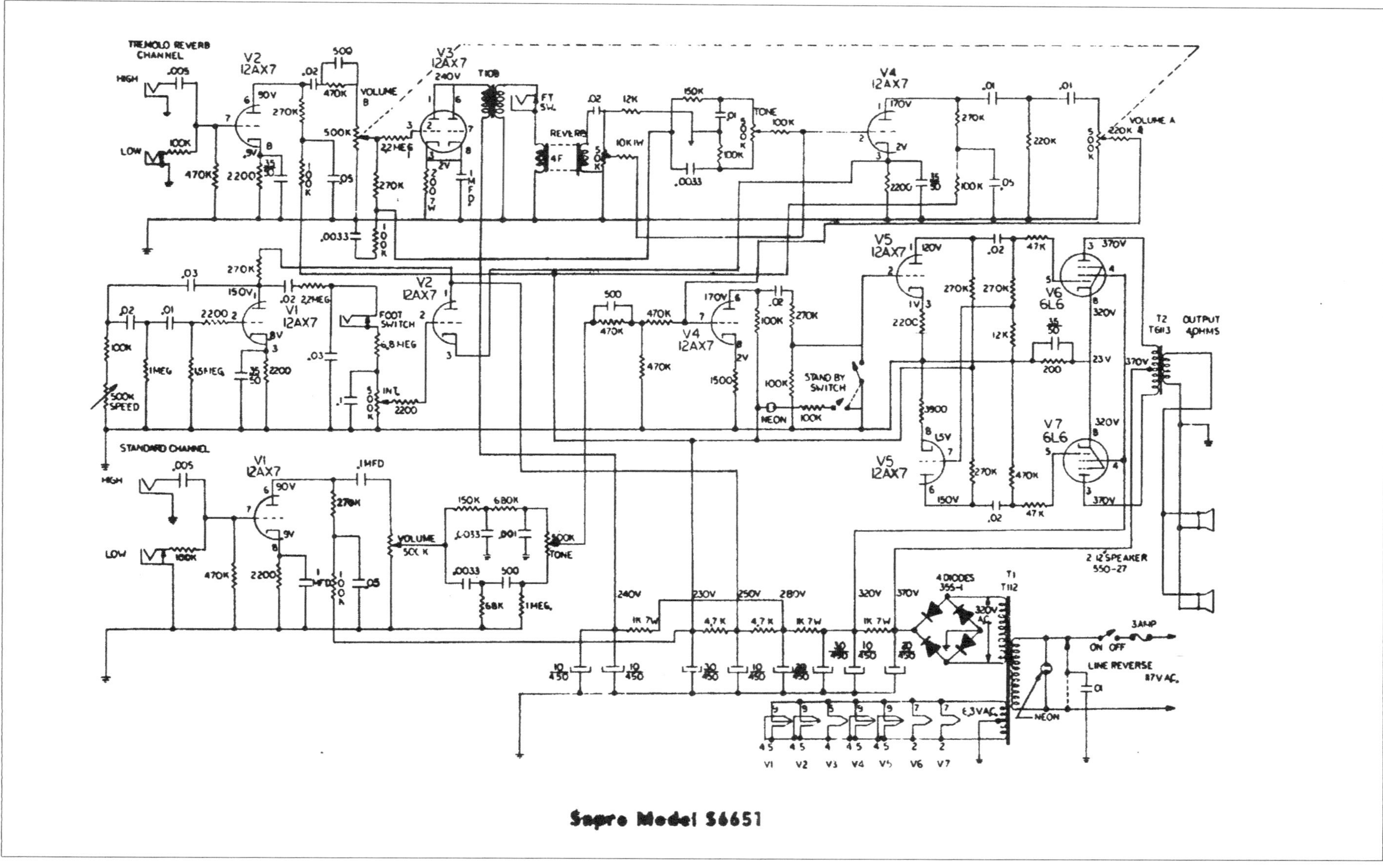

Supro Model S6651

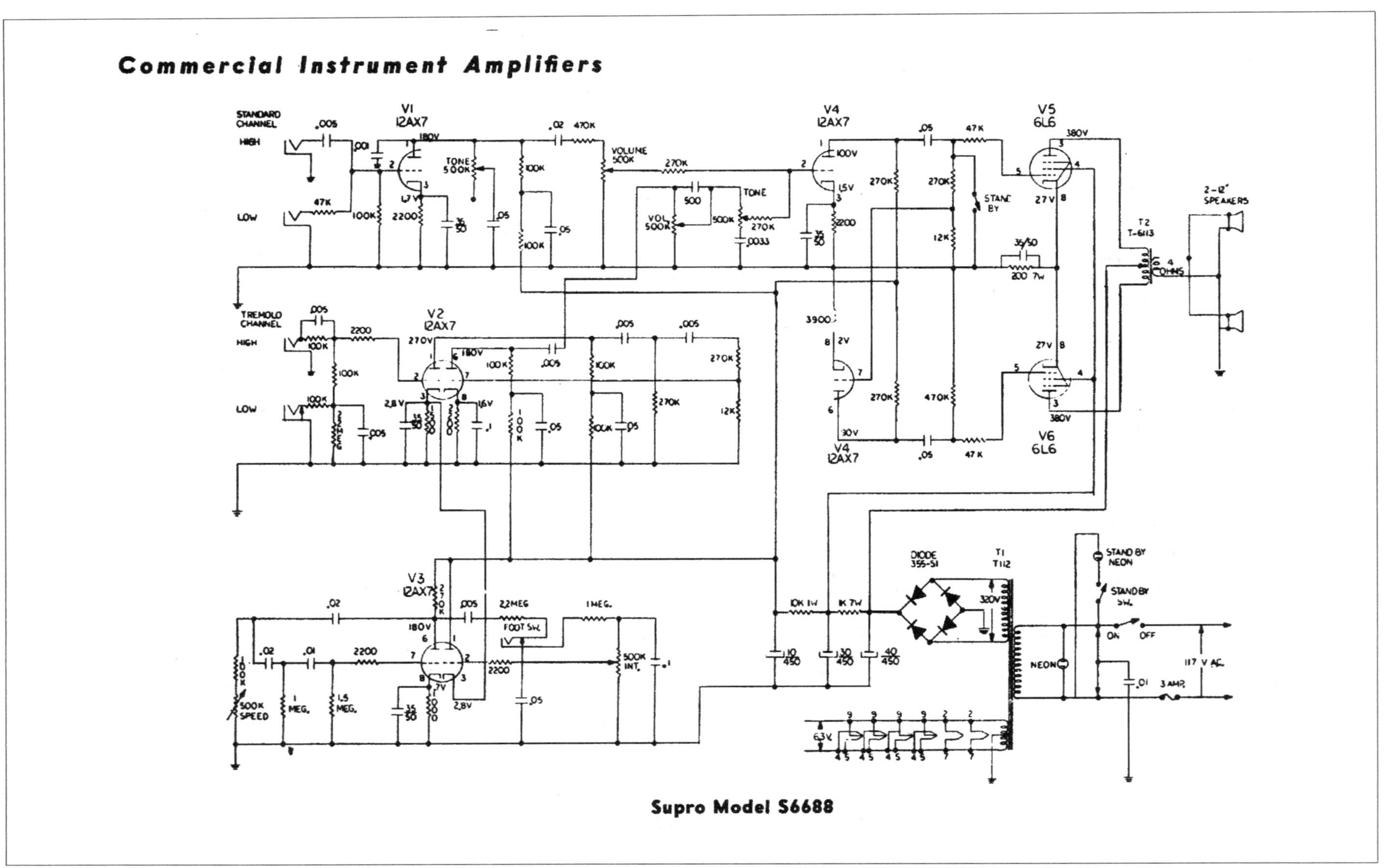

Supro Model S6688

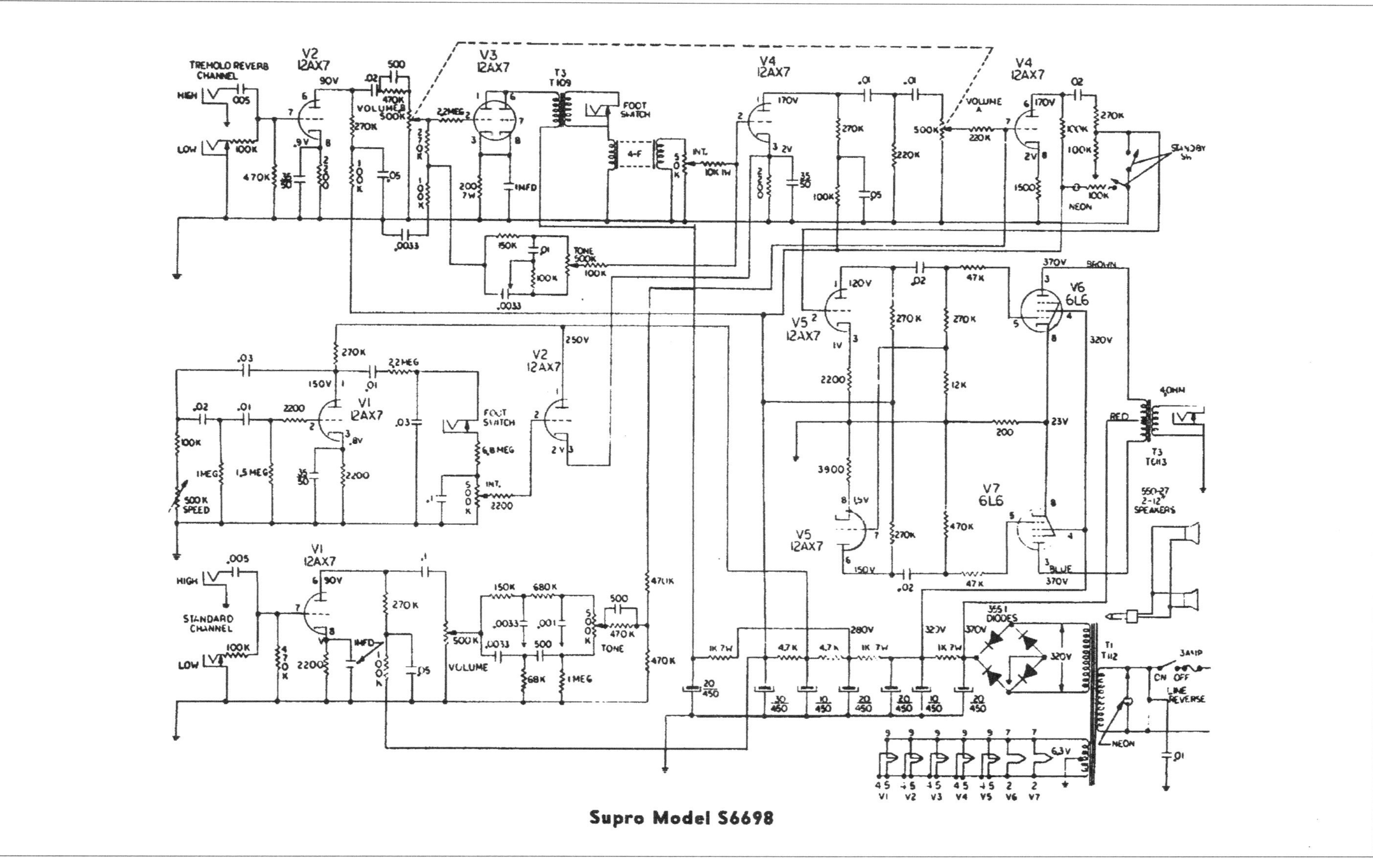

Supro Model S6698

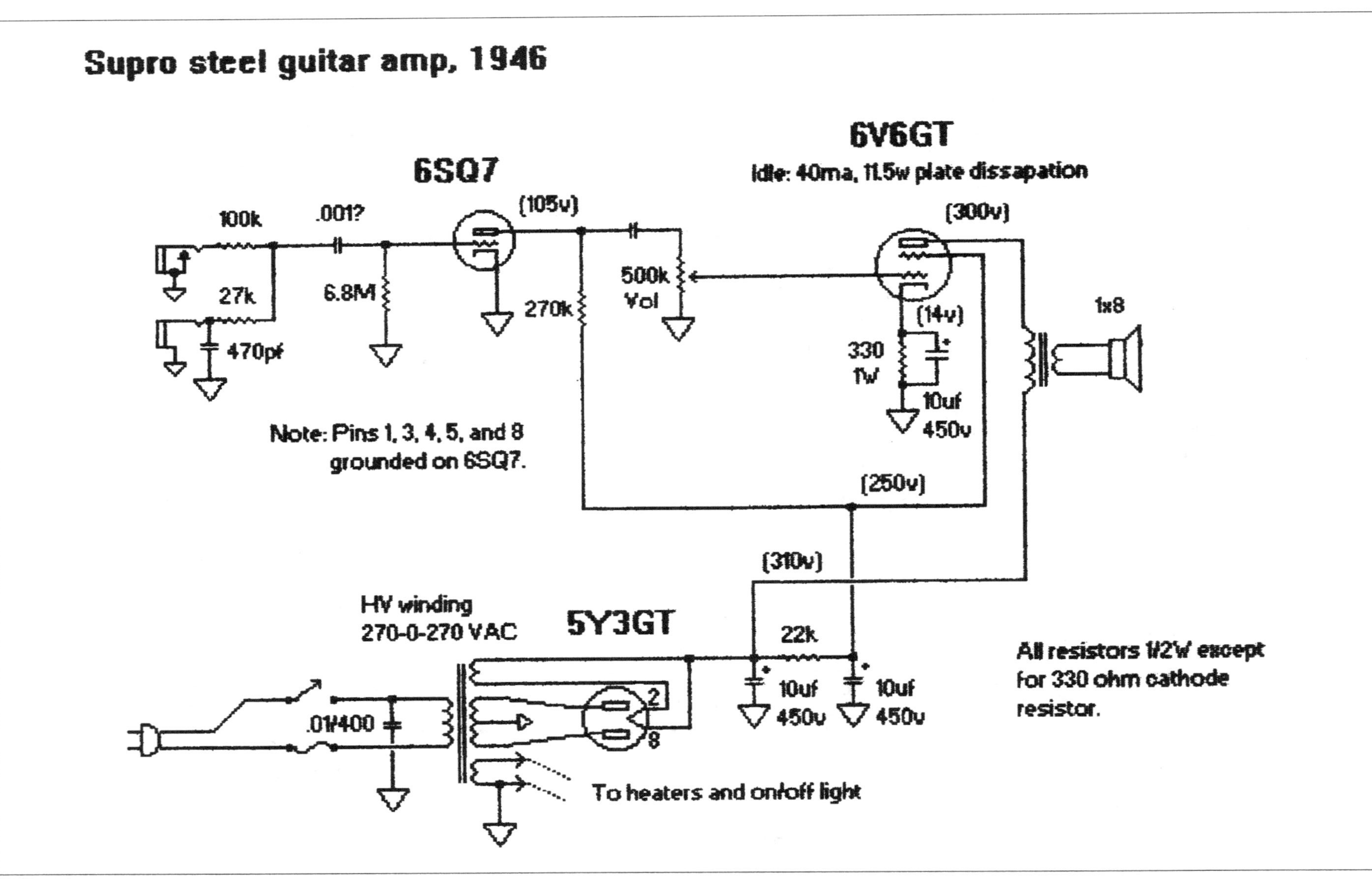
Supro steel guitar amp, 1946
6SQ7
6V6GT
Idle: 40ma, 11.5w plate dissapation
100k
.001?
(105v)
(300v)
27k
6.8M
500k
Vol
270k
470pf
(14v)
330
1W
1x8
10uf
450v
Note: Pins 1, 3, 4, 5, and 8
grounded on 6SQ7.
(250v)
(310v)
HV winding
270-0-270 VAC
5Y3GT
22k
.01/400
2
8
10uf
450v
10uf
450v
All resistors 1/2W except
for 330 ohm cathode
resistor.
To heaters and on/off light

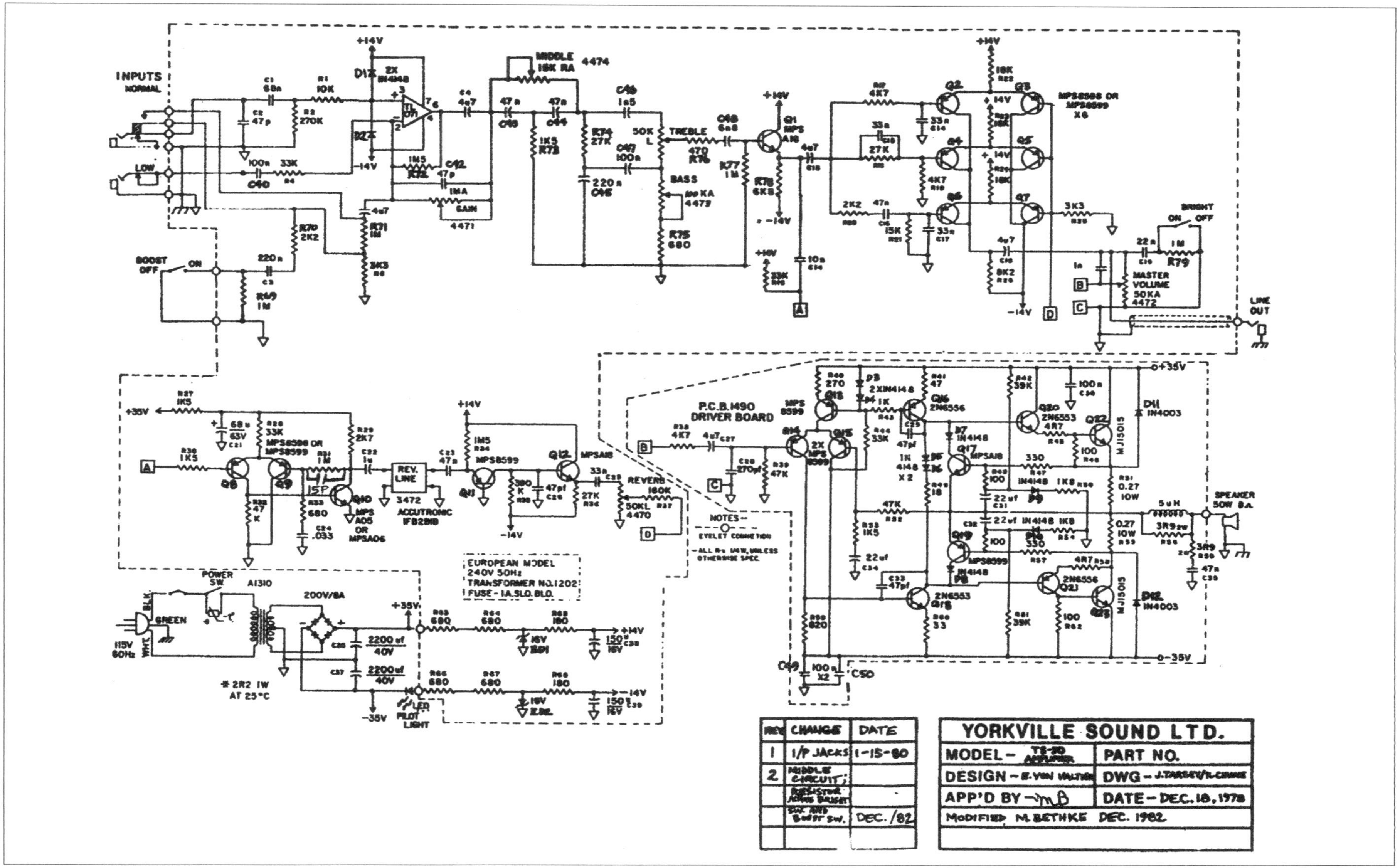

TS 50

NOTES –
EYELET CONNECTION
– ALL R's 1/4W, UNLESS OTHERWISE SPEC.

EUROPEAN MODEL
240V 50Hz
TRANSFORMER NO.1202
FUSE - 1A.SLO.BLO.

REV	CHANGE	DATE
1	I/P JACKS	1-15-80
2	MIDDLE CIRCUIT; RESISTOR ADDED; BOOST SW.	DEC./82

YORKVILLE SOUND LTD.	
MODEL – TS-50 AMPLIFIER	PART NO.
DESIGN – E.VON VALTIER	DWG – J.TARSEY/L.CRANE
APP'D BY – MB	DATE – DEC.18, 1978
MODIFIED M.BETHKE DEC. 1982	

YBA-1 BASS MASTER

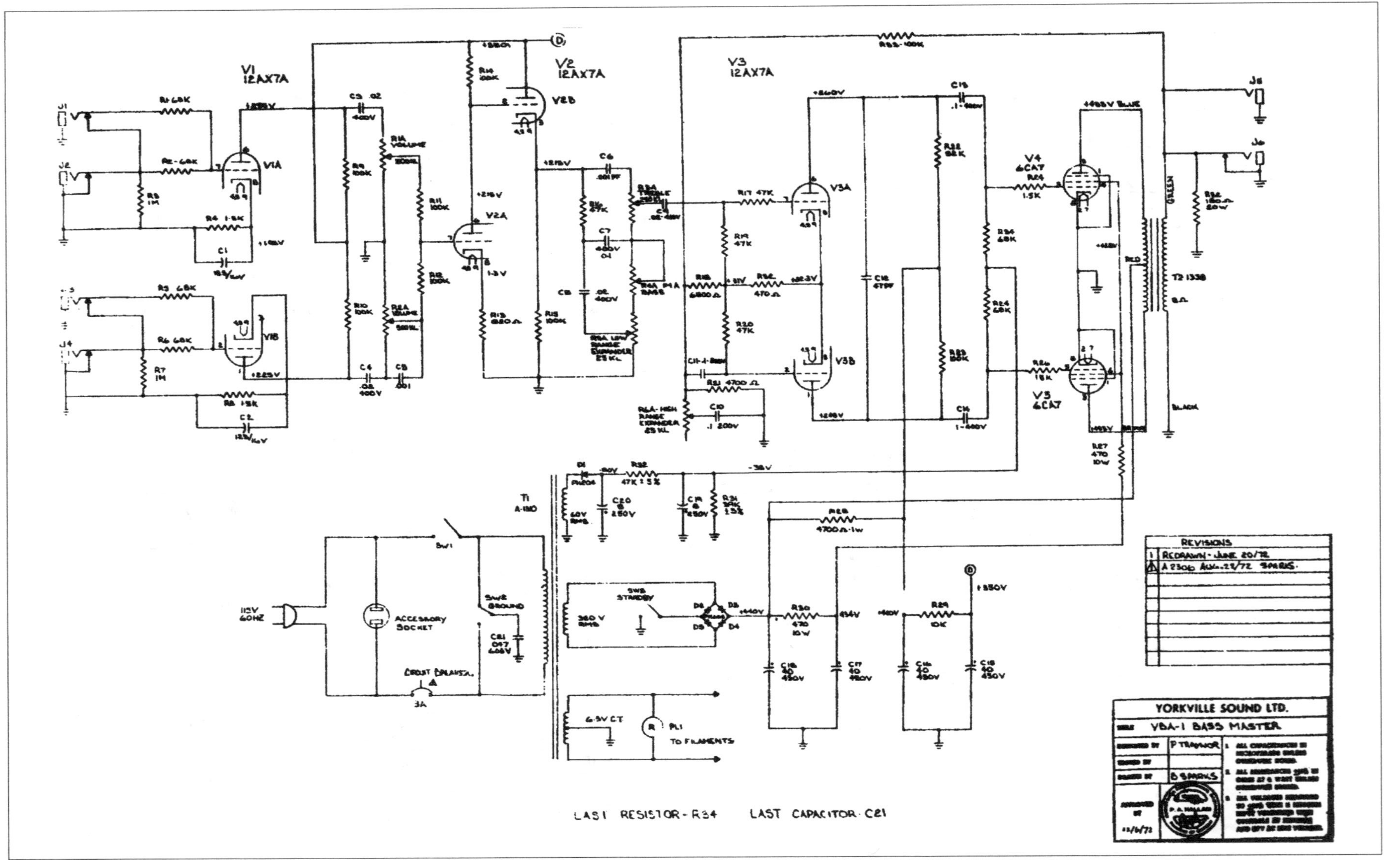

YBA-1 **BASS MASTER**

Traynor

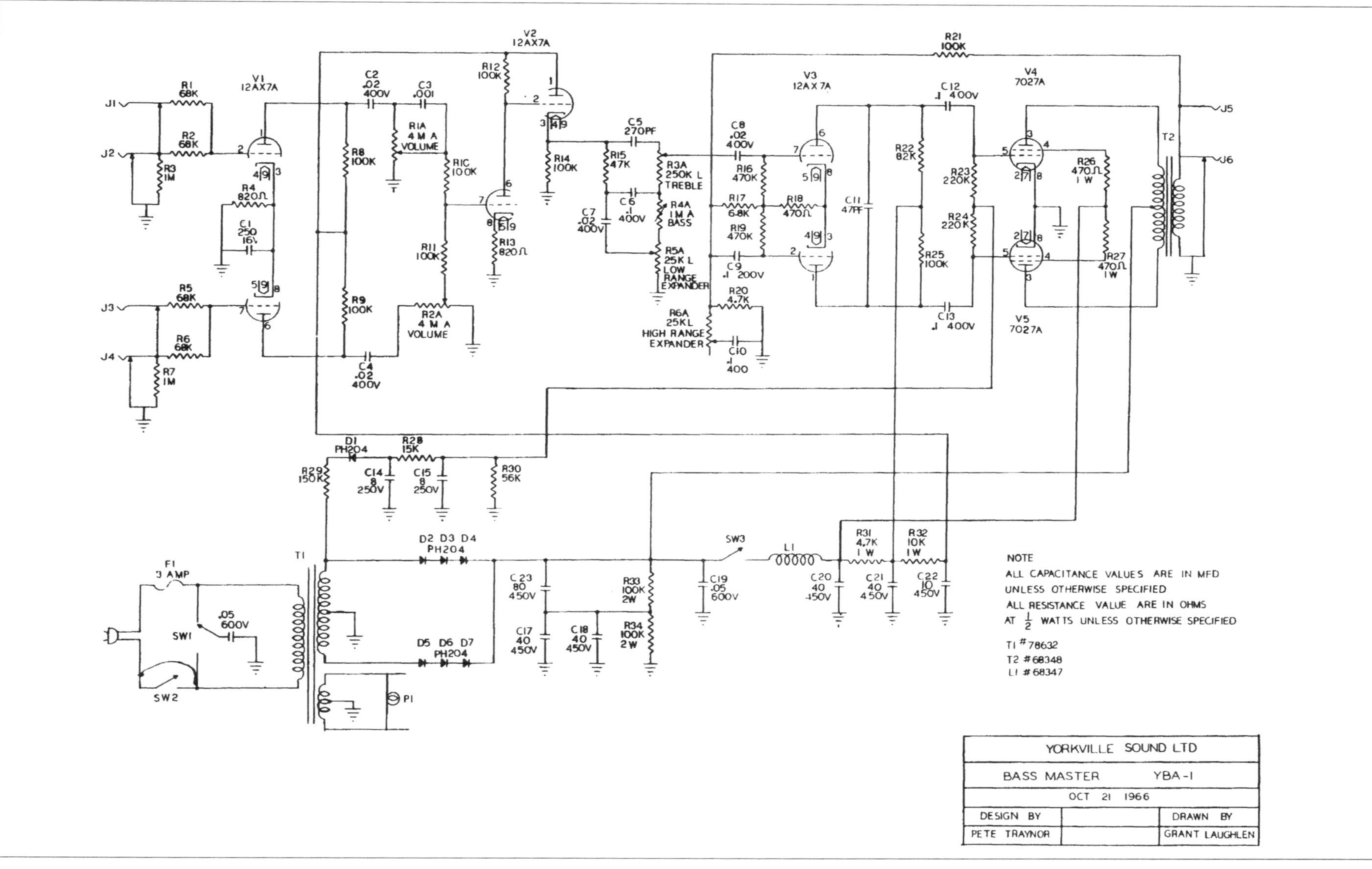

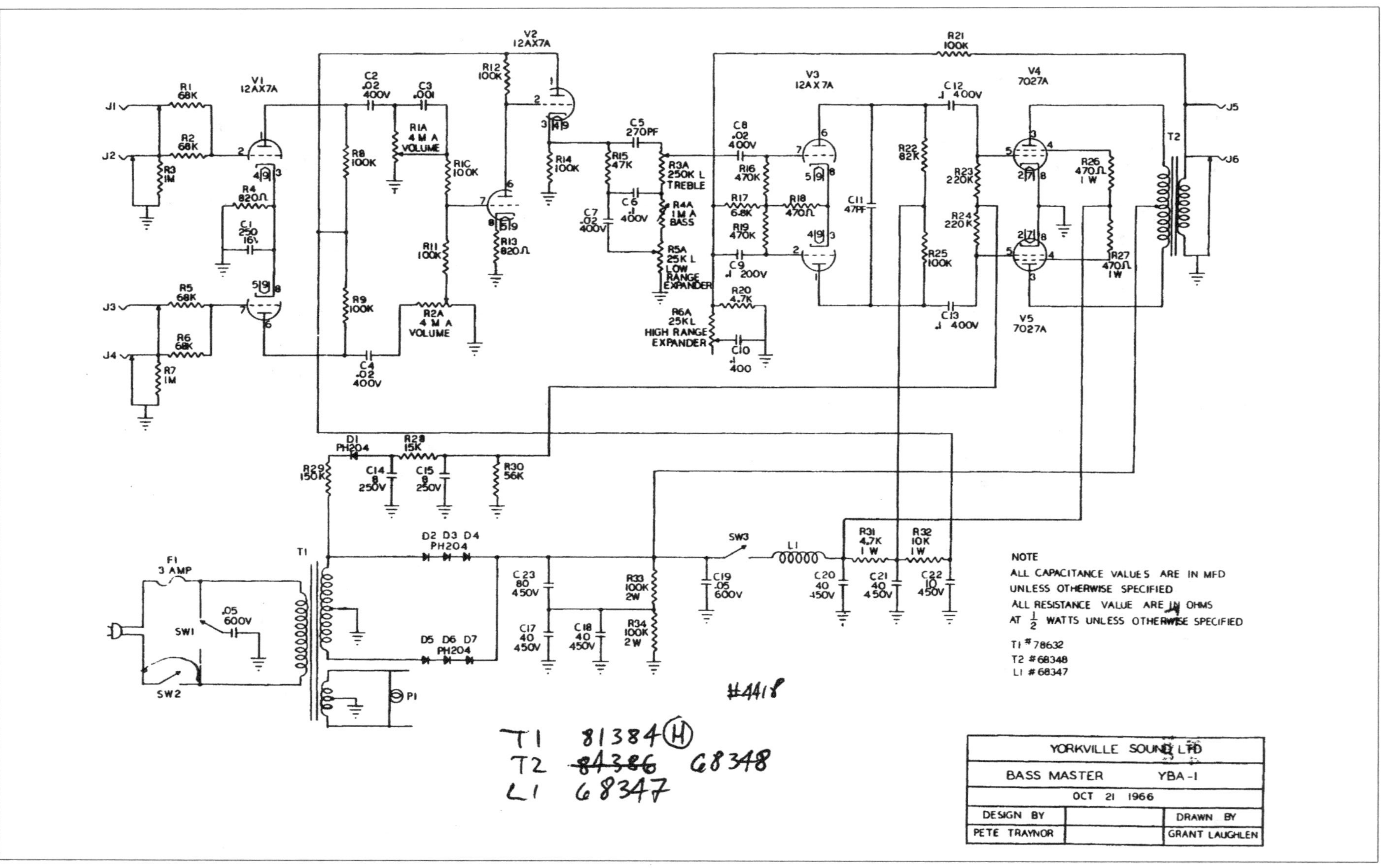
NOTE
ALL CAPACITANCE VALUES ARE IN MFD
UNLESS OTHERWISE SPECIFIED
ALL RESISTANCE VALUE ARE IN OHMS
AT 1/2 WATTS UNLESS OTHERWISE SPECIFIED
T1 #78632
T2 #68348
L1 #68347
YORKVILLE SOUND LTD
BASS MASTER
YBA-1
OCT 21 1966
DESIGN BY
PETE TRAYNOR
DRAWN BY
GRANT LAUGHLEN
#4418
T1 81384 (H)
T2 84386 68348
L1 68347

YBA-1 (LATEST VERSION)

Traynor

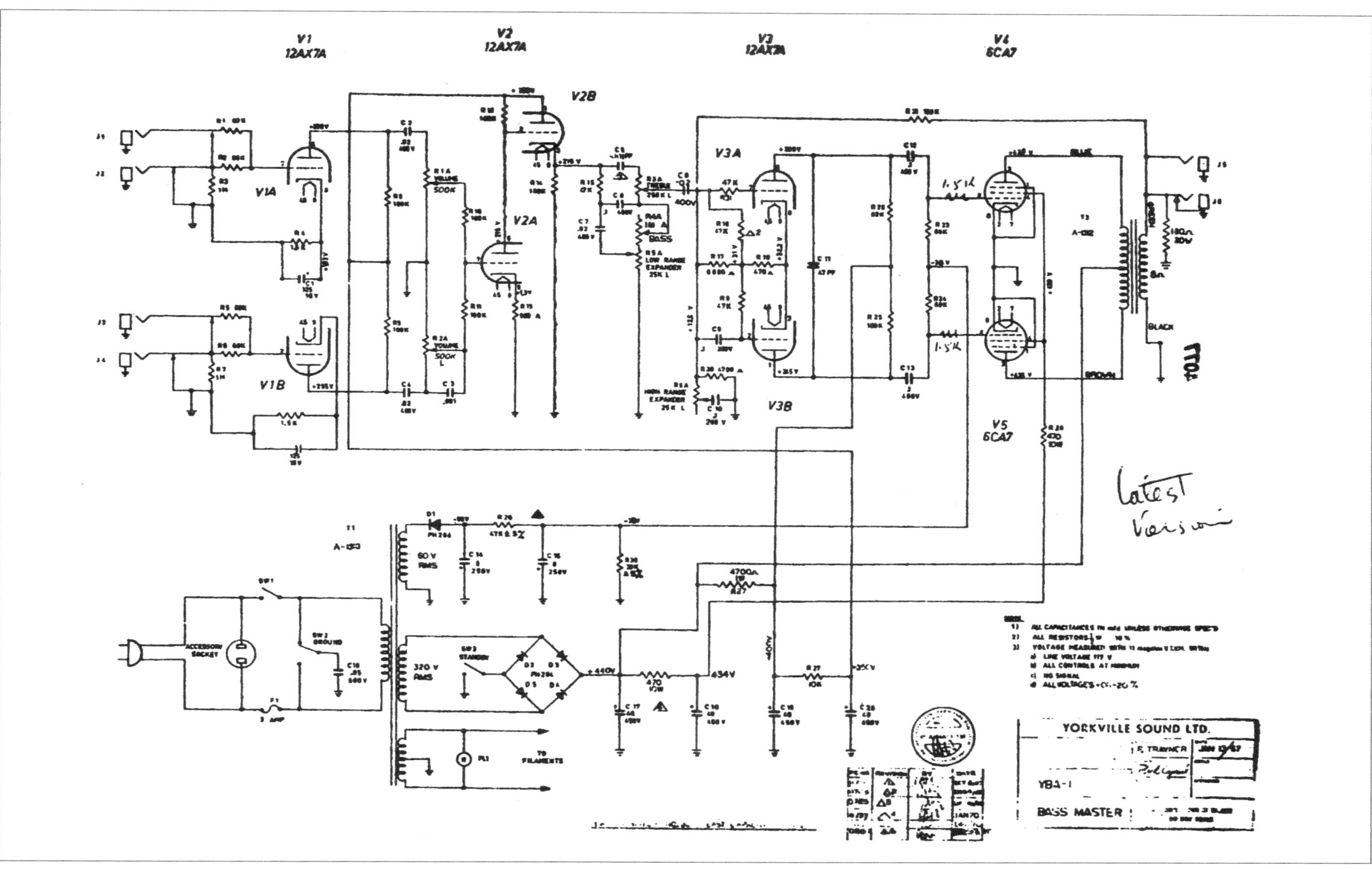

YBA-1 Power Supply

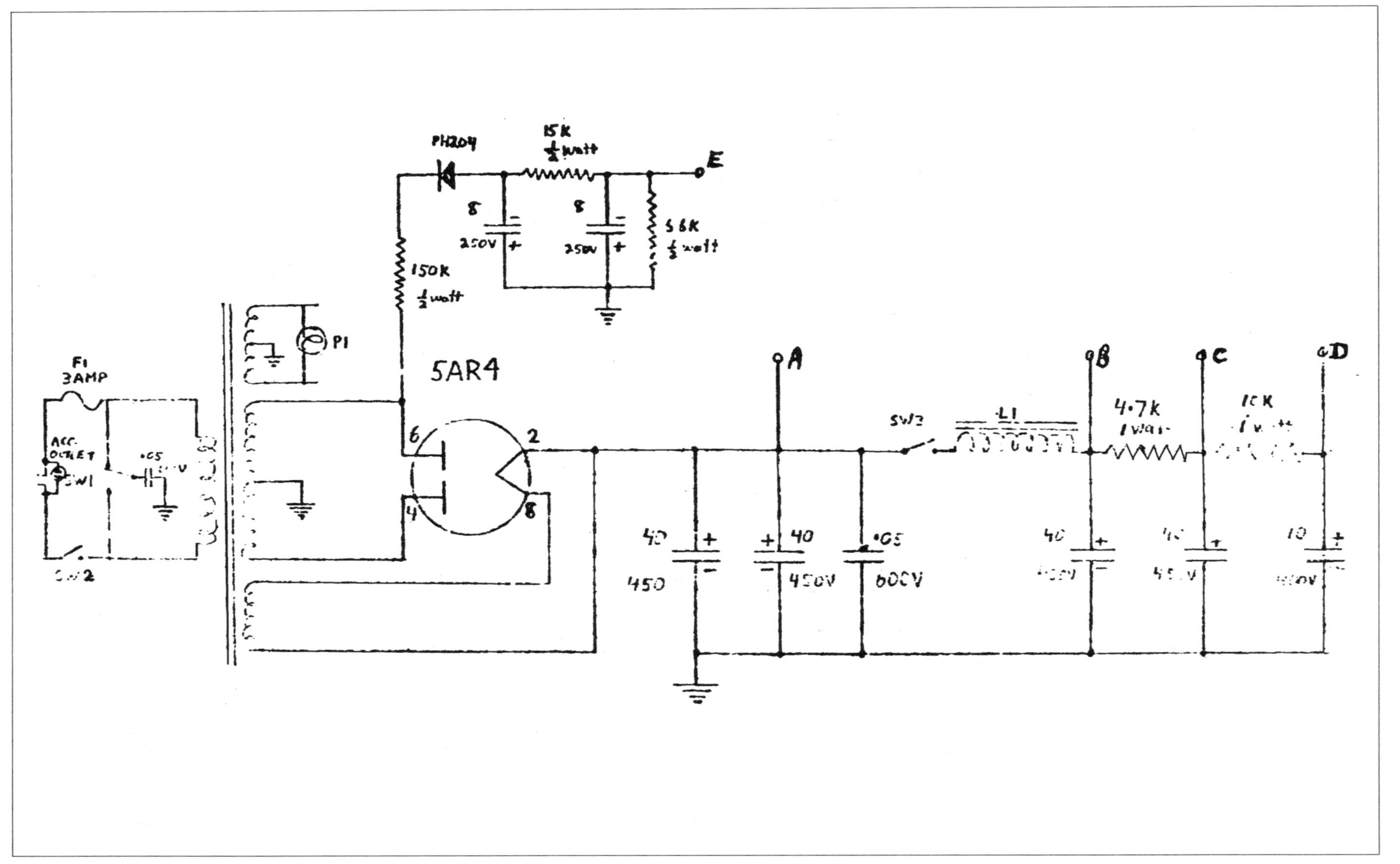

YBA-1 & YBA-4

Traynor

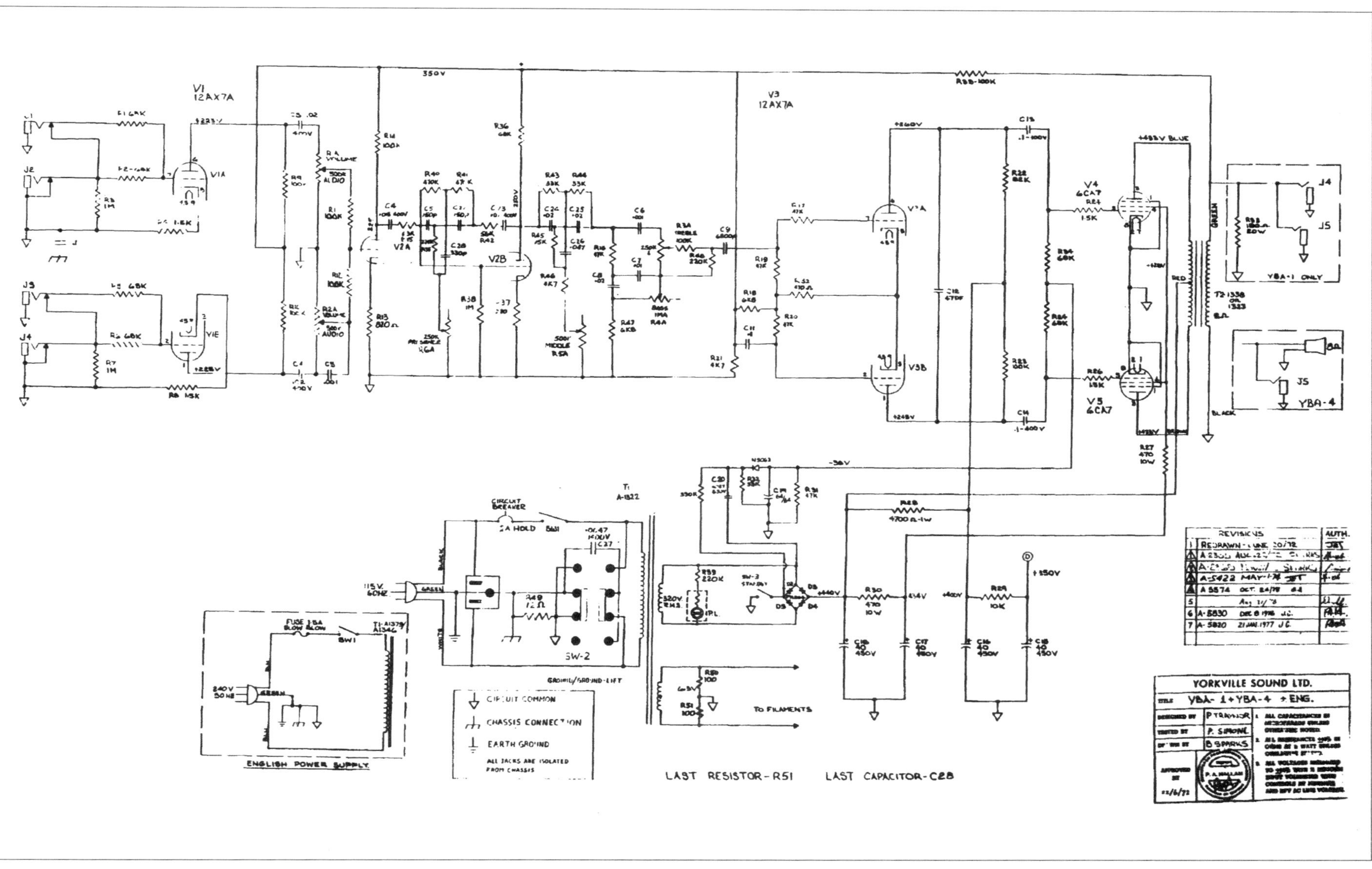

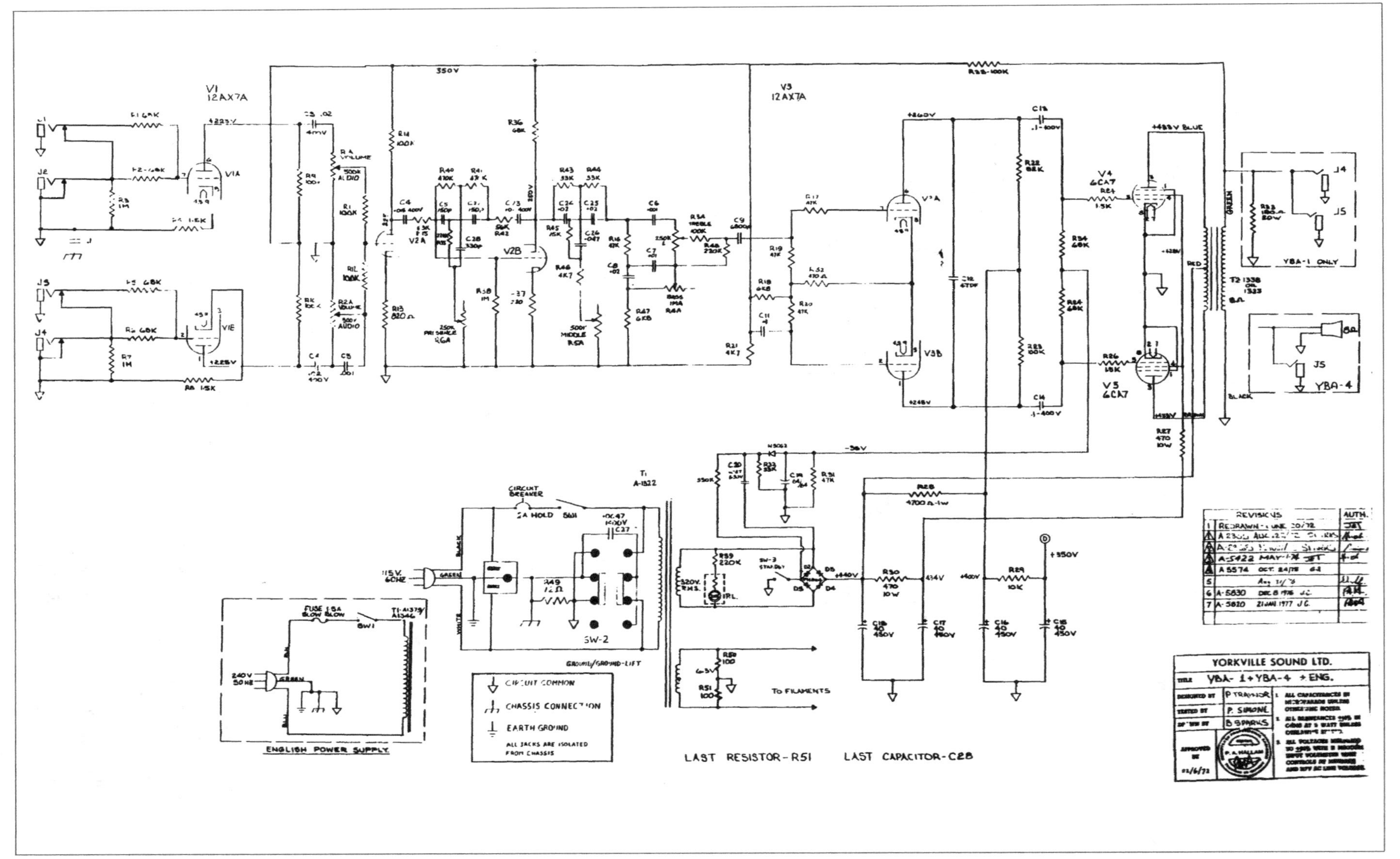
V1
12AX7A
V3
12AX7A
V4
6CA7
V5
6CA7
V2A
V2B
V1A
V1B
V3A
V3B
350V
YBA-1 ONLY
YBA-4
T2 1338
T1
A-1822
SW-2
CIRCUIT COMMON
CHASSIS CONNECTION
EARTH GROUND
ALL JACKS ARE ISOLATED FROM CHASSIS
ENGLISH POWER SUPPLY
TO FILAMENTS
LAST RESISTOR-R51
LAST CAPACITOR-C28
REVISIONS
AUTH.
YORKVILLE SOUND LTD.
YBA-1 + YBA-4 + ENG.
P. TRAYNOR
P. SIMONE
B. SPARKS

YBA-1A **BASS MASTER MK II**

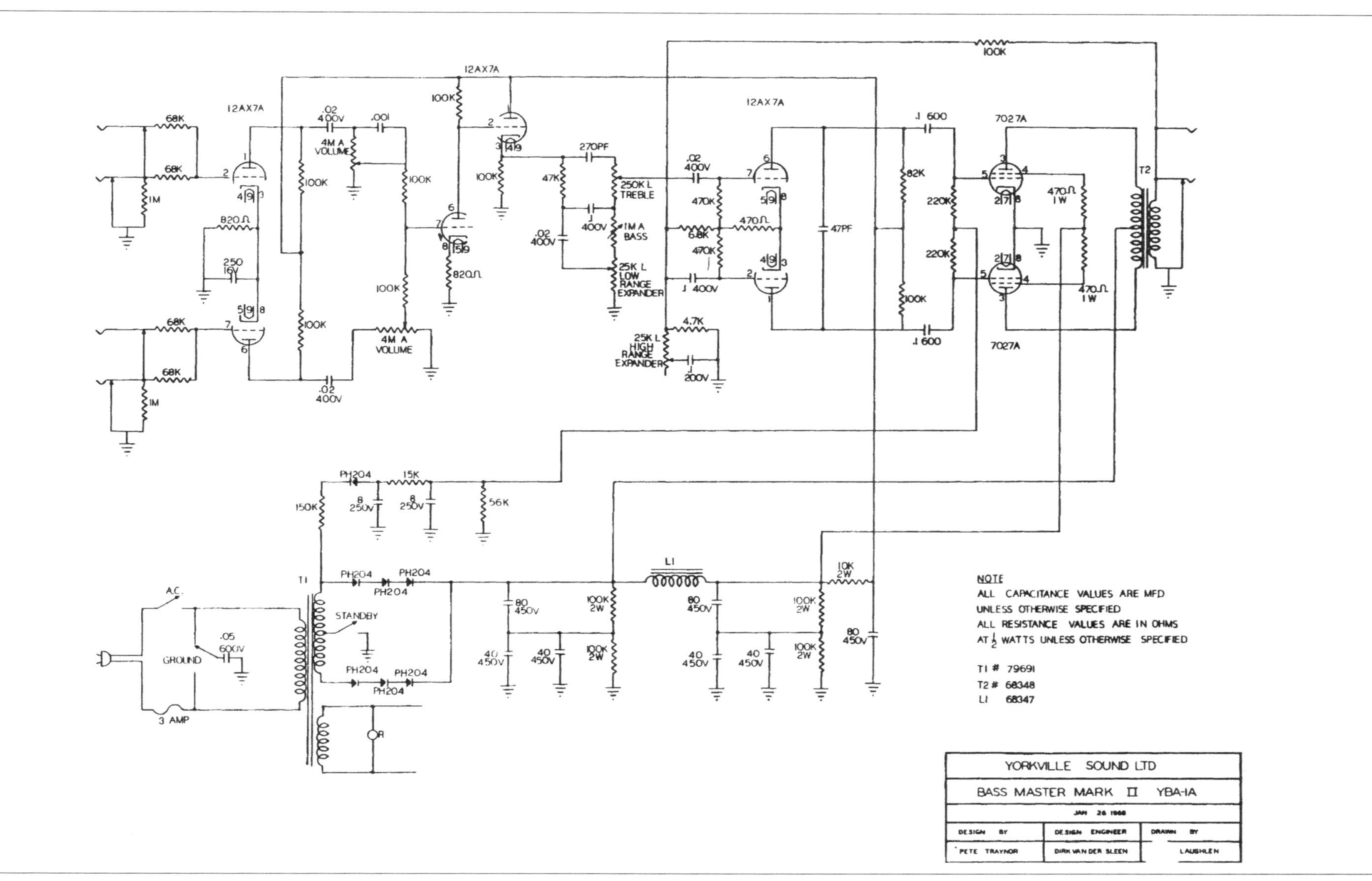

YBA-1A & ENG BASS MASTER MK II

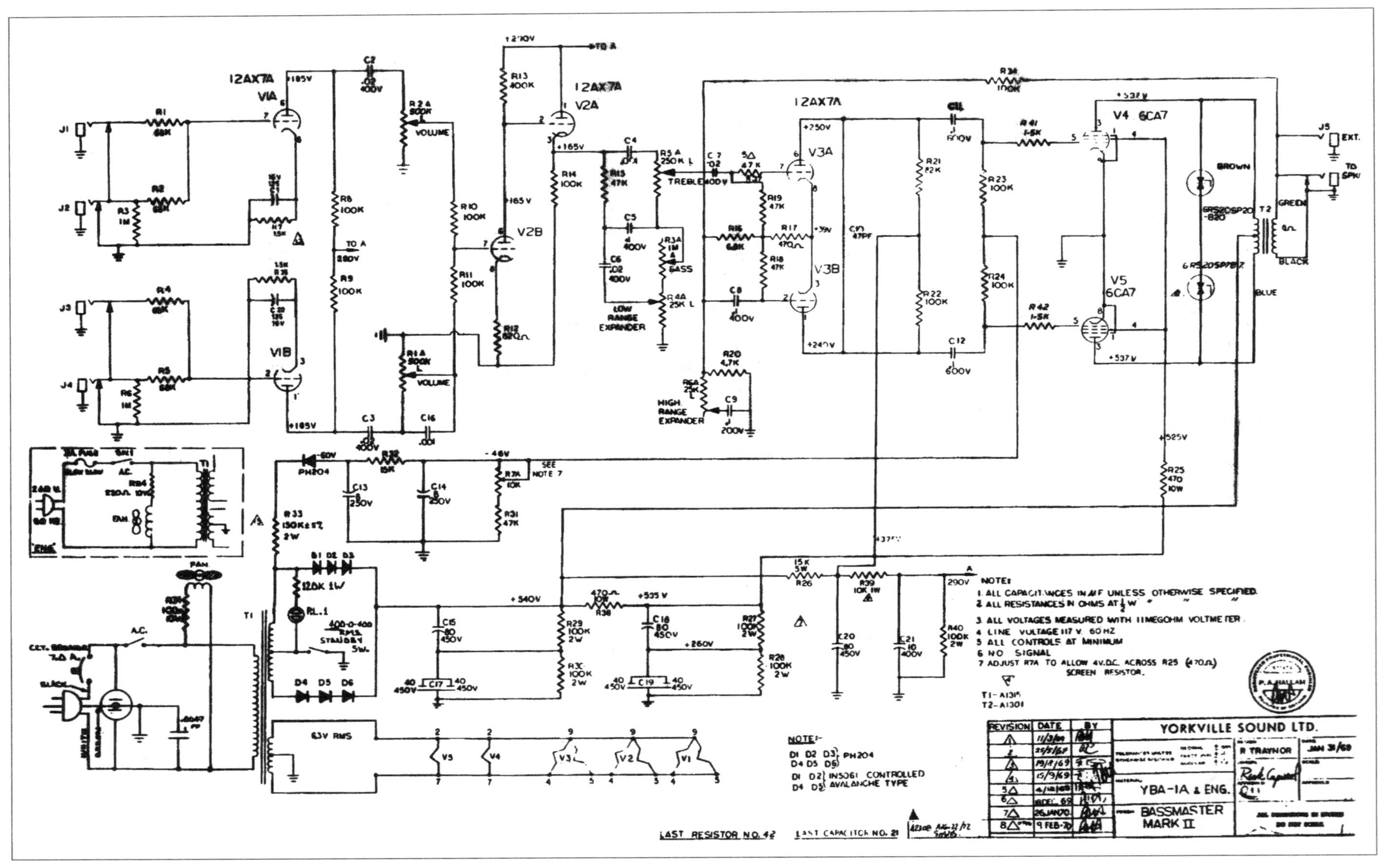

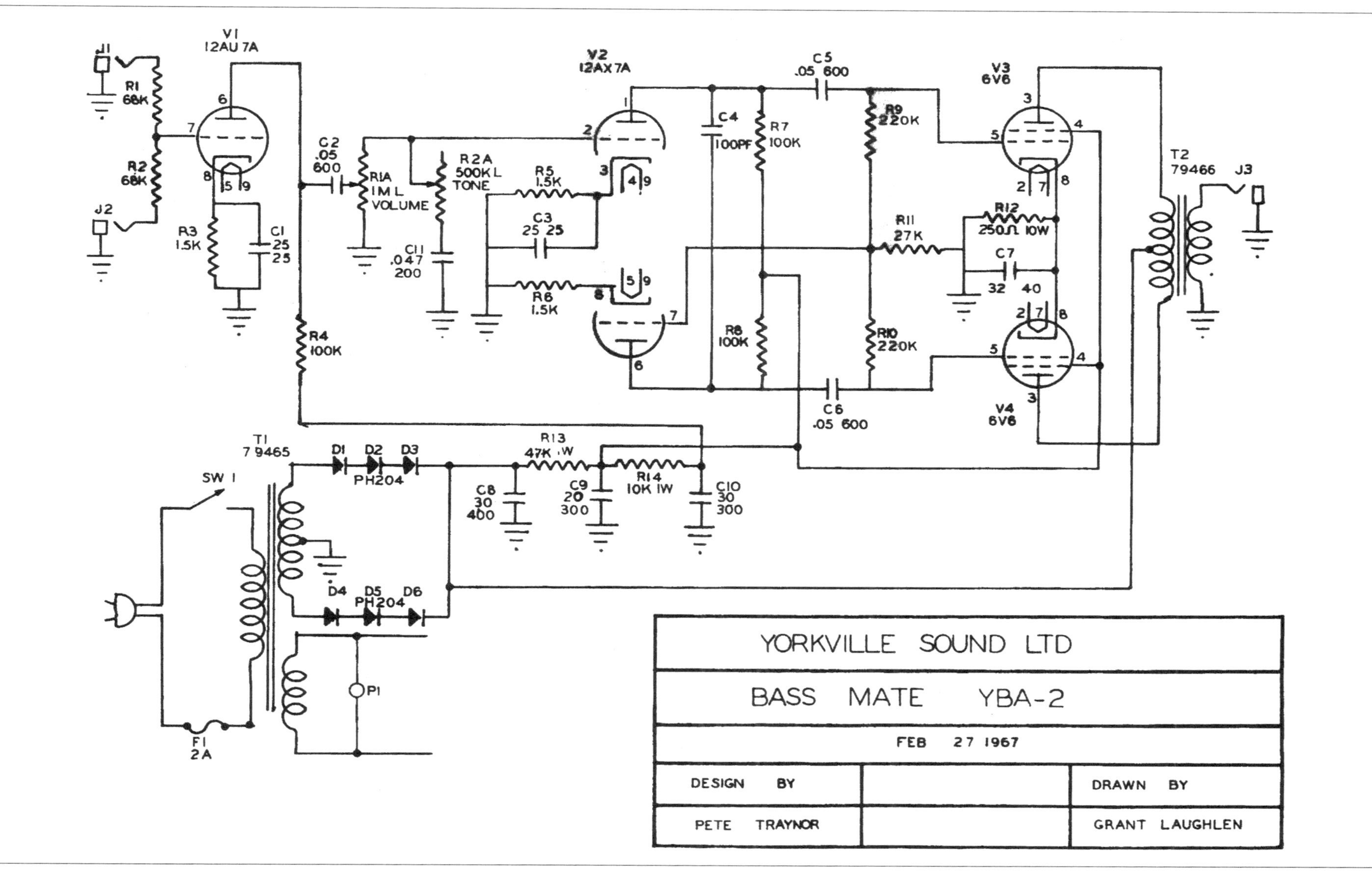
J1
R1 68K
R2 68K
J2
V1 12AU 7A
R3 1.5K
C1 25 25
C2 .05 600
R1A 1M L VOLUME
R2A 500K L TONE
C11 .047 200
R4 100K
R5 1.5K
C3 25 25
R6 1.5K
V2 12AX7A
C4 100PF
R7 100K
R8 100K
C5 .05 600
C6 .05 600
R9 220K
R10 220K
R11 27K
R12 250Ω 10W
C7 32 40
V3 6V6
V4 6V6
T2 79466
J3
T1 7 9465
SW 1
F1 2A
D1 D2 D3 PH204
D4 D5 D6 PH204
P1
R13 47K 1W
R14 10K 1W
C8 30 400
C9 20 300
C10 30 300
YORKVILLE SOUND LTD
BASS MATE YBA-2
FEB 27 1967
DESIGN BY
PETE TRAYNOR
DRAWN BY
GRANT LAUGHLEN

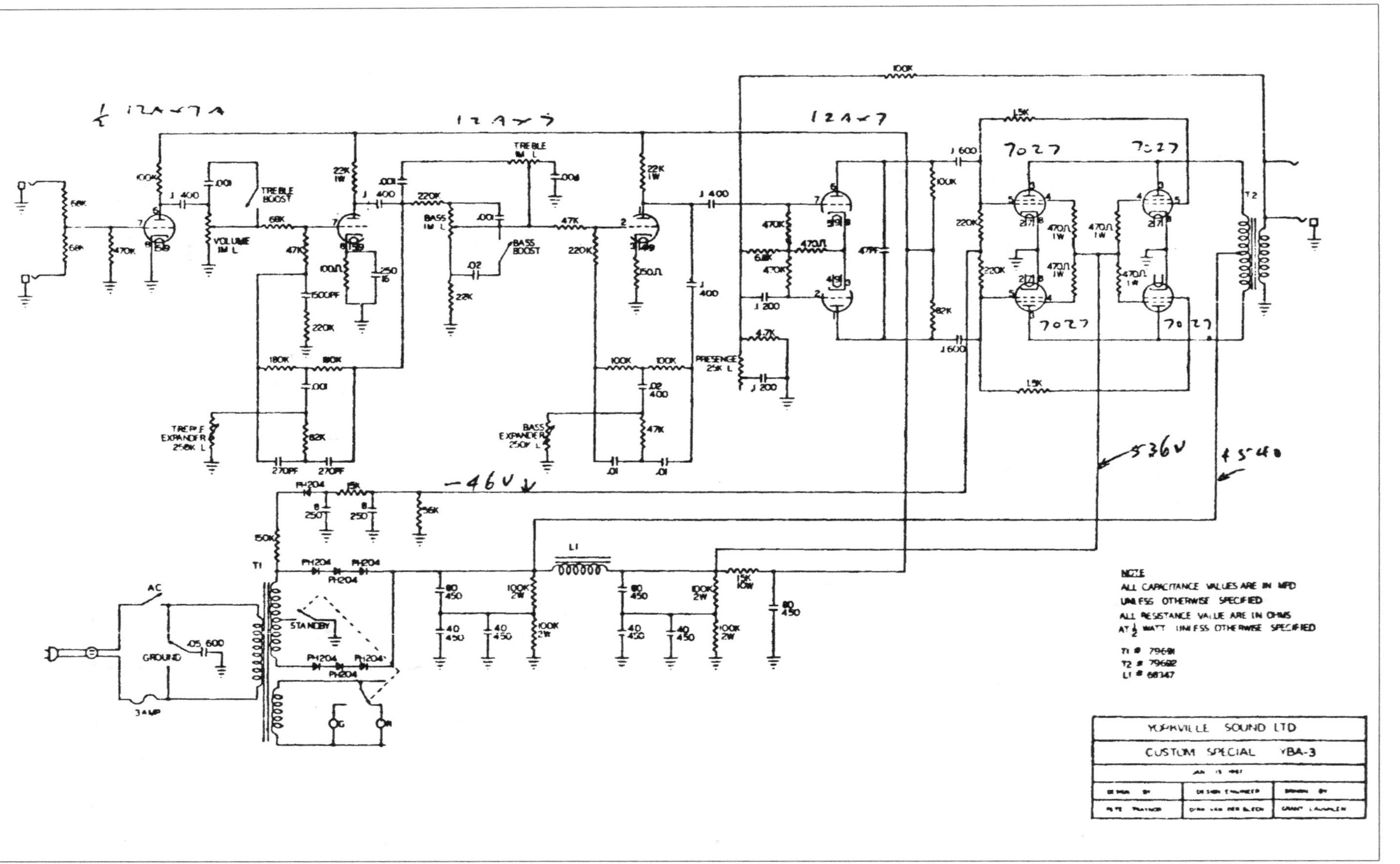
12AX7
12AX7
7027
7027
7027
7027
TREBLE BOOST
VOLUME 1M L
TREBLE 1M L
BASS 1M L
BASS BOOST
PRESENCE 25K L
TREBLE EXPANDER 250K L
BASS EXPANDER 250K L
-46V
536V
AC
GROUND
STANDBY
3 AMP
T1
T2
L1
NOTE
ALL CAPACITANCE VALUES ARE IN MFD
UNLESS OTHERWISE SPECIFIED
ALL RESISTANCE VALUE ARE IN OHMS
AT ½ WATT UNLESS OTHERWISE SPECIFIED
T1 # 79691
T2 # 79692
L1 # 68347
YORKVILLE SOUND LTD
CUSTOM SPECIAL YBA-3

YBA-3 & ENG **CUSTOM SPECIAL**

Traynor

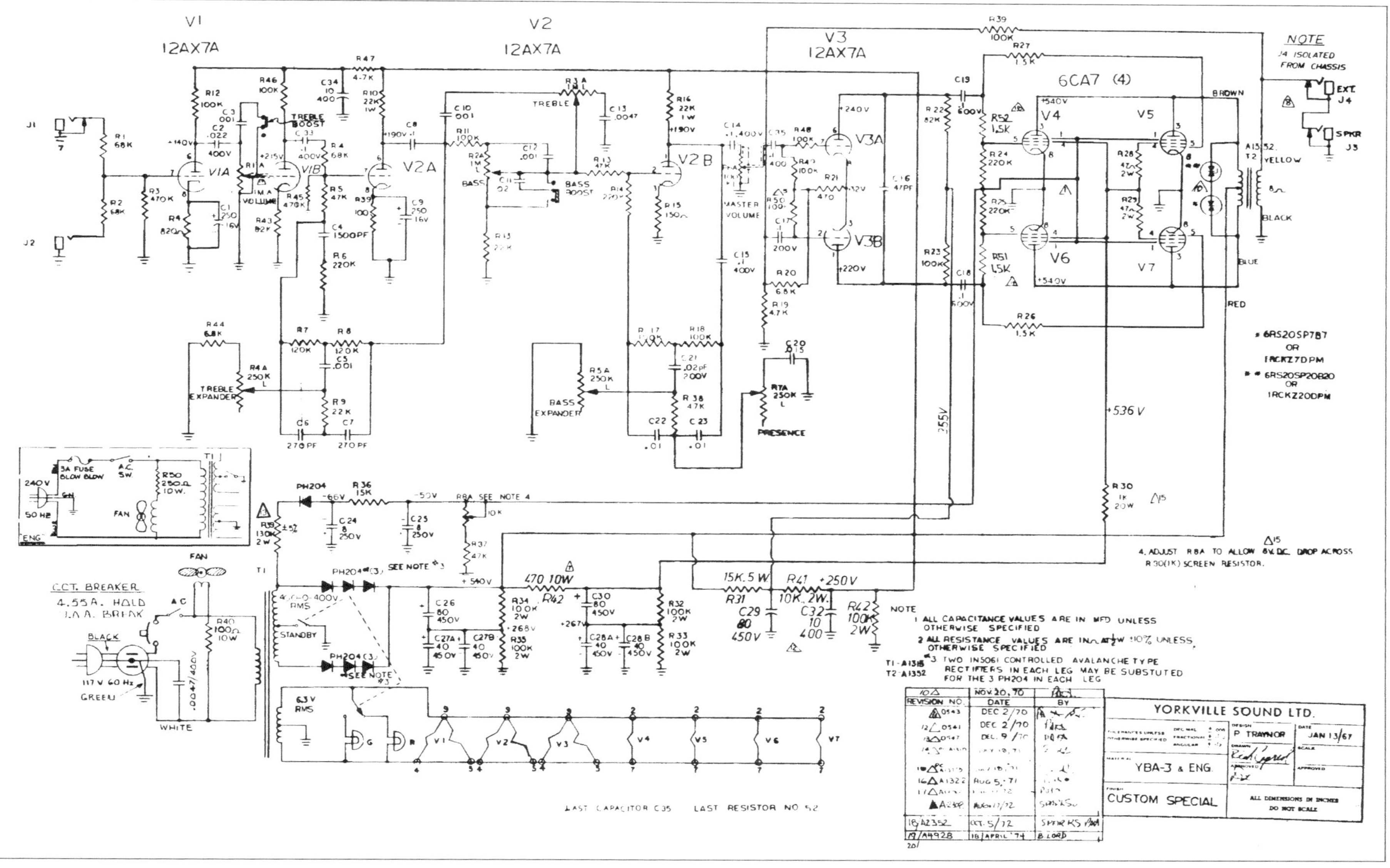

YBA-3 CUSTOM SPECIAL (REVISED)

Traynor

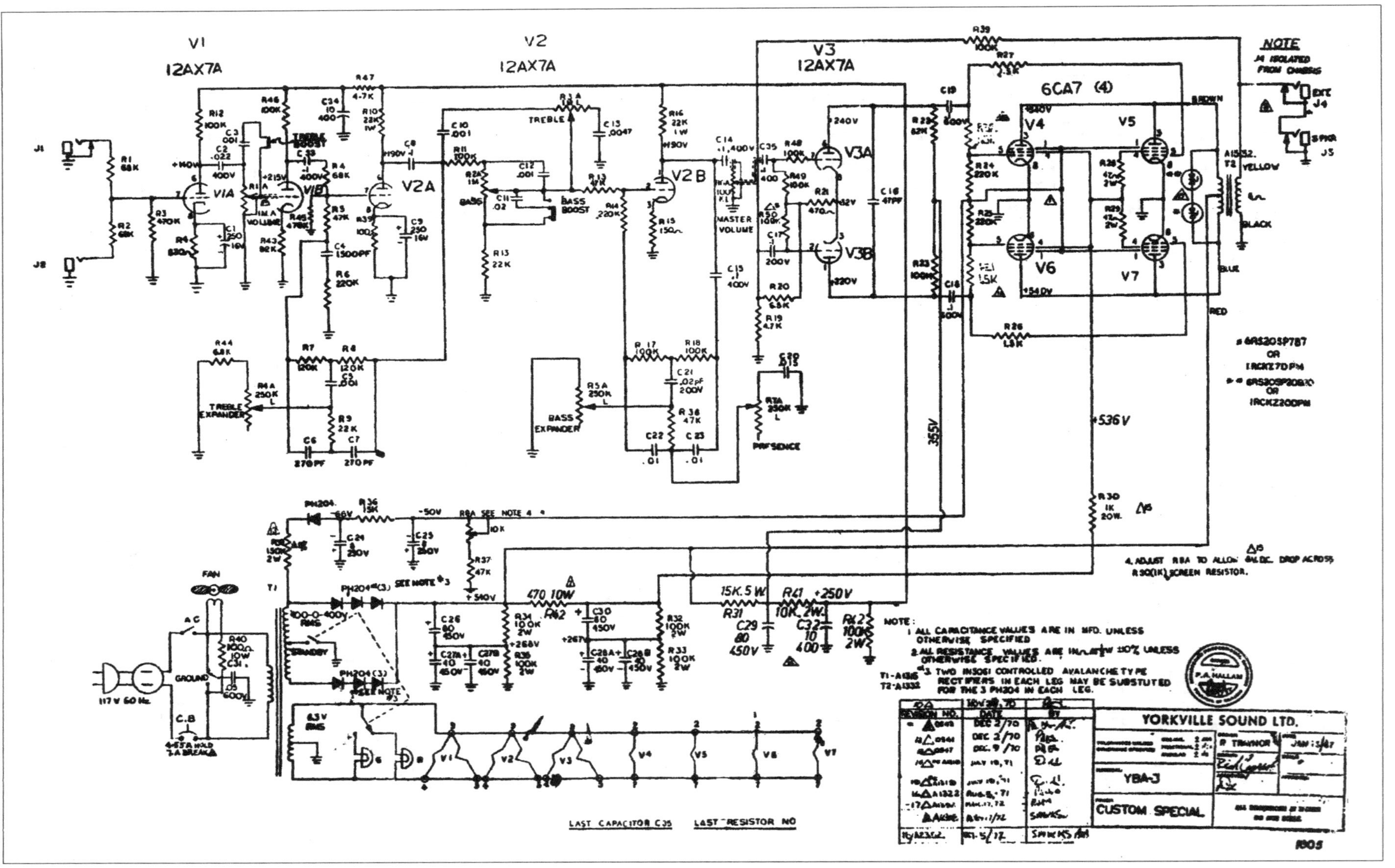

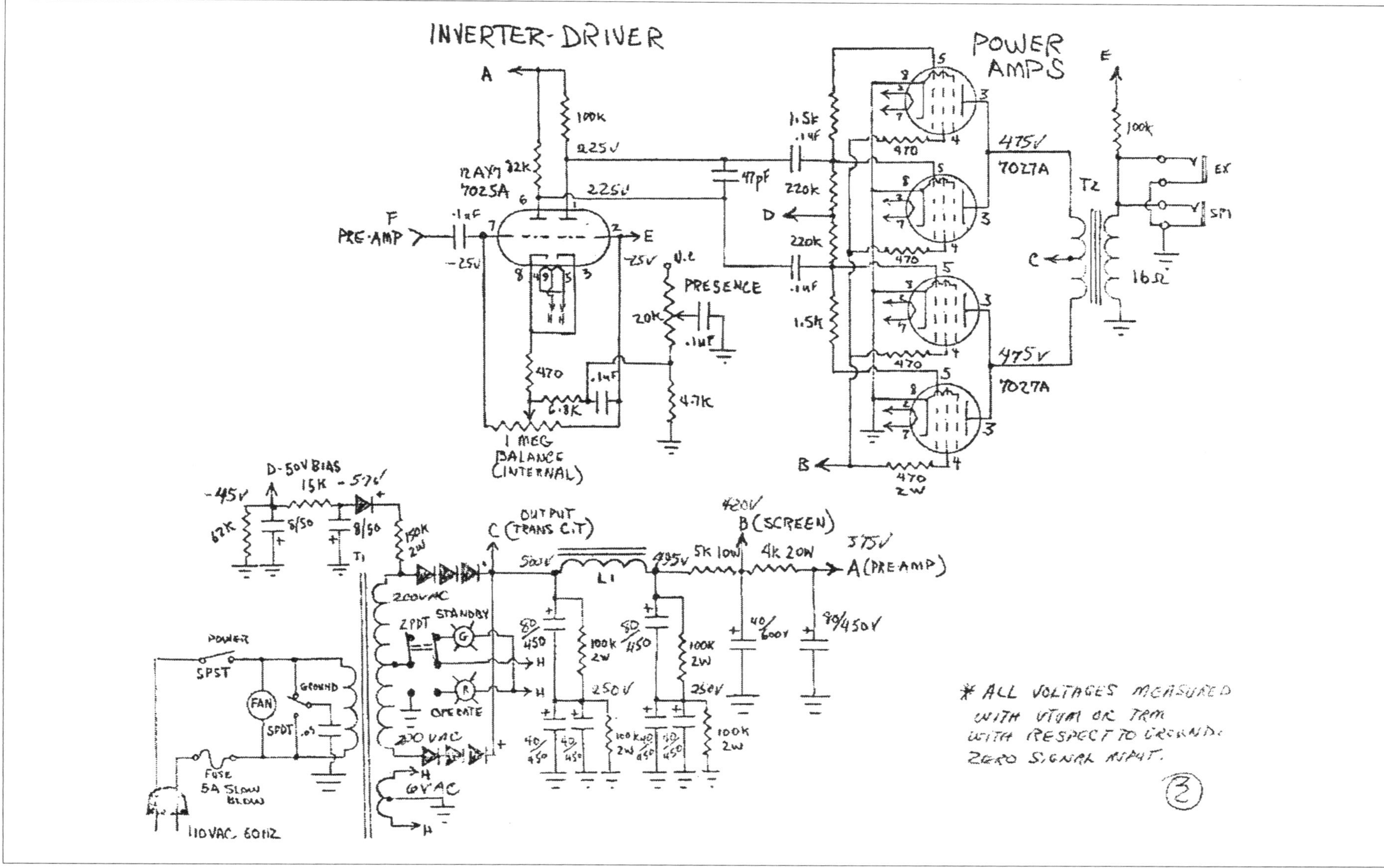
INVERTER-DRIVER
POWER AMPS
12AY7
7025A
PRE-AMP
PRESENCE
1 MEG BALANCE (INTERNAL)
7027A
T2
16Ω
EX
SPK
D-50V BIAS
OUTPUT C (TRANS C.T)
B (SCREEN)
A (PRE-AMP)
L1
T1
STANDBY
OPERATE
POWER SPST
GROUND SPDT
FAN
FUSE 5A SLOW BLOW
110VAC 60HZ
200VAC
6VAC
* ALL VOLTAGES MEASURED WITH VTVM OR TRM WITH RESPECT TO GROUND. ZERO SIGNAL INPUT.
3

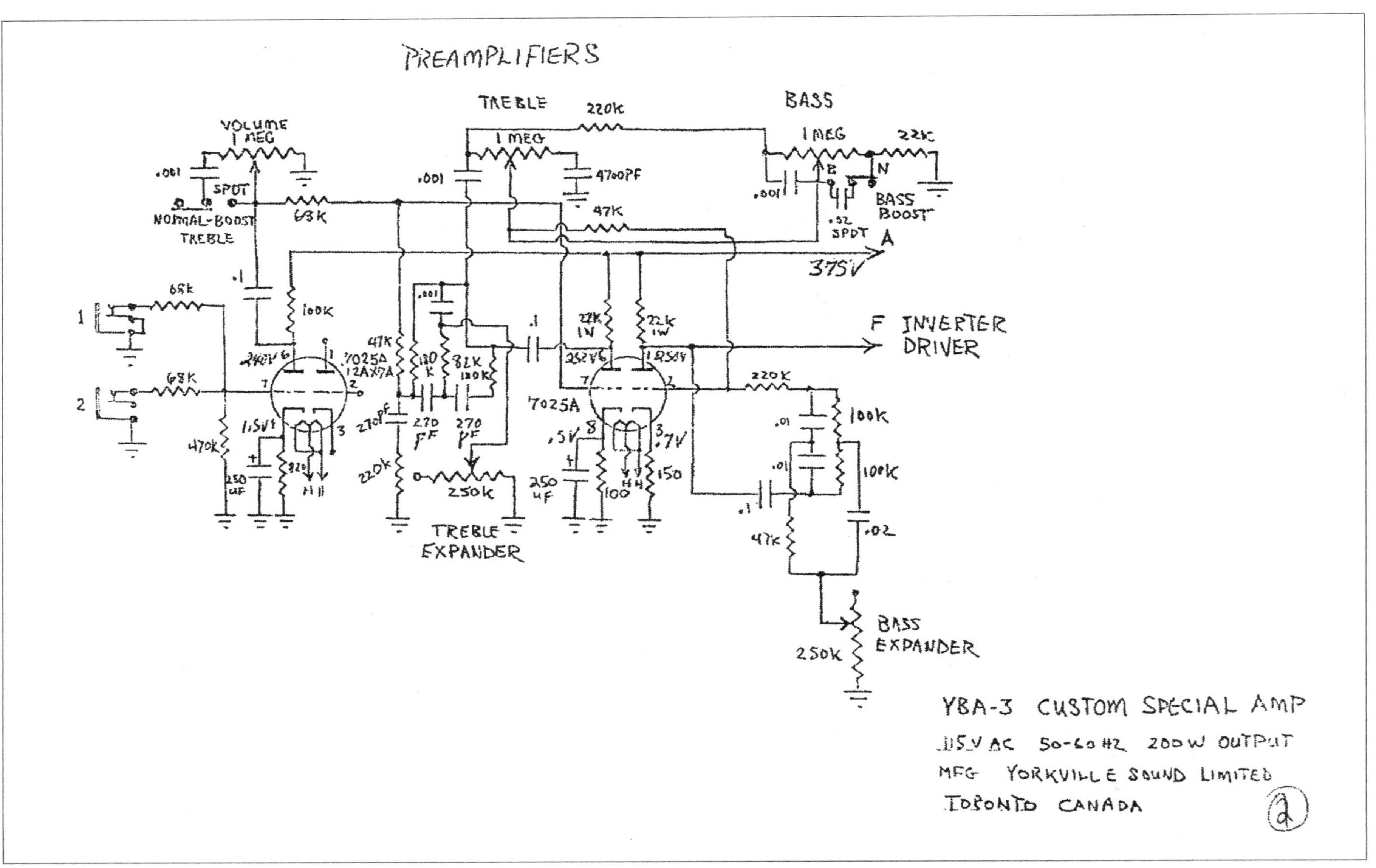
PREAMPLIFIERS
VOLUME 1 MEG
.001
SPDT
NORMAL-BOOST TREBLE
68K
.1
100K
68K
68K
470K
250 UF
7025A 12AX7A
47K
180 K
82K
180K
270 PF
270 PF
270 PF
220K
250K
TREBLE EXPANDER
TREBLE
1 MEG
.001
4700PF
47K
220K
.1
22K 1W
22K 1W
7025A
250 UF
100
150
BASS
1 MEG
22K
.001
.02 SPDT
BASS BOOST
A
375V
F INVERTER DRIVER
220K
100K
100K
.01
.01
.1
47K
.02
250K
BASS EXPANDER
YBA-3 CUSTOM SPECIAL AMP
115 V AC 50-60 HZ 200W OUTPUT
MFG YORKVILLE SOUND LIMITED
TORONTO CANADA
2

YBA-4

Traynor

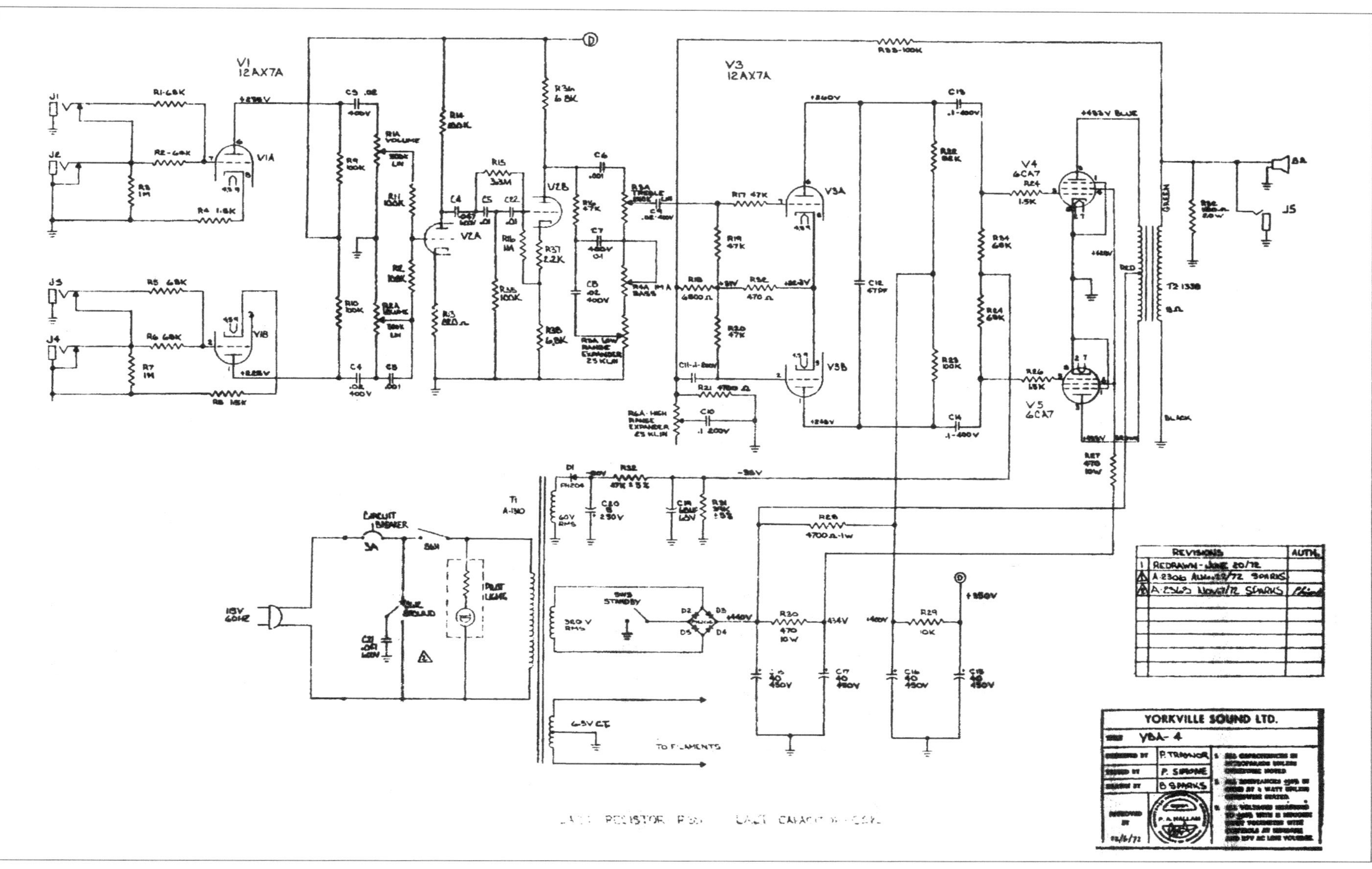

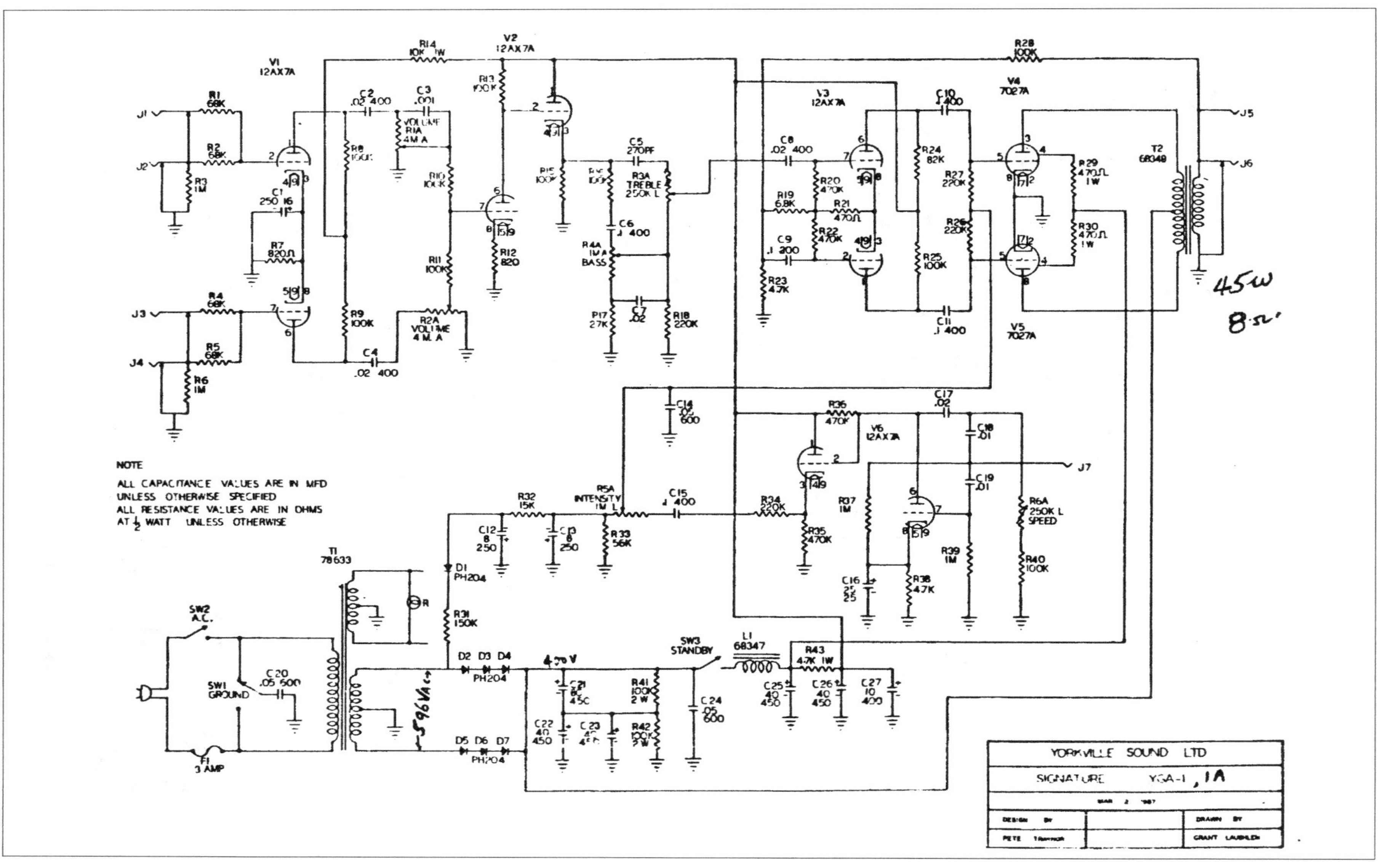
V1 12AX7A
V2 12AX7A
V3 12AX7A
V4 7027A
V5 7027A
V6 12AX7A
R14 10K 1W
R28 100K
C2 .02 400
C3 .001
VOLUME R1A 4M A
R2A VOLUME 4M A
C4 .02 400
C5 270PF
R3A TREBLE 250K L
C6 .1 400
R4A 1MA BASS
C8 .02 400
C10 .1 400
C11 .1 400
T2 68048
45w
8Ω
R29 470Ω 1W
R30 470Ω 1W
R5A INTENSITY 1M L
C15 .1 400
R6A 250K L SPEED
J7
NOTE
ALL CAPACITANCE VALUES ARE IN MFD
UNLESS OTHERWISE SPECIFIED
ALL RESISTANCE VALUES ARE IN OHMS
AT ½ WATT UNLESS OTHERWISE
T1 78633
SW2 A.C.
SW1 GROUND
C20 .05 600
F1 3 AMP
D1 PH204
R31 150K
D2 D3 D4 PH204
D5 D6 D7 PH204
SW3 STANDBY
L1 68347
R43 4.7K 1W
R41 100K 2W
R42 100K 3W
YORKVILLE SOUND LTD
SIGNATURE YGA-1, 1A
DESIGN BY
PETE TRAYNOR
DRAWN BY
GRANT LAUGHLEN

YGL-3 **MK III**

Traynor

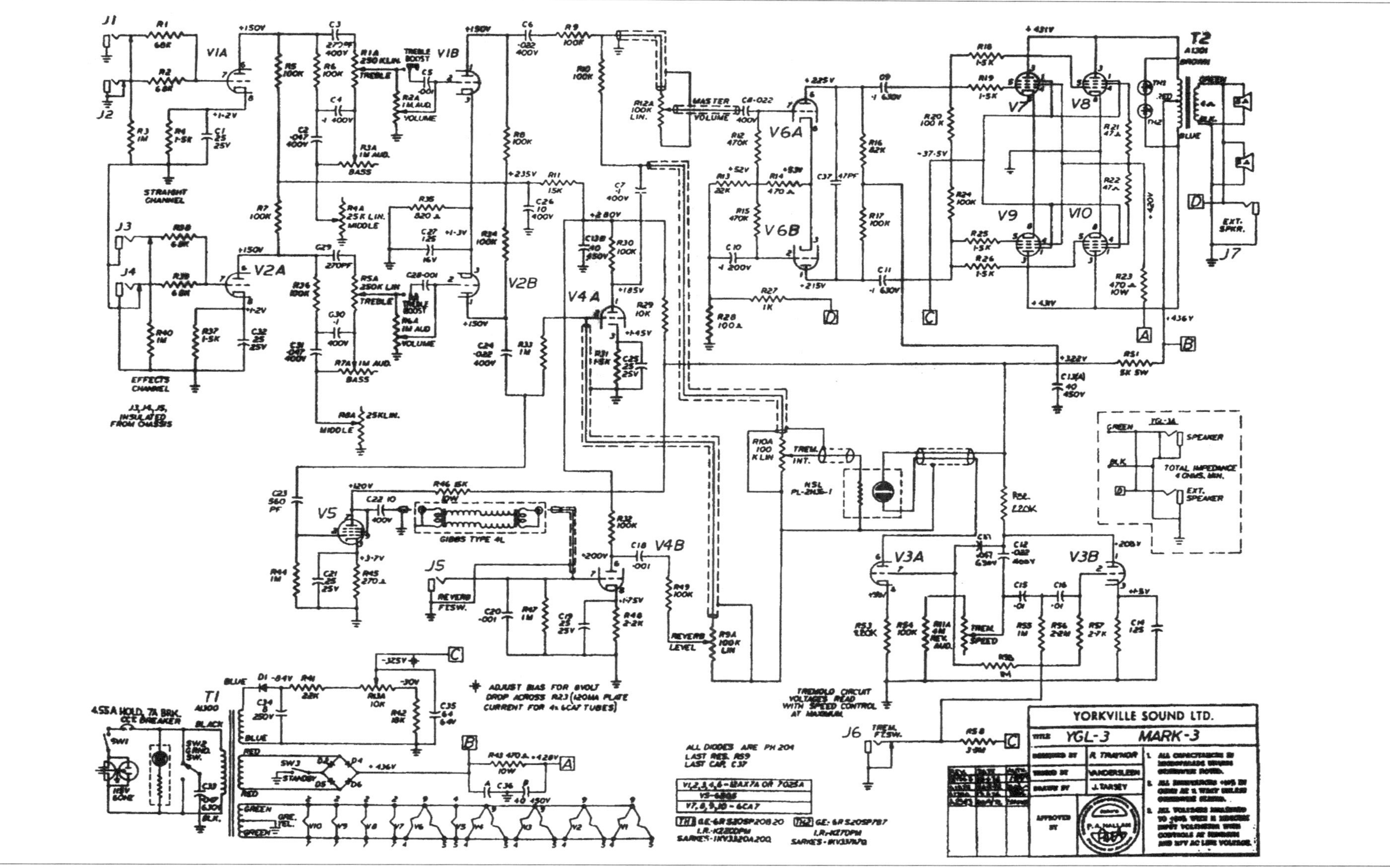

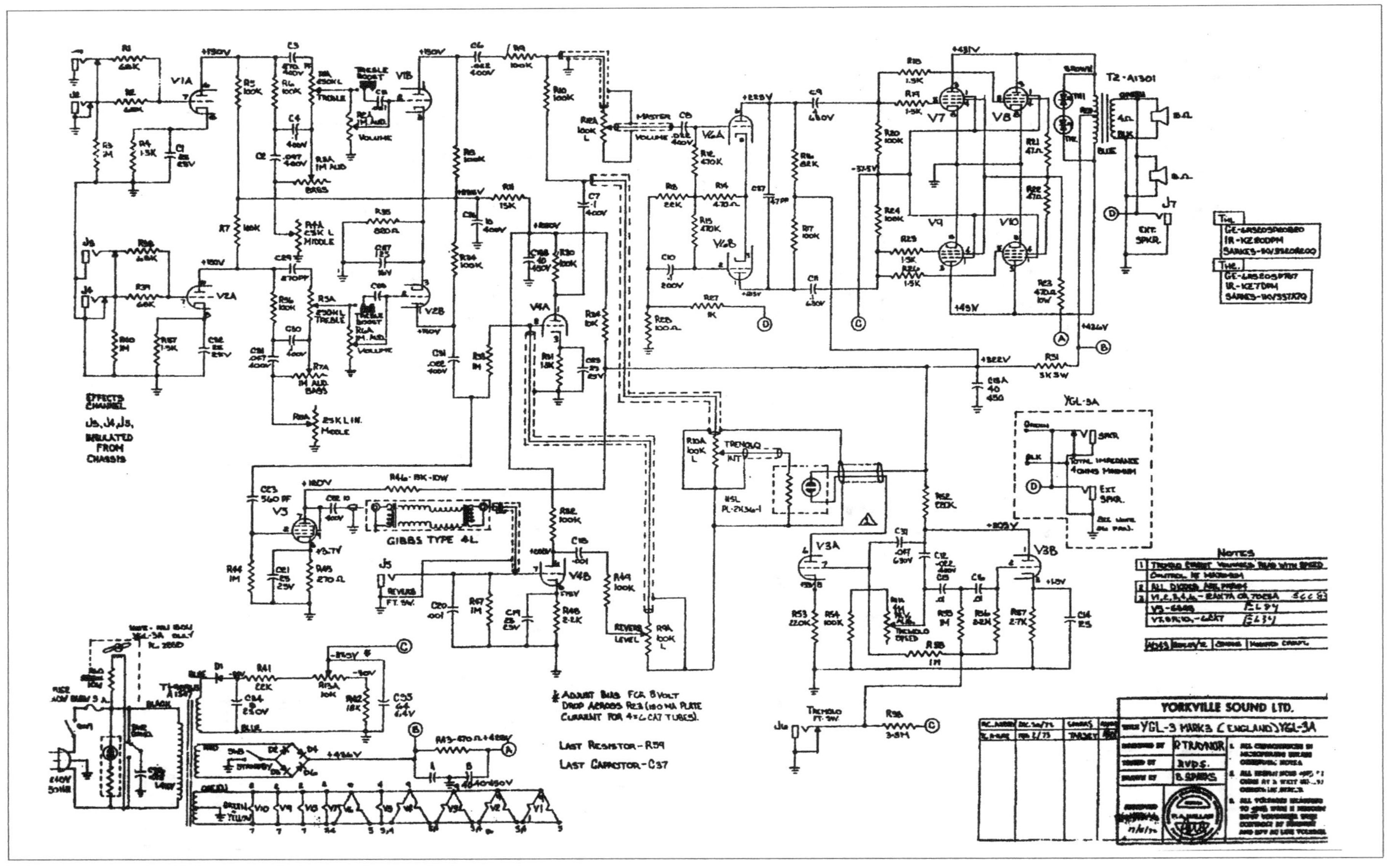
V1A
V1B
V2A
V2B
V3
V3A
V3B
V4A
V4B
V6A
V6B
V7
V8
V9
V10
T2-A1301
MASTER VOLUME
VOLUME
TREBLE
BASS
MIDDLE
GIBBS TYPE 4L
REVERB FT. SW.
REVERB LEVEL
TREMOLO FT. SW.
TREMOLO INT
TREMOLO SPEED
EXT. SPKR.
EFFECTS CHANNEL J3, J4, J5, INSULATED FROM CHASSIS
YGL-3A
TOTAL IMPEDANCE 4 OHMS MINIMUM
NOTES
ADJUST BIAS FOR 8 VOLT DROP ACROSS R23 (180 MA PLATE CURRENT FOR 4 × 6CA7 TUBES).
LAST RESISTOR - R59
LAST CAPACITOR - C37
YORKVILLE SOUND LTD.
YGL-3 MARK 3 (ENGLAND) YGL-3A
R TRAYNOR
240V 50HZ

YGM-2 **GUITAR MATE**

Traynor

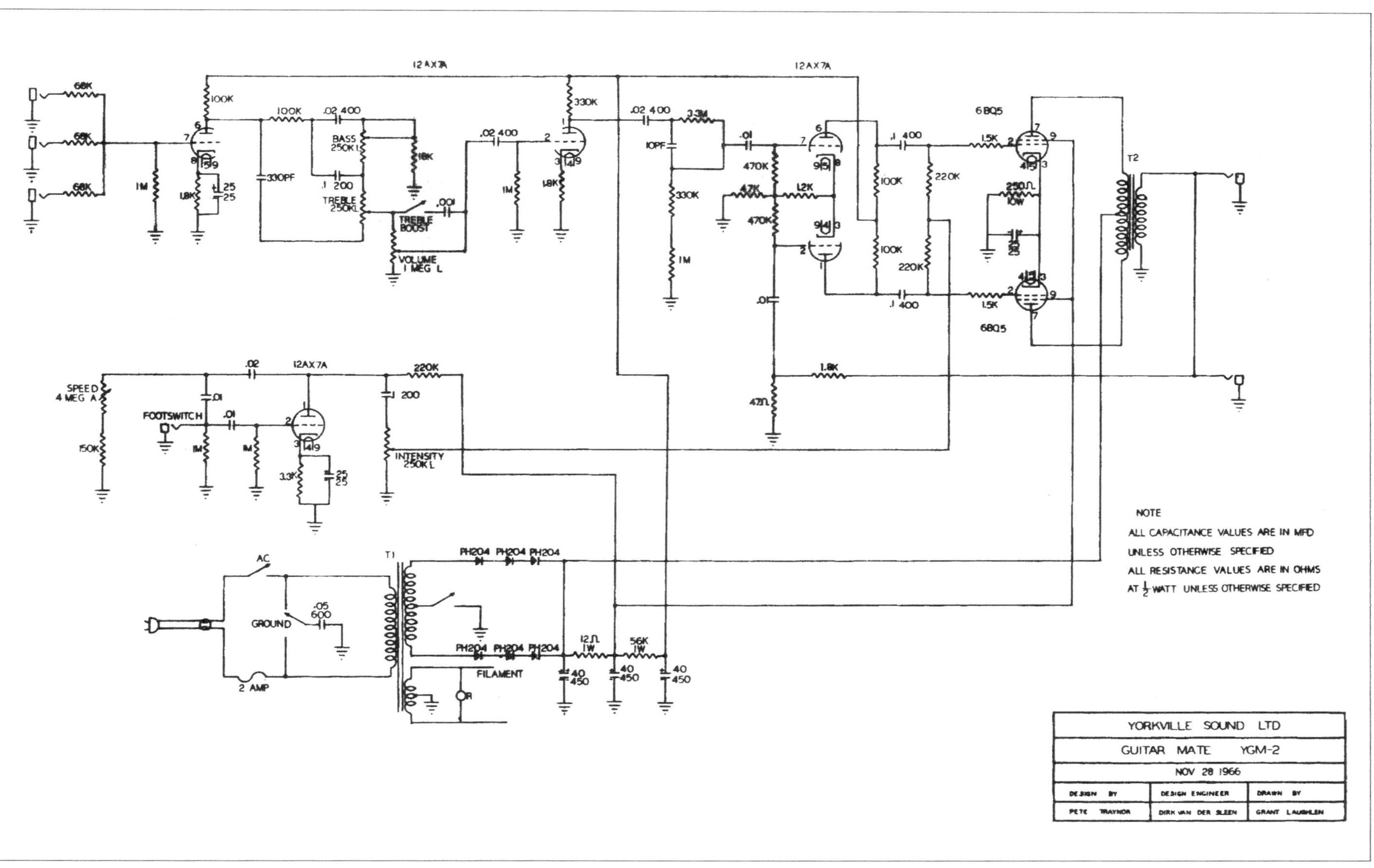

YGM-3 & 4

Traynor

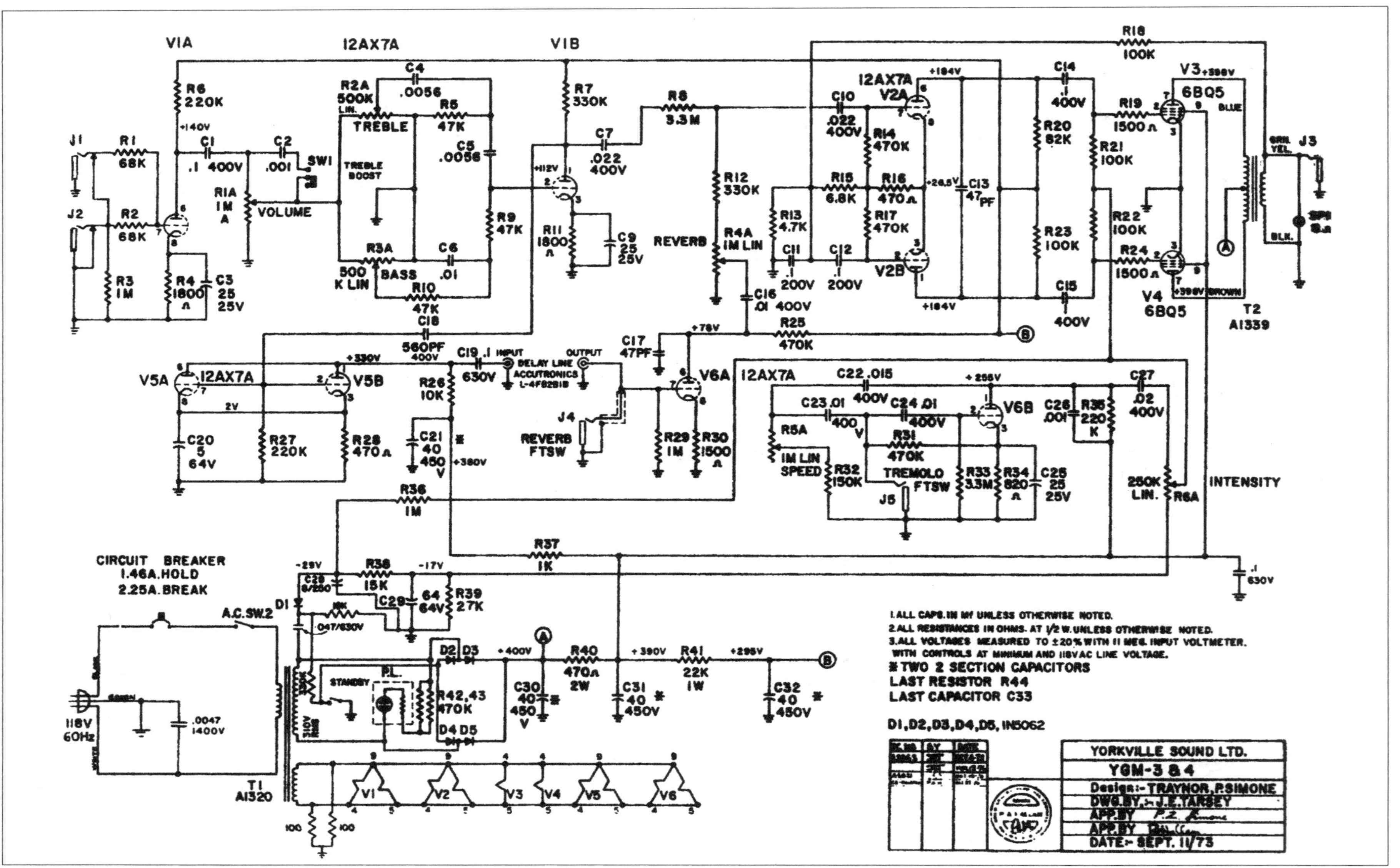

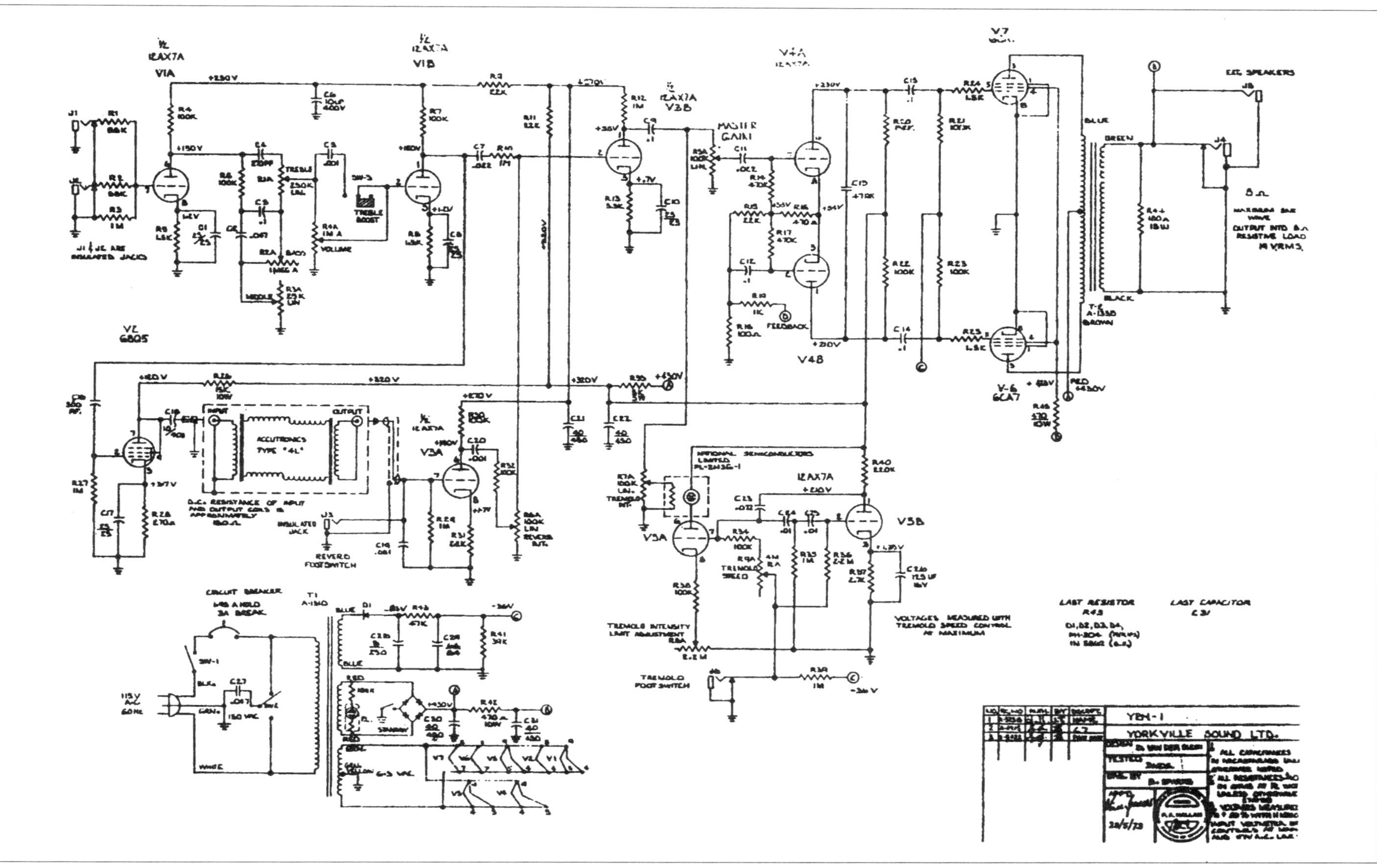
V1A
V1B
V3B
V4A
V4B
V3A
V5A
V5B
MASTER GAIN
TREBLE
BASS
MIDDLE
VOLUME
TREBLE BOOST
ACCUTRONICS TYPE "4L"
REVERB FOOTSWITCH
TREMOLO FOOTSWITCH
TREMOLO SPEED
EXT. SPEAKERS
FEEDBACK
LAST RESISTOR R45
LAST CAPACITOR C31
YRM-1
YORKVILLE SOUND LTD.

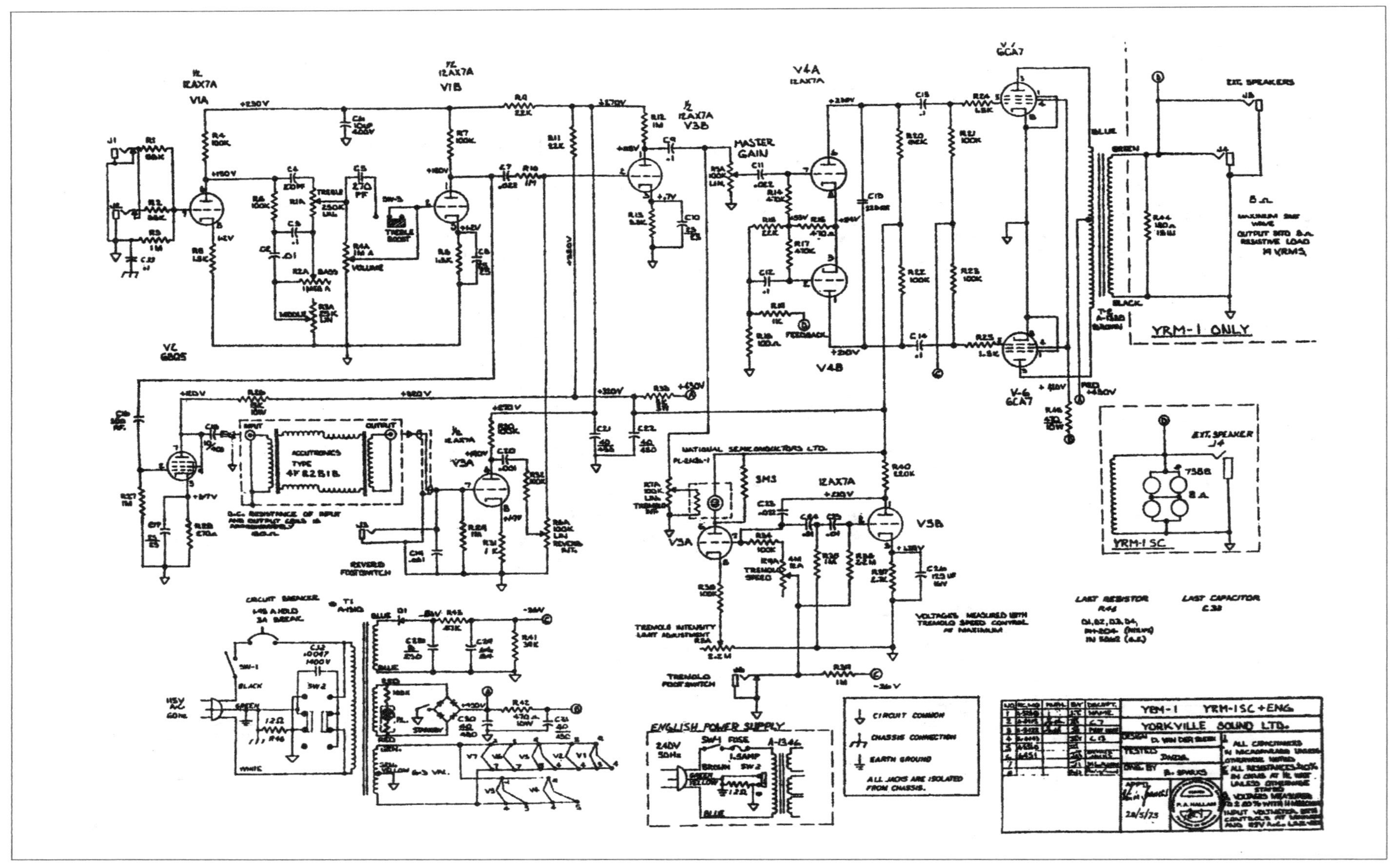
V1A
V1B
V3B
V4A
V4B
V3A
V5B
MASTER GAIN
YRM-1 ONLY
YRM-1 SC
ENGLISH POWER SUPPLY
YRM-1
YRM-1SC+ENG
YORKVILLE SOUND LTD.

YSR-1

Traynor

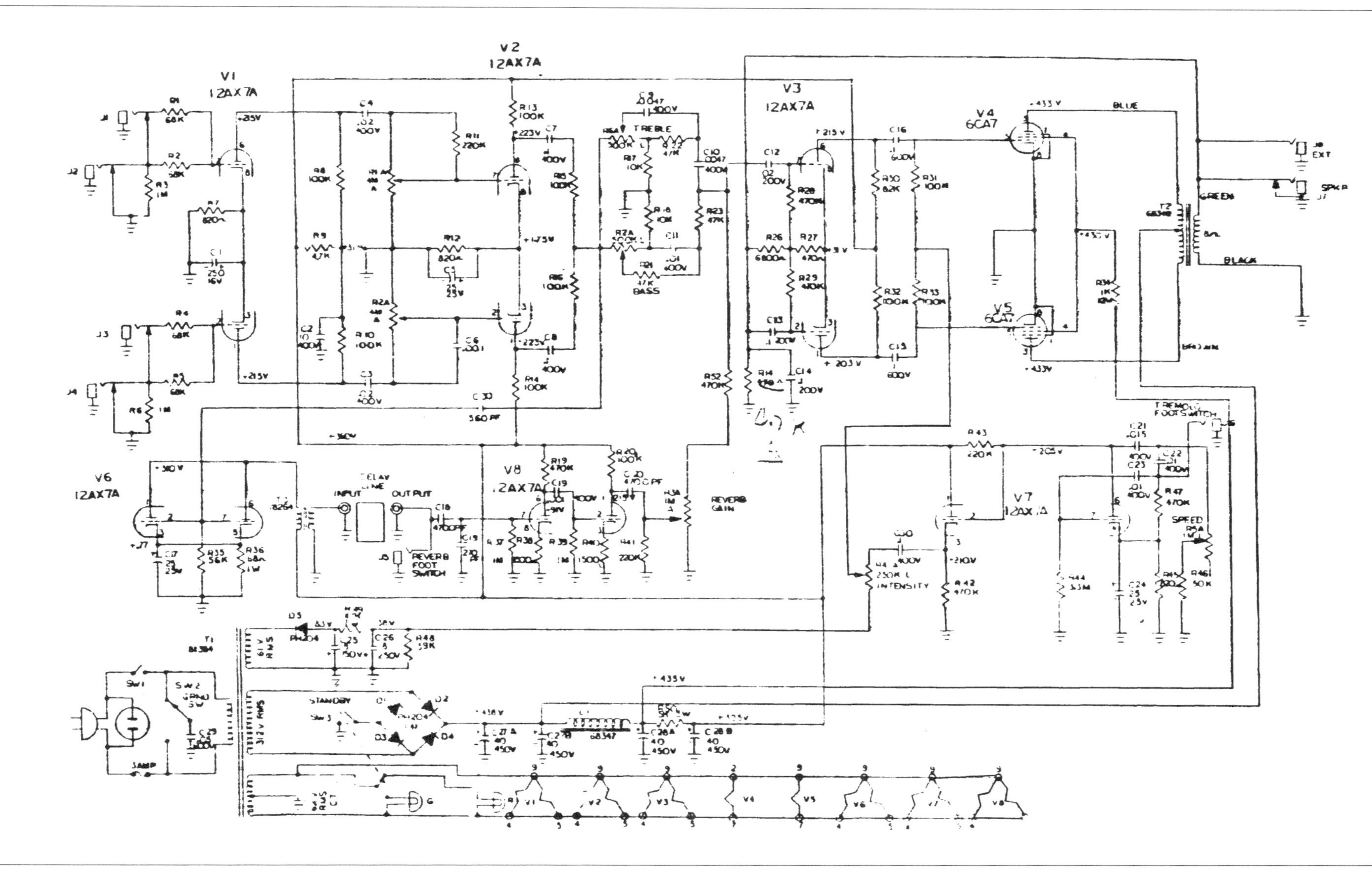

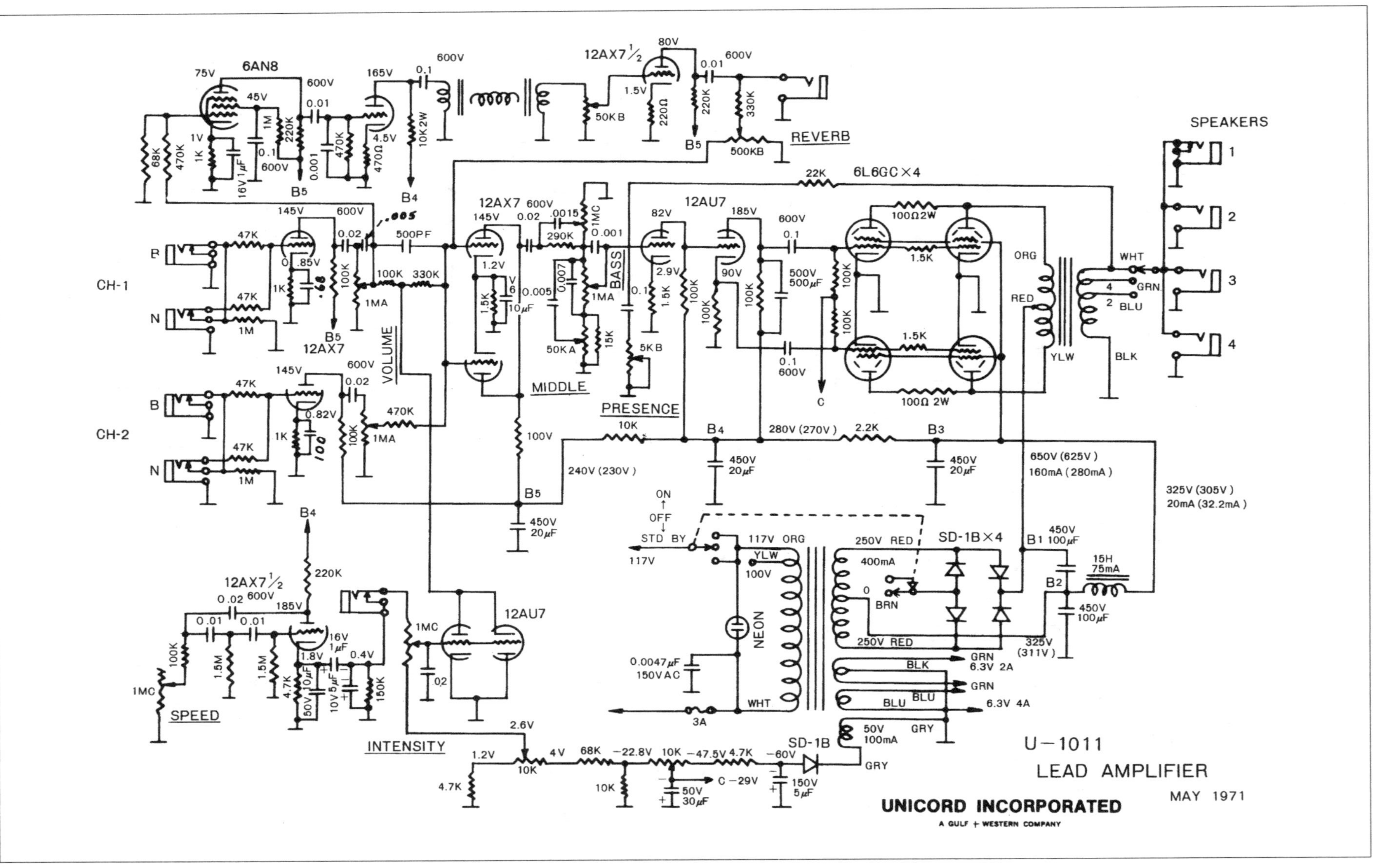
6AN8
12AX7½
REVERB
500KB
22K
6L6GC×4
SPEAKERS
1
2
3
4
CH-1
CH-2
B
N
12AX7
12AU7
VOLUME
MIDDLE
PRESENCE
BASS
5KB
50KA
15K
1MA
1MC
WHT
GRN
BLU
BLK
ORG
RED
YLW
100Ω2W
100Ω 2W
1.5K
B4
B3
B5
280V (270V)
2.2K
240V (230V)
650V (625V)
160mA (280mA)
325V (305V)
20mA (32.2mA)
ON
OFF
STD BY
117V
117V ORG
100V
NEON
0.0047μF
150VAC
3A
250V RED
400mA
BRN
SD-1B×4
B1 100μF
B2
15H
75mA
325V
(311V)
GRN
6.3V 2A
6.3V 4A
GRY
50V
100mA
SD-1B
SPEED
INTENSITY
220K
1.2V
2.6V
4V
68K
−22.8V
10K
−47.5V
4.7K
−60V
C −29V
50V
30μF
150V
5μF
U−1011
LEAD AMPLIFIER
MAY 1971
UNICORD INCORPORATED
A GULF + WESTERN COMPANY

U-1226 **LEAD AMP**

Univox

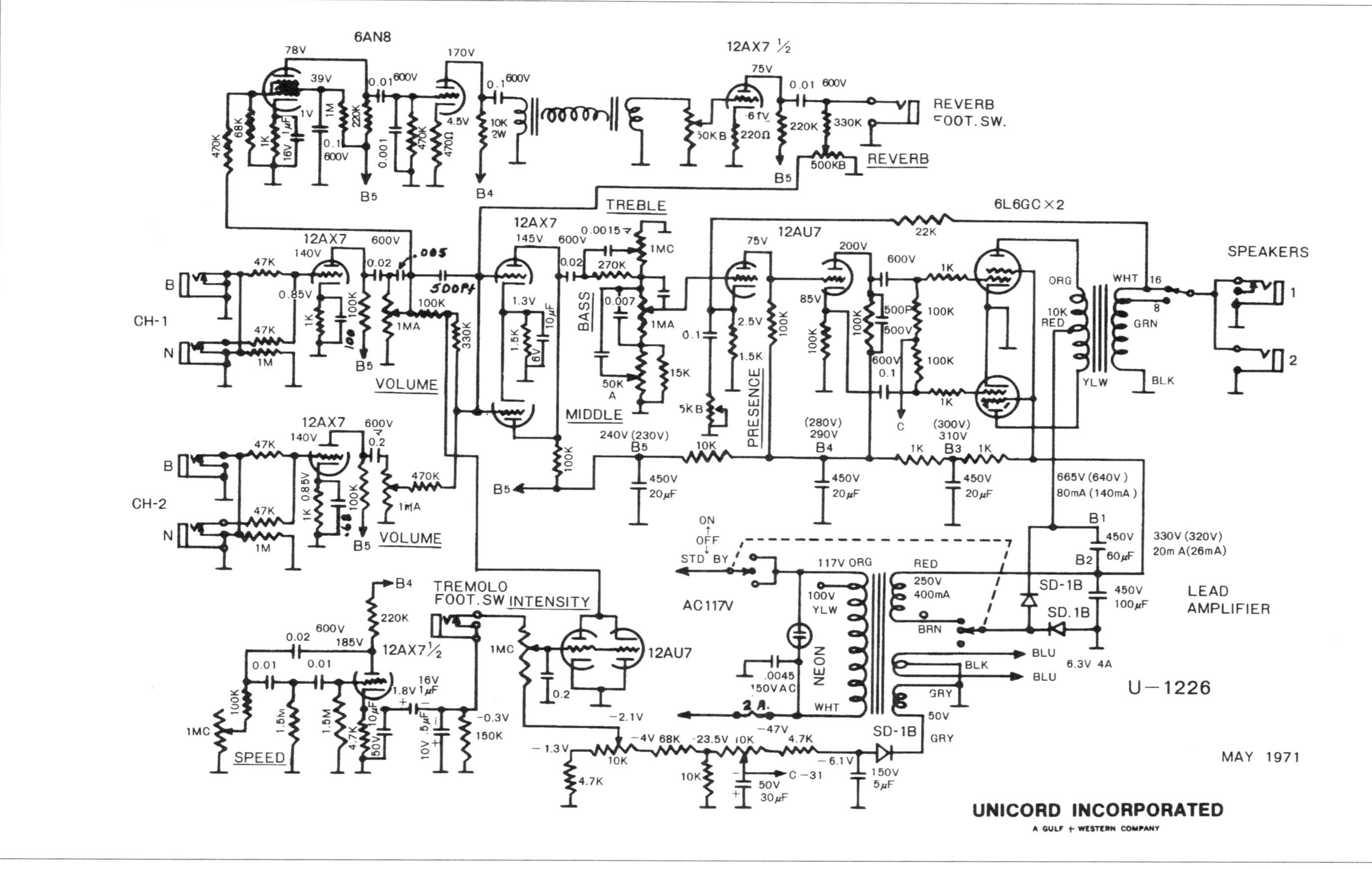

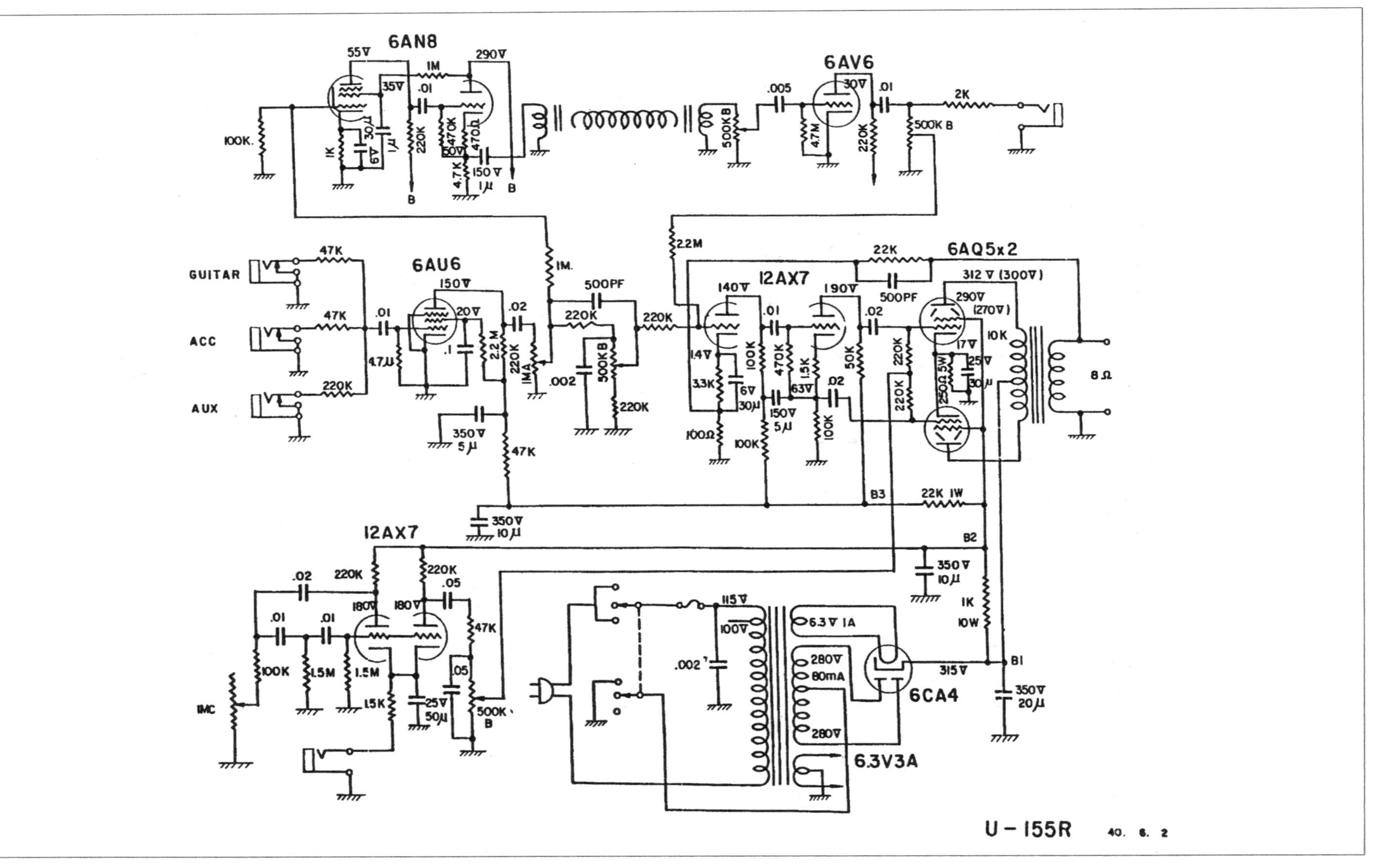
6AN8
55V
35V
1M
290V
.01
100K.
1K
30µ
6V
1µ
220K
470K
50V
470Ω
4.7K
150V
B
6AV6
.005
30V
.01
2K
500K B
4.7M
220K
GUITAR
ACC
AUX
47K
47K
220K
6AU6
150V
20V
.01
47µ
.1
2.2M
.02
220K
1MA
1M.
500PF
220K
.002
500K B
220K
350V 5µ
47K
350V 10µ
2.2M
220K
12AX7
140V
190V
1.4V
3.3K
100Ω
100K
6V 30µ
470K
1.5K
63V
150V 5µ
100K
.02
50K
22K
500PF
6AQ5x2
312V (300V)
290V (270V)
10K
17V
25V 30µ
250Ω 5W
8Ω
B3
22K 1W
B2
350V 10µ
1K 10W
B1
315V
350V 20µ
6CA4
6.3V 1A
280V
80mA
280V
6.3V3A
115V
100V
.002
12AX7
.02
220K
220K
.05
180V
180V
.01
.01
100K
1.5M
1.5M
1.5K
25V 50µ
.05
47K
500K B
1MC
U-155R
40. 6. 2

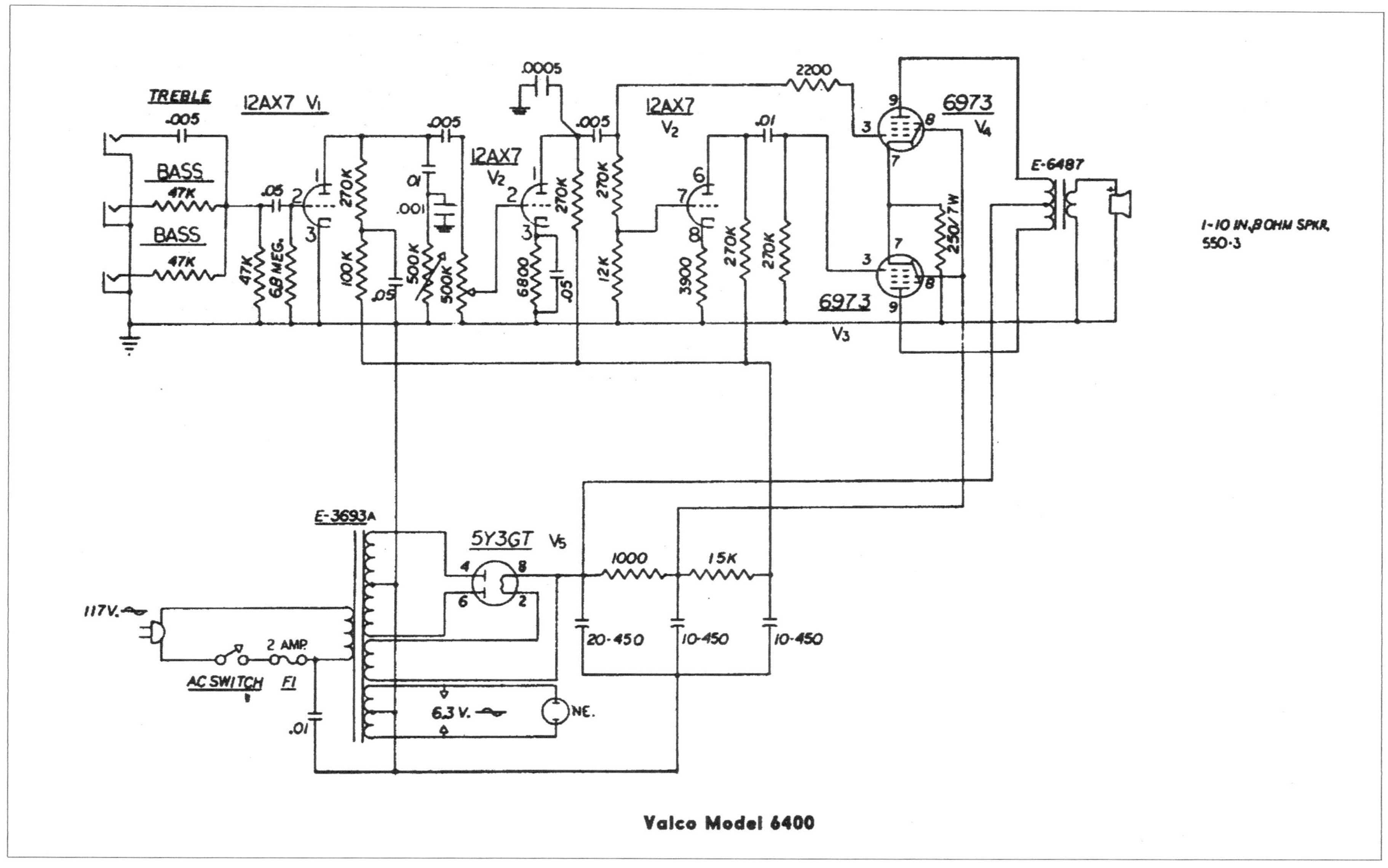

Valco Model 6400

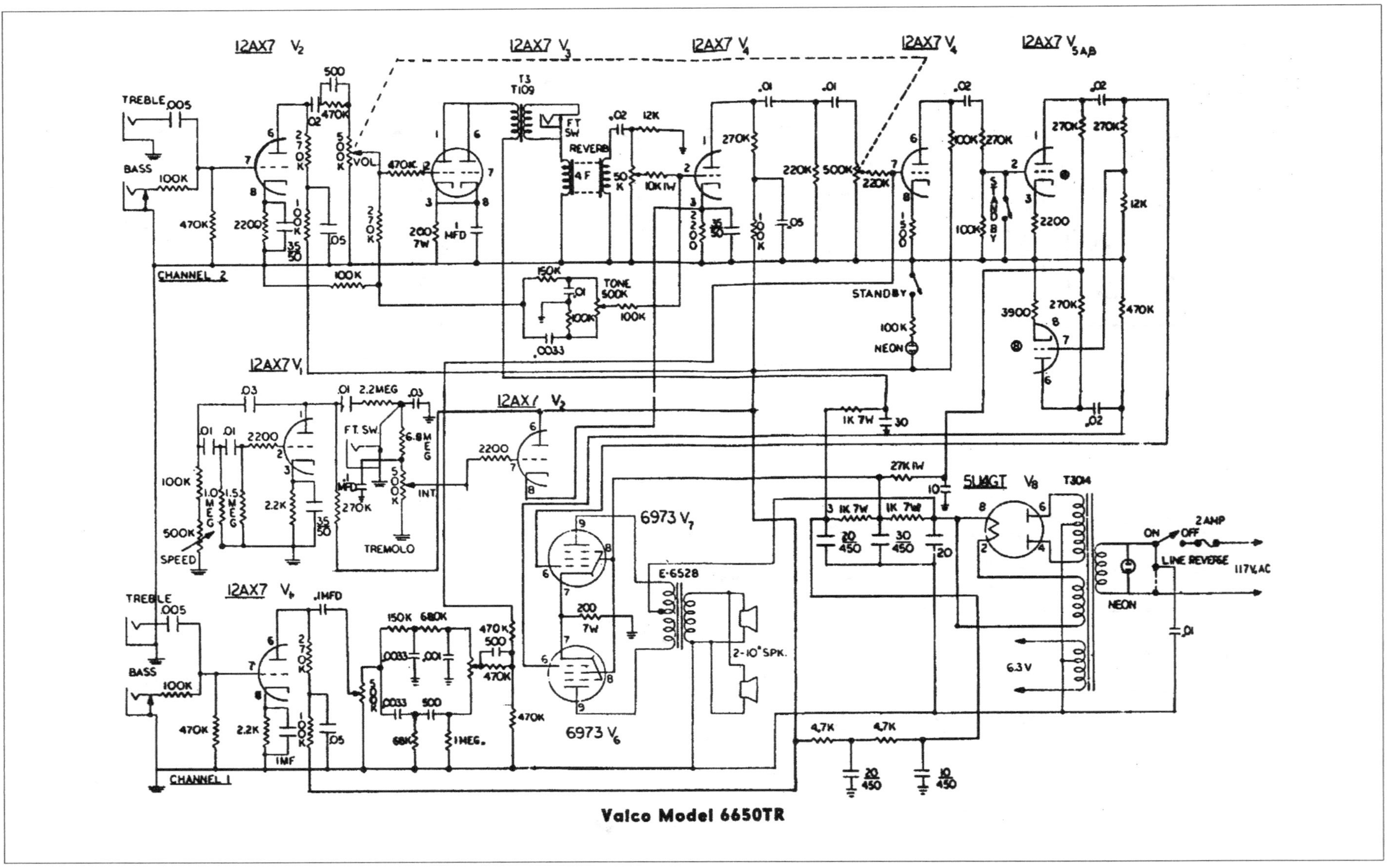

Valco Model 6650TR

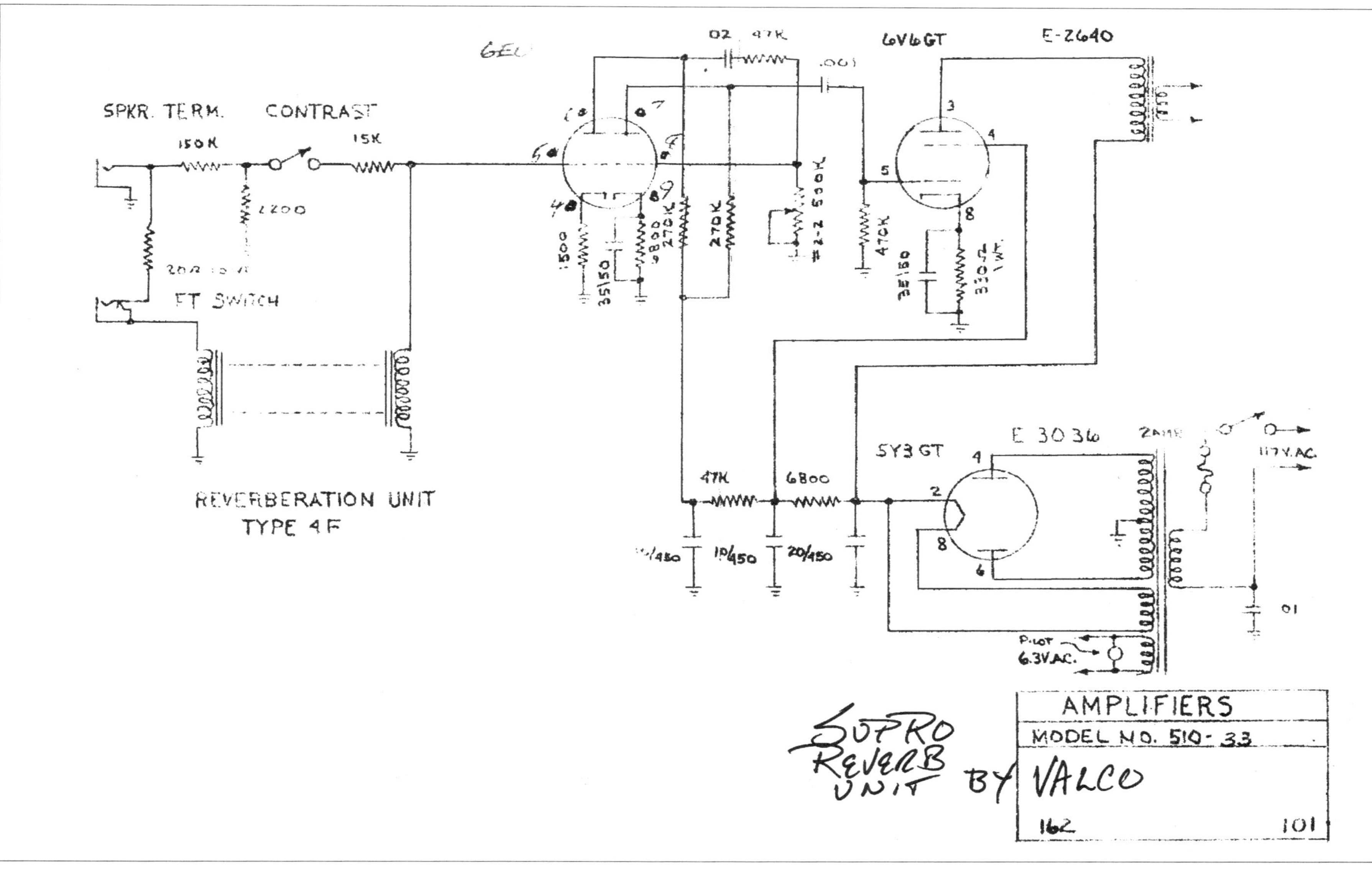
SPKR. TERM.
CONTRAST
150K
15K
2200
FT SWITCH
REVERBERATION UNIT
TYPE 4F
.02
47K
.001
6V6GT
E-2640
1500
35/50
6800
270K
270K
#2-2 500K
470K
330Ω
1WT.
47K
6800
10/450
20/450
5Y3GT
E 3036
2AMP.
117V.AC.
.01
PILOT
6.3V.AC.
AMPLIFIERS
MODEL NO. 510-33
SUPRO
REVERB
UNIT
BY VALCO
162
101

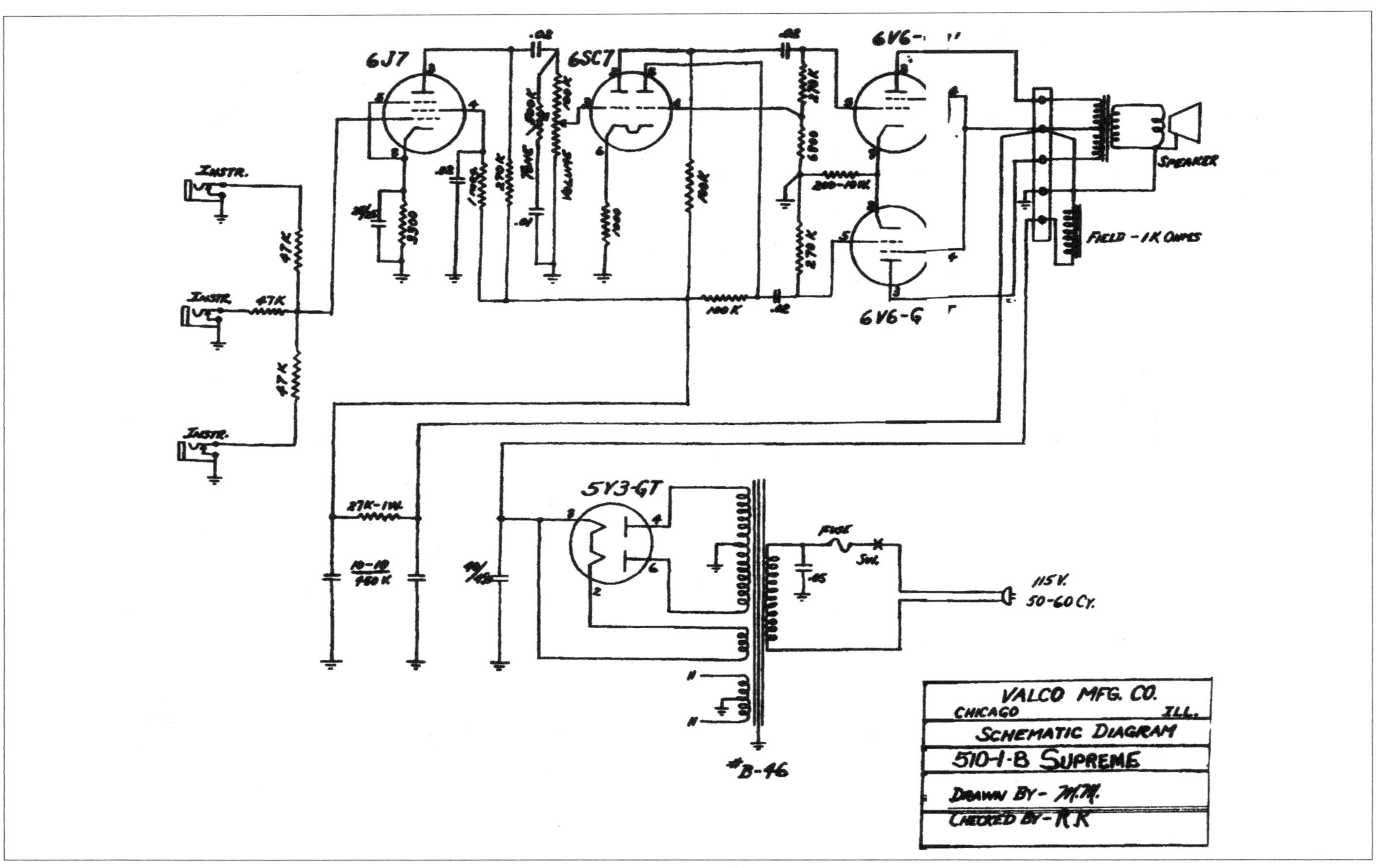
6J7
6SC7
6V6-G
6V6-G
5Y3-GT
INSTR.
INSTR.
INSTR.
47K
47K
47K
3300
270K
100K
6800
270K
27K-1W
10-10
450 V
.05
TONE
VOLUME
SPEAKER
FIELD - 1K OHMS
FUSE
SW.
115 V.
50-60 CY.
#B-46
VALCO MFG. CO.
CHICAGO ILL.
SCHEMATIC DIAGRAM
510-1-B SUPREME
DRAWN BY- M.M.
CHECKED BY- R.K

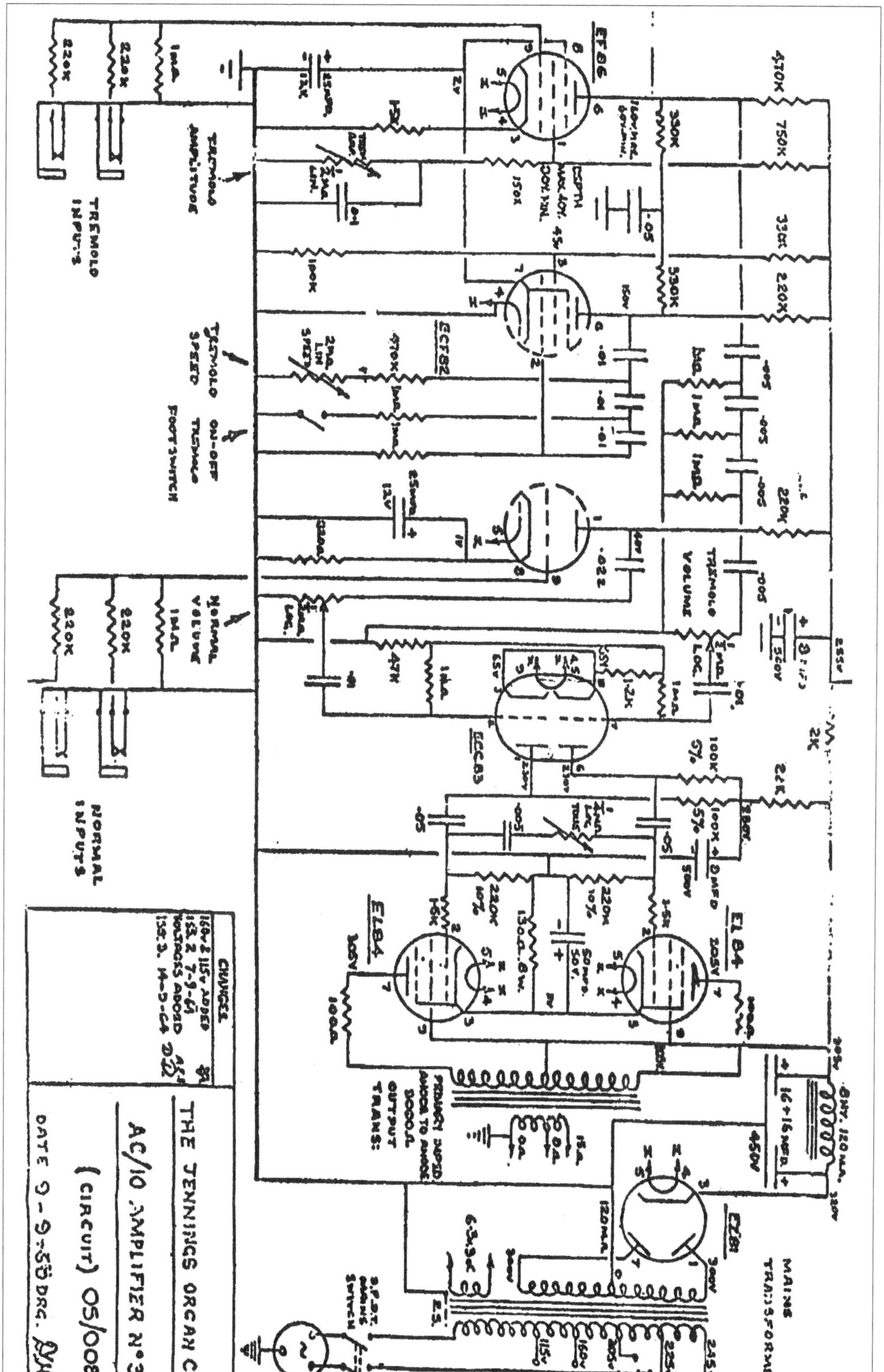

EF86 · ECF82 · ECC83 · EL84 · EL84 · EZ81

TREMOLO AMPLITUDE · TREMOLO SPEED · ON-OFF TREMOLO FOOTSWITCH · NORMAL VOLUME · TREMOLO VOLUME

TREMOLO INPUTS · NORMAL INPUTS

MAINS TRANSFORMER · PRIMARY IMPED. ANODE TO ANODE 8000 Ω OUTPUT TRANS: · 8HY. 120 MA. · 16+16 MFD. 450V · S.P.D.T. MAINS SWITCH

CHANGES	
160v & 115v ADDED 153.2 7-9-61	
VOLTAGES ADDED 153.3. 14-9-64	

THE JENNINGS ORGAN Co.

AC/10 AMPLIFIER N°3

(CIRCUIT) 05/008

DATE 9-9-58 DRG.

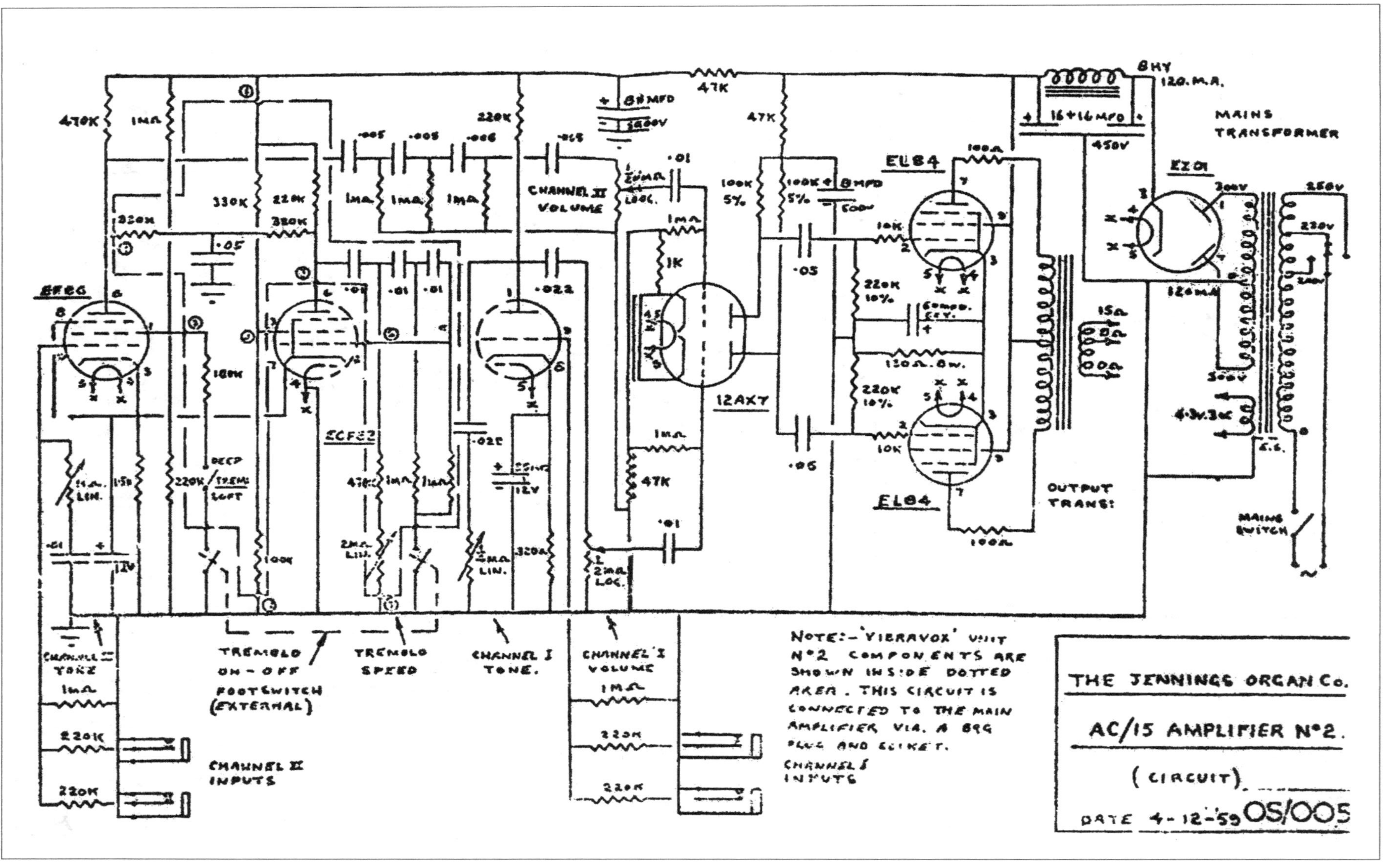
MAINS TRANSFORMER
OUTPUT TRANS:
EL84
EL84
EZ81
12AX7
ECF82
EF86
8HY 120.M.A.
16+16 MFD 450V
CHANNEL II VOLUME
CHANNEL II TONE
CHANNEL II INPUTS
TREMOLO ON-OFF FOOTSWITCH (EXTERNAL)
TREMOLO SPEED
CHANNEL I TONE.
CHANNEL I VOLUME
CHANNEL I INPUTS
MAINS SWITCH
NOTE:- 'VIBRAVOX' UNIT N°2 COMPONENTS ARE SHOWN INSIDE DOTTED AREA. THIS CIRCUIT IS CONNECTED TO THE MAIN AMPLIFIER VIA. A 8 PIN PLUG AND SOCKET.
THE JENNINGS ORGAN Co.
AC/15 AMPLIFIER N°2.
(CIRCUIT)
DATE 4-12-59 OS/005

AC15 **MK III (1960)**

Vox

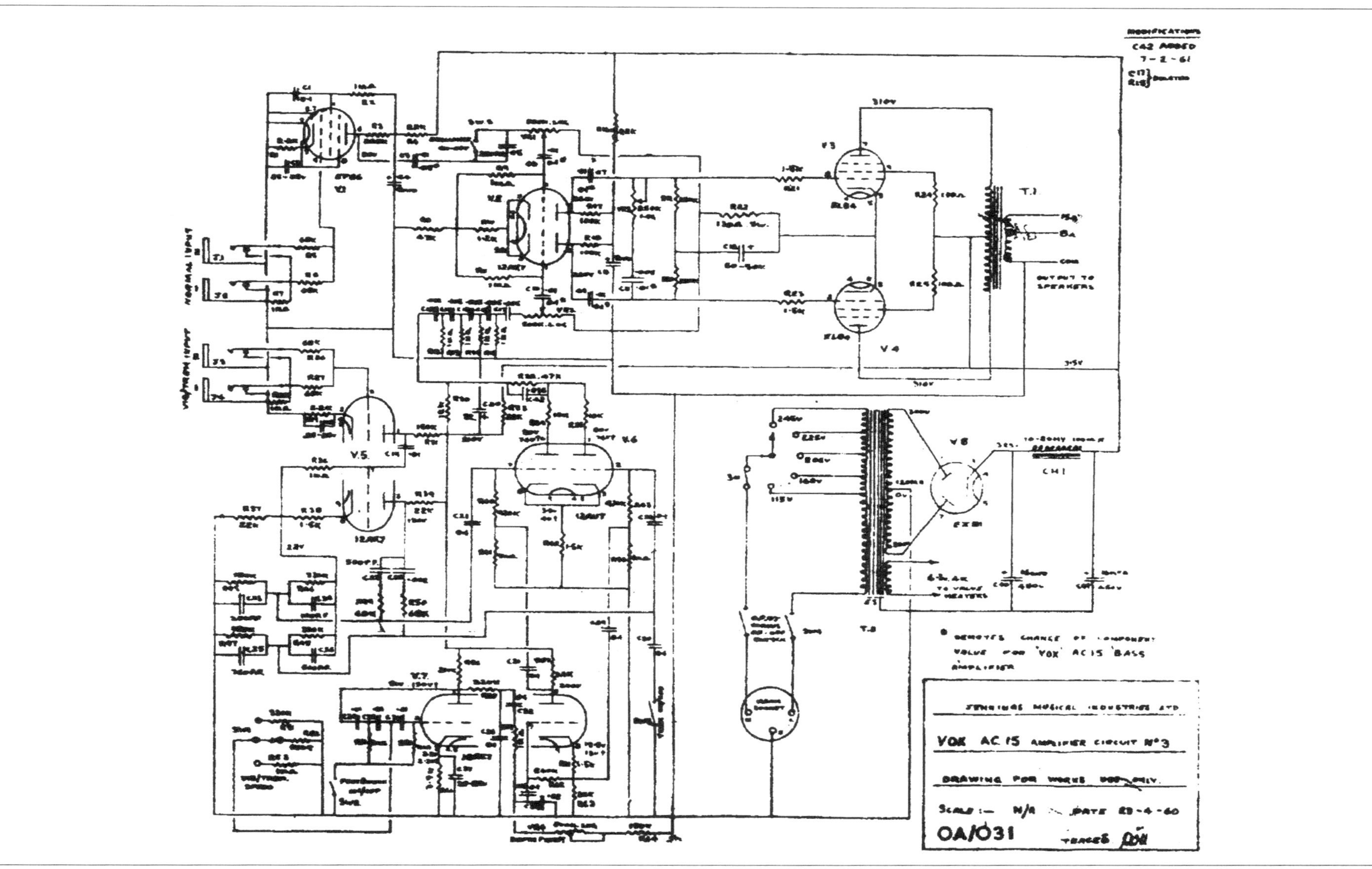

AC30 REVERB (1978)

Vox

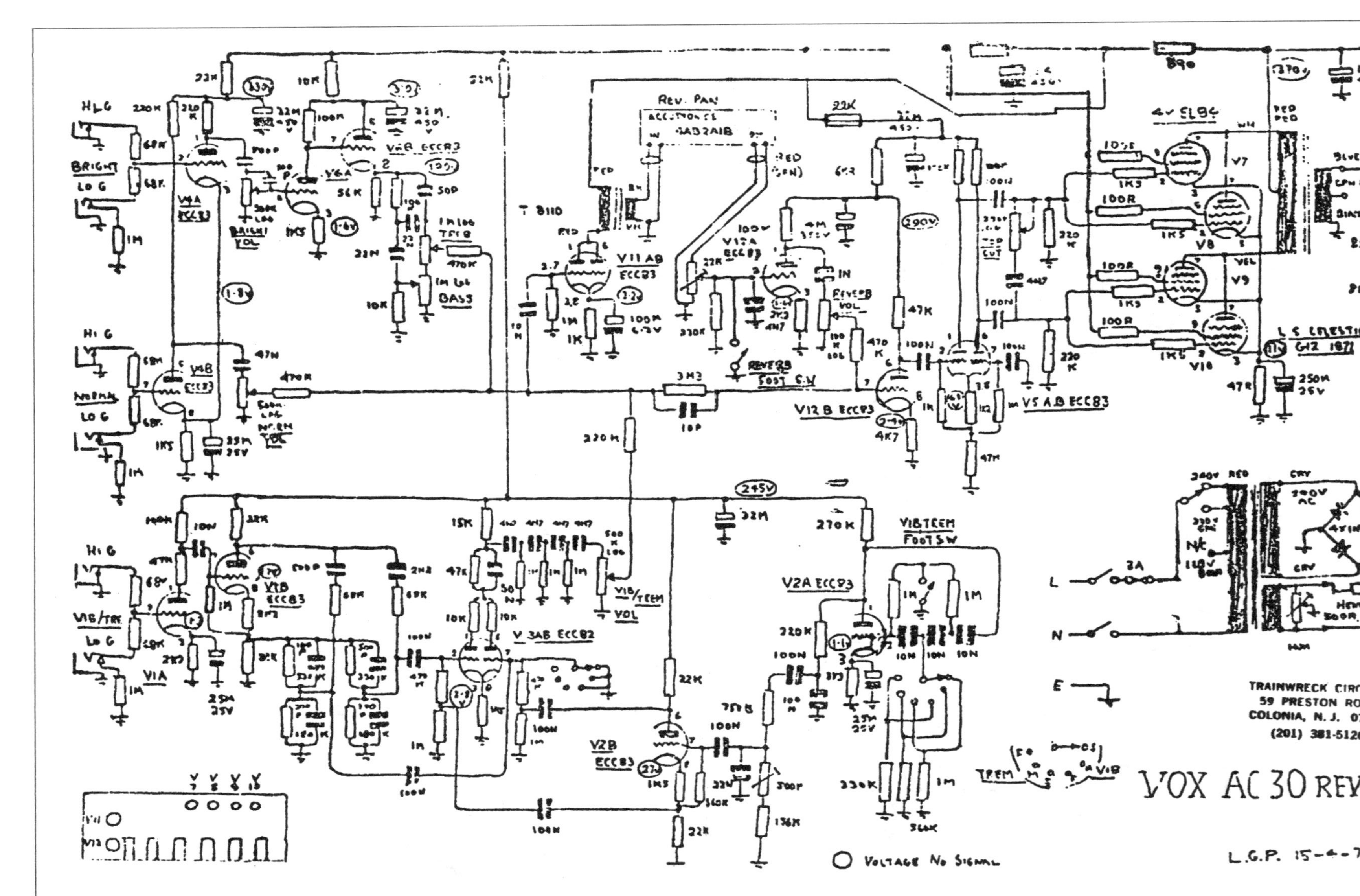

AC30 SILVER JUBILEE (VOX LTD)

Vox

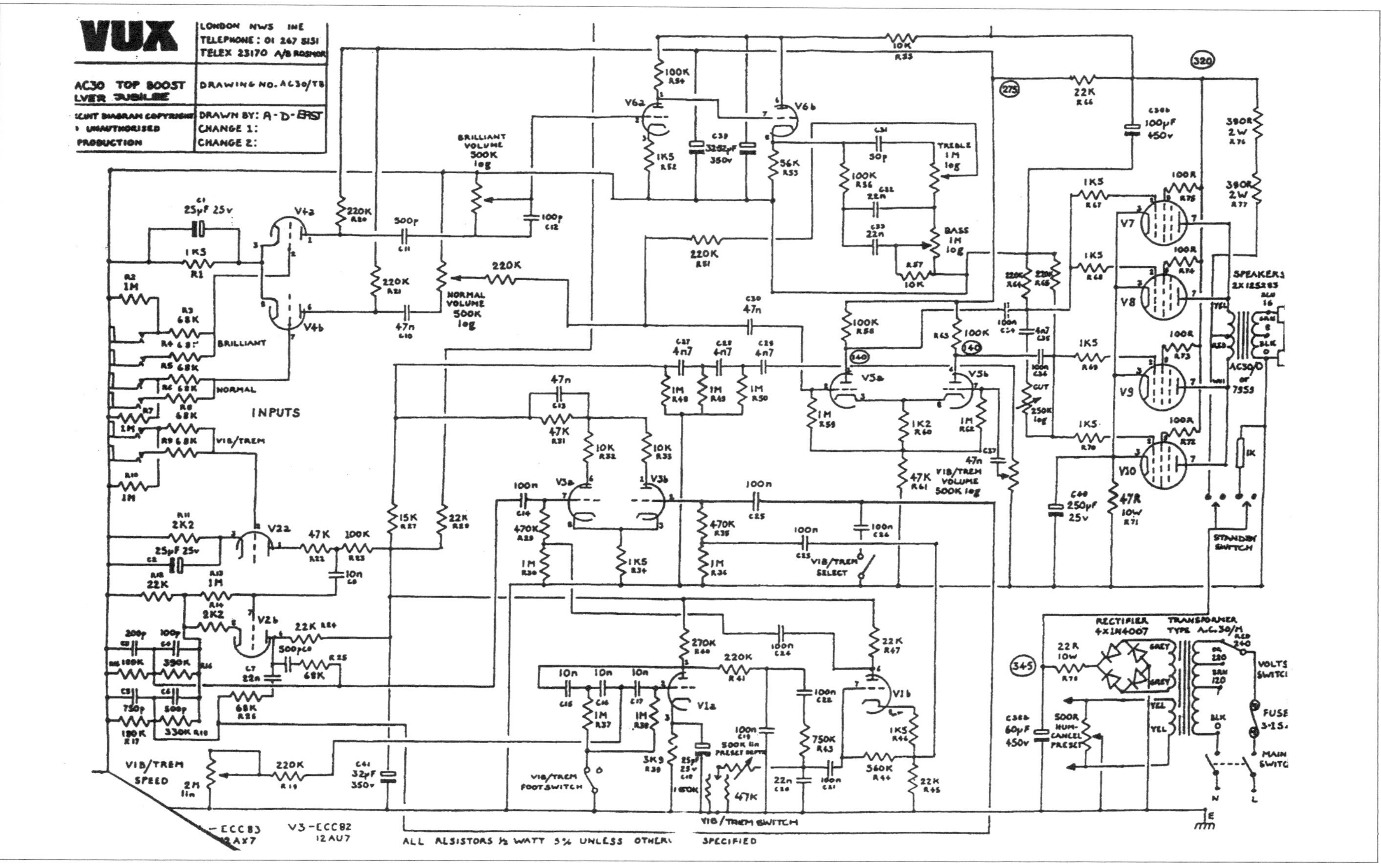

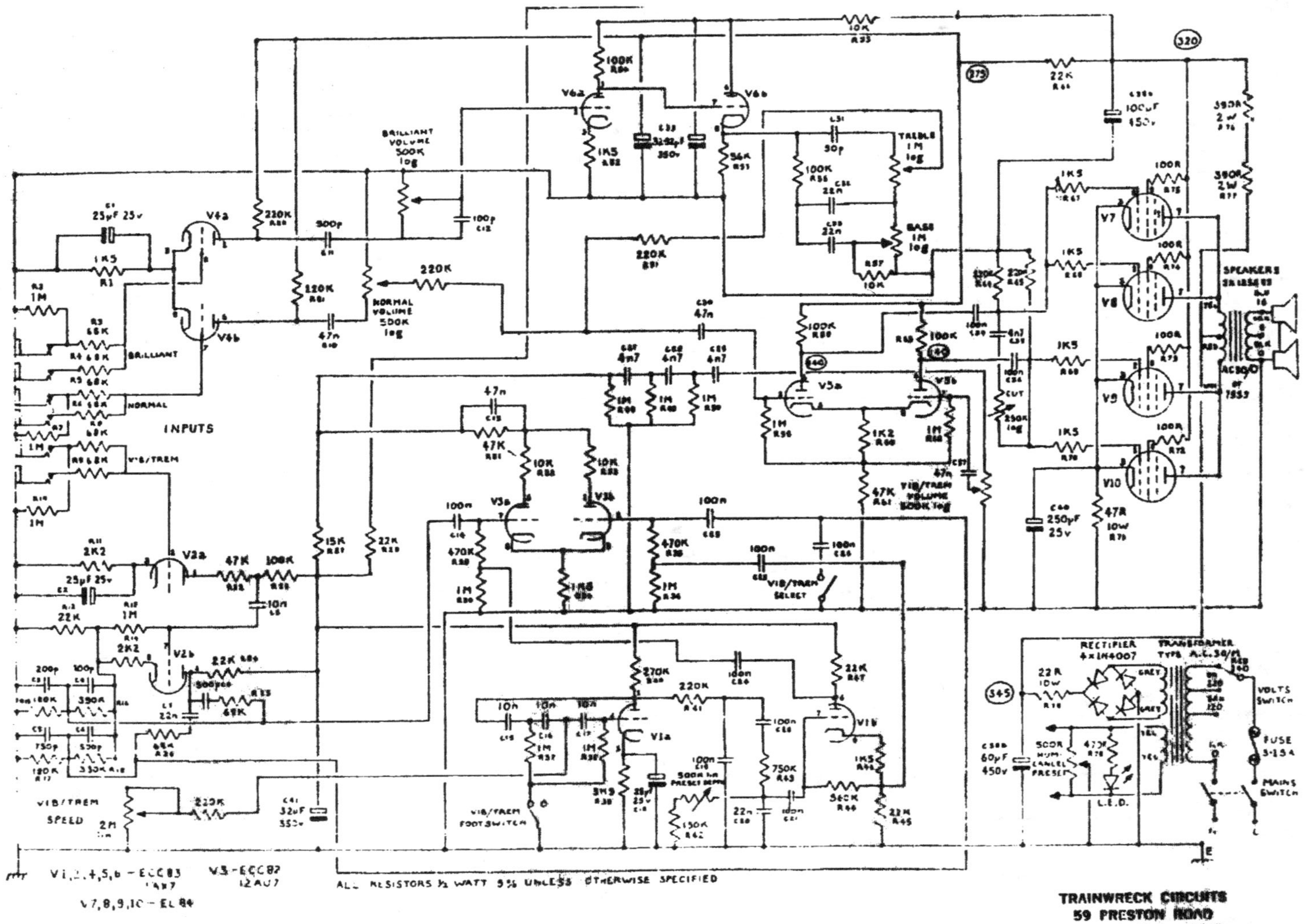

AC30 (S.S. RED)

V1,2,4,5,6 – ECC83 12AX7
V3 – ECC82 12AU7
V7,8,9,10 – EL84

ALL RESISTORS ½ WATT 5% UNLESS OTHERWISE SPECIFIED

TRAINWRECK CIRCUITS
59 PRESTON ROAD
COLONIA, N. J. 07067
(201) 381-8128

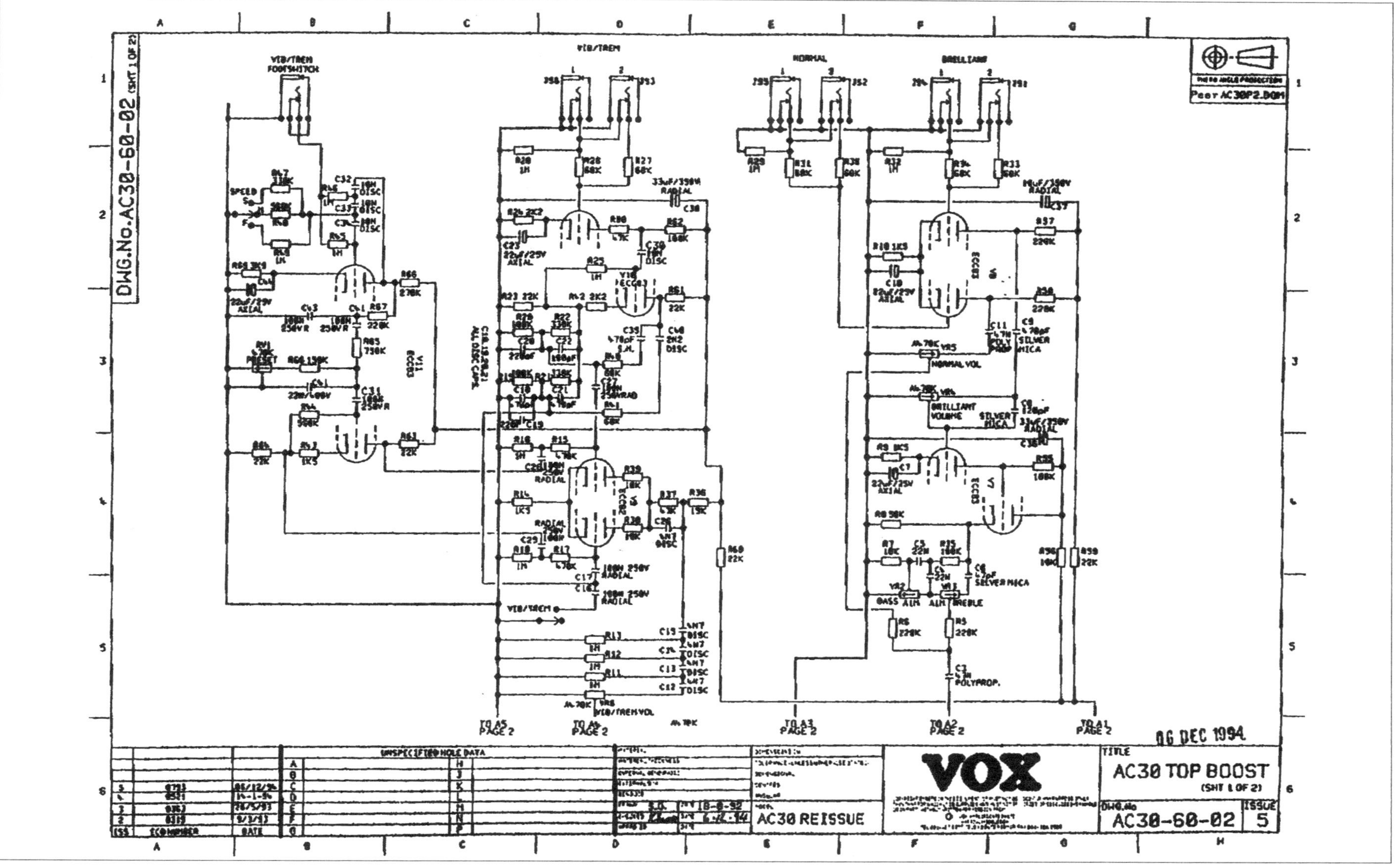
DWG.No.AC30-60-02 (SHT 1 OF 2)
VIB/TREM FOOTSWITCH
VIB/TREM
NORMAL
BRILLIANT
ECC83
ALL DISC CAPS. C18,19,20,21
NORMAL VOL
BRILLIANT VOLUME
BASS
TREBLE
POLYPROP.
SILVER MICA
VIB/TREM VOL
TO A5 PAGE 2
TO A4 PAGE 2
TO A3 PAGE 2
TO A2 PAGE 2
TO A1 PAGE 2
06 DEC 1994
UNSPECIFIED HOLE DATA
VOX
AC30 REISSUE
TITLE
AC30 TOP BOOST
(SHT 1 OF 2)
DWG.No
AC30-60-02
ISSUE
5

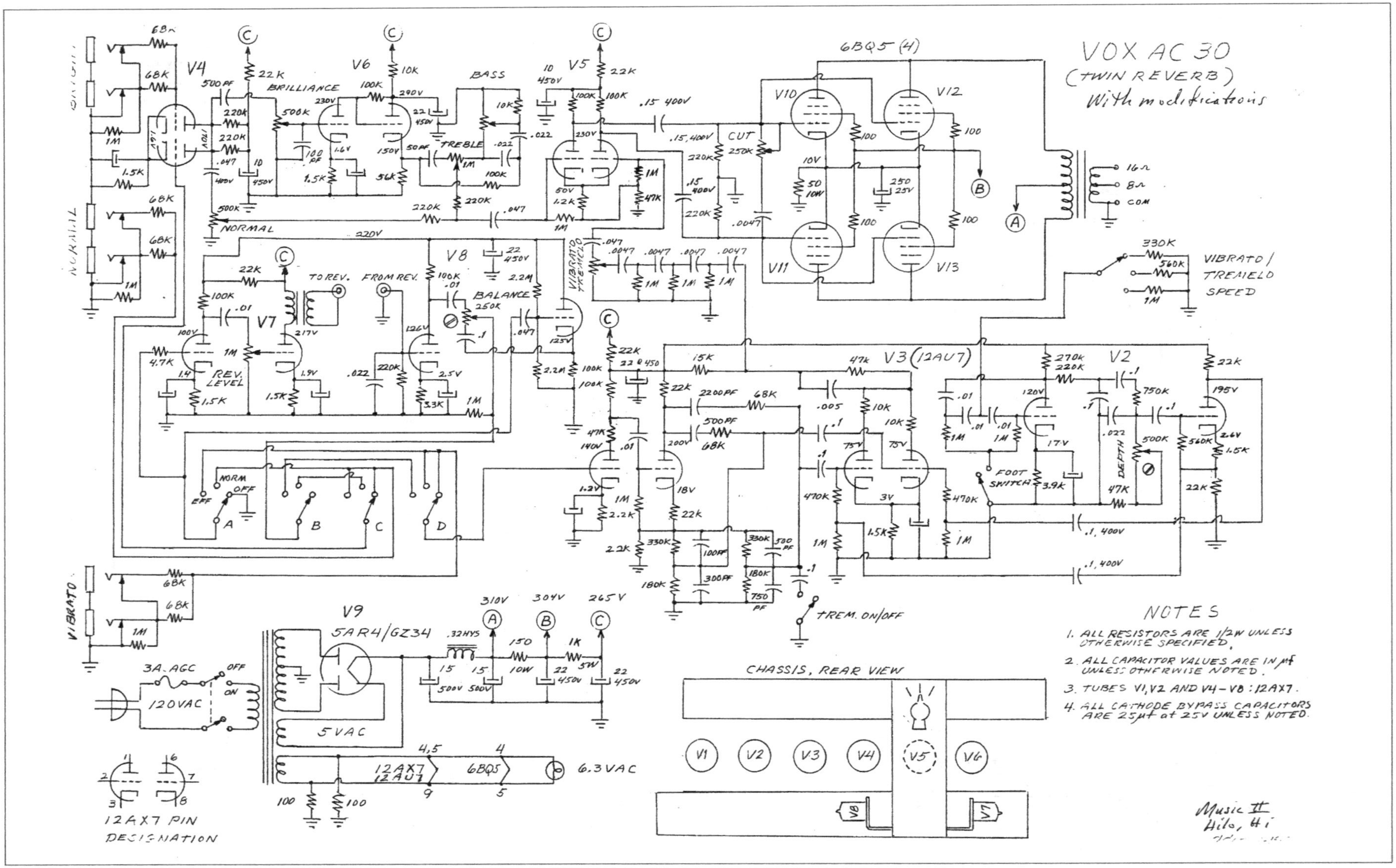

AC30 TWIN REVERB (W/MOD)

Vox

VOX AC 30
(TWIN REVERB)
With modifications

NOTES

1. ALL RESISTORS ARE 1/2W UNLESS OTHERWISE SPECIFIED.
2. ALL CAPACITOR VALUES ARE IN µf UNLESS OTHERWISE NOTED.
3. TUBES V1, V2 AND V4 – V8 : 12AX7.
4. ALL CATHODE BYPASS CAPACITORS ARE 25µf at 25V UNLESS NOTED.

Music II
Hilo, Hi

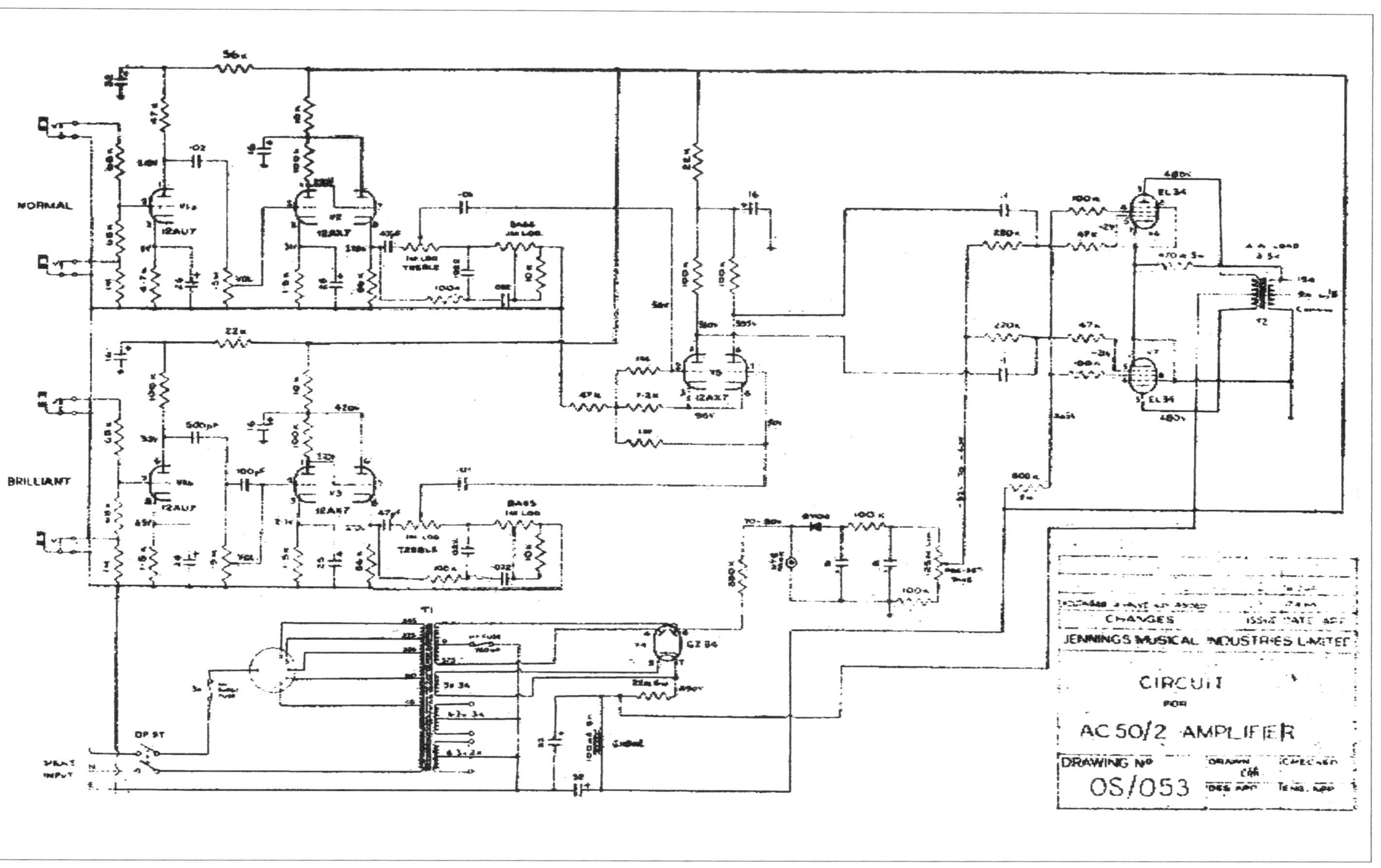
NORMAL
BRILLIANT
12AU7
12AX7
EL34
GZ34
DP ST
MAINS INPUT
T1
T2
BASS 1M LOG
TREBLE 1M LOG
VOL
CHANGES
ISSUE DATE APP
JENNINGS MUSICAL INDUSTRIES LIMITED
CIRCUIT
FOR
AC 50/2 AMPLIFIER
DRAWING No
OS/053
DRAWN
CHECKED
DES APP
TECH APP

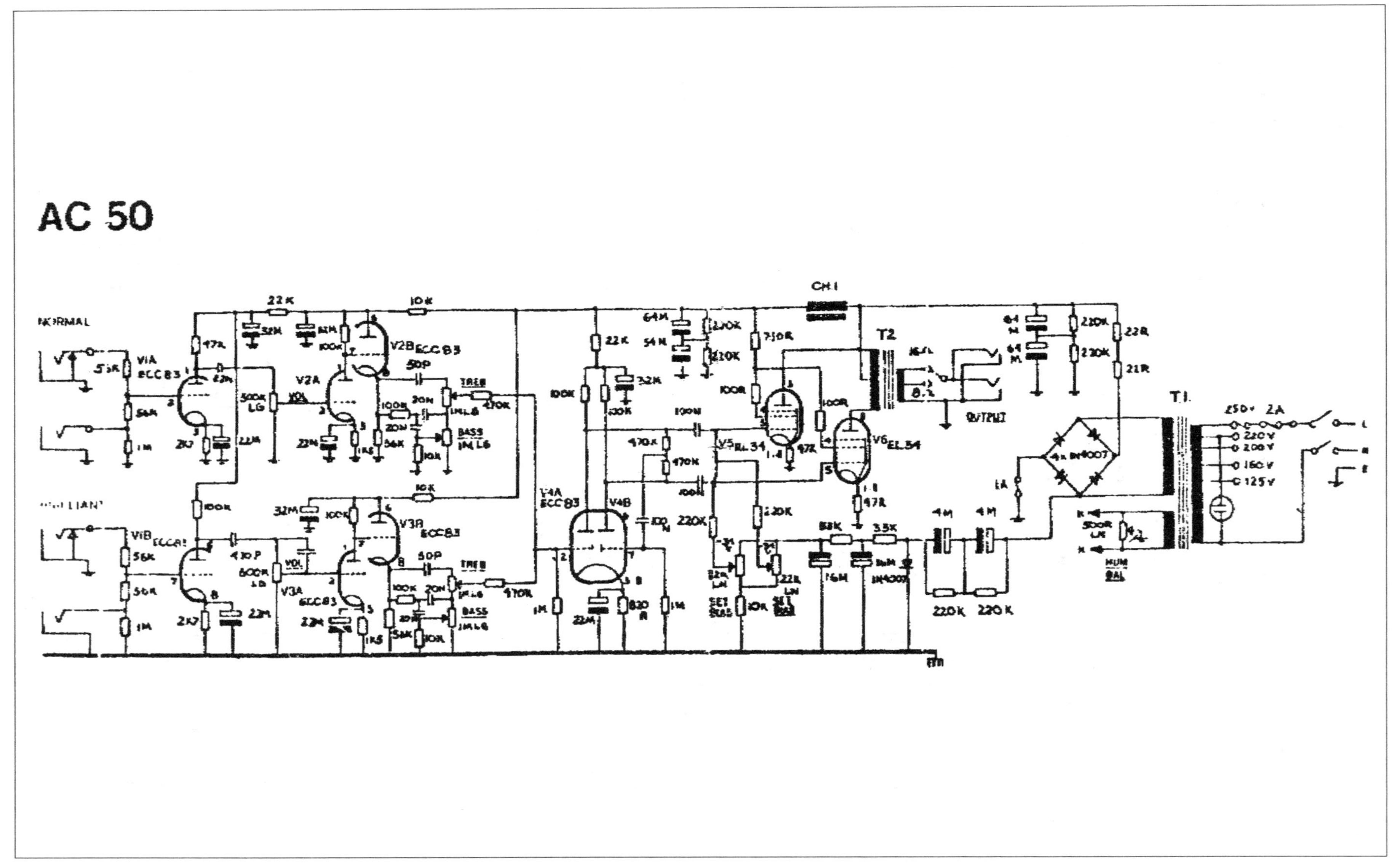
AC 50
NORMAL
V1A ECC83
V1B ECC83
V2A
V2B ECC83
V3A ECC83
V3B ECC83
V4A ECC83
V4B
V5 EL34
V6 EL34
CH1
T2
T1
OUTPUT
TREB
BASS
VOL
SET BIAS
HUM BAL
4 x IN4007
250V 2A
220V
200V
160V
125V

AC50/4 **MK III**

Vox

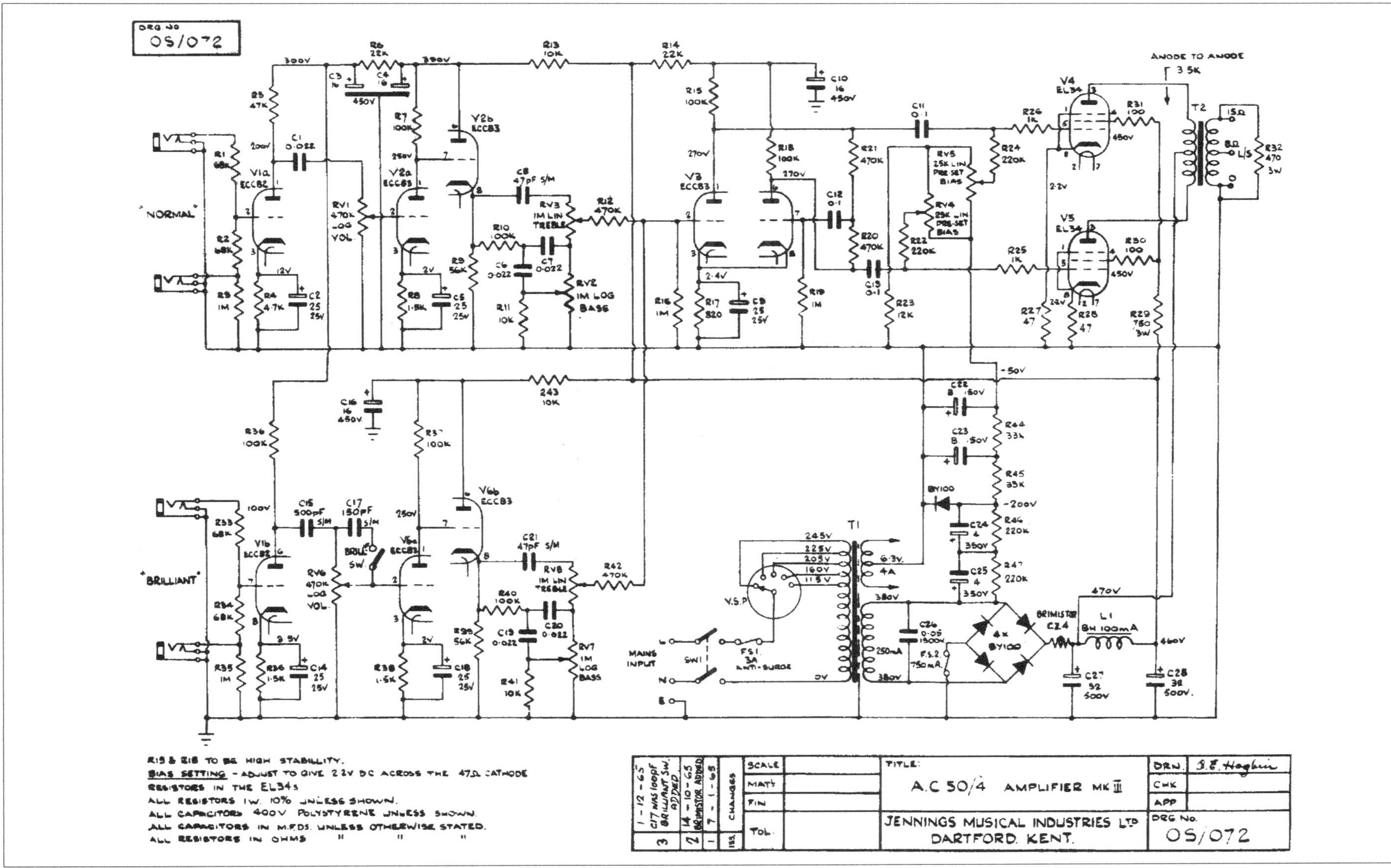

BERKELY II SUPER REVERB V-8

Vox

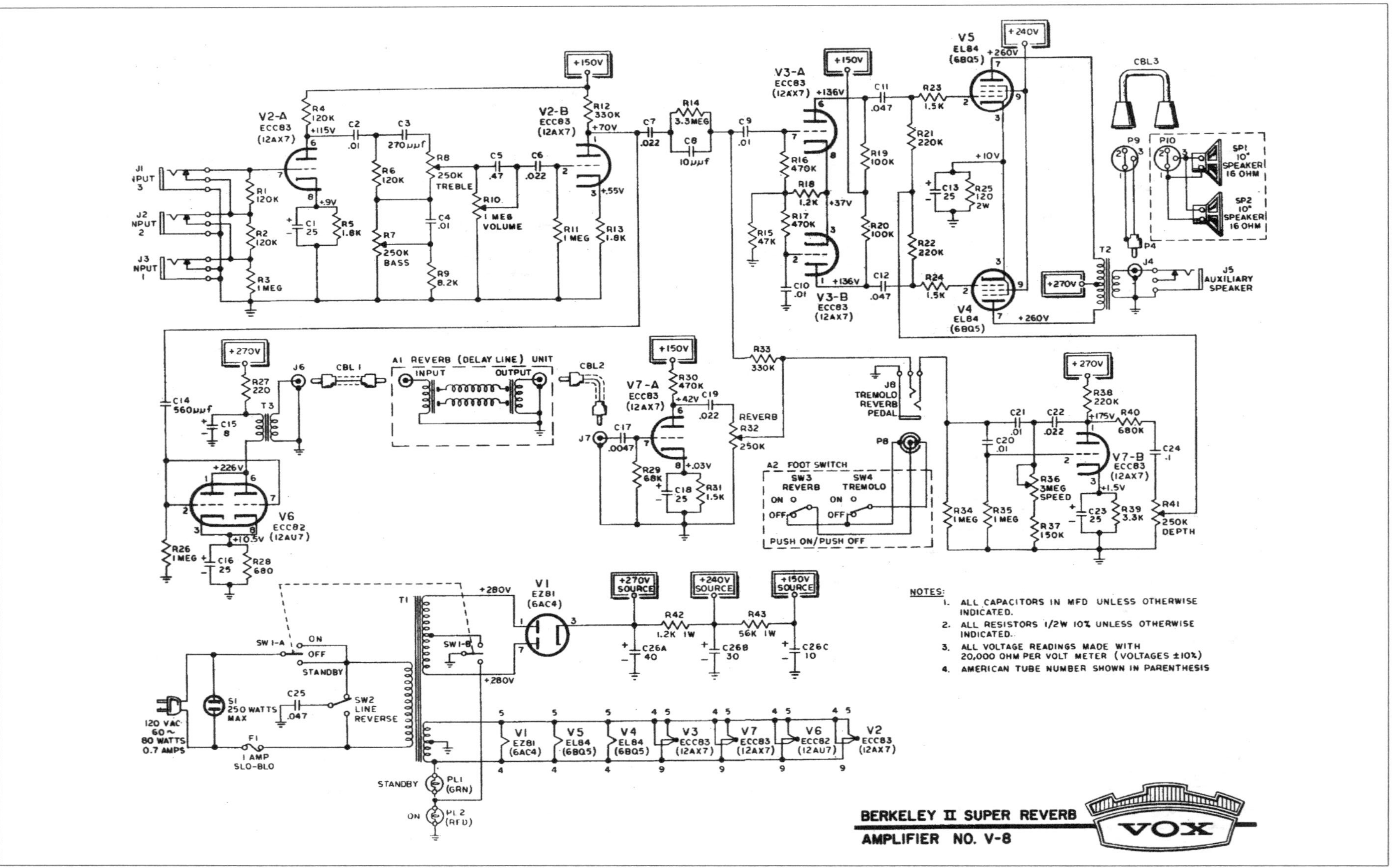

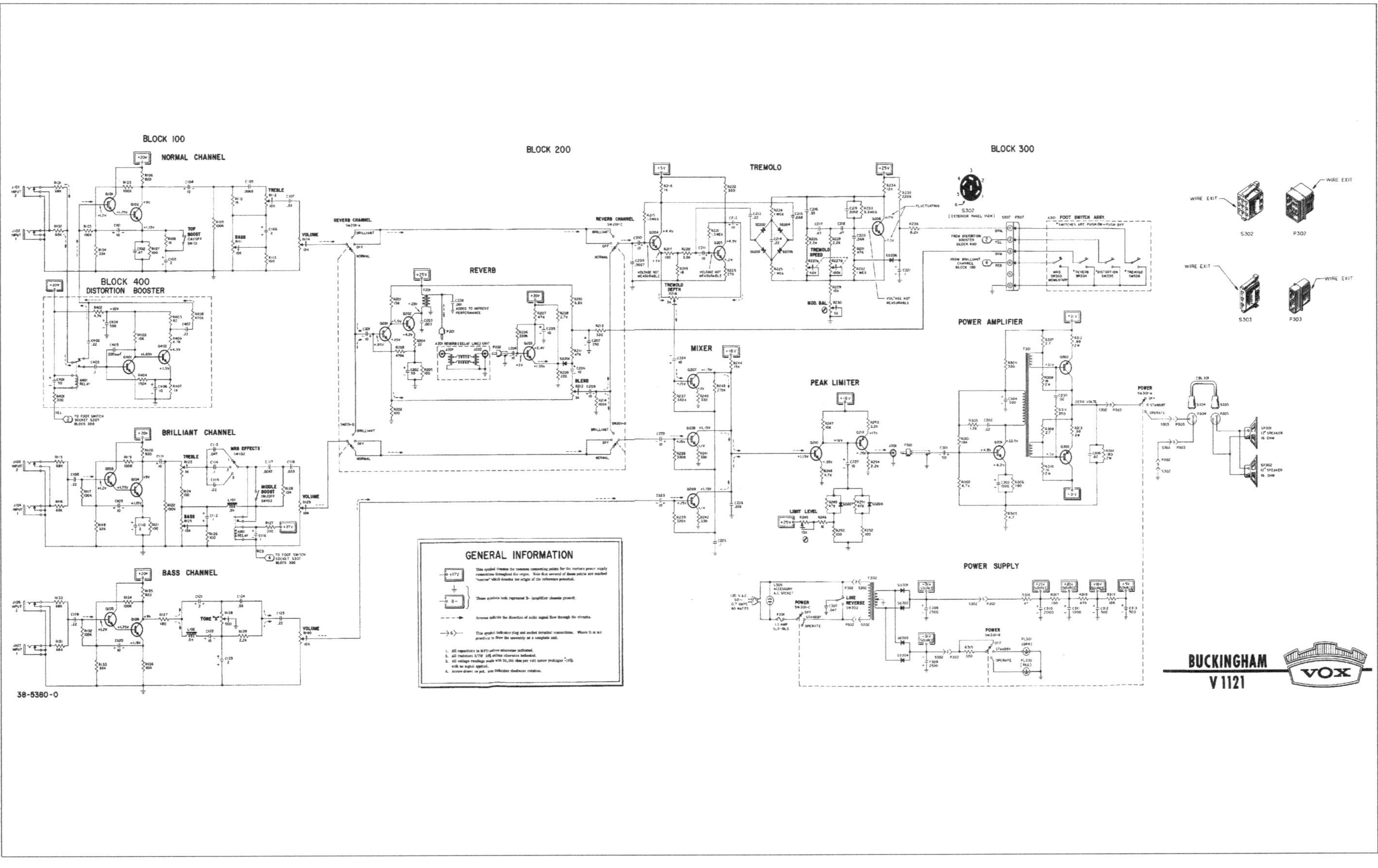
BLOCK 100
NORMAL CHANNEL
BLOCK 400
DISTORTION BOOSTER
BRILLIANT CHANNEL
BASS CHANNEL
BLOCK 200
REVERB
TREMOLO
MIXER
PEAK LIMITER
BLOCK 300
POWER AMPLIFIER
POWER SUPPLY
GENERAL INFORMATION
38-5360-0
BUCKINGHAM
V 1121
VOX

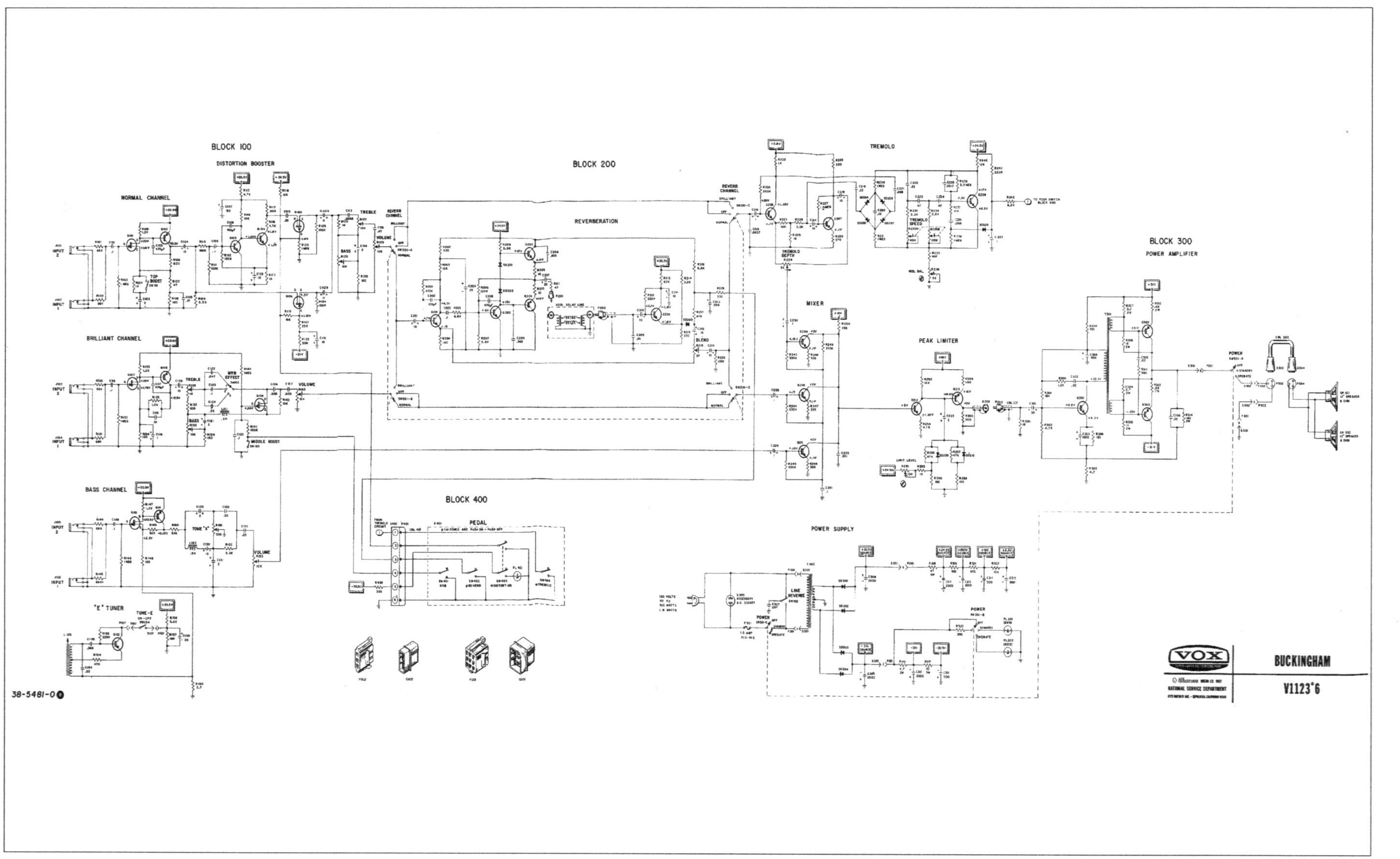
BLOCK 100
DISTORTION BOOSTER
NORMAL CHANNEL
BRILLIANT CHANNEL
BASS CHANNEL
"E" TUNER
BLOCK 200
REVERBERATION
TREMOLO
MIXER
PEAK LIMITER
BLOCK 300
POWER AMPLIFIER
BLOCK 400
PEDAL
POWER SUPPLY
VOX
BUCKINGHAM
V1123*6
38-5481-0

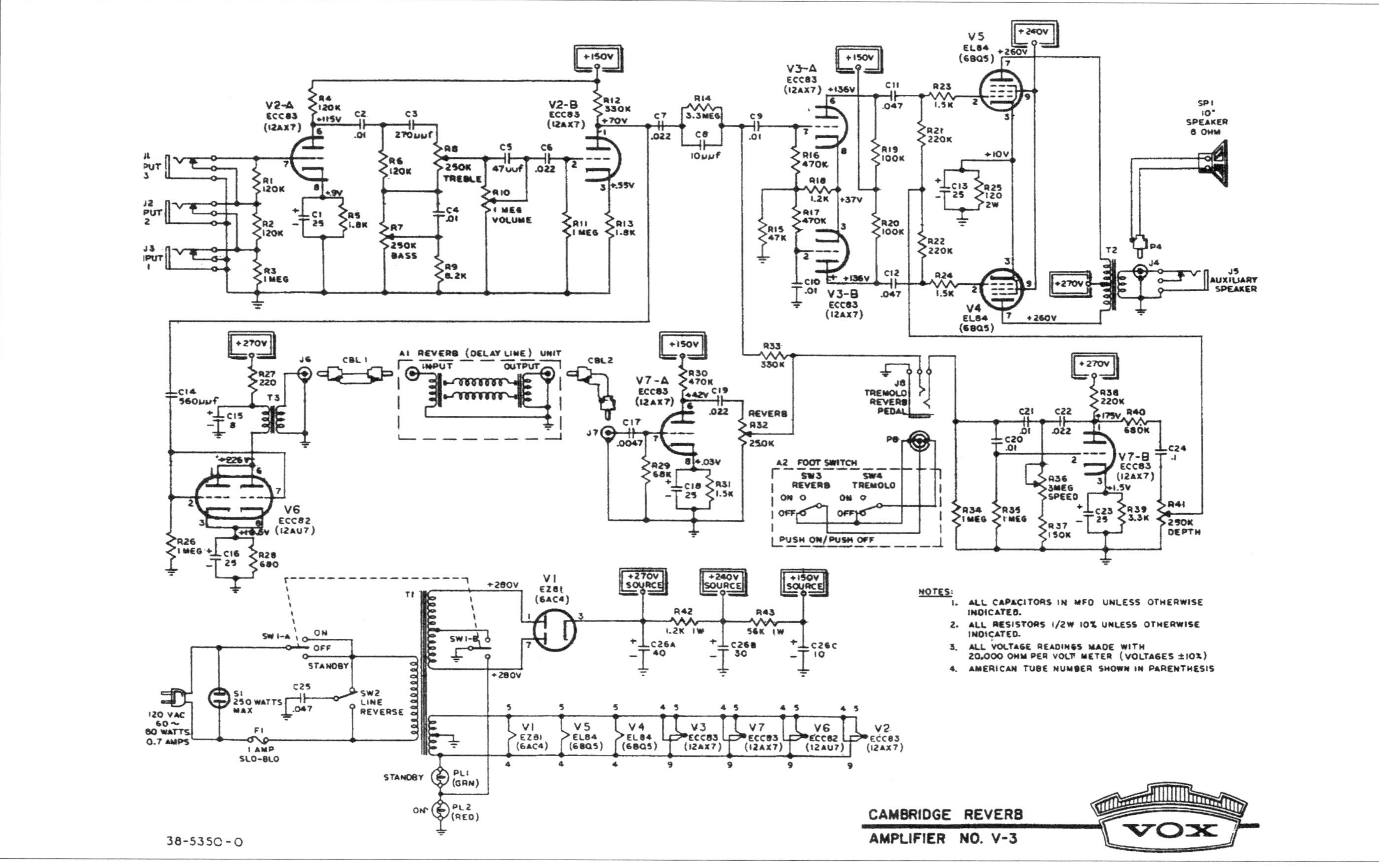
V2-A ECC83 (12AX7)
V2-B ECC83 (12AX7)
V3-A ECC83 (12AX7)
V3-B ECC83 (12AX7)
V4 EL84 (6BQ5)
V5 EL84 (6BQ5)
V6 ECC82 (12AU7)
V7-A ECC83 (12AX7)
V7-B ECC83 (12AX7)
V1 EZ81 (6AC4)
J1 INPUT 3
J2 INPUT 2
J3 INPUT 1
R7 250K BASS
R8 250K TREBLE
R10 1 MEG VOLUME
A1 REVERB (DELAY LINE) UNIT
INPUT
OUTPUT
CBL 1
CBL 2
REVERB R32 250K
R36 3MEG SPEED
R41 250K DEPTH
SP1 10" SPEAKER 8 OHM
J5 AUXILIARY SPEAKER
J8 TREMOLO REVERB PEDAL
A2 FOOT SWITCH
SW3 REVERB
SW4 TREMOLO
PUSH ON/PUSH OFF
+150V
+270V
+240V
+270V SOURCE
+240V SOURCE
+150V SOURCE
SW1-A
ON
OFF
STANDBY
SW2 LINE REVERSE
S1 250 WATTS MAX
120 VAC 60~ 80 WATTS 0.7 AMPS
F1 1 AMP SLO-BLO
PL1 (GRN)
PL2 (RED)
NOTES:
1. ALL CAPACITORS IN MFD UNLESS OTHERWISE INDICATED.
2. ALL RESISTORS 1/2W 10% UNLESS OTHERWISE INDICATED.
3. ALL VOLTAGE READINGS MADE WITH 20,000 OHM PER VOLT METER (VOLTAGES ±10%)
4. AMERICAN TUBE NUMBER SHOWN IN PARENTHESIS
CAMBRIDGE REVERB
AMPLIFIER NO. V-3
VOX
38-5350-0

PERCUSSION KEY **V829**

Vox

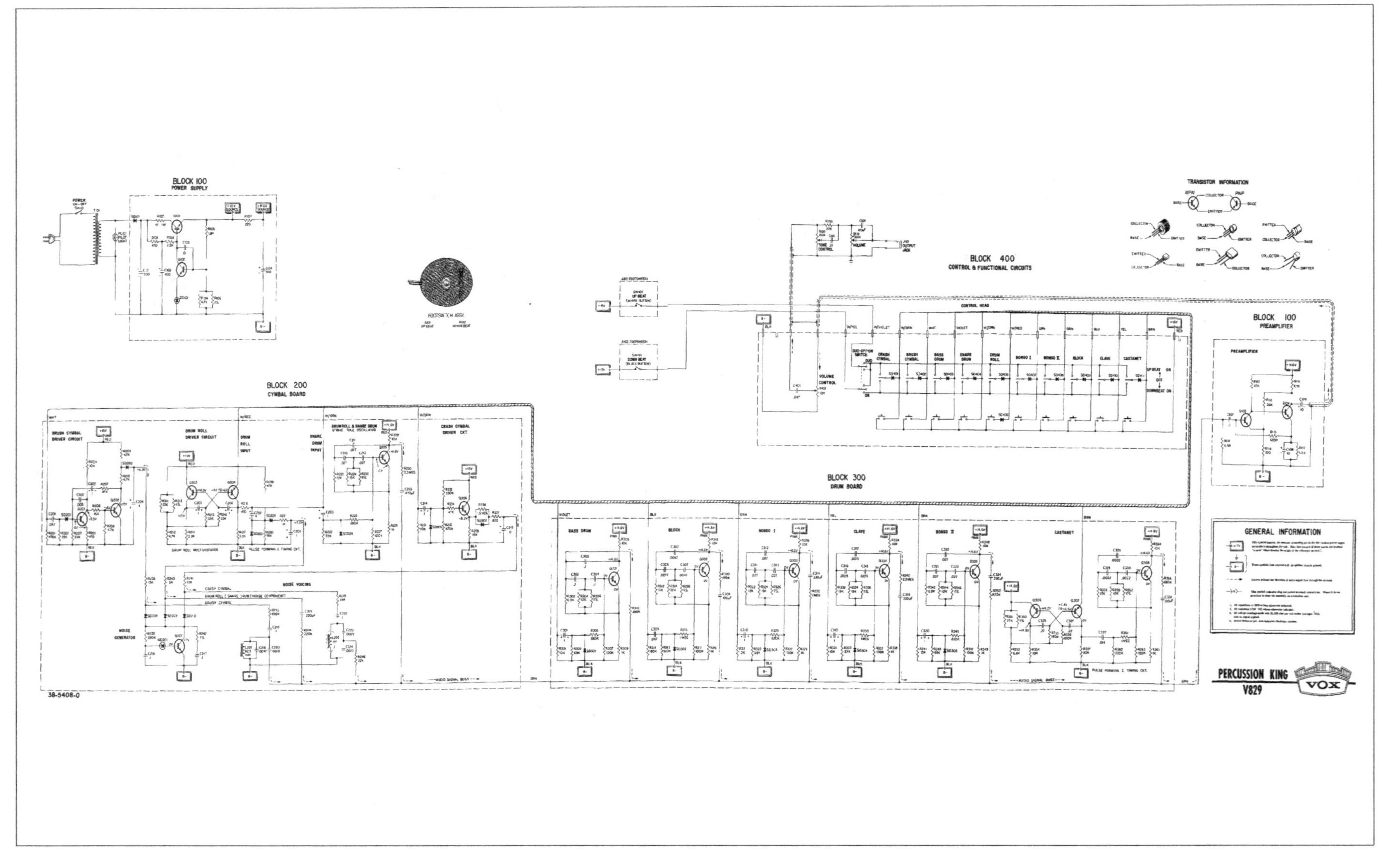

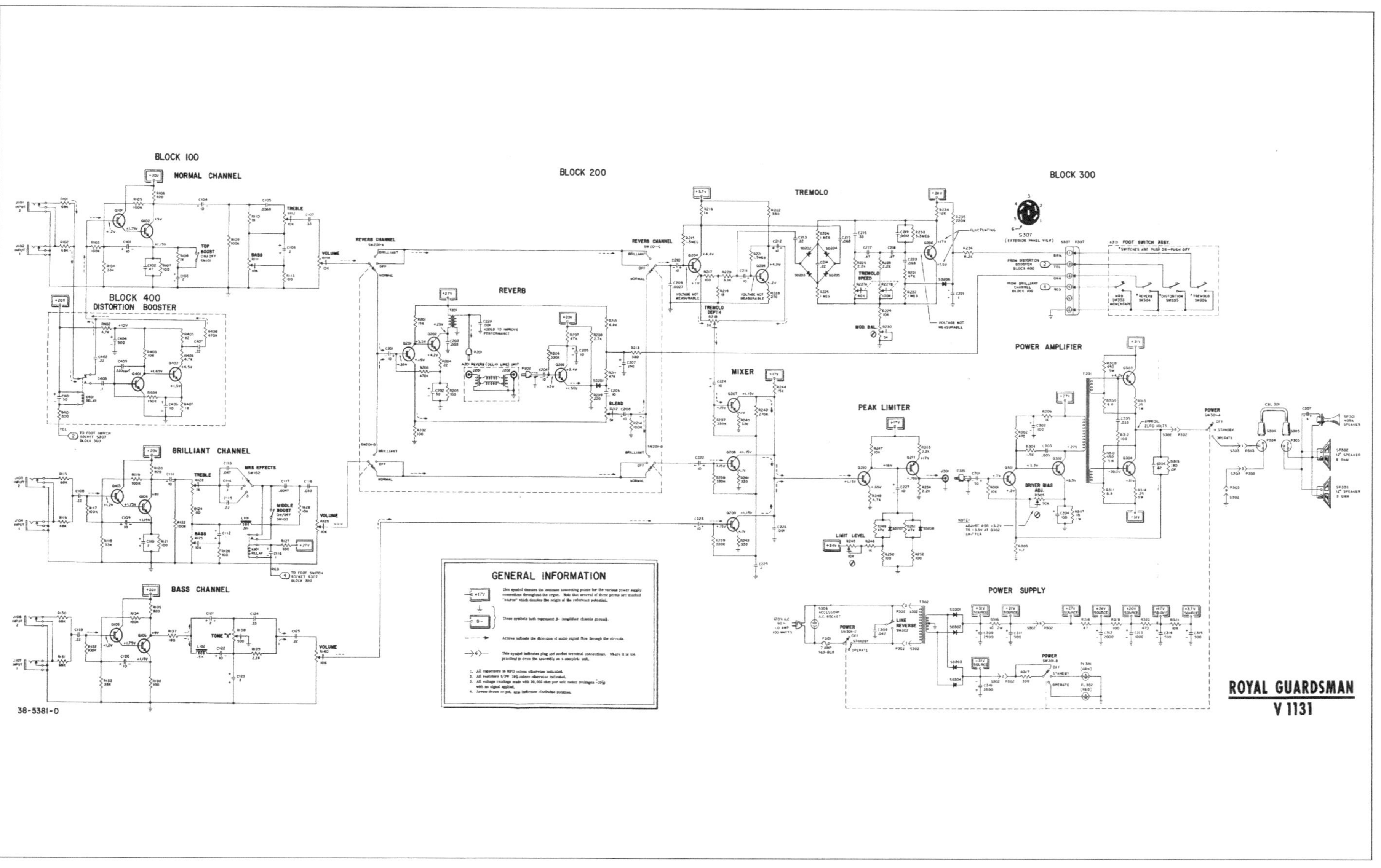
BLOCK 100
NORMAL CHANNEL
BLOCK 200
TREMOLO
BLOCK 300
BLOCK 400
DISTORTION BOOSTER
REVERB
MIXER
POWER AMPLIFIER
BRILLIANT CHANNEL
PEAK LIMITER
BASS CHANNEL
GENERAL INFORMATION
POWER SUPPLY
ROYAL GUARDSMAN
V 1131
38-5381-0

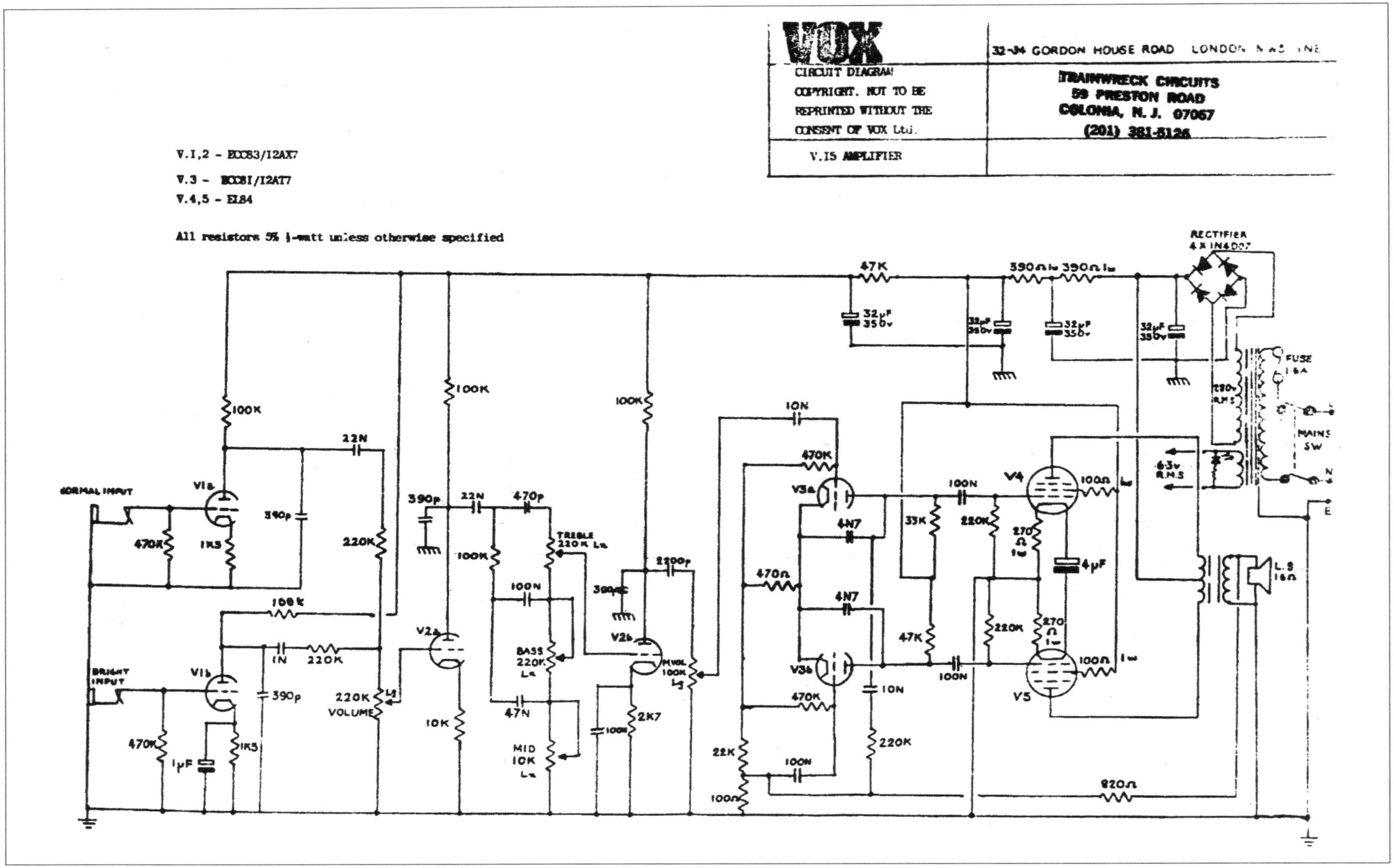

V15 (VOX LTD)

VOX

CIRCUIT DIAGRAM

V.15 AMPLIFIER

32-34 GORDON HOUSE ROAD LONDON N.W.5 1NE

TRAINWRECK CIRCUITS
59 PRESTON ROAD
COLONIA, N. J. 07067
(201) 381-5126

V.1,2 - ECC83/12AX7

V.3 - ECC81/12AT7

V.4,5 - EL84

All resistors 5% ½-watt unless otherwise specified

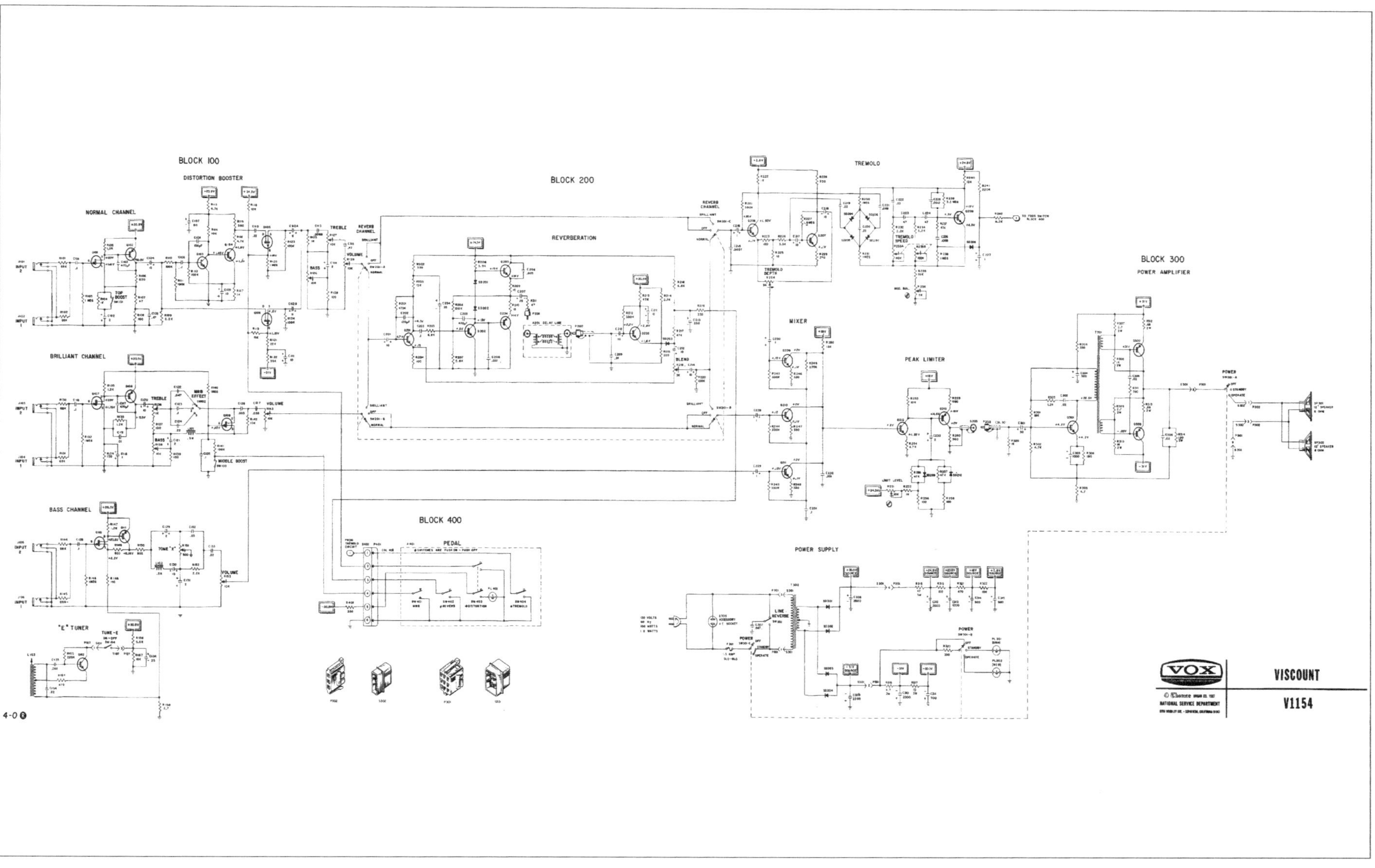
BLOCK 100
DISTORTION BOOSTER
NORMAL CHANNEL
BRILLIANT CHANNEL
BASS CHANNEL
"E" TUNER
BLOCK 200
REVERBERATION
BLOCK 400
PEDAL
TREMOLO
MIXER
PEAK LIMITER
POWER SUPPLY
BLOCK 300
POWER AMPLIFIER
VOX
NATIONAL SERVICE DEPARTMENT
VISCOUNT
V1154
4-0

WESTMINSTER V118

Vox

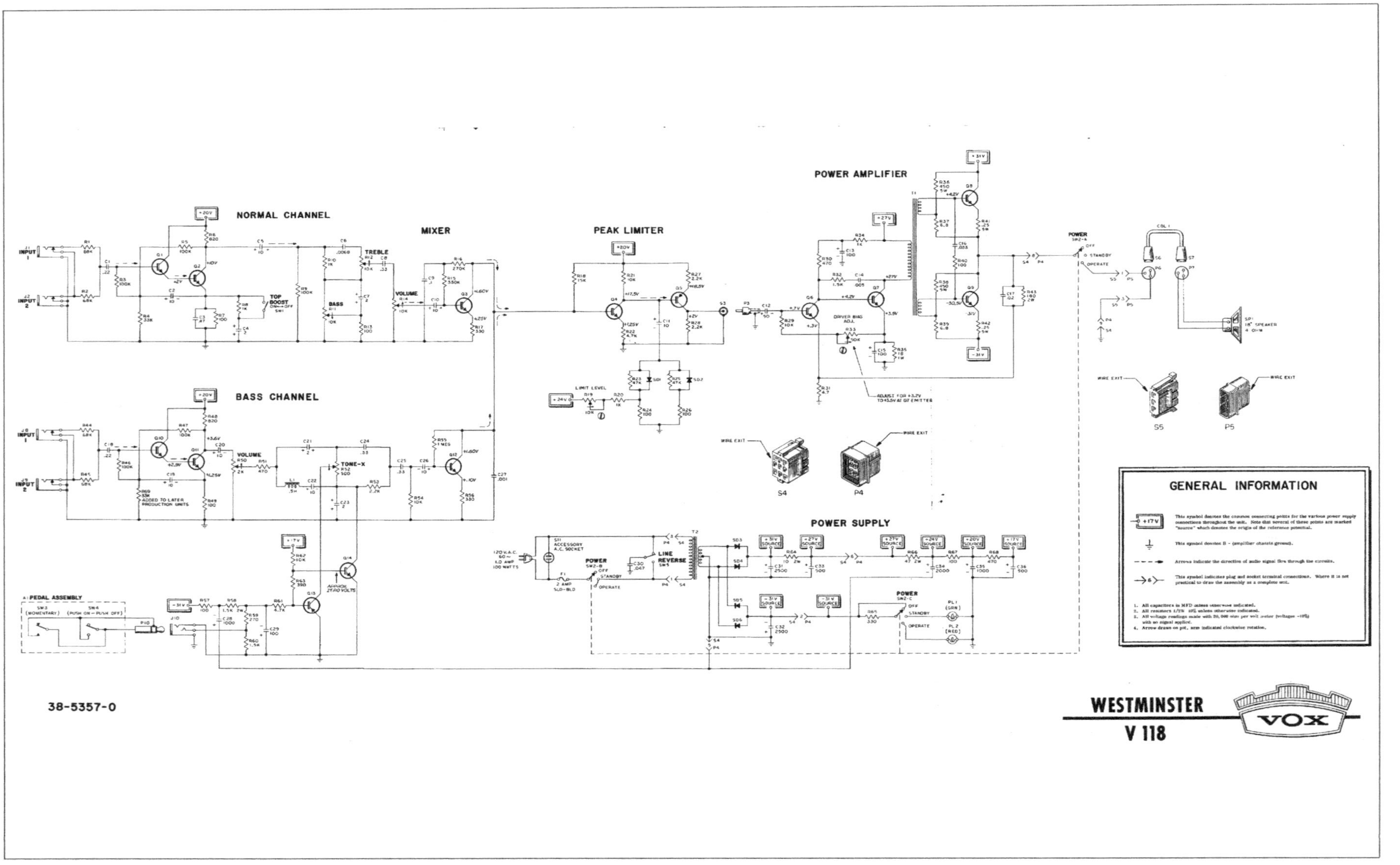

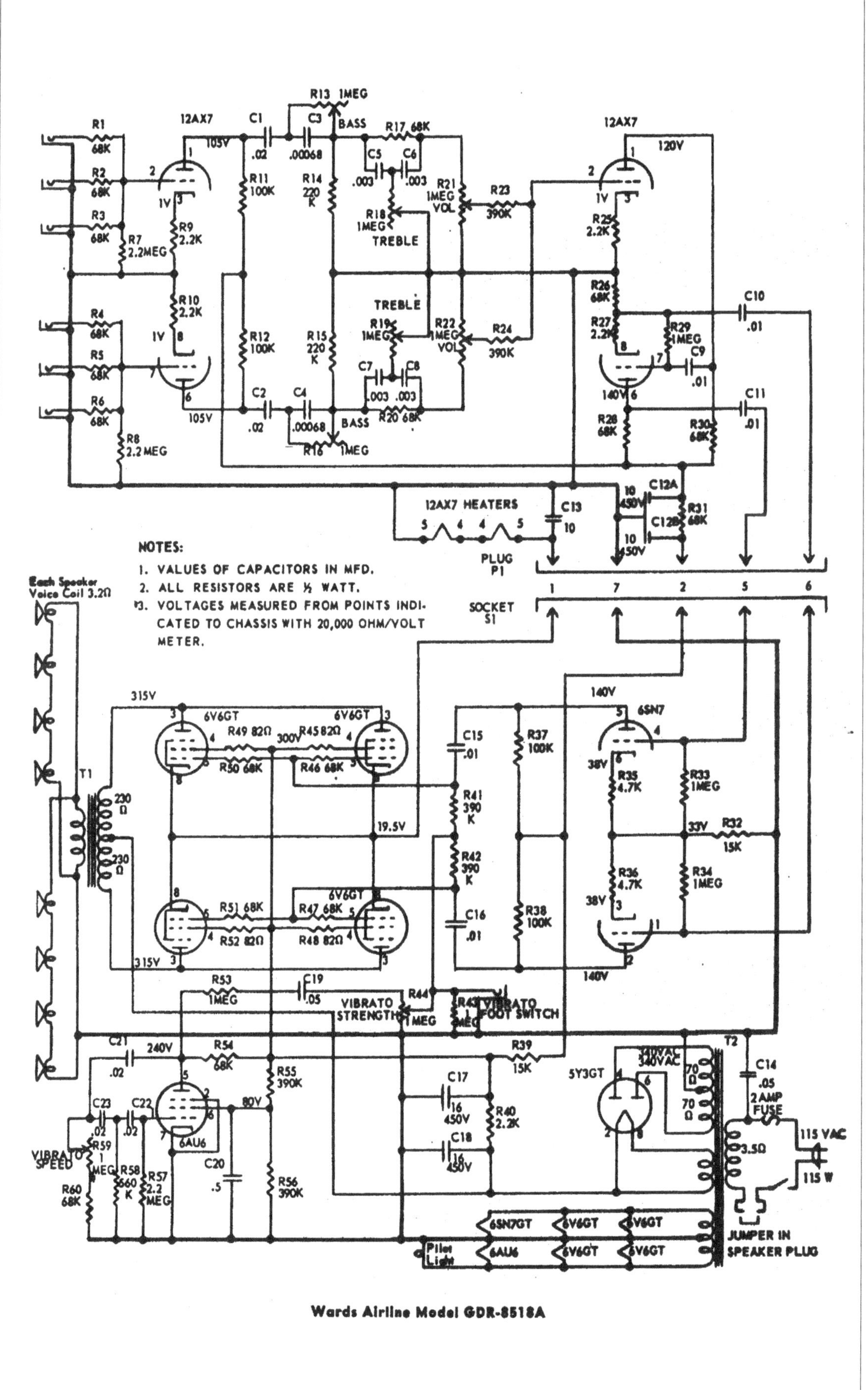
NOTES:
1. VALUES OF CAPACITORS IN MFD.
2. ALL RESISTORS ARE ½ WATT.
*3. VOLTAGES MEASURED FROM POINTS INDICATED TO CHASSIS WITH 20,000 OHM/VOLT METER.
Each Speaker Voice Coil 3.2Ω
12AX7 HEATERS
PLUG P1
SOCKET S1
VIBRATO STRENGTH
VIBRATO FOOT SWITCH
VIBRATO SPEED
JUMPER IN SPEAKER PLUG
Pilot Light
115 VAC
115 W
2 AMP FUSE
TREBLE
BASS
VOL
Wards Airline Model GDR-8518A

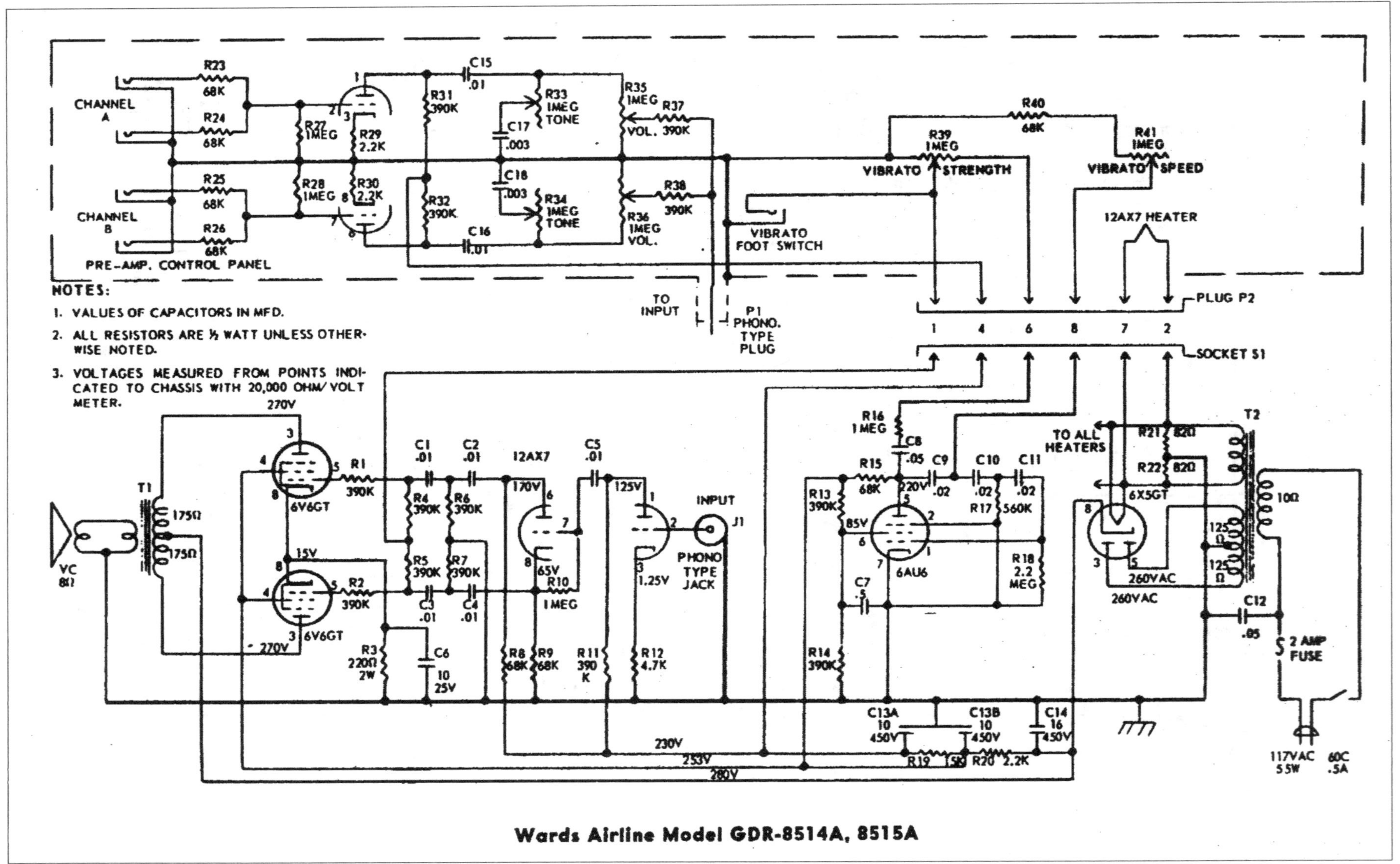

Wards Airline Model GDR-8514A, 8515A

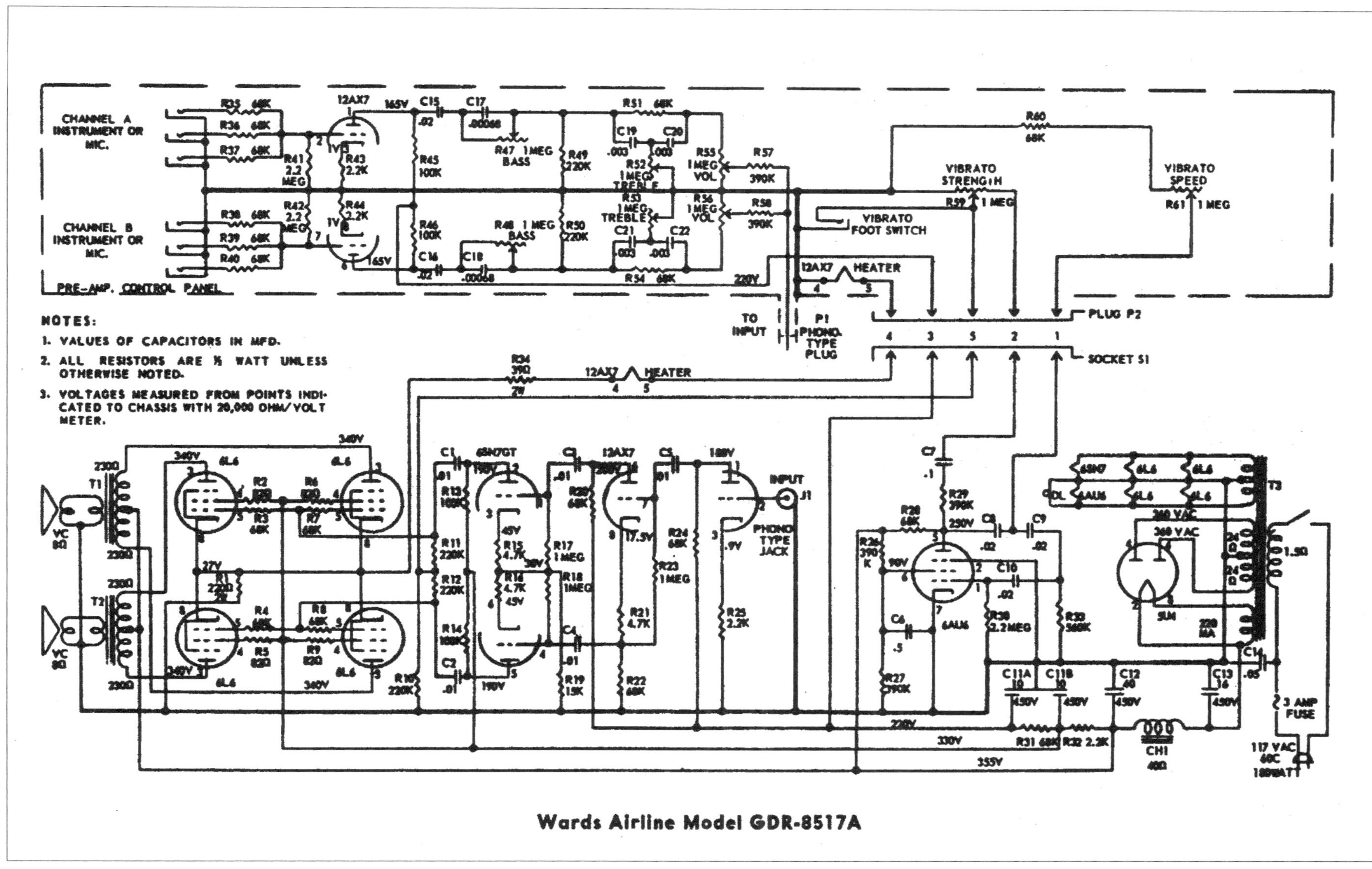

Wards Airline Model GDR-8517A

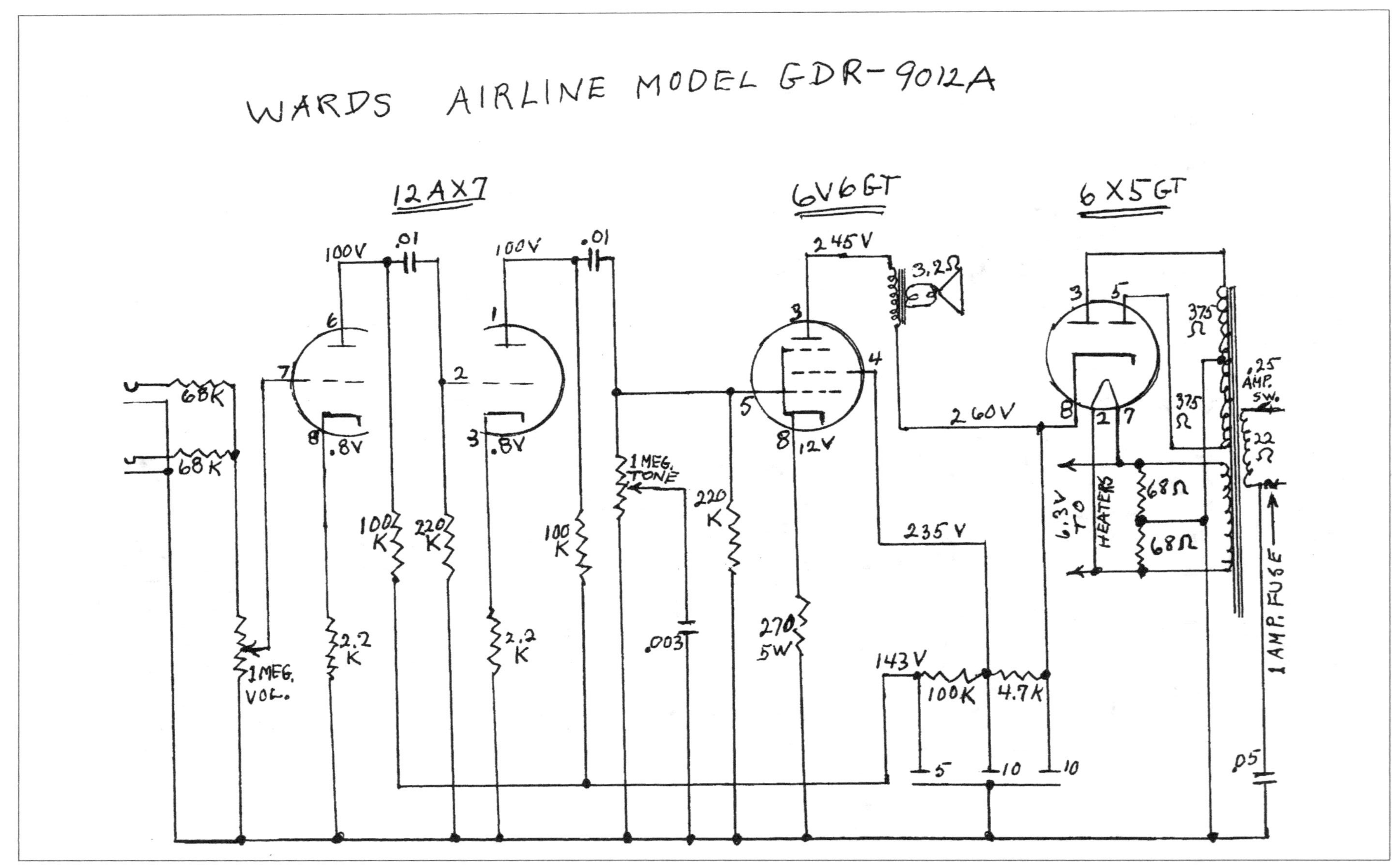
WARDS AIRLINE MODEL GDR-9012A
12AX7
6V6GT
6X5GT
100V
.01
100V
.01
245V
3.2Ω
68K
68K
.8V
.8V
1 MEG. TONE
220 K
100 K
220 K
100 K
2.2 K
2.2 K
1 MEG. VOL.
.003
270 5W
12V
260V
235V
143V
100K
4.7K
5
10
10
375 Ω
375 Ω
.25 AMP. SW.
22 Ω
68Ω
68Ω
6.3V TO HEATERS
1 AMP. FUSE
.05

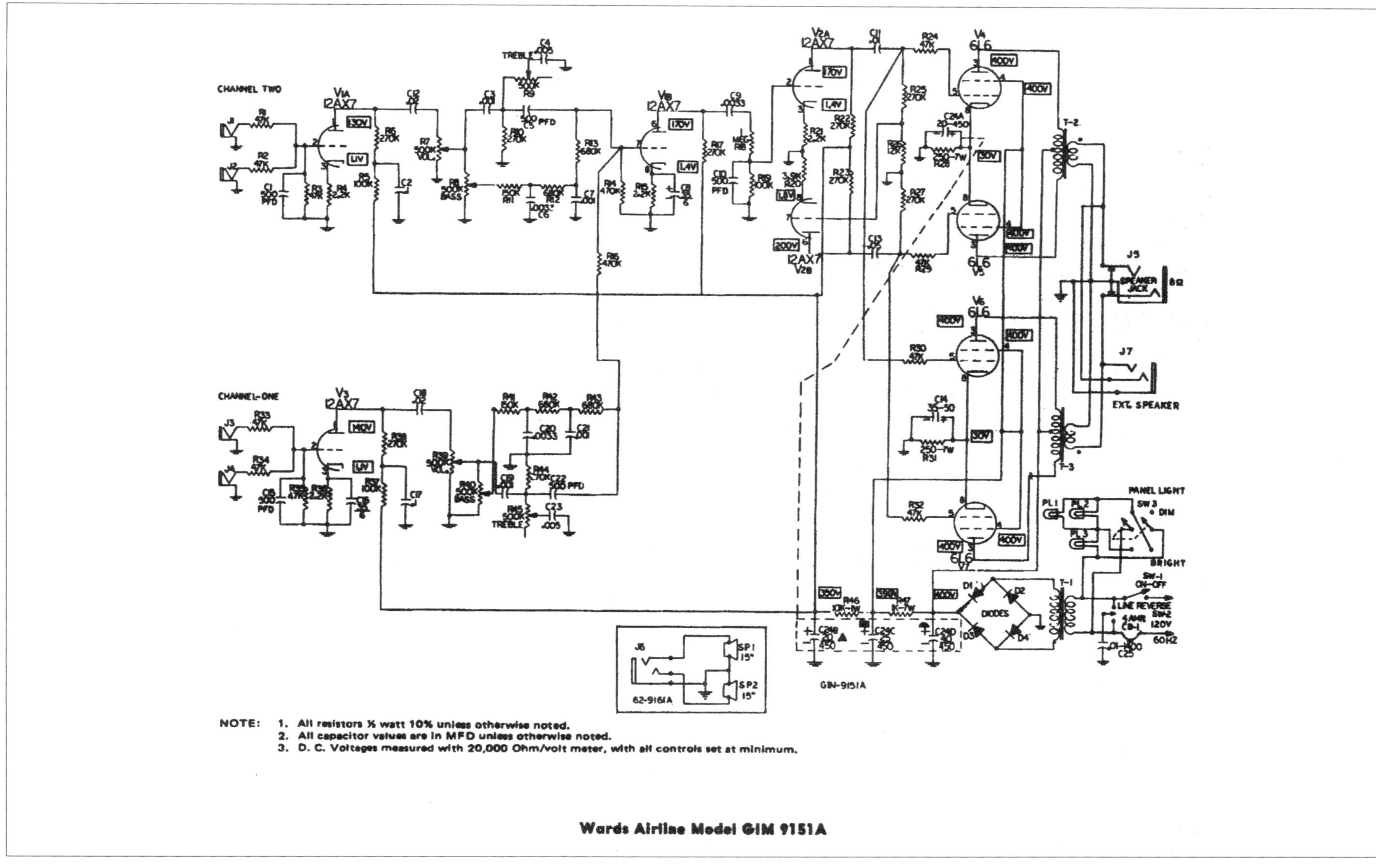

NOTE:
1. All resistors ½ watt 10% unless otherwise noted.
2. All capacitor values are in MFD unless otherwise noted.
3. D. C. Voltages measured with 20,000 Ohm/volt meter, with all controls set at minimum.

Wards Airline Model GIM 9151A

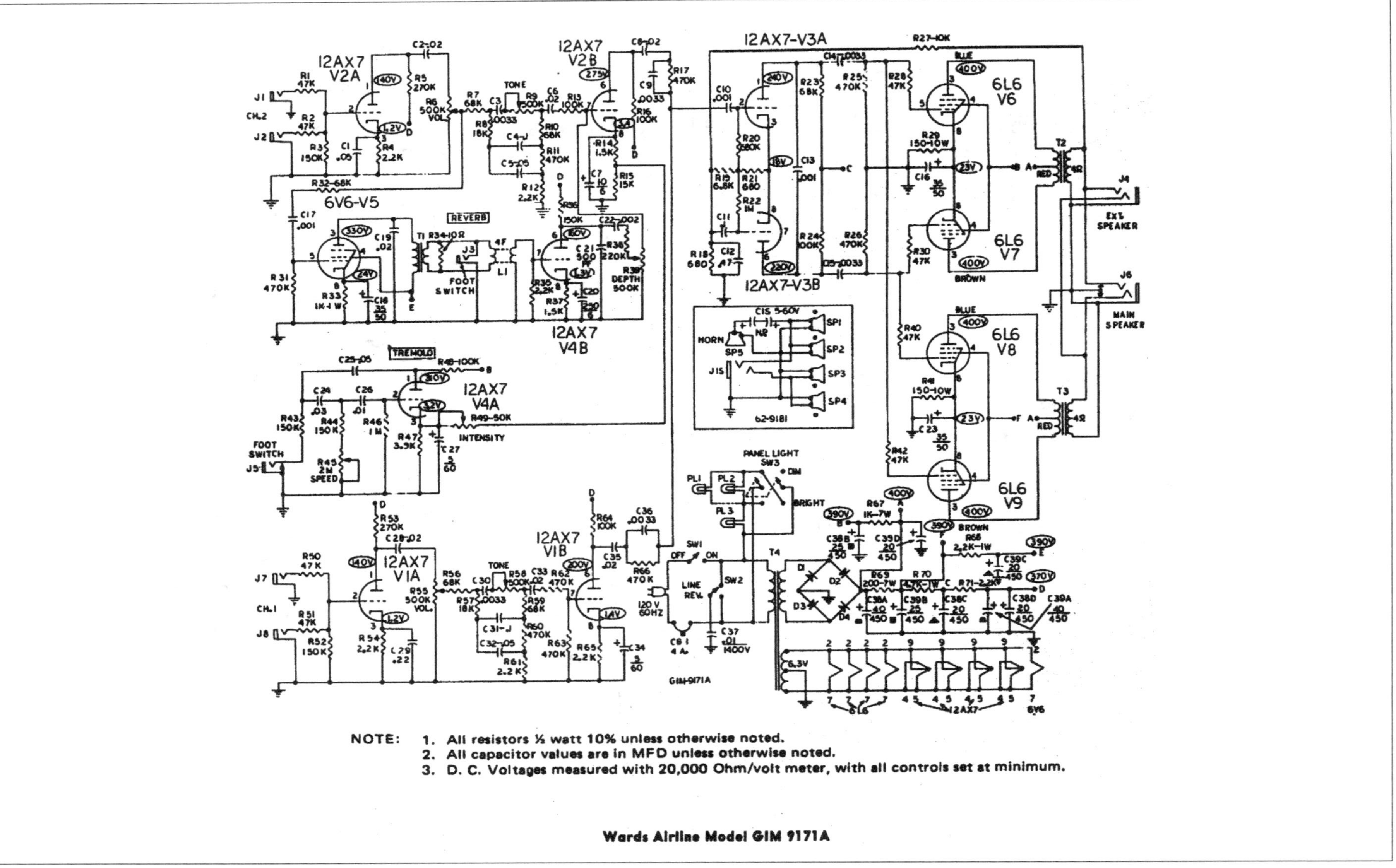

NOTE:
1. All resistors ½ watt 10% unless otherwise noted.
2. All capacitor values are in MFD unless otherwise noted.
3. D. C. Voltages measured with 20,000 Ohm/volt meter, with all controls set at minimum.

Wards Airline Model GIM 9171A

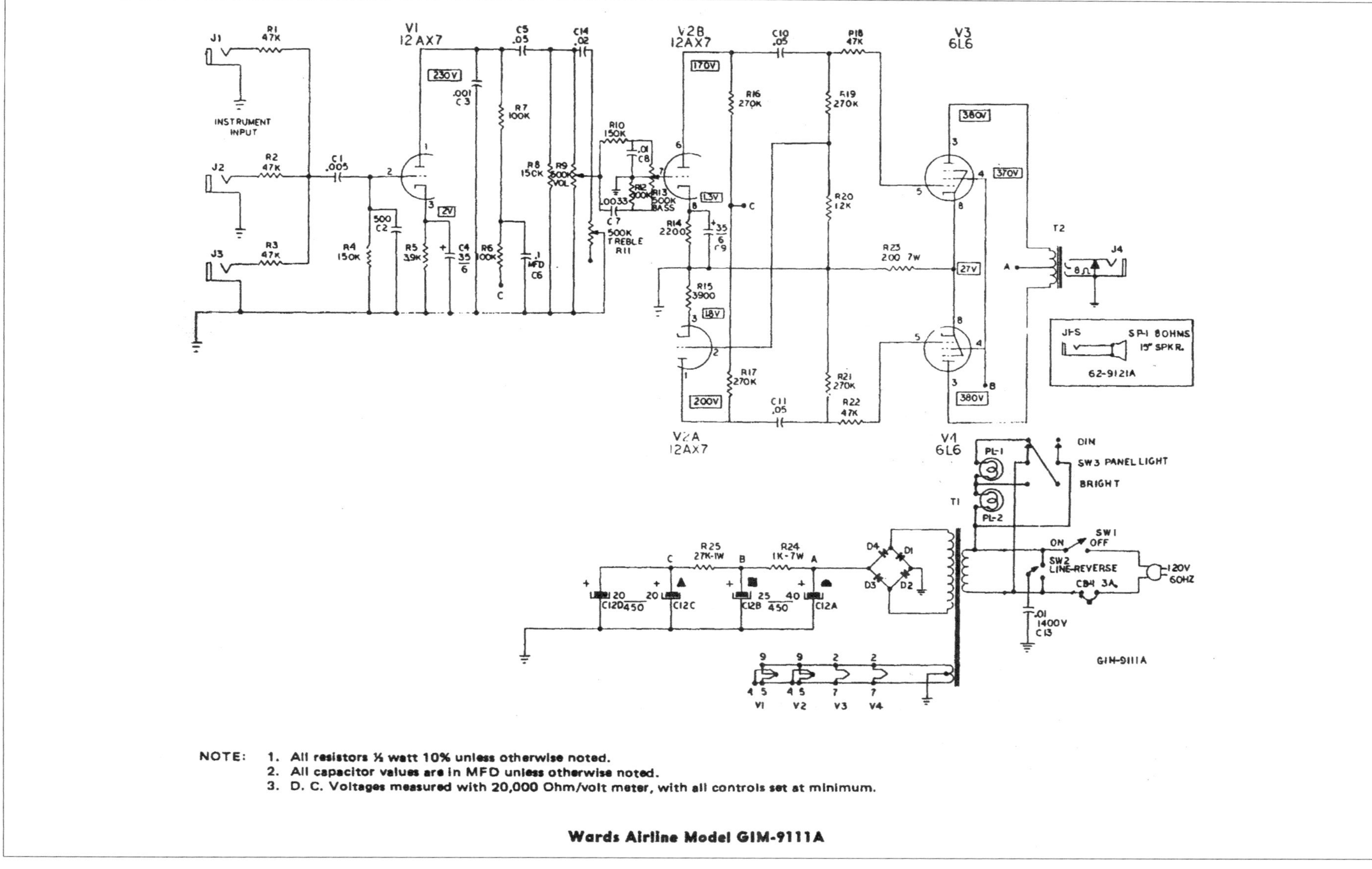

NOTE:
1. All resistors ½ watt 10% unless otherwise noted.
2. All capacitor values are in MFD unless otherwise noted.
3. D. C. Voltages measured with 20,000 Ohm/volt meter, with all controls set at minimum.

Wards Airline Model GIM-9111A

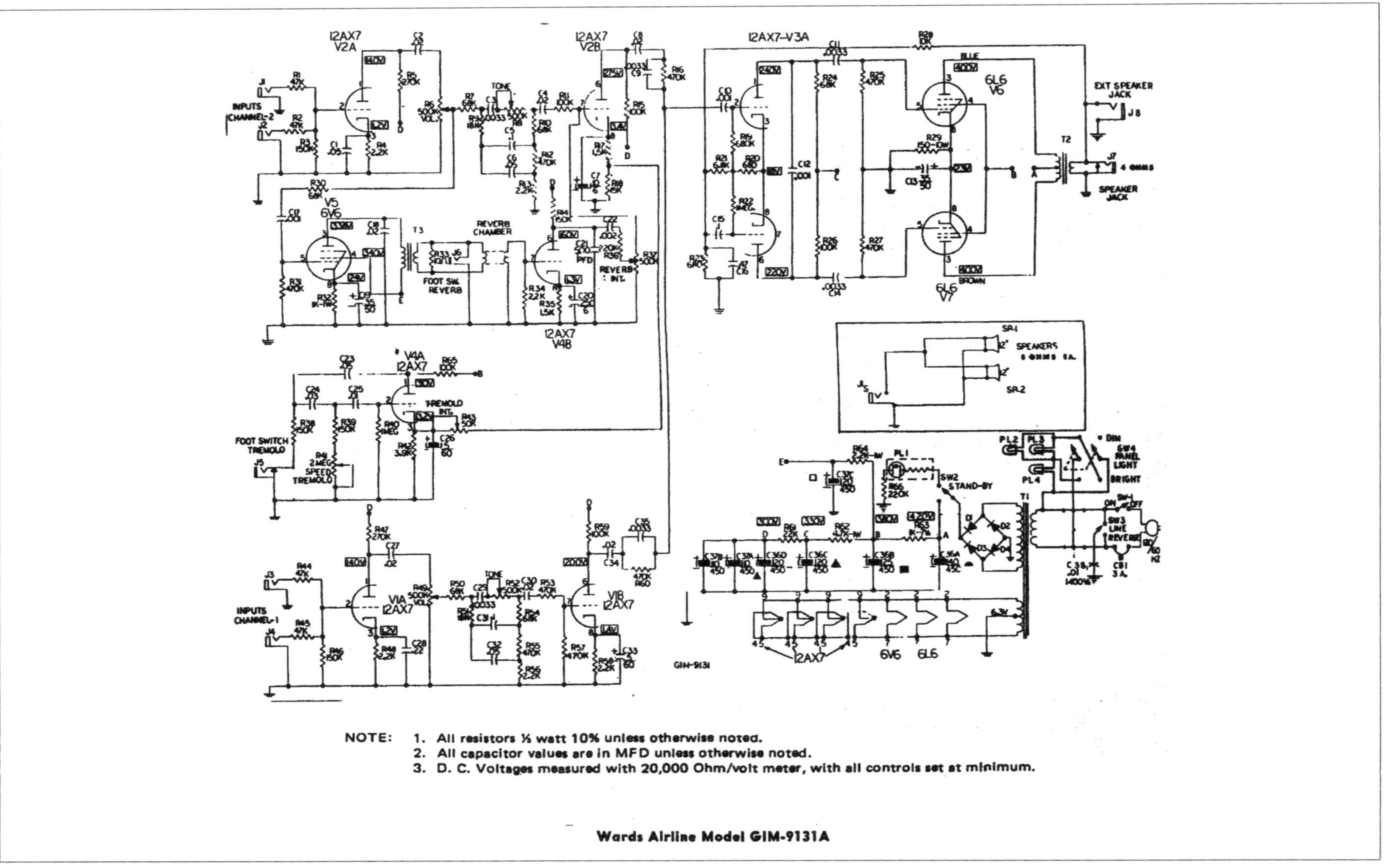

NOTE:
1. All resistors ½ watt 10% unless otherwise noted.
2. All capacitor values are in MFD unless otherwise noted.
3. D. C. Voltages measured with 20,000 Ohm/volt meter, with all controls set at minimum.

Wards Airline Model GIM-9131A

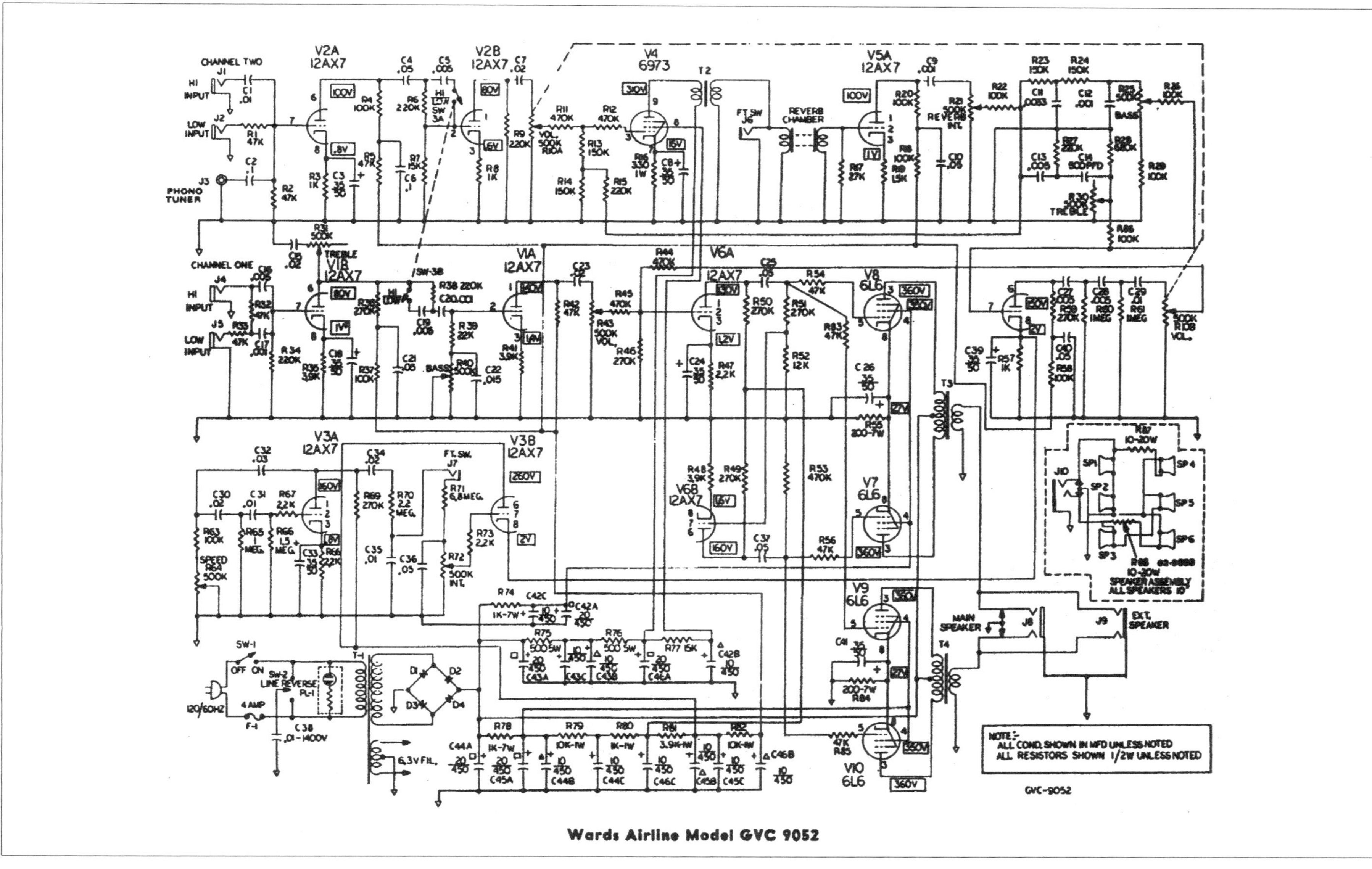

Wards Airline Model GVC 9052

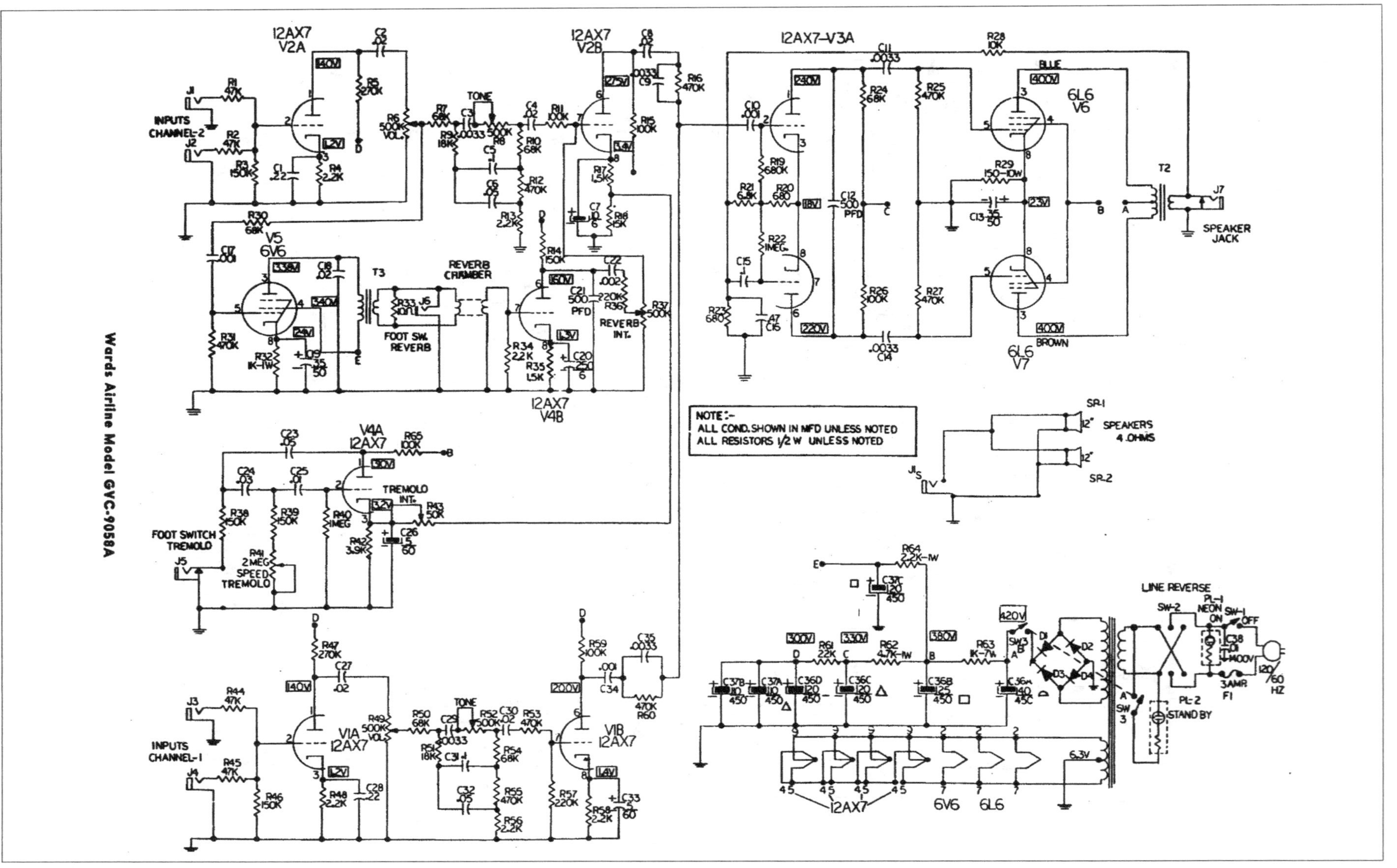

Wards Airline Model GVC-9058A

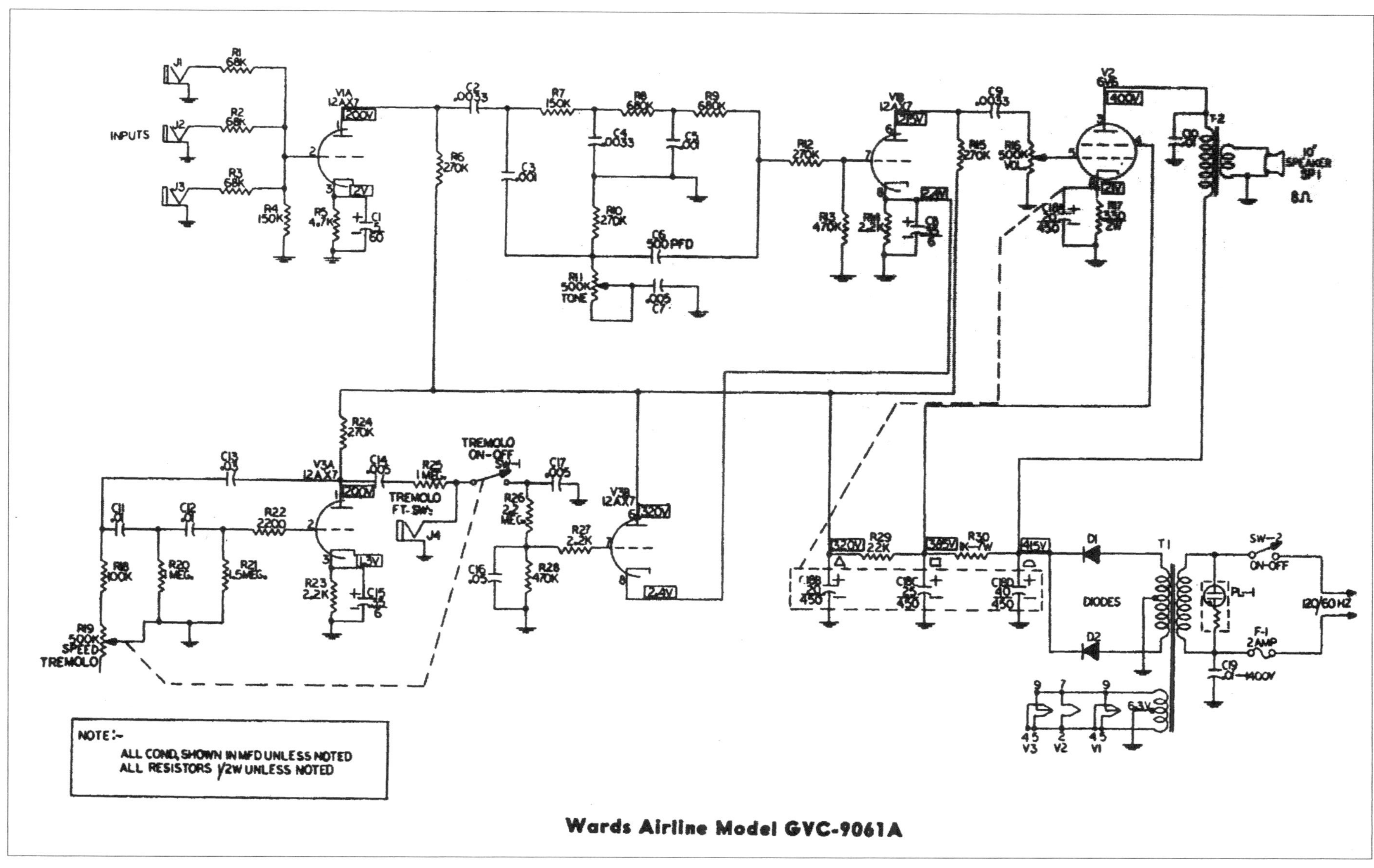

Wards Airline Model GVC-9061A

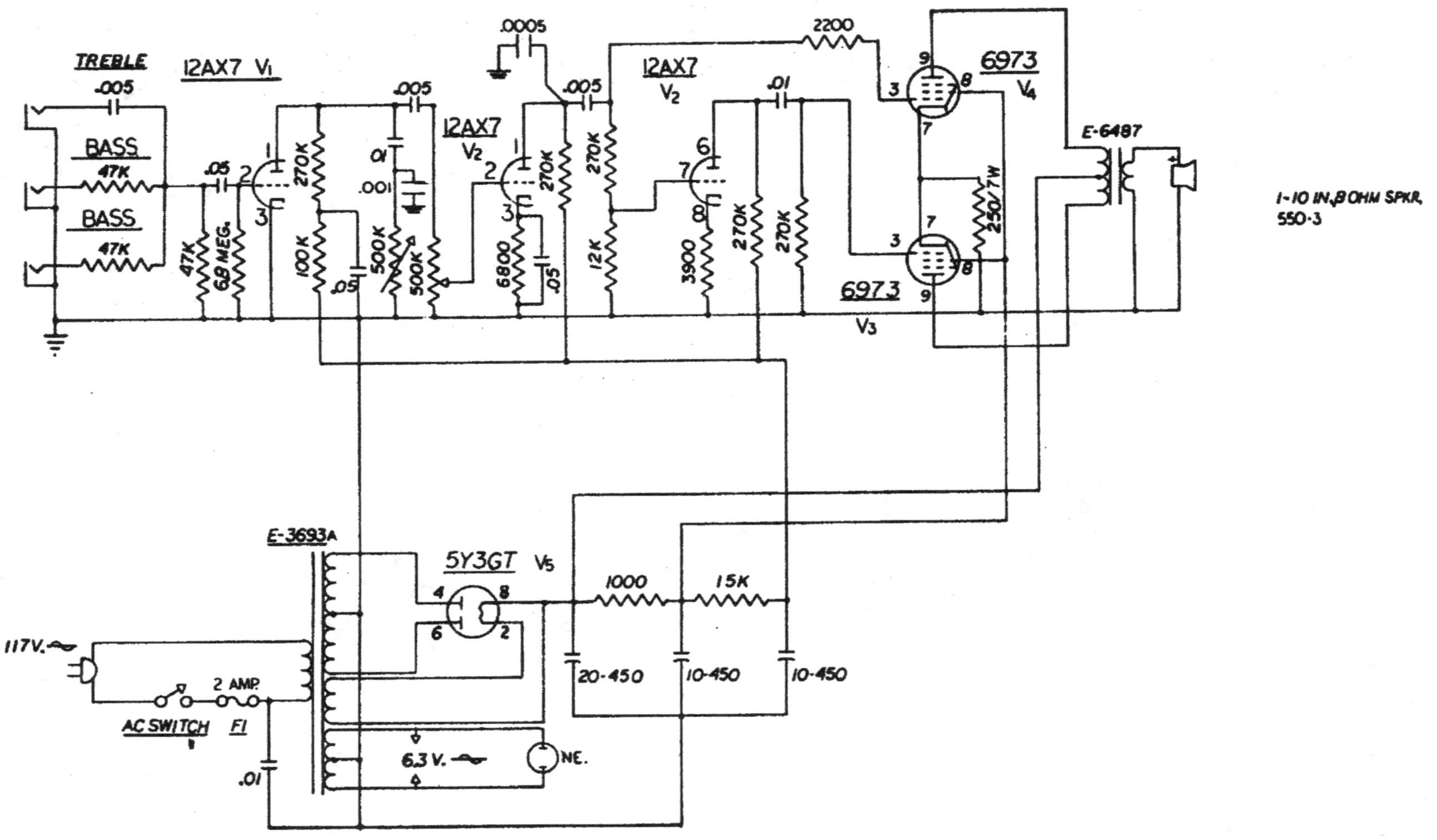

Valco Model 6400

1959 COPICAT **MKII**

Watkins

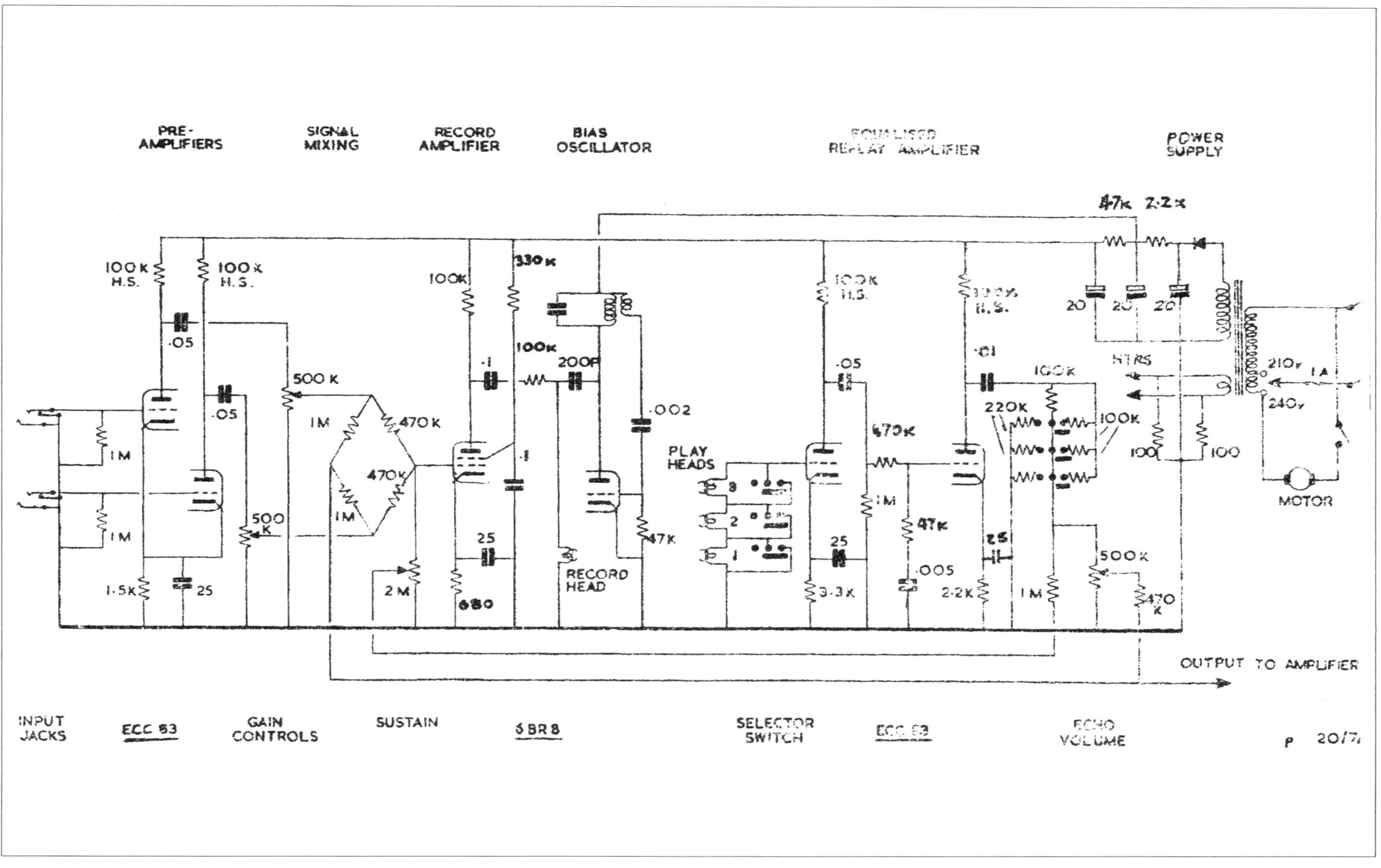

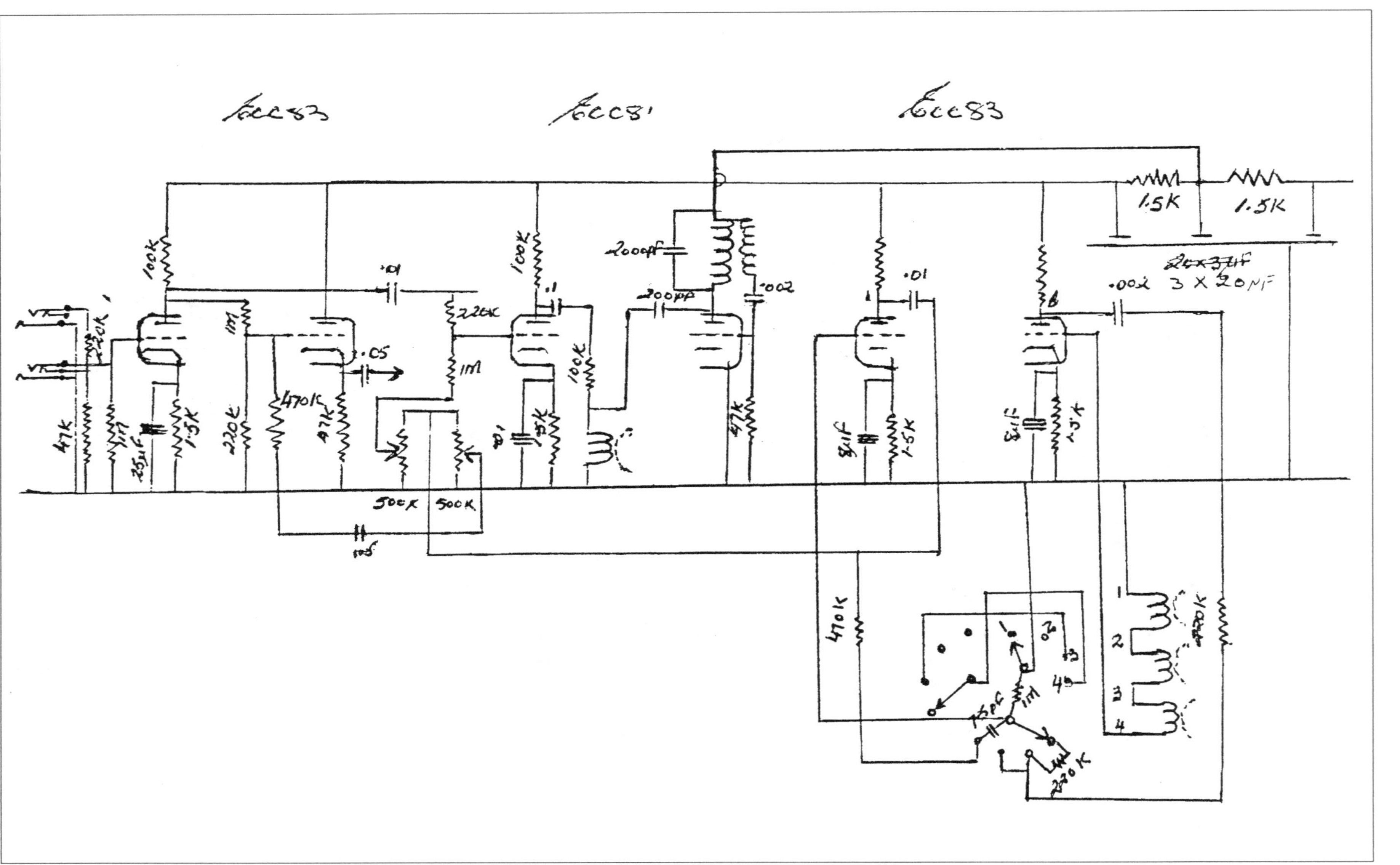
ECC83
ECC81
ECC83
1.5K
1.5K
3 X 20MF
.002
.01
100K
220K
1M
47K
1.5K
25µF
.05
470K
500K
500K
.1
2000pF
200pF
.002
470K
8µF
220K

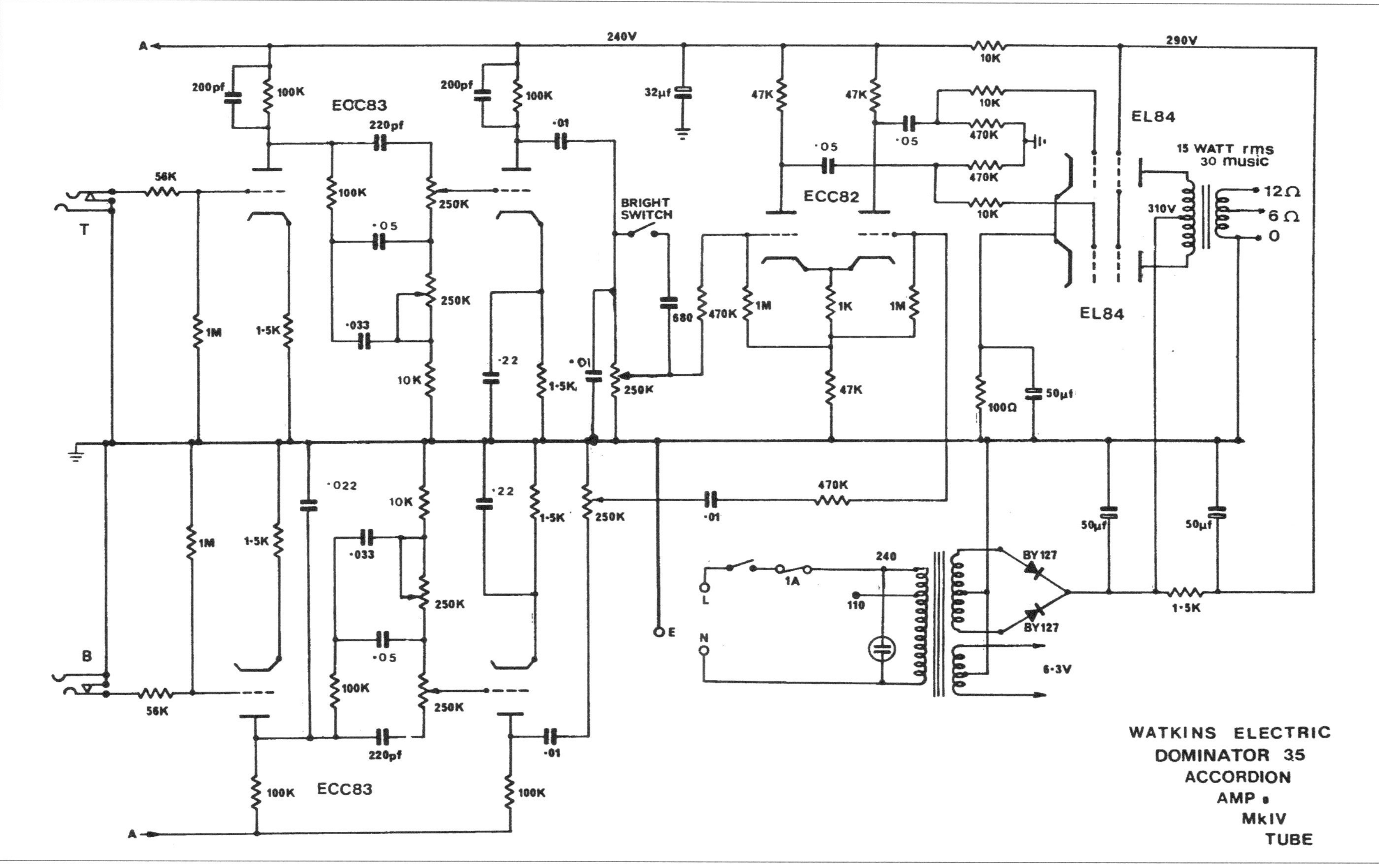
WATKINS ELECTRIC
DOMINATOR 35
ACCORDION
AMP
MkIV
TUBE
ECC83
ECC82
EL84
240V
290V
BRIGHT SWITCH
15 WATT rms
30 music
12Ω
6Ω
0
310V
BY127
6·3V
1A
240
110
T
B
E
L
N
A

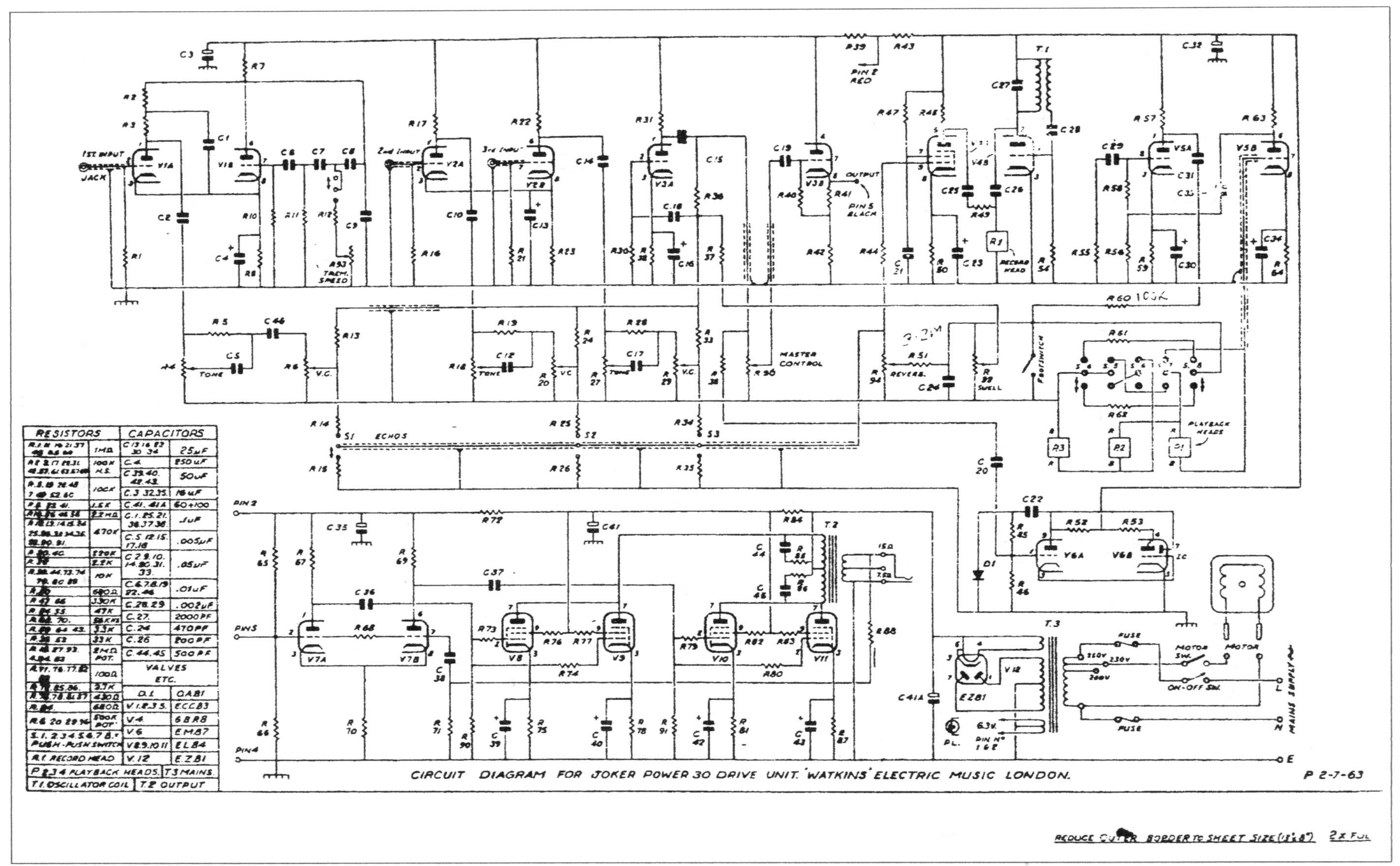

RESISTORS	
R.1. [illegible] 19 21 37 [illegible]	1MΩ
R2 [illegible] 17 22.31 [illegible]	100K H.S.
R.3. [illegible] 26.48 7 [illegible] 52.60	100K
R.[illegible] 41.	1.5K
R.[illegible]	2.2MΩ
R.[illegible] 13.14.15.24 25.[illegible] 90. 91.	470K
R.[illegible] 40	220K
R.[illegible]	2.2K
R.[illegible] 44.73.74 [illegible] 80 88	10K
R.[illegible]	680Ω
R.[illegible] 66	330K
R.[illegible] 35.	47K
R.[illegible] 70.	[illegible]K
R.[illegible] 64 43.	3.3K
R.[illegible] 53	33K
R.[illegible] 27.93. 4.[illegible]	2MΩ POT.
R.71. 76. 77. [illegible]	100Ω
R.[illegible] 85.86.	2.7K
R.[illegible] 78. 81. 87	430Ω
R.[illegible]	680Ω
R.6 20 29.96	500K POT.
S.1. 2.3.4.5.6.7.8. PUSH-PUSH SWITCH	
R.1. RECORD HEAD	
P.2.3.4 PLAYBACK HEADS.	
T.1. OSCILLATOR COIL	

CAPACITORS	
C 13 16 23 30 34	25μF
C.4	250uF
C 39.40. 42.43.	50uF
C.3 32.35.	16uF
C.41. 41A	60+100
C.1. 25. 21. 36. 37. 38.	.1uF
C.5 12.15. 17.18	.005μF
C 2.9.10. 14.20.31. 33.	.05μF
C.6.7.8.19 22.46	.01uF
C.28.29	.002uF
C.27.	2000PF
C.24	470PF
C.26	200PF
C.44.45	500PF
VALVES ETC.	
D.1.	OA81
V.1.2.3.5.	ECC83
V.4	6BR8
V.6	EM87
V.8.9.10.11	EL84
V.12	EZ81
T3 MAINS.	
T.2 OUTPUT	

CIRCUIT DIAGRAM FOR JOKER POWER 30 DRIVE UNIT. 'WATKINS' ELECTRIC MUSIC LONDON.

P 2-7-63

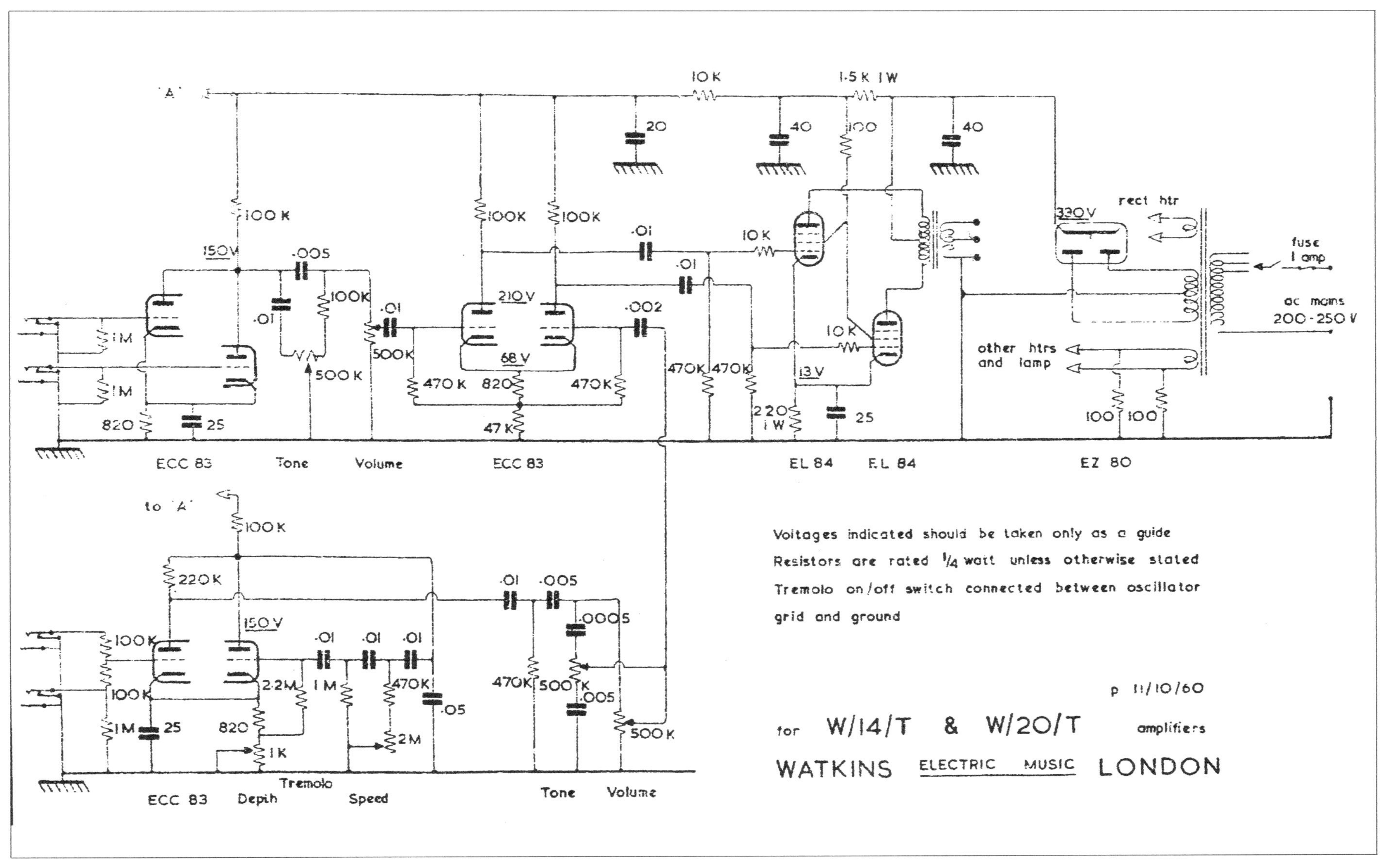
'A'
10 K
1.5 K 1W
20
40
100
40
100 K
100K
100K
.01
10 K
.01
rect htr
330V
fuse
1 amp
150V
.005
.01
100K
.01
500 K
500K
210V
68V
.002
ac mains
200-250 V
1M
1M
820
25
470 K
820
470K
47 K
470K
470K
10 K
13 V
220
1 W
25
other htrs
and lamp
100
100
ECC 83
Tone
Volume
ECC 83
EL 84
EL 84
EZ 80
to 'A'
100 K
220K
150V
100K
100K
1M
25
.01
.01
.01
2.2M
1 M
470K
.05
820
1K
2M
.01
.005
.0005
470K
500K
.005
500 K
ECC 83
Depth
Tremolo
Speed
Tone
Volume
Voltages indicated should be taken only as a guide
Resistors are rated 1/4 watt unless otherwise stated
Tremolo on/off switch connected between oscillator
grid and ground
p 11/10/60
for W/14/T & W/20/T amplifiers
WATKINS ELECTRIC MUSIC LONDON

CLUBMAN **MK8**

Watkins/WEM

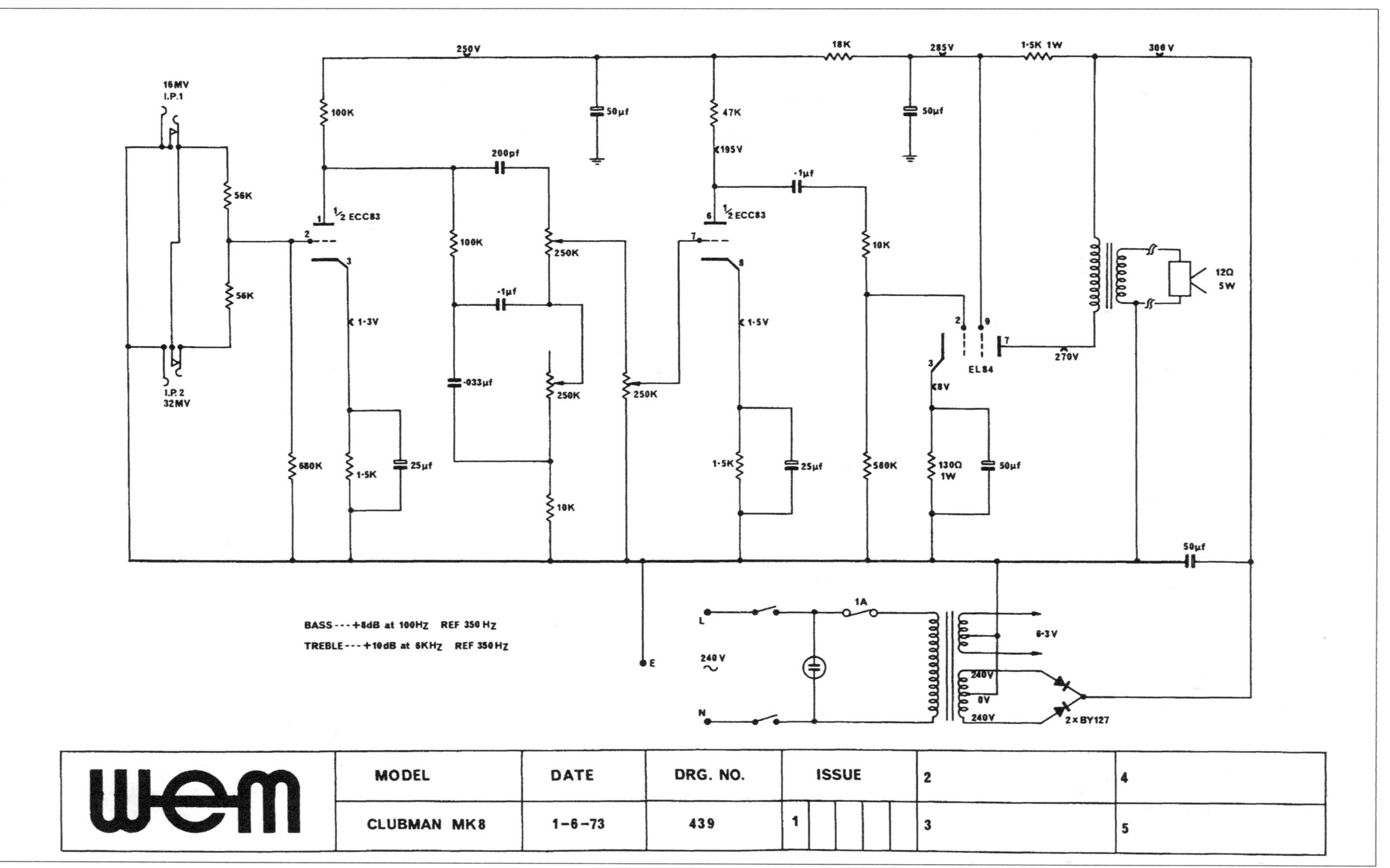

DOMINATOR 50

Watkins/WEM

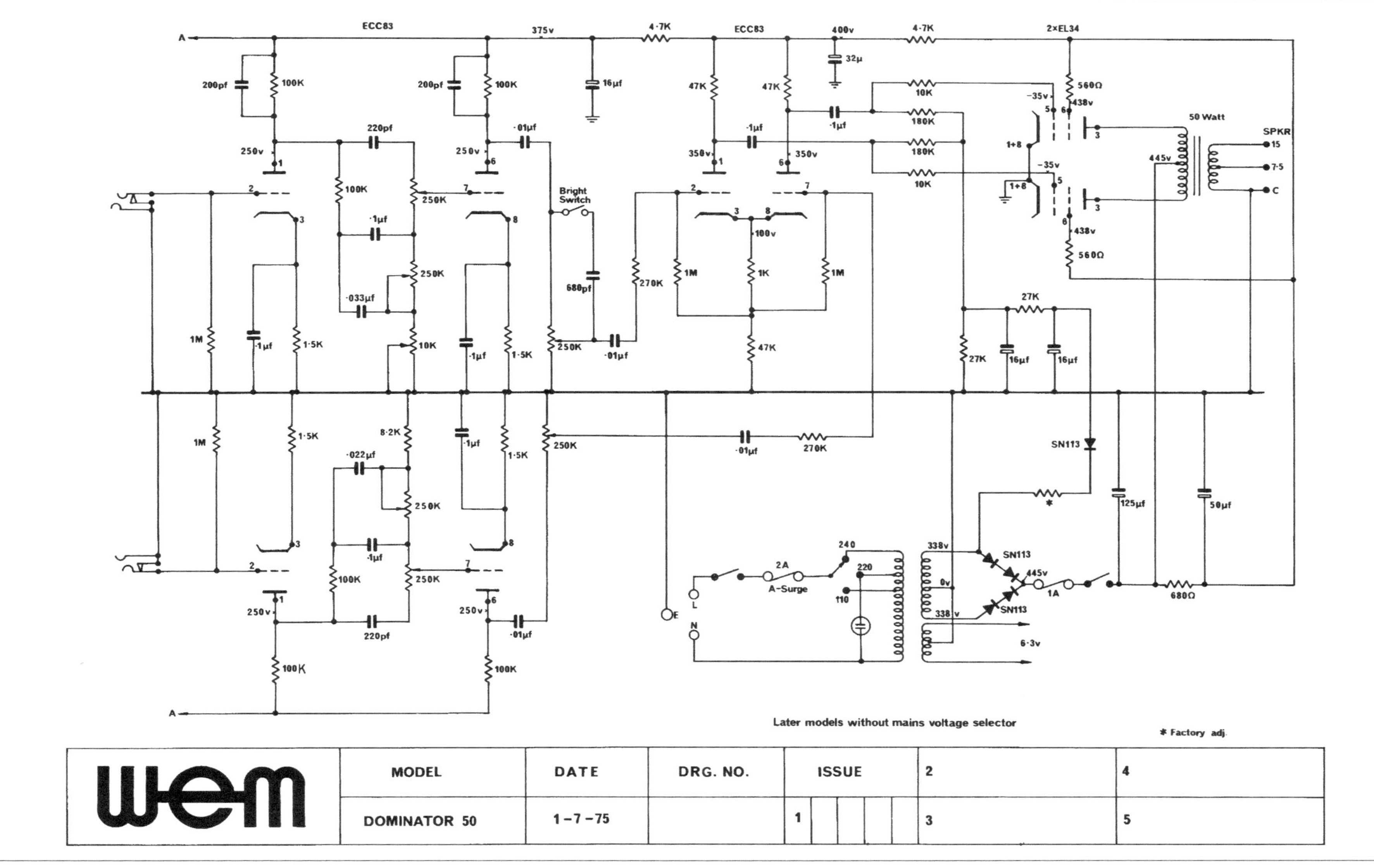

WESTMINSTER **MK9**

WatkinsWEM

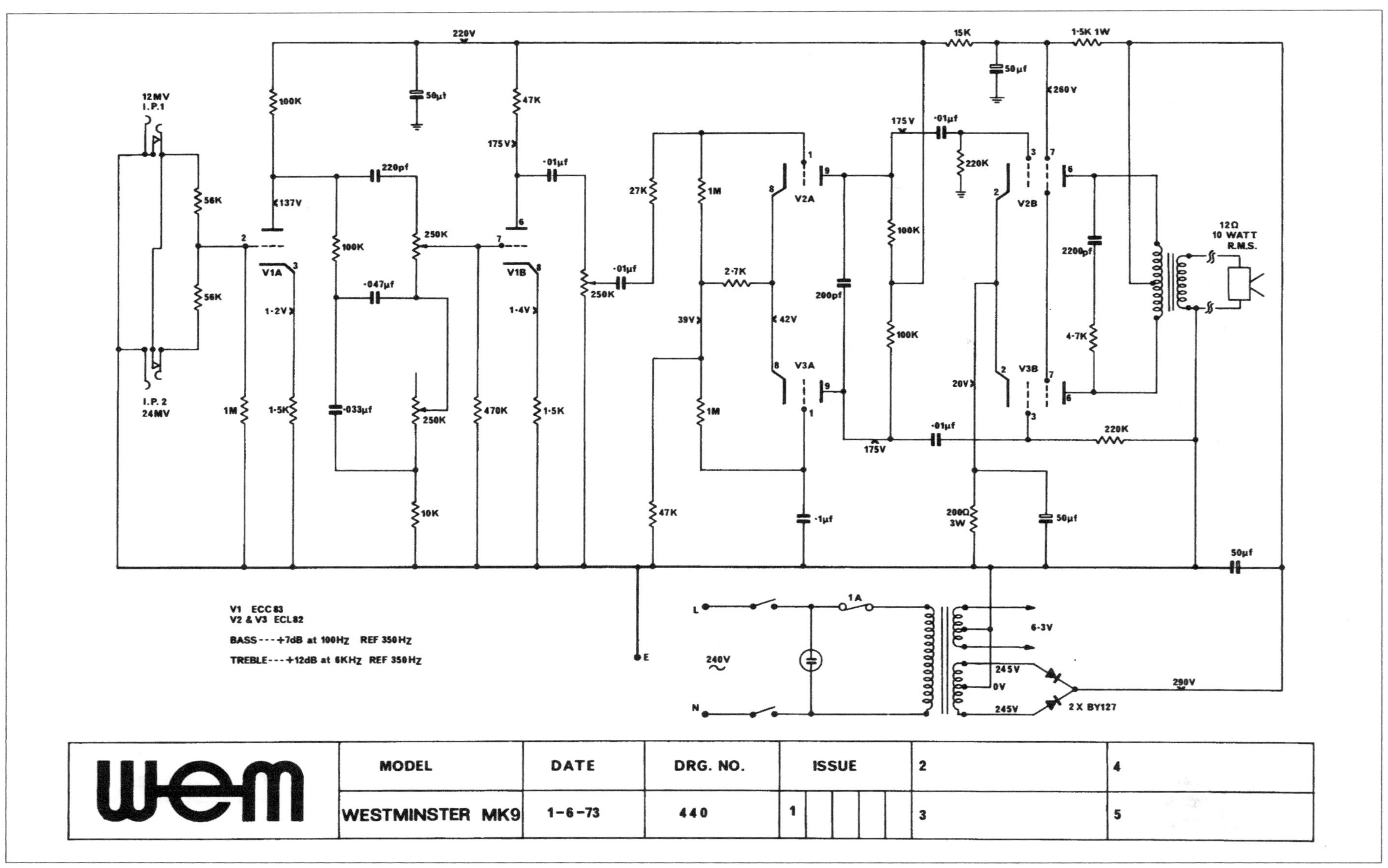

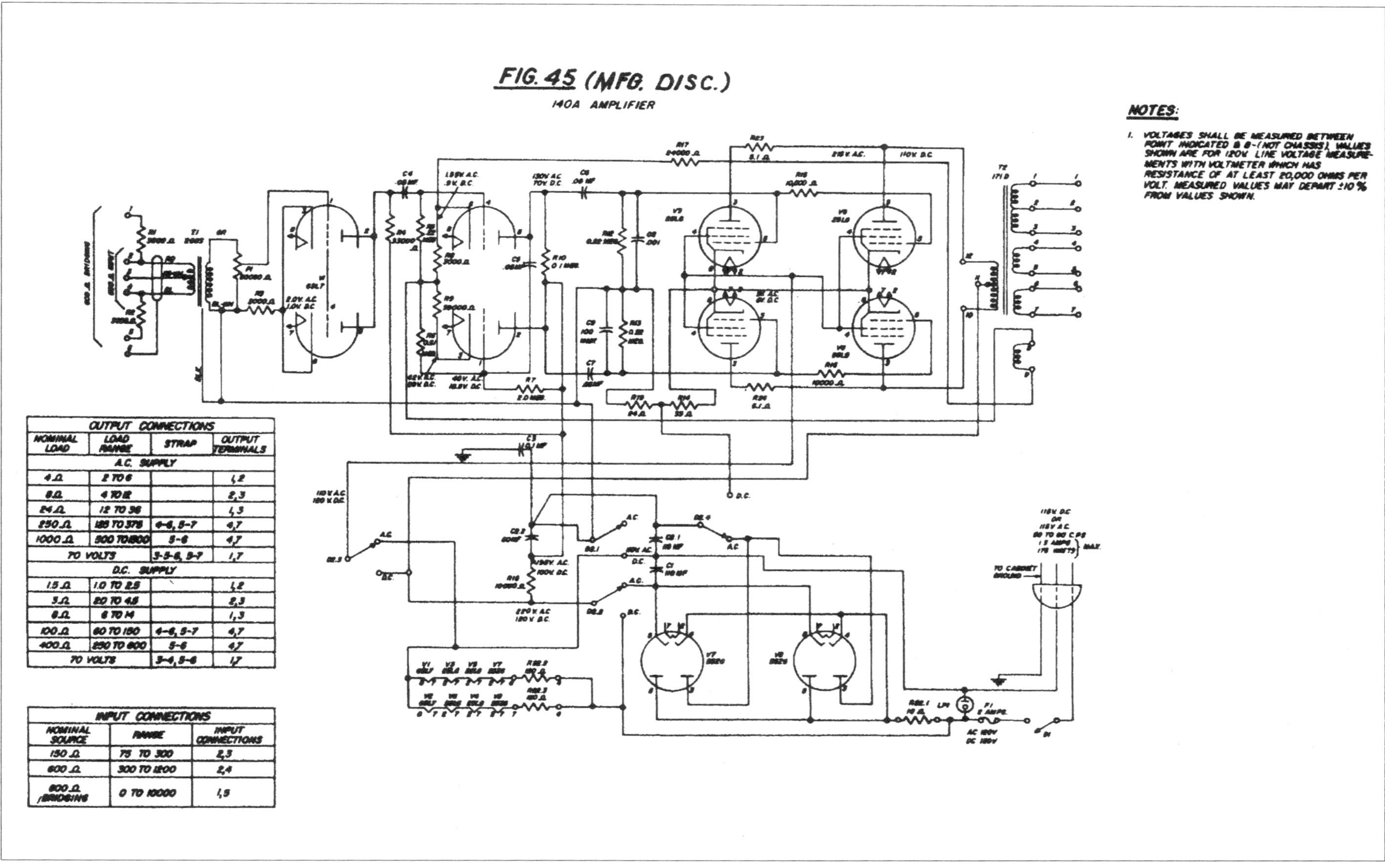

NOTES:
1. VOLTAGES SHALL BE MEASURED BETWEEN POINT INDICATED & B-(NOT CHASSIS). VALUES SHOWN ARE FOR 120V. LINE VOLTAGE MEASUREMENTS WITH VOLTMETER WHICH HAS RESISTANCE OF AT LEAST 20,000 OHMS PER VOLT. MEASURED VALUES MAY DEPART ±10% FROM VALUES SHOWN.

FIG. 45 (MFG. DISC.)
140A AMPLIFIER

OUTPUT CONNECTIONS			
NOMINAL LOAD	LOAD RANGE	STRAP	OUTPUT TERMINALS
A.C. SUPPLY			
4 Ω	2 TO 6		1,2
8 Ω	4 TO 12		2,3
24 Ω	12 TO 36		1,3
250 Ω	125 TO 375	4-6, 5-7	4,7
1000 Ω	500 TO 1500	5-6	4,7
70 VOLTS		3-5-6, 5-7	1,7
D.C. SUPPLY			
1.5 Ω	1.0 TO 2.5		1,2
3 Ω	2.0 TO 4.5		2,3
8 Ω	6 TO 14		1,3
100 Ω	60 TO 150	4-6, 5-7	4,7
400 Ω	250 TO 600	5-6	4,7
70 VOLTS		3-4, 5-6	1,7

INPUT CONNECTIONS		
NOMINAL SOURCE	RANGE	INPUT CONNECTIONS
150 Ω	75 TO 300	2,3
600 Ω	300 TO 1200	2,4
600 Ω /BRIDGING	0 TO 10000	1,5

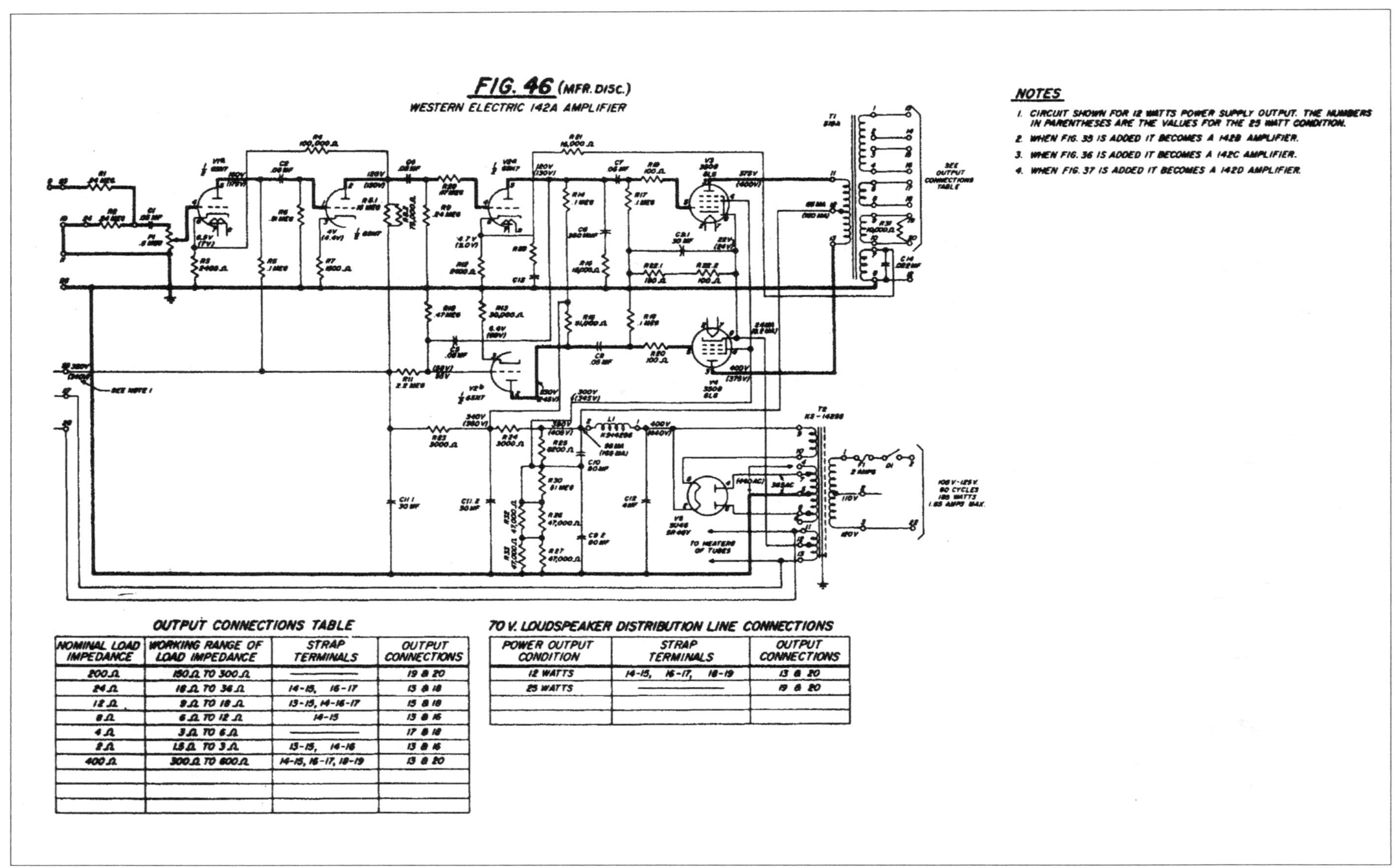

NOTES

1. CIRCUIT SHOWN FOR 12 WATTS POWER SUPPLY OUTPUT. THE NUMBERS IN PARENTHESES ARE THE VALUES FOR THE 25 WATT CONDITION.
2. WHEN FIG. 35 IS ADDED IT BECOMES A 142B AMPLIFIER.
3. WHEN FIG. 36 IS ADDED IT BECOMES A 142C AMPLIFIER.
4. WHEN FIG. 37 IS ADDED IT BECOMES A 142D AMPLIFIER.

OUTPUT CONNECTIONS TABLE

NOMINAL LOAD IMPEDANCE	WORKING RANGE OF LOAD IMPEDANCE	STRAP TERMINALS	OUTPUT CONNECTIONS
200 Ω	150 Ω TO 300 Ω	———	19 & 20
24 Ω	18 Ω TO 36 Ω	14-15, 16-17	13 & 18
12 Ω	9 Ω TO 18 Ω	13-15, 14-16-17	15 & 18
8 Ω	6 Ω TO 12 Ω	14-15	13 & 16
4 Ω	3 Ω TO 6 Ω	———	17 & 18
2 Ω	1.5 Ω TO 3 Ω	13-15, 14-16	13 & 16
400 Ω	300 Ω TO 600 Ω	14-15, 16-17, 18-19	13 & 20

70 V. LOUDSPEAKER DISTRIBUTION LINE CONNECTIONS

POWER OUTPUT CONDITION	STRAP TERMINALS	OUTPUT CONNECTIONS
12 WATTS	14-15, 16-17, 18-19	13 & 20
25 WATTS	———	19 & 20

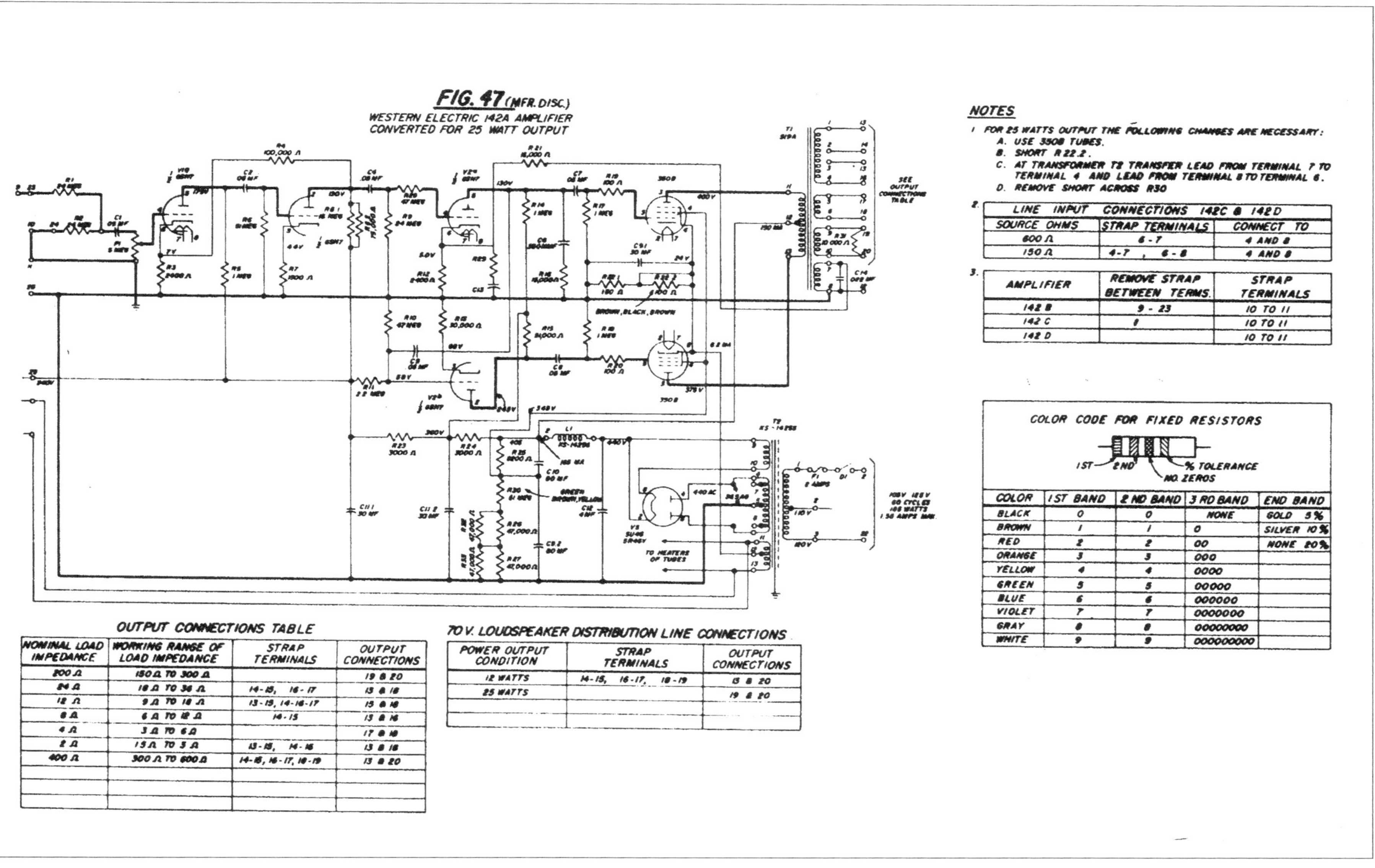

FIG. 47 (MFR. DISC.)
WESTERN ELECTRIC 142A AMPLIFIER
CONVERTED FOR 25 WATT OUTPUT

NOTES

1 FOR 25 WATTS OUTPUT THE FOLLOWING CHANGES ARE NECESSARY:
- A. USE 350B TUBES.
- B. SHORT R 22.2.
- C. AT TRANSFORMER T2 TRANSFER LEAD FROM TERMINAL 7 TO TERMINAL 4 AND LEAD FROM TERMINAL 8 TO TERMINAL 6.
- D. REMOVE SHORT ACROSS R30

2.

LINE INPUT CONNECTIONS 142C & 142D		
SOURCE OHMS	STRAP TERMINALS	CONNECT TO
600 Ω	6 - 7	4 AND 8
150 Ω	4-7 , 6 - 8	4 AND 8

3.

AMPLIFIER	REMOVE STRAP BETWEEN TERMS.	STRAP TERMINALS
142 B	9 - 23	10 TO 11
142 C	8	10 TO 11
142 D		10 TO 11

COLOR CODE FOR FIXED RESISTORS

1ST — 2ND — % TOLERANCE — NO. ZEROS

COLOR	1ST BAND	2 ND BAND	3 RD BAND	END BAND
BLACK	0	0	NONE	GOLD 5%
BROWN	1	1	0	SILVER 10%
RED	2	2	00	NONE 20%
ORANGE	3	3	000	
YELLOW	4	4	0000	
GREEN	5	5	00000	
BLUE	6	6	000000	
VIOLET	7	7	0000000	
GRAY	8	8	00000000	
WHITE	9	9	000000000	

OUTPUT CONNECTIONS TABLE

NOMINAL LOAD IMPEDANCE	WORKING RANGE OF LOAD IMPEDANCE	STRAP TERMINALS	OUTPUT CONNECTIONS
200 Ω	150 Ω TO 300 Ω		19 & 20
24 Ω	18 Ω TO 36 Ω	14-15, 16-17	13 & 18
12 Ω	9 Ω TO 18 Ω	13-15, 14-16-17	15 & 18
8 Ω	6 Ω TO 12 Ω	14-15	13 & 16
4 Ω	3 Ω TO 6 Ω		17 & 18
2 Ω	15 Ω TO 3 Ω	13-15, 14-16	13 & 18
400 Ω	300 Ω TO 600 Ω	14-15, 16-17, 18-19	13 & 20

70 V. LOUDSPEAKER DISTRIBUTION LINE CONNECTIONS

POWER OUTPUT CONDITION	STRAP TERMINALS	OUTPUT CONNECTIONS
12 WATTS	14-15, 16-17, 18-19	13 & 20
25 WATTS		19 & 20

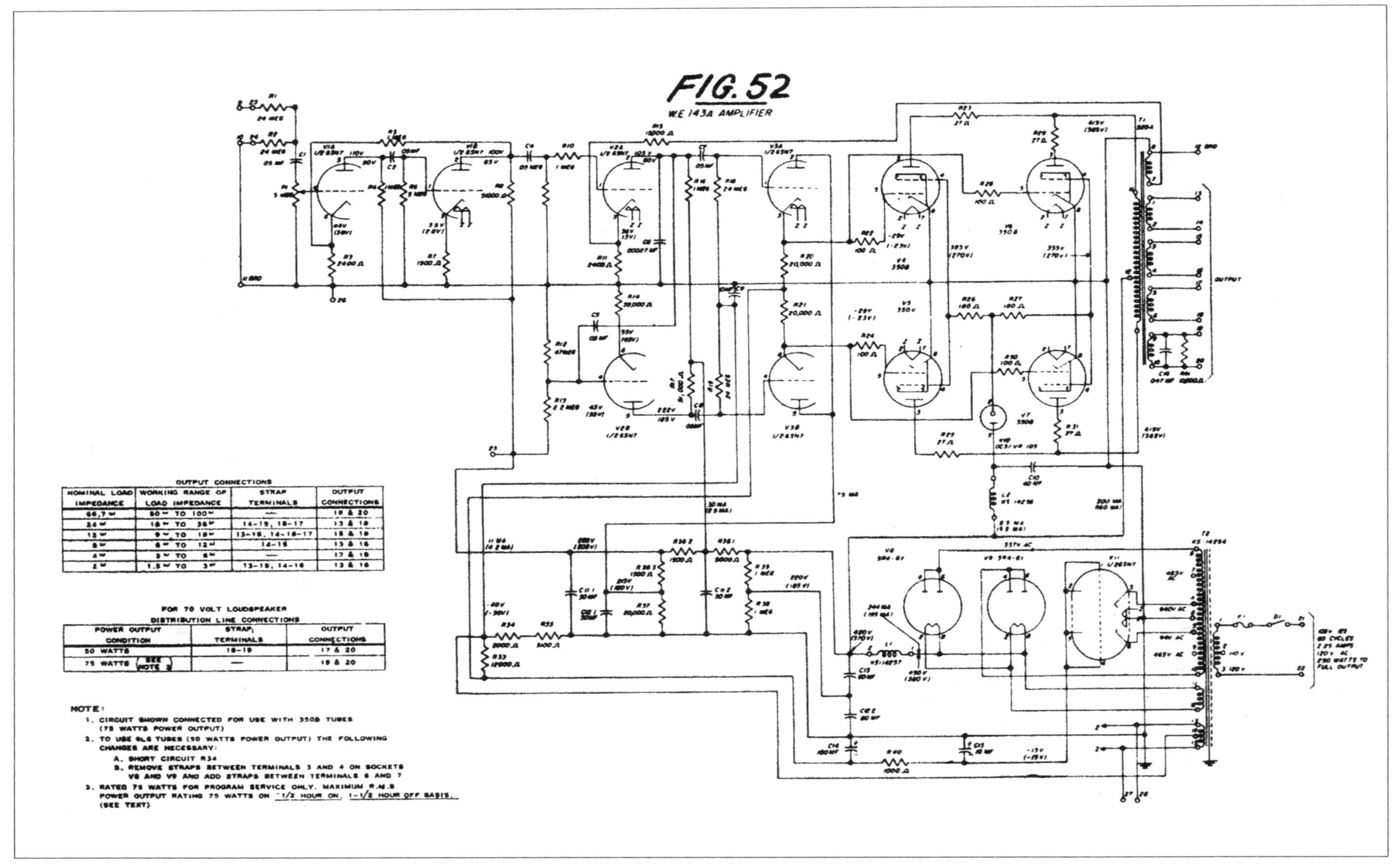

OUTPUT CONNECTIONS

NOMINAL LOAD IMPEDANCE	WORKING RANGE OF LOAD IMPEDANCE	STRAP TERMINALS	OUTPUT CONNECTIONS
66.7Ω	50Ω TO 100Ω	—	19 & 20
24Ω	18Ω TO 36Ω	14-19, 18-17	13 & 18
12Ω	9Ω TO 18Ω	13-18, 14-16-17	18 & 19
8Ω	6Ω TO 12Ω	14-15	13 & 16
4Ω	3Ω TO 6Ω	—	17 & 19
2Ω	1.5Ω TO 3Ω	13-15, 14-16	13 & 16

FOR 70 VOLT LOUDSPEAKER DISTRIBUTION LINE CONNECTIONS

POWER OUTPUT CONDITION	STRAP, TERMINALS	OUTPUT CONNECTIONS
50 WATTS	18-19	17 & 20
75 WATTS (SEE NOTE 3)	—	19 & 20

NOTE:

1. CIRCUIT SHOWN CONNECTED FOR USE WITH 350B TUBES (75 WATTS POWER OUTPUT)
2. TO USE 6L6 TUBES (50 WATTS POWER OUTPUT) THE FOLLOWING CHANGES ARE NECESSARY:
   A. SHORT CIRCUIT R34
   B. REMOVE STRAPS BETWEEN TERMINALS 3 AND 4 ON SOCKETS V8 AND V9 AND ADD STRAPS BETWEEN TERMINALS 6 AND 7
3. RATED 75 WATTS FOR PROGRAM SERVICE ONLY. MAXIMUM R.M.S POWER OUTPUT RATING 75 WATTS ON 1/2 HOUR ON, 1-1/2 HOUR OFF BASIS. (SEE TEXT)

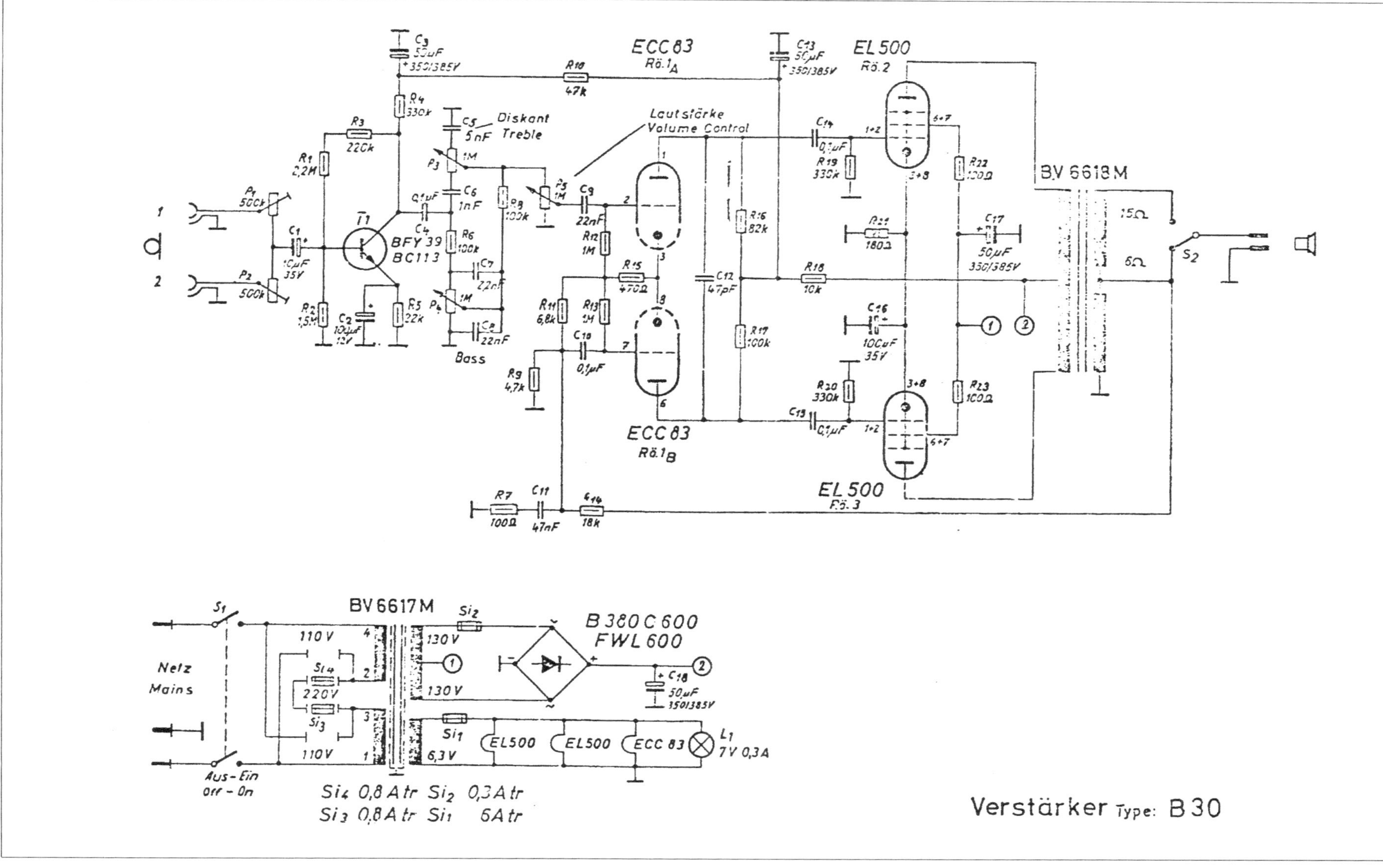

Verstärker Type: B 30

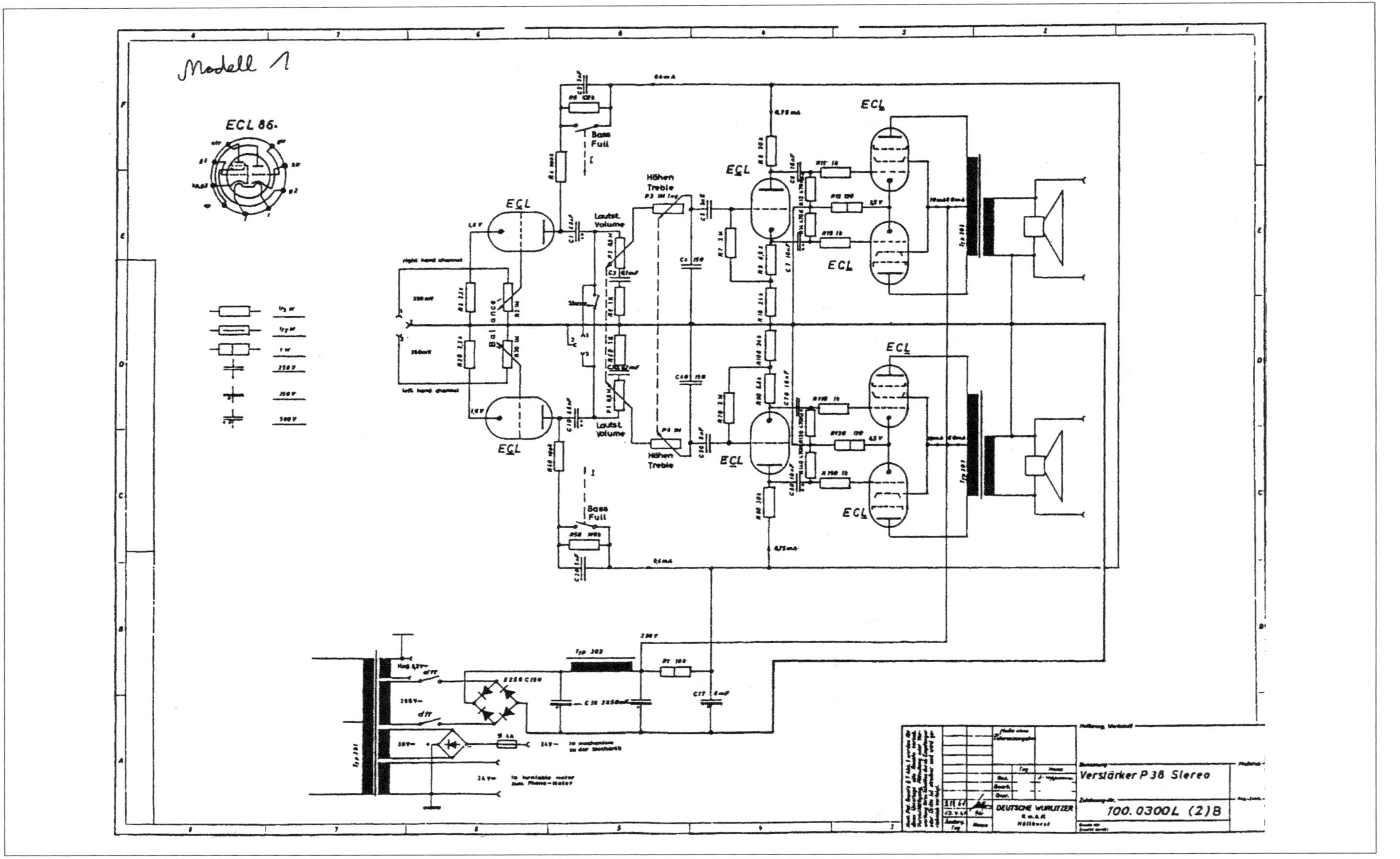
Modell 1
ECL 86
ECL
Lautst. Volume
Höhen Treble
Bass Full
Balance
right hand channel
left hand channel
Verstärker P 38 Stereo
100.0300L (2)B
DEUTSCHE WURLITZER

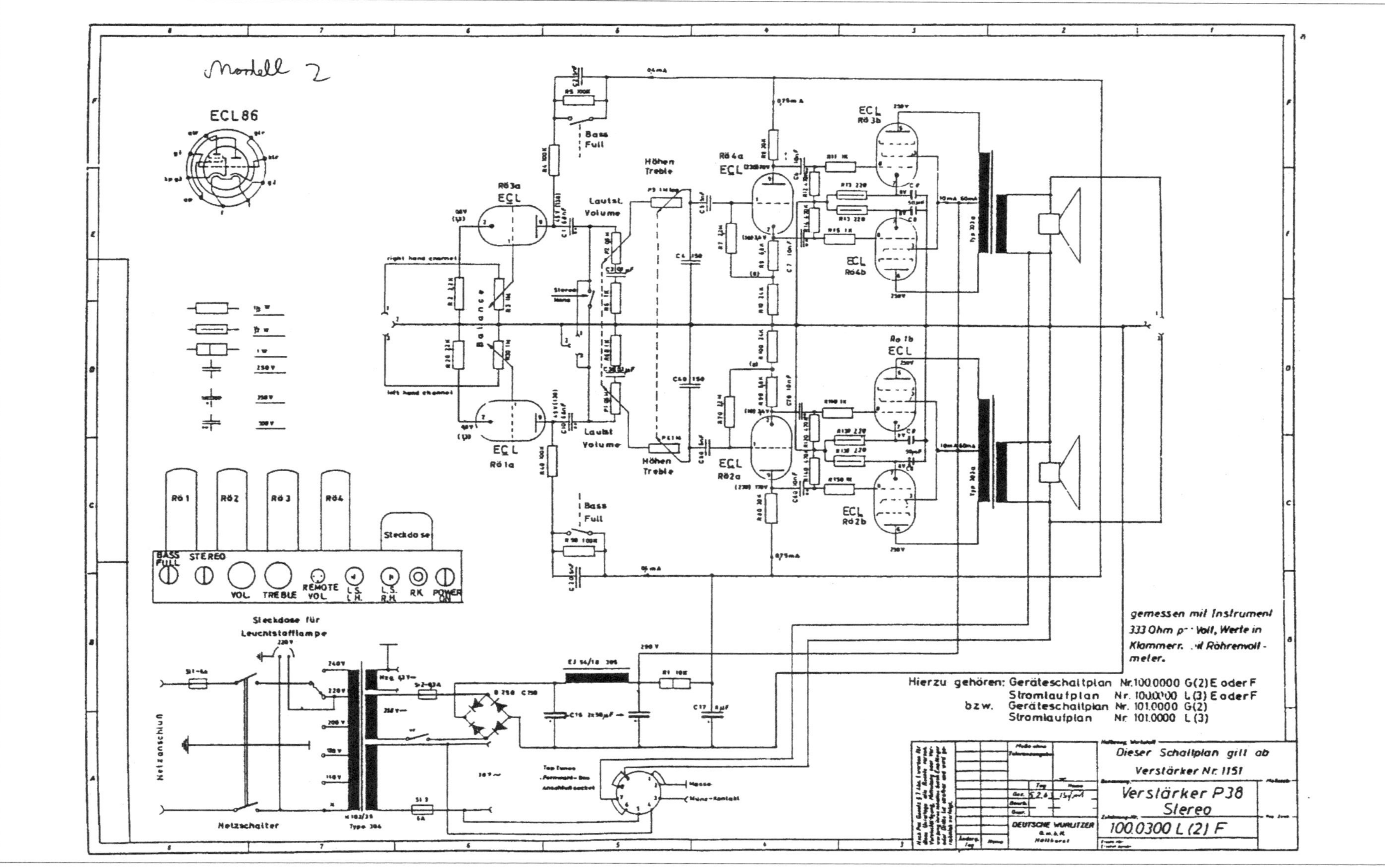
Modell 2
ECL86
Höhen
Treble
Lautst.
Volume
Bass
Full
Balance
right hand channel
left hand channel
Stereo
Mono
Rö 1
Rö 2
Rö 3
Rö 4
Steckdose
BASS FULL
STEREO
VOL.
TREBLE
REMOTE VOL.
L.S. L.H.
L.S. R.H.
RK
POWER ON
Steckdose für Leuchtstofflampe
Netzanschluß
Netzschalter
gemessen mit Instrument 333 Ohm pro Volt, Werte in Klammern mit Röhrenvoltmeter.
Hierzu gehören: Geräteschaltplan Nr. 100.0000 G(2) E oder F
Stromlaufplan Nr. 100.0000 L(3) E oder F
bzw. Geräteschaltplan Nr. 101.0000 G(2)
Stromlaufplan Nr. 101.0000 L(3)
Dieser Schaltplan gilt ab Verstärker Nr. 1151
Verstärker P38 Stereo
100.0300 L (2) F
DEUTSCHE WURLITZER

2 X 7886 MASTER VOLUME **MOD**

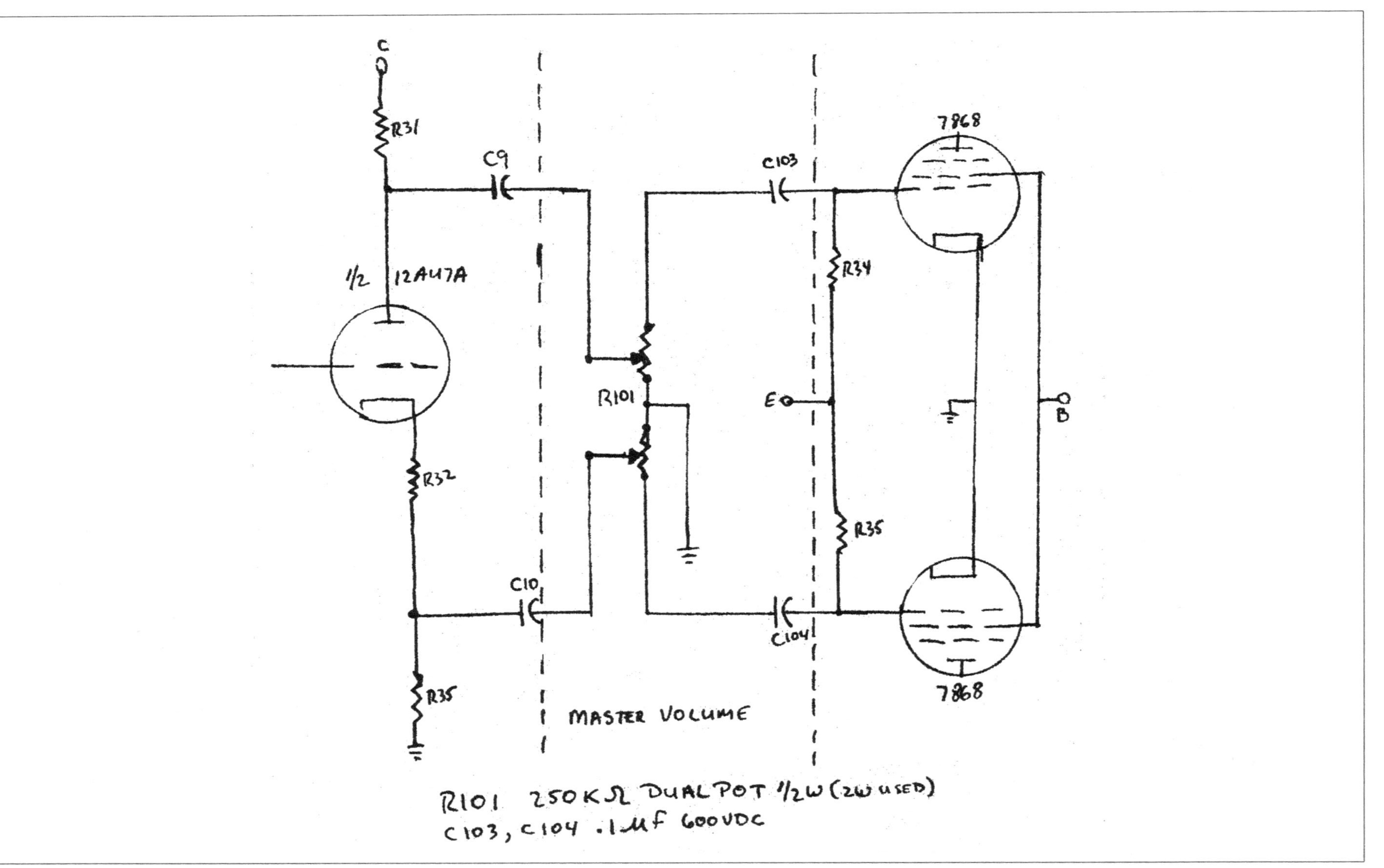

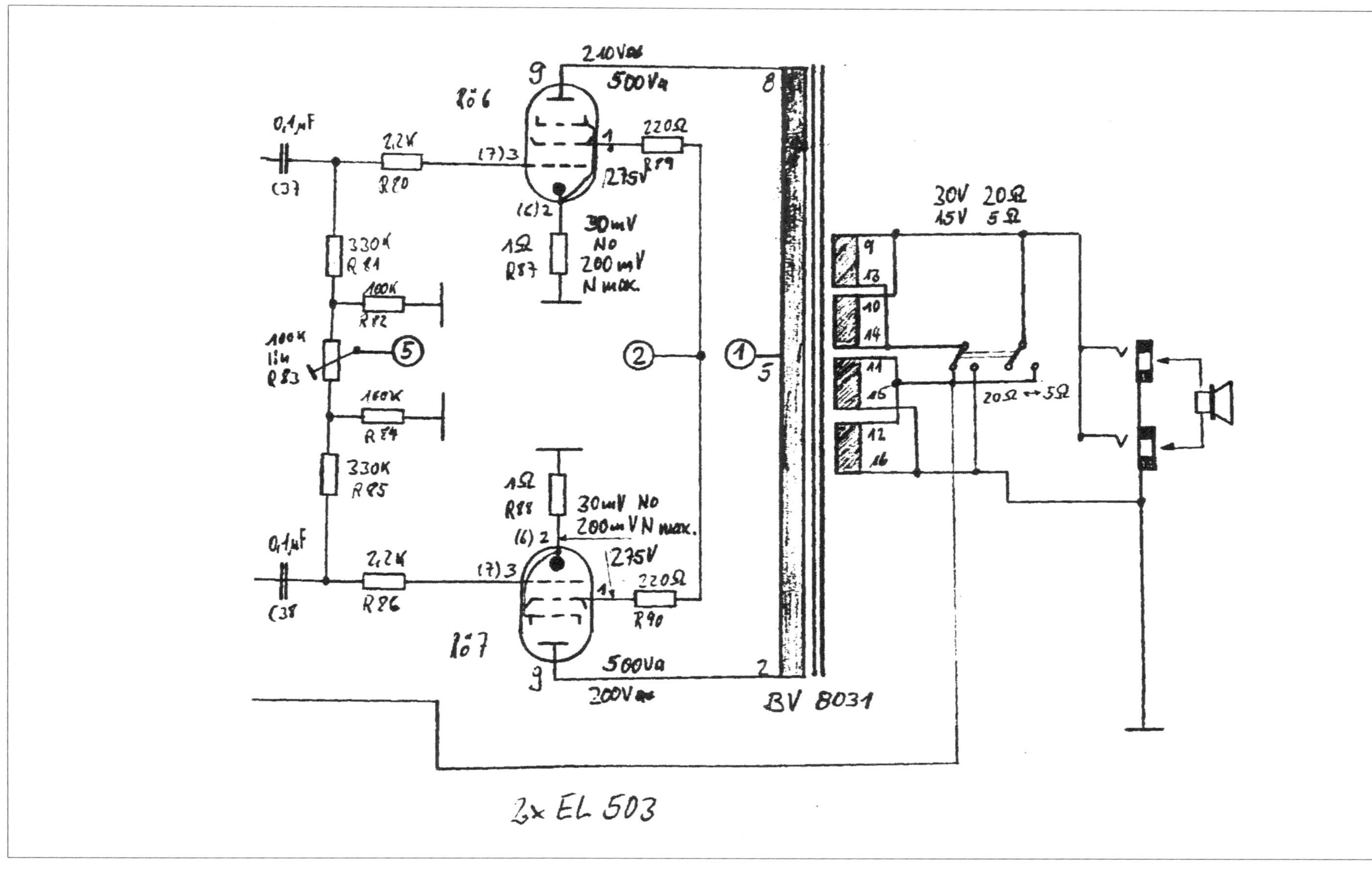

240Vac
500Va
Rö6
0,1µF
2,2K
(7) 3
220Ω
C37
R80
275V
R79
(6) 2
30mV
NO
200mV
N max.
1Ω
R87
330K
R81
100K
R82
100K
lin
R83
5
2
1
5
8
100K
R84
330K
R85
1Ω
R88
30mV NO
200mV N max.
(6) 2
275V
0,1µF
2,2K
(7) 3
220Ω
C38
R86
R90
Rö7
500Va
200Vac
2
BV 8031
30V 20Ω
15V 5Ω
9
13
10
14
11
15
12
16
20Ω ↔ 5Ω
2x EL 503

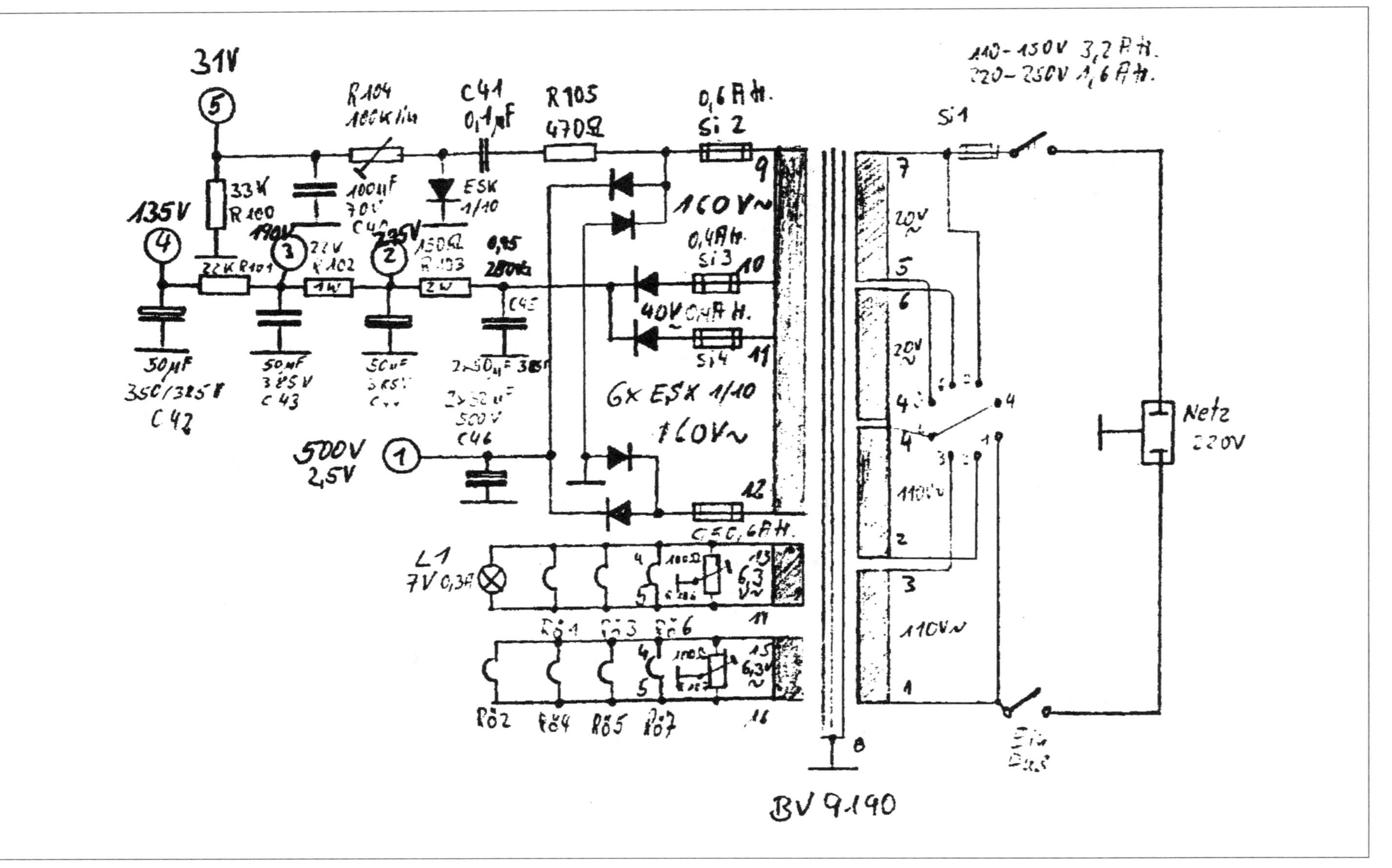
31V
135V
500V
2,5V
R 104
C 41
0,1µF
R 105
470Ω
0,6 A th.
Si 2
Si 1
110-150V 3,2 A th.
220-250V 1,6 A th.
Netz
220V
33K
R 100
22K R 101
R 102
1W
R 103
2W
ESK
1/10
160V~
0,4 A th.
Si 3
Si 4
6x ESK 1/10
160V~
50µF
350/385V
C 42
50µF
385V
C 43
C 46
L1
7V 0,3A
Rö 2
Rö 4
Rö 5
Rö 7
110V~
20V
BV 9.190

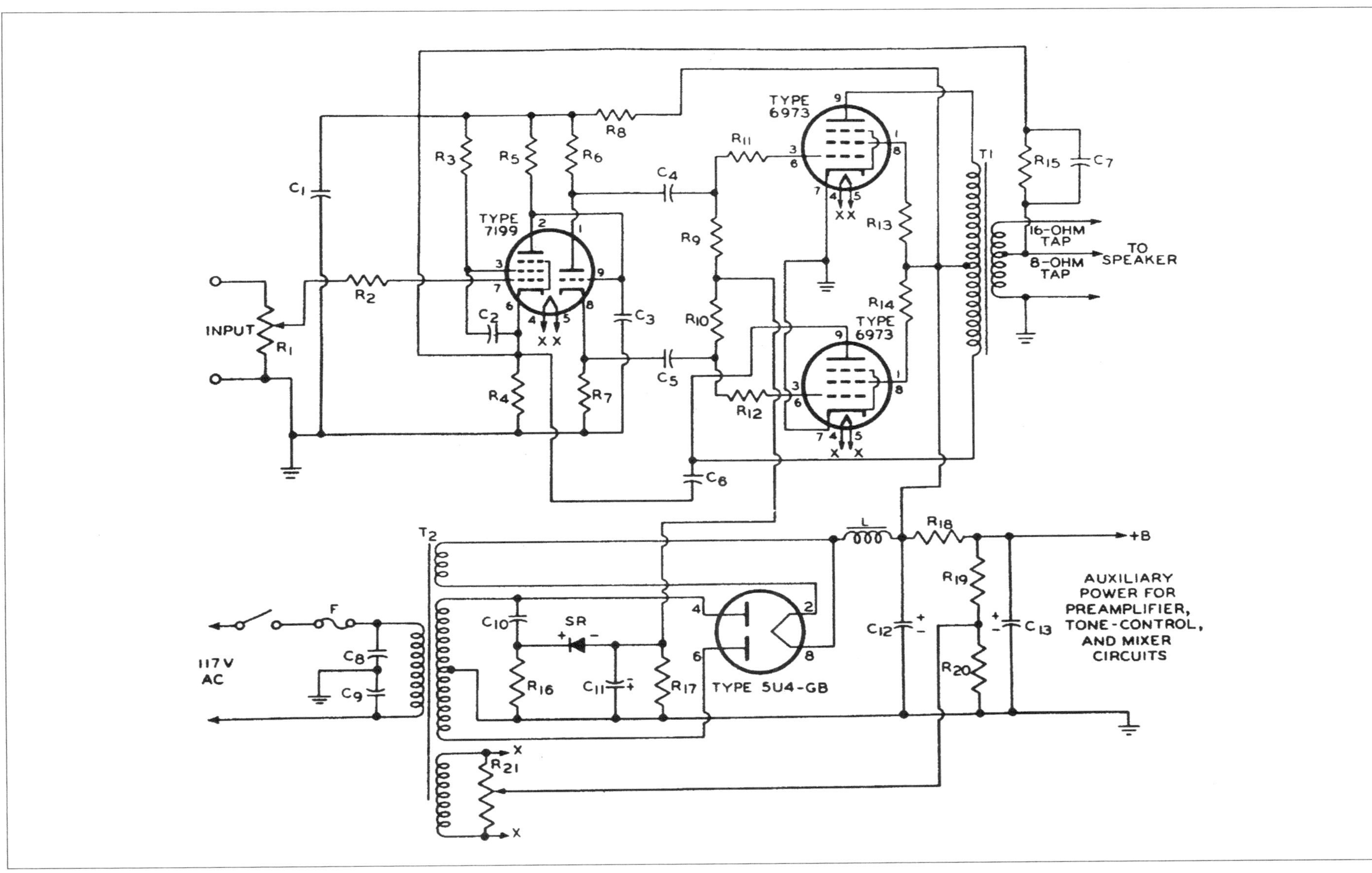
INPUT
TYPE 7199
TYPE 6973
TYPE 6973
TYPE 5U4-GB
16-OHM TAP
8-OHM TAP
TO SPEAKER
117V AC
+B
AUXILIARY POWER FOR PREAMPLIFIER, TONE-CONTROL, AND MIXER CIRCUITS

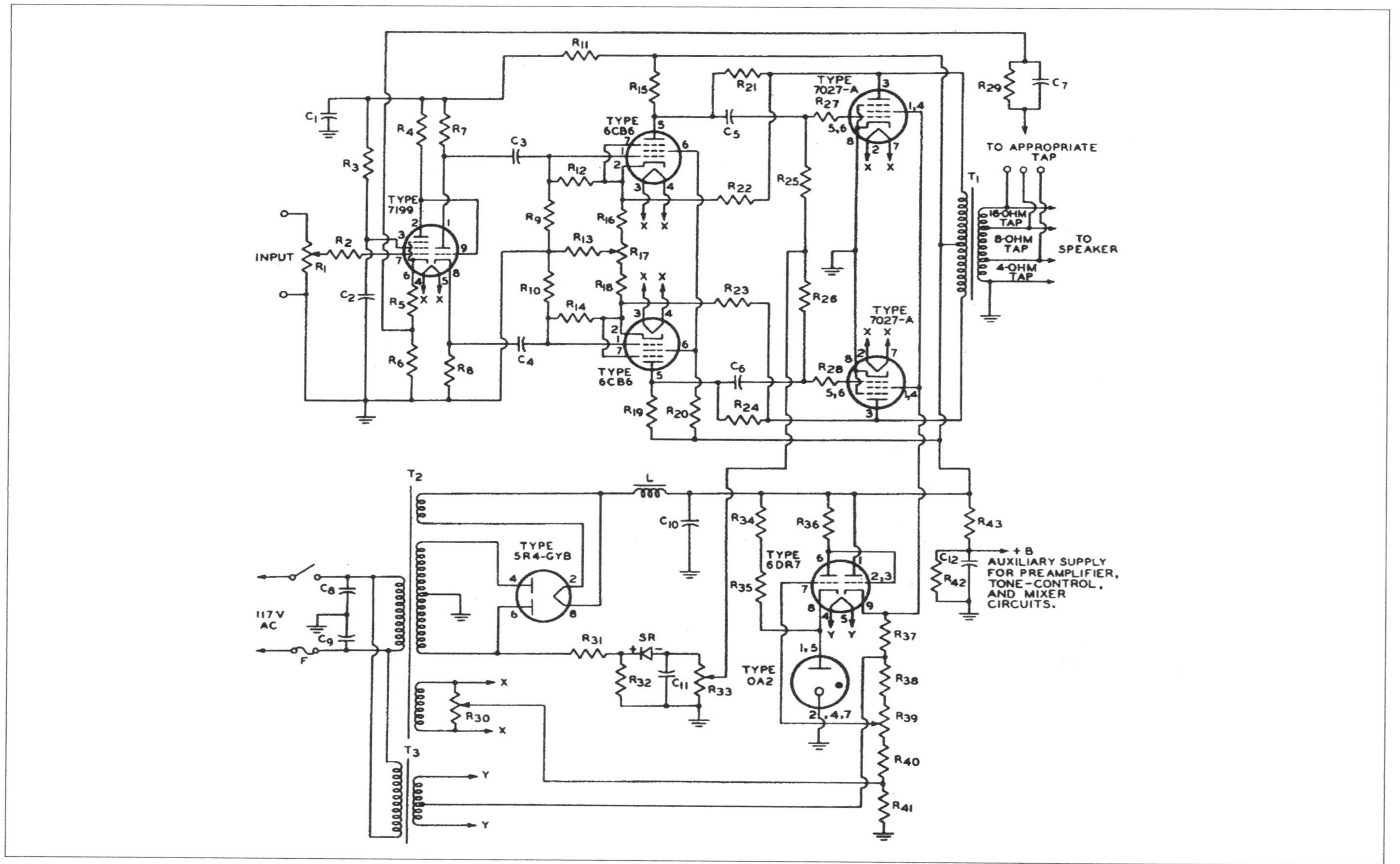
INPUT
TYPE 7199
TYPE 6CB6
TYPE 6CB6
TYPE 7027-A
TYPE 7027-A
TO APPROPRIATE TAP
16-OHM TAP
8-OHM TAP
4-OHM TAP
TO SPEAKER
TYPE 5R4-GYB
TYPE 6DR7
TYPE OA2
117V AC
+B
AUXILIARY SUPPLY FOR PREAMPLIFIER, TONE-CONTROL, AND MIXER CIRCUITS.

POWER SUPPLY

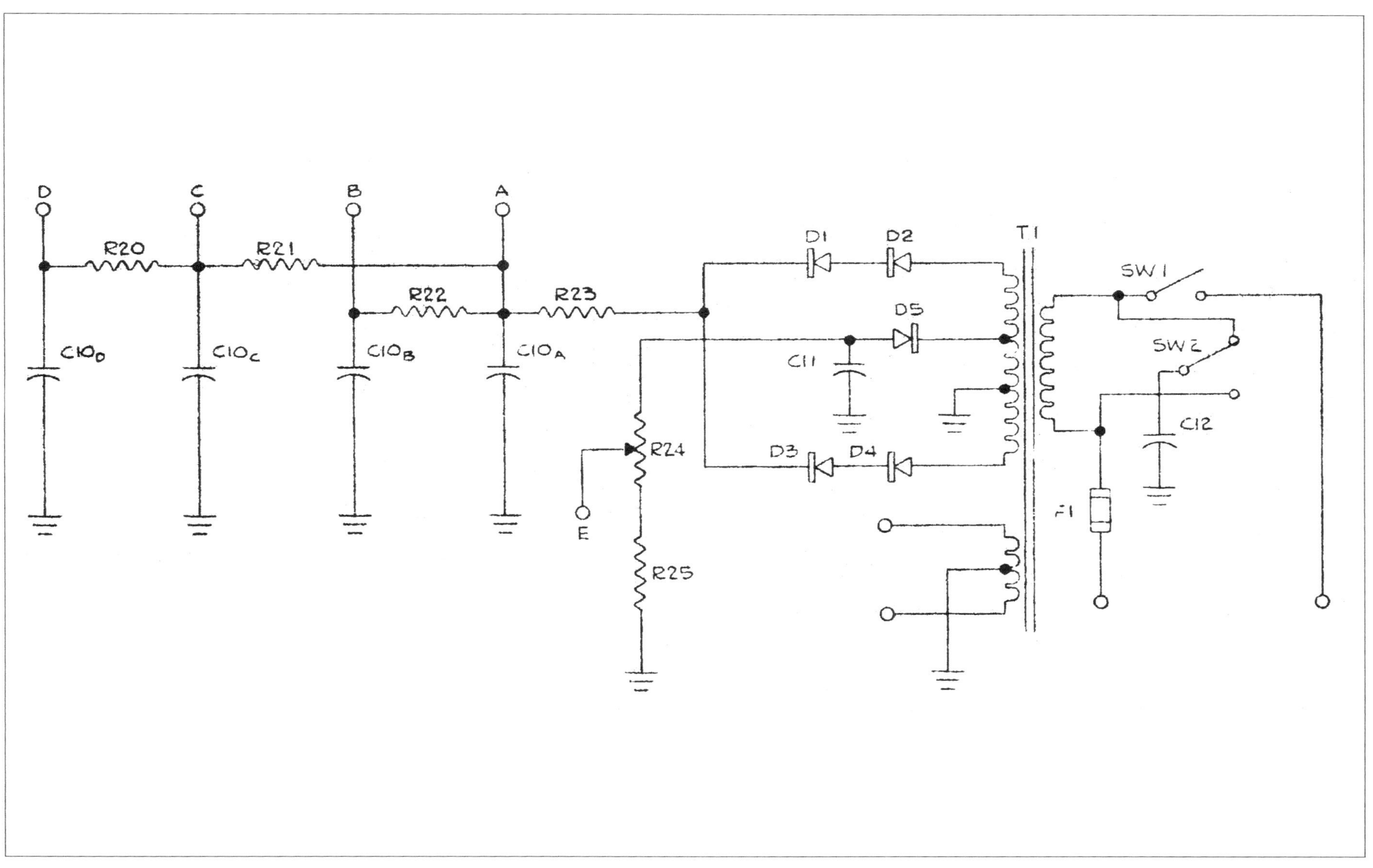

www.ingramcontent.com/pod-product-compliance
Lightning Source LLC
Chambersburg PA
CBHW081427170526
45166CB00008B/2124

9781500582456